**Handbook of
Combinatorial Chemistry**
Drugs, Catalysts, Materials
*Edited by K. C. Nicolaou,
R. Hanko, and W. Hartwig*
Volume 1

Further Titles of Interest

W. Bannwarth, E. Felder (Eds.)

Combinatorial Chemistry

2000, ISBN 3-527-30186-0

G. Jung (Ed.)

Combinatorial Chemistry

Synthesis, Analysis, Screening

1999, ISBN 3-527-29869-X

F. Zaragoza Dörwald

Organic Synthesis on Solid Phase

Supports, Linkers, Reactions

2000, ISBN 3-527-29950-5

A. Beck-Sickinger, P. Weber

Combinatorial Strategies in Biology and Chemistry

2002, ISBN 0-471-49726-6

Handbook of Combinatorial Chemistry

Drugs, Catalysts, Materials

Volume 1

Edited by K. C. Nicolaou, R. Hanko, and W. Hartwig

CHEM
o 11930627

Editors

Prof. Dr. K. C. Nicolaou
Department of Chemistry and the Skaggs
Institute for Chemical Biology
The Scripps Research Institute
10550 North Torrey
Pines Road
La Jolla, CA 92037
USA

Dr. R. Hanko
Bayer AG
Chemical Division
Head of BUFine Chemicals
51368 Leverkusen
Germany

Prof. Dr. W. Hartwig
Bayer AG
Geschäftsbereich Pharma
Leiter Forschung International
42096 Wuppertal
Germany

Library of Congress Card No.: applied for
A catalogue record for this book is available
from the British Library.
Die Deutsche Bibliothek – CIP
Cataloguing-in-Publication Data
A catalogue record for this publication is
available from Die Deutsche Bibliothek

Printed in the Federal Republic of
Germany.
Printed on acid-free paper.

Typesetting Asco Typesetters, Hong Kong
Printing Strauss Offsetdruck GmbH,
69503 Mörlenbach
Bookbinding J. Schäffer GmbH & Co.,
KG, 67269 Grunstadt
ISBN 3-527-30509-2

ac

Contents

List of Authors

Markus Albers
Bayer AG
Central Research
D-51368 Leverkusen
Germany

Valery V. Antonenko
Affymax Research Institute
3410 Central Expressway
Santa Clara
CA 95051
USA

Bill Archibald
H3-Technologies Inc.
1021 Grebe Street
Foster City
CA 94404
USA

István Bágyi
ComGenex Inc.
Bem rkp. 33-34
H-1024 Budapest
Hungary

Carmen Baldino
Department of Chemistry
ArQule Inc.
19 Presidential Way
Woburn
MA 01801
USA

Marcus Bauser
Bayer AG
Pharma Research
PH-R-CR VII
D-42096 Wuppertal
Germany

Stefan Bräse
Kekulé-Institut für Organische Chemie und
    Biochemie
Universität Bonn
Gerhard-Domagk-Straße 1
D-53121 Bonn
Germany

Gabriele Bräunlich
Bayer AG
Pharma BPA
Leverkusen
Germany

O. Brümmer
Symyx Technologies Inc.
3100 Central Expressway
Santa Clara
CA 95051
USA

Roger M. Brunne
Bayer AG
Pharma Research Centre
42096 Wuppertal
Germany

David L. Coffen
Discovery Partners International
San Diego
CA 92121
USA

Christopher P. Corrette
Array Biopharma Inc.
1885 33rd Street
Boulder
CO 80301
USA

Stefan Dahmen
Kekulé-Institut für Organische Chemie und
  Biochemie
Universität Bonn
Gerhard-Domagk-Straße 1
D-53121 Bonn
Germany

Ferenc Darvas
ComGenex Inc.
Bem rkp. 33-34
H-1024 Budapest
Hungary

M. Devenney
Symyx Technologies Inc.
3100 Central Expressway
Santa Clara
CA 95051
USA

Roland E. Dolle
Department of Chemistry
Adolor Corporation
371 Phoenixville Pike
Malvern
PA 19355
USA

György Dormán
ComGenex Inc.
Bem rkp. 33-34
H-1024 Budapest
Hungary

Markus Eckert
Bayer AG
Chemicals Division
Business Unit Fine Chemicals
Research and Development
D-51368 Leverkusen
Germany

Demosthenes Fokas
Department of Chemistry
ArQule Inc.
19 Presidential Way
Woburn
MA 01801
USA

A. Ganesan
Department of Chemistry
University of Southampton
Highfield
Southampton SO17 1BJ
UK

Christoph Gerdes
Bayer AG
Pharma Research
Pharmacology
42096 Wuppertal
Germany

D. Giaquinta
Symyx Technologies Inc.
3100 Central Expressway
Santa Clara
CA 95051
USA

S. Gorer
Symyx Technologies Inc.
3100 Central Expressway
Santa Clara
CA 95051
USA

Philipp Grosche
Institute for Organic Chemistry
University of Tübingen
Auf der Morgenstelle 18
D-72076 Tübingen
Germany

Rainer Haag
Freiburger Materialforschungszentrum und
Institut für Makromolekulare Chemie
Albert-Ludwigs-Universität Freiburg
Stefan-Meier-Straße 21
D-79104 Freiburg
Germany

R. Hanko
Bayer AG
Chemical Division
Head of BUFine Chemicals
D-51368 Leverkusen
Germany

Jan-Gerd Hansel
Bayer AG
Central Reseach
D-51368 Leverkusen
Germany

Michael W. Härter
Business Group Pharma
Building 470
D-42096 Wuppertal
Germany

Wolfgang Hartwig
Bayer AG
Geschäftsbereich Pharma
Leiter Forschung International
D-42096 Wuppertal

Tamio Hayashi
Department of Chemistry
Graduate School of Science
Kyoto University
Sakayo, Kyoto 606-8502
Japan

André Hebel
Freiburger Materialforschungszentrum und
Institut für Makromolekulare Chemie
Albert-Ludwigs-Universität Freiburg
Stefan-Meier-Str. 21
D-79104 Freiburg
Germany

Gerhard Hessler
Bayer AG
Central Research
D-51368 Leverkusen
Germany

Berthold Hinzen
Bayer AG
Pharma Research
Medicinal Chemistry
D-42096 Wuppertal
Germany

Amir H. Hoveyda
Department of Chemistry
Merkert Chemistry Center
Boston College
2609 Beacon Street
Chestnut Hill
MA 02467
USA

Conrad Hummel
Array Biopharma Inc.
1885, 33rd Street
Boulder
CO 80301
USA

Bernd Jandeleit
XenoPort Inc.
2631 Hanover Street
Palo Alto
CA 94304
USA

Stephan Jordan
Bayer AG
Central Reseach
D-51368 Leverkusen
Germany

Günther Jung
Institute for Organic Chemistry
University of Tübingen
Auf der Morgenstelle 18
D-72076 Tübingen
Germany

Christopher Kallus
Bayer AG
Central Research
D-51368 Leverkusen
Germany

Tamás Karancsi
ComGenex Inc.
Bem rkp. 33-34
H-1024 Budapest
Hungary

Jörg Keldenich
Business Group Pharma
Building 470
D-42096 Wuppertal
Germany

Thomas Krämer
Bayer AG
Pharma Research
Medicinal Chemistry
D-42096 Wuppertal
Germany

Jochen Krüger
Bayer AG
Business Group Pharma
PH-R CR-MC III
D-42096 Wuppertal
Germany

Nicolay V. Kulikov
Affymax Research Institut
3410 Central Expressway
Santa Clara, CA 95091
USA

Joachim E. A. Luithle
Bayer AG
Business Group Pharma
Medicinal Chemistry
D-42096 Wuppertal
Germany

Klemens Lustig
Bayer AG
Pharma Research
Pharmakokinetics
42096 Wuppertal
Germany

Thorsten Meyer
Bayer AG
Central Research
D-51368 Leverkusen
Germany

Reza Mortezaei
Affymax Research Institut
3410 Central Expressway
Santa Clara, CA 95091
USA

Ingo Muegge
Bayer Research Center
400 Morgan Lane
West Haven
CT 06516
USA

Tamás Nagy
ComGenex Inc.
Bem rkp. 33-34
H-1024 Budapest
Hungary

K. C. Nicolaou
Department of Chemistry and the Skaggs
    Institute for Chemical Biology
The Scripps Research Institute
10550 North Torrey Pines Road
La Jolla
CA 92037
USA
and
Department of Chemistry and Biochemistry
University of California
San Diego
9500 Gilman Dr
La Jolla, CA 92093
USA

Ulrich Nielsch
Bayer AG
Pharma Research
Pharmacology
42096 Wuppertal
Germany

Ulrich Notheis
Bayer AG
Chemicals Division
Business Unit Fine Chemicals
Research and Development
D-51368 Leverkuten
Germany

Josef Pernerstorfer
Bayer AG
Pharma Research
Medicinal Chemistry
D-42096 Wuppertal
Germany

Jeffrey A. Pfefferkorn
Pharmacia Corp.
Combinatorial and Medicinal Chemistry
7000 Portage Rd
Kalamazoo, MI 49001
USA

Jörg Rademann
Institute for Organic Chemistry
University of Tübingen
Auf der Morgenstelle 18
D-72076 Tübingen
Germany

Walter Schmitt
Business Group Pharma
Building 470
D-42096 Wuppertal
Germany

Pamela Sears
Department of Chemistry
The Scripps Research Institute
10550 N. Torrey Pines Rd
La Jolla
CA 92037
USA

Mukund P. Sibi
Department of Chemistry
North Dakota State University
Fargo
ND 58105-5516
USA

Adrian L. Smith
Amgen Inc.
One Amgen Center Drive
Thousand Oaks, CA 91320
USA

Hubertus Stakemeier
Bayer AG
Central Research
ZF-LSc
D-51368 Leverkusen
Germany

Henning Steinhagen
Bayer AG
Business Group Pharma
PH-R CR-MC II
D-42096 Wuppertal
Germany

Jean-François Stumbé
Freiburger Materialforschungszentrum und
Institut für Makromolekulare Chemie
Albert-Ludwigs-Universität Freiburg
Stefan-Meier-Str. 21
D-79104 Freiburg
Germany

Sean V. Taylor
Laboratorium für Organische Chemie
ETH Hönggerberg-HCI
CH-8093 Zürich
Switzerland

Arounarith Tuch
Bayer AG
Central Research
ZF-LSc-KC
D-51368 Leverkusen
Germany

Tetsuo Uno
Genomics Institute of the Novartis Research
   Foundation
3115 Merryfield Row Suite 200
San Diego
CA 92121
USA

Yasuhiro Uozumi
Laboratory of Complex Catalysis
Institute for Molecular Science
Nishigonaka 38
Myodaiji
Okazaki 444-8585
Japan

Stefan Wallé
Bayer AG
Central Research
ZF-LSc-KC
D-51368 Leverkusen
Germany

W. H. Weinberg
Symyx Technologies Inc.
3100 Central Expressway
Santa Clara
CA 95051
USA

T. Weskamp
Symyx Technologies Inc.
3100 Central Expressway
Santa Clara
CA 95051
USA

Chi-Huey Wong
Department of Chemistry and Skaggs
   Institute for Chemical Biology
The Scripps Research Institute
10550 N. Torrey Pines Rd
La Jolla
CA 92037
USA

Tobias Wunberg
Bayer AG
Business Group Pharma
PH-R CR-MC II
D-42096 Wuppertal
Germany

Florencio Zaragoza
Novo Nordisk A/S
Novo Nordisk Park
DK-2760 Måløv
Denmark

Part I
General Aspects

1
Combinatorial Chemistry in Perspective

K. C. Nicolaou, R. Hanko, and W. Hartwig

1.1
Introduction

A fundamental aspect of organic chemistry is its ability to create new carbon-containing substances. This endeavor, called organic synthesis, made huge contributions to society by delivering a myriad of synthetic materials. Until recently the science of organic synthesis and its branches, such as medicinal chemistry, practiced the construction of organic molecules by targeting one molecule at a time. During the last decade of the twentieth century, however, a new concept took hold by which a collection of molecules is targeted, simultaneously producing a library of compounds instead of a single product. This strategy of concurrently synthesizing large numbers of compounds is called combinatorial chemistry or combinatorial synthesis. Although the roots of combinatorial chemistry were already in place, this philosophy took on a special meaning and assumed high priority in the 1990s because of pressures within the pharmaceutical industry to speed up the drug discovery process and because of the advent of high-throughput screening.

Although combinatorial chemistry was initially met with resistance from some medicinal and other chemists, it rapidly became mainstream, and its practice today is widespread in academia and industry. It is utilized to synthesize libraries of compounds either as mixtures or as single compounds. The latter approach is preferred by most practitioners, and as new synthetic technologies, high-speed purification techniques and characterization strategies are developed, combinatorial chemistry will assume an increasingly prominent position within the armamentarium of the synthetic chemist. Combinatorial chemistry is currently being applied, in addition to natural product synthesis and drug discovery, to agricultural chemistry, chemical biology, catalyst discovery and material science.

Combinatorial synthesis may be performed either in solution or on solid support. Each having their own advantages and disadvantages, both methods have been widely used in the construction of various compound libraries. Although solid-phase chemistry is currently not as well developed as solution-phase chemistry, particularly with regard to small organic molecules, the former method has distinct advantages over the latter. First, in solid-phase synthesis, large excesses of reagents can be used to accelerate reactions and to drive them to completion; these reagents can then conveniently be removed at the end by filtration and washing. Furthermore, and because of easy separation, solid-phase chemistry can be automated through robotics to a higher degree than solution chemistry can. Most importantly, solid-phase chemistry can be applied to the elegant and powerful "split-and-pool" synthesis strategy for combinatorial chemistry. Despite its dramatic contribution to increasing efficiency over traditional methods, high-throughput parallel synthesis remains a laborious task. Thus, combinatorial chemists quickly recognized the benefits of automation as a crucial component of combinatorial chemistry. Besides the synthetic sequence, purification and characterization of compounds are also important aspects of combinatorial chemistry. The development of high-throughput chromatography methods to support automated parallel synthesis and of high-performance liquid chromatography–mass spectrometry (HPLC/MS) systems to direct the collection of the desired products are only the beginning of such high-throughput methods. Progress in this area is moving quickly and will be indispensable to any serious combinatorial chemistry effort.

1.2
Brief History of Combinatorial Chemistry

While combinatorial synthesis is a relatively new field of chemistry, Nature has been utilizing the same principles since the beginning. Although biologists had previously recognized the power of combinatorial chemistry, its application to problems relating to chemistry did not emerge until recently. Chemists' confidence in rational design has previously kept them away from systematic explorations in chemical synthesis.

The rather dramatic developments in molecular biology and high-throughput screening increased the demand for large numbers of small organic molecules to be screened against the ever-increasing biological targets. A solution to these challenges came from the peptide and oligonucleotide chemists, who could conveniently implement combinatorial chemistry strategies given the ease with which the amide and phosphate bonds could be constructed from the readily available building block libraries of amino acids and nucleotides respectively.

Solid-phase chemistry was pioneered by Merrifield [1] and applied to the peptide and oligonucleotide fields quite effectively. In the early 1970s developments had already occurred in solid-phase synthesis of nonpeptide and nonoligonucleotide molecules. For example, the groups of Leznoff [2], Fréchet [3], Camps [4], Patchornik [5] and Rapoport [6] all reported early results on solid-phase synthesis. Camps et al. even applied solid-phase synthesis to the pharmaceutically relevant benzodiazepine system [7].

In Germany in the 1980s Frank and coworkers synthesized collections of oligonucleotides and, later, peptides on circles of cellulose paper [8]. Geysen et al. in

Australia prepared a library of peptides [9] on functionalized polypropylene pins by immersing them sequentially into various solutions of activated amino acids held in the wells of a microtiter plate. Houghten at The Scripps Research Institute in La Jolla synthesized a library of 260 peptides [10] in polypropylene mesh containers encapsulating polystyrene resin, a process that came to be known as the "tea-bag" strategy. Both the pin and the tea-bag techniques went on to gain wide popularity and led to new generations of improved technologies for combinatorial chemistry. Researchers at Affymax reported very large spatially addressable libraries on glass chips using photolithographic techniques in conjunction with photolabile protecting group chemistry [11]. In parallel with the chemical approaches to peptide diversity, phages were being exploited to display very large libraries of peptides [12].

In 1992, Bunin and Ellman reported another synthesis of a benzodiazepine library [13] using the "multi-pin" technology pioneered by Geysen. At about the same time, a group of scientists at Parke–Davis reported the construction of hydantoins and benzodiazepines using a semiautomated robotic synthesizer [14]. In addition, a Chiron group reported the synthesis of a library of peptoids [oligo(N-substituted glycine)] and a robotic synthesizer of such compounds [15].

In the meantime, an elegant and ingenious strategy for combinatorial synthesis was proposed and demonstrated. This strategy called "split synthesis" or "split and pool" was introduced by Furka and coworkers at two European symposia in 1988; this work was published in 1991 [16]. The groups of Lam [17] and Houghten [18] independently developed the same technique and also published their results in 1991. These strategies led to the concept of "one bead–one compound" and promised the delivery of millions of compounds synthesized simultaneously on beads and with unprecedented rapidity. As elegant as it is, this method left much to be desired in terms of structure deconvolution and quantity of material produced. To solve the first problem, a number of encoding strategies were developed based on technologies ranging from DNA sequences to polychlorinated aromatics as well as nonchemical encoding methods such as radiofrequency tagging and two-dimensional (2D) bar-coding (for further discussion of library encoding, see Chapter 5).

From the early 1990s onwards, the chemical literature exploded with reports addressing all aspects of combinatorial synthesis, including solid-phase chemistry, encoding strategies and molecular diversity.

In the late 1990s alternative strategies were investigated, and an interesting compromise between solid-phase and solution-phase chemistry was found with polymers which are soluble in certain solvents but can be precipitated efficiently in others [19]. Thus the reactions on such polymers are carried out in homogeneous solution while the convenience of purification via a simple filtration is maintained. In a highly efficient extension of this principle, Curran and coworkers [20] have developed a number of fluorous tags which allow extraction of tagged compounds into a three-phase separation system (aqueous, organic, and fluorinated).

Today, many well-known solution-phase reactions have been demonstrated to perform equally well on solid phase [21] and a plethora of reagents have been immobilized on solid supports [22]. Such techniques lead to high-speed purification procedures and often to higher yields of targeted products, which in turn lead to an increase in efficiency and productivity.

While the peptide and oligonucleotide chemists may have opened the field of combinatorial chemistry, it was left to those chemists concerned with small organic molecules to make the methods widely applicable to more "lead-like" and "drug-like" structures. Of particular interest were new solid-phase synthetic strategies, new linkers for solid-phase chemistry [23], and new polymer-bound reagents [24].

1.3
Applications of Combinatorial Chemistry

With the advent of combinatorial chemistry, the traditional *one molecule at a time* approach to drug discovery was severely shaken. The initial euphoria of the early 1990s, however, was based to a considerable extent, on faulty grounds. The idea of synthesizing a myriad of compounds randomly, often as mixtures, sounded like a dream to many chemists of biotechnology and pharmaceutical companies. Soon thereafter, however, the principles of combinatorial chemistry for small organic molecules crystallized on a more pragmatic platform. The prevailing approach today is that based on both solution-phase and solid-phase chemistry applied in parallel or split-and-pool formats and directed at discrete and high-purity compounds [25, 26]. Initial-phase combinatorial chemistry is applied to discover lead compounds rapidly which are then subjected to lead optimization to produce drug candidates. The last part of the process is the domain of the medicinal chemists, who may also practice combinatorial strategies to achieve their goals. Thus, smaller focused libraries are carefully designed and synthesized, either in parallel or by the split-and-pool strategy using solution- or solid-phase chemistry. Combinatorial chemistry has, therefore, penetrated the laboratories of medicinal chemists who recognized its power in delivering the targeted compounds in a much faster way, and in acceptable quantities and purities. In similar ways, academic laboratories have adopted and refined combinatorial techniques in their quest for libraries of compounds needed for chemical biology studies [27].

The capability of combinatorial chemistry to produce large numbers of compounds rapidly is a powerful tool not only for chemical biology and drug discovery but also for a host of other research endeavors. Indeed, this philosophy and these combinatorial processes have been successfully applied to reaction optimization, the discovery of new materials [28] and the development of new catalysts [29]. In pioneering work, Fuchs and coworkers [30] reported in 1984 the use of automation to optimize reaction conditions with multiple variables. Reddington and Sapienza [31] reported in 1998 results from a highly parallel, optical screening method to discover novel electrocatalysts. Such practices are currently gaining wide popularity in industry for the optimization of process chemistry. The first report of a combinatorial approach to new high-technology materials came from Schultz and coworkers [32], who prepared a spatially addressable array of potential superconducting materials. Similar techniques were then applied to a number of studies including ferroelectric materials [33] and phosphorescent materials [34]. Combinatorial techniques have also been applied to the development of chiral separation methods [35, 36] and optimization of protein catalysts through DNA-shuffling techniques [37]. Most significantly, combinatorial chemistry has proven itself to be useful in the discovery of new catalytic systems. Early examples include Liu and Ellman's synthesis of 2-pyrrolidinemethanol ligands intended for enantio-selective additions [38], Burgess and coworkers' diazo-compound library for a C–H insertion [39], Hoveyda and coworkers' library of dipeptide Schiff base ligands for enantioselective addition of trimethylsilylcyanide to epoxides [40], and Sigman and Jacobsen's Schiff base catalysts for the Strecker reaction [41].

The payoff of combinatorial chemistry to drug discovery is already becoming obvious to the industry in terms of a significant increase in the number of drug candidates and of decreases in time from target identification to drug candidates and manpower employed per drug candidate. Similar benefits are beginning to emerge in process chemistry, catalyst discovery and material science, where combinatorial chemistry techniques have also been implemented.

1.4
Outline of the Book

The following chapters in this series will address the various aspects of combinatorial chemistry in order to facilitate further advances in the field as well as to aid the reader in his or her practice of combinatorial chemistry.

The book is divided into three parts. The first chapters (1–5) serve as the introductory part and build the foundation on which the next chapters are based. The following part (Chapters 5–20) deals with basic reaction mechanisms. The aim is to describe thoroughly the repertoire of combinatorial chemistry in solution and on solid support and to provide the reader with a critical overview of the best methods and conditions found so far. As an introduction, each chapter briefly discusses the applicability of the reactions dealt with for library synthesis in general and for solution-phase or solid-phase chemistry in particular. This entry is followed by the detailed reflection of each reaction type including the most recent developments/achievements. Based on mechanistic considerations, emphasis is put on the suitability of a given reaction for library synthesis using either solid-phase or solution-phase chemistry.

Generally, synthetic examples from all techniques are presented (solution-phase parallel synthesis, solid-phase chemistry, solid-supported reagents). In most cases examples originate from syntheses of large libraries.

The last part of the book (Chapters 21–35) describes special topics such as applications, design and instrumentation as well as aspects beyond pure synthetic organic chemistry. In addition, since combinatorial chemistry is a tool aiming at specific applications, for example medicinal chemistry, material research and catalysis, selected examples of breakthroughs in these applications are discussed.

Acknowledgments

This book would not have been possible without the dedicated work of the editing team, especially Berthold Hinzen and Tobias Wunberg. We would also like to thank P. Gölitz and his team for their conceptual contributions and their support to ensure the ambitious scheduling of this book.

References

1 R. B. MERRIFIELD, *J. Am. Chem. Soc.* **1963**, *85*, 2149–2154.

2 a) C. C. LEZNOFF, J. Y. WONG, *Can. J. Chem.* **1972**, *50*, 2892–2893; b) C. C. LEZNOFF, *Acc. Chem. Res.* **1978**, *11*, 327.

3 J. M. J. FRÉCHET, *Tetrahedron* **1981**, *37*, 663–683.

4 F. CAMPS, J. CASTELLS, M. J. FERRANDO, J. FONT, *Tetrahedron Lett.* **1971**, *20*, 713–1714.

5 A. PATCHORNIK, M. A. KRAUS, *J. Am. Chem. Soc.* **1970**, *92*, 7587–7589.

6 a) J. I. CROWLEY, H. RAPOPORT, *J. Am. Chem. Soc.* **1970**, *92*, 6363–6365; b) J. I. CROWLEY, H. RAPOPORT, *Acc. Chem. Res.* **1976**, *9*, 135–144.

7 F. CAMPS, J. CASTELLS, J. PI, *An. Quim.* **1974**, *70*, 848–849.

8 R. FRANK, W. HEIKENS, G. HEISTERBERG-MOUTSIS, H. BLÖCKER, *Nucl. Acids Res.* **1983**, *11*, 4365–4377.

9 H. M. GEYSEN, R. H. MELOEN, S. J. BARTELING, *Proc. Natl. Acad. Sci. USA* **1984**, *81*, 3998–4002.

10 R. A. HOUGHTEN, *Proc. Natl. Acad. Sci. USA* **1985**, *82*, 5131–5135.

11 a) S. P. A. FODOR, R. J. LEIGHTON, M. C. PIRRUNG, L. STRYER, A. T. LU, D. SOLAS, *Science* **1991**, *251*, 767–773; b) C. Y. CHO, E. J. MORAN, S. R. CHERRY, J. C. STEPHANS, S. P. A. FODOR, C. L. ADAMS, A. SUNDARAM, J. W. JACOBS, P. G. SCHULTZ, *Science* **1993**, *261*, 1303–1305.

12 J. K. SCOTT, G. P. SMITH, *Science* **1990**, *249*, 386–390.

13 B. A. BUNIN, J. A. ELLMAN, *J. Am. Chem. Soc.* **1992**, *114*, 10997–10998.

14 S. H. DEWITT, J. S. KIELY, C. J. STANKOVIC, M. C. SCHROEDER, D. M. R. CODY, M. R. PAVIA, *Proc. Natl. Acad. Sci. USA* **1993**, *90*, 6909–6913.

15 R. N. ZUCKERMANN, J. M. KERR, M. A. SIANI, S. C. BANVILLE, *Int. J. Pept. Protein Res.* **1992**, *40*, 497–506.

16 Á. FURKA, F. SEBESTYÉN, M. ASGEDOM, G. DIBÓ, Highlights of Modern Biochemistry, Proceedings of the 14th International Congress of Biochemistry, Prague, Czechoslovakia, 1988, VSP, Ultrecht, The Netherlands, **1988**, *13*, 47; b) Á. FURKA, F. SEBESTYÉN, M. ASGEDOM, G. DIBÓ, *Int. J. Peptide Prot. Res.* **1991**, *37*, 487–493.

17 K. S. LAM, S. E. SALMON, E. M. HERSH, V. J. HRUBY, W. M. KAZMIERSKI, R. J. KNAPP, *Nature* **1991**, *354*, 82–84.

18 R. A. HOUGHTEN, C. PINILLA, S. E. BLONDELLE, J. R. APPEL, C. T. DOOLEY, J. H. CUERVO, *Nature* **1991**, *354*, 84–86.

19 D. J. GRAVERT, K. D. JANDA, *Chem. Rev.* **1997**, *97*, 489–510.

20 A. STUDER, S. HADIDA, R. FERRITTO, S.-Y. KIM, P. JEGER, P. WIPF, D. CURRAN, *Science* **1997**, *275*, 823–826.

21 a) P. H. H. HERMKENS, H. C. J. OTTENHEIJM, D. REES, *Tetrahedron*

1996, *52*, 4527–4554; b) P. H. H. HERMKENS, H. C. J. OTTENHEIJM, D. REES, *Tetrahedron* **1997**, *53*, 5643–5678.

22 S. V. LEY, *J. Chem. Soc. Perkin Trans 1*, **2000**, 1235.

23 For a review of linkers, see: I. W. JAMES, *Tetrahedron* **1999**, *55*, 4855–4946

24 For a recent review, see: S. BHATTACHARYYA, *Comb. Chem. High Throughput Screening* **2000**, *3*, 65–92.

25 H. N. WELLER, M. G. YOUNG, S. J. MICHALCZYK, G. H. REITNAUER, R. S. COOLEY, P. C. RAHN, D. J. LOYD, D. FIORE, S. J. FISHMAN, *Mol Diversity* **1997**, *3*, 61–70

26 L. ZENG, L. BURTON, K. YUNG, B. SHUSHAN, D. B. KASSEL, *J. Chromat. A* **1998**, *794*, 3–13.

27 For examples, see: a) S. FENG, J. K. CHEN, H. YU, J. A. SIMON, S. L. SCHREIBER, *Science* **1994**, *266*, 1241–1247; b) K. C. NICOLAOU, N. WINSSINGER, J. PASTOR, S. NINKOVIC, F. SARABIA, Y. HE, D. VOURLOUMIS, Z. YANG, T. LI, P. GIANNAKAKOU, E. HAMEL, *Nature* **1997**, *387*, 268–272; c) N. S. GRAY, L. WODICKA, A.-M. W. H. THUNNISSEN, T. C. NORMAN, S. KWON, F. H. ESPINOZA, D. O. MORGAN, G. BARNES, S. LECLERC, L. MEIJER, S.-H. KIM, D. J. LOCKHART, P. G. SCHULTZ, *Science* **1998**, *281*, 533–538.

28 a) B. JANDELEIT, D. J. SCHAEFER, T. S. POWERS, H. W. TURNER, W. H. WEINBERG, *Angew. Chem. Int. Ed.* **1998**, *38*, 2494–2532; b) P. G. SCHULTZ, X.-D. XIANG, *Curr. Opin. Solid State Mater. Sci.* **1998**, *3*, 153–158.

29 For recent reviews, see: a) K. D. SHIMIZU, M. L. SNAPPER, A. H. HOVEYDA, *Chem. Eur. J.* **1998**, *4*, 1885–1889; b) T. BEIN, *Angew. Chem. Int. Ed.* **1999**, *38*, 323–326.

30 A. R. FISBEE, M. H. NANTZ, G. W. KRAMER, P. L. FUCHS, *J. Am. Chem. Soc.* **1984**, *106*, 7143–7145.

31 E. REDDINGTON, A. SAPIENZA, *Science* **1998**, *280*, 1735.

32 X.-D. XIANG, X.-D. SUN, G. BRICENO, Y. LOU, K. A. WANG, H. CHANG, W. G. WALLACE-FREEDMAN, S.-W. CHEN, P. G. SCHULTZ, *Science* **1995**, *268*, 1738–1740.

33 H. CHANG, C. GAO, I. TAKEUCHI, Y. YOO, J. WANG, P. G. SCHULTZ, X.-D. XIANG, R. P. SHARMA, M. DOWNES, T. VENKATESAN, *Appl. Phys. Lett.* **1998**, *72*, 2185–2187.

34 E. DANIELSON, J. H. GOLDEN, E. W. MCFARLAND, C. M. REAVES, W. H. WEINBERG, X. D. WU, *Nature* **1997**, *389*, 944–948.

35 K. SADA, K. YOSHIKAWA, M. MIYATA, *Chem. Commun.* **1998**, 1763–1764.

36 P. MURER, K. LEWANDOSKI, F. SVEC, J. M. J. FRÉCHET, *Chem. Commun.* **1998**, 2559–2560.

37 W. P. C. STEMMER, *Nature* **1994**, *370*, 389–391.

38 G. C. LIU, J. A. ELLMAN, *J. Org. Chem.* **1995**, *60*, 7712–7713.

39 K. BURGESS, H.-J. LIM, A. M. PORTE, G. A. SULIKOWSKI, *Angew. Chem. Int. Ed. Engl.* **1996**, *35*, 220–222.

40 B. M. COLE, K. D. SHIMIZU, C. A. KRUEGER, J. P. A. HARRITY, M. L. SNAPPER, A. H. HOVEYDA, *Angew. Chem. Int. Ed. Engl.* **1996**, *35*, 1668–1671.

41 M. S. SIGMAN, E. N. JACOBSEN, *J. Am. Chem. Soc.* **1998**, *120*, 4901–4902.

2

Introduction to Combinatorial Chemistry

David L. Coffen and Joachim E. A. Luithle

2.1
Combinatorial Chemistry in Drug Discovery – a Perspective

The categorical imperatives of modern drug discovery are to produce better clinical candidates that are less prone to failure at a late-stage, and to do this more rapidly than the industry performance standards of the past two decades would predict and at a cost that is responsive to social and political pressures on drug prices. Combinatorial chemistry is nothing less than a cornerstone technology in the realization of these imperatives. Its function is basically that of a high-output engine, providing very large numbers of well-designed, well-made, high-quality compounds for high-throughput evaluation as potential drug candidates. In fact, this ability to leverage productivity has prompted researchers in other fields, such as catalysis and material science, to adopt and adapt this engine to their needs as well.

The purpose of this chapter is to provide an overview of the basic principles of combinatorial chemistry and to outline the operating principles associated with its most widely practiced forms. Subsequent chapters provide more detailed accounts of how the various types of synthetic chemistry are utilized, of specific combinatorial chemistry technologies, and of specific applications.

By way of background one can compare how a typical medicinal chemist of 1975 would set about making a compound for testing with the way that the same chemist would approach the task today. First of all, each decision about what compound to make was typically embedded in the program category in which the chemist was a participant. The program description frequently included a structural component, such as 'muscarinic cholinergics', 'cephalosporin antibiotics', 'benzodiazepine anxiolytics', or 'tricyclics', and the chemist would have had considerable expertise in the synthetic aspects, structure–activity relationships, the literature and patents, and, to some extent at least, the pharmacology of the particular compound class. Each new hypothesis was based on a blend of past experience, recent papers and/or patents in the field, and on the latest biological testing results, and was then refracted through knowledge of how to make the compounds of interest and what was available to make them from. If the synthesis worked, a

1–5-g sample (huge by today's standards) would be made and rigorously purified by column chromatography or recrystallization. Meticulous analyses using, as a minimum, thin-layer chromatography (TLC), mass spectrometry (MS), infrared (IR), nuclear magnetic resonance (NMR), ultraviolet (UV), and combustion analysis determined if the sample was suitable for registration and biological testing. A compound data sheet that included analytical results, melting point, solubility data, and other information would typically be prepared, reviewed, approved, and filed before a sample could be sent for testing. Using this process, a chemist with one or two assistants was deemed productive if 50–200 new compounds a year were forthcoming, depending on the complexity of the structural class.

In contrast, the drug designer of year 2000 is most likely a hybrid medicinal, combinatorial, computational, and analytical chemist – with a working knowledge of molecular and cellular biology. He or she spends little time in the library, lots of time at a computer, and much of the remaining time meeting with the project team to review results and plan activities. Also, in comparison with the handful of laboratory appliances considered essential 25 years ago, today's chemistry laboratory is replete with complex (and expensive) productivity-enhancing equipment, making it possible for the 200 compounds of yesteryear to be made in an automated synthesizer during the hour spent in a meeting.

The decision as to which compound to make next has been replaced by a design process regarding which set (or library) of compounds to make next. Routine synthesis of single compounds now occurs mainly in the later stages of lead optimization. The design process flows not from a compound class associated with a particular drug discovery program, but from the biomolecular targets associated with the program. This fact underlies the critical relationship between modern medicinal chemistry and structural biology/computational chemistry. As an illustration, a 1970s' program on influenza drugs might be centered on aminoadamantane derivatives (antiviral compound *class*), whereas a modern program might be focused on inhibitors of neuraminidase (viral protein *target*). The discovery phase of the program is driven by the target and by knowledge of its structure and natural ligand in the case of a receptor or by its mechanism of catalysis and natural substrate in the case of an enzyme.

Overall, it can be stated that the integration of biomolecular target-driven computational design methods, combinatorial techniques for compound production, and high-throughput biological screening technologies has, in principle, resulted in a huge increase in productivity in medicinal chemistry. The role of combinatorial chemistry in modern pharmaceutical research and development (R&D) is perhaps best shown in Figs. 2.1 and 2.2, which depict as triads the major components of both the overall R&D process and its discovery phase.

2.2
Key Issues

The basic principles of combinatorial chemistry are often tied to, and sometimes confused with, the origins of combinatorial chemistry. The founding (but flawed)

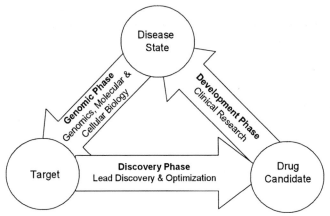

Fig. 2.1. The essential elements of modern pharmaceutical research and development.

principle was that, given a sufficiently large and diverse set of compounds to test, the discovery of an ideal drug for any given disease state would be statistically unavoidable. This principle was frequently popularized in metaphorical terms, such as the likelihood that any lock could be opened if a sufficiently large number of keys are tried, or that the pharmacological richness of natural products could be easily accessed and expanded through the laboratory equivalent of a rain forest. The anticipated surge in overall pharmaceutical R&D productivity has not materialized and this optimistic view of combinatorial chemistry has largely been abandoned. The contemporary view is more pragmatic and generally conforms to the principles outlined below.

1 *The synthesis of many compounds simultaneously is more efficient than the synthesis of a single compound.* In terms of simple time utilization, this principle was

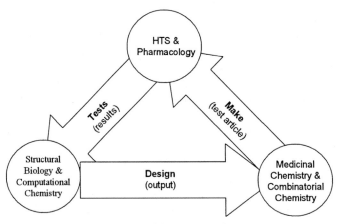

Fig. 2.2. The essential elements of the discovery phase.

intuitive to generations of chemists who set up two or more reactions in a hood each night. In the combinatorial context it has far greater significance as the number of compounds made from combinations of reagent sets increases geometrically with the number of reagents in these sets.

2 *Any synthesis scheme can be executed in a manner that affords multiple products if individual reagents are replaced by complementary reagent sets used in different combinations.* This perception is eponymous with combinatorial chemistry and lies at its heart. It can be readily illustrated with a Diels–Alder reaction (Fig. 2.3).

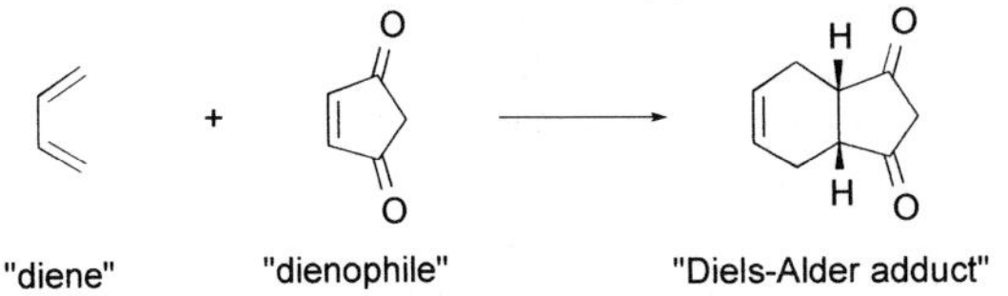

"diene"   "dienophile"   "Diels-Alder adduct"

Fig. 2.3. Combinatorial chemistry illustrated with a Diels–Alder reaction.

A chemist could easily execute this synthesis in a single workday. However within this same day, that chemist could weigh and prepare stock solutions of ten dienes and ten dienophiles. Then by making equimolar combinations of each diene with each of the ten dienophiles and running the corresponding Diels–Alder reactions, a total of 100 Diels–Alder adducts could be synthesized, thus achieving a 100-fold productivity gain. When a synthesis scheme provides for three reagent sets (incorporated into the products), the use of ten members in each set affords 1000 products. Similarly the use of four sets of ten reagents distributed over one (e.g. Ugi reaction) or several stages of a synthesis affords 10,000 possible combinations. The number of products increases geometrically ($A \times B \times C$) whereas the number of reagents increases arithmetically ($A + B + C$) and in this simple reality lies the true power of combinatorial chemistry.

3 *Rigorous design of products and synthesis is critical and large numbers cannot compensate for poor design.* Combinatorial chemistry, often viewed and applied solely as a tool to leverage serendipity, frequently fails to meet expectations. However when used to leverage the output of sound experimental designs based on synthetic and medicinal chemistry knowledge, the outcome is generally more satisfactory. There is probably a more fundamental principle behind this empirically derived view and it relates to the fact that the pharmacological effects of bioactive small molecules are derived from their ability to bind to and modulate the function of macromolecular targets, usually proteins. Proteins are built from a bounded set of amino acids. Their biological and biochemical functions have structural determinants derived from a bounded set of secondary structure motifs and these are usually associated with small molecule binding. Thus the number and types of small molecules that are likely to exhibit biological activity also represent bounded sets and combinatorial chemistry applied within this domain is far more likely to succeed than a totally random approach.

4 *While synthetic organic chemistry evolved primarily in terms of serial processing, its productivity can be greatly enhanced by introducing parallel processing.* Strong emphasis needs to be put on the importance of parallel processing as an explicit component of experimental design. The synthesis plan must comprehend the fact that hundreds or thousands of compounds are being synthesized simultaneously using hundreds or thousands of reagent combinations in each synthetic step. Each unit operation and each operating parameter – such as solvent used, temperature, and level of agitation – should encompass the conditions needed to bring each reagent combination to complete reaction while minimizing side-reactions and product decomposition. Sample tracking and in-process control data must also be managed at very high volume.

5 *Laboratory automation, robotics, and mechanical devices enabling the simultaneous performance of multiple tasks are essential to combinatorial chemistry.* The range of enabling tools goes from simple devices such as multichannel hand pipettes to high-performance, fully integrated systems such as the Irori NanoKan system. This issue will be revisited, but the basic principle is that the successful practice of combinatorial chemistry requires some level of commitment to and investment in laboratory automation.

6 *Robust process chemistry is required to assure the desired outcome.* In the pharmaceutical world, the importance of process chemistry lies at the two extremes. For single compound synthesis in a medicinal chemistry laboratory, a good process may be on the "nice to have" list, but as long as the synthesis scheme affords some of the desired compounds, intermediate and endstage purifications will overcome deficiencies in the process. For bulk pharmaceutical production, the process is everything: it must be extremely efficient, fully validated, and conform to regulated manufacturing practices. The importance of process chemistry in combinatorial synthesis is closer to the bulk production context.

   For large compound libraries, the opportunity for product purification is very limited (although technologies for doing this are evolving nicely). Library members with poor-quality or outright failures are highly undesirable as they waste time and resources in biological testing laboratories. The best resolution of this dilemma is to develop the synthesis scheme to be used for production of a library into a robust, well-defined process. This requires careful optimization of reaction conditions, validating the individual members of the reagent sets, developing reliable analytical methods, and defining the process in something equivalent to a 'standard operating procedure'.

7 *Electronic tracking and control systems are critical components of combinatorial chemistry.* Every chemist has been exposed to the tedium of labeling samples by hand, filling out analytical request forms, and completing compound data sheets. Preserving the identity of samples and information about them by linkage to laboratory notebook pages is a cumbersome system that works when a few hundred samples per year are involved but it cannot function when individual experiments are producing thousands of individual compounds. The data management aspect of combinatorial chemistry requires access to electronic tracking and control systems for sorting and labeling samples (e.g. with

machine-readable barcodes), for generating structure lists (SD files) that link sample codes to compound structures, and for collecting and processing analytical data and linking these data to sample codes and structure files. In addition to these sample and data management applications, electronic control systems embedded in the operating software of automated synthesis systems are necessary to ensure that each in-process material is in the right place at the right time for each step in order to ensure that the library plan is faithfully executed.

8 *Analysis and quality control procedures are just as important in combinatorial chemistry as in other forms of synthetic chemistry.* The shortcomings of early expressions of the combinatorial concept included the misperception that quality was not important. In fact many libraries were deliberately prepared as compound mixtures. It was taken for granted that biologically active samples could be separated, deconvoluted, or in some way dealt with after hits were detected. Two things resulted. The fact that synthesis without analysis is a prescription for poor science was reaffirmed and confidence in combinatorial chemistry as a productive discovery tool developed rather slowly. At its current state of development and acceptance, combinatorial chemistry is expected to provide samples of individual compounds with a purity level [by high-performance liquid chromatography (HPLC)] of at least 80%. The identity of compounds is routinely verified by mass spectrometry, using electronic comparison of probable molecular ion peaks with calculated molecular weights. Combined liquid chromatography/mass spectrometry (LC/MS) is preferable as it will confirm or deny that the major peak in an HPLC trace corresponds to the design intent for that particular library member. An aspect of quality control which needs further development is the application of in-process control procedures (IPCs) which are basic and routine in multistep synthesis of single compounds.

9 *Since the purpose of combinatorial chemistry is to facilitate the discovery of useful compounds, combinatorial syntheses must be reproducible and scalable.* This is not to say that the reaction sequence and mode of synthesis (solid or solution phase) must eventually serve to provide multikilogram supplies, but the compounds in a library have little value if interesting compounds cannot be conveniently resynthesized in small amounts for closer examination.

10 *As a final basic principle it should be understood that combinatorial chemistry is a productivity-enhancing tool* for chemists engaged in pharmaceutical research, agrochemicals, catalysis, and materials science – any field where the preparation and testing of new compositions of matter are the essential elements of discovery. However it does not displace or supplant established fields of synthetic chemistry such as medicinal chemistry – its best use is to leverage the productivity of existing fields.

2.3
Combinatorial Synthesis

As mentioned before libraries of mixtures and of single compounds are accessible by combinatorial chemistry. Furthermore the syntheses can either be performed in solution or on solid support. In the following, both methods will be discussed in detail and the underlying principles will be described. To be able to compare the advantages and disadvantages of these technologies the following aspects will be emphasized: reaction conditions, equipment, and possible degree of automation.

2.3.1
Solid-phase Combinatorial Synthesis

Solid-phase combinatorial chemistry relies on the fact that the molecule under construction is attached via a linker to a polymeric carrier (bead) (Fig. 2.4). This immobilization allows a simple separation of intermediates and finally of the product from reagents and soluble byproducts. To guarantee the efficacy of this principle, the bead and linker have to be stable under the reaction conditions. Furthermore, the bond between the target product and the linker has to be cleaved selectively under mild conditions without destroying the product. Thus, after simple filtration, nothing but the liberated pure product AB is obtained.

Fig. 2.4. Solid-phase chemistry. After reaction the product is liberated and filtered off.

2.3.1.1 Reagents and Conditions
A key feature of solid-phase synthesis is the possibility of driving reactions to completion with excess reagent. In normal solution-phase chemistry such forcing conditions would have devastating effects on product quality, whereas in solid-phase chemistry the excess reagents are simply removed by filtration. Another important effect is the pseudo dilution caused by spatial separation of reactive sites. Difunctional compounds such as diamines or diacids react selectively in a monofunctional manner with only one of the two groups resulting in a highly effective desymmetrization. The spatial separation of the functional groups also renders the macrocyclization reaction a comparatively effective processe. Solution-phase chemistry, on the other hand, often affords statistical mixtures of products under these conditions.

The positive features mentioned before are opposed by some drawbacks: adapting a standard solution-phase reaction to solid phase frequently entails problems associated with finding suitable, robust and versatile linkers – points of attachment in the starting material. These inherent features limit the choice of possible reaction conditions. Furthermore, the selection of usable solvents can be quite restricted. While crosslinked polystyrene resin beads may be insoluble, polyethylene or polypropylene used in containers, pins, etc. will deform or dissolve in many solvents at elevated temperatures. Another restriction stems from the swelling/shrinking characteristics of crosslinked polystyrene beads. Dichloromethane, tetrahydrofuran (THF), dimethylformamide (DMF), toluene, and (marginally) dimethyl sulfoxide (DMSO) are all suitable in this respect, but ether, methanol, ethyl acetate, acetonitrile, and water are not. Tentagel resins are compatible with more solvents but introduce several more restrictions. Also the temperature that can be applied is restricted: $+100$ °C is generally at or beyond the upper limit for most solid-phase systems – room temperature to 80 °C being the 'comfort zone'. It should be noted, however, that very low temperature operations involving, for example, reactive enolates generated at -78 °C are more easily handled in a directed split-and-pool solid-phase mode (fewer batches). Chemistry involving heterogeneous catalysts such as palladium on charcoal or solid reagents such as manganese dioxide are intrinsically incompatible with solid-phase synthesis methods. Furthermore, the solid-phase chemistry literature is still limited, and only a relatively small proportion of the synthetic solution-phase repertoire has been adapted to solid phase. Therefore the development of a suitable methodology for the synthesis of large libraries can be time-consuming (several months). Another drawback is the difficulty to analyze the outcome of a given reaction. On-bead analysis is still not satisfying, although NMR and IR as well as MS techniques have been improved. In most cases, reaction products have to be cleaved from the support and are analyzed by normal methods (HPLC, NMR).

2.3.1.2 **Automation**
As pointed out before, the filtration of the reaction mixture, e.g. the separation of solid-supported products from starting materials and reagents, is the key process in solid-phase chemistry. Because of its simplicity, this process can easily be automated and run in parallel. Thus, the first robotic synthesizers were those for solid-phase chemistry: many reactions were set up in parallel by automated dispensing stations and the normally very tedious work-up procedure was reduced to simple filtrations. This worked fine as long as the library size was not too large (<1000 compounds). Larger libraries are difficult to prepare by this process. A much more efficient way to prepare large libraries is the so-called split-and-combine protocol.

2.3.1.3 **Split and Combine**
As depicted in Fig. 2.5, a polymer support is reacted with three different substances – A, B, and C – in separate reaction vessels. Subsequently, the beads are combined and again split into three portions. Each of them is submitted to a reaction with one of the substances D, E, or F. This process can be repeated, and with each iteration the number of compounds produced is multiplied by a factor of 3. It has to be emphasized that every bead bears only one compound. This is especially true because reactions can be driven to completeness by excess reagent, which is finally washed away together with impurities. After a sequence of two iterations nine compounds, and after three iterations 27 compounds, are produced. However, to prepare single compounds and not mixtures, several encoding technologies had to be developed.

Structure elucidation is possible employing chemical encoding technologies. In addition to the reagent, a small quantity (e.g. 1%) of a chemical tag is reacted with the bead, which makes the tracking of the history of a bead, and thus the identification of the product, unambiguously possible. The tagging process should neither interfere with the reaction nor consume too much of the bead capacity. After the

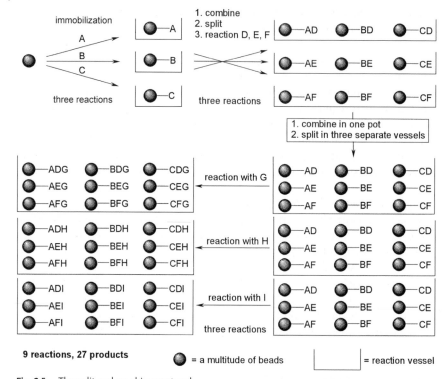

Fig. 2.5. The split-and-combine protocol.

reaction the product is selectively cleaved from the bead. The identity of the product can subsequently be determined by conventional analytical methods analyzing the tagged bead. W. Clark Still invented in 1993 the first binary encoding using haloaromatic reagents for tagging (see Chapter 5). Several more possibilities, such as tagging with oligonucleotides or peptides, can be used for the chemical encoding of single beads.

When encapsulated in porous containers, aliquots of beads traverse the split-and-pool protocol with the same outcome as a single bead – all the beads in each container carry the same compound at every stage. Today radiofrequency tagging is the most prominent encoding technology. This method is based on a small radio transponder, which is attached to the synthetic platform (e.g. a polypropylene reactor). Upon activation this device emits a given radiofrequency, thus making the identification possible. Development of chemically passive tagging systems for solid-phase synthesis vastly enhanced the utility of the split-and-pool protocol. The enormous gain-in-function offered by split and pool can now be used to produce huge libraries of 'pure', single compounds in multimilligram quantities with minimal synthetic effort. For example, the production of a 160,000-compound library can be achieved in just 80 discrete synthetic operations if the library design embraces four sets of building blocks with 20 members in each set. Automated sorting and washing stations have been built, but handling became so easy that small libraries can be prepared with minimal automation.

2.3.1.4 Cost

Modest capital investments are required for tag-reading/sorting devices, apparatus for cleavage and delivery of compounds to plates, and centrifugal evaporation. The instrumentation requirements for product quality control (QC) are, of course, the same as in solution-phase chemistry. Solid-phase methods foster profligate use of reagents to drive reactions to completion, as mentioned before. Added to these considerations is the fact that speciality resins, 'hi-tech' linkers, and single-use resin containers can add substantial material costs to library production.

2.3.1.5 The Products from Solid-phase Chemistry

Each compound in a library produced by solid-phase methods will generally have a structural feature, usually a functional group, associated with its covalent linkage to a resin bead. This feature becomes an obligatory part of every compound in the library – they will all be amines, phenols, acids, etc. depending on the type of linker used. Such features are not always desirable, and considerable effort is being invested in the development of 'traceless' linkers – often silicon based (see Chapter 4). Another approach utilizes intramolecular displacement cleavage strategies in which a (desirable) new ring is formed at the cleavage stage.

2.3.2
Solution-phase Combinatorial Synthesis

When combinatorial chemistry first emerged solution-phase reactions were not considered to be suitable for preparation of larger libraries (> 1000 compounds). Only a few reactions lead to quantitative conversion using equimolar amounts of reagents. Solution-phase methods encourage parsimony in the use of reagents as any excess can become a product impurity. The use of excess reagents in solution phase can only be contemplated when facile removal of the excess is possible. However, purification and isolation procedures were tremendously improved in the last year, allowing the purification of hundreds of compounds per day by preparative HPLC methods. Furthermore, more recently, a number of technologies have emerged making combinatorial chemistry in solution phase competitive to solid-phase reactions.

2.3.2.1 Reagents and Conditions

In principle, any chemistry and any reaction can be employed in solution-phase methods, including complex organometallic reagents, biocatalysis, etc. Few, if any, of the reactions and procedures documented in compendia of 'organic syntheses' and 'organic reactions' could not be adapted to solution-phase combinatorial chemistry if suitable time and effort were applied. Adapting viable reactions to solution-phase combinatorial chemistry entails a set of problems which can be routinely solved in a month or two. The problems are fairly standard and address aspects such as control of stoichiometry, capturing a broad range of reactivity within a single set of reaction conditions, and product quality assurance.

Complex molecule synthesis relies heavily on protecting groups. However their use in solution-phase library synthesis is restricted to situations where their removal gives volatile byproducts, e.g. Boc or Cbz groups. Certain synthetic transformations that produce nonvolatile coproducts are also a problem in solution. For example, Mitsunobu reactions, which produce phosphine oxide and hydrazide coproducts, work very well in solid phase but require major adaptation of procedures for use in solution. Multistep synthesis conducted in normal solution-phase parallel synthesis leads to rapid deterioration of product quality because of incomplete reactions and accumulating byproducts.

Compatibility with automated solution-handling devices is the only solvent and reagent restriction for solution-phase methods. Problems that may exist, such as loss of accuracy in dispensing small volumes of highly volatile solutions in ether or dichloromethane, can usually be overcome with less volatile alternatives such as dioxane or tetrachloroethylene.

Solution-phase systems equipped to prevent condensation of water vapor (or icing up) at the low end, or equipped to condense/reflux solvent vapor at the high end, can easily operate in the −20 °C to +150 °C range.

2.3.2.2 Scavenger Resins, Polymer-supported Reagents and Fluorous Tags

Solution-phase combinatorial chemistry suffers from one main disadvantage. To drive reactions to completion, more than one equivalent of reagent is frequently necessary, but the use of excess reagents is often prohibitive because their removal causes in most cases severe problems. Two new techniques were established to overcome these difficulties: polymer-supported reagents and scavenger resins. The latter are functional group-specific, reactive resins, e.g. polystyrene-bound isocyanates which react selectively with primary and secondary amines. Thus excess reagents are first used to drive reactions to completion. Subsequently, scavenger resins are added to the reaction mixture. After selective coupling of the resin to the excess starting material the insoluble material is removed by filtration and the product remains in pure form (Fig. 2.6).

Also the second approach to simplify solution-phase chemistry relies on the separation of insoluble material from the reaction mixture and combines the benefits of solid- and solution-phase chemistry (Fig. 2.7): instead of immobilizing the starting material on the support, only the employed reagents are polymer supported. Again, the work-up and purification are reduced to a simple filtration.

Furthermore, different polymer-supported reagents do not interfere with each other, making one-pot reactions possible that would not work in classic solution

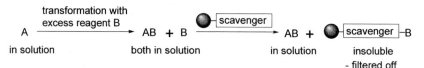

Fig. 2.6. Excess reagents can easily be removed using scavengers.

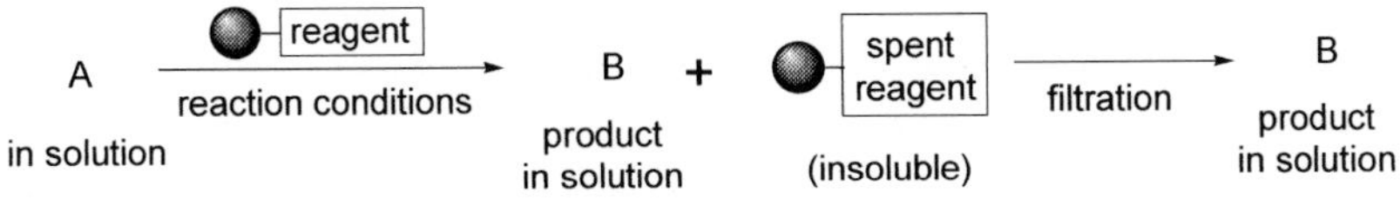

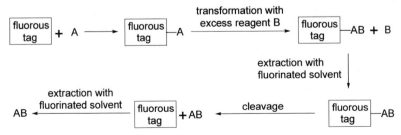

Fig. 2.7. The principle of polymer-supported reagents.

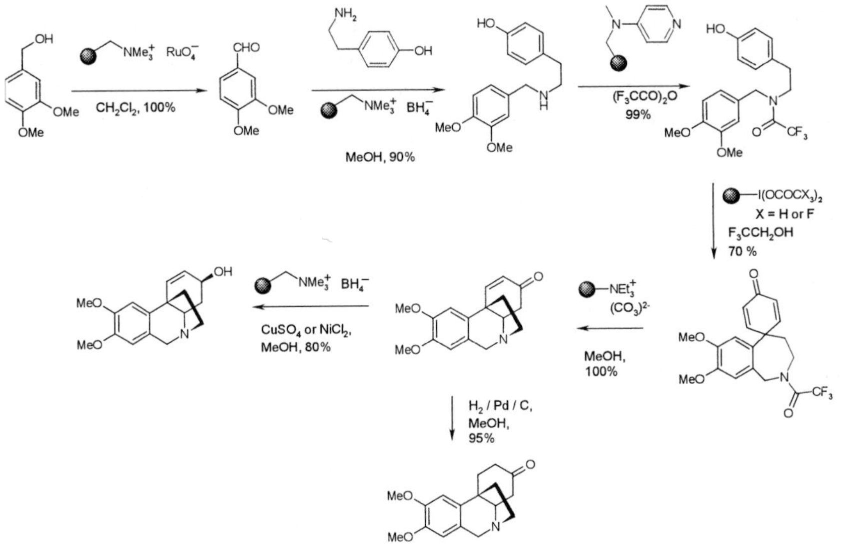

Fig. 2.8. The total synthesis of (±)-oxomaritidine and (±)-epimaritidine.

chemistry (Fig. 2.8). The versatility of this approach has been proved by Ley et al. in the total synthesis of two natural products, (±)-oxomaritidine and (±)-epimaritidine.

Another possibility to deal with the problem of excess reagents in solution-phase combinatorial chemistry is the use of fluorous tagging and extraction with fluorinated solvents (Fig. 2.9). Substrate A is attached to a moiety which is polyfluorinated. Because of the fluorous tag the product of the reaction is exclusively soluble in the fluorous solvents. After reaction with excess quantities of reagent B an extraction with a polyfluorinated solvent is performed. Reagent B is more solu-

Fig. 2.9. Fluorous tagging – a powerful tool for solution-phase chemistry.

ble in the organic phase and can be separated from the product. Finally the fluorous tag is cleaved off, removed by another extraction and the pure product can be isolated.

2.3.2.3 Equipment and Costs

Automation in solution-phase chemistry is more difficult than in solid-phase chemistry. Especially the purification of products requires large systems for preparatory HPLC. For *highly automated* solution-phase parallel synthesis on the scale of >100,000 compounds per year, an investment of several million dollars may be required to build and equip a suitable facility.

2.4
Conclusion

Solid-phase chemistry has made tremendous progress during recent years, and today highly complex natural products can be prepared on solid supports (see Chapter 3). Nevertheless, method development is slow and often fails to produce the compounds needed, e.g. in medicinal chemistry optimization programs. Thus, careful validation of the possible outcome of solid-phase chemistry has to be given the highest priority before the development of methodology is started. However, for the preparation of large libraries with only minor time constraints, solid-phase chemistry remains the method of choice. On the other hand, solution-phase chemistry is much more flexible and quicker. However, the production of large libraries is very tedious owing to the difficult purification of the products.

Polymer-supported reagents and scavengers represent a versatile addition to solid-phase organic synthesis and parallel solution-phase chemistry. The combination of these reagents offers exciting possibilities. The methods described in this chapter are interesting and fascinating but, so far, they do not enable the chemist to exploit all the synthetic routes that he or she might think of. Thus, there is a need for more solid-supported reagents, linkers, scavengers, and, especially, solid-phase methodology.

Bibliography

1 J. N. ABELSON, *Combinatorial Chemistry*, Academic Press, San Diego, **1996**.

2 G. JUNG, *Combinatorial Peptide and Nonpeptide Libraries: A Handbook*, VCH, Weinheim, **1996**.

3 I. M. CHAIKEN, K. D. JANADA, *Molecular Diversity and Combinatorial Chemistry: Libraries and Drug Discovery*, ACS, Washington, DC, **1996**.

4 A. W. CZARNIK, S. H. DEWITT, *A Practical Guide to Combinatorial Chemistry*, American Chemical Society, Washington, DC, **1997**.

5 G. B. FIELDS, *Solid-Phase Peptide Synthesis*, Academic Press, San Diego, **1997**.

6 B. A. BUNIN, *The Combinatorial Index*, Academic Press, San Diego, **1998**.

7 E. M. GORDON, J. F. KERWIN, *Combinatorial Chemistry and Molecular*

Diversity in Drug Discovery, John Wiley & Sons, New York, **1998**.

8 K. BURGESS, *Solid-Phase Organic Synthesis*, John Wiley & Sons, New York, **2000**.

9 D. OBRECHT, *Solid-Supported Combinatorial and Parallel Synthesis of Small-Molecular-Weight Compound*, Pergamon Press, Oxford, **1998**.

10 W. BANNWARTH, E. FELDER, *Combinatorial Chemistry. A Practical Approach*, Wiley-VCH, Weinheim, New York, **2000**.

11 M. R. PAVIA, T. K. SAWYER, W. H. MOOS, *Bioorg. Med. Chem. Lett.* **1993**, *3*, 387–396.

12 W. H. MOOS, G. D. GREEN, M. R. PAVIA, *Annu. Rep. Med. Chem.* **1993**, *28*, 315–324.

13 K. S. LAM, M. LEBL, *Methods: A Companion to Methods in Enzymology* **1994**, *6*, 372–380.

14 K. D. JANDA, *Proc. Natl. Acad. Sci. USA* **1994**, *91*, 10779–10785.

15 M. A. GALLOP, R. W. BARRETT, W. J. DOWER, S. P. A. FODOR, E. M.

GORDON, *J. Med. Chem.* **1994**, *37*, 1233–1251.

16 E. M. GORDON, R. W. BARRETT, W. J. DOWER, S. P. A. FODOR, M. A. GALLOP, *J. Med. Chem.* **1994**, *37*, 1385–1401.

17 G. LOWE, *Chem. Soc. Rev.* **1995**, *24*, 309–382.

18 Dedicated issue, *Account. Chem. Res.* **1996**, *29*, 111–170.

19 F. BALKENHOHL, C. VON DEM BUSSCHE-HÜNNEFELD, A. LANSKY, C. ZECHEL, *Angew. Chem. Int. Ed. Engl.* **1996**, *35*, 2288–2337.

20 P. H. H. HERMKENS, H. C. J. OTTENHEIJM, D. REES, *Tetrahedron* **1996**, *52*, 4527–4554.

21 R. E. DOLLE, *Mol. Diversity* **1997**, *2*, 223–236.

22 P. H. H. HERMKENS, H. C. J. OTTENHEIJM, D. REES, *Tetrahedron* **1997**, *53*, 5643–5678.

23 R. E. DOLLE, *Mol. Diversity* **1998**, *3*, 199–233.

24 D. P. CURRAN, *Angew. Chem. Int. Ed.* **1998**, *37*, 1175–1196.

25 R. E. DOLLE, K. H. NELSON, JR, *J. Comb. Chem.* **1999**, *1*, 235–282.

Valuable Internet Links

http://www.5z.com/divinfo/
http://www.combinatorial.com/

3
Solid Phase and Soluble Polymers for Combinatorial Synthesis

Rainer Haag, André Hebel, and Jean-François Stumbé

3.1
Introduction

Polymeric supports have revolutionized organic synthesis in the past decade and have become a major driving force for laboratory automation and combinatorial chemistry in general. While polystyrene (PS)-based ion exchange resins have been known since the 1950s, the first solid-phase peptide synthesis on modified PS microbeads was reported by Merrifield in 1963 [1]. Until the late 1980s, when the first nonpeptidic molecule libraries were reported [2], polymeric supports were mainly used for special applications, such as peptide and oligonucleotide synthesis. In these cases the polymeric support had to be stable only toward two repetitive reaction conditions (coupling and deprotection). Therefore, some of these original supports are of limited use for parallel multistep synthesis of small organic molecules, generally known as combinatorial chemistry. In the last decade a rapidly increasing number of new polymeric supports, crosslinked (insoluble) [3–5] and noncrosslinked (soluble) polymers [6–10], have been described and used for combinatorial synthesis. Although solid-phase synthesis on PS-based resins exhibits a number of problems due to the heterogeneous nature of reactions and the low concentration of functional groups (typically ≤ 1.5 mmol substrate/g polymer) [11], no other polymeric support has yet found the same broad application for combinatorial synthesis [12]. However, the reader should be aware of the fact that there is no polymeric support for general application in organic synthesis. Every polymer has its drawbacks (e.g. chemical stability, polarity) and hence is stable only within a certain range of reaction conditions.

In this chapter, we will describe the structure and the properties of polymeric supports as well as the effects of different spacers (Fig. 3.1). Spacer molecules, compared with linker molecules, are used to provide more accessible linker functionalities and to modify the properties of the polymer backbone (e.g. polarity, swelling characteristics). Linker functionalities, which also influence the materials' properties, will be the subject of the next chapter. In the following sections the two major classes of polymeric supports – solid and soluble polymers – will be further classified by their chemical structures and the polymer topologies. A special focus

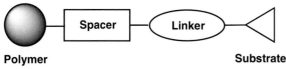

Polymer   **Substrate**

Fig. 3.1.   General structure of a polymeric support.

will be put on the polymer properties, such as stability, swelling property, reactive site distribution, loading capacity and the range of applications in organic synthesis.

3.2
Solid-phase Supports

3.2.1
Polystyrene-based Resins

Functionalized polystyrenes are available as linear noncrosslinked and as crosslinked polymers. The latter, generally referred to as PS resins, are among the most commonly used solid supports for solid-phase organic synthesis (SPOS), solid-phase-supported peptide (SPPS) synthesis and nucleotide synthesis. Nowadays, numerous types of polystyrene resins are commercially available and a nonexhaustive list of the most frequently used resins is given in Table 3.1. They are available in different sizes, with different loading capacities (amount of functionalization) and degrees of crosslinking.

   Macroporous and microporous polystyrene resins are typically prepared by suspension polymerization. The basis of this process is the dispersion of an organic phase (consisting of a monomer, a radical initiator, a crosslinking agent and potentially a comonomer) into an aqueous phase [13, 14]. The size of the initial droplets is adjusted by emulsifying the organic phase under stirring in the presence of a polymeric surfactant, which governs the final size and the final size distribution of the beads after polymerization. The different bead sizes are then separated by a multiple sieving process. This is why the size of the beads is usually given in mesh (number of sieve holes per inch); however, for most chemists the μm size of the final bead is more relevant for handling (Fig. 3.2). Resin beads used for combinatorial synthesis are spherical particles typically in the range 50–500 μm which can be easily handled (weighting, filtering and drying).

3.2.1.1 General Aspects

Crosslinkers   In order to obtain insoluble resins, given amounts of crosslinking agent have to be used for the synthesis of the beads by suspension polymerization (see also Sect. 3.2.1.2 and Sect. 3.2.1.3). A crosslinking agent is generally a bi- or multifunctional molecule that can be incorporated into two or more growing

Tab. 3.1. Features and applications of frequently used solid-phase resins.

Type	Common features	Representatives	Applications[a]	Loading [mmol g^{-1}]
Gel (microporous)	High capacity	1% or 2% crosslinked PS (Merrifield resin)	A, B, D, E	0.2–5.0
	Different solvent swelling			
	Reagent access by diffusion	PEGA, PAP	A, F	0.4 0.7
	Uniform sites	POEPOP SPOCC POEPS-3	A	0.4–0.7 0.4–1.2 0.2–0.3
	Many unsuitable for continuous flow synthesis	CLEAR	A, F	0.2–0.3
Hybrid	High capacity	ROMP	D	3.0
	Different solvent swelling (but less than microporous gels)	Rasta silanes	D	1.6–3.8
Encapsulated gel	Suitable for continuous flow synthesis	PolyHIPE (acrylamide resin encapsulated in PS "shell")	F	1.0–5.0
	Unstable to agitation	Pepsyn-K (similar resin encapsulated in inorganic scaffold)	A, F	–
Graft copolymers	Lower capacity	PEG-PS	D, E	1.2–2.9
	More uniform swelling	Tentagel®	A, B, D	0.25–0.6
	Pressure stable	Argogel®	A, B, D	0.4–0.5
	Suitable for continuous flow synthesis	NovaGel®	A, B, D	0.5–0.7
Rigid macrorporous supports	Capacity and efficiency depend on specific surface area, mean pore volume, pore distribution, and mean pore size	Highly crosslinked PS	E, F	<1.0
		Silica	A, E, F	0.006–0.06
		CPG	A, B, E, F	0.1

[a] Applications: A, peptide synthesis; B, oligonucleotide synthesis; C, oligosaccharide synthesis; D, synthesis of small organic molecules; E, polymer supported reagents; F, continuous flow synthesis.

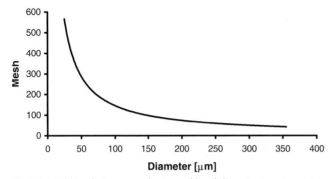

Fig. 3.2. Relation between mesh size and bead diameter in micrometers.

chains during the polymerization process, leading to interconnected chains. The most popular crosslinking agent, used in the presence of styrenic monomers, is divinylbenzene (DVB).

Crosslinked polystyrene resins can be further modified in many ways, such as Friedel–Crafts acylations [15, 16], electrophilic substitutions [17, 18], lithiations and so forth. The resulting beads can be used in further reactions to produce resins with various linker functionalities. However, for the synthesis of standard Merrifield resins, the functional group (e.g. chloromethyl styrene) is typically incorporated as a comonomer during the polymerization process.

Structure, morphology, and reactive sites distribution of crosslinked polymer beads
For chemists working with polymeric supports, it is of great importance to have a good knowledge of the internal structure and morphology of the microbeads because they strongly influence the physical properties and, as a consequence, the reactivity of the functional sites. For example, if active sites are located in highly crosslinked microdomains, as can be the case in macroporous resins, they will remain inaccessible for reactions and the effective loading will be lower than the theoretical one [11].

Even in microporous resins, one can expect some heterogeneity due to the different reactivities of the monomers and crosslinking mixtures involved in the suspension polymerization process [19, 20]. For example, if styrene is copolymerized with DVB and chloromethylstyrene, the reaction mixture consists of at least six monomer entities: styrene, chloromethylstyrene, *m*-DVB (**1**), *p*-DVB (**2**), and (*m*, *p*)-ethylvinylbenzenes (**3**).

For a better understanding of the polymer properties, several groups have studied the structure, morphology and reactive site distribution of Merrifield resin and its derivatives in more detail. The analytical techniques include autoradiography [21], confocal fluorescence spectroscopy (CFS) [22], confocal raman spectroscopy [23], and scanning secondary ion mass spectrometry [24]. For Merrifield resin, an inhomogeneous site distribution and reactivity have recently been discussed in the literature on the basis of whole-bead CFS [22]. However, the unexpected results described in this work were recently revised by further studies using confocal raman spectroscopy [23] and confocal fluorescence spectroscopy of thin slices in order to avoid fluorescence quenching [25]. It is now obvious that the distribution of active sites is essentially uniform throughout polystyrene resins, as shown already by Merrifield [21]. However, for large beads, diffusion processes can certainly play an important role and lead to an unequal reactivity of functional groups located closer to the core and the shell of the individual bead. Based on geometrical considerations, one should also be aware of the fact that, in microbeads with a diameter of 100 μm, 50% of the active sites are within the first 10 μm of the outer shell [23].

3.2.1.2 Macroporous Resins

Macroporous resins are generally highly (>5%) crosslinked polystyrene microbeads [12]. The term "macroporous" refers to their inner skeleton, which is made up of a permanent porous structure even in the dry state (see Scheme 3.1c). Historically, functionalized macroporous resins have mainly been used for ion exchange. Nowadays, many new applications, especially in the field of polymer-supported reagents [26, 27], have been developed (see also Section end).

Macroporous resins are prepared by suspension polymerization of monomers such as styrene, vinyl pyridine, acrylamide, or glycidyl methacrylate with a porogen agent (Scheme 3.1) [28, 29]. Thus, a mixture of monomer with potentially a comonomer and a crosslinking agent are copolymerized after dispersion in an aqueous medium in the presence of the porogen, which, remains within the beads during the polymerization and acts as a template for the formation of the permanent internal porous structure of the final resin. Porogen agents can be of different natures (e.g. solvents, noncrosslinked polymers). After completion of the polymerization the porogens are removed by processes dependent on their characteristics and a hard opaque bead with a rough surface remains. The opacity of the macroporous resins, compared with the glassy appearance of the microporous beads, is due to their heterogeneous structure, which is made up of highly crosslinked polymeric microdomains and pores that are devoid of polymer [20, 30].

Scheme 3.1. Synthesis and structure of macroporous resins: (a) polymer network forming; (b) porogen phase acts as pore template; (c) dry macroporous resin with large interconnected pores. (1) Porogen and network start to phase separate; (2) porogen phase removed to yield pores (hatched area, crosslinked polymer; dots, porogen phase).

As mentioned above the generation of pores can take place in two ways: noncrosslinked polymers as well as organic solvents [28, 31]. For instance, if linear

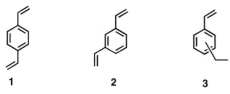

1   **2**   **3**

Fig. 3.3. Different chemical species present in the technical grade DVB mixture.

and noncrosslinked macromolecules are dissolved in the starting organic phase, they can be removed from the final polymerized beads by Soxhlet extraction. In this process, these linear soluble macromolecules act as templates by creating permanent pores after their removal [32, 33]. The size and the distribution of the pores can be controlled by the characteristics of the linear polymer (i.e. molecular weight, concentration) [29]. Another possibility is suspension polymerization in the presence of an organic solvent that is a poor solvent, or even a nonsolvent, of the final polymer. Phase separation that occurs within the beads during the polymerization process leads to the formation of globular entities that are free of polymer with characteristics that depend on the nature and on the concentration of the porogen [20]. The amount of crosslinking agent plays a major role in the formation of the macroporous structure and has to be carefully considered. For example, during polymerization conducted with relatively large amounts of crosslinkers (up to 20–25 mol% compared to the monomer), the formation of highly DVB-crosslinked microdomains ensures a rapid phase separation [20, 32]. The most commonly used crosslinking agents for the preparation of macroporous resins are divinylbenzenes (DVB) (**1**, **2**). The technical grade DVB typically used is a mixture of three chemical species: 27% *m*-DVB (**1**), 53% *p*-DVB (**2**), and 20% ethylvinylbenzenes (**3**) (Fig. 3.3).

Other frequently used crosslinkers for macroporous resins include ethyleneglycol dimethacrylate (EGDMA) (**4**) and methylene bisacrylamide (**6**) (Fig. 3.4). They are selected depending on the nature and reactivity of the monomer engaged in the polymerization.

Structure and physical properties of macroporous resins   As a consequence of their mode of preparation, macroporous microbeads consist, on one side, of a permanent macroporous internal structure and, on the other side, of highly crosslinked areas (Scheme 3.1c). The porous areas are made up of numerous interconnected cavities of different sizes leading to a large internal surface available for functionalization, whereas, the crosslinked areas provide the rigidity for such structures [20, 28]. This high internal surface area – typically ranging from 50 to 1000 $m^2\ g^{-1}$ (determined by N_2 BET-isotherm) – is accessible even in the dry state [20].

In general macroporous resins show very low swelling in organic solvents because of the very highly crosslinked areas. For this reason, macroporous beads remain unaffected by changes in the direct environment, even in the presence of good solvents. Another consequence is that the pores can accommodate a large variety of solvents, including polar solvents such as water and low-molecular-

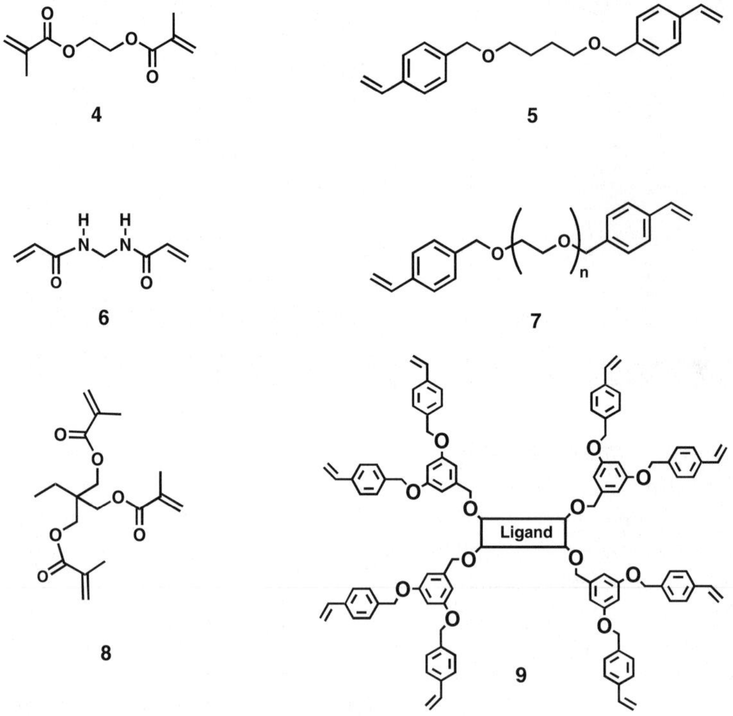

Fig. 3.4. Different crosslinking agents for PS beads.

weight alcohols. The pore size and the presence of channels interconnecting these cavities allow solvents to diffuse quickly in and out of these pores.

One drawback of these heterogeneous structures is the very low accessibility of the solvents and reagents to the highly crosslinked areas, leading to limited loading capacities with a typical range of 0.8–1.0 mmol g^{-1} [34, 35]. However, some commercial ion exchange resins, also used for organic reagents, have loading capacities up to 4.5 mmol g^{-1} [26]. Generally, macroporous resins display lower reactivities than microporous swollen beads. In contrast to microporous resins, they show high resistance toward osmotic shock (see Fig. 3.5b) [20], but they can be brittle when not manipulated carefully [13].

Applications of macroporous resins The most extensively used macroporous resins are polystyrene-based ionic exchange resins. They are made of poly(styrene-co-divinylbenzene) copolymers and are subsequently modified to arylsulfonic acids, quarternary ammonium salts or other derivatives mainly located on the internal surface of the pores [28, 36]. This renders them accessible to numerous organic solvents, including water and alcohols. Recently, these ion exchange resins have had a revival for the immobilization of ionic reagents in automated synthesis [26, 27]. Macroporous beads have also been used for the immobilization of cata-

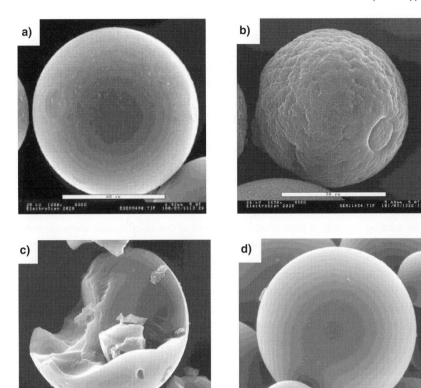

Fig. 3.5. ESEM pictures demonstrating the mechanical stability of 2% crosslinked Merrifield resins under different conditions: (a) after swelling in THF and drying; (b) after osmotic shock (swelling in THF, washing with MeOH and drying); (c) after magnetic stirring of a swollen resin (24 h, 300 rpm, 20 °C, DMF); (d) after mechanical shaking of a swollen bead (7 days, 60 rpm, 20 °C, DMF). The ESEM pictures were taken by Dr. Rüdiger Landers, Materials Research Center, University of Freiburg, Germany.

lysts [28] and as adsorbents [37]. Other interesting applications involve their use as chromatographic materials and for continuous flow synthesis in columns [28, 38, 39].

3.2.1.3 Microporous Resins

Microporous beads are weakly crosslinked resins obtained by suspension polymerization of undissolved styrene and divinylbenzene in the absence of any porogen agent. This process leads to the formation of a homogeneous network evidenced by a glassy and transparent appearance. The size of the beads is controlled by the stirring conditions and by the nature and the amount of the stabilizing agent, e.g.

a partially hydrolyzed polyvinylalcohol (PVA) (85–87%) [40]. The most commonly used supports for solid-phase organic synthesis are styrene–divinylbenzene co-polymers crosslinked with 1–2% DVB. Numerous derivatives of these copolymers are commercially available. Three synthetic routes are used for the preparation of functionalized polystyrene microbead derivatives (i.e. Merrifield resins):

1 chemical transformation of the unfunctionalized poly(styrene-co-divinylbenzene) [1, 41, 42];
2 copolymerization of a mixture of styrene, divinylbenzene with 4-(chloromethyl)-styrene [43, 44];
3 copolymerization of the mixture styrene, divinylbenzene with α-methylstyrene and its consecutive transformation by chlorination [45, 46].

The loading capacity is controlled by the yield of the electrophilic aromatic sub-stitution in method 1 or by the amount of 4-(chloromethyl)-styrene incorporated in the starting organic phase for copolymerization in method 2. Typical loading values are between 0.2 and 4.0 mmol g^{-1}, and a loading capacity of 1.5 mmol g^{-1} (for the most commonly used Merrifield resins) corresponds approximately to 20% substituted aromatic groups. Higher loading Merrifield resins (4 mmol g^{-1}) have also been used in organic synthesis [47]; however, the highest possible loading 6.55 mmol g^{-1}, which corresponds to 100% chloromethylstyrene, would not be useful in practice [12].

For microporous resins the exact degree of crosslinking and the nature of the crosslinker are even more important than for macroporous resins, owing to the severe effect on the swelling properties. The most common microporous resins are 1–2% crosslinked, but resins with less crosslinkage have also been studied [48]. They are mechanically weak and consequently easily subject to damage. However, increased reaction rates have been observed for these more flexible polymer net-works. It is important to keep in mind that the DVB used for crosslinking is usu-ally a technical grade product with a composition that can vary from one batch to another (see Fig. 3.3), which influences the properties of the beads. The con-sequences are that varying amounts of crosslinking agents are incorporated in resins depending on the different polymerization batches, which generally leads to relatively high error values (±0.5%) in the degree of crosslinking indicated in the catalogues of many suppliers.

Other crosslinkers are ethyleneglycol dimethacrylate (EGDMA) (4), *N,N*-methylene bisacrylamide (6) (MBA), trimethylolpropane trimethacrylate (8) (TRIM), and, more recently, novel crosslinkers have been introduced such as 1,4-bis(vinylphenoxy)-butane (5) [49], and bis(vinylphenoxy)-polyethylene glycol (PEG) (7) [50, 51] which present the advantage of having a strong influence on the swell-ing properties due to the increased flexibility between the two crosslinking units (Fig. 3.4).

In addition, styryl-terminated dendrimers have been introduced as novel poly-mer crosslinkers (9). They consist of 8–16 peripheral styryl units attached to aryl end branches of dendritic TADDOL or BINAP ligands and these were copolymerized with styrene by suspension polymerization [52]. However, it is not yet clear whether such highly functional crosslinkers are of any advantage for practical use.

Physical properties of microporous resins In contrast to macroporous resins, microporous beads have a low internal surface area in the dry state of less than 10 $m^2 g^{-1}$ (determined by N_2 BET-isotherm) [20], owing to their more homogeneous structure that does not allow the diffusion of gases or "bad" solvents into the polymeric network. As polystyrene is hydrophobic and nonpolar, swelling of microporous polystyrene resins will occur in nonprotic solvents such as dioxane, dichoromethane, dimethylformamide (DMF), tetrahydrofuran (THF), or toluene, but not in polar protic solvents (e.g. water, alcohols) and apolar aprotic solvents (e.g. alkanes) [53].

However, the swelling of a resin depends not only on the nature of the solvent but also on the degree of crosslinking and the spacer molecules [48, 54]. It is indeed obvious that the percentage of swelling is inversely proportional to the crosslinker–monomer ratio. Thus, 1% DVB polystyrene swells 4–6 times its volume in dichloromethane, whereas in contrast 2% DVB polystyrene swells 2–4 times its volume in dichloromethane. The transformation of Merrifield microbeads by the conversion of the chloromethyl groups or the grafting of spacer groups can lead to dramatic changes in the swelling properties, as illustrated in Table 3.2. In order to anticipate the swelling properties of a modified resin, the chemist can compare the Hildebrand solubility parameters of solvents and of the polymer [55]. Thus, the calculation of the solubility parameter for crosslinked polystyrene gives values of 9.1 $(Cal mL^{-1})^{0.5}$ and this value can then be compared with the solubility parameters of different solvents [49]. It is generally accepted that polymers and solvents are miscible when the difference of solubility parameters is not higher than 1. For example, the comparison of the values for the solubility parameter of polystyrene and of different solvents allow for the selection of "good" solvents for swelling the resin beads such as chloroform, DMF, dioxane, toluene, or THF, which was confirmed experimentally (Table 3.2) [49].

Mechanical and chemical stability of microporous resins Care has to be taken when working with microporous resins in different solvents. Osmotic shock can occur when a preswollen bead (in a "good" solvent) is introduced into a "bad" solvent. The beads start to shrink rapidly under expulsion of the "good" solvent and are subjected to stress. This leads to mechanical damage or at least to non-negligible modifications of the structure, as shown in Fig. 3.5b (compare with the surface of the original Merrifield resin in Fig. 3.5a).

Another problem that has to be encountered when conducting a reaction on swollen microporous resins is that they can break if agitation conditions are too vigorous. Figure 3.5c shows an example of beads after a solid-phase reaction conducted with a magnetic stirrer. The scanning electron microscope (SEM) picture reveals that most of the beads are broken compared with the beads collected after a reaction conducted on a mechanical shaker (Fig. 3.5d). The consequence for broken resin beads is the clogging of the filters used during their purification.

Tab. 3.2. Swelling of selected solid phase resins (mL g^{-1}).

Solvent	Merrifield resins (1% crosslinked)	Tentagel S	ROMP spheres	PS dendrimers		PAP	CLEAR	POEPOP	SPOCC	POEPS-3
				Polyamide dendrimer	Arylether dendrimer			M_n(PEG) = 1500 g mol^{-1}		
CF$_3$CH$_2$OH	–	–	–	–	–	14.1	12.0	–	–	–
Water	1.6	3.6	–	3.2	3.5	10.5	8.0	14.0	10.5	8.0
MeOH	1.7	3.6	1.0	–	3.3	9.5	7.0	–	–	–
EtOH	–	2.9	–	4.0	–	8.1	–	–	–	–
2-Propanol	–	–	–	–	–	7.3	–	–	–	–
DMF	5.6	4.7	1.8	4.0	5.8	6.4	8.0	15.5	11.0	9.5
DMSO	–	–	1.3	5.4	–	–	–	–	–	–
MeCN	3.2	4.2	1.5	–	–	1.9	6.5	12.0	7.9	6.0
EtOAc	–	–	–	–	–	–	5.0	–	–	–
THF	8.8	5.0	2.2	3.2	4.7	–	6.5	12.5	8.1	7.0
Dioxane	7.8	5.4	–	2.8	–	–	–	–	–	–
Acetone	–	–	–	–	–	1.5	–	–	–	–
CH$_2$Cl$_2$	8.3	6.3	2.2	3.7	4.6	6.4	10.0	22.8	15.5	12.0
CHCl$_3$	–	–	–	4.0	–	6.8	–	–	–	–
Et$_2$O	4.0	1.9	1.5	–	–	–	–	–	–	–
Toluene	8.5	4.8	–	3.0	–	–	5.0	–	–	–
Hexane	–	–	–	2.4	–	–	3.0	–	–	–

Concerning the chemical stability of Merrifield resins and their derivatives, it has been shown that they are relatively stable toward weak oxidants, strong bases, and acids. In fact, reactions that are known to proceed on alkyl substituted aromatic compounds, especially electrophilic substitutions, will also occur on crosslinked polystyrene [12]. Strong oxidants at elevated temperatures and electrophilic reagents should therefore be avoided [12, 13].

3.2.2
Polystyrene Hybrid Supports

3.2.2.1 PEGylated Resins
Merrifield resins and their derivatives are still the most commonly used resins for the synthesis of small molecules, but one of their limitations is the poor swelling in polar protic solvents. For instance, Merrifield resins cannot be applied for synthesis of polar compounds in water or alcohols. This problem, however, can be overcome by designing "amphiphilic" resins made of a 1% crosslinked polystyrene matrix onto which polyethylene oxide chains are grafted [56, 57]. These resins are commercially available as TentaGel® (Fig. 3.6). Typically, their composition is 70 wt% of polyethylene glycol (PEG) grafts (average molecular weight of 3000 g mol^{-1}) and 30 wt% of PS. These resins can be prepared by two different pathways:

1 grafting of PEG chains onto Merrifield resins [58, 59];
2 anionic ring opening polymerization of oxiranes initiated from the active sites of a hydroxy-terminated polystyrene resin [56, 57, 60].

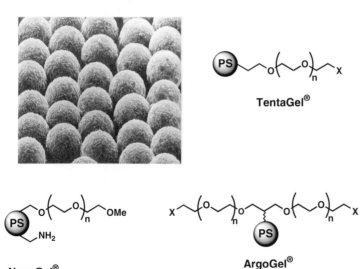

TentaGel®

NovaGel®

ArgoGel®

Fig. 3.6. Picture of Tentagel® resin and chemical structures of TentaGel®, ArgoGel® and NovaGel® resins.

The main drawback of the first technique is the possibility of additional cross-linking due to the fact that PEG macromolecules are bifunctional and consequently that the two end chains can both react with the resin [56]. This, however, can be overcome if monoprotected PEGs are used.

PEGylated supports are far more polar than Merrifield resins and hence they swell in a broad range of solvents from apolar aprotic to polar protic solvents (Table 3.2) [53]. However, the loading of TentaGel® resins is relatively low (loading values range from 0.15 to 0.40 mmol g^{-1} for PEG$_{35}$ chains) compared with the higher loading capacities that can be reached for Merrifield resins. A similar type of PEGylated resin known as ArgoGel® has slightly higher loading capacities (up to 0.5 mmol g^{-1} for PEG$_{35}$ chains) because it has two PEG chains that are grafted onto every active site [61, 62].

PS–PEG hybrid resins opened many new possibilities in combinatorial synthesis by allowing the use of protic solvents. Even on bead, screening is possible with these resins. However, the user should be aware of some drawbacks, such as bleeding, which is defined as the loss and the release of PEG grafts when treated, for example, with strong acids [63, 64]. It has been reported for instance that trifluoroacetic acid (TFA), which is commonly used for cleavage of protecting groups, can also cleave the benzylic–CH_2O–PEG bond [65]. This problem can be overcome by using a TentaGel® resin with the PEG spacer attached to the polystyrene backbone via an alkyl linkage. This hybrid resin is not sensitive to acids or bases [34, 57, 62, 64]. Other drawbacks of such PS–PEG hybrid resins originate from the nature of the PEG polymer, which is hygroscopic. Hence, the presence of large amounts of PEG (up to 70%) render the beads more sticky and more difficult to manipulate in some cases than regular Merrifield resins [20].

In order to overcome the decrease of loading capacity due to bleeding of the PEG spacers, a similar type of resin has been designed (e.g. NovaGel®) with active sites located on the PS backbone [64] rather than at the end of the PEG-grafted chains as in TentaGel® and ArgoGel® resins. This results not only in a good swelling behavior in both apolar and polar solvents due to the PEG chains (PEG content ~ 50%) but also in a slightly higher loading capacity (0.7 mmol g^{-1}). In addition, eventual bleeding of the PEG grafts does not decrease the loading capacity and only a slight modification of the swelling properties will occur. However, the advantage of improved kinetics and better on-bead analytics (e.g. nuclear magnetic resonance) due to the highly flexible polyether spacers are lost in this approach. It should be pointed out again that *all* PEGylated resins are sensitive toward oxidation and hence bleeding will occur if not properly handled.

3.2.2.2 High-loading PS Hybrid Supports
A general problem in solid-phase combinatorial chemistry is low loading capacity of the commonly used resins. Often, large quantities of resin are required in order to obtain substantial amounts of products.

Ring-opening-metathesis-polymer hybrids: ROMP spheres A recent approach toward high-loading solid-phase supports uses cross metathesis between vinyl polystyrene beads and norbornene derivatives [66]. In a first step the immobilization of the ruthenium catalyst (Grubbs catalyst) on vinyl polystyrene (**11**) can be achieved via insertion of ruthenium into the styryl double bond (Scheme 3.2). The PS-supported ruthenium alkylidene (**12**) shows good stability under normal atmospheric conditions when dried. In the presence of solvents (e.g. dichloromethane), however, the catalyst becomes inactive within a period of 5 h. Treatment of the PS-supported catalyst (**12**) with an excess of a norbornene derivative (**13**) yields a ROMP-based polymer (**14**), so-called ROMP spheres (Scheme 3.2).

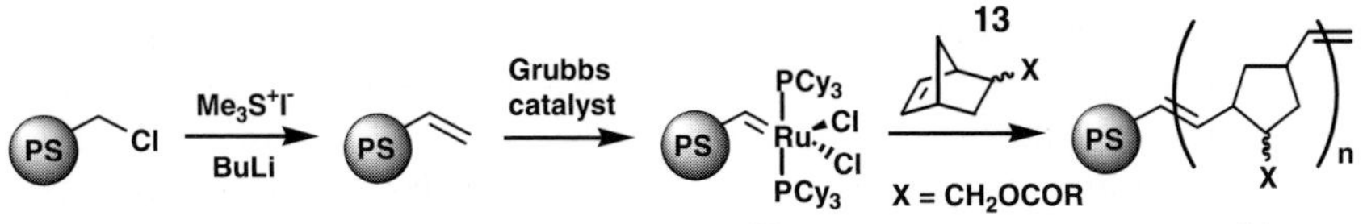

Scheme 3.2. Cross metathesis of the supported Grubbs catalyst and a norbornene derivate to yield ROMP spheres (**14**).

This resin shows high loading capacities (up to 3.0 mmol g^{-1}), but the swelling properties are lower than for conventional resins, "good" solvents include THF and DCM (see Table 3.2). The utility of these high-loading supports was exemplified by a palladium-catalyzed coupling reaction between a resin-bound bromobenzoate and aryl zincates in a THF/MeOH (4:1) mixture. Also, crosslinked ROMP-based polymers have been used as supports for catalysts [67] and high-loading amine scavengers [68].

Rasta silanes Silyl linkers (see Chapter 4.3.5) offer a broad range of mild and chemospecific cleavage conditions and are frequently used in SPOS. A high-loading silyl-functionalized PS hybrid resin, so-called 'Rasta silane', has recently been introduced for solid-phase chemistry [69]. The required silyl substituted styrene monomer (**16**) can be prepared by lithium–halogen exchange of *p*-bromostyrene (**15**) and subsequent quenching with the appropriate dialkylchlorosilane. A living free radical polymerization initiated by heating TEMPO-methyl resin (**17**) with the silyl styrene (**16**) under solvent-free conditions at 130 °C leads to the formation of Rasta silanes (**18**) (Scheme 3.3).

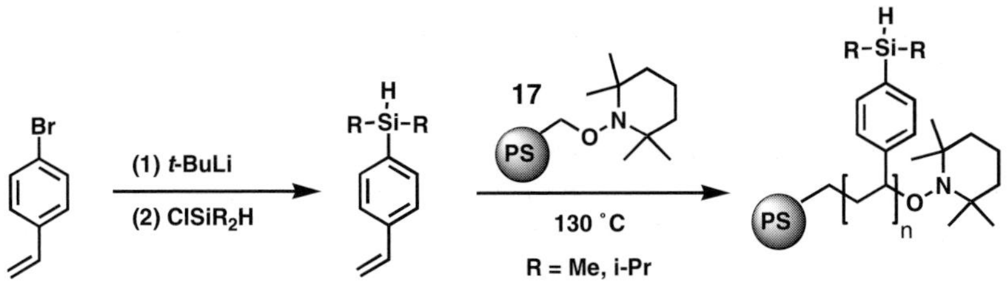

Scheme 3.3. Formation of high-loading Rasta silane resins (**18**) by living free radical polymerization.

Rasta resins are characterized by high loading levels between 1.6 and 3.8 mmol g^{-1} with regard to the silicon linker. These values are significantly higher than for conventional silicon-linker resins, which show loading capacities between 0.5 and 1.6 mmol g^{-1}. Easy conversion of the silyl units to reactive silyl chloride or triflate by standard synthetic methods allow the immobilization of alcohols and phenols for solid-phase synthesis. It has been shown that these Rasta silanes allow fast diffusion of reagents throughout the resin beads [70].

Polystyrene dendrimer hybrids Another approach to high-loading resins are polystyrene dendrimer hybrids [71]. Dendrimers are highly branched macromolecules built by a stepwise approach from a central core. The dendritic spacer molecules provide a rapid and efficient method of increasing the loading capacity. For example, the naturally occurring amino acid lysine has been used as a building block in creating an inert dendrimeric scaffold [72]. The tris-Boc-protected amino acid (**20**) can be attached to a preswollen aminomethylated PS resin (**19**) (Scheme 3.4) and thus creates the first generation of this dendrimer (**21**).

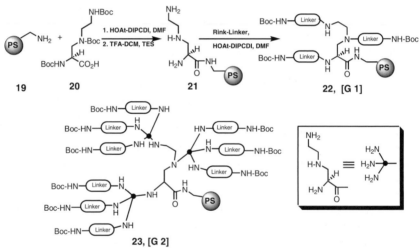

Scheme 3.4. Synthesis of dendritic polyamide PS hybrid resins by stepwise attachment of a tris-Boc amino acid (**20**) to an amino-methylated PS resin (**19**) (0.8 mmol g^{-1}) via the DIPCDI-HOAt-mediated amide coupling leads to the first generation [G1] resin (**22**) (Fmoc loading capacity 0.9 mmol g^{-1}) and [G2] **23** (Fmoc loading capacity 1.2 mmol g^{-1}).

After coupling of a Rink amide linker [73] onto resin (**21**), the derivatized PS dendrimer resin (**22**) with a loading capacity of 0.9 mmol g^{-1}, with regard to the Fmoc (9-fluorenylmethoxycarbonyl) amino groups, is obtained. Repeating this procedure yields the second generation [G2] (**23**). This exhibits a slightly higher loading capacity of 1.17 mmol g^{-1}. It is important to mention that the loading per gram of resin does not increase substantially because of the mass of the attached linker. However, the loading per bead increases geometrically from one generation

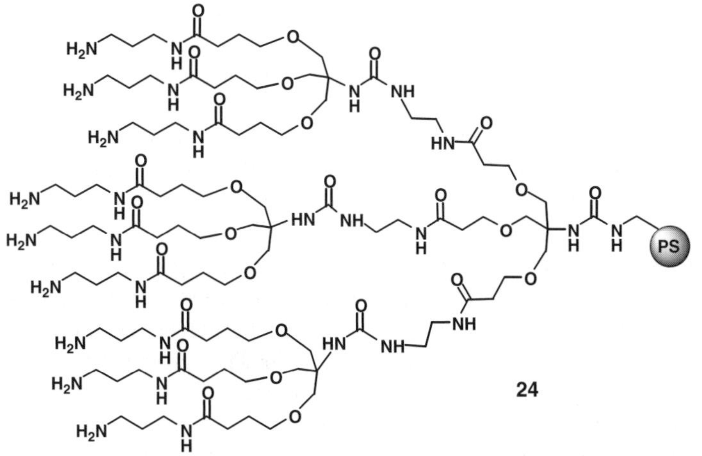

Fig. 3.7. [G2] of a triple branched symmetrical PS dendrimer resin (**24**) (loading capacity 0.8 mmol g^{-1}, 30.6 nmol per bead).

to the next. These PS dendrimer resins have successfully been tested in the synthesis of small peptides [72].

The synthesis of symmetrical tri-branching dendrimers on aminomethyl polystyrene macrobeads has also been described [74]. These [G2] dendritic resins (Fig. 3.7) offer loadings of 30.6 nmol per macrobead (0.8 mmol g^{-1}). A split-and-mix synthesis (e.g. SPPS, Suzuki-coupling) on a gram of these resin beads ($\sim$ 27,000 beads g^{-1}) provides the number of compounds and the amounts necessary for a single bead screening approach (including ^{1}H-NMR!). The obtained beads, especially the PS–[G2] dendrons (**24**), show higher swelling values in polar solvents such as methanol (MeOH) and dimethyl sulfoxide (DMSO) than the originally used aminomethyl resin (see Table 3.2).

Some aryl ether dendrimers can be synthesized directly on hydroxymethyl polystyrene (**25**). Mitsunobu reaction with 3,5-bis(acetoxymethyl)-phenol (**26**) gives the branched precursor (**27**) (Scheme 3.5) [75]. After deprotection and repetition of the sequence a [G3] dendrimer resin was obtained with a loading capacity of 2.85 mmol g^{-1} (3 nmol per bead). In unpolar solvents such as DCM and THF the dendrimer resin (**28**) swells better than in MeOH or water.

The development of high-loading immobilized catalysts is another interesting application for PS dendrimer resins. An example is given by the rhodium complex

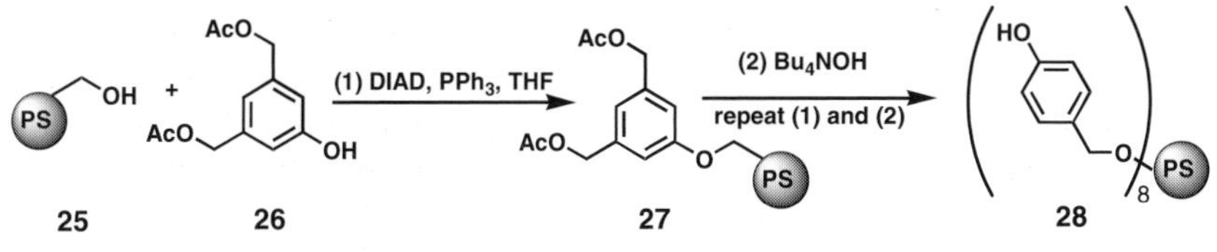

Scheme 3.5. Synthesis of PS–arylether dendrimer hybrid (**28**) (loading capacity 2.85 mmol g^{-1}).

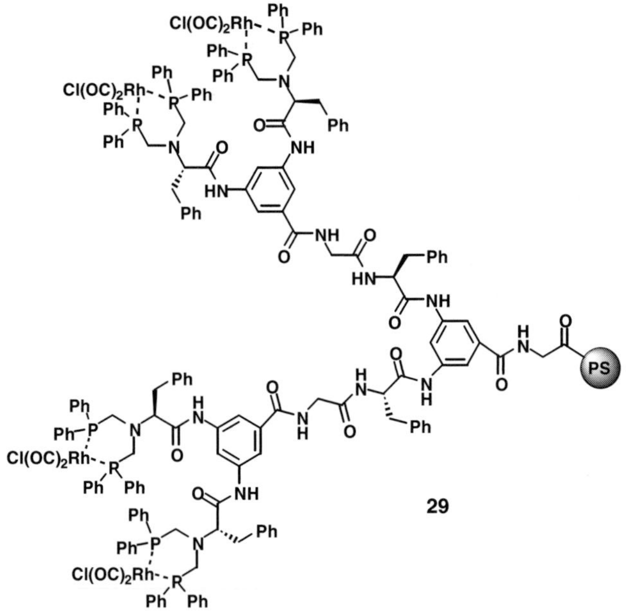

Fig. 3.8. PS–[G2] dendrimer hybrid (**29**) supporting a tetravalent catalyst for heterogeneous hydroformylation reactions.

with dendritic phosphine ligands that are anchored onto a PS resin (Fig. 3.8) [76]. Different dendrimer generations have been prepared and tested as catalysts for the hydroformylation reaction of several olefins. The PS–[G2] dendrimer (**29**) shows higher reactivities than the first generation and does not decrease for up to five cycles. The origin of this increased reactivity might be due to the higher density of ligands on the outer core and cooperative effects.

High internal phase emulsion polymers: PolyHIPE PolyHIPE structures are formed by polymerization of a *h*igh *i*nternal *p*hase *e*mulsion with styrene and divinylbenzene [77]. The resulting material consists of an extremely porous and rigid matrix, which provides a scaffold onto which functional polymers (e.g. polyacrylamide gels) can be chemically bound (Scheme 3.6). The polyacrylamide gel is prepared

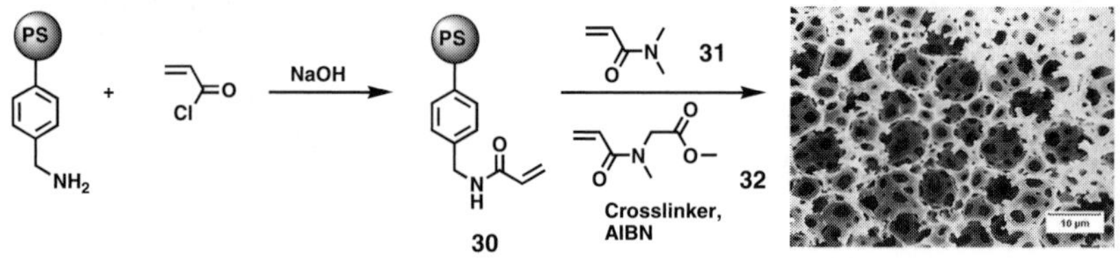

SEM of a PolyHIPE support

Scheme 3.6. Preparation of highly porous 'PolyHIPE' PS–acrylamide composite resins.

from *N,N'*-dimethylacrylamide (**31**), acryloylsarcosine methyl ester (**32**), a cross-linker and a radical initiator that is grafted onto the highly porous support (**30**). The polymer matrix (**30**) can be obtained from aminomethylated PS and acrylic chloride. Although these polyacrylamide grafts are of limited use for SPOS, very high loading capacities (up to 5 mmol g^{-1}) can be achieved by this approach [78].

The high internal surface area and the limited swelling allow the usage of these materials in column reactors for automated synthesis. Applications in low-pressure continuous-flow solid-phase peptide synthesis have already been described [78]. In principle these highly porous PS structures (see Scheme 3.6) should also be suitable as supports for reagents or catalysts in SPOS.

3.2.3
Other Crosslinked Polymeric Supports

3.2.3.1 Crosslinked Acrylamides
Solid-phase crosslinked acrylic amides possess a more polar character than the frequently used polystyrene supports. Thus, they are rather compatible in respect to solvation properties with a growing peptide chain. A wide range of polar and protic solvents can freely permeate and swell these materials. This class of resins, however, has been specifically developed for peptide and oligonucleotide synthesis. The limited chemical stability of the amide bonds causes these resins to be less suitable for broad application in SPOS with only some limited specific applications.

Pepsyn-K resins   Crosslinked and functionalized polydimethylacrylamide gels which are kept inside the pores of an inert, rigid and macroporous matrix are well suited for continuous flow applications. Compared with PolyHIPE materials, in which crosslinked polystyrene is used as matrix material, Pepsyn-K resins contain a highly porous kieselguhr matrix as the scaffold [79]. A mixture of dimethylacrylamide (**31**), ethylene bisacrylamide (**33**), acryloylsarcosine methyl ester (**32**) in DMF and water is soaked into kieselguhr particles and then polymerized at room temperature (Scheme 3.7). This support is a hybrid between an inorganic and an organic support. So far it has been used only for peptide synthesis.

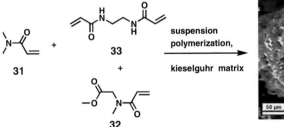

suspension polymerization,

kieselguhr matrix

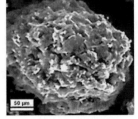

SEM of a Pepsyn-K support

Scheme 3.7.   Synthesis of Pepsyn-K resins from dimethylacrylamide (**31**), ethylene bisacrylamide (**33**), and acryloylsarcosine methyl ester (**32**) on a highly porous kieselguhr matrix.

Poly(ethylene glycol)-copoly(N,N'-dimethylacrylamide) (PEGA) Another flow-stable, highly polar solid support is poly(ethylene glycol)-copoly(N,N'-dimethylacrylamide) (PEGA). This resin is accessible by an inverse suspension copolymerization of N,N'-dimethyl acrylamide (**31**), acryloylsarcosine ethyl ester (**34**) and bis-2-acrylamidoprop-1-yl-PEG$_{1900}$ (**35**) in silicon oil as a dispersing medium (Scheme 3.8) [80, 81].

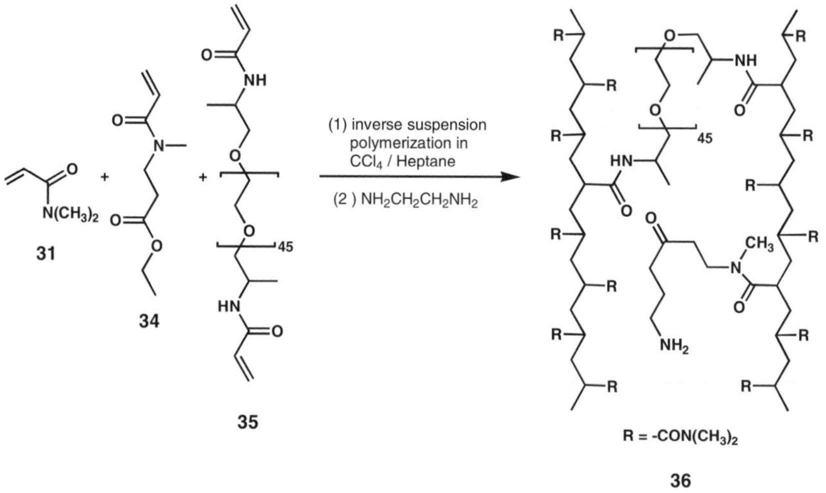

Scheme 3.8. Synthesis and structure of the high-capacity PEGA resin (**36**), suitable for continuous flow synthesis.

The obtained PEGA resin (**36**) is characterized by good mechanical stability as well as by high swelling properties in both organic solvents and aqueous buffers. The loading capacity is 0.4 mmol g^{-1}. Furthermore, it has been shown that larger peptides and other macromolecules can easily diffuse through the PEGA polymer matrix [82]. It is, therefore, suitable for solid-phase synthesis using both the batch method and the continuous flow synthesis [83]. In addition, it was demonstrated that biological screening is possible on the resin-bound substrate without further cleavage [84].

Poly(N-acrylylpyrrolidine) (PAP) It has been reported that acrylamide polymers swell very well in polar solvents such as water, DMF, and certain alcohols but less well with CH$_2$Cl$_2$ and CHCl$_3$ [85]. Development of poly(N-acrylylpyrrolidine) (PAP) allows peptide-coupling reactions in organic media of high and low polarity as well as in aqueous solution. The synthesis of PAP (**39**) can be achieved by inverse suspension polymerization. An aqueous solution made up of the monomer N-acrylpyrrolidine (**37**), the functionalization agent N-acryl-1,6-diaminohexane hydrochloride (**38**) and the crosslinker N,N'-bis(acrylyl)-1,2-diaminoethane (**33**) is dispersed in a mixture of hexane and carbon tetrachloride. The polymerization is then initiated with a redox initiator system consisting of ammonium peroxydisulfate and N,N,N',N'-tetramethyl-1,2-diaminoethane (Scheme 3.9) [86].

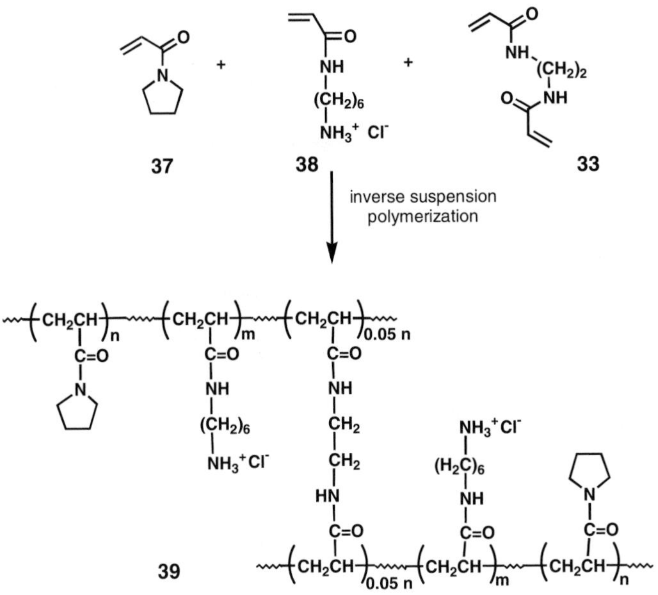

Scheme 3.9. Preparation and structure of poly(N-acrylyl-pyrrolidine) (PAP) resins (**39**).

The obtained PAP resin (**39**) has a loading capacity of 0.7 mmol g^{-1} (amino groups), which can be used for SPPS. In either its protonated or acylated form, this resin exhibits high swelling values in water, methanol, CH_2Cl_2 and $CHCl_3$. Several applications of PEGA resins have been reported for SPPS [87].

Crosslinked-ethoxylate acrylate resins (CLEAR) Another class of highly crosslinked polymeric supports for SPPS are crosslinked ethoxylate acrylate resins (CLEARs) [88]. The CLEAR supports are prepared from branched crosslinker (**8, 40**) that is used in high molar ratios for copolymerization with different monomers (**41–44**) (Fig. 3.9). The resulting polymers show a polyethylene glycol-like character, although individual chains are quite short compared with chains of already known PEG-containing resins such as PEGA or TentaGel®.

Both the oxyethylene chains and the ester functionalities are responsible for the hydrophilic character of this class of resins. In contrast to many other resins, amino groups can be introduced directly onto the resin. It does not require additional transformations of other functional groups or deprotection of a protected amino polymer. CLEAR supports have good swelling properties, in both hydrophilic and hydrophobic solvents (see Table 3.2). This magnifies the range of chemistry that could be carried out using these resins. The loading capacities are typically between 0.20 and 0.29 mmol NH$_2$ g^{-1}, which is significantly lower than those of PAP (**39**), POEPOP (**51**), and SPOCC (**52**) (see below). CLEAR supports have been successfully tested in batchwise and continuous-flow peptide synthesis

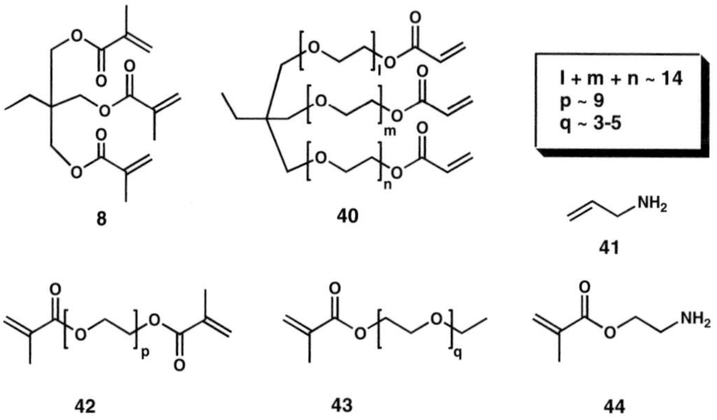

Fig. 3.9. CLEAR supports consist of these monomers which can be combined in different ways, yielding resins that are suitable for batchwise and continuous-flow modes.

but they should also be suitable for limited application in SPOS, especially when protic solvents are required.

3.2.3.2 Crosslinked PEGs

This class of PEG-based supports is chemically more stable than the polyacrylamide-based supports. However, their applications in SPOS have not yet been extensively demonstrated. Interesting features of these resins include the magnitude of swelling in all kinds of solvents and a relatively high loading capacity when compared with regular PS–PEG hybrid resins (see Sect. 3.2.2.1).

Polyoxyethylene-polyoxypropylene (POEPOP) Polyethylene glycol derivatives [e.g. polyethylene glycol polyacrylamide (PEGA **36**)], as a result of their inert character, offer a wide range of applications in polymer chemistry and biomedical sciences. Endgroup-modified PEGs (see Section 3.2.2.1) are often used as polymeric supports in peptide and combinatorial synthesis [89]. In the case of PEGA, a limiting factor is its amide backbone. An exchange of the amide linker in PEGA for the more stable ether linkages offers a larger diversity for application in organic reactions. These properties are represented in polyoxyethylene-polyoxypropylene (POEPOP **51**), which can be obtained by anionic polymerization between end-modified PEG macromonomers (**46**, **47**) and epichlorhydrin (**48**) (Scheme 3.10) [90]. The resulting POEPOP (**51**) shows an optimized balance between hydrophilic and hydrophobic properties and has a high mechanical and chemical stability [91]. However, due to the physical properties of the PEG chains these beads might be sticky in some cases [20].

One prerequisite for a successful polymeric support is that good swelling properties are available both in organic solvents and in enzyme buffers. The POEPOP resins swell best in CH_2Cl_2, less in DMF and water, and much less in THF and MeOH (see Table 3.2) [90]. However, swelling is strongly influenced by the length of the PEG chains. The longer the chain the higher the swelling, but at the same

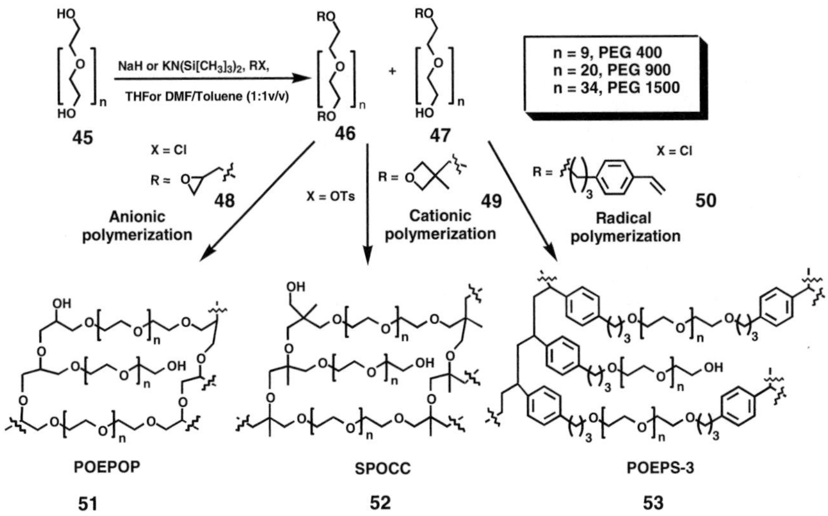

Scheme 3.10. Preparation of PEG macromonomers and polymerization to provide the hydrophilic POEPOP (**51**), SPOCC (**52**) and POEPS-3 (**53**) resins.

time the loading capacity drops. Typical loading capacities for POEPOP resins are between 0.4 and 0.7 mmol g^{-1}. In comparison with typical polystyrene-based TentaGel® resins (see Table 3.1), this is still a relatively high value [90]. POEPOP supports have been successfully employed for a solid-phase synthesis of peptide isoesters by nucleophilic reactions [91].

Polyoxyethylene-polyoxethane (SPOCC) As a result of the secondary ether bonds that are formed during the polymerization reaction, POEPOP (**51**) is unstable under extremely strong acidic or basic conditions. Cationic polymerization of end-modified PEG macromonomers (**46, 47**) with 3-methyl-oxetan-3-yl-4-methyl toluenesulfonate (**49**) leads to polyoxyethylene-polyoxetanes (SPOCC **52**; Scheme 3.10) [89]. It contains only primary ethers and alcohols and exhibits, therefore, an excellent chemical stability. No changes in the SPOCC structure were observed by treatment with 37% HCl or neat anhydrous hydrogen fluoride. In addition, heating with a large excess of thionyl chloride in toluene did not change the resin structure. The same conditions, however, dissolve POEPOP within a few minutes. Loading capacities of SPOCC are typically in the range 0.4–1.2 mmol g^{-1}. The swelling behavior is similar to that of POEPOP in the same solvents, but SPOCC resins swell slightly more than POEPOP resins with equal PEG chain length [90]. Easy conversion of the primary alcohols into amine or thiol functionalities is possible and therefore allows different types of chemistry. Solid-phase glycosylation of peptides as well as enzymatic reactions have been performed on this resin [89].

Polyoxyethylene-polystyrene (POEPS-3) Another class of crosslinked polyethylene glycols are polyoxyethylene-polystyrene resins (POEPS-3 **53**). These supports can be obtained via radical polymerization of an end-modified PEG macromonomer (**46, 47**) with chloro-propyl-styrene (**50**) (Scheme 3.10). In order to obtain beads this reaction is performed by inverse suspension polymerization [90]. Loading capacities are similar to TentaGelSOH and are in the range 0.2–0.3 mmol g^{-1}. POEPS-3 (**53**) shows similar swelling properties to POEPOP and SPOCC. With regard to the chemical stability of these compounds, POEPS-3 is less stable than SPOCC but more stable than PEGA due to its polyether backbone. As shown by fluorescence quenching diffusion experiments in DMF, diffusion processes are slower than in TentaGel and ArgoGel. In water, however, all three crosslinked PEG-based resins show faster diffusion rates than the PS–PEG hybrid resins [90]. So far POEPS-3 resins have been used exclusively for SPPS [92].

3.2.4
Inorganic Supports

In the following section, we give a short overview of some inorganic supports that are suitable for SPOS. A good summary was published recently [4, 26]. Different forms of silica gel are very often used for SPOS. They offer a number of advantages and are now commercially available in various derivatized forms. These supports are rigid and do not swell or contract in the solvents employed in synthetic cycles. Reagents and products can easily diffuse into and out of the pores. Typical pore sizes are between 2 and 10 nm, and the surface area is between 200 and 800 m^2 g^{-1}. Although higher coupling rates in peptide synthesis were obtained on silica than on polystyrene [93], silica gel did not become an established support for solid-phase organic synthesis. In the meantime, silica gel has been partially replaced by controlled pore glass (CPG) as an inorganic support [38, 94–98]. The main advantages of CPG, compared with silica gel, are its more regular particle size and shape, and greater mechanical stability. Functionalization with long-chain alkylamine groups leads to exceptional properties: rigid, nonswelling, mechanical stability within a series of polar solvents, and chemical stability over the whole pH range. Furthermore, CPG is unreactive to a broad variety of nucleophiles and electrophilic reagents. Its stability during heating and its excellent solvation and flow properties make it the support of choice for automated solid-phase synthesis. Typical loading capacities, however, are rather low (0.1 mmol g^{-1}).

3.3
Soluble Polymeric Supports

In contrast to solid supports, soluble noncrosslinked polymers give rise to homogeneous reaction conditions [6–8, 10]. Many of the problems of crosslinked polymers arising from heterogeneous reaction conditions, such as nonlinear kinetic behavior, reactive site distribution and solvation problems, can be overcome by using soluble polymeric supports. Unlike solid-phase supports, soluble supports enable the use of insoluble reagents or catalysts and also allow for the characterization of intermediates by standard analytical techniques (e.g. NMR, UV), in the homogeneous phase. On the other hand, the isolation and separation of these soluble materials is not as straightforward as for solid-phase resins. Although soluble polymeric supports have been used in organic synthesis since the 1970s, it is only recently that several groups have started to investigate automated separation techniques that enable the use of these soluble supports for parallel synthesis.

Soluble polymeric supports can be classified according to their topology, which strongly influences the physical properties of these materials. Linear polymers with one or two functional endgroups, polyfunctional linear structures with many functional groups along the chain and highly branched dendritic polymers will be discussed in the following sections.

3.3.1
Separation Techniques for Soluble Polymeric Supports

It is generally believed that soluble polymeric supports, neglecting their advantages, are difficult to separate from the reaction mixture. For many years the technique of precipitation and subsequent filtration seemed to be the most promising method for separating soluble polymers from low-molecular-weight species. However, for automation several other separation techniques appear to be more promising [10]. Especially in the field of homogeneous catalysis, membrane filtration techniques have found application and show great potential for industrial use [99].

Therefore it seems appropriate to summarize briefly the multiple commercially available techniques for the separation of soluble polymers from the reaction mixture, i.e. low-molecular-weight compounds (Table 3.3). There are several methods which have been used for the separation of macromolecules by size, i.e. dialysis [100, 101], membrane filtration [102], preparative size exclusion chromatography (SEC) [103] and filtration through a silica cartridge [68]. All of them are suitable for automation. However, little effort has been undertaken to develop multiparallel automation in this area compared with the progress achieved in solid-phase synthesis. For efficient and fast separations by size, soluble polymeric supports should have medium molecular weights (5000–10,000) and narrow molecular weight distributions (< 1.5). In addition, macromolecules with a more globular branched structure are preferable to a linear polymer structure.

Other separation techniques for soluble polymers, such as precipitation/ filtration and liquid–liquid phase separation rely on polymer properties rather than on their hydrodynamic volume. Precipitation is frequently used in polymer chemistry to purify the respective polymer from low-molecular-weight impurities. This method works especially well when the polymer is crystalline and the T_g is above ambient temperature. It has also been used in the separation of functionalized PEG supports [6–8, 10]. Precipitation, however, is unsuitable for multistep syntheses because impurities often remain trapped in the polymer. In addition, large solvent volumes are required to perform quantitative precipitations and hence automation of the process is difficult. Another relatively simple separation technique, which is suitable for the separation of organic molecules from water-

Tab. 3.3. Separation techniques for soluble polymeric supports.

Parameter	Dialysis	Ultrafiltration	SEC	Filtration through a silica cartridge	Precipitation/filtration	Liquid–liquid phase separation
Separation by	Hydrodynamic volume	Hydrodynamic volume	Hydrodynamic volume	Hydrodynamic volume, polarity	Solubility	Phase distribution coefficients
Minimum MW of polymer	>1000 g mol^{-1}	>1000 g mol^{-1}	–	–	>3000 g mol^{-1}	–
Typical sample volume	10 mL to 1 L	1–100 mL	<1 mL	0.1–10 mL	1–100 mL	10 mL to 1 L
Commercially available	Yes	Yes	Yes	Yes	–	Yes
Suitable for automation	Yes	Yes	Yes	Yes	No	Yes
Suitable for high throughput	No	Yes	Yes	Yes	No	No
Limitations	Unsuitable for final cleavage step	–	–	Suitable only for final cleavage step or removal of polymeric reagents	Unsuitable for multistep syntheses	Different solubilities required

soluble polymers, is based on liquid–liquid phase separation between an organic phase which contains the cleaved organic product and an aqueous phase which contains the water-soluble polymer [104]. This separation technique, however, is limited to systems with different distribution coefficients of the components in the two phases. Yet another separation technique that is used for soluble polymeric reagents is filtration through a silica cartridge [68]. This technique has already been established for automated synthesis in many laboratories and does not require additional equipment. However, it can only serve for purification of the product after the final cleavage step or the removal of a polymeric reagent.

3.3.2
Terminal Functionalized Linear Polymeric Supports

Linear polymers that carry functionalities only on their chain ends (e.g. PEGs), have been frequently used as soluble polymeric supports and are listed in Table 3.4 (Entries 1 and 2) [6–10]. Because of the limited number of functional groups (one or two per chain) these materials have rather poor loading capacities (typically <0.5 mmol g^{-1}) and products coupled to the polymeric support are more difficult to analyze (e.g. by NMR) than higher loading supports. On the other hand, these soluble polymers do not dramatically change their physical properties upon functionalization. This is extremely important if precipitation is used for separation.

The most widely used soluble polymeric support in organic synthesis is mono-methylated polyethylene glycol (typically MPEG 5000), which is soluble in many organic solvents and easily precipitated in nonpolar solvents (e.g. diethyl ether) [7, 105]. It was introduced by Bayer and Mutter in the 1970s for peptide synthesis and was rediscovered by Janda and others as a soluble support for organic synthesis in the 1990s [6, 8, 9]. Owing to the linear topology of this polyether, it contains only one reactive functionality and hence exhibits a rather poor loading capacity (0.2 mmol g^{-1}). In order to increase the loading capacity, bifunctional PEGs endcapped with benzylether dendrons have been prepared and used for the synthesis of β-lactams [106, 107]. Another approach is the coupling of PEG arms onto a multifunctional core to give a PEG-star polymer. Such multiarm PEG-star polymers have been recently introduced as supports for organic synthesis based on pentaerythritol [108], hyperbranched polyglycerol (see Section 3.3.4) [109] and cyclotriphosphazene [110]. The advantages of these polyethers are their high chemical stability and the good reactivity of the functional groups in homogeneous phase. The achieved loading capacities of these multifunctional PEG polymers (≤ 1.0 mmol g^{-1}), however, are only marginal better than the commercially available mono- and bifunctional PEG derivatives (0.2 and 0.4 mmol g^{-1} respectively).

3.3.3
Polyfunctional Linear Polymeric Supports

Polyfunctional linear polymeric supports that carry reactive side-groups on every monomer unit are summarized in Table 3.4 (Entries 3–10). These high-loading

Tab. 3.4. Selected linear soluble polymeric supports for organic synthesis.

Entry	Structure	Name	Loading[a]	Applications[b]
1		Polethylene glycol (PEG)	0.2–0.5 (MPEG)	A, B, C, D, E
2		Polypropylene oxide (PPO)	0.2–0.4	A
3		Polystyrene non-crosslinked (PS)	(6.6)	A, B, C
4		Polyacrylic acid	(13.9)	A
5		Polyacrylamide	(14.1)	C
6		Polyvinylalcohol	10.0 (22.7)	A, B, C, D
7		Polynorbornene ROMP gel	3.3–5.8 (9.4)	E
8		Cellulose	(18.5)	B
9		Poly(vinyl alcohol-b-1-vinyl-2-pyrrolidone)	(6.5)	A, B
10		Poly(styrene-b-vinyl-alcohol)	(6.8)	D

[a] Loading capacity: experimental (theoretical) [mmol g^{-1}].
[b] Applications: A, petide synthesis; B, oligonucleotide synthesis; C, oligosaccharide synthesis; D, synthesis of small organic molecules; E, polymer-supported reagents.

polymers have been used for various organic synthesis applications as well as for polymer-supported reagents. However, the physical properties (i.e. solubility) of these materials change dramatically upon functionalization and can be problematic in some cases for broad application in combinatorial synthesis.

The most frequently used examples are polyvinylalcohol, polyacrylamide, non-crosslinked polystyrene and more recently ROMP-based polymers. For example, commercially available polyvinylalcohol (PVA) (Table 3.4) has been used as a soluble support for oligonucleotide synthesis, and a loading of up to 10 mmol g^{-1} of mononucleotide has been reported [111–113]. However, polymeric supports with a lower loading capacity and better solubility in polar organic solvents were recommended by the authors. Another plausible reason for the limited use of PVA as a soluble polymeric support is its poor solubility in organic solvents and the need for harsh reaction conditions for functionalization [10]. Therefore block copolymers of PVA with other monomers, such as styrene, have also been used as multifunctional polymeric supports with improved solubilities [114].

Another high-loading linear polymeric support is polyacrylamide [115, 116]. This very polar resin has better solubilities than PVA and allows the copolymerization of functional monomers. Several applications, for example synthesis of oligonucleotides, have been reported for this soluble support [6, 7]. A limiting factor for the general use of polyamide supports, however, is their reduced chemical stability compared with polystyrene or polyether supports, especially for reactions with organometallic reagents and strong bases. A soluble form of Merrifield resin (non-crosslinked PS/chloromethyl-PS copolymer) has recently been used for combinatorial solution-phase synthesis [9]. Even though this resin has been used for peptide chemistry before, this was the first application in the synthesis of natural products (e.g. prostaglandins). In this application the support was isolated by precipitation.

More recently high-loading ROMP-based polymers (Table 3.4) have been introduced as polymer-supported reagents. As these "ROMPgels" have so far been used only as polymeric reagents, in this case the limited chemical stability of the polynorbornene backbone is of minor concern. For example, "ROMPgels" have been used for Horner–Emmons olefinations, as acylation reagents and for the synthesis of oxadiazoles [117–119]. These polymers have a gel-like behavior when solvated in organic solvents and can be removed from the reaction mixture by filtration through a silica cartridge. An advantage in the preparation of ROMP-based reagents is the great tolerance of the ruthenium (Grubbs) catalyst toward many functional groups.

3.3.4
Dendritic Polymeric Supports

The disadvantages of linear polymers, such as limited solubility in many organic solvents, gel formation and problematic thermal behavior (high melting points and T_g values) in some cases, can be overcome by the use of dendritic polymer architectures. An extreme in terms of tree-like branching are the perfectly branched dendrimers [120]. These well-defined macromolecules are soluble in many organic solvents (depending on their end functionalities) and possess a maximum capacity of functional groups in their periphery. However, the relatively high price and limited chemical stability of the commercially available polyamidoamine or polyamine dendrimers might well be the reasons for their limited use as supports in organic synthesis. Despite these problems, polyamidoamine dendrimers have been successfully used as high-loading supports in the synthesis of indoles [121]. In this case, the separation of the macromolecules from low-molecular-weight species was performed by size exclusion chromatography. More recently, polysilane dendrimers have been used as high-loading soluble polymeric supports in organic synthesis [122].

A very promising class of high-loading polymeric supports are dendritic aliphatic polyethers as highly branched analogues of PEG [123, 124]. They are chemically stable, soluble in many organic solvents, show good accessibility and reactivity of the functional groups, and possess a high concentration of OH groups (up to 14 mmol g^{-1}). However, the general drawback of any dendrimer is the tedious multistep preparation of higher generations (molecular weight exceeding 1500 g mol^{-1}), which is the lower limit for dialysis and ultrafiltration procedures (see Table 3.3). Yet another problem of high-generation dendrimers appears to be steric hindrance and site–site interaction at the outer functional shell. This problem might be overcome by using randomly branched polymer structures as supports [125]. Nevertheless, in some cases the high degree of catalytically active sites on a dendrimer surface can result in cooperative effects and enhanced reactivities [126].

In contrast to perfect dendrimers, hyperbranched polymers are easily available in one reaction step. They contain dendritic, linear and terminal units in their skeleton and hence can be considered as intermediates between linear polymers [degree of branching (DB) = 0%] and dendrimers (DB = 100%) with an approximate DB of between 50% and 75% [127]. The high potential loading capacities of these hyperbranched polymers are similar to those for dendrimers (5–14 mmol g^{-1}) and some hyperbranched polymers are commercially available [128]. The use of these commercial materials as supports for organic synthesis, however, is limited because of the chemical stability of the respective polymer backbone (e.g. polyamines, polyesters) and the relatively broad molecular weight distributions (typically >2).

Hyperbranched polymers (Fig. 3.10) have recently been introduced as soluble supports in organic synthesis [103, 104]. A commercial polyester support (Boltorn **54**) containing 1,3-diols as terminal units was used for the synthesis of disaccharides (Scheme 3.6) [103]. Theoretically, these polyesters have a relatively high loading capacity (8.8 mmol OH g^{-1}), but the experimentally achieved loading with monosaccharides attached to a photo-labile linker was 0.8 mmol g^{-1}, additionally reduced by the weight of the linker functionality. The general use of this hyperbranched polymer (**54**) as a soluble polymeric support, however, is limited as a result of the chemical sensitivity and the low molecular weights of the polyester backbone [129]. This is especially problematic for the separation by dialysis or ultrafiltration with a minimum molecular weight cut-off (MWCO) of 1000 g mol^{-1} (see Table 3.3).

Recently, the controlled synthesis of well-defined hyperbranched polyglycerols (**55**) (R = H) has been achieved, by using both racemic and enantiomerically pure glycidol monomers [130, 131]. These polyether polyols are conveniently prepared in a one-step synthesis on a kilogram scale [132], and possess molecular weights (M_n) up to 30.000 g mol^{-1} with molecular weight distributions typically below 1.5.

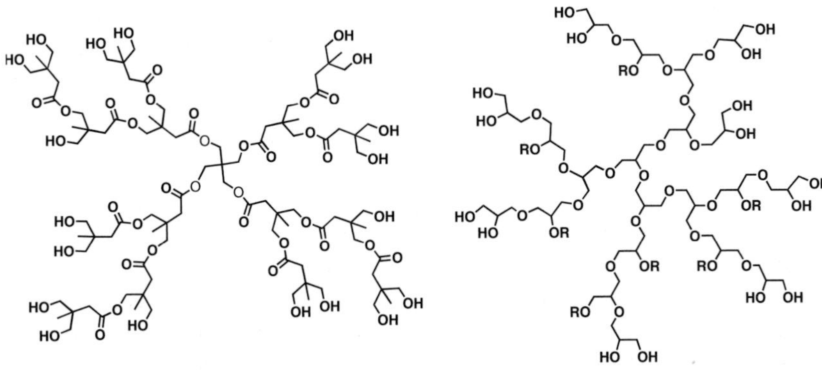

Fig. 3.10. Dendritic polyester (**54**) and polyether (**55**) (R = H, alkyl, glyceryl) as high-loading soluble polymeric supports.

The dendrimer-like structure of the hyperbranched polyglycerol (**55**) is characterized by exactly one core unit with multiple hydroxyl groups randomly incorporated as linear (OH groups) and terminal groups (1,2-diols). The total density of functional groups in polymer **55** (R = H) is 13.5 mmol OH g^{-1} polymer, of which approximately 60% (8.2 mmol OH g^{-1}) are terminal 1,2-diols. These terminal diols can be used directly as linker functionalities for many applications in organic synthesis [10].

The complete derivatization of the terminal diols in polyglycerol (**55**) (R = H), with, for example, acetals, leaves about 40% of the OH groups unaffected (see Scheme 3.3). These remaining OH groups might limit the scope of this new polymeric support for some synthetic applications. For the preparation of a chemically inert polyether support (**55**) (R = alkyl) these residual OH groups can be selectively alkylated by using phase transfer conditions to obtain dendritic polymers with exclusively diol linkers [133]. This approach also permits the introduction of a second type of functional group and tunes the solubility of the polymer in various organic solvents. In order to increase the loading capacity of the dendritic polyglycerol support further, the linear glycerol units can also be converted into terminal 1,2-diols [124]. This strategy increases the capacity of the terminal 1,2-diol units from 4.1 mmol g^{-1} for the hyperbranched polymer (**55**) (R = H) to 7.1 mmol g^{-1} for the dendritic structure (**55**) (R = glyceryl) and preserves all advantages of the polyether scaffold.

For a more general application of polyglycerol (**55**) in organic synthesis and in order to increase the scope of possible reactions on this support, the conversion of the hydroxyl groups into various other linker functionalities by postsynthetic transformations has been explored [134]. In one or two synthetic steps after the polymerization several reactive linker functionalities, such as aldehydes, alkenes, amines, carbonic acids and esters, are accessible in good yields and with high loading [135]. Dendritic polyglycerols (**55**) are readily soluble in many organic solvents and can easily be separated from low-molecular-weight compounds by dialysis or ultrafiltration with a typical MWCO of 1000 g mol^{-1}. Because of the high flexibility of the branches, the terminal 1,2-diol groups show excellent accessibility and can be used directly as linker groups for various applications in organic synthesis [104]. In addition the characterization of these dendritic polyethers by standard analytical techniques (e.g. NMR) is much more rapid than the monofunctionalized PEG derivatives owing to the high loading capacity of functional groups [104].

3.3.5
Microgels

Recently, Wulff et al. introduced "microgels" as a new type of polymeric supports for use in parallel synthesis [136]. Microgels are intramolecularly crosslinked polymer molecules that build stable solutions in suitable solvents [137]. Oxazaborolidine functionalized PS-microgels have been prepared and used as catalysts in enantioselective reduction of prochiral ketones [136]. In contrast to gel-type polymer supports (see Section 3.2.1.3), microgels offer the advantage of higher loading capacities and homogeneous reaction conditions. Like dendritic polymeric supports (see Section 3.3.4), they have a very low solution viscosity, which simplifies the handling of the reaction mixture. Also, separation can be performed by ultrafiltration and the process can be performed in an automated or continuous mode.

3.4
Conclusions

After more than 10 years of intensive research on the preparation and the application of new polymeric supports in combinatorial chemistry many new polymeric materials have been developed and evaluated. However, there is still no polymeric support for general application in combinatorial chemistry. Every polymer has its drawbacks (e.g. chemical stability, polarity, loading capacity). Therefore, a polymeric support has to be carefully selected for the synthetic problem that needs to be solved. In solid-phase organic synthesis several new supports with better swelling properties in a wider range of organic solvents and higher loading capacities have been introduced. Among these are several new PS hybrid resins as well as a new family of crosslinked PEGs. In terms of material stability, inorganic supports, i.e. controlled pore glass, cannot be beaten. However, the width of synthetic applications is very limited so far and will depend on the chemical stability of the spacer and linker molecules used. In addition, these supports suffer a heterogenic nature and rather low loading capacities.

Soluble polymeric supports, like solid supports, have had a similar revival over the past decade. In terms of stability, aliphatic polyethers and noncrosslinked polystyrene are among the most promising candidates. Dendritic and linear polyfunctional soluble polymers have by far the highest loading capacities and show great potential as supports for reagents and catalysts in combinatorial synthesis because of their homogeneous reaction conditions. The use of these soluble polymers for

SPOS, however, requires further progress in automation of solution-phase separation techniques. Nevertheless, the current trend in the pharmaceutical industry to move back to combinatorial solution-phase chemistry renders these high-loading soluble polymers as potential homogeneous supports for solution-phase synthesis.

References

1 R. B. MERRIFIELD, *J. Am. Chem. Soc.* **1963**, *85*, 2149–2154.

2 S. R. WILSON, A. W. CZARNIK, *Combinatorial Chemistry*, Wiley, New York, 1997.

3 D. HUDSON, *J. Comb. Chem.* **1999**, *1*, 333–360.

4 D. HUDSON, *J. Comb. Chem.* **1999**, *1*, 403–457.

5 J. W. LABADIE, *Curr. Opin. Chem. Biol.* **1998**, *2*, 346–352.

6 D. J. GRAVERT, K. D. JANDA, *Chem. Rev.* **1997**, *97*, 489–509.

7 K. E. GECKELER, *Adv. Polymer Sci.*, **1995**, *121*, 31–79.

8 P. WENTWORTH, K. D. JANDA, *Chem. Commun.* **1999**, 1917–1924.

9 P. H. TOY, K. D. JANDA, *Acc. Chem. Res.* **2000**, *33*, 546–554.

10 R. HAAG, *Chem. Eur. J.* **2001**, *7*, 327–335.

11 P. HODGE, *Chem. Soc. Rev.* **1997**, *26*, 417–424.

12 F. ZARAGOSA-DÖRWALD, *Organic Synthesis on Solid Phase*, Wiley-VCH, Weinheim, 2000.

13 A. AKELAH, A. MOET, *Functionalized Polymers and their Applications*, London, 1990.

14 X. HOHENSTEIN, US Patent No. 2524627, **1950**.

15 T. MIZAGUCHI, *Chem. Pharm. Bull.* **1970**, *18*, 1465–1474.

16 R. B. SCARR, M. A. FINDEIS, *Pept. Res.* **1990**, *3*, 238–241.

17 M. J. FARRALL, J. M. J. FRÉCHET, *J. Org. Chem.* **1976**, *41*, 3877–3882.

18 H. SELIGER, *Makromol. Chem.* **1973**, *169*, 83–93.

19 R. H. WILEY, *Pure Appl. Chem.* **1975**, *43*, 57–75.

20 D. C. SHERRINGTON, *Chem. Commun.* **1998**, 2275–2286.

21 V. K. SARIN, S. B. H. KENT, R. B. MERRIFIELD, *J. Am. Chem. Soc.* **1980**, *102*, 5463–5970.

22 S. R. MCALPINE, S. L. SCHREIBER, *Chem. Eur. J.* **1999**, *5*, 3528–3532.

23 J. KRESS, A. ROSE, J. G. FREY, W. S. BROCKLESBY, M. LADLOW, G. W. MELLOR, M. BRADLEY, *Chem. Eur. J.* **2001**, *7*, 380–383.

24 S. B. ROSCOE, J. M. J. FRECHET, J. F. WALZER, A. J. DIAS, *Science* **1998**, *280*, 270–273.

25 J. RADEMANN, M. BARTH, R. BROCK, G. JUNG, H.-J. EGELHAAF, **2001**, to be published.

26 S. V. LEY, I. R. BAXENDALE, R. N. BREAM, P. S. JACKSON, A. G. LEACH, D. A. LONGBOTTOM, M. NESI, J. S. SCOTT, R. I. STORER, S. J. TAYLOR, *J. Chem. Soc. Perkin I* **2000**, 3815–4195.

27 A. KIRSCHNING, H. MONENSCHEIN, R. WITTENBERG, *Angew. Chem., Int. Ed. Engl.* **2001**, *4*, 650–679.

28 F. SVEC, J. M. J. FRÉCHET, *Science* **1996**, *273*, 205–211.

29 J. SEIDL, J. MALINSKY, K. DUSEK, W. HEITZ, *Adv. Polym. Sci.* **1967**, *5*, 113–213.

30 R. KUNIN, E. MEITZNER, N. BORTNICK, *J. Am. Chem. Soc.* **1962**, *84*, 305–306.

31 J. R. MILLAR, D. G. SMITH, W. E. MARR, T. R. E. KRESMAN, *J. Chem. Soc.* **1963**, 218–224.

32 I. M. ABRAMS, J. R. MILLAR, *React. Funct. Polym.* **1997**, *35*, 7–22.

33 A. GUYOT, *Synthesis and Separations Using Functional Polymers*, Chap. 1, Wiley, Chichester, UK, 1988.

34 J. W. LABADIE, *Curr. Opin. Chem. Biol.* **1998**, *2*, 346–352.

35 M. HORI, D. J. GRAVERT, P. WENTWORTH, K. D. JANDA, *Bioorg. Med. Chem. Lett.* **1998**, *8*, 2363–2368.

36 A. Guyot, *Pure Appl. Chem.* **1988**, *60*, 365–376.

37 J. M. J. Fréchet, *Makromol. Chem., Macromol. Symp.* **1993**, *289*, 70–71.

38 H. Köster, F. Cramer, *Liebigs Ann. Chem.* **1974**, 946–958.

39 C. McCollum, A. Andrus, *Tetrahedron Lett.* **1991**, *32*, 4069–4072.

40 T. Balakrishnan, W. T. Ford, *J. Appl. Polym. Sci.* **1982**, *27*, 133–138.

41 K. W. Pepper, H. M. Paisley, M. A. Young, *J. Chem. Soc.* **1953**, 4097–4105.

42 R. P. Pinnel, G. D. Khune, N. A. Khatri, S. L. Manatt, *Tetrahedron Lett.* **1984**, *25*, 3511–3514.

43 R. Arshasy, G. W. Kenner, A. Ledwith, *Makromol. Chem.* **1976**, *177*, 2911–2918.

44 W. T. Ford, S. A. Yacoub, *J. Org. Chem.* **1981**, *46*, 819–821.

45 S. Mohanraj, W. T. Ford, *Macromolecules* **1986**, *19*, 2470–2472.

46 Q. Sheng, H. D. H. Stöver, *Macromolecules* **1997**, *30*, 6712–6714.

47 S. P. Raillard, G. Ji, A. D. Mann, T. A. Baer, *Org. Process Res. Dev.* **1999**, *3*, 177–183.

48 S. Rana, P. White, M. Bradley, *J. Comb. Chem.* **2001**, *3*, 9–15.

49 A. R. Vairo, K. D. Janda, *J. Comb. Chem.* **2000**, *2*, 579–596.

50 S. Itsuno, I. Moue, K. Ito, *Polym. Bull.* **1989**, *21*, 365–370.

51 M. Renil, M. Meldal, *Tetrahedron Lett.* **1996**, *37*, 6185–6188.

52 P. B. Rheiner, H. Sellner, D. Seebach, *Chem. Eur. J.* **2000**, *6*, 3692–3705.

53 R. Santini, M. C. Griffith, M. Qi, *Tetrahedron Lett.* **1998**, *39*, 8951–8954.

54 S. Pickup, F. D. Blum, W. T. Ford, M. Periyasami, *J. Am. Chem. Soc.* **1986**, *108*, 3987–3990.

55 D. W. V. Krevelen, *Properties of Polymers*, 3rd edn, Elsevier, New York, 1990.

56 E. Bayer, *Angew. Chem., Int. Ed. Engl.* **1991**, *30*, 113–128.

57 B. D. Park, H. I. Lee, S. J. Ryoo, Y. S. Lee, *Tetrahedron Lett.* **1997**, *38*, 591–594.

58 H. Hellermann, H. W. Lucas, J. Maul, V. N. R. Pillai, M. Mutter, *Makromol. Chem.* **1983**, *184*, 2603–2617.

59 S. A. Kates, B. F. McGuiness, C. Blackburn, G. W. Griffin, N. A. Solé, G. Barany, F. Albericio, *Biopolymers* **1998**, *47*, 365–380.

60 P. Wright, D. Lloyd, W. Rapp, A. Andrus, *Tetrahedron Lett.* **1993**, *34*, 3373–3376.

61 O. W. Gooding, S. Baudart, T. L. Deegan, K. Heisler, J. W. Labadie, W. S. Newcomb, J. A. Porco Jr, P. V. Eikeren, *J. Comb. Chem.* **1999**, *1*, 113–122.

62 J. A. Porco, T. Deegan, W. Devonport, O. W. Gooding, K. Heisler, J. W. Labadie, B. Newcomb, C. Nguyen, P. V. Eikeren, J. Wong, P. Wright, *Mol. Diversity* **1997**, *2*, 197–206.

63 V. Smali, N. J. Wells, G. J. Langley, M. Bradley, *J. Org. Chem.* **1997**, *62*, 4902–4903.

64 J. H. Adams, R. H. Cook, D. Hudson, V. Jammalamadaka, M. H. Lyttle, M. F. Songster, *J. Org. Chem.* **1998**, *63*, 3706–3716.

65 E. Bayer, W. Rapp in: Poly(ethylene glycol) chemistry: Biotechnical and Biomedical Applications, Harris, M. (ed.), Plenum Press, New York, 1992, p. 325.

66 A. G. M. Barrett, S. M. Cramp, R. S. Roberts, *Org. Lett.* **1999**, *1*, 1083–1086.

67 M. R. Buchmeiser, K. Wurst, *J. Am. Chem. Soc.* **1999**, *121*, 11101–11107.

68 T. Arnauld, A. G. M. Barrett, S. M. Cramp, R. S. Roberts, F. J. Zécri, *Org. Lett.* **2000**, *2*, 2663–2666.

69 C. W. Lindsley, J. C. Hodges, G. F. Filzen, B. M. Watson, A. G. Geyer, *J. Comb. Chem.* **2000**, *2*, 550–559.

70 S. R. McAlpine, C. W. Lindsley, J. C. Hodges, D. M. Leonard, G. F. Filzen, *J. Comb. Chem.* **2001**, *3*, 1–5.

71 V. Swali, N. J. Wells, G. J. Langley, M. Bradley, *J. Org. Chem.* **1997**, *62*, 4902–4903.

72 A. Mahajan, S. R. Chabra, W. C. Chan, *Tetrahedron Lett.* **1999**, *40*, 4909–4912.

73 H. Rink, *Tetrahedron Lett.* **1987**, *28*, 3787–3790.

74 C. Fromont, M. Bradley, *Chem. Commun.* **2000**, 283–284.

75 A. Basso, B. Evans, N. Pegg, M. Bradley, *Chem. Commun.* **2001**, 697–698.

76 A. Arya, N. V. Rao, J. Singkhonrat, *J. Org. Chem.* **2000**, *65*, 1881–1885.

77 N. R. Cameron, D. C. Sherrington, *Adv. Pol. Sci.* **1996**, *126*, 163–214.

78 P. W. Small, D. C. Sherrington, *Chem. Commun.* **1989**, *21*, 1589–1591.

79 E. Atherton, E. Brown, R. C. Sheppard, *Chem. Commun.* **1981**, 1151–1152.

80 M. Meldal, *Tetrahedron Lett.* **1992**, *33*, 3077–3080.

81 M. Renil, M. Meldal, *Tetrahedron Lett.* **1995**, *36*, 4647–4650.

82 M. Meldal, F. I. Auzanneau, O. Hindsgaul, M. M. Palcic, *Chem. Commun.* **1994**, 1849–1850.

83 M. Meldal, F. I. Auzanneau, O. Hindsgaul, M. M. Palcic, *Chem. Commun.* **1994**, 1949–19850.

84 P. M. S. Hilaire, M. Meldal, *Angew. Chem.* **2000**, *112*, 1210–1228.

85 E. Atherton, D. L. J. Clive, R. C. Sheppard, *J. Am. Chem. Soc.* **1975**, *97*, 6584–6585.

86 G. L. Stahl, R. Walter, C. W. Smith, *J. Am. Chem. Soc.* **1979**, *101*, 5383–5394.

87 C. W. Smith, G. L. Stahl, R. Walter, *Pept. Prot. Res.* **1979**, *13*, 109–112.

88 M. Kempe, G. Barany, *J. Am. Chem. Soc.* **1996**, *118*, 7083–7093.

89 J. Rademann, M. Grøtli, M. Meldal, K. Bock, *J. Am. Chem. Soc.* **1999**, *121*, 5459–5466.

90 M. Grøtli, C. H. Gotfredsen, J. Rademann, J. Buchardt, A. J. Clark, J. Ø. Duus, M. Meldal, *J. Comb. Chem.* **2000**, *2*, 108–119.

91 J. Rademann, M. Meldanl, K. Bock, *Chem. Eur. J.* **1999**, *5*, 1218–1225.

92 M. Renil, M. Meldal, *Tetrahedron Lett.* **1996**, *37*, 6185–6188.

93 E. Bayer, G. Jung, I. Halász, I. Sebastian, *Tetrahedron Lett.* **1970**, 4503–4505.

94 R. T. Pon, K. K. Ogilvie, *Tetrahedron Lett.* **1984**, *25*, 713–716.

95 J. W. Engels, W. Uhlmann, *Angew. Chem. Int. Ed.* **1989**, *28*, 716–734.

96 P. K. Ghosh, P. Kumar, K. C. Gupta, *J. Indian Chem. Soc.* **1998**, *75*, 206–218.

97 K. K. Ogilvie, M. J. Nemer, *Tetrahedron Lett.* **1980**, *21*, 4159–4162.

98 H. Köster, J. Biernat, J. McManus, A. Wolter, A. Stumpe, C. K. Narang, N. D. Sinha, *Tetrahedron Lett.* **1984**, *40*, 103–112.

99 For a recent example, see: K. D. Smet, S. Aerts, E. Ceulemans, I. F. J. Vanklecom, P. A. Jacobs, *Chem. Comm.* **2001**, 597–598.

100 H. Köster, *Tetrahedron Lett.* **1972**, 1535–1538.

101 D. Paul, *Chem. i. u. Zeit* **1998**, *32*, 197–205.

102 H. Determann, K. Lampert, *Mitt. Dtsch. Pharmaz. Ges.* **1970**, *40*, 117–134.

103 A. B. Kantchev, J. R. Parquette, *Tetrahedron Lett.* **1999**, *40*, 8049–8053.

104 R. Haag, A. Sunder, A. Hebel, S. Roller, *J. Combinational Chem.*, **2001**, in press.

105 M. Mutter, E. Bayer, *Angew. Chem.* **1974**, *86*, 101–102.

106 M. Benaglia, R. Annunziata, M. Cinquini, F. Cozzi, S. Ressel, *J. Org. Chem.* **1998**, *63*, 8628–8629.

107 R. Annunziata, M. Benaglia, M. Cinquini, F. Cozzi, *Chem. Eur. J.* **2000**, *6*, 133–138.

108 J. Chang, O. Oyelaran, C. K. Esser, G. S. Kath, G. W. King, B. G. Uhrig, Z. Konteatis, R. M. Kim, K. T. Chapman, *Tetrahedron Lett.* **1999**, *40*, 4477–4480.

109 R. Knischka, P. Lutz, A. Sunder, R. Mülhaupt, H. Frey, *Macromolecules* **2000**, *33*, 315–320.

110 N. N. Reed, K. D. Janda, *Org. Lett.* **2000**, *2*, 1311–1313.

111 H. Schott, F. Brandstetter, E. Bayer, *Makromol. Chem.* **1973**, *173*, 247–251.

112 H. Schott, *Angew. Chem.* **1973**, *85*, 263–264.

113 H. Schott, F. Brandstetter, E. Bayer, *Makromol. Chem.* **1975**, *176*, 2163–2175.

114 J. J. V. Eynde, D. Rutot, *Tetrahedron* **1999**, *55*, 2687–2694.

115 D. A. Wellings, A. Williams, *Reactive Polymers* **1987**, *6*, 143–157.

116 E. RANUCCI, G. SPAGNOLI, L. SARTORE, P. FERRUTI, P. CALICETI, O. SCHIAVON, F. VERONESE, *Makromol. Chem. Phys.* 1994, *195*, 3469–3479.

117 A. G. M. BARRETT, S. M. CRAMP, R. S. ROBERTS, F. J. ZECRI, *Org. Lett.* 1999, *1*, 579–582.

118 A. G. M. BARRETT, S. M. CRAMP, R. S. ROBERTS, F. J. ZECRI, *Org. Lett.* 2000, *2*, 261–264.

119 A. G. M. BARRETT, S. M. CRAMP, R. S. ROBERTS, F. J. ZÉCRI, *Comb. Chem. High Throughput Screening* 2000, *3*, 131–133.

120 G. R. NEWKOME, C. N. MOOREFIELD, F. VÖGTLE, *Dendrimers and Dendrons: Concepts, Syntheses, Perspectives*, Wiley-VCH, Weinheim, 2001.

121 R. M. KIM, M. MANNA, S. M. HUTCHINS, P. R. GRIFFIN, N. A. YATES, A. M. BERNICK, K. T. CHAPMAN, *Proc. Natl. Acad. Sci. USA* 1996, *93*, 10012–10017.

122 N. J. HOVESTAD, A. FORD, J. T. B. H. JSTRZEBSKI, G. V. KOTEN, *J. Org. Chem.* 2000, *65*, 6338–6344.

123 S. M. GRAYSON, M. JAYARAMAN, J. M. FRÉCHET, *Chem. Commun.* 1999, 1329–1330.

124 R. HAAG, A. SUNDER, J.-F. STUMBÉ, *J. Am. Chem. Soc.* 2000, *122*, 2954–2955.

125 C. SCHLENK, A. W. KLEIJ, H. FREY, G. V. KOTEN, *Angew. Chem.* 2000, *112*, 3587–3589.

126 R. BREINBAUER, E. N. JACOBSEN, *Angew. Chem.* 2000, *112*, 3750–3753.

127 A. SUNDER, J. HEINEMANN, H. FREY, *Chem. Eur. J.* 2000, *6*, 2499–2505.

128 H. FREY, R. HAAG, in Encyclopedia of Materials, Science and Technology. MÜLHAUPT, R., KRAMER, E. (eds), Elsevier, New York, 2001.

129 A. BURGATH, A. SUNDER, H. FREY, *Macromol. Chem. Phys.* 2000, *201*, 782–791.

130 A. SUNDER, R. HANSELMANN, H. FREY, R. MÜLHAUPT, *Macromolecules* 1999, *32*, 4240–4246.

131 A. SUNDER, R. MÜLHAUPT, R. HAAG, H. FREY, *Macromolecules* 2000, *33*, 253–254.

132 For further information, see: www.hyperpolymers.com.

133 R. HAAG, J. F. STUMBÉ, A. SUNDER, H. FREY, A. HEBEL, *Macromolecules* 2000, *33*, 8158–8166.

134 A. SUNDER, H. TÜRK, R. HAAG, H. FREY, *Macromolecules* 2000, *33*, 7682–7692.

135 A. SUNDER, R. MÜLHAUPT, R. HAAG, H. FREY, *Adv. Mater.* 2000, *12*, 235–239.

136 C. SCHUNICHT, A. BIFFIS, G. WULFF, *Tetrahedron* 2000, *56*, 1693–1699.

137 For a review see: M. ANTONIETTI, *Angew. Chem. Int. Ed. Engl.* 1988, *27*, 1743–1747.

4
Linkers for Solid-phase Synthesis

Stefan Bräse and Stefan Dahmen

4.1
Introduction

Solid-phase organic chemistry is one of the key tools in combinatorial chemistry [1–6] used to synthesize large compound libraries of potential new drugs and other biologically active compounds, especially in automated synthesis. With the aid of linkers that are capable of attaching building blocks or intermediates onto solid support as well as facilitating their ultimate release into solution, the synthetic gap between solid and liquid phase is diminished [7–10].

Linkers and their associated synthesis strategies, therefore, play a pivotal role in the successful implementation of solid-phase organic synthesis and its application to combinatorial chemistry [11].

> ■ Linker: *bifunctional chemical moiety attaching a compound to a solid support or soluble support which can be cleaved to release compounds from the support. A careful choice of the linker allows cleavage to be performed under appropriate conditions compatible with the stability of the compound and assay method [12].*

In solid-phase synthesis, a starting material is attached reversibly to a linker, which is bound again directly or over a spacer (Sect. 4.2.2, for example a polyethylene glycol chain such as that shown in Fig. 4.1) to the actual resin (usually with divinylbenzene crosslinked polystyrene, see Chapter 3). Anchoring groups for the use of soluble polymers [MeO-polyethylene glycol (PEG)] are in principle the same as for the use of insoluble polymers. Therefore, throughout this chapter they will not be distinguished.

While the group attached to the solid support is in general unchanged upon cleavage, the anchoring bond to the compound is sensitive to certain conditions, which then lead to bond breakage and release of the final compounds. Traditionally, linkers were designed to release one functional group and hence acted more or less as bulky protecting groups (see Sect. 4.3). Therefore, the release of carbox-

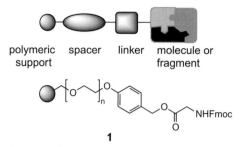

polymeric   spacer   linker   molecule or
support   fragment

1

Fig. 4.1. Linker construct.

ylic acids and amines, which are essential for peptide synthesis, has been studied extensively. However, the synthesis of nonoligomeric compound libraries requires more versatile linkers (see Sect. 4.5) [13].

A particular linker should resist the entire synthesis sequence without bias to the diversity or structure of the target compound library and without limiting the chemical methodology. None of the reaction conditions used for the construction of the building blocks should lead to a premature cleavage (orthogonality principle; Sect. 4.3.8). Beyond that, the linker must be cleavable off the resin as mildly as possible to insure that the product will not be affected.

In this chapter, linker types classified according to their structures, and cleavage strategies are presented. In general, linkers for all kinds of building blocks (peptides, oligosaccharides, nucleic acids and small molecules) are covered.

4.2
General Linker Structures

4.2.1
Immobilization of Molecules

In general, the anchoring of molecules to a resin can be realized by two different strategies. Direct loading is clearly the most straightforward technique for the set-up of a solid-phase organic synthesis. A molecule with a reactive or potentially reactive functional group is coupled directly to the preformed linker. This strategy is useful if the linker and the building block can be coupled efficiently. Coupling rates greater than 90% are essential at this point. Successful examples are, for instance, the formation of an amide bond, reductive aminations, alkylation reactions (including Mitsunobu reactions), and olefin metathesis. Since the building blocks can be used without any additional purification step, this method is especially suitable for the anchoring of libraries of starting materials and/or automated synthesis. The attachment of molecules to a particular resin is strongly dependent on the nature of the linker and will be discussed for each linker separately. While in

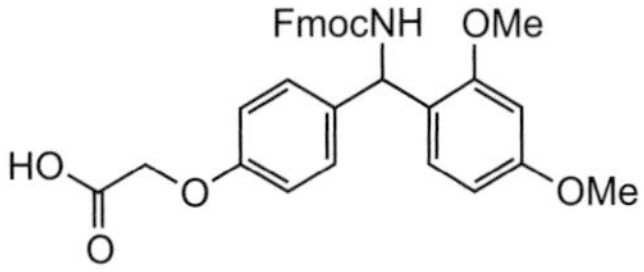

2

Fig. 4.2. The unloaded Knorr linker (**2**) as a handle.

some cases a simple and rapid mixing of reagents is sufficient to drive the attachment to completion, in other cases tedious monitoring is necessary. Additionally, in some cases the attachment proceeds under similar conditions as the detachment, for example the formation of ketals, where excess reagents are required to drive the reaction to completion.

In a second general method for attachment, the building block can be coupled to give the preformed linker in solution. The fragment thus formed, which is called the handle (see for example Fig. 4.2), can then be activated for attachment onto the resin using the coupling strategies outlined above. This method requires an additional step in solution, which can be advantageously used for an increase of purity. Especially in cases where the activated linker tends to decompose (e.g. silyl chlorides) or can be formed only in moderate yields and purities, this handle approach can be favorable. However, in automated synthesis, the required solution-phase step is clearly a drawback.

4.2.2
Spacers

The linker either can be attached directly to the resin or can be located further from the polymeric backbone using spacers. These bifunctional constructs either can be built sequentially or can be attached via a handle approach. The spacer acts as a connection to give the building block more mobility and hence the kinetics for a given reaction are superior to the corresponding resin without spacers [14]. Furthermore, larger spacers considerably change the physical properties, such as swelling behaviors. This has been demonstrated with Tentagel resin [15]. However, one must take into account both that the use of spacers requires an additional synthetic step usually connected with a decrease of loading capacity and that the spacer has to be as robust as the linker toward the reaction conditions performed on the bead; Tentagel resin, in particular, leads to bleeding.

For the characterization of compounds attached to a polymeric support by nuclear magnetic resonance (NMR) spectroscopy, however, long spacers are an advantage, because they increase the mobility of the substrate and reduce the line broadening usually observed in NMR spectra of polymers.

In most cases, a clear distinction between linker and spacer is not easy. The linker is the minimal part of the resin required for the functional cleavage (for the silyl linkers this is the silyl group, for the trityl linkers this is the triphenylmethyl moiety, for the triazene linker this is the 1-aryltriazenyl group, and so on).

The spacer therefore is the part between the linker and the resin as depicted in Fig. 4.1 and described in Chapter 3 of this book.

4.2.3
Functionalized Linkers as Analytical Constructs

Functionalized spacers or linkers can play an important role for the determination of loading and reaction conversion. The NMR-active fluorinated linkers are especially suitable in this regard [16].

4.3
Linker Families

Usually, linkers are categorized according to the kind of functional group or substrate class they are able to selectively immobilize (linkers for carboxylic acids, alcohols, amines, and so on). As there are various types of linkers available for solid-phase synthesis, many of them belong to certain well-established classes of protecting groups (Table 4.1) and therefore can be grouped into families. The members of each family have certain reactivity patterns in common.

While a linker presents the chemical structure essential for the loading and cleavage of a particular functional group, a linker system provides the whole protocol for the attachment to and cleavage from the resin.

However, because of an increased demand for the flexible anchoring of molecules, other families of linkers such as sulfur linkers, triazenes linkers, among others, have emerged. Nevertheless, the largest number of linkers developed so far are based on benzylic-type groups.

Tab. 4.1. Common protecting groups and the analogous linker families.

Protecting group in liquid phase	Functional group protected	Linkers or linker families
Benzyl	Alcohols, esters	Benzyl-type linkers: Sect. 4.3.1, Table 4.2
Allyl	Amines	Allyl-type linkers: Sect. 4.3.2
Cbz (Z)	Amines	Carbamate-based linkers: Sect. 4.3.4 [17]
Alloc	Amines	Allyl-type linkers: Sect. 4.3.2
Boc	Alcohols	Boc-type linker [18]
Silyl ether	Alcohols	Silyl-type linkers: Table 4.6. [19, 20]
Alkyl ester	Carboxylic esters	Sect. 4.3.4
SEM (trimethylsilylethoxymethyl)	Alcohols	SEM linker (**99**) [21]

4.3.1
Benzyl-type Linkers Including Trityl and Benzhydryl Linkers

Benzyl-type linkers are the most common anchoring groups for various kinds of functionalities. In particular, esters, amides, amines, alcohols, and thiols can be immobilized by this linker family.

As the pioneering works of Merrifield [22] and Wang [23] are based on this linker type, they represent the starting point of modern linker development. Benzylic linkers are typically cleaved by strong acids (e.g. trifluoroacetic acid, TFA), which causes a protonation and subsequent elimination. A nucleophilic scavenger usually quenches the resonance-stabilized cation thus formed. An increase of acid lability can thus be achieved by stabilization of this intermediate by, for example, *o*- and *p*-substitution of methoxy groups onto the ring [24]. This has been demonstrated in the development of the SASRIN resin (**11**; *super acid-sensitive resin*: 1% TFA cleavable) [25, 26] with one additional alkoxy group related to the Wang resin and the HAL linker (**13**; *hypersensitive acid labile*: 0.1% TFA cleavable) [27] having two additional alkoxy groups. In addition, benzyl-type linkers might be cleaved by ammonolysis [28], light (Sect. 4.4.3; see Table 4.15), metal salts [29] (see Scheme 4.8), and oxidation reagents (e.g. Wang resin with H_2O_2 [30] or dichlorodicyano-benzoquinone (DDQ) [31]). The introduction of nitro groups onto benzyl-type linkers leads to photolabile systems (see Sect. 4.4.3).

The prototype of a functional group with an appropriate breakable bond is the Wang resin (**6**), which contains a 4-hydroxybenzyl alcohol linker moiety. The benzyl alcohol hydroxyl group can be functionalized using either electrophilic or nucleophilic substrates (Scheme 4.1) to give a benzylic linkage. It is very stable in a whole set of reactions, but can be cleaved by acids such as trifluoroacetic acid or HF. Acids, alcohols, esters, and amides can be obtained as products after cleavage.

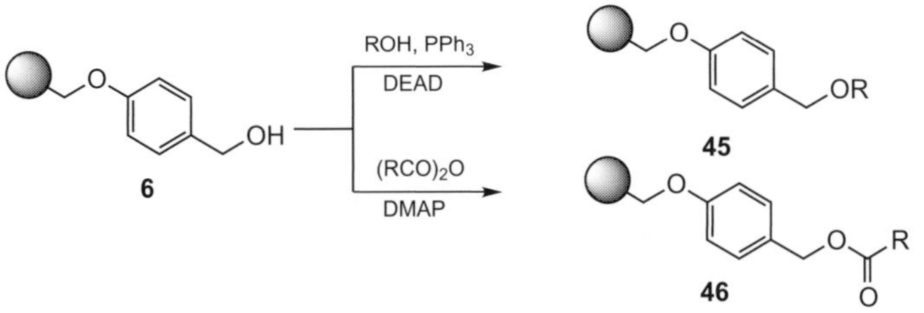

Scheme 4.1. Loading of the Wang resin (**6**). DEAD, diethyl azodicarboxylate. DMAP, 4-dimethylaminopyridine.

The Rink resin is particularly useful for the attachment of various functionalities such as primary amines (see for example [82]; Scheme 4.2) and is commercially available from various sources. The loading can be achieved via the Rink chloride (**29**) or triflate (**30**) (Table 4.2).

A linker that is particularly suitable for peptide amides and cyclopeptides is the PAL linker and the backbone amide linker (BAL) concept [83] (Scheme 4.3).

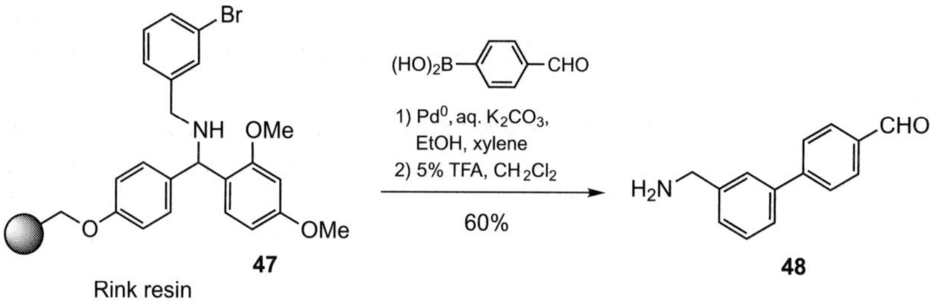

Scheme 4.2. The Rink resin as a linker for primary amines according to Garigipati [82].

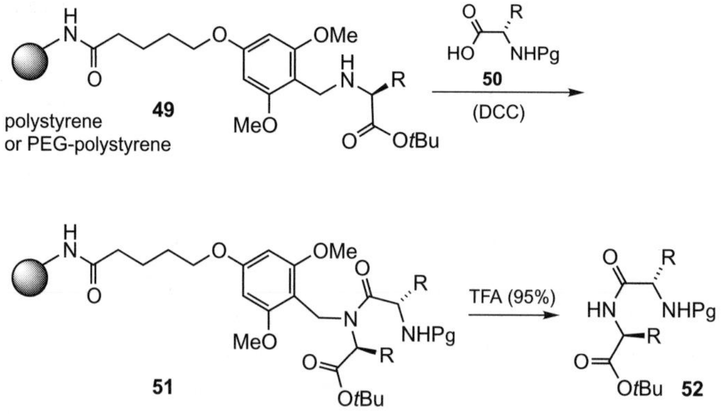

Scheme 4.3. Backbone amide linkage using the PAL linker by Barany and coworkers [83]. Pg, Protecting Group.

Trityl resins (**38–43**) (Table 4.2) are especially suitable for the immobilization of nucleophilic substrates such as acids, alcohols, thiols, and amines. They are quite acid sensitive and can be cleaved with acetic acid, for example; this feature is useful in cases where acid-labile protecting groups are used. The stability of trityl resin can be tailored using substituted arene rings, as demonstrated with the chlorotrityl resin, which furnishes a more stable linker than the trityl resin itself. Furthermore, the steric hindrance prohibits the formation of diketopiperazines during the synthesis of peptides. The orthogonality towards allyl-based protective groups was demonstrated in the reverse solid-phase peptide synthesis of oligopeptides [84] (Scheme 4.4).

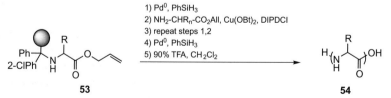

Scheme 4.4. Reverse (N → C) directed solid-phase peptide synthesis with the 2-chlorotrityl resin according to Thieriet et al. [84].

Tab. 4.2. Structures and properties of benzyl-type linkers including trityl and benzhydryl linkers (excluding photolabile linkers: see Sect. 4.4.3, Table 4.15).

Generic name of the resin	Structure	Cleavage	Comments
Merrifield resin (X = Cl: **3**) [22]; AM PS (Aminomethyl polystyrene) (X = NH$_2$: **4**); hydroxymethyl polystyrene (X = OH: **5**)	polystyrene	HF	Standard resin
Wang linker (X = OH: **6**) [23], Boba resin (X = NH$_2$: **7**) [32]		Cleavage for thiols: HF [33]; for esters: CF$_3$SO$_3$H [34]; 95% TFA [23]; H$_2$O$_2$ [30]	Standard resin; very robust; for Boc strategy
PAM (**8**) (phenylacetamido-methyl) [35–37]		Cleavage for esters: 90% HF, anisole [35]; stable in 50% TFA, CH$_2$Cl$_2$ [35]	Very robust
HMPA (**9**) (hydroxymethyl-phenoxyacetic acid) [38]		Cleavage for esters: 95% TFA [28, 39]; 20% TFA, CH$_2$Cl$_2$ (incomplete) [40]	
HMPP (**10**) (hydroxymethyl-phenoxypropionic acid) [40]		Cleavage for esters: 20% TFA, CH$_2$Cl$_2$ [40]	
SASRIN (**11**) (super acid-sensitive resin) [25]		Cleavage for esters: 1% TFA, CH$_2$Cl$_2$, 20 °C, 5 min [25]	Very sensitive
R = Me: AMEBA (**12**) (acid-sensitive methoxy-benzaldehyde) [41, 42]; MALDRE (α-methoxybenz-aldehyde resin) [43]; [44]; R = H: AHB linker [45]		Precursor of the SASRIN linker (R = Me)	
HAL (**13**) (hypersensitive acid labile) [27]		Cleavage for esters: 0.1% TFA (CH$_2$Cl$_2$), 25 °C, 1 h	Very sensitive
PAL (peptide amide linker) (**14**) [46, 47]	polystyrene or PEG polystyrene	Cleavage for amides: TFA [48]; TFA or HF [49] (quinoxalinone)	
BAL (backbone amide linker) (**15**) [50]	polystyrene or PEG polystyrene	Cleavage for esters: [14]	

Tab. 4.2. *(continued)*

Generic name of the resin	Structure	Cleavage	Comments
Extended HAL linker (**16**) [51]		Cleavage for amides: 25% TFA, CH$_2$Cl$_2$ [51]	
Ramage linker (**17**) [52, 53]		Cleavage for esters: nBu$_4$NF	
HMB (**18**) (Sheppard) [28]		Cleavage for amides: NH$_3$ [28]	
19 [54]	Synphase crowns	Cleavage for amides: 20% TFA, CH$_2$Cl$_2$ [54]	
SAL linker (**20**) [55] (*silyl amide linker*)		Cleavage for amides: 90% TFA, scavenger [55]	
SAC linker (**21**) [56] (*silyl acid linker*)		Cleavage for acids: TBAF [56]	
BHA (**22**) [57] (*benzylhydryl amin*)			Starting resin for various linkers
MBHA (**23**) (*methylbenzyl-hydryl amin*) [58]		Cleavage for amides: HF; CF$_3$SO$_3$H [58], HF/PhOMe (sec. amide) [59]	
MAMP linker (**24**) (*Merrifield, α-methoxybenzyl*) [60, 61]		Cleavage for amides: TFA/CH$_2$Cl$_2$/H$_2$O (9:90:1) [60]	
p-Acyloxy BHA (**25**) [62]		Cleavage for amides: HF, cresol (9:1) [62, 63]	

Tab. 4.2. *(continued)*

Generic name of the resin	Structure	Cleavage	Comments
SCAL (26) (safety-catch acid-sensitive linker) (Sect. 4.5.1) [64] (cf. [65])		Cleavage for amides: TFA/(EtO)$_2$(PS)SH [64]; Me$_3$SiBr/ PhSMe/TFA [64]	
Rink acid (X = OH: 27); Rink amide (X = NH$_2$: 28) [66, 67]; Rink chloride (X = Cl: 29); Rink triflate (X = OTf: 30)	polystyrene	Cleavage for esters: 10% AcOH, CH$_2$Cl$_2$, 20 °C, 1.5 h; 1% TFA [66]; cleavage for amides: TFA [66]; 20% TFA, CH$_2$Cl$_2$ [68]	Widely used anchor for amines
Rink amide AM (31) (RAM) (rink amide aminomethyl)			Similar to the Rink resin above
32 [69, 67]		Cleavage for esters: TBAF or Cs$_2$CO$_3$ [69]	
Sieber XAN (33) [70–72]		Cleavage for amides: [70]	
2-XAL (34), 3-XAL (35) (xanthyl acid-labile linker)	3-XAL$_4$	Cleavage for amides: TFA [71]	
CHA (36), CHE (37) (dibenzo[a,d]cyclohepta(e)ne) [73]	Tentagel CHA (CHE)	Cleavage for amides: 10% TFA, CH$_2$Cl$_2$ [73]	More sensitive than Rink or PAL [73]
Trityl (38) [74, 75]		Cleavage for esters: 1% TFA/AcOH [76, 77]	Quite unstable for esters
2-Chlorotrityl (39)		Cleavage for esters: (CF$_3$)$_2$CHOH (HFIP) (20%), CH$_2$Cl$_2$ [48]; AcOH/CF$_3$CH$_2$OH/ CH$_2$Cl$_2$	Increased stability compared to trityl linker

Tab. 4.2. *(continued)*

Generic name of the resin	Structure	Cleavage	Comments
4-Methyltrityl (**40**)	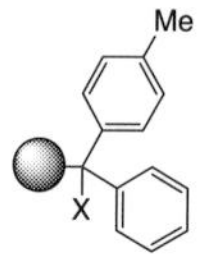	Cleavage for esters: AcOH/CF$_3$CH$_2$OH/ CH$_2$Cl$_2$ (1:1:8)	Increased stability compared to 2-chlorotrityl linker
4-Methoxytrityl (**41**)	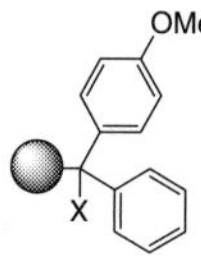	Cleavage for esters: AcOH/CF$_3$CH$_2$OH/ CH$_2$Cl$_2$ [78]	Increased stability compared to 4-methyltrityl linker
4-Fluoro(chloro)-tritylcarboxyl linker (**42**)	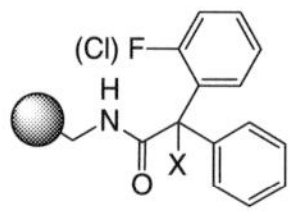	Cleavage for esters: 0.1% TFA [79], AcOH/CF$_3$CH$_2$OH/ CH$_2$Cl$_2$ (1:1:8)	
4-Cyanotrityl (**43**)	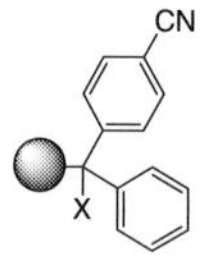	Cleavage for esters: [80]	
(**44**) [81]	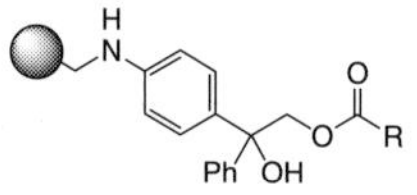	Cleavage for esters: TFA [81]	

Besides benzyl-type linkers, other arylmethyl moieties can serve as linkers. Recently, a new backbone amide linker has also been devised using indole chemistry [85] (Scheme 4.5).

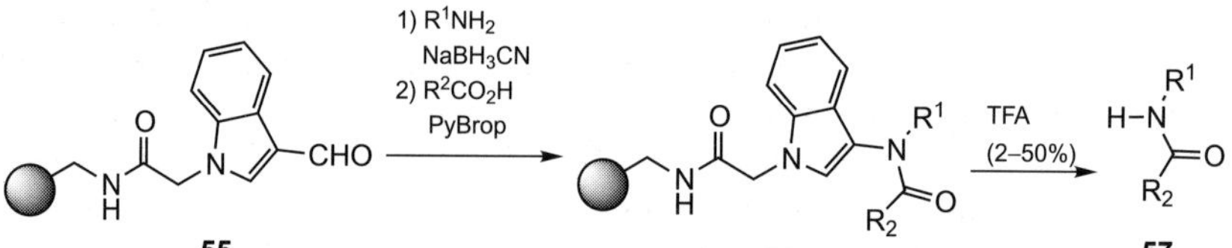

Scheme 4.5. The indole linker according to Estep et al. [85].

4.3.2
Allyl-based Linkers

A second, to the benzyl linkers related class is the family of the allyl-based linkers (Table 4.3). They have been used for the attachment of carboxylic acids, which can

Tab. 4.3. Overview of allylic linkers.

Structure	Reference
HYCRAM **58**	[86]
HYCRAM with β-Ala-AMPS **59**	[86]
HYCRON with β-Ala-AMPS **60**	[90, 91, 98]
Cellulose **61**	[93]
62	[94]

be detached using metal catalysis. The advantages of linker cleavage under palladium catalysis are the mild reaction conditions and the orthogonality (Sect. 4.3.8) to various protecting groups. Kunz et al. [86, 87] developed the first and simplest linker to use the π-allyl detachment strategy. Starting from 2-bromocrotonic acid, attachment to an amino group on a resin and further reaction with the cesium salt of an appropriate protected amino acid or peptidic structure yields the HYCRAM (*hydroxycrotonylamide*) resin [88]. The allylic cleavage proceeds with Pd(PPh$_3$)$_4$ and morpholine or hydroxybenzotriazole [89]. The readily available (hydroxycrotyl-oligoethyleneglycol-*n*-alkanoyl) HYCRON linker [90–92] is based on a similar concept; however, in this case, a handle comprising an amino acid and a preformed linker has been used to minimize the risk of racemization upon cleavage. A higher stability towards unwanted nucleophilic cleavage was achieved than that observed with the HYCRAM linker. The incorporation of β-alanine facilitates easier monitoring of the reaction. Several other similar constructs have been used for comparable purposes [93–96]. Recently, the semisynthesis of vancomycin on solid support was accomplished using an allylic linker [97].

An example for the use of the allyl linkers has been provided in the synthesis of peptide nucleic acids (PNAs) [91] (Scheme 4.6).

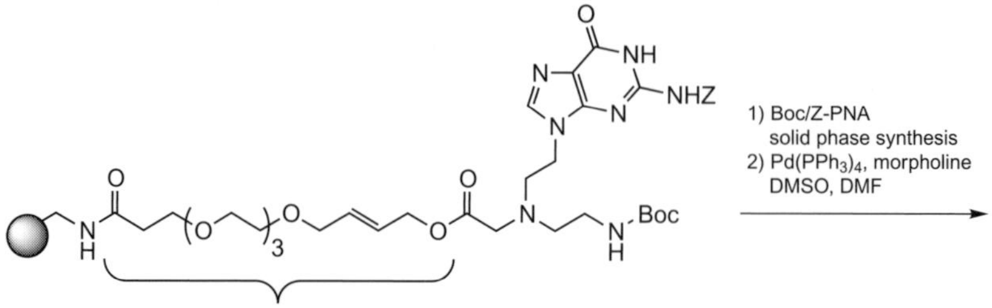

1) Boc/Z-PNA
   solid phase synthesis
2) Pd(PPh₃)₄, morpholine
   DMSO, DMF

HYCRON with AMPS   **63**

Ac-Gᶻ-Cᶻ-Cᶻ-Aᶻ-Cᶻ-Gᶻ-Gᶻ-OH

64

Scheme 4.6. Synthesis of PNAs using the HYCRON linker according to Seitz [91].

4.3.3
Ketal/acetal-based Linkers

Ketals and their corresponding sulfur analogs are well-established protecting groups in solution-phase synthesis. However, only a few constructs have been used in solid-phase organic synthesis (SPOS) as linkers (Table 4.4).

The most versatile ketal linker is the tetrahydropyranyl THP linker developed by Thompson and Ellman [99]. This linker allows the attachment of alcohols, phenols, and nitrogen functionalities in the presence of pyridinium toluene sulfonate. Both the loading and the detachment proceed under acidic conditions. Similarly, other acetal linkers have also been used for the attachment of alcohols [100, 106]. All these linkers are stable toward bases and organometallic reagents. However, a drawback is certainly the formation of diastereomers as a result of the chirality of these linkers.

Tab. 4.4. Overview of acetal/ketal-based linkers.

Structure	Reference	Comments
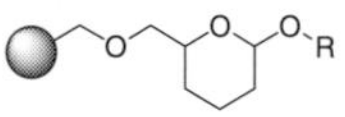 **65**	THP resin [99]	Most common linker for alcohols
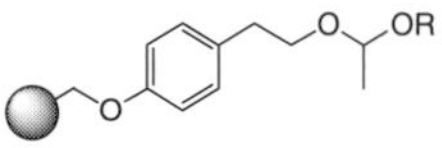 **66**	[100]	New linker for alcohols
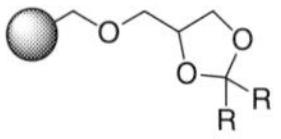 **67**	[101–104]	Common linker for ketones and aldehydes
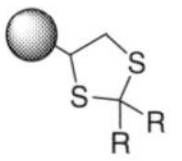 **68**	[105]	Linker for sterically hindered ketones

An example of the synthesis of complex molecules using THP-type linkers was provided by Chen and Janda [107] in their synthesis of prostaglandins on non-crosslinked polystyrene (Scheme 4.7). Further successful examples demonstrate the generality of the THP ketal-type linkers [108–114].

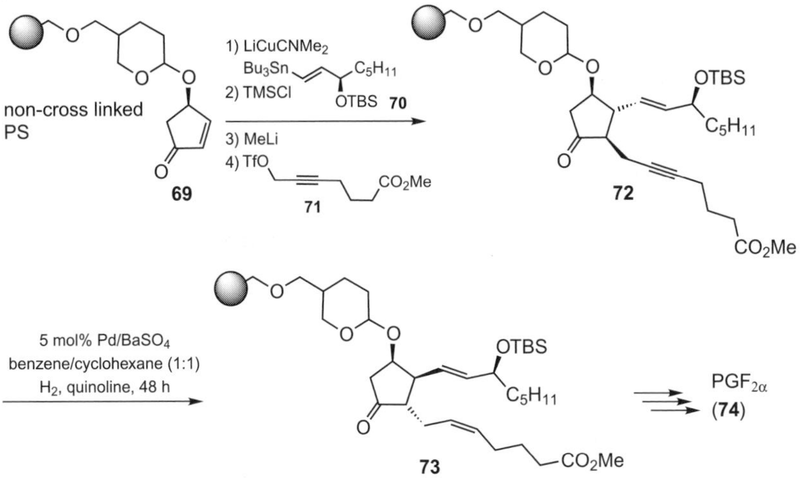

Scheme 4.7. Prostaglandin synthesis on solid phase according to Chen and Janda [107]. TBS, tert-butyldimethylsilyl.

Another class of linkers incorporating a ketal moiety are the Leznoff diol-linked aldehydes [101, 102] and ketones [103, 104]. A similar linker using dithianes is also suitable for ketones [105] (Sect. 4.3.7.1).

4.3.4
Ester-, Amide-, and Carbamate-based Linkers

Ester and amide moieties are, apart from being used in the benzyl and allyl linker types, also suitable for attachment. Various examples have demonstrated the useful applications of these functional groups (Table 4.5). Basically, two different strategies can be used (Fig. 4.3). For example, May et al. [115] recently introduced thioesters as linkers for alcohols, ketones, and lactones.

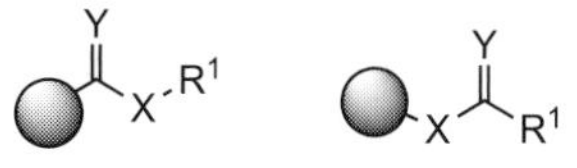

$X = O, NR^2, N=N; Y = O, S, NR^2$

75   **76**

Fig. 4.3. Ester linkers: general structures.

Tab. 4.5. Overview of some ester- and amide-based linkers (see also Table 4.16).

Structure	Reference	Possible structures achievable
77 **78**	e.g. [116]	Esters: cleavage with methoxide; alcohols: cleavage by Grignard reagents [117] Alcohols: cleavage with methoxide
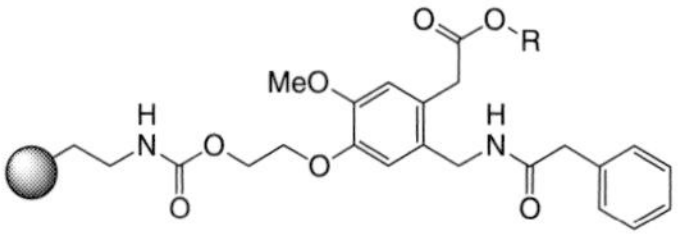 **79**	[118]	Alcohols: cleavage with enzymes
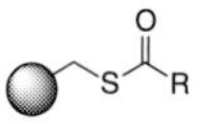 **80**	[115]	Ketones, amides: cleavage with reducing reagents
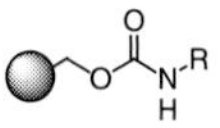 **81**	[17]	Amines: cleavage with Pd(OAc)$_2$, H$_2$ (45 psi), DMF, rt, 16 h
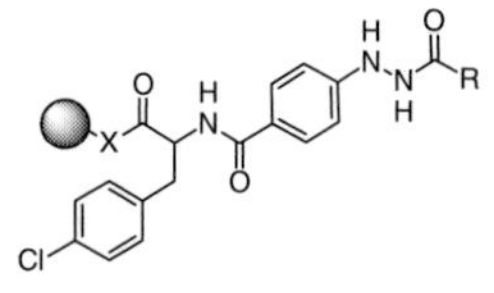 **82** TentaGel	[119]	Amides, esters: cleavage with Cu(II) salts
83	[120]	Methyl arenes: cleavage under photolytic conditions

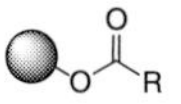

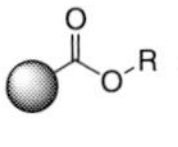

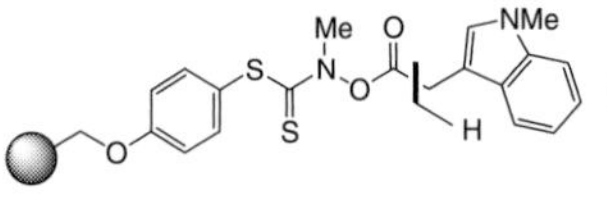

Benzyl carbamates are also useful linkers for the synthesis of amines since they are readily cleavable by palladium salts [17] (Scheme 4.8).

Carboxylic esters have been released by the action of alkoxides on ester resins. In most cases, the cleavage has been performed using methoxide in methanol (e.g. [116]; Scheme 4.9). Drawbacks to this method are the need to remove excess metal salts and/or the aqueous work-up. Alternatively, after cleavage from solid support, postsynthesis of methyl esters with diazomethane is also possible [121] (Scheme 4.10).

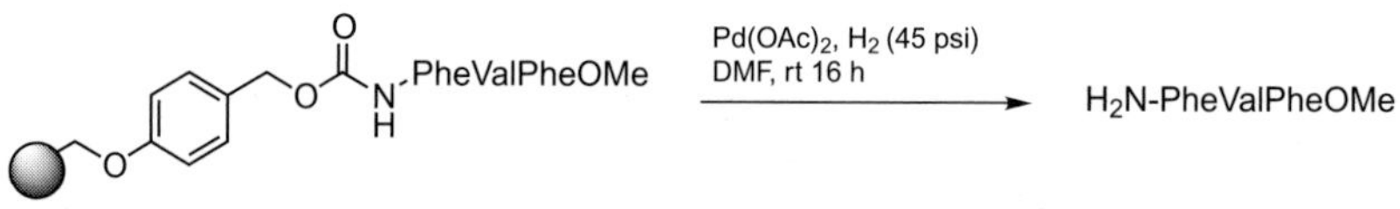

84
polystyrene or TentaGel

Pd(OAc)$_2$, H$_2$ (45 psi)
DMF, rt 16 h
→ H$_2$N-PheValPheOMe

85

Scheme 4.8. Detachment of peptides from polymeric benzyl-type protecting groups [17].

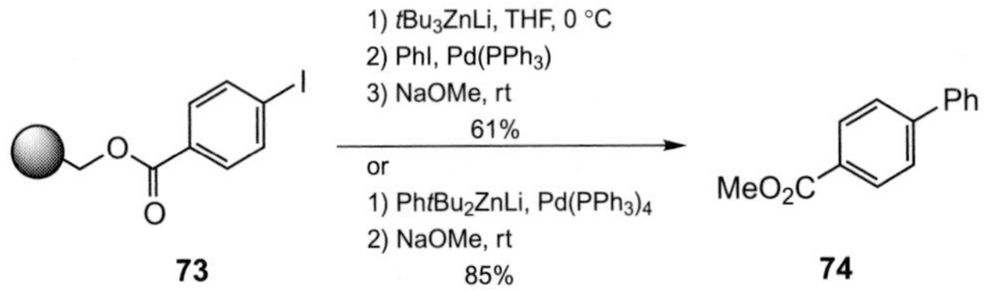

Scheme 4.9. The use of esters as linkers for benzoic acid derivative by Kondo et al. [116].

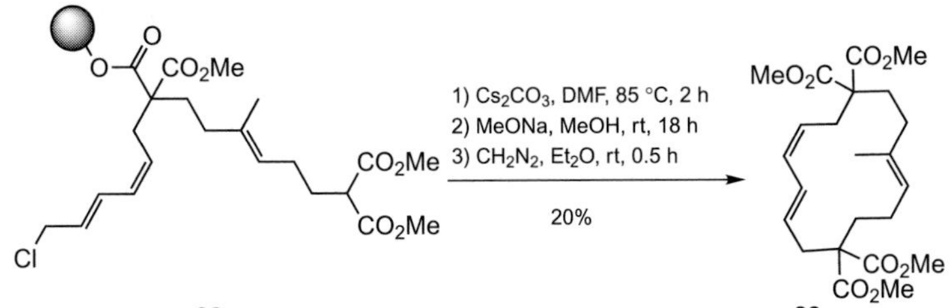

Scheme 4.10. Synthesis of macrocycles by Soucy et al. [121].

Arylhydrazides can serve as safety-catch linkers for C-terminal carboxylic acid, amide, or ester functionalities. The cleavage proceeds via an oxidation with copper(II) and subsequent cleavage of the diazenyl moiety by means of a nucleophile (Scheme 4.11) [119].

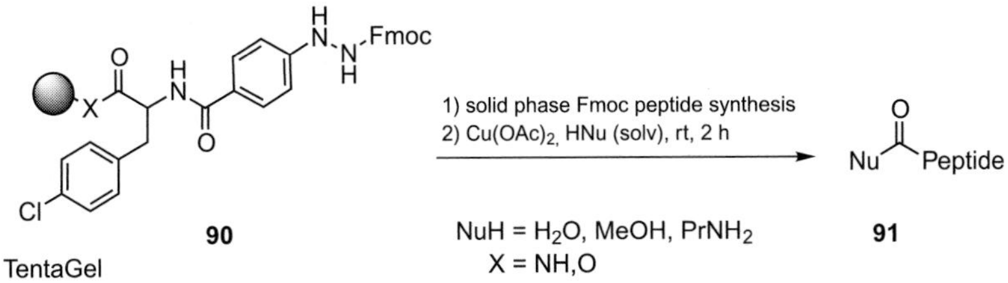

Scheme 4.11. Detachment of peptides from hydrazide resins by Millington et al. [119].

4.3.5
Silyl Linkers

The electronic and steric properties of silicon compounds have been used in many applications for the design and use of new linker types (Table 4.6).
The different applications can be divided as follows:

1 Direct attachment of building blocks on silylated resins: linker for alcohols [19, 20], traceless linkers for arenes (Sect. 4.5.5).
2 Use of the β-silicon effect for elimination reactions: e.g. the SEM (2-trimethylsilylethoxymethyl) linker [21].
3 Silylated benzhydryl linkers: [69, 122].

Tab. 4.6. Overview of silyl linkers.

Structure	Reference	Possible structures achievable
Si O R nBu nBu **92**	[123]	Alcohols; cleavage with (Tetrabutylammonium fluoride) TBAF
(structure) Bn OR H $SiMe_3$ **93**	[21]	Alcohols
(structure) O Si H, X **94**	[124]	Traceless linking of arenes (Scheme 4.60)
(structure) Si —H, Br **95** X	[125]	Traceless linking of arenes
(structure) O O Si —H,I,Br **96** X	[126]	Traceless linking of arenes
(structure) O Si R R —H **97** X	[127–129]	Traceless linking of arenes; cleavage with fluoride
(structure) NHR $SiMe_3$ SAL linker (**20**)	[55]	Amides; cleavage with 90% TFA, scavenger
(structure) OR $SiMe_3$ SAC linker (**21**)	[56]	Acids; cleavage by TBAF [56]
(structure) Ph tBu Si O X Pbs linker ("silico Wang linker") (**98**) BHA or methyl polystyrene	[122]	Carboxylic acids; cleavage with TBAF
(structure) $SiMe_3$ X Ramage linker (**15**)	[52, 53]	Carboxylic acids; cleavage with TBAF
(structure) X O $SiMe_3$ SEM linker (**99**)	[21]	Alcohols; cleavage with TBAF
(structure) X $SiMe_3$ **32**	[67, 69]	Estes; cleavage with TBAF or Cs_2CO_3 [69]

The robustness of silicon linkers against basic and organometallic reagents makes them especially suitable for solid-phase organic synthesis. Cleavage can be affected by electrophiles such as protons (trifluoroacetic acid; TFA). A special feature of silyl linkers is their sensitivity to fluoride ions, which makes them ideally orthogonal (Sect. 4.3.8) to various other functionalities present in the molecule. The fine-tuning of electronic and steric properties is possible by using different substituents on the silicon atom (trimethylsilyl vs. *tert*-butyldimethylsilyl).

The first traceless linkers (Sect. 4.5.5) for arenes were described independently by Ellman and coworkers [125, 130] and Chenera et al. [124] (see Scheme 4.60) in the 1990s using silyl linkers. This linker type was used in Ellman's synthesis of a benzodiazepine library, which is a milestone in the solid-phase synthesis of small organic molecules (Scheme 4.12). The synthesis of the linker involves a lithium–halogen exchange and, after chlorosilane attachment, coupling with an aryl halide. As the silyl arene might be cleaved in the unwanted direction to give silylated arenes, further improvement led to the development of a germanium linker [131], which gives rise to the formation of pure material (Sect. 4.5.5, see Scheme 4.61). In addition, cleavage of these linkers can be accomplished by electrophiles other than protons to yield halogenated residues, thus rendering this linker into a multiple cleavage linker system [see 379] (Sect. 4.5.6). The range of electrophiles is limited, since only small, reactive electrophiles (e.g. iodine, bromine, and chlorine) react to give the desired products.

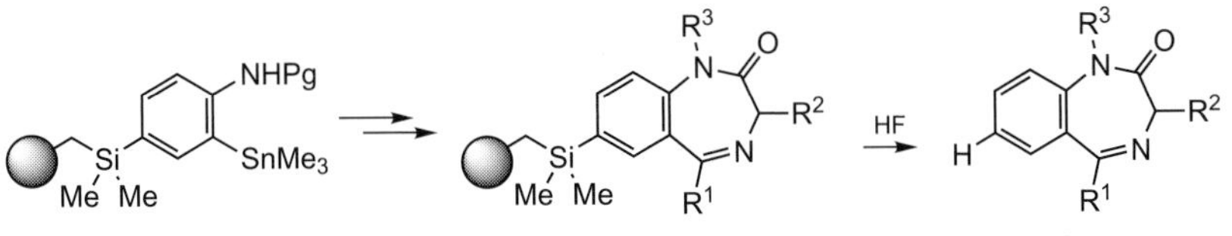

Scheme 4.12. Synthesis of a benzodiazepine library using silyl linker [131].

The silicon-based linkers, which are commercially available [132], were developed further by several groups [133] and have been used in the synthesis of various systems for the traceless detachment of various arenes and heteroarenes [126–128, 133–143]. They are also suitable as traceless linkers for allyl silanes to give alkenes [144] (Sect. 4.5.5).

A recent traceless application was demonstrated in the synthesis of chromenones (**104**) [145]. In this case, a mild cyclization method was used to circumvent a premature cleavage for the support (Scheme 4.13). Similarly, heteroarenes are also accessible [128] (Scheme 4.14).

Other linkers having silyl fragments are the silyl acid linker (SAC linker **21**) [56], the silyl amide linker (SAL linker **20**) [55], the Pbs linker ("silico Wang linker") (**98**) [122], the Ramage silyl linker (**15**) [52], the SEM linker (**99**) [21], and silylated benzhydryl linkers (**32**) [69].

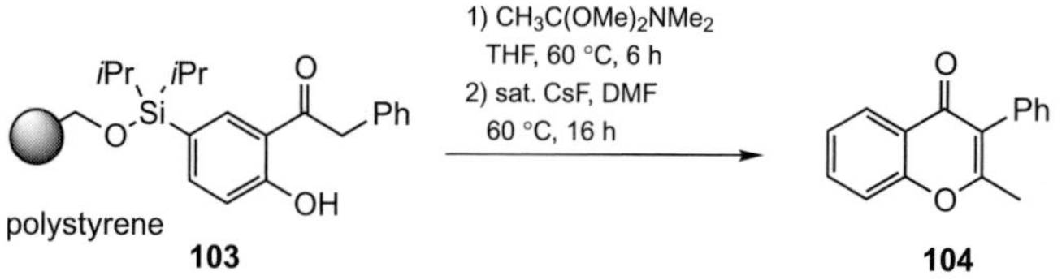

Scheme 4.13. Silicon linker for the synthesis of chromenones **104** by Harikrishnan and Showalter [145].

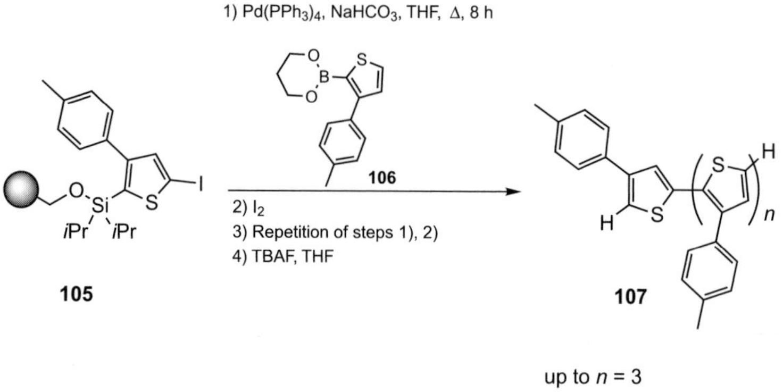

up to $n = 3$

Scheme 4.14. Synthesis of oligo 3-arylthiophenes **107** by Briehn et al. [128].

4.3.6
Boronate Linkers

Boronates have been used in various linker types either as linkers for diols [146] or as precursors for metal-mediated cleavage (Table 4.7).

A boronic acid ester, which contains an aryl iodide moiety attached by an appropriate tether, can act as an intramolecular arylation agent. Thus, Li and Burgess [148] developed a polymer-bound precursor, the ensuing cleavage of which furnished a macrocyclic constrained β-turn peptide mimic via biaryl coupling (Scheme 4.15).

Tab. 4.7. Overview of boronate linkers.

Structure		Reference	Possible structures achievable
	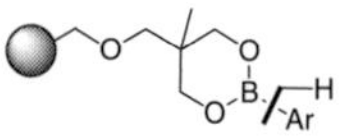 108	[147]	Arenes; cleavage by Ag^+
	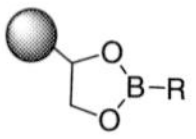 109	[148]	Biaryls by Suzuki crosscoupling
	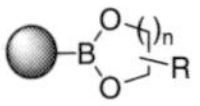 110	[146]	Diols

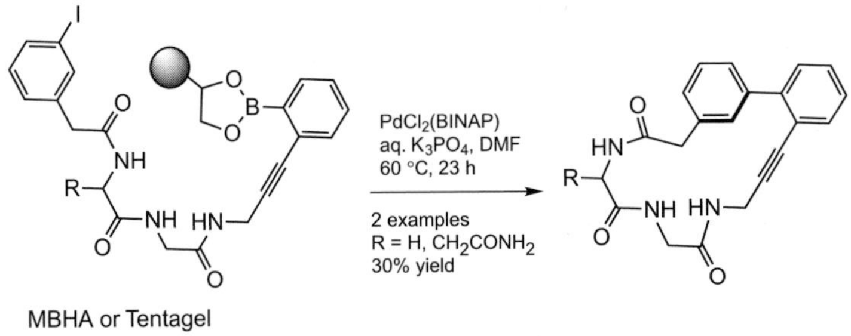

Scheme 4.15. Intramolecular cleavage Suzuki coupling by Li and Burgess [148].

Immobilized aryl boronic esters (**113**) can be cleaved to the corresponding hydrogen-substituted products (**115**) (traceless cleavage, Sect. 4.5.5) using aqueous silver nitrate solution [147] (Scheme 4.16).

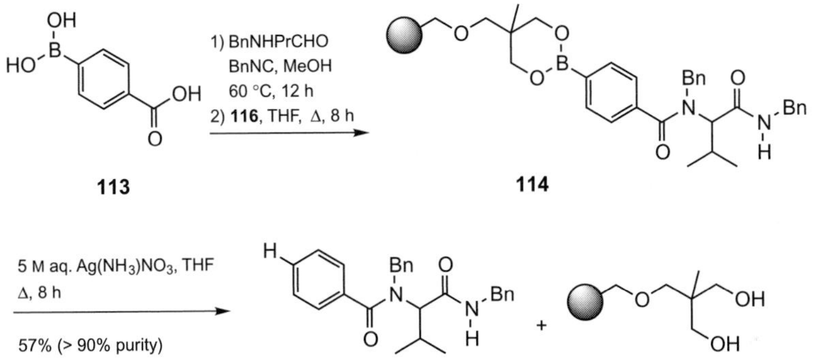

Scheme 4.16. Traceless cleavage of boronic acid derivatives according to Pourbaix et al. [147].

4.3.7
Sulfur, Stannane- and Selenium-based Linkers

A set of modern linkers (Table 4.8) based on sulfur, stannane, and selenium chemistry can be found in the literature. Their popularity obviously stems from the fact that these elements can favorably be tailored for use as fragile points of attachment.

4.3.7.1 Sulfur-based Linkers
Sulfur has been used in linkers as thioethers, sulfoxides, sulfones, sulfonic acids, and their corresponding derivatives. The relatively weak carbon–sulfur bond can be cleaved under reductive conditions (see Scheme 4.62) [155, 156], photolytic conditions, or in the presence of strong bases [160]. Since thiols can be oxidized to the corresponding sulfoxides and/or sulfones, various safety-catch linkers [155, 156] (Sect. 4.5.1) have benefited from this fact.

Tab. 4.8. Overview of sulfur, stannane and selenium linkers.

Structure	Reference	Possible structures achievable
117 (NHR², methylation)	[149] (Scheme 4.42)	Amides
118 (oxidation)	[150] (Scheme 4.43)	Carboxylic acids
119 (oxidation)	[151] (Scheme 4.20)	Amines
80	[115]	Ketones, amides; cleavage with reducing reagents
68	[105]	Linker for sterically hindered ketones
120	[152, 153]	a) R = Aryl: Arenes (Scheme 4.17); cleavage with palladium/formate; b) R = α-carbonylalkyl: heterocycles
121	[154]	Diaryl methanes (Scheme 4.18); cleavage with alkylation/Suzuki coupling
122 (MeO-PEG-O)	[155–159]	Alkanes (Scheme 4.62); cleavage with Na/Hg
123	[160]	Benzofurans (Scheme 4.56); cleavage with bases
124 (Tentagel, MeO)	[161]	Methylarenes; photolytic cleavage

Tab. 4.8. *(continued)*

Structure	Reference	Possible structures achievable
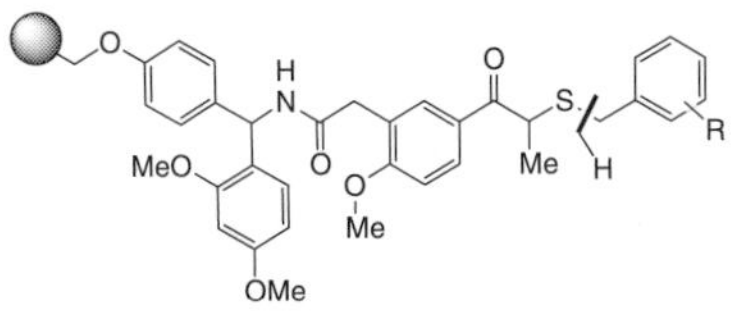 **125**	[162]	Methylarenes; cleavage under photolysis
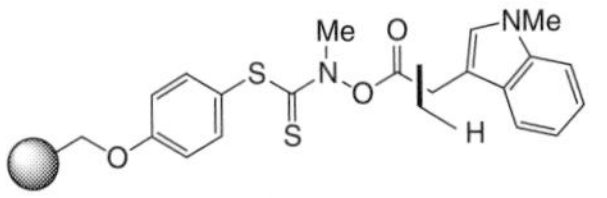 **126**	[120]	Methylarenes; photolytic cleavage
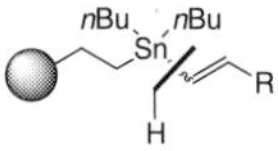 **127**	[163]	Alkenes (Scheme 4.35, Scheme 4.36); cleavage with oxidants
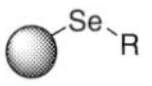 **128**	[164, 165]	Alkanes, alkenes

Aryl sulfonates can be used as linkers for arenes, as shown by Jin et al. [152]. These can be cleaved under reducing conditions to give the corresponding hydrocarbons (**130**). In addition, this linker might be suitable as a multifunctional anchor, as proposed in a patent [166] (Scheme 4.17).

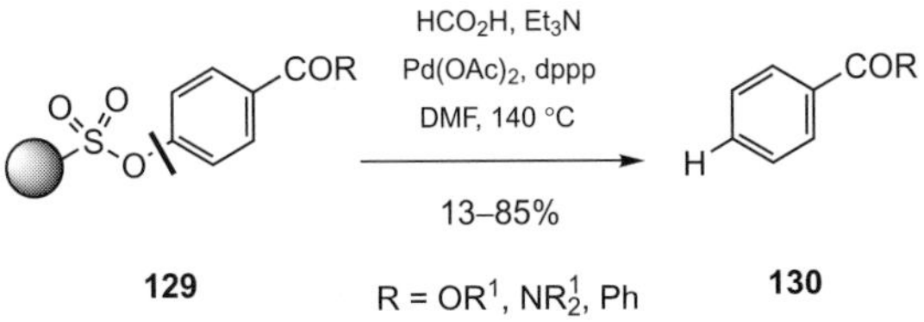

HCO$_2$H, Et$_3$N
Pd(OAc)$_2$, dppp
DMF, 140 °C

13–85%

129   R = OR^1, NR$_2^1$, Ph   **130**

dppp = 1,3-bis(diphenylphosphino)propane

Scheme 4.17. Cleavage-hydrogenation reaction with a sulfonate linker (**129**) according to Jin et al. [152].

Arylmethyl(homobenzyl)ethylsulfonium salts are also appropriate substrates for Suzuki-type coupling reactions. In this reaction performed on a polymer-bound sulfonium tetrafluoroborate, the benzyl fragment on the sulfur is transferred to the boronic acid residue. The sulfonium salt was prepared from an alkylthiol resin by alkylation with a substituted benzyl halide to give thioether (**131**) and subsequent alkylation with triethyloxonium tetrafluoroborate. Reaction with a boronic acid derivative yielded a diaryl methane (**132**) [154] (Scheme 4.18).

The cleavage of sulfonamides on substituted thiophenes deposited on platinum electrodes can be conducted by an electrochemical cleaving step [167] (Scheme 4.19).

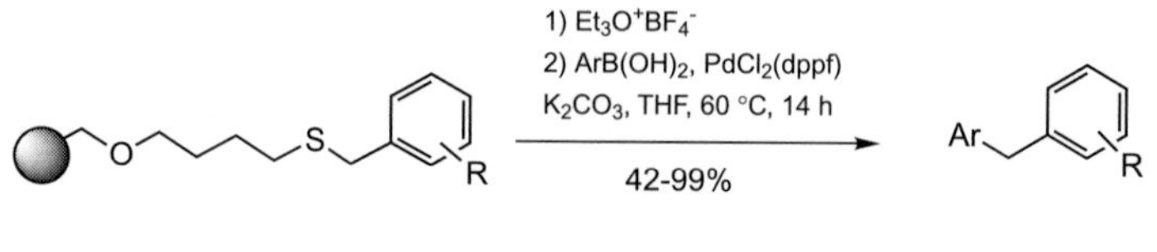

131 **132**

Scheme 4.18. Cleavage Suzuki coupling approach using sulfonium salts by Vanier et al. [154].

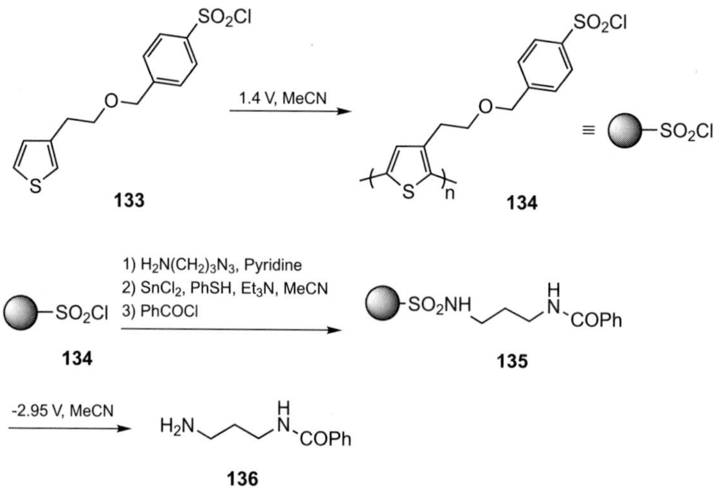

133 **134**

134 **135**

136

Scheme 4.19. The use of an electrocleavable anchoring by Marchand et al. [167].

Photolabile sulfur linkers are based either on the relatively weak carbon–sulfur bond in thioethers [161] or on the photolytic decarboxylation of thiohydroxamic acids to give methylindoles [120]. Also, a safety-catch linker (Sect. 4.5.1) for amines is based on 2-(thiobenzyl)ethylcarbamates [151, 168]. The linkage is performed with preformed handles containing ethenyloxycarbonyl-protected amines (**138**). Attachment to thiomethylated polystyrene (**139**) was performed under conditions involving radicals. The cleavage was carried out with an oxidizing agent, which forms the retro-Michael substrate [151] (Scheme 4.20).

The utility of a thioacetal-based anchor (**68**) as a chemically robust linker for the immobilization of ketones employed the commercially available (±)-alpha-lipoic acid. The products were easily cleaved from solid support by treatment with [bis(trifluoroacetoxy)iodo]-benzene [105] (Table 4.4).

Recently, Nicolaou et al. [153] have shown that the reaction of alkenes in the presence of dimethyldioxirane could be used for loading onto polystyrene sulfonic acid resin. Subsequent cleavage with nucleophiles proceeded smoothly to give a vast array of heterocycles.

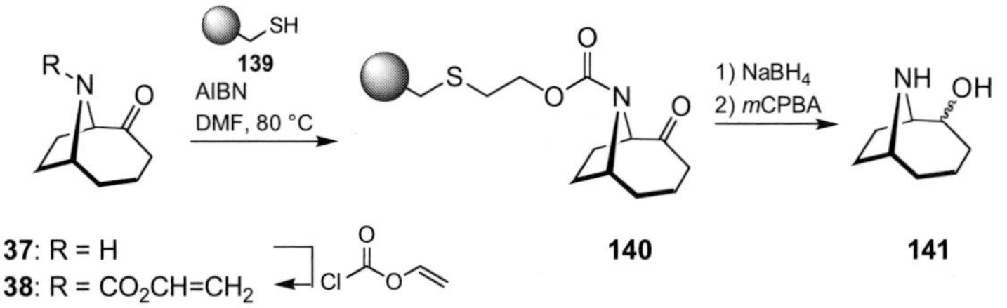

137: R = H
138: R = CO₂CH=CH₂

Scheme 4.20. Thiobenzylethylcarbamates as linkers for amines according to Timar and Gallagher [151]. AIBN, Azobis-isobutyronitrile.

4.3.7.2 Stannane-based Linkers

Stannanes have become prominent members in the area of multifunctional anchoring groups. A polymer-bound tin hydride (**142**) has been used to hydrostannylate alkynes under palladium catalysis to give polymer-bound alkenylstannanes (**143**). Alternatively, the latter can be prepared from a polymer-bound tin chloride and an alkenyl lithium or magnesium halide reagent [163] (Scheme 4.21). These alkenyl stannanes were employed in intermolecular [169] as well intramolecular Stille reactions. Alkenylstannanes can also undergo protonation to give alkenes (**145**) in a traceless fashion (Sect. 4.5.5). Therefore, this linker is able to operate in a multifunctional mode (see Fig. 4.5) (Sect. 4.5.6).

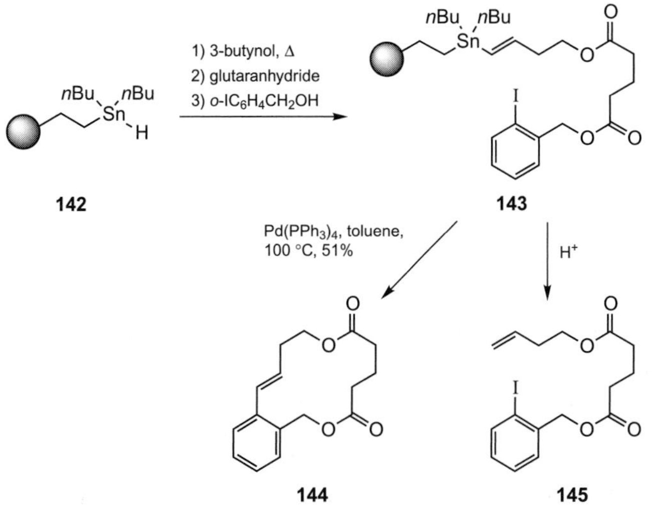

Scheme 4.21. The stannane linker for Stille reactions according to Nicolaou et al. [163].

4.3.7.3 Selenium-based Linkers

The selenium–carbon bond is, because of its weakness ($E = 217$ kJ mol^{-1}), prone to undergo homolytic cleavage, thus producing radicals. This fact was first recognized and used for solid-phase synthesis by Michels et al. in 1976 [170]. More recently, Nicolaou et al. [164] and Ruhland et al. [165] independently developed more efficient methods for the preparation of selenium-containing supports in their development of traceless linkers (Sect. 4.5.5). Starting from polystyrene, various steps including selenation with selenium powder or MeSeSeMe give rise to the formation of selenium resins, which can then be alkylated to give selenoethers (**148**). The traceless cleavage that yields alkanes (**149**) can be conducted by reduction with tributyltin hydride, while the formation of alkenes (**150**) can be observed after mild oxidation (Scheme 4.22). This linker holds promise for numerous applicability since the starting materials (alkenes, alkyl halides) are readily available, although the toxicity of the reagents and starting materials has to be considered.

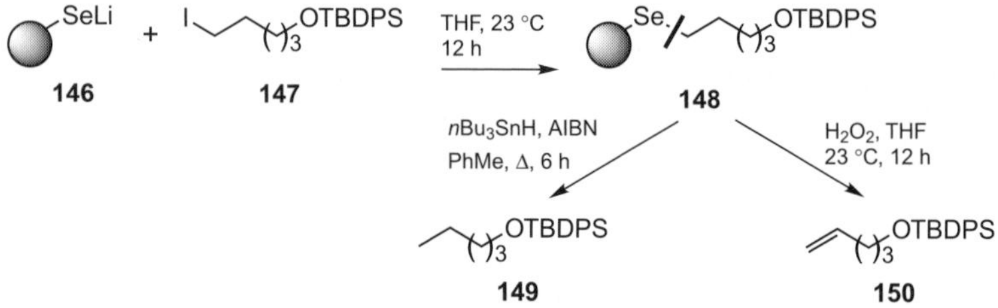

Scheme 4.22. The selenium linker for alkanes and alkenes according to Nicolaou et al. [164]. TBDPS, *tert*-butyldiphenyl-silyl.

Based on a similar concept, a selenium linker has been described that is loaded via a preformed handle [171]. Oxidation and thermal elimination give rise to the formation of alkenes (**153**) (Scheme 4.23). Interestingly, the selenoxides (**152**) decompose at room temperature, whereby the corresponding sulfoxides fragment at 100 °C.

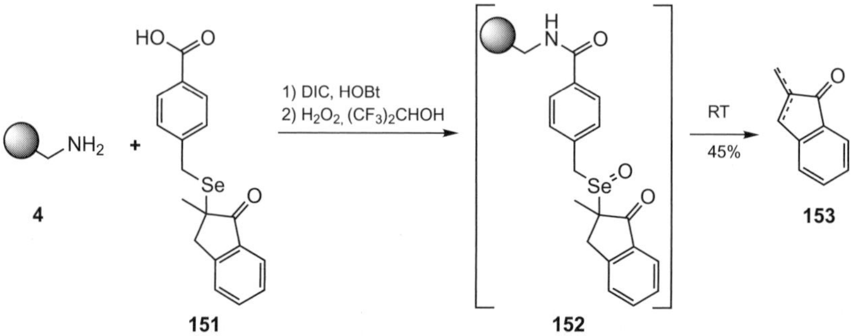

Scheme 4.23. The selenium linker according to Russell et al. [171].

The selenium bromide linker (**154**) (Scheme 4.24) is also the starting point for a library of natural products and analogs of the benzopyran type through reaction of prenylphenols [172–174] (Scheme 4.24). Furthermore, this linker enables the synthesis of medicinally interesting molecules [175].

Besides phenols, allyl anilines can be used in this sequence to produce indolines. Depending on the substitution pattern, either traceless (→ **162**) (Sect. 4.5.5)

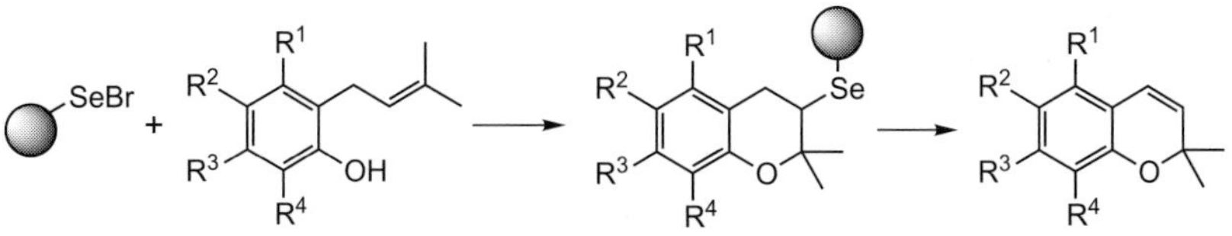

154 **155** **156** **157**

Scheme 4.24. The selenium linker in chromene synthesis according to Nicolaou et al. [172].

or cyclative (→**160**) cleavage can be realized [176] (Scheme 4.25). The seleno linker has also been used in the synthesis of 2-deoxy glycosides, orthoesters, and allyl orthoesters [177]. Finally, the SEM linker for the attachment and detachment of alcohols takes advantage of the substitution of phenylselenyl groups at seleno acetal resins [21].

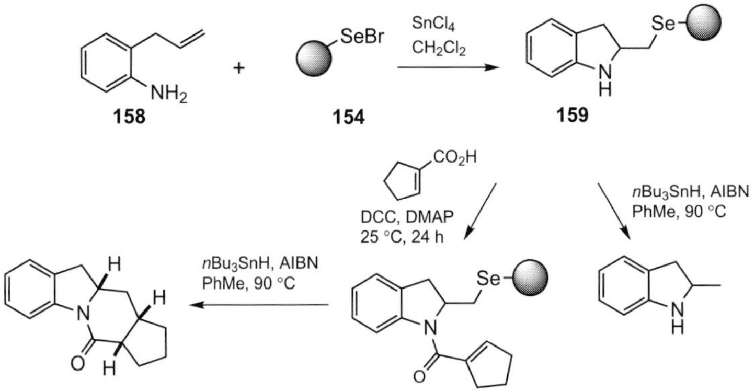

158 **154** **159**

160 **161** **162**

Scheme 4.25. The selenium linker in the indoline synthesis by Nicolaou et al. [176].

4.3.8 Triazene-based Linkers

Inspired by the use of triazenes in the total synthesis of vancomycin [178] and the pioneering work of Moore and coworkers [179, 180] and Tour and coworkers [181] in the synthesis of triazenes on a solid support and the final detachment to give iodoarenes (**180-I**), a whole set of triazene-based linkers (Table 4.9) has been developed [182].

The chemistry of diazonium salts provides tremendous opportunities for the construction of a wide range of aromatic compounds. Triazenes, which have been used as traceless linkers for arenes as shown in Sect. 4.5.5, provide both interesting new possibilities for activation of the *ortho*-position of the arenes and are ideal synthons for diazonium salts. Triazenes are stable toward light, moisture, and bases; however, they are cleaved by Brønsted acids and certain Lewis acids to give diazonium salts and amines.

Tab. 4.9. Overview of triazene linkers.

Structure	Possible structures achievable
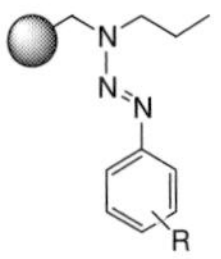 **163**	Iodoarenes [179]
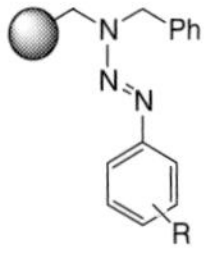 **164**	T1 resin: "traceless" linker [182–185]; synthesis of phenols [186], biaryls, alkyl arenes [187, 188], azides [189], aromatic hydrazines, halides [190, *cf.* 180, 181], ester, azo compounds, cinnolines [191], benzotriazoles [192]
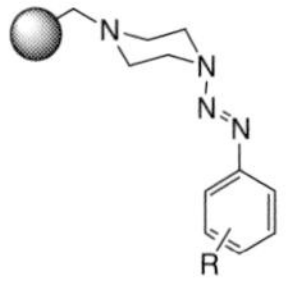 **165**	As above
166 **167**	T2 resin: synthesis of substituted amines [193], amides (peptides) [194], (thio)ureas [194, 195], hydrazines, alcohols, esters [183, 196, 197], guanidines [195], alkyl halides [183, 196, 197], sulfoximines T2* resin: as above [198]; scavenger for amines and phenols

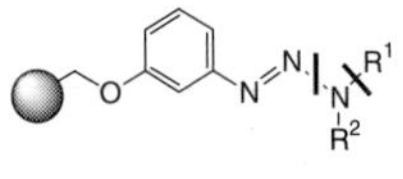

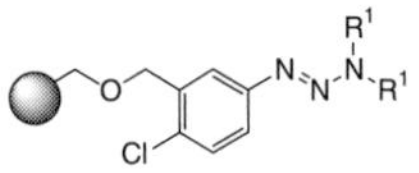

Two linkers based on triazene chemistry have been developed. While the T1 linker system consists of 3,3-dialkyl-1-aryl triazene bound to a support via the alkyl chain (Scheme 4.26), the T2 linker family is based on immobilized aryl diazonium salts.

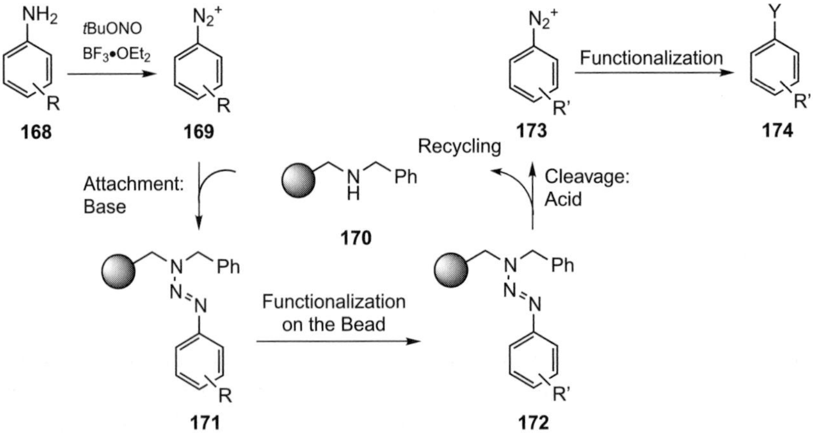

Scheme 4.26. Concept of the T1 linker [183].

The triazene T1 linker has been successfully used as a linker for arenes (see also Sect. 4.5.5). Until now, approximately 100 different anilines (**168**) have been immobilized. Functionalization on the bead has been demonstrated extensively. These immobilized diazonium derivatives are stable towards various reaction conditions, such as alkyl lithium reagents, reducing agents, and oxidizing reagents. However, acids cleave the triazenes to give the amine resin (**170**) and the modified aryl diazonium salts (**173**).

The latter can be transformed into various different products giving modified arenes in high yields and purities [purities > 90–95% according to gas chromatography (GC), NMR, and high-performance liquid chromatography (HPLC)] directly at the cleavage step [183] (Scheme 4.27).

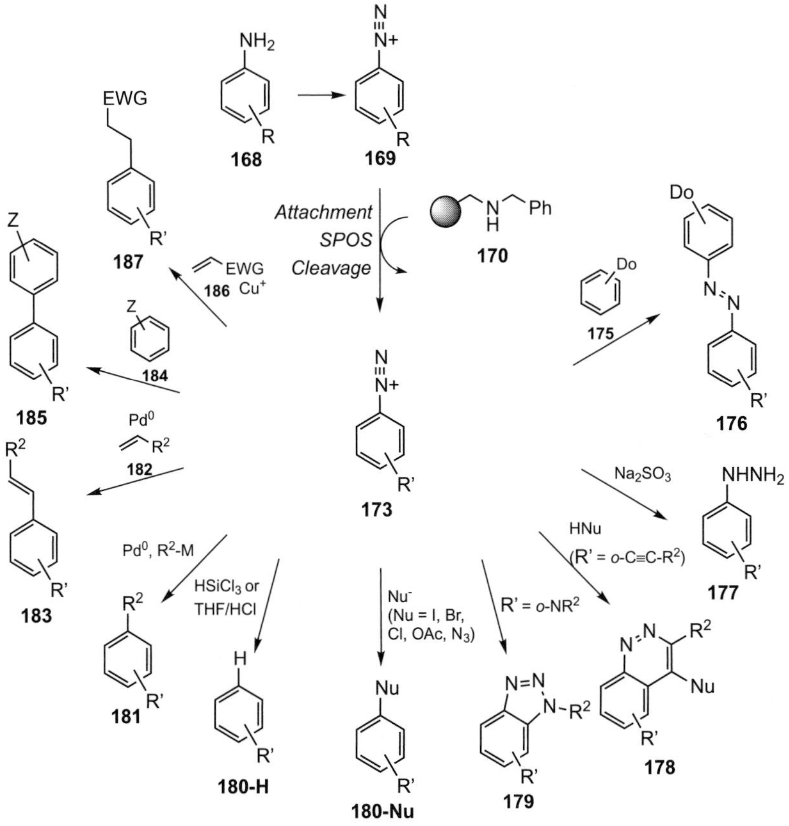

Scheme 4.27. Possibilities of the T1 triazene linker [183].

The diazonium salts, for example, can be reduced to the hydrocarbon (**180-H**) in THF with the aid of ultrasound [182]. The latter facilitates this reduction due to a radical pathway. A new reagent for this reduction was found to be trichlorosilane [184]. This is not only a source of traces of hydrochloric acid, which cleaves the triazene moiety, but, as a hydride donor, it is also able to reduce diazonium ions cleanly (Scheme 4.27 and see Scheme 4.64).

As already shown by Moore and coworkers [180] and Tour and coworkers [181], addition of methyl iodide to a triazene resin at elevated temperature (110 °C) gives rise to aryliodides (**180-I**) (Nu = I) in excellent yields. Furthermore, aryl halides (**180-X**) (X = Cl, Br, I) are readily available by the action of lithium halides in the presence of an acidic ion exchange resin or with the corresponding trimethylsilyl halide at room temperature [190]. A mixture of acetic anhydride and acetic acid produces phenol acetates (**180-OAc**) [186]. Although quite flexible in the range of possible electrophiles that may be employed, the most striking feature was the development of a cleavage cross-coupling strategy [187]. Starting from modified triazene resins, a one-pot cleavage cross-coupling reaction was conducted with two equivalents of trifluoroacetic acid in MeOH at 0 °C to give a diazonium ion. *In situ* coupling with various alkenes (**182**) in the presence of catalytic amounts (5 mol%) palladium(II) acetate furnished the corresponding products (**183**) in excellent yield and purities. Using palladium on charcoal as the catalyst has the advantage of decreasing palladium contamination, as well as providing a subsequent hydrogenation option [187]. Multicomponent Heck reactions (domino Heck Diels–Alder reaction) are possible in this context and lead to further diversification [187]. In the examples above, the diazonium group, upon cleavage from the resin, is lost as dinitrogen. However, a suitable nucleophilic *ortho*-substituent favors cyclization to give heterocyclic structures. Benzotriazoles, for example, are accessible from *o*-aminoaryl-substituted triazenes [192]. Other heterocyclic systems such as cinnolines (**178**) are available by a cleavage Richter reaction strategy, which starts from *o*-alkynylaryl triazenes. Cleavage was conducted with aqueous hydrogen chloride or hydrogen bromide in acetone or dioxane at room temperature to produce various cinnolines (**178**) in a library format with up to 95% yield and with a range of purities between 60% and 95% [191]. In addition to these methods, reduction of diazonium salts gives rise to the formation of hydrazines (**177**) [199], which are important building blocks for the synthesis of other heterocyclic compounds.

Whereas the T1 linker involves the immobilization of a diazonium salt on an amine resin, the T2 linker represents the reverse of this concept. Thus, an immobilized diazonium salt (**190**) was prepared from Merrifield resin (**3**) in two steps, and subsequent additions of primary and secondary amines generated triazenes (**191**). In addition, attachment of hydroxylamines, hydrazines, sulfoximines, and phenols (to give azo coupling products) proceeds equally well (Scheme 4.28).

Secondary amines can be cleaved directly [193] or after modification from the resin. Primary amines can be derivatized on the free N–H functionality and can therefore be modified to an array of products [194, 195, 198]. Thus, ureas (**198**) [194], thioureas (**197**) [195], guanidines (**196**) [195], and carboxamides (**199**) [194] were prepared in excellent yields (Scheme 4.29).

While the cleavage of trisubstituted triazenes gives rise to the formation of secondary amines in excellent yields [193], the cleavage of disubstituted triazene (**194**) gives rise to aliphatic diazonium salts [183, 196, 197]. The diazonium ion thus formed undergoes substitution with the nucleophile present in the reaction mix-

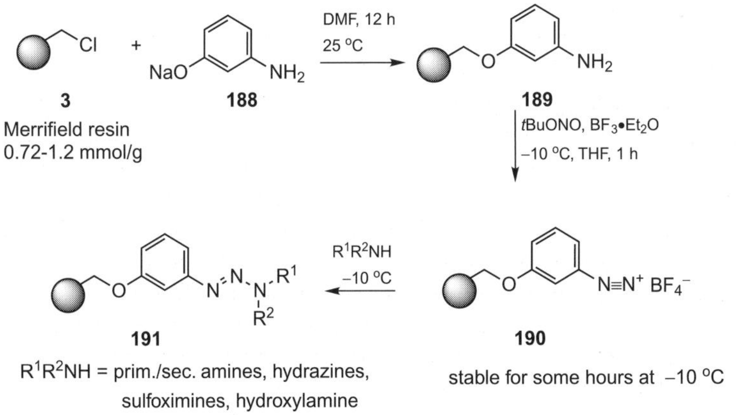

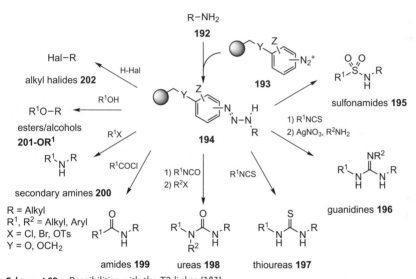

Scheme 4.28. Synthesis of the T2 linker [193].

Scheme 4.29. Possibilities with the T2 linker [183].

ture. Therefore, alkyl halides (**202**), alcohols (**201-OH**), and alkyl esters (**201-OCOR**) can be formed by cleavage with trimethylsilyl halides (X = I, Br, Cl) or carboxylic acids (X = for example OAc, OTfa) [200] [183, 196, 197]. The regio-selectivity of the cleavage can be explained by the presence of one tautomer of the triazene in which the hydrogen atom is next to the arene ring. Overall, this reaction sequence provides a substitution of an amino group with oxygen or a halogen (Cl, Br, I).

Various immobilized diazonium salts have been prepared [201], thermo-analytically characterized [202], and used for the linkage of nucleophiles. The structure of the salts clearly influences the stability of the diazonium moiety. The thermally

Tab. 4.10. Orthogonality of linker families.[a]

Cleavage reagents	Benzyl-type linkers	Ketal/acetal linkers	Esters/amide linkers	Silyl linkers	Triazene linkers	Selenium/sulfur/stannyl linkers
Electrophiles	++	++	++	++	++	++
Nucleophiles	0	0	++	0	0	0
Fluoride	0	0	0	++	0	0
$h\nu$	(++)	0	0	0	+	0
Oxidative conditions	0	0	0	0	0/+	++
Reductive conditions	0	0	+	0	0/+	++

[a] (++): specially designed linker; ++: cleavage; +: partial cleavage;
0: no cleavage.

stable diazonium ion (**194a**) (Z = Cl, Y = CH$_2$O) [$t_{1/2}$ (25 °C) > 100 days] is also capable of scavenging various nucleophiles (amines, phenols, and anilines) [198].

4.3.9
Orthogonality Between Linkers

The orthogonality of linkers is important for the design and execution of both simple and complex reaction sequences performed on a solid support. Recently, an entire set of innovative linkers and cleavage strategies has been disclosed, which enables the full set of orthogonality to be produced.

As discussed above, each linker family is sensitive toward a certain spectrum of cleavage conditions and is therefore stable to dissimilar conditions. Since most of the linkers are based on well-established protecting groups, Table 4.10 can be used for the determination of orthogonality. For example, benzyl-type linkers, which are mostly cleaved by electrophiles and are stable toward nucleophiles, can be combined with ester-based protective groups.

4.4
Cleavage

In this section, various methods and reagents for the cleavage of linkers are presented. In most cases, cleavage of linkers is conducted with protons. However, other electrophiles, photones, oxidizing and reducing reagents, and nucleophiles can be used in many cases.

4.4.1
Electrophilic and Nucleophilic Cleavage

Cleavage of linkers can be conducted with various kinds of electrophiles and nucleophiles (Tables 4.11 and 4.12). The most popular cleaving reagent is trifluoro-

Tab. 4.11. Typical electrophiles and Lewis acids used for detachment.

Electrophiles (concentration)	Solvent	Additive	Example for a suitable linker	Product
HF			[203–205]	
HF			Wang linker [33]	Thiols
HF		Anisole	PAM resin [35]	Carboxylic acids
HF		Cresol	p-Acyloxy BHA resin [62]	Amides
CF$_3$SO$_3$H			[206]	
TFA (0.1% to neat)	CH$_2$Cl$_2$		Various linkers	
TFA		p-Cresol	[205, 207, 208]	
TFA		p-Cresol, Me$_2$S	[205]	
TFA		Anisole (PhOMe)		
TFA (25%)		Et$_3$SiH	[209]	
TFA		Ethanedithiol (EDT)	[210]	
TFA		iPr$_3$SiH	[211]	
TFA		Et$_3$SiH	[212, 213]	
TFA		PhSMe	[214]	
HCl	Dioxane		Ketals [103]	Ketones
CF$_3$SO$_2$OSiMe$_3$			[215]	
HBr/Ac$_2$O			[216, 217]	
AcOH	CH$_2$Cl$_2$		[66]	
AlCl$_3$	CH$_2$Cl$_2$/MeNO$_2$		[218]	
Et$_2$AlCl			[219]	
Me$_3$SiCl			[194]	

Tab. 4.12. Typical nucleophiles used for detachment.

Nucleophile (concentration)	Solvent	Additive	Suitable linker	Products(s)
RMgX			Carboxyl linker	Ketones, alcohols
RMgX	THF		Thioester	Alcohols [115]
R$_2$CuLi	THF		Thioester	Ketones [115]
NaOH			Carboxyl linker	Carboxylic acids
KOH	THF/H$_2$O		Reissert complex	Isoquinolines [351]
NaOMe			Carboxyl linker	Esters
F$^-$			Silyl linkers	Hydrocarbons, alcohols (Sect. 4.3.5)
HSCH$_2$CH$_2$OH (2-mercaptoethanol)	DMF	NMM, AcOH	Dinitroaryl linker	Thiols [229]
nPrNH$_2$ (10%)	DMF		Dde [228]	Primary amines
N$_2$H$_4$	DMF		Dde [226]	Primary amines
NH$_3$	(vapor)		[230]	

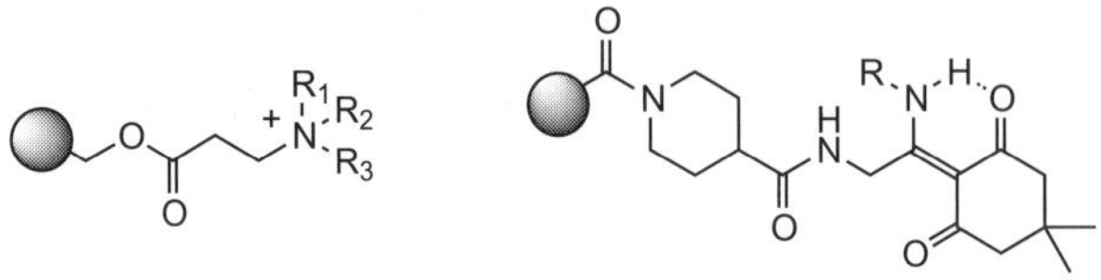

REM linker (203)   Dde linker (204)

Fig. 4.4. The REM and the Dde linker.

acetic acid in various solvents and concentrations. Because of its low boiling point, removal is readily achieved. Besides this reagent, various other acids have been used. Anhydrous HF, quite a toxic reagent, or triflic acid are required for more stable linkers. A mild reagent is trimethylsilyl chloride, which solvolyzes slowly to HCl and hexamethyldisiloxane.

Typical nucleophilically cleavable linkers are the REM (regenerated Michael acceptor) linker (203) [220–225] and Dde (Dimethyldioxocyclohexylidene)ethyl group (204) (ADCC anchor) [226–228], which are linkers for tertiary and primary amines, respectively (Fig. 4.4).

4.4.2
Oxidative/Reductive Methods

Besides electrophiles and nucleophiles, several linkers are designed to be cleaved by oxidative or reductive methods (Tables 4.13 and 4.14). Besides the feature of

Tab. 4.13. Some reducing agents used for detachment.

Reducing agent	Solvent	Suitable linker	Product
nBu_3SnH	Toluene	Selenium linker	Hydrocarbons (Scheme 4.25) [176]
$NaBH_4$	THF/H_2O	Amide	Alcohols [231]
$LiBH_4$	THF	Thioester	Alcohols [115]
Na/Hg	MeOH	Sulfon linker	Alkanes [155] (Scheme 4.62)
Phosphines		Disulfide linker	Thiols [232–235]

Tab. 4.14. Some oxidizing agents used for detachment.

Oxidant	Solvent	Suitable linker	Product
mCPBA	CH_2Cl_2	Thiol-based safety-catch linkers	Secondary amines [151]
mCPBA	CH_2Cl_2	Selenium linkers	Alkenes [177]
[Bis(trifluoroacetoxy)iodo] benzene		Thioketal-based linker	Ketones [105] (Sect. 4.3.7.1)
DDQ		Wang resin	Alcohols [31]
H_2O_2		Wang resin	Carboxylic acids [30]
$Cu(OAc)_2$	Pyr, MeCN	Hydrazide linker	Amides (Scheme 4.48) [236]
$Cu(OAc)_2$		Hydrazide linker	Arenes [237]

orthogonality with other cleaving methods, a drawback for the use of oxidative or reductive reagents is the necessity to remove excess reagents or byproducts.

4.4.3
Photocleavable Linkers

Light-induced cleavage offers new possibilities for orthogonal use of linkers (Sect. 4.3.8) and acid- or base-labile protecting groups (Table 4.15).

The first photolabile linker (**215**), which was based on the o-nitrobenzyl protecting group, was developed by Rich and Gurwara [238, 247] for the synthesis of protected peptides (Scheme 4.30). This linker was developed further because in the synthesis of the original linker, a nitration of Merrifield resin was involved, thus leading to nitration of excess phenyl rings. Therefore, mostly preformed handles have been used. In all cases, upon ultraviolet (UV) photolysis the photo byproduct, a nitrosobenzaldehyde, is also photoactive and causes a reduction in cleavage yield from the support. To circumvent this problem, an additional methyl group was introduced to give linkers, which lead to the photoreactive nitrosoactophenone system [248]. Moreover, introduction of methoxy groups *para* to the nitro groups (vanilline-type linker) [249] improves cleavage properties and the compounds are typically released within 3 h in >90% yield and >95% purity under neutral conditions [241, 250, 251]. Therefore, these linkers are biocompatible and are suitable for Fmoc solid-phase peptide synthesis (Fmoc-SPPS) (however, see [252]). Various modifications and improvements [239, 241, 248, 253–258] and applications [259] of this linker type have been reported [260–263].

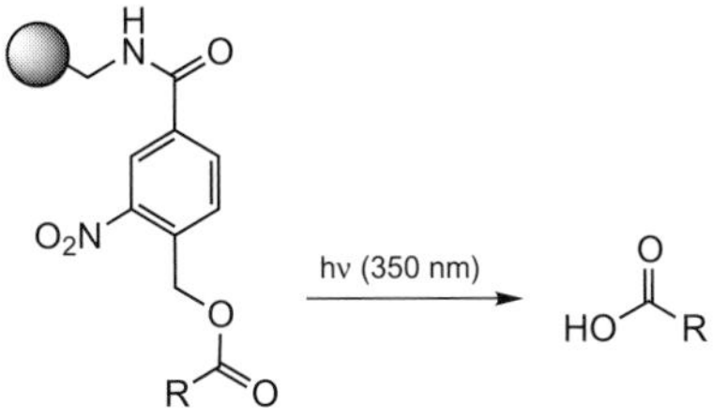

Scheme 4.30. The prototype of a photolabile linker according to Rich and Gurwara [238].

Benzoins [264] and related *β*-keto systems (phenacyl esters) [245, 246, 265] have been used as linkers for over 25 years now. A novel safety-catch linker (**217**) (Sect. 4.5.1) based on the benzoin-protecting group has been utilized to anchor carboxylic acids. The use of a dithiane-protected 3-alkoxy-benzoin allows for elaboration of molecules, linked as esters to the secondary hydroxy function of the benzoin, prior to deprotection of the dithiane and photolytic cleavage [244, 266] (Scheme 4.31).

A new photolabile linker (**126**) based on a thiohydroxamic acid has been shown to be an efficient "traceless" linker yielding an aliphatic CH bond upon photolysis

Tab. 4.15. Overview of photolabile linkers.

Structure	Reference	Possible structures achievable
205	[238]	Carboxylic acids
206	[239]	Carboxylic acids
207	[240]	Amines
208	[241]	Amines
209	[242]	Carboxylic acids
210	[243]	Alcohols
211	[244]	Carboxylic acids
124	[161]	Methylarenes

Tab. 4.15. *(continued)*

Structure	Reference	Possible structures achievable
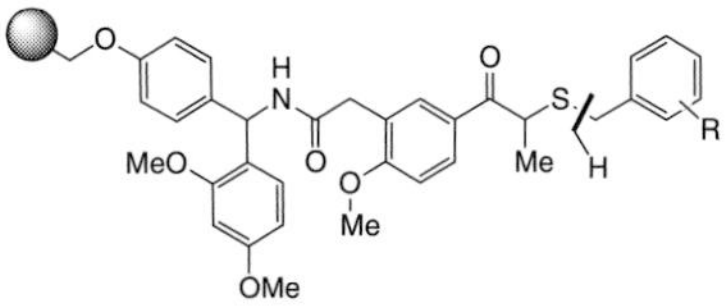 125	[162]	Methylarenes
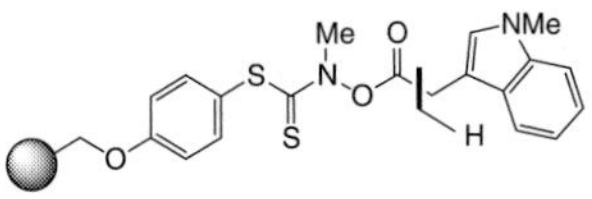 126	[120]	Methylarenes
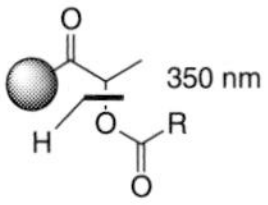 212	[245, 246]	Carboxylic acids

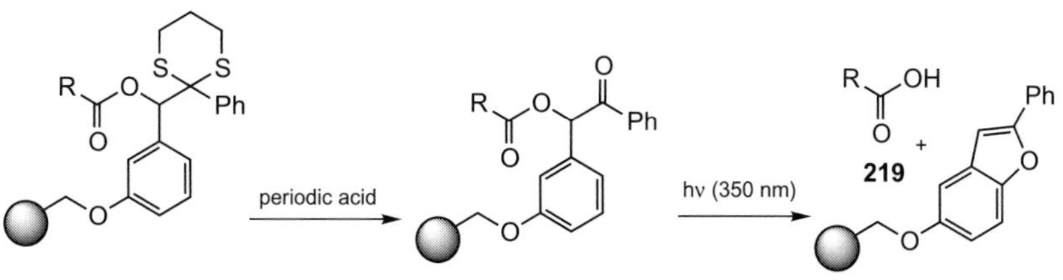

Scheme 4.31. A benzoin safety-catch photolabile linker according to Balasubramanian and coworkers [244, 266].

at 350 nm [120]. Alternatively, the Methoxynitrophenyldithiooxopropylphenyl-acetamide NPSSMPact handle is suitable for the detachment of benzyl-type structures (**124, 125**) for the traceless synthesis of methylarenes [161, 162] (see Scheme 4.59).

Besides these structures, various other linkers, such as those (**209, 210**) introduced by Giese and coworkers [242, 243] that are cleavable by C–C fragmentation reaction, and other systems have been developed [267–272].

4.4.4
Metal-assisted Cleavage

Cleavage mediated or catalyzed by (transition) metals is particularly interesting for several reasons. First, this type of cleavage is in most cases orthogonal (Sect. 4.3.8) to other procedures, thus enabling various types of transformations. Second, reactive intermediate organometallics can be suitable for further transformations.

In particular, the cleavage of substrates from a solid support using palladium-promoted or -catalyzed reactions has some advantages over other cleavage methods. Since most protecting groups and functionalities are resistant towards palladium complexes, a selective surgical removal is frequently possible. In addition, intermediate π-allyl– and σ-aryl–palladium complexes can in principle be used for further derivatization with the use of appropriate linker types.

Boronates can be reduced to the corresponding hydrocarbons using silver salts, making this linker as a traceless linker (Sect. 4.5.5) for arenes [147]. Similarly, aryl sulfonates are reduced by the action of palladium or a reducing medium [166].

4.4.4.1 Cleavage with Ensuing Allylic Substitution or Cross-coupling Reactions

The detachment of molecules with a concomitant cross-coupling or allylic substitution is an elegant method for the increase of diversity upon cleavage (see also Chapter 19). A common drawback of most methods is the contamination with transition metal catalysts and organometallic byproducts. However, various methods are available for the sequestering of transition metals from the products. The same holds true the removal of other byproducts.

Allylic substitution reactions   The cleavage of polymer-bound allyl esters with palladium catalysts provides a general access to π-allyl complexes, which can react with various nucleophiles. This approach has been used in the development of π-allyl-based linkers (Sect. 4.3.2).

Schürer and Blechert [273] reported on an ene–yne cross metathesis (see also Sect. 4.4.4.2) and a subsequent cleavage in the presence of various nucleophiles to yield the corresponding functionalized dienes (**224**) (Scheme 4.32).

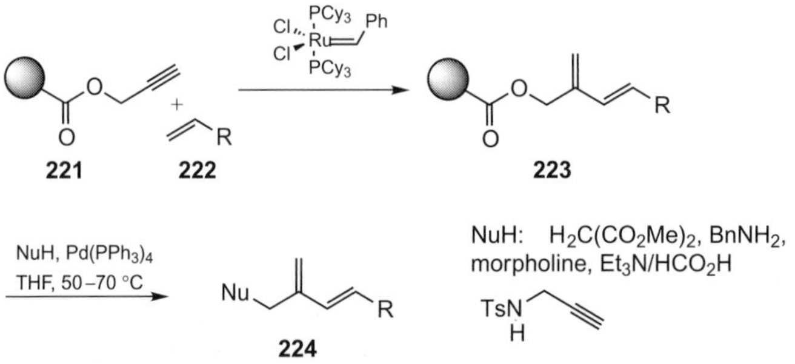

Scheme 4.32.   Cleavage via formation of π-allyl intermediates according to Schürer and Blechert [273].

Similarly, polymer-bound 1-alkenylcyclobutylsulfones (**225**) can be reacted with suitable nucleophiles (**226**) and palladium catalysts to give the corresponding cyclobutylidene derivatives (**227**) [274] (Scheme 4.33). The latter linker employed in an allylic substitution reaction can be regarded as a multifunctional linker (Sect. 4.32).

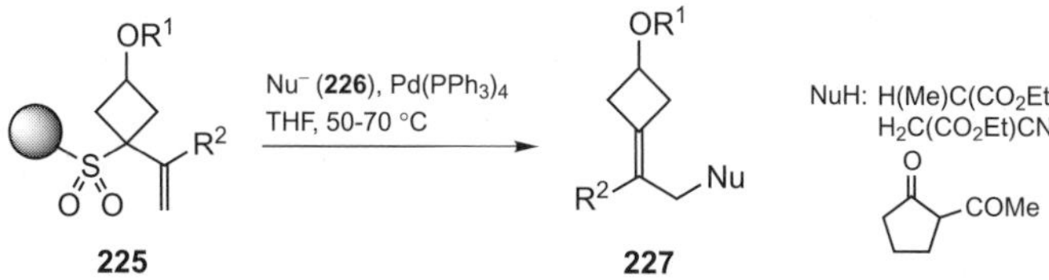

Scheme 4.33. Cleavage of allylsulfones according to Cheng et al. [274].

Heck reactions Cleavage by an ensuing Heck reaction was developed utilizing the T1 triazene linker [187]. Upon cleavage with trifluoroacetic acid, a diazonium ion is formed which can couple to an alkene under palladium catalysis (Scheme 4.34). The coupling proceeds well with simple terminal alkenes, styrenes, as well as di- and even trisubstituted alkenes. The coupling with 1,3-cyclohexadiene eventually yields a biaryl, apparently by a facile dehydrogenation of the primary coupling product. The advantage of this process is clearly the possibility of using volatile alkenes (and alkynes) without contamination of any salt or other less volatile by-products, particularly with the use of palladium on charcoal as the catalyst. In this case, a subsequent hydrogenation is also possible [187].

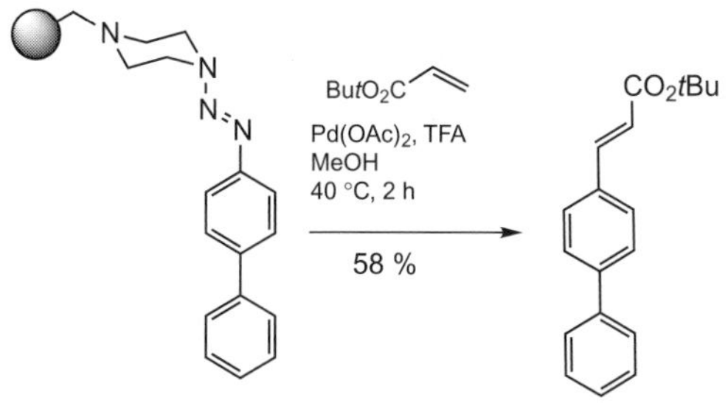

Scheme 4.34. Cleavage with ensuing Heck coupling using the triazene linker by Bräse and Schroen [187].

Stille couplings The intermolecular Stille reaction of aryl halides with immobilized stannanes (Scheme 4.21) provide coupling products in good yields, as demonstrated by Kuhn and Neumann [169]. In addition, the stannylated resin produced in the cleavage coupling can be recycled. Although the products obtained were not contaminated by any stannane, they were separated from an excess of the reactive electrophiles that had to be applied in the cleavage-coupling step. The intramolecular variant, which was used by Nicolaou et al. [163] to produce macrocyclic ring systems such as the natural product (S)-zearalenone (**231**), does not have this drawback (Schemes 4.35 and 4.36).

Suzuki couplings Suzuki couplings following a cleavage reaction are potentially applicable in a multifunctional sense (Sect. 4.5.6). However, owing to the tendency

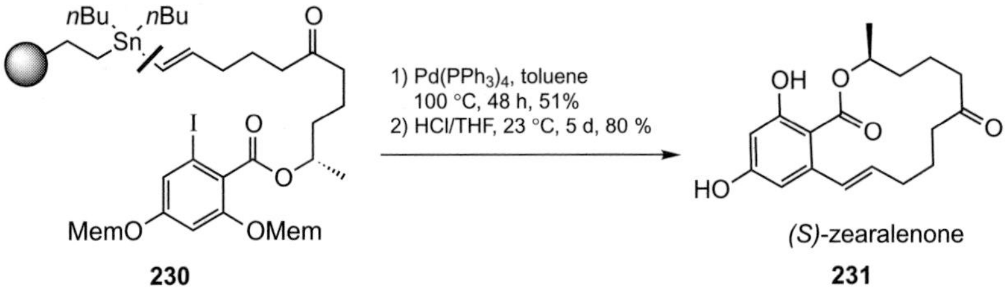

Scheme 4.35. A cleavage Stille strategy using a stannane linker for the synthesis of zearalenone (**231**) by Nicolaou et al. [163]. Mem, methoxyethoxymethyl.

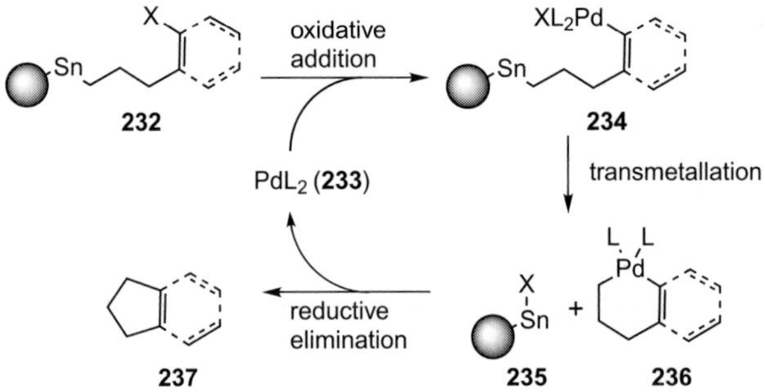

Scheme 4.36. Schematic reaction for the Stille cleavage reaction.

of the boronic acid derivative to give homocoupling products, the need to apply additional ligands, and the low volatility of the boronic acid derivative, a more or less tedious work-up is required after these types of transformations. A few studies have proven that certain functionalities, when generated during cleavage, may act as leaving groups for a subsequent Suzuki reaction. One of these is the diazonium group, which can be generated by cleavage of the triazene T1 linker. While the Heck-type coupling with alkenes gives good yields of the desired products [187] (Scheme 4.34), the analogous reaction with phenylboronic acid appears to be difficult because of work-up problems [187] (Scheme 4.37).

Arylmethyl(homobenzyl)ethylsulfonium salts [154] and aryl boronates [148] have been used as precursors for a cleavage/Suzuki approach (Schemes 4.15 and 4.20).

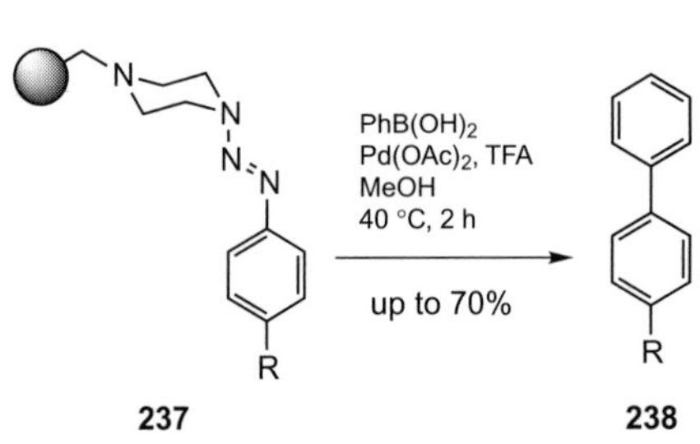

Scheme 4.37. Cleavage with subsequent Suzuki coupling according to Bräse and Schroen [187].

Sonogashira-type couplings The coupling of alkynes with diazonium salts has been reported in the context of the T1 linker. Here, the product (**240**) was isolated in moderate yields, and it had to be separated by chromatography from alkyne homodi- and trimers [187] (Scheme 4.38).

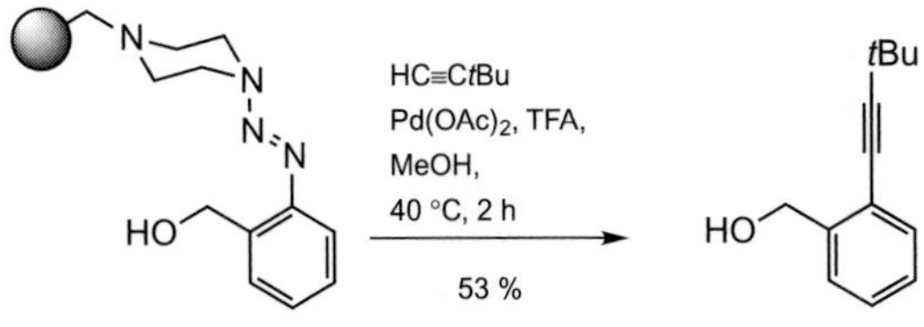

Scheme 4.38. Sonogashira coupling associated with the cleavage according to Bräse and Schroen [187].

4.4.4.2 Cleavage via Alkene Metathesis

The cleavage via alkene metathesis is particularly useful since a clean and selective scissoring of molecules is possible (see also Chapter 20). The cleavage by meta-thesis (Scheme 4.39) can be performed by cyclization during cleavage [275–281] (ring-closing metathesis, RCM), by intermolecular metathesis [144, 281] (cross-metathesis) (Sect. 4.5.6), or by intramolecular metathesis [275]. Successful examples for this cleavage/cyclization method are the synthesis of epothilone A [277] and medium-sized heterocycles [278–282].

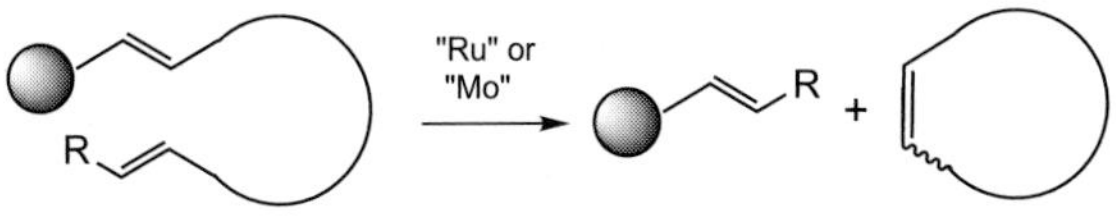

Scheme 4.39. The concept of metathetic cleavage by Blechert and coworkers [275, 276], Nicolaou et al. [277], Piscopio et al. [278–280], and Van Maarseveen et al. [281].

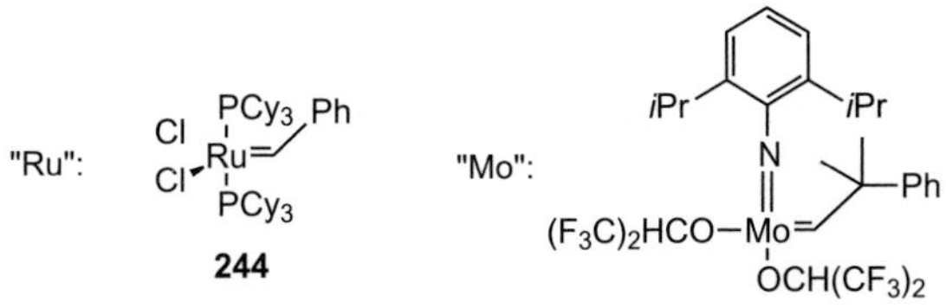

The intermolecular cross-metathesis of alkenes and alkynes provides a general access to 1,3-dienes [273] (see also Sect. 4.4.4, Scheme 4.32). A new anchoring group for the solid-phase synthesis of oligosaccharides has been described by Melean et al. [283]. Alkenyl units are not suitable for glycosidation reactions in the presence of strong electrophilic activators. However, using the 4,5-dibromooctane (DBOD) anchor, an iodide-mediated elimination reaction provides the active linker, which can be cleaved under metathesis conditions [283] (Scheme 4.40).

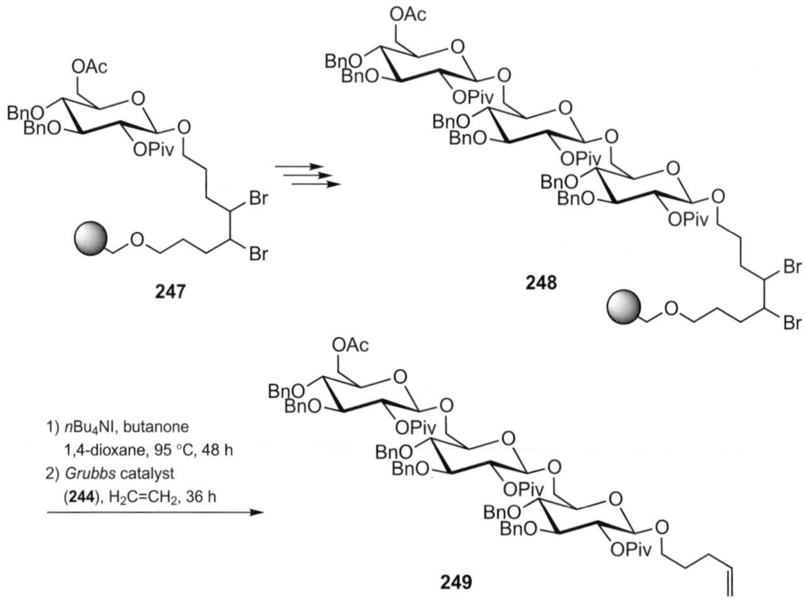

Scheme 4.40. Oligosaccharide synthesis on solid support using a linker cleavable by metathesis according to Melean et al. [283].

4.4.5
Unusual Cleavage Methods

While most linkers are cleavable with electrophilic or nucleophilic reagents, or under photolytic conditions, some more or less unusual cleaving conditions have occasionally been used.

An enzyme-labile safety-catch linker (Sect. 4.5.1) was reported by Grether and Waldmann [118] (Scheme 4.41). They used an acyl-protected amine, which was deprotected by an acylase that triggered the release of the ester-bound substrate. A number of other enzyme-labile linkers have also been reported [284, 285].

4.5
Linker and Cleavage Strategies

Apart from simple monofunctional cleavage, various different linker strategies have been developed in recent years. In particular, new concepts based on safety-

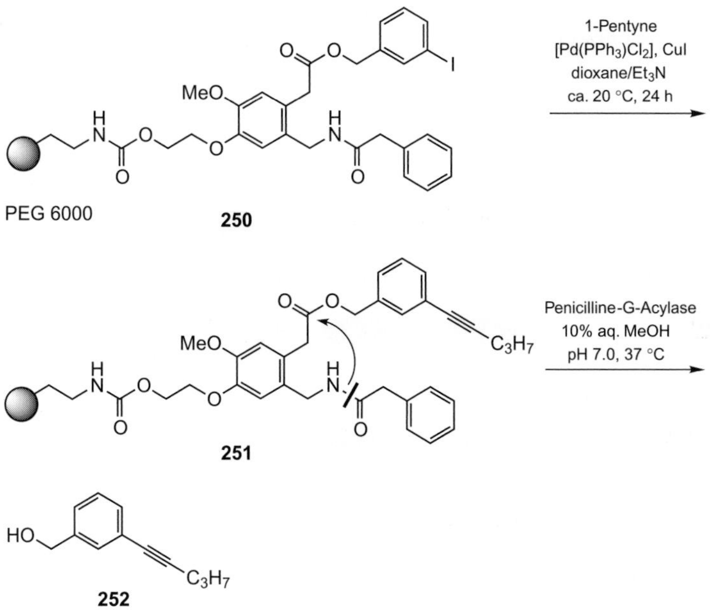

Scheme 4.41. An enzyme-labile linker by Grether and Waldmann [118].

catch linkers (Sect. 4.5.1), cyclative cleavage strategies, and fragmentation reactions have been presented.

Cleavage of linkers might be monofunctional or with functionalization of the linking site, whatever is required (Fig. 4.5). In the latter case, which is also known as the multifunctional cleavage strategy (see Sect. 4.5.6), the membership of library compounds is multiplied by the number of building blocks or functional groups that can be incorporated into the cleavage step. Hence, an anchoring group capable of functionalization and traceless linking is a versatile tool for enhancing diversity in a given system.

4.5.1
Safety-catch Linkers

> ■ Safety-catch linker: *a linker which is cleaved by performing two different reactions instead of the normal single step, thus providing greater control over the timing of compound release [12].*

The 'safety-catch' consists of a linker that, during synthesis, is inert towards the cleavage conditions and has to be activated (Table 4.16). Ellman and coworkers [286–288] have used this strategy in various applications, such as in the sulfonamide linker proposed by Kenner et al. [289]. Since 'safety-catch' means the activa-

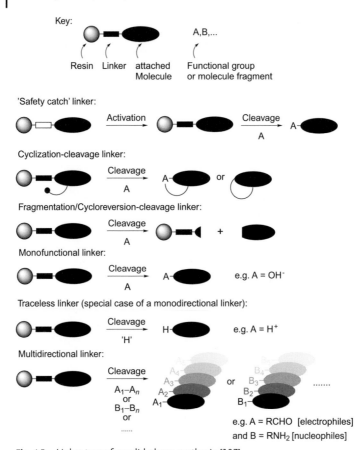

Fig. 4.5. Linker types for solid-phase synthesis [197].

tion of the linker before cleavage, such a system can be applied to monodirectional linkers, such as traceless linkers (Sect. 4.5.5) or to multifunctional linkers (Sect. 4.5.6), as well as to cleavage-cyclization strategies.

Recent applications of the Kenner linker have been shown in the synthesis of vinylsulfones [149] (Scheme 4.42) or in the synthesis of amides [294, 295].

The oxidation of sulfides to sulfones [150] or the reduction in the opposite direction [64, 65] are popular methods for the design of a safety-catch protocol. A resin-bound thioether can be activated by oxidation to insure the nucleophilic cleavage of a phenoxide moiety [150] (Scheme 4.43).

A safety-catch π-allyl-cleavable linker was developed for the synthesis of DNA on solid support. Starting from a resin carrying an Alloc-protected amino group fragment, conventional phosphoramidite chemistry was carried out to build up the desired nucleotide (**267**). Removal of the Allyloxycarbonyl (Alloc) group under palla-

Tab. 4.16. Overview of safety-catch linkers.

Structure	Reference	Possible structures achievable
253	[289]	Amides, carboxylic acids, hydrazides
117	[149] (Scheme 4.42)	Amides
254	[290]	Amides
118	[150] (Scheme 4.43)	Carboxylic acids
119	[151] (Scheme 4.20)	Amines
SCAL linker (**26**)	[64] (cf. [65])	Amines; cleavage with TFA/(EtO)$_2$(PS)SH [64]; Me$_3$SiBr/PhS Me/TFA [64]
255	[65]	Esters
256	[296]	Nucleic acids

Tab. 4.16. *(continued)*

Structure	Reference	Possible structures achievable
Dpr(Phoc) Linker / Phoc linker **257**	[291, 292]	Amides
79 hydrolysis	[118]	Alcohols; cleavage with enzymes
258	[293]	Amides
259	[236]	Diketopiperazines
123	[160]	Benzofurans (Scheme 4.56); cleavage with bases

dium catalysis and neutral conditions produces a polymer-bound intermediate (**268**) with a free amino group that can intramolecularly attack the activated phosphonates and liberate the nucleotide (**269**) from the solid support [296] (Scheme 4.44).

Other safety-catch linkers have been developed using the selenium (Scheme 4.22), amide [293] and ester linkers [297], the Phenyloxycarbonyl (Phoc) linker [291, 292], intramolecular catalysis by an imidazole residue [298], hydrazide linkers [236], sulfone linkers [156], and other structures [290, 299–303].

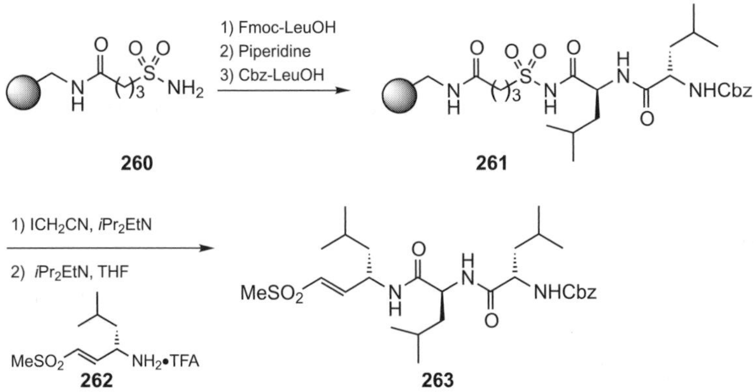

Scheme 4.42. The Kenner safety-catch linker according to Overkleeft et al. [149].

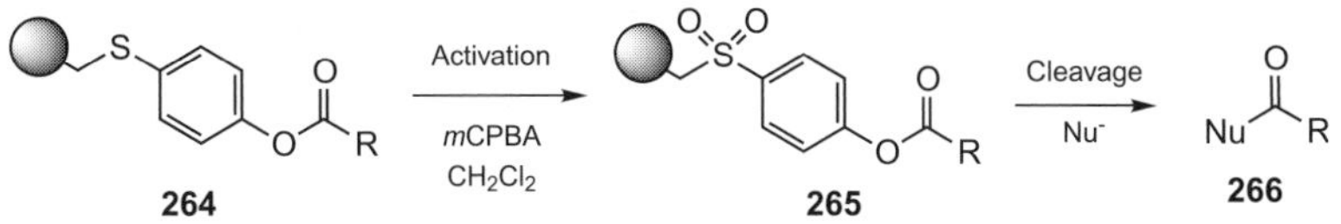

Scheme 4.43. Sulfide safety-catch linker by Marshall and Liener [150].

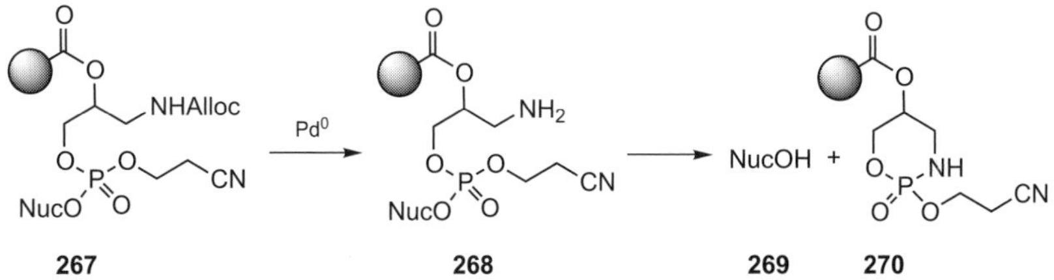

Scheme 4.44. A safety-catch palladium activated linker by Lyttle et al. [296]. Nuc = nucleotide.

4.5.2
Cyclative Cleavage (Cyclorelease Strategy)

> ■ Cyclative cleavage: *cleavage resulting from intramolecular reaction at the linker which results in a cyclized product. The cleavage may also act as a purification if resin-bound side-products are incapable of cyclizing, and thus remain attached to the solid support on release of the desired material [12].*

The cyclization-cleavage strategy (cyclative cleavage or cyclorelease strategy) is a typical example of the reaction for the synthesis of cyclic structures on solid support. It uses the characteristics of quasi-high-dilution kinetics on solid support and thus offers advantages that are not found in solution-phase chemistry.

In general, the starting material for cyclative cleavage is anchored to the resin via a leaving group (Scheme 4.45). An internal nucleophile provides the ring closure by displacement of this leaving group either directly or after activation. Apart from nucleophilic attack, cyclative cleavage can be achieved, for example, by Stille (Scheme 4.35) or Wittig–Horner reactions [304, 305].

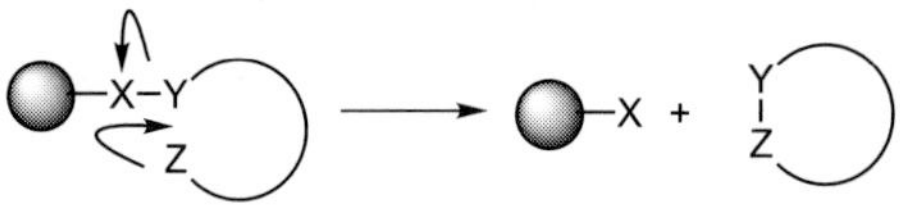

Scheme 4.45. General scheme for the cyclative cleavage.

Since the intramolecular reaction is by far faster than any intermolecular step, this strategy provides an additional purification step since only the cyclized structures are detached from the bead. Incomplete building blocks will remain on the solid support.

Cyclative cleavage has to be distinguished from cases in which the cyclization occurs in solution after cleavage because unsuccessful cyclization precursors remain in the liquid phase (Sect. 4.5.3).

In most cases, the precursor for the cyclization cleavage is linked via an ester bond to solid support whereby the nucleophile is based on an amine functionality. The product thus formed is therefore a cyclic amide or analog. Indeed, one of the first examples of this type was the pioneering benzodiazepine synthesis by Camps and Castells in 1974 [306]. In this case, the benzodiazepine ring formation proceeded by simultaneous cleavage from the bead. A vast list of examples has appeared since that suggests various kinds of heterocyclic systems. Examples are lactams, hydantoins [307–319], thiohydantoins [315], oxazolidinones [320, 321] (Scheme 4.46), diketopiperazines [322–331], benzodiazepines and benzodiazepinones [306, 318, 332–334], pyrazolones [335, 336], diketomorpholines [323], tetramic acids [337–340], quinazolinediones [341], dihydropyrimidine-2,4-diones [342], quinolinones [343], tetrahydrocarbolines [326], thiazoles [317], perhydrodiazepinones [327], sulfahydantoins [344], and benzimidazoles [345].

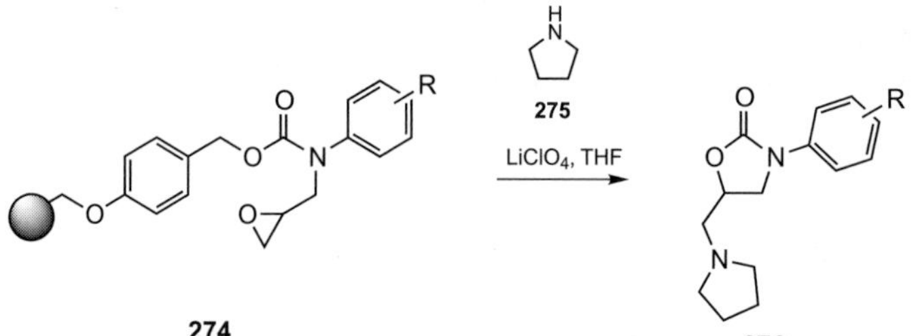

Scheme 4.46. Synthesis of oxazolidines (**276**) by cyclative cleavage according to Buchstaller [320].

In addition to nitrogen functionalities, oxygen nucleophiles can also act effectively. This concept was demonstrated by an approach to 3,5-disubstituted 1,3-oxazolidinones (**276**) via a ring-opening cyclization-cleavage step [320] (Scheme 4.46). Lactones are also accessible in a similar way [346].

The intramolecular Wittig reaction provides elegant access to cyclic alkenes. Here, an intermediate ylide is formed which interacts with an internal carbonyl functionality. Release from the bead is achieved via final elimination or cycloreversion. Thus, starting from the appropriate substituted phosphonium salts that have regional amide functionality, treatment with a base provides indoles when this reaction is performed under anhydrous conditions [364] (Scheme 4.47). Even phosphonates can undergo cyclative cleavage, as shown by Nicolaou et al. [304] in their synthesis of α,β-unsaturated macrolactones using the Horner–Wadsworth alkene synthesis.

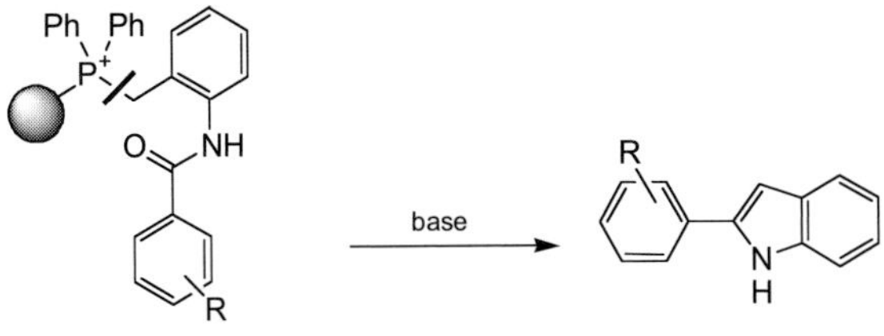

277   **278**

Scheme 4.47. The phosphonium linker for the synthesis of indoles by Hughes [363].

A new safety-catch linker (Sect. 4.5.1), which is based on the lability of the diazenyl carbonyl derivative, was developed for the synthesis of monoketopiperazines (**285**) (piperazinones). After elaboration of the anchoring system, the dimethylbenzyl group was removed by TFA and the hydrazide thus formed was oxidized by copper(II) acetate furnishing a diazenyl group. This motif was attacked intramolecularly by the primary amine functionality, which resulted in cleavage of the product from the resin [236] (Scheme 4.48). The potential of this linker class for the synthesis of heterocyclic amidic structures is discussed in Chapter 22.

A cyclative approach for the synthesis of thiazoles (**288**) begins with the acylated Rink amide resin (**286**) which has been transformed into the thioamide using Lawesson reagent. The S-alkylation with α-bromo ketones (**287**) proceeds with concomitant cleavage from the resin [347] (Scheme 4.49).

The intramolecular Stille reaction proceeds similarly under cyclorelease conditions (see above; see also Schemes 4.35 and 4.36).

4.5.3
Cleavage-cyclization Cases

The cleavage-cyclization reaction has to be distinguished from cases where a cyclization proceeds after cleavage (and not at the same time). This is true for most acidic cleavage conditions. In these cases, larger quantities of non-cyclized bypro-

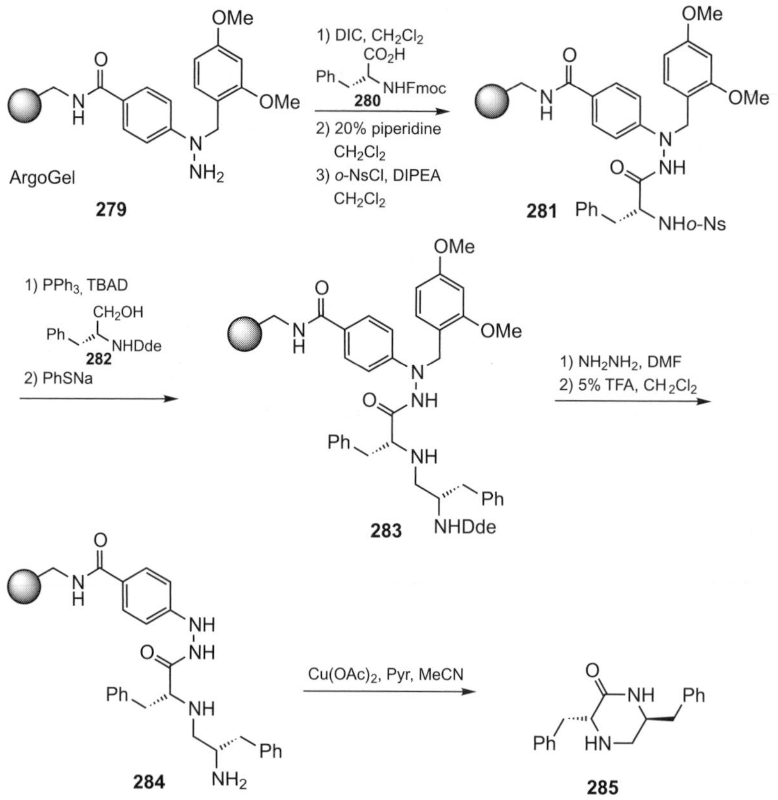

Scheme 4.48. A safety-catch linker for the cyclative cleavage by Murray and coworkers [236]. TBAD, *tert*-butyl azodicarboxylate. DiPEA, diisopropylethylamine.

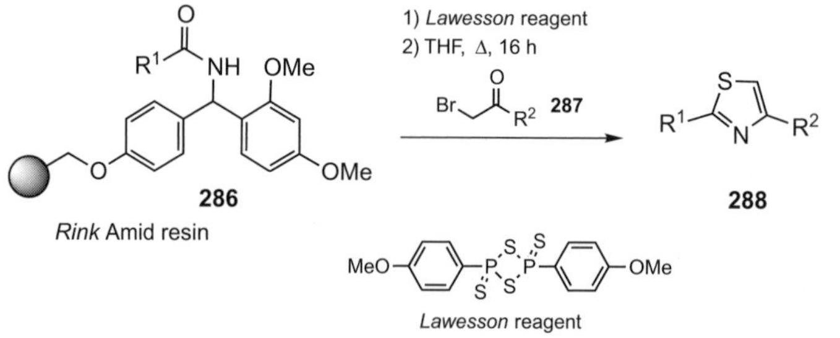

Scheme 4.49. Cyclorelease strategy for the synthesis of thiazoles by Brookfield and coworkers [347].

ducts can be obtained, e.g. the synthesis of benzofurans. The alkylidenylation of esters with thioacetals (**290**) and titanium complexes by the Takeda process proceeds without any problems [348]. The resulting enol ethers can be cleaved from the solid support, resulting in the formation of benzofurans (**291**) (Scheme 4.50).

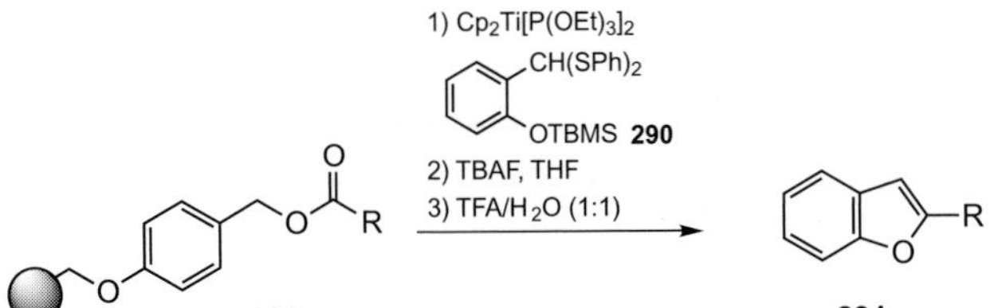

Scheme 4.50. Alkylidenylation and subsequent cleavage-cyclization reaction for the synthesis of benzofurans (**291**) by Hartley and coworkers [348].

Another example is the synthesis of imidazoquinoxalinones (**293**) (Scheme 4.51) [209, 349]. Likewise, diketopiperazines are formed using the T2 linker and immobilized glycinamide (**294**) [350] (Scheme 4.52). Finally, the synthesis of certain heterocycles belongs to this strategy (Chapter 22; Sect. 4.6.1.11).

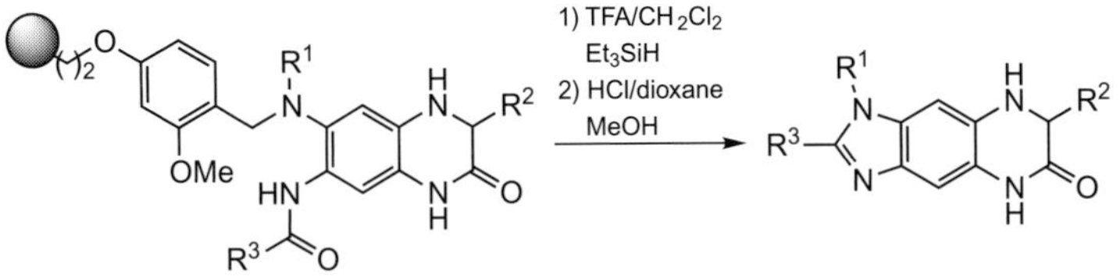

Scheme 4.51. Cleavage and subsequent cyclization in the synthesis of heteroanellated benzimidazoles (**293**) by Mazurov [209].

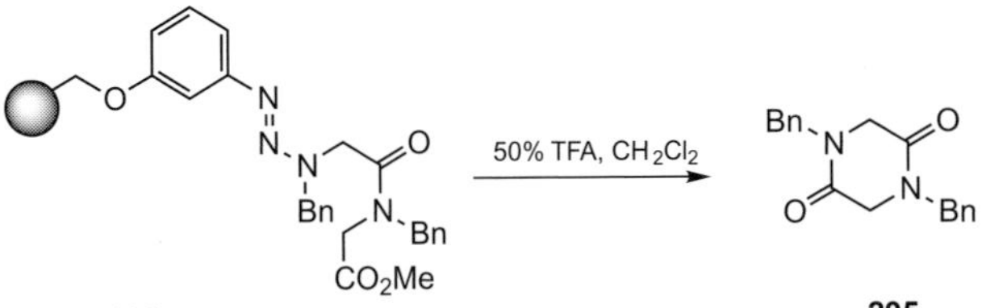

Scheme 4.52. Cleavage of diketopiperazines (**295**) from triazene T2 resin by Bräse and Lazny [350].

4.5.4
Fragmentation Strategies

The fragmentation strategy is related to the traceless anchoring groups (defined in Sect. 4.5.5) and also contains strategies which can be considered as retro-cyclo-addition cleavage, cycloelimination, or cyclofragmentation reactions. Here, a double or triple bond results from 1,n-elimination processes. Occasionally, a retro-cyclo-addition should also be considered as a fragmentation cleavage.

The only example so far of an attachment of heteroarenes via an addition/elimination strategy has been described by Kurth and coworkers [351, 352]. While arenes are more or less resistant toward addition, heteroaromatic systems such as isoquinolines (296) are prone to the addition of nucleophiles. Subsequent reactions with the addition of electrophiles results in the so-called Reissert compounds (298). These are stable compounds, which for example can be alkylated. In the case of solid-phase synthesis, the electrophile chosen was a polymer-based acid chloride. The detachment can be carried out by simple addition of hydroxide ions (Scheme 4.53).

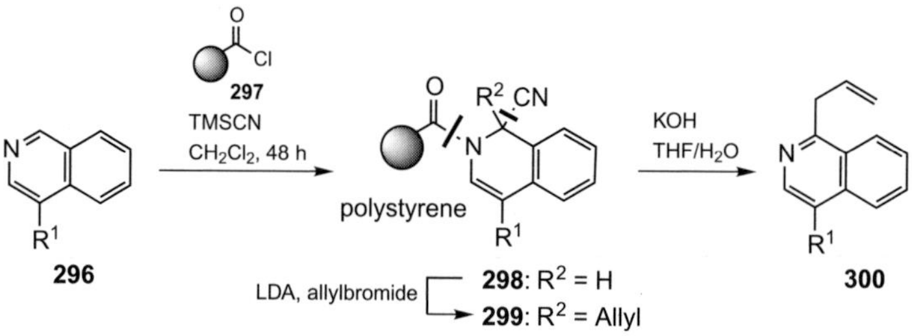

Scheme 4.53. The Reissert complex strategy used by Kurth and coworkers [351].

Recently, Gibson et al. demonstrated that arenes can be attached to a solid support using chromium arene complexes (301) [353; cf. 354] (Scheme 4.54). This method even allows, at least from a theoretical point of view, the complete variation of the arene backbone; however, modification of the arene ring system might be limited to a certain extent. It has already been shown that various other metals or unsaturated molecules (e.g. alkynes with dicobaltoctacarbonyl fragments) can lead to interesting examples for the design of new linkers [355]. Linkers based on arene complexes have been patented [356, 357].

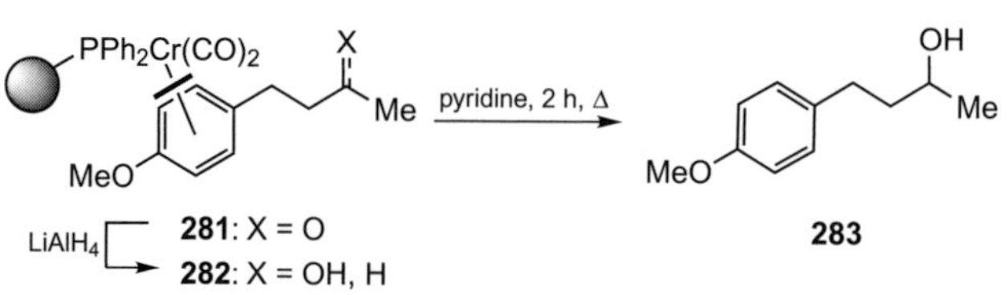

Scheme 4.54. The π-complexation for the linkage of arenes by Gibson et al. [353].

The fragmentation-cleavage strategy has been used for the synthesis of pyrazoles (308) [358]. An α-silylated N-nitrosamide (304) has been rearranged via a thermal silyl shift to an azomethine ylide (306), which then reacted with the dipolarophile (305); subsequent fragmentation yielded the target pyrazoles (308) (Scheme 4.55).

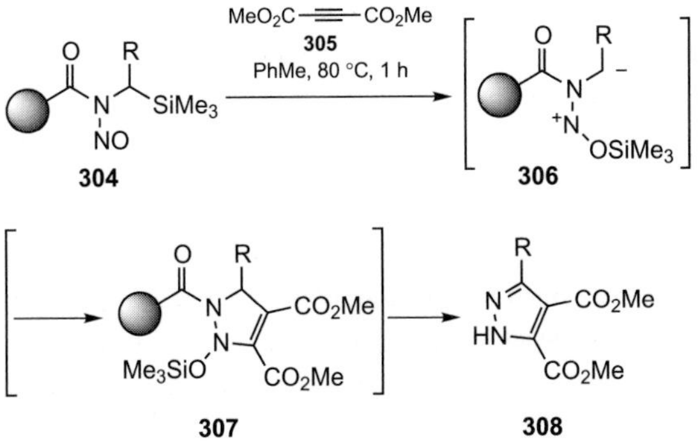

Scheme 4.55. The synthesis of pyrazole (**308**) according to Komatsu and coworkers [358].

A recent synthesis of benzofurans was based on cyclofragmentation. An appropriately substituted sulfone was used as a nucleophile in an intramolecular ring opening of an epoxide, wherein the resulting molecule lost a sulfinate and formaldehyde. By immobilization of the sulfone precursor to a resin, this sequence can be used for the cleavage of benzofurans from solid support, as demonstrated by Nicolaou et al. [160] (Scheme 4.56).

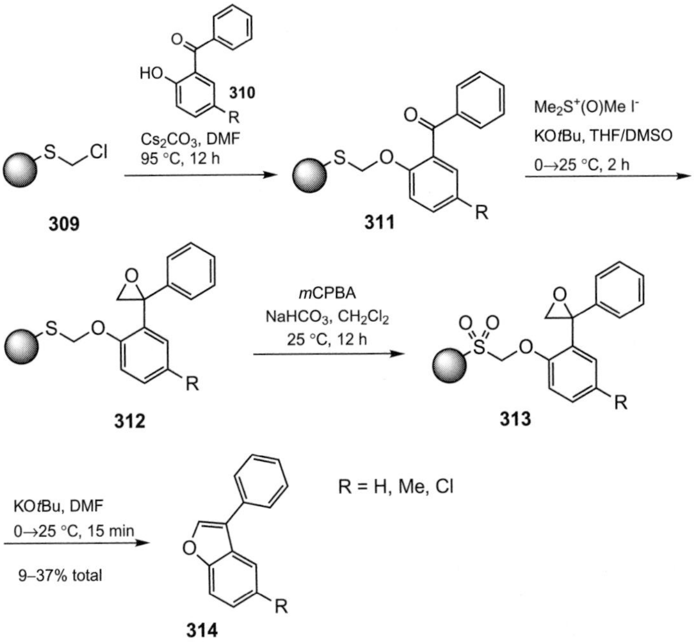

Scheme 4.56. Benzofuran synthesis according to Nicolaou et al. [160].

Further examples of the fragmentation strategy have also been published [359–362]. The semisynthesis of vancomycin clearly demonstrates the advantage of this kind of linker strategy [97].

4.5.5
Traceless Linkers

> ■ Traceless linker: *type of linker which leaves no residue on the compound after cleavage, i.e. replaced by hydrogen [12].*

The term "traceless linker" has led to ambiguous interpretations in the past. Many authors have claimed their linkers to be traceless because the term has, in the past, been quite fashionable even when the reported linker was used to immobilize and release amines (which upon cleavage carry a hydrogen).

"Traceless linking" is nowadays considered to be "leaving no functionality," meaning for arenes and alkanes that only a C–H bond remains at the original position of attachment (Scheme 4.57). A broadening of this definition to OH or NH groups is not useful, because otherwise every linker derived from polymeric protecting groups would have to be regarded a traceless linker.

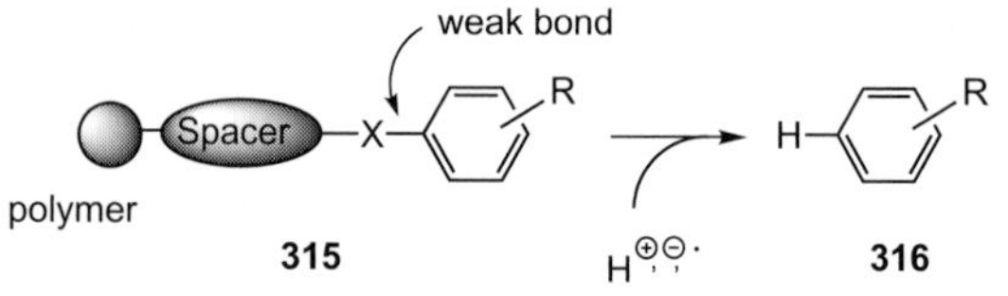

X = B, Si, Ge, N=N, Se, OSO$_2$

Scheme 4.57. Principle of a traceless linker.

When designing a traceless linker, one has to start from a heteroatom–carbon bond, which is labile toward protogenolytic, hydrogenolytic, or hydridolytic cleavage. Since most heteroatom–carbon single bonds are less stable than a carbon–carbon bond, traceless linkers can be synthesized based on nearly all heteroatoms. However, the enthalpies of C–X bonds are only relevant for homolytic bond scission. Many linkers are cleaved heterolytically, and the kinetic stability toward heterolytic bond cleavage is decisive in these cases.

The first traceless linker was developed by Kamogawa and coworkers as early as 1983 [367]. Starting from a polymer-bound sulfonylhydrazine, formation of a sulfonylhydrazone resin (**325**) was achieved by reaction with ketones or aldehydes. The cleavage step was conducted either by reduction with borohydride or alanate to yield alkanes (**326**), or by treatment with a base to give the corresponding alkenes (**327**) (Bamford–Stevens reaction) (Scheme 4.58).

In addition, one of the first papers dealing with traceless linkers was published in 1994 by Sucholeiki [161] and describes the use of thioethers (**328**) that are attached via an aromatic core that enhances the photolytic cleavage (Sect. 4.4.3). Irradiation at 350 nm gives rise to the formation of hydrocarbon **329** [161, 162]

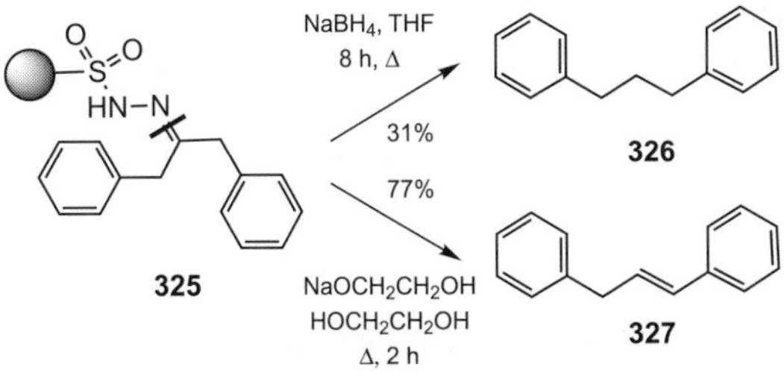

Scheme 4.58. The first traceless linker by Kamogawa et al. [367].

(Scheme 4.59). So far, this linker has not been fully explored and is limited in its range of functionalized arenes, since a phenyl substitution instead of biphenyl results in the formation of disulfides.

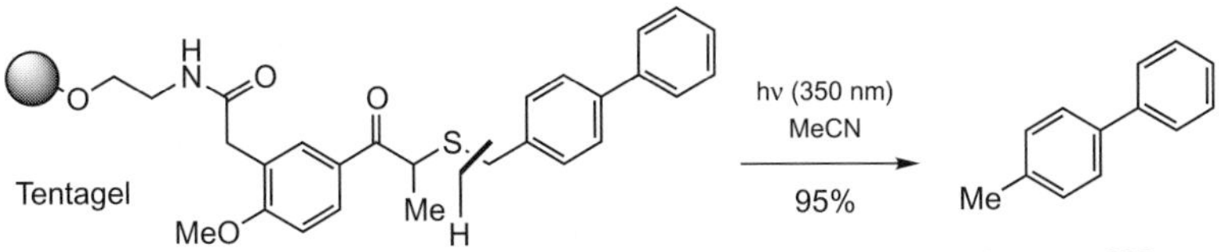

Scheme 4.59. Photolabile traceless linkage by Sucholeiki [161].

However, the most prominent anchors for traceless linkage for arenes (Table 4.17) are based on silyl linkers [124, 125, 369, 370] (Scheme 4.60) (Sect. 4.3.5). The generation of a diverse benzodiazepine library by Plunkett and Ellman [131] has shown clearly the advantages of this type of detachment since no additional functionalities were retained in the final molecules, which might bias the library. Starting from an immobilized stannane, palladium-catalyzed coupling with acid chlorides, deprotection of the aniline protecting group, acylation of the aniline with a series of Fmoc amino acids, Fmoc deprotection, and cyclization afforded resin-bound benzodiazepines, which were cleaved from the support using trifluoroacetic acid. Improvements in the chemoselectivity of the cleavage step – the silyl linker produces a substantial amount of the silyl arene upon cleavage – were accomplished using a germanium linker, which is more labile towards acids [131,

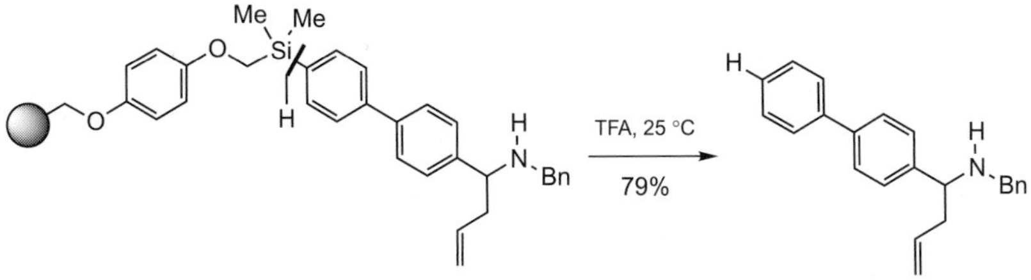

Scheme 4.60. The traceless silicon linker according to Veber and coworkers [124].

Tab. 4.17. Overview of traceless linkers for arenes.

Structure	Reference	Cleavage
94	[124]	Scheme 4.60; cleavage with fluoride
95	[125]	Cleavage with TFA
96	[126]	
97	[127, 128]	Cleavage with TBAF
319	[131, 363]	Scheme 4.61
120	[152]	Scheme 4.17; cleavage with palladium/formate
164	[182–185]	T1 resin: cleavage with HCl/THF or HSiCl$_3$
165	[182]	T1 resin: cleavage with HCl/THF or HSiCl$_3$
320	[237]	Scheme 4.65
108	[147]	Arenes; cleavage by Ag$^+$ (Scheme 4.16)

363] (Scheme 4.61). For example, the synthesis of the silyl linker has been optimized using preloaded handles to assist the coupling of the product to the resin. Other arylsilyl linkers have also been used (Table 4.6) to facilitate loading, synthesis, and/or detachment from the support [126–128, 130, 134–136].

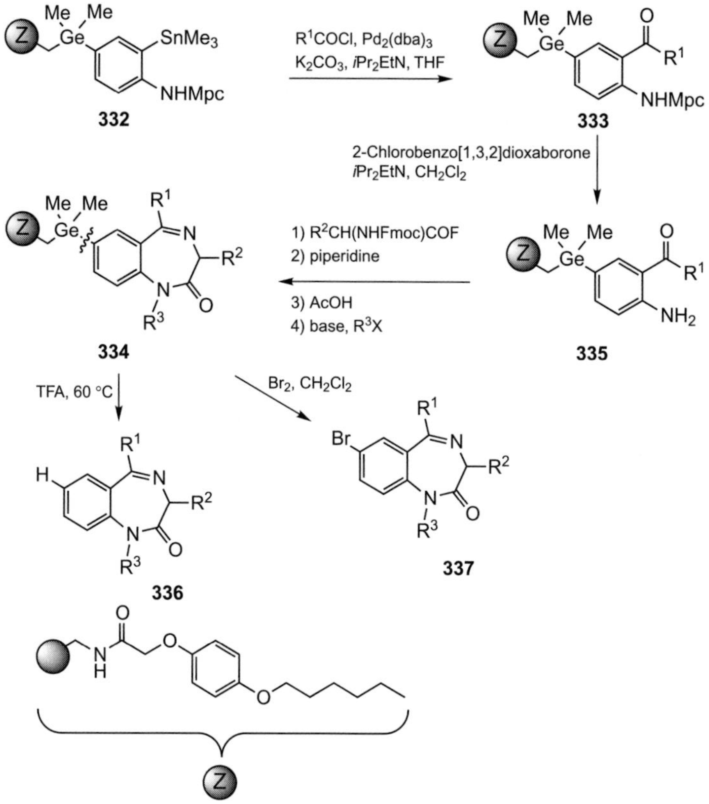

Scheme 4.61. Synthesis of a benzodiazepine library with the aid of a germanium linker by Plunkett and Ellman [131].

Furthermore, silicon linkers can be used for the attachment for allylsilanes, which can be cleaved to alkenes in a traceless fashion [144].

The use of sulfones as suitable anchoring groups for alkanes in soluble polymer chemistry has been reported previously [155, 156] (Table 4.18). After oxidation of a sulfide to a sulfone (**338**), treatment of the latter with sodium/mercury gives rise to the formation of the parent hydrocarbon (**339**) in high yields. However, aqueous work-up is necessary to provide the pure product (**339**) [155] (Scheme 4.62).

Another traceless linker type was independently developed by the groups of Nicolaou et al. [164] and Ruhland et al. [165]. Starting from lithiated selenium polystyrene (**146**), readily available from metallated polystyrene and selenium reagents, reaction with iodoalkanes led to the smooth formation of alkylated compounds. The cleavage can be conducted in such a way as to give alkenes (**150**)

Tab. 4.18. Overview of traceless linkers for alkenes and alkanes.

Structure	Reference	Possible structures achievable
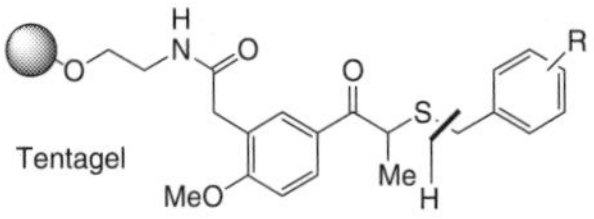 Tentagel **124**	[161]	Methylarenes; photolytic cleavage
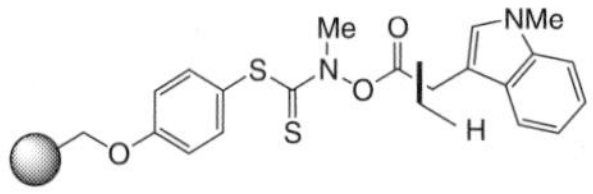 **126**	[120]	Methylarenes; photolytic cleavage
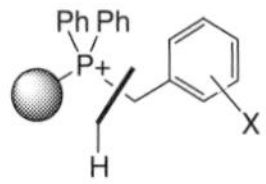 **321**	[364, 365]	Scheme 4.63
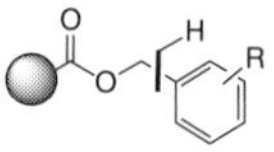 **322**	[366]	Scheme 4.67
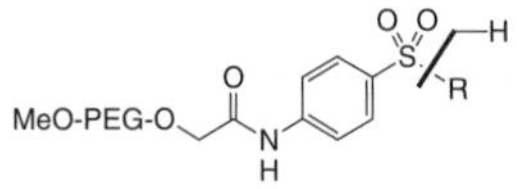 MeO-PEG-O **122**	[156, 155]	Alkanes; cleavage with Na/Hg
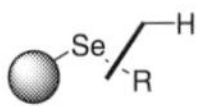 **128**	[164, 165]	Alkanes
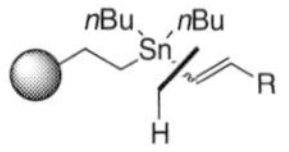 **127**	[163]	Alkenes
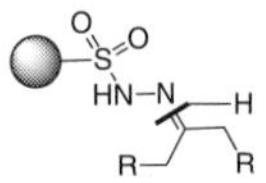 **323**	[367]	Alkenes and alkanes; cleavage by reducing media or bases (Scheme 4.58)
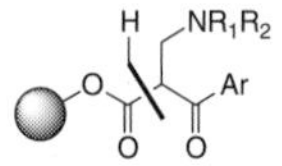 **324**	[368]	Ketones

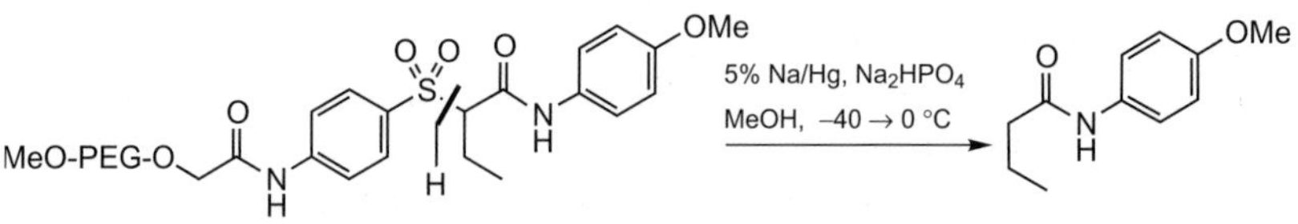

Scheme 4.62. The sulfone traceless linker for alkanes according to Janda and coworkers [156].

upon treatment with hydrogen peroxide. Alternatively, tin hydride reduction leads to the generation of alkanes (**149**) (Scheme 4.22).

Electron-poor aryl sulfonates are suitable candidates for oxidative palladium insertion. Hence, immobilized phenol sulfonates (**129**) have been employed in a palladium-catalyzed reductive cleavage using formic acid to yield arenes (**130**) with overall traceless cleavage [152] (Scheme 4.17). It might be anticipated that this type of linker is also suitable for functionalization (see below).

The phosphorus–carbon bond in phosphonium salts is readily cleavable by the aid of a base in the absence of an aldehyde. Hence, the polymer-bound phosphonium salt (**340**) gives direct access to methylarenes (**341**). An interesting feature of this linker is the fact that carbonyl compounds can be olefinated, which leads to a cleavage-olefination linker system [364, 365] (Scheme 4.63).

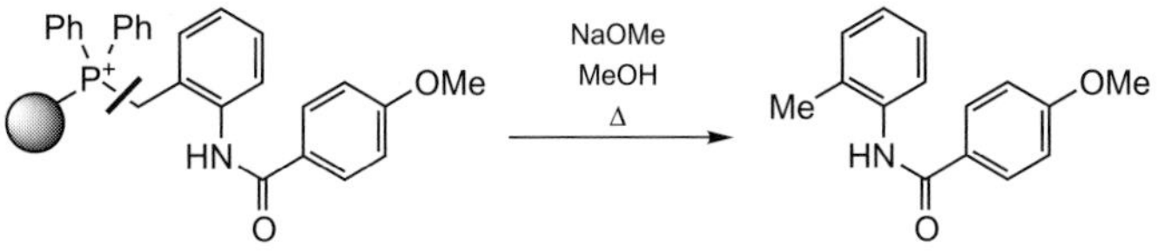

Scheme 4.63. The phosphonium linker for methylarenes according to Hughes [364].

Tin hydride reagents are versatile tools for the functionalization of alkenes and alkynes. Based on this concept, Nicolaou and coworkers [163] developed a polymer-bound tin hydride (**142**), which reacts in a hydrostannylation reaction with alkynes to give alkenylstannanes (**143**). After further transformation, the latter undergo proteolytic traceless cleavage to yield unsubstituted alkenes (**145**) [163] (Scheme 4.21).

The decarboxylation of appropriately substituted arenes [371] and alkanes [343, 368, 372–375] has been used to generate the parent hydrocarbons. Since the neighboring group effect is essential, limitation to special substrates is required.

One possible method for converting functionalized arenes into the corresponding hydrocarbons is the reduction of diazonium compounds [182]. Hence, the synthetic utility of the triazene linker as a traceless anchor for arenes has been demonstrated by Bräse et al. [182] using short reaction sequences. Thus, cinnamic esters were synthesized in a sequence starting from the iodoarene resin (**343**). Heck coupling with acrylates using palladium catalysis affords an immobilized cinnamate. This can be detached either directly or by a sequence of transformations yielding to an allyl amine in a traceless fashion either using trichlorosilane [184] or a HCl/THF mixture [182], to give the products **342** and **344**, respectively, in high yields and without further purification or aqueous work-up (Scheme 4.64).

Acyl aryl diazenes are known to fragment upon treatment with nucleophiles, a strategy which was used for a linker for carboxylic acid derivatives [119] (Scheme 4.11). Waldmann and coworkers recently developed a traceless linker for arenes based on this methodology [237]. Starting from a hydrazide resin (**345**), which

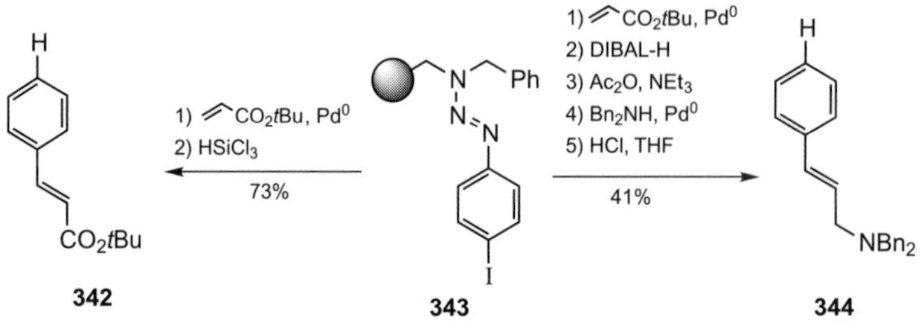

Scheme 4.64. The T1 linker for traceless cleavage [182, 184].

is converted into an activated species by oxidation, detachment of the molecule is carried out by the addition of nucleophiles (Scheme 4.65). This safety-catch principle – the activated linker is generated before cleavage (Sect. 4.5.1) – shows promising stability in palladium-catalyzed reactions. Furthermore, arylboronic acids can be used for the traceless synthesis of arenes [147] (Scheme 4.16).

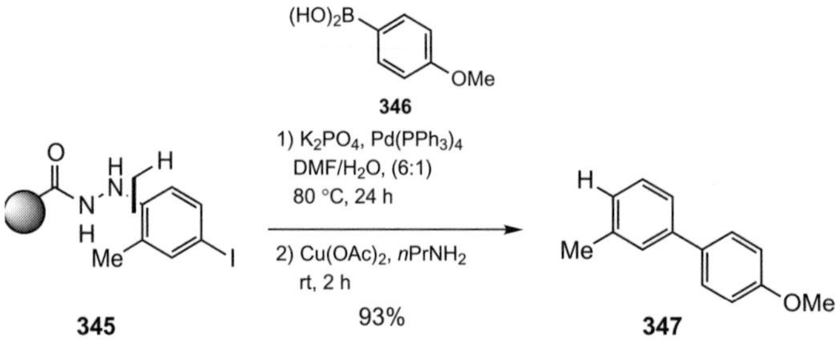

Scheme 4.65. The hydrazide linker according to Waldmann and coworkers [237].

A new traceless photolabile linker has been published using a hydroxamic acid derivative (**83**) [120].

The cleavage of specially designed polymeric benzyl-type protecting groups has been achieved using heterogeneous palladium black. In these cases, the catalytic hydrogenation furnishes methyl-substituted arenes as side products or targets. An early example takes advantage of the properties of the MeO-PEG-type support for the synthesis of di- and higher oligosaccharides (Scheme 4.66). It is interesting to note that the DOX linker enables the cleavage of the PEG structure, leaving the *p*-methylbenzyl group attached under certain conditions [376] (but see [29]) (Scheme 4.66).

Similarly, cleavage from polystyrene resins was achieved using homogeneous palladium catalysis (palladium acetate) either with formate reduction [366] (Scheme 4.67) or under an atmosphere of hydrogen [377] to yield methyl arenes.

The chemistry of traceless linkers is a fast-emerging field in the intensively investigated area of solid-phase organic synthesis [for reviews, see 183, 197, 378]. Although some confusion about the definition or classification has been related to

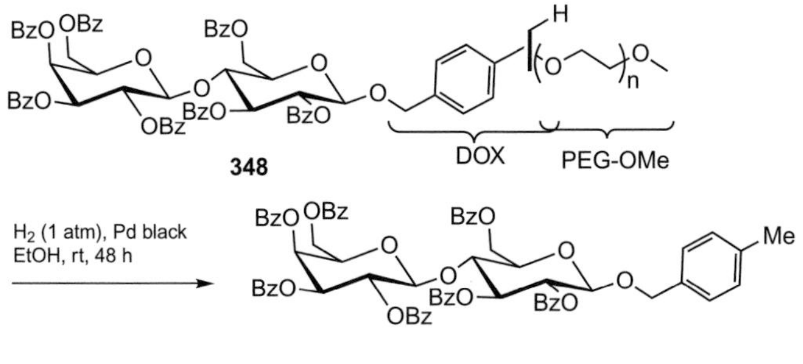

349

Scheme 4.66. Syntheses of methylarenes on polymeric support [376].

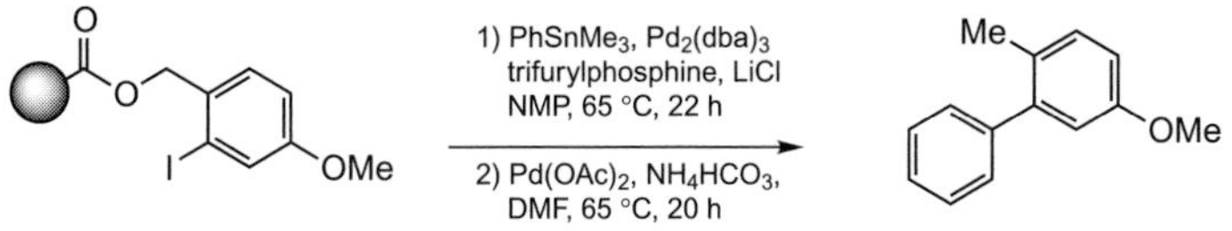

350 351

polystyrene with magnetite core

Scheme 4.67. Syntheses of methylarenes (**351**) on solid support [366].

this linker type, and therefore a careful designation has to be made, it is now clear that this anchoring mode will play an important role in the design and syntheses of drug-like molecules.

4.5.6 Multifunctional Cleavage

Traceless cleavage (Sect. 4.5.5) provides an efficient access to hydrocarbon-like molecules. However, monofunctional linkers (Table 4.19) provide only one type of compound in a library. The so-called multifunctional cleavage [379] offers an important opportunity to incorporate additional diversity upon cleavage (Scheme 4.68). Hence, the number of new functionalities (Fig. 4.5) can multiply the number of produced compounds. If the linker is amenable to various types of building blocks (e.g. nucleophile [A] and electrophiles [B]) incorporated during cleavage, a substantial library of novel molecules can be prepared from one immobilized compound [380].

However, when considering using a multifunctional linker, one must take into account the nature of the cleavage reagent and the cleavage step. A cleavage consisting of, for example, addition of a Grignard reagent to an ester with a huge excess of the organometallic component requires an aqueous work-up and hence potential annihilation of valuable material. Thus, the excess reagents need to be easily removable (volatile, low or very high solubility in certain solvents, easy to

Tab. 4.19. Overview of multifunctional linkers.

Structure	Reference	Possible structures achievable
355	[273]	Scheme 4.32
321	[364]	Alkenes by Wittig reaction
127	[163]	Alkanes by crosscoupling (Scheme 4.35)
164		T1 resin: "traceless" linker [182–184]; synthesis of phenols [186], biaryls, alkyl arenes [187, 188], azides, aromatic hydrazines, halides [190, cf. 181, 180], ester, azo compounds; cinnolines [191], benzotriazoles [192]
166		T2 resin: synthesis of alcohols, esters [196, 197, 183], alkyl halides [196, 197, 183]
167		T2*-resin: as above [198]
356	[381] (Scheme 4.69)	Amines, thiols [381], halides [382], azides [382], acetates [382] by nucleophilic displacement
357	[274] (Scheme 4.32)	Alkenes by palladium-catalyzed cleavage
121	[154] (Scheme 4.18)	Diaryl methanes by activation and subsequent palladium-catalyzed cleavage
120 polystyrene	[153]	Heterocycles by cleavage with nucleophiles
358	[277, 154] (Scheme 4.70)	Alkenes by cleavage through metathesis

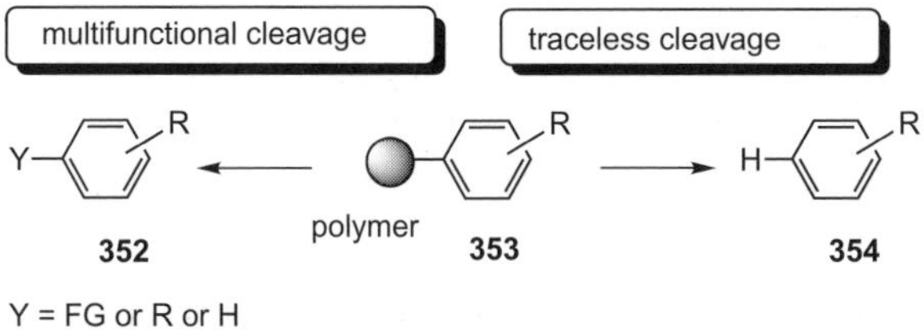

Scheme 4.68. Traceless vs. multifunctional cleavage.

eliminate or to be removed by for example scavenger resins, etc.) and should not interfere with the functionalities of the library compounds.

The addition of nucleophiles such as amines, alkoxides, thiolates, and carbon nucleophiles to carbonyl groups leading to modified carbonyl moieties has been used widely for multifunctional cleavage. Hence, anchoring a Weinreb amide to the resin and cleaving with Grignard reagents leads to the formation of ketones [383]. Alternatively, thioesters provide access to amides or ketones, whereas sulfonate esters such as **359** provide access to amines and thiols (**360**) [381] (Scheme 4.69), halides [382], azides [382], or acetates [382].

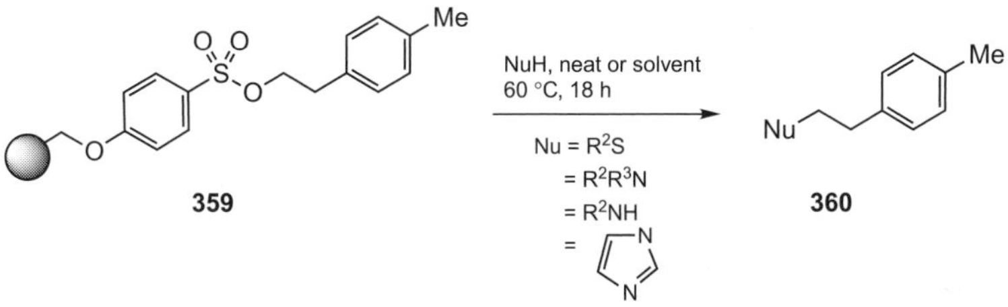

Scheme 4.69. Synthesis of tertiary amines, thioethers or imidazoles using a nucleophilic substitution cleavage [381].

Aromatic sulfones, synthesized by oxidation of thiopyrimidines (safety-catch linker, Sect. 4.5.1), have been used for the synthesis of aminopyrimidines by displacement with amines [384]. Silyl linker **96**, described previously [126], may be used for a smooth multifunctional cleavage. Hard electrophiles such as chlorine, bromine, iodine, or nitrosyl replace the silicon at the *ipso*-position of the silyl arene, leading to a variety of arenes upon cleavage. The same strategy has been demonstrated with the germanium linker.

Cleavage by metathesis has been used by various groups to accomplish either detachment by cyclization (ring-closing metathesis, RCM) [281], intermolecular metathesis (cross-metathesis) [385] (Scheme 4.70), or intramolecular metathesis [275] (see also Sect. 4.4.4.2). One advantage of cleavage by metathesis is the possibility of introducing fragments with additional functionalities. The successful examples include the total synthesis of epothilone by this cleavage strategy [277].

The nucleophilic substitution of allylic fragments provides a general access by multifunctional anchoring. Staring from dienes, prepared by en-yne cross-

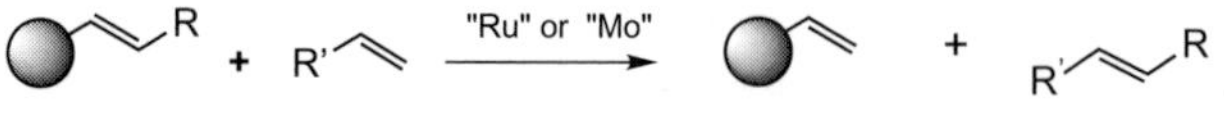

Scheme 4.70. The concept of cleavage by cross-metathesis.

metathesis, nucleophilic substitution catalyzed by palladium via π-allyl complexes provides access to functionalized dienes [273] (Scheme 4.33). Similarly, 1-alkenyl-cyclobutylsulfones can be used to yield cyclobutylidene derivatives [274] (Scheme 4.32).

Besides the possibility of conducting traceless cleavage (Sect. 4.5.5) from tri-azene linkers, a cleavage cross-coupling can be achieved using palladium catalysis [187, 188] (Scheme 4.34). In addition, functionalization of triazene resins upon cleavage leads to the formation of aryl halides (**180-Nu**, Nu = halogen) [191], phe-nols, and aryl ethers (**180-Nu**, Nu = OAc), azo compounds (**176**), biaryls (**185**) by a Gomberg–Bachmann reaction, Meerwein alkylation products (**187**), benzotriazoles (**177**) [192], and Richter products such as cinnolines (**178**) [191] (Scheme 4.27).

Disubstituted triazenes (**194**) prepared on the T2 resin undergo cleavage to yield aliphatic diazonium ions, which in turn solvolyze in the presence of nucleophiles to give alkyl halides (**202**), alcohols (**201-OH**) and alkyl esters (**201-OCOR**) [see 200] [183, 196, 197] (Scheme 4.29). It is possible, but not fully established, that silyl linkers can be cleared in a multifunctional fashion to enable the synthesis of arenes and aryl halides (see Scheme 4.61).

A flexible tool for the synthesis of heterocycles is the cleavage of α-sulfonylated ketones attached to a solid support [153]. Since the loading proceeds smoothly from alkenes or epoxides onto polystyrene sulfonic acid resin, this method pro-vides ample possibilities for the functionalization of simple compounds.

4.5.7
Linkers for Asymmetric Synthesis

The use of enantiomerically pure drugs has increased within the pharmaceutical industry over recent years. It is assumed that the market was well over US$120 billion in the year 2000 [386, 387], and approximately 60% of all marketed drugs are enantiomerically pure. Chiral building blocks have been used extensively in solid-phase synthesis (Fig. 4.6) and have been incorporated into modern small-molecule compound libraries.

However, asymmetric synthesis based on linking strategies, with either diaster-eoselective or enantioselective methods, has been more or less neglected. The ob-vious advantage of a linker, which induces stereoselectivity, is the ease with which the auxiliary can be removed and recycled [388].

Pioneering work on asymmetric synthesis on an insoluble support was de-scribed in 1972 when Kawana and Emoto [389, 390] reported the synthesis of an atrolactic acid on a polymer containing the sugar 1,2-o-cyclohexylidene-α-D-xylofur-anose (**365**). Shortly after this, Leznoff and coworkers [391, 392] demonstrated the synthesis of α-chiral cyclohexanone derivatives using polymer-bound imines (**366**).

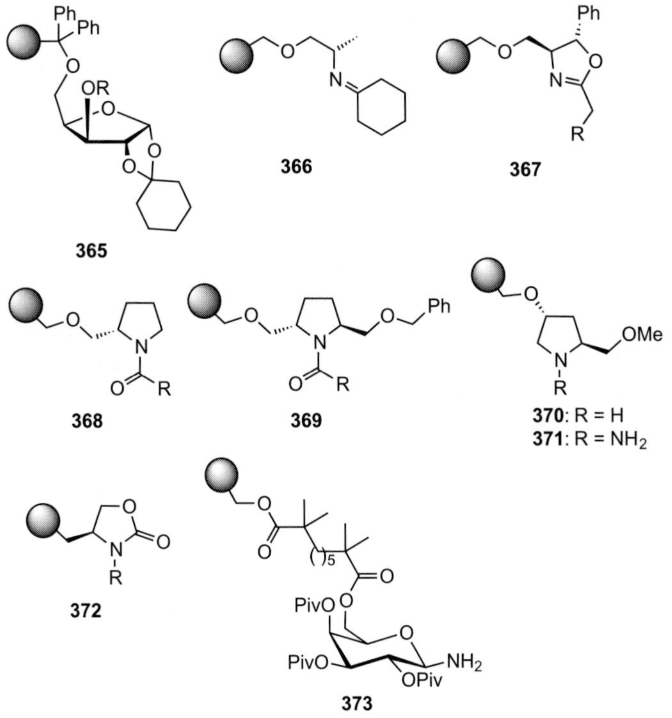

Fig. 4.6. Overview of chiral auxiliaries on solid support.

These early examples have demonstrated that polymer-bound chiral auxiliaries are suitable both for anchoring of organic molecules and for the induction of asymmetry. Furthermore, the recycling of chiral auxiliaries can be confirmed.

In recent years, a series of auxiliaries has been immobilized on solid support, including oxazolidinones (**372**) according to Evans [231, 393–398], oxazoles (**367**) [399], mono- (**368**) [400] and bisalkoxymethylpyrrolidines (**369**) [401], as well as SMP/SAMP auxiliaries (**370**, **371**) by Enders and coworkers [402, 403]. Alternatively, polymer-bound chiral sulfinamides [404], sulfoximines by Hachtel and Gais [405], imines [406], and amines (as galactosylamine) (Scheme 4.71) have been used.

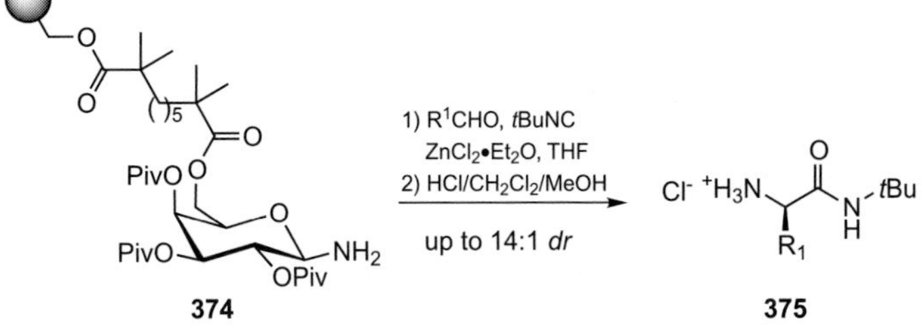

Scheme 4.71. Stereoselective Ugi-3CC according to Kunz and coworkers [407].

A recent example is the diastereoselective Ugi reaction for the synthesis of α-amino acids by Kunz et al. [407]. Starting from a galactosylamine (**374**), a three-component reaction of aldehydes and isocyanides produced amides (**375**) in good diasteroselectivities (Scheme 4.71) after removal of the auxiliary.

4.6
Linkers for Functional Groups

The requirement for diverse compound libraries by means of solid-phase synthesis has led to the development of linkers for most functional groups found in organic synthesis. The number of linkers developed for a specific group also reflects the distribution of pharmacophoric groups (Fig. 4.7) present in natural products and other bioactive compounds. In this section, linkers for functional groups are highlighted. In all cases, the functional group that is attached to the solid support, whether it remains unchanged upon cleavage or is formed during cleavage, acts as the guide through this section. For example, a benzodiazepine attached via the aromatic core would be found in the section for the linking of arenes (Sect. 4.6.7.2), whereas attachment via the nitrogen atom (amide functionality) will be described in Sect. 4.6.2.4.

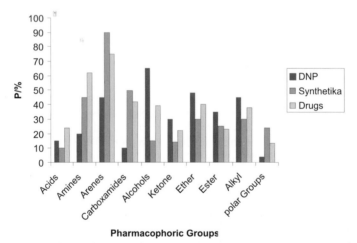

Pharmacophoric Groups

Fig. 4.7. Distribution of pharmacophoric groups of compounds in three different databases according to Henkel et al. [408]. DNP, *Dictionary of Natural Products*; drugs, internal database of the Bayer company; synthetika, screening pool of the Bayer company.

4.6.1
Linkers for Nitrogen Functionalities

Nitrogen-containing structures are mandatory structural units in biologically active compounds. Therefore, it is not surprising that a vast number of linkers for various nitrogen functionalities has been developed.

4.6.1.1 Linkers for Amines

The solid-phase synthesis of primary amines is of great importance to various fields of organic and medicinal chemistry, since these compounds are both valuable synthetic intermediates and often interesting target molecules. The generation of libraries of commercially unavailable amines has received considerable interest within combinatorial chemistry as a large number of primary amines show biological activity and the primary amine moiety is incorporated into various molecules encountered in medicinal chemistry. Moreover, these amines can serve as building blocks in the preparation of further libraries. Consequently, linkers for amines have been used excessively. Among these are the Rink linker, carbamate linkers [301], and Rink carbamates [409].

4.6.1.2 Linkers for Primary Amines

For the solid-phase synthesis of primary amines miscellaneous linker types have been developed [85, 94, 316, 410–425] which are resistant to various reaction conditions. The standard linker family suitable for the detachment of amines is the benzyl-type linkers (Sect. 4.3.1) [32, 82, 426–438], such as the Rink linker [82] (Scheme 4.72), the trityl linkers [76], the indole linker [85], the BAL linker [14], and others [226, 439]. Carbamates are suitable linkers for primary amines, which can be released by diluted TFA [94, 268, 440–446].

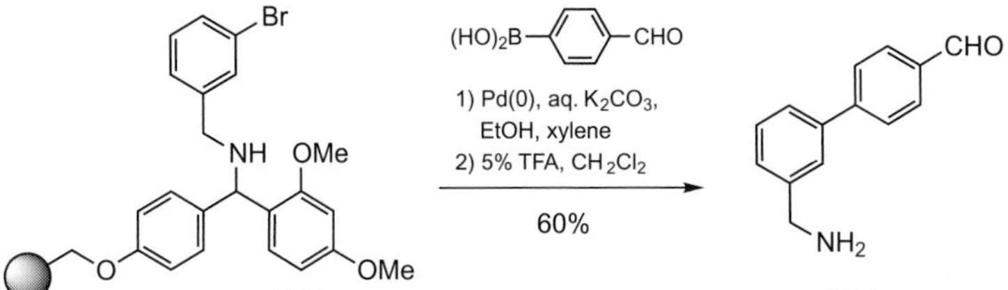

Scheme 4.72. The Rink resin as a linker for primary amines [82].

Alternatively, benzyl carbamates can be attached to the solid support, and hydrogenolytic cleavage can be used to detach the molecules, which are then usually left with a nitrogen functionality (cleavage of C–O bond) [29, 447]. The polymers in these cases are formally immobilized Benzyloxycarbonyl Cbz (Z) groups. Interestingly, TentaGel and polystyrene give the products in similar yields under identical conditions [17] (Scheme 4.73). Benzylic linkers can also be used advantageously in the presence of other benzylic protecting groups, since they can be removed in the same step [29].

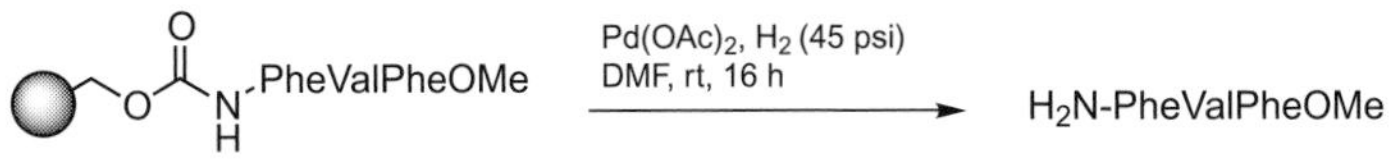

Wang or TentaGel Wang resin

Scheme 4.73. Detachment of peptides from polymeric benzyl-type protecting groups [17].

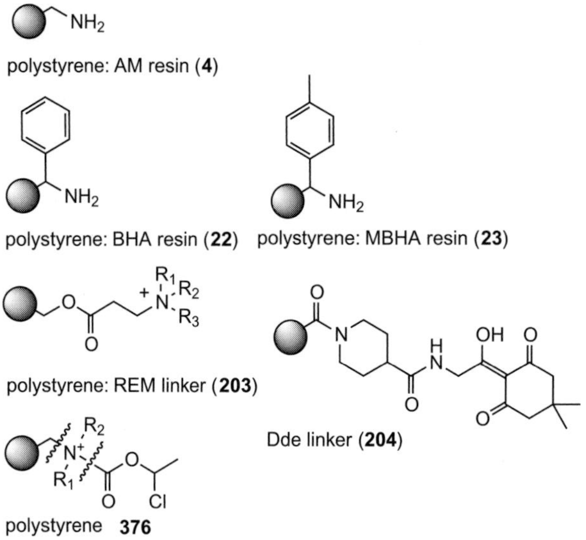

polystyrene: AM resin (**4**)

polystyrene: BHA resin (**22**)   polystyrene: MBHA resin (**23**)

polystyrene: REM linker (**203**)

Dde linker (**204**)

polystyrene **376**

Fig. 4.8. Overview of amine linkers.

An anchor for primary amines, which is cleavable under basic conditions (Table 4.12), is the dimedone-based Dde group (**204**) [226, 228, 425] (Fig. 4.8).

Anilines are released from the PAL linker, the Rink linker [82], the PhFl linkers [437, 448], the *p*-benzyloxybenzylamine (BOBA) resin [32], and carbamate linkers. The latter type has also been used as a safety-catch version [168] (Sect. 4.5.1).

A novel dialkylhydrazine linker (**381**), which is stable toward organometallic reagents, has recently been reported. A series of α-branched primary amines was synthesized by attachment of various aldehydes, the resulting hydrazones (**384**) were modified via 1,2-addition of organolithium reagents furnishing trisubstituted hydrazines (**385**) which in turn were cleaved from the solid support to yield acylated α-branched primary amines [449] (Scheme 4.74).

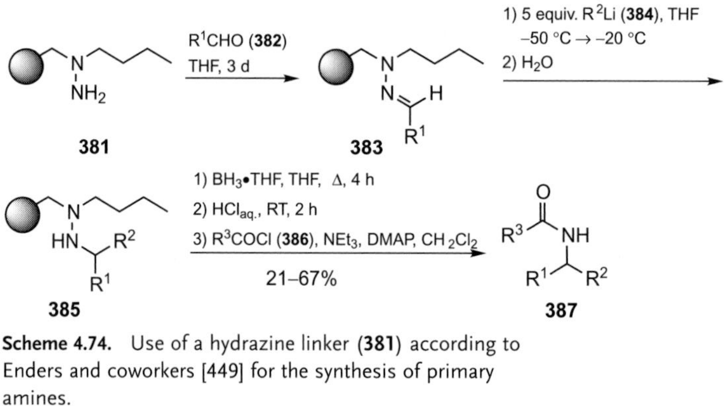

Scheme 4.74. Use of a hydrazine linker (**381**) according to Enders and coworkers [449] for the synthesis of primary amines.

The cleavage of primary sulfonamides to give primary amines can be performed with electrochemical cleavage, when carried out on a conducting polymer [167] (Scheme 4.19).

4.6.1.3 Linkers for Secondary Amines

Secondary amines have been detached from solid support using the trityl, chlorotrityl, and other linkers [69, 193, 340, 416, 418, 450] (Fig. 4.9). Secondary amines can be detached from benzyl-type resins if they are activated to the corresponding α-chlorocarbamates 376 [417].

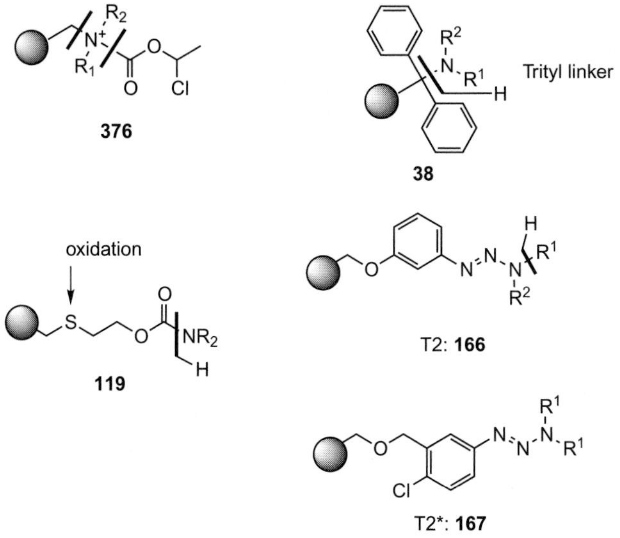

Fig. 4.9. Overview of linkers for secondary amines.

Carbamates are suitable linkers for secondary amines, which can be released by diluted TFA [17, 345, 451, 452], as demonstrated in a Hantzsch dihydropyridine synthesis by Breitenbucher and Figliozzi [409] (Scheme 4.75).

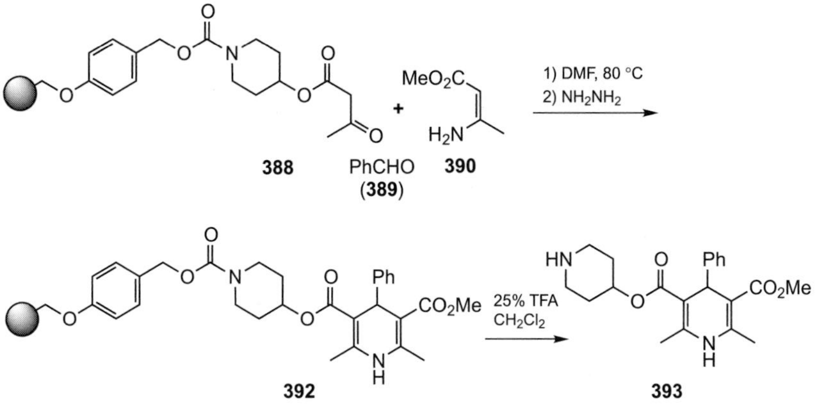

Scheme 4.75. Carbamates as linkers for amines by Breitenbucher et al. [409].

The triazene T2 linker is capable of clean detachment of secondary amines. Since disubstituted triazenes can be alkylated with electrophiles in the presence of a strong base, this method provides a clean procedure for the conversion of primary to secondary amines [198] (Sect. 4.3.8) (Scheme 4.76).

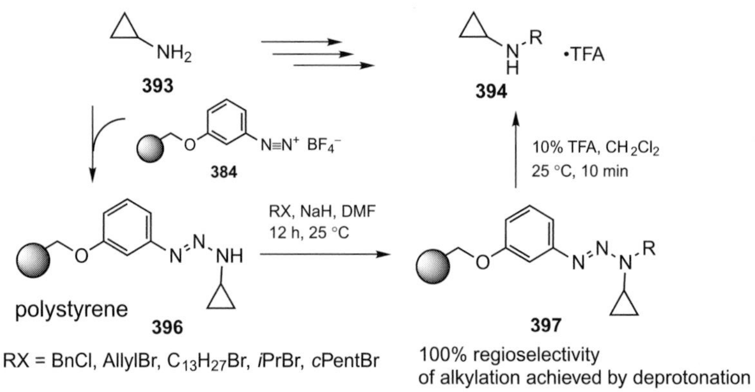

Scheme 4.76. The T2 linker for the synthesis of secondary amines on solid support [198].

Other linkers for secondary amines are the thiobenzylethylcarbamate linker [151] (Scheme 4.20), PAL-type linkers [453] (dihydropyridines) [453] (Scheme 4.51) [454], and other systems [455–457].

4.6.1.4 Linkers for Tertiary Amines

The cleavage of tertiary amines from solid support is somehow different from the analogous primary and secondary amines, since no apparent linking site is available. However, the ease of formation of the tetravalent nitrogen moiety can be favorably used for the linking of tertiary amines (Fig. 4.10).

The prototype is the REM resin, which is based on a Michael addition/alkylation and elimination sequence on an immobilized acrylate [220–225] (Scheme 4.77). Besides acrylates, vinyl sulfones might serve the same purpose [458].

The cleavage of dialkylaryl triazenes with methyl iodide was investigated by Moore and coworkers [180] and Tour and coworkers [181] during their synthesis of aryl iodides on solid support (Sect. 4.14.3.8). Using the piperazine resin (**394**), cleavage with methyl iodide in dichloromethane yielded the tertiary amines with good purities. A quarternization of the nitrogen was not detected [459] (Scheme 4.78).

Similarly, the hydroxylamine linker (**396**) can be used for the synthesis of tertiary amines [460], as demonstrated in the synthesis of the analgetic (±)-Tramadol® (**401**) on solid support [461] (Scheme 4.79).

Carbamates can be cleaved by means of lithium aluminum hydride in THF to give *N*-methylalkylamines with high purities [446, 462, 463]. While the latter

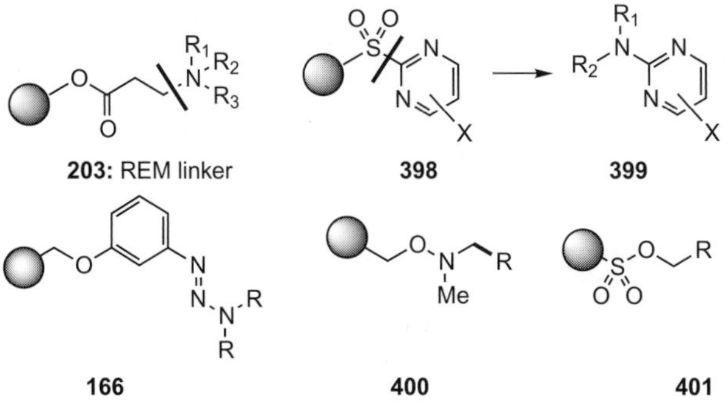

Fig. 4.10. Overview of linkers for tertiary amines.

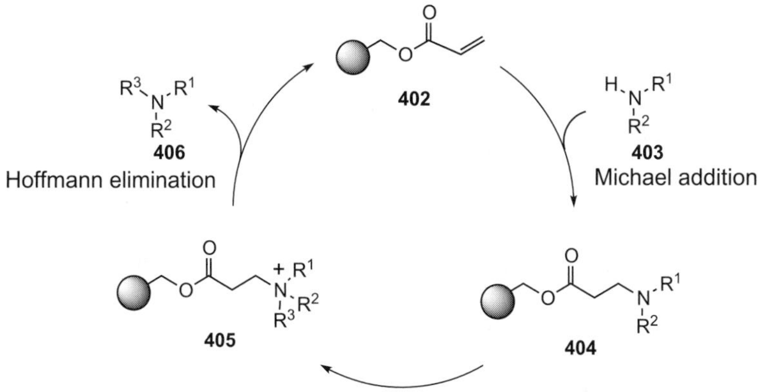

Scheme 4.77. The REM linker for the synthesis of tertiary amines [223].

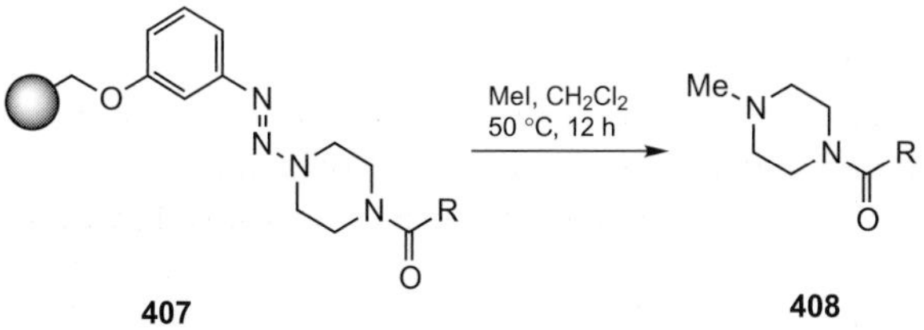

Scheme 4.78. Alkylative cleavage of triazenes [459].

linker provides methylamines, the reaction of secondary amines with electrophiles consisting of polymer-bound leaving groups enables the synthesis of higher substituted amines. However, the removal of the excess of secondary amine can be a drawback to this method [153, 381] (Scheme 4.80).

Besides aliphatic electrophiles, certain heteroaromatic structures can also be

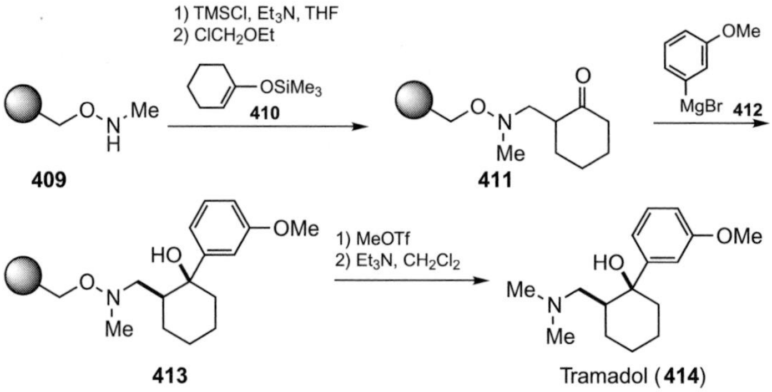

Scheme 4.79. Tramadol® synthesis according to Grigg and coworkers [461].

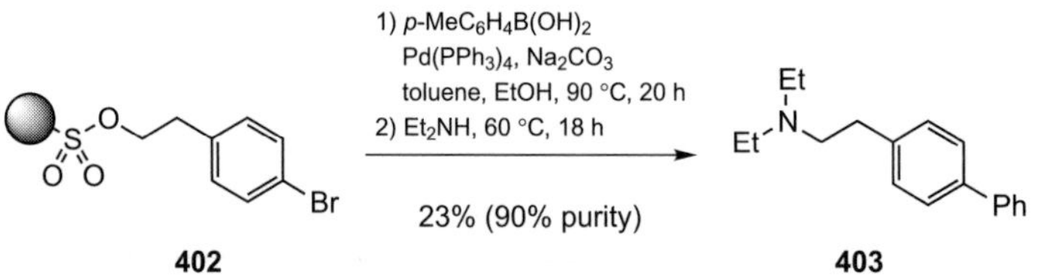

Scheme 4.80. Synthesis of tertiary amines using a sulfonate linker [464].

used [384]. Furthermore, various other linkers are suitable for tertiary amines [224, 225, 465].

4.6.1.5 Linkers for Hydrazines, Hydrazones and Hydroxylamines
While hydrazines can be released from trityl linkers, the detachment of disubstituted hydrazones (**418**) is possible with the triazene linker [466] (Scheme 4.81). The hydrazones (**417**) are readily available from polymer-bound hydrazines and subsequent condensation with aldehydes.

4.6.1.6 Linkers for Diazonium Salts
The triazene T1 linker is a suitable anchoring group for the detachment of diazonium salts. Upon cleavage with dilute acids, the diazonium salts can be obtained

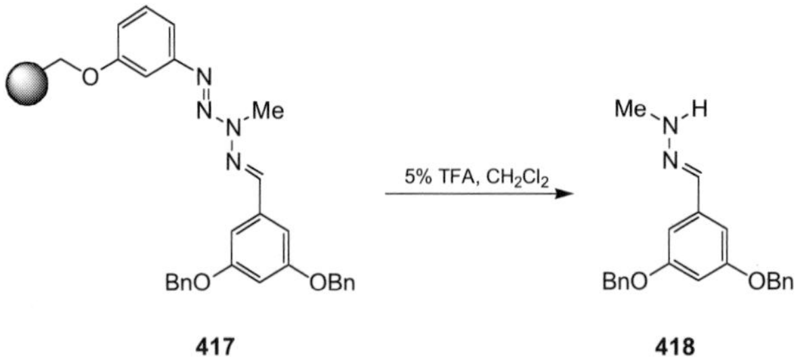

Scheme 4.81. Synthesis of hydrazones [466].

in high yields. However, owing to their instability, the diazonium ion is mostly transformed *in situ* to a new functional group [467] (Scheme 4.82) (see also Scheme 4.27).

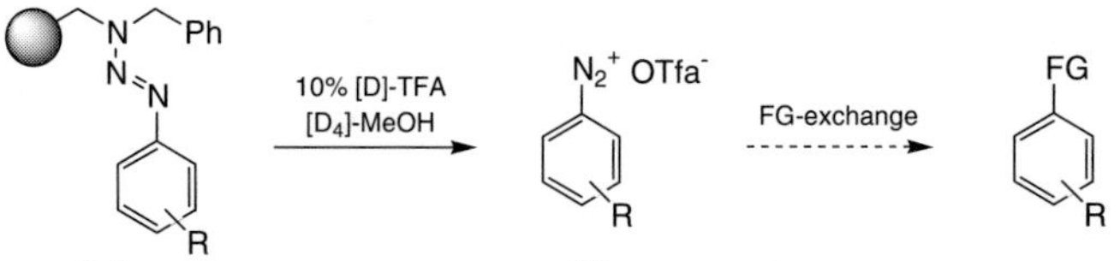

Scheme 4.82. *In situ* cleavage of triazenes for the synthesis of diazonium salts [467].

4.6.1.7 Linkers for Azides
Aromatic and aliphatic azides are interesting targets and intermediates in solid-phase organic synthesis. Aryl diazonium salts react with azide ions without a catalyst to give aryl azides. In contrast to the classical Sandmeyer reaction, this transformation proceeds without cleavage of the carbon–nitrogen bond. This reaction can be used for the synthesis of libraries of aryl azides. Starting from the triazene T1 resin, cleavage with trifluoroacetic acid in the presence of trimethylsilyl azide furnished, after simple removal of the solvent, the aryl azides in good yields and excellent purities [189] (Scheme 4.27). This synthesis can also be applied to the synthesis of heterocycles such as benzofuroxanes [189]. Aliphatic azides have been released from solid support using a sulfonate linker [382, 465, 468].

4.6.1.8 Linkers for Nitro Compounds
The reaction of arenes attached to silyl-type linkers with nitronium ions might lead to the formation of nitroarenes, as proposed in [128].

4.6.1.9 Linkers for Azo Compounds
Although important structures for the synthesis of dyes, the synthesis of azo compounds on solid support has scarcely being described. The transfer of diazonium salts to nucleophiles such as phenols can be used for the synthesis of azo compounds. This approach has been demonstrated either with diazonium salts immobilized on an ion exchange resin [469] (Scheme 4.83) or starting from triazene T1 resins [199] (Scheme 4.84).

4.6.1.10 Linkers for Nitriles
Polymeric sulfonylhydrazines react readily with aldehydes and ketones to form hydrazones, which can release the corresponding nitriles [367].

4.6.1.11 Linkers for N-Heterocycles
Various heterocyclic nitrogen compounds can be detached from solid support (for a review, see [470]). Basically, two options are conceivable. The removal of heterocycles bound via the nitrogen atom to solid support provides a general entry to this

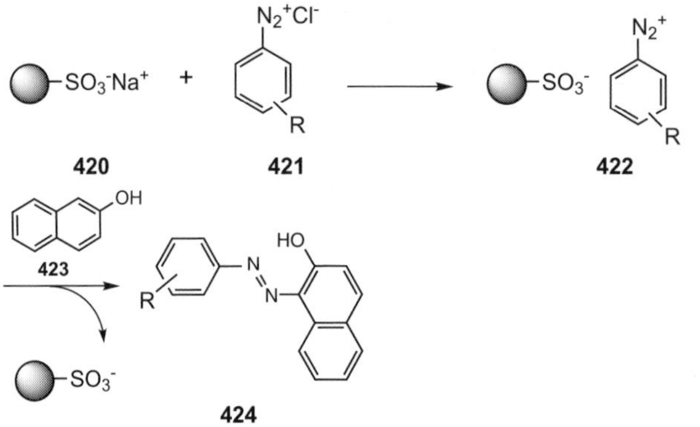

420 421 422 423 424

Scheme 4.83. Azo coupling through immobilized diazonium salts according to Das et al. [469].

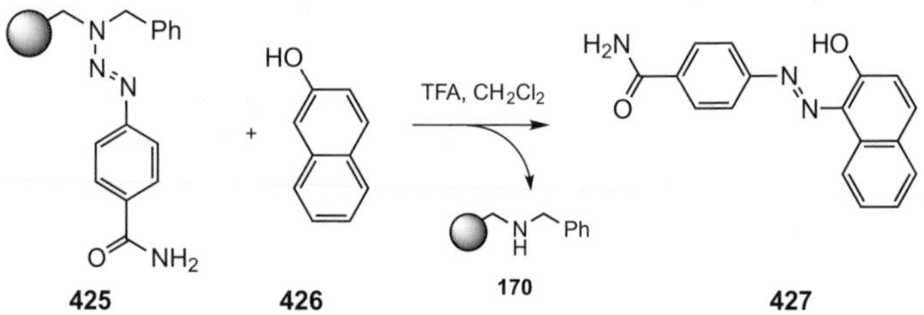

425 426 170 427

Scheme 4.84. Cleavage of a triazene with concomitant azo coupling [199].

class of compounds. Otherwise, a cleavage-cyclization technique can be considered to build up the nitrogen heterocycle as the very last step. Both strategies have been used extensively.

Tetrazoles, for example, can be attached to a THP linker. Cleavage is best achieved either with 3% HCl or with TFA [112] (Scheme 4.85).

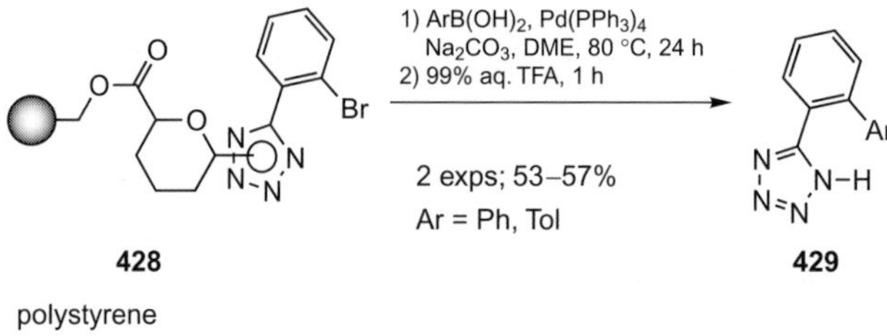

428 429

Scheme 4.85. Cleavage of tetrazoles from a THP-type linker [112].

The THP (Table 4.4) [99] and other acetal linkers [100] are also suitable for the attachment of indoles. Alternatively, heterocyclic amines with a free N–H functionality can be linked via an acetal, which can be cleaved first under acidic and then under basic conditions [456] (Scheme 4.86).

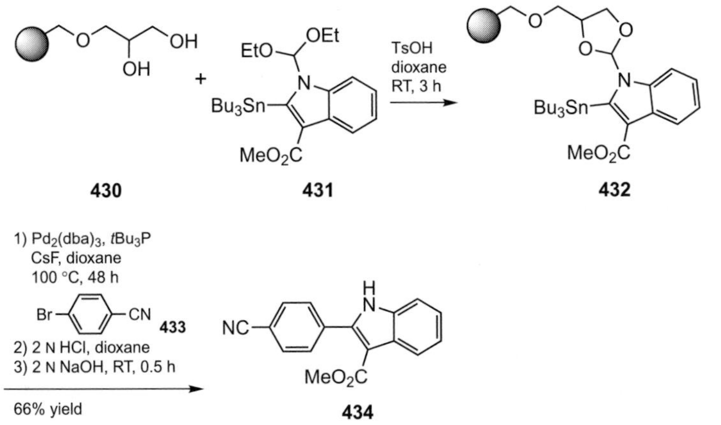

Scheme 4.86. Synthesis of indoles (**434**) using a diol linker by Gmeiner and coworkers [456].

Indole synthesis can be realized on solid support via a sulfone linker. Advantageously, the heterocyclic core was installed in this case via a palladium-catalyzed process [302] (Scheme 4.87). Similarly, benzimidazoles can be released via a retro-Michael addition from sulfonate resins [471] (Scheme 4.88).

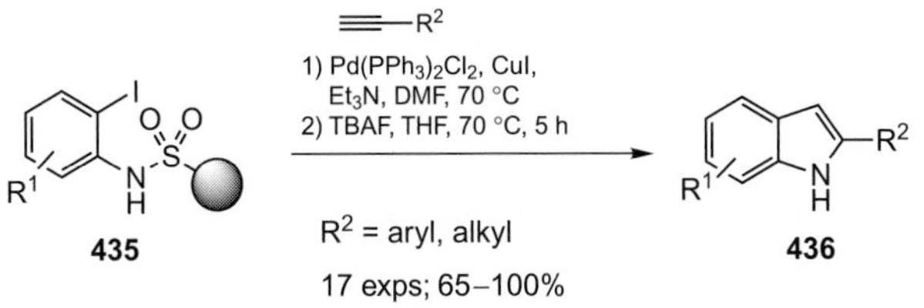

Scheme 4.87. Cleavage of indoles from sulfonate resins [302].

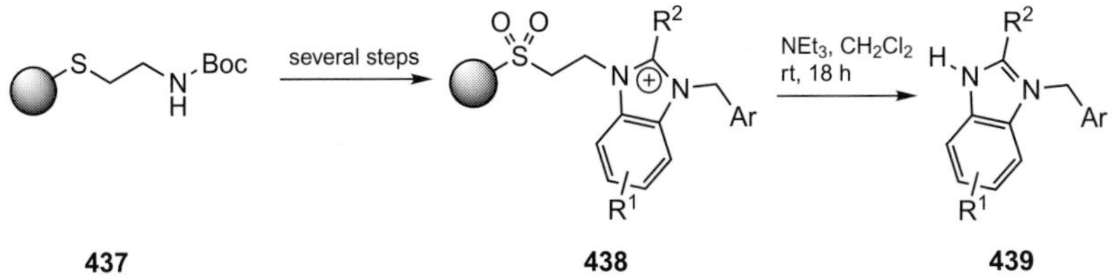

Scheme 4.88. Cleavage of benzimidazoles [471].

Other examples for the synthesis of aromatic nitrogen heterocycles include the synthesis of pyridines and tetrahydropyridines [452], benzimidazoles [471], isoxazolines [472], isoquinolines via Reissert complexes (Scheme 4.53) [351, 352], quinoxalinones [49], benzisoxazoles [473], and imidazoles on trityl linkers [78].

The second strategy is dedicated to cleavage conditions, which result in the installation of the heterocyclic core; this strategy obviously belongs to the cleavage-cyclization technique (Sect. 4.5.3). An example of this detachment uses diazonium salts, which can be cyclized when a nucleophilic *ortho*-substituent is present (Scheme 4.89). This reaction yields cinnolines [191] (Scheme 4.90), benzotriazoles (Scheme 4.91) [192], benzothiadiazoles, indazoles, benzo[c]cinnolines, and other structures. Suitable precursors are the triazene T1 resins (Sect. 4.14.3.8), which are cleavable by mild acids. Azides are also precursors for a similar cyclization. This process has been demonstrated in a synthesis of benzofuroxanes on solid support [189] (Scheme 4.92). Finally, certain arylsulfonyl hydrazones cyclized to the corresponding thiadiazoles (**453**) when treated with thionyl chloride [474] (Scheme 4.93).

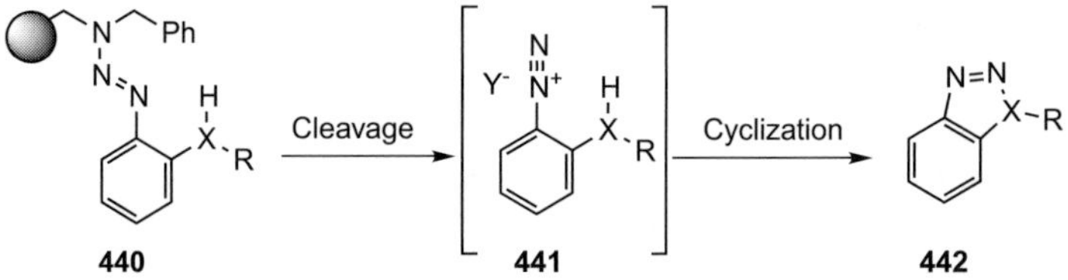

Scheme 4.89. General synthesis of heterocycles via cyclization of diazonium salts generated from the T1 linker [191].

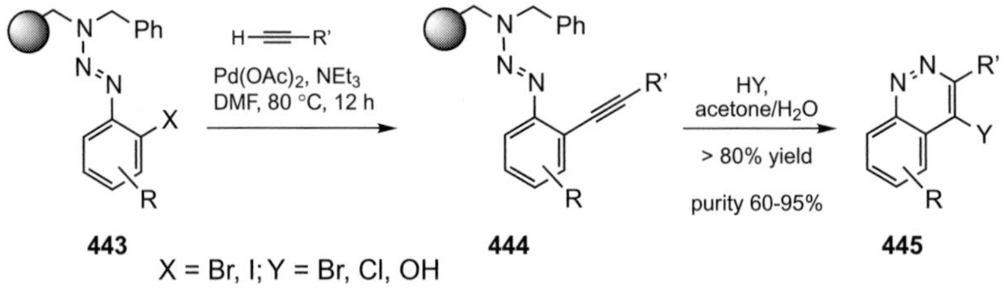

X = Br, I; Y = Br, Cl, OH

Scheme 4.90. Synthesis of cinnolines (**445**) [191].

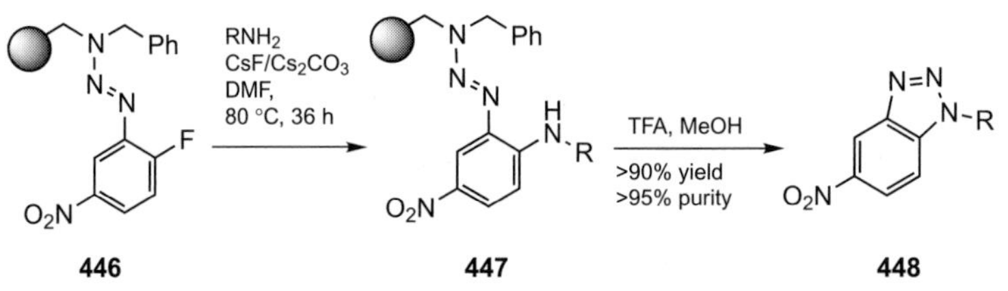

Scheme 4.91. Synthesis of benzotriazoles (**448**) [192].

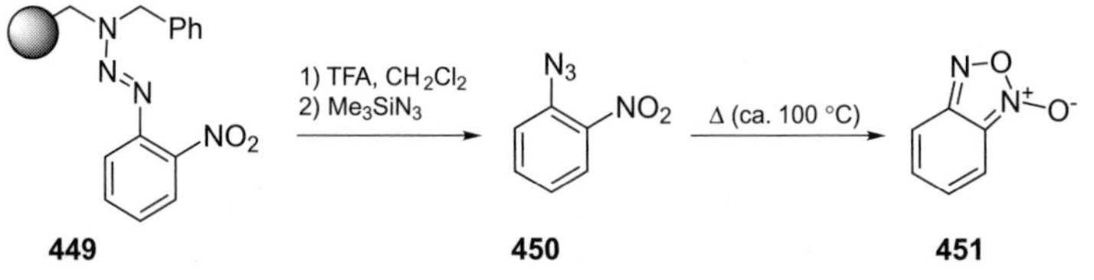

Scheme 4.92. Synthesis of benzofuroxanes (**438**) [189].

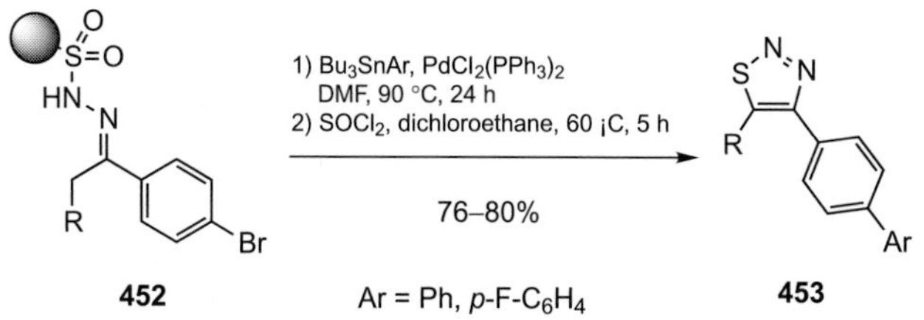

Scheme 4.93. Synthesis of thiadiazoles [474].

4.6.2
Linkers for Carbonyl Functionalities

4.6.2.1 Linkers for Carboxylic Acids

The prominent number of linkers is certainly dedicated to the anchoring and detachment of carboxylic acid. This is due to the widespread application of solid-phase peptide synthesis. However, a large number of pharmacologically active, low-molecular-weight compounds contain the carboxylic acid functionality and therefore new and innovative linkers have been developed [475]. In general, most syntheses rely on the Wang resin [476] or related benzyl, trityl, and benzhydryl systems, which are cleavable by trifluoroacetic acid [23, 34, 48, 66, 77] (see Table 4.2) or HF/anisole [35]. Other examples include acetal resins, sulfones [477], β-thioethylesters [168], fluoride-labile linkers [52, 69, 122], Kaiser oxime resin [478], π-allyl linkers, photo linkers [238, 241, 244, 245, 260, 266, 479], the safety-catch ketal linker [293], and other linkers [18, 442, 443, 480, 481].

A new anchoring group based on phenanthridines has been developed [482], which can be acylated by conventional methods. It shows stability toward acidic, basic, and reductive conditions; however, in the presence of a suitable oxidant, detachment of the carboxylate takes place (Scheme 4.94).

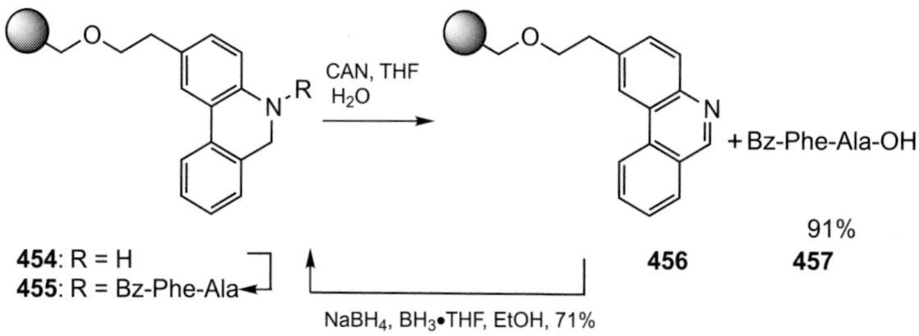

Scheme 4.94. Anchoring and detachment of carboxylic acids by Li et al. [482].

The well-optimized and established BMPSE (β-dimethylphenylsilylethyl) protocol for esters has been adapted by Kurth and coworkers [483] for the immobilization of carboxylates, which can be cleaved under mild conditions (Scheme 4.95).

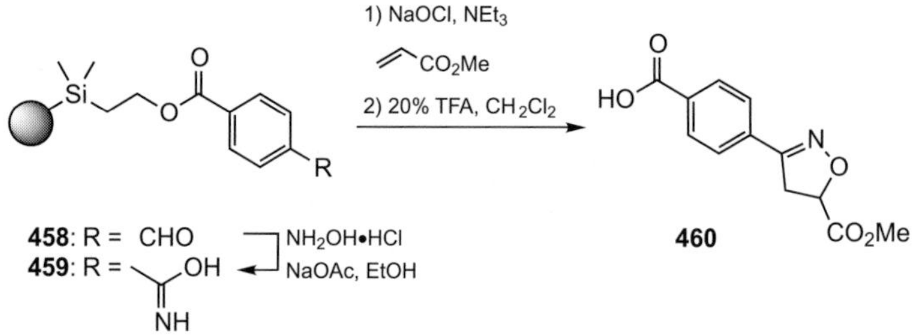

Scheme 4.95. The solid-phase BMPSE group by Kurth and coworkers [483].

Other reports describe the use of arylhydrazides as linkers [119, 484] and also other linkers [372, 373, 485–489].

4.6.2.2 Linkers for Carboxylic Esters, Anhydrides and Lactones

Carboxylic esters have been released mostly by the action of alkoxides onto ester resins. In most cases, the cleavage has been performed using methoxide in methanol (Scheme 4.96). A drawback is certainly the necessity to remove excess metal salts.

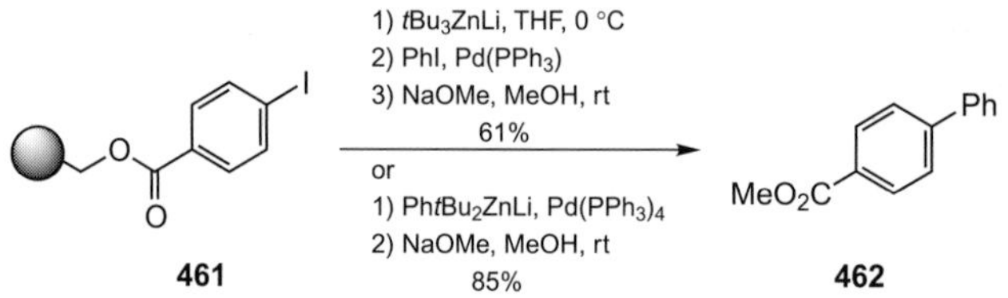

Scheme 4.96. The use of esters as linkers for benzoic acid derivative by Kondo et al. [116].

A new safety-catch linker based on the activation of an amide bond by cyclization to an indolyl carboxamide is suitable for the release of methyl esters [293]. Similarly, the *tert*-butoxy group of polymer-bound 2-(*tert*-butoxy)phenyl carboxylates may be removed with trifluoroacetic acid, and subsequently the ester cleaved by primary or secondary amines [121]. Alternatively, postsynthesis with diazomethane is also possible [121] (Scheme 4.97).

Polymer-bound arylhydrazides have been used in the synthesis of peptide carboxylic esters in a safety-catch strategy (Sect. 4.5.1) [119, 484].

A direct synthesis of carboxylic esters based on the cleavage of triazene T1 resins in the presence of a palladium catalyst under carbon monoxide atmosphere (1 bar) to yield methyl benzoate (**466**) in 87% yield (92% purity) [187] (Scheme 4.98).

Arylalkyl triazenes with a free NH moiety can also be used for the synthesis of carboxylic esters. Starting from amines attached to the T2 linker [183, 196, 197] or the analogous *para*-substituted resin [200], cleavage with different carboxylic acids proceeds smoothly to give esters (Scheme 4.99).

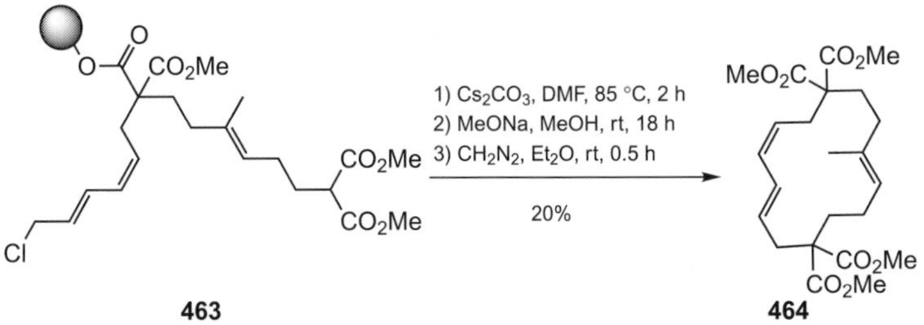

463   **464**

Scheme 4.97. Synthesis of macrocycles by Deslongchamps and coworkers [121].

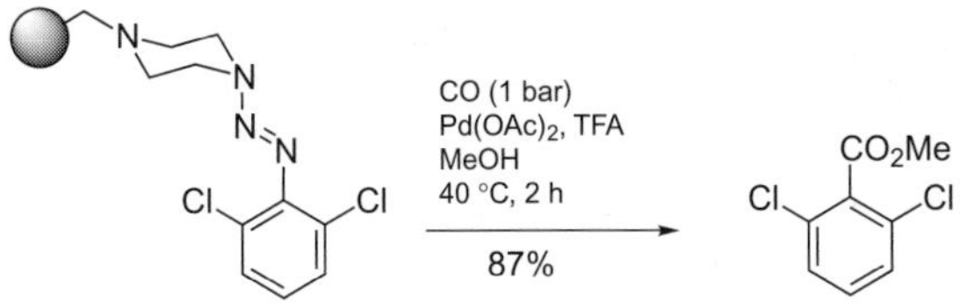

465   **466**

Scheme 4.98. Cleavage and carbonylation of the triazene T1 resin (**465**) [187].

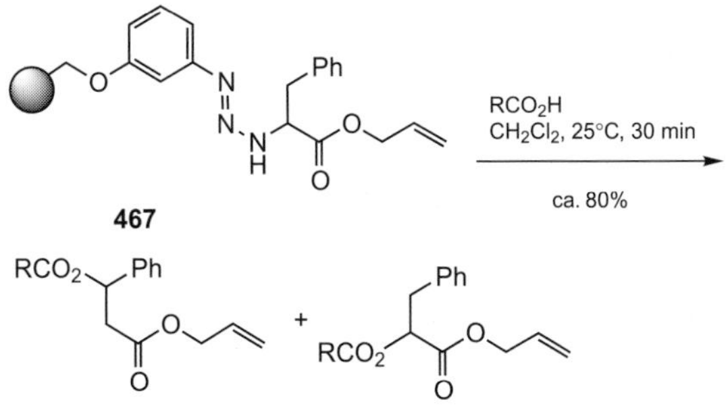

468   80:20 – 65:35   **469**

R = Me, CF$_3$

Scheme 4.99. Synthesis of carboxylic esters via triazene T2 resins [183, 196, 197].

Lactones are accessible via thioesters [115], esters [346, 400], or amides [490] with a suitably positioned internal nucleophile or electrophile. In addition, alkyl acetates are also accessible from sulfonates upon cleavage with sodium acetate [382].

4.6.2.3 Linkers for Thiocarboxylic Acids and Esters
A benzhydryl linker has been used to attach thioesters, while the cleavage can be conducted with HF [491]. Alternatively, treatment of an oxime linker with bistri-ethylthioether results in the fission of the product to give the thio carboxylic acid [478, 492].

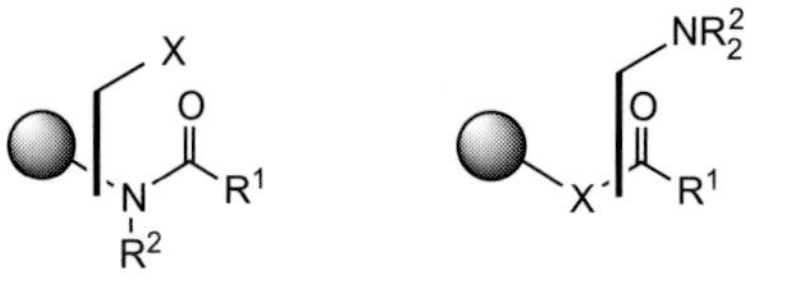

X = H, R
470

X = O, S, N=N
471

Fig. 4.11. Strategies for amide linkers.

4.6.2.4 Linkers for Carboxamides and Related Structures

The solid-phase synthesis of amide-containing structures is important to various fields of organic and bioorganic chemistry. The solid-phase peptide synthesis (SPPS) has provided the chemical community with various solutions toward linking, reaction, and detachment of amide structures. In general, these protocols involve the attachment of amine derivatives by their carbon backbone or, in case of amino acids, by their carboxy functionality. The coupling with carboxyl derivatives proceeds via the free amine, usually using Fmoc-protected amino acids.

Basically, two strategies have been used for the synthesis on solid support (Fig. 4.11). In most cases, an amine has been attached to solid support and then acylated on the bead to give secondary amides (**470**). The latter can be cleaved under acidic conditions. Alternatively, carboxyl functionalities (**471**) can be cleaved with ammonia or with primary or secondary amines. A synthesis of secondary and tertiary amides is based on Lewis acid-mediated cleavage of the ester resin [493].

Safety-catch approaches (Sect. 4.5.1) in the synthesis of peptide carboxamides use polymer-bound arylhydrazides [119, 236, 484] (see Scheme 4.48) or indolyl carboxamides [293].

Linkers for primary amides   Primary amides can be synthesized on solid support using, for example, the Rink resin [68, 494]. This support is suitable for various reaction conditions, as outlined in Schemes 4.100 and 4.101.

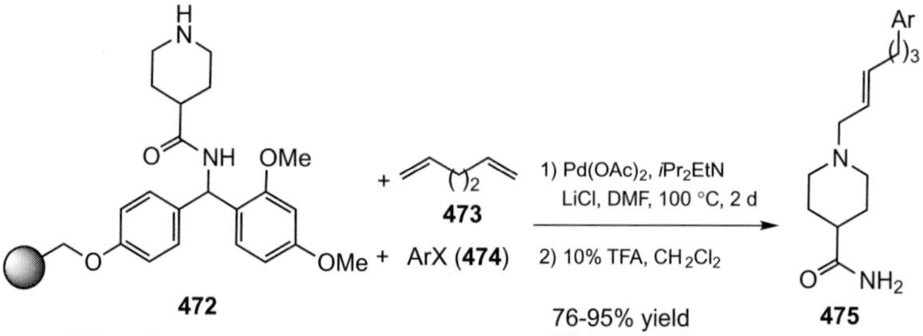

Scheme 4.100. Detachment of primary amides from a Rink resin by Wang et al. [494].

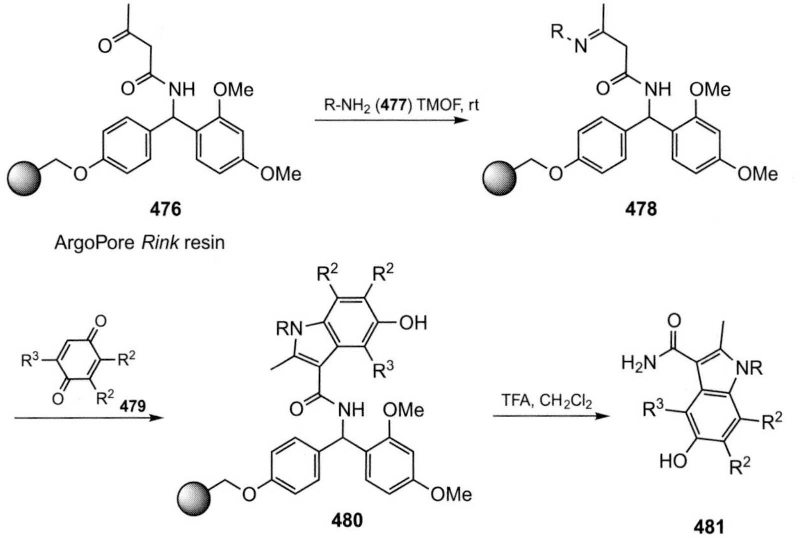

Scheme 4.101. Synthesis of Nenitzescu products according to Wilson and coworkers [495].

Linkers for secondary amides Linking by the N–H of a secondary amide bond has been developed, hence leading to so-called backbone amide linkers (BAL). Originally designed for the N–H protection of amide bonds to circumvent β-turns and other problems during peptide synthesis, the Hydroxymethylbenzoic acid Hmb group [496, 497] or the Silylmethylmethoxyhydroxybenzyl SiMB group [498] can also serve as linkers for SPPS.

Although important in the synthesis of cyclopeptides, they also play an important role in the synthesis of small molecule libraries containing amide moieties. Barany and coworkers have described an application of a backbone amide linker for the synthesis of oligopeptides [40, 83, 424, 499, 500] (Scheme 4.102), glycopep-

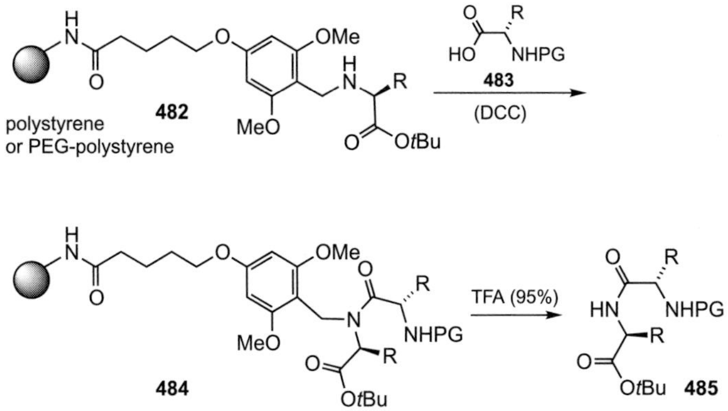

Scheme 4.102. Backbone amide linkage using the PAL linker by Barany and coworkers [83].

tides, and peptide aldehydes [501] based on the peptide amide linker (PAL) concept. Attachment of amines furnishes alkylated products (**482**), which in turn can be acylated under peptide-coupling reaction conditions to give amidic structures (**484**). The latter can be cleaved from the resin using acidic media to give amides (**485**). This concept was used for the synthesis of benzodiazepine libraries by Ellman and coworkers [47] (Scheme 4.103), hapalosin mimetics [502], and further refined to yield an alternative, TFA-stable benzylamine linker [503]. Benzylamine-based linkers have encountered some problems during cleavage which also caused the fragmentation of the whole linker. Furthermore, the reactivity is decreased because of steric hindrance [14, 241].

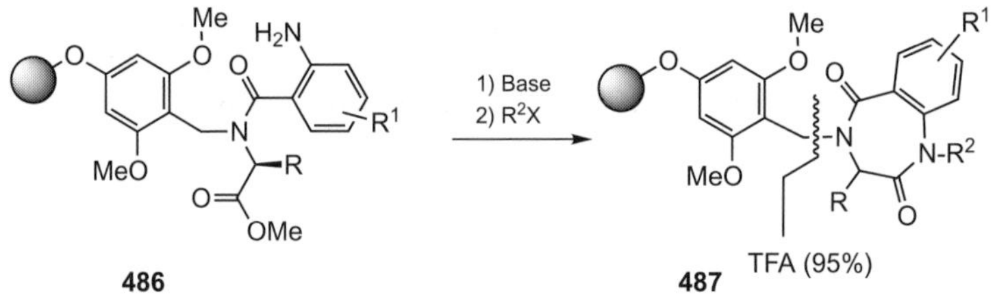

Scheme 4.103. Synthesis of benzodiazepines according to Ellman and coworkers [47].

Recently, it has been shown that the BAL linker is also suitable as a safety-catch linker (Sect. 4.5.1) for the synthesis oligosaccharides [454] (Scheme 4.104). Starting from an aldehyde, reductive amination of a protected glucosamine and subse-

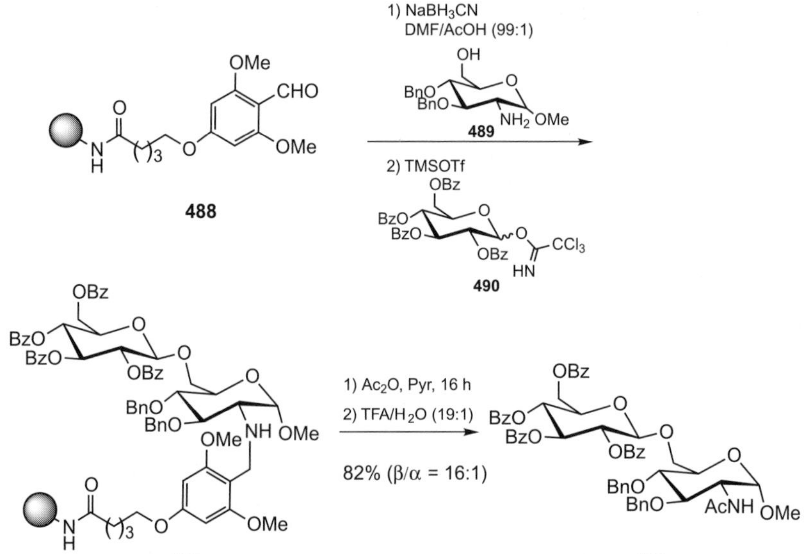

Scheme 4.104. Oligosaccharide synthesis via the BAL structure according to Jensen and coworkers [454].

quent glycosidation with a glucopyranosyltrichloroacetimidate yielded the corresponding disaccharide (**492**). The cleavage was performed after acylation of the amine functionality, which is required to suppress premature decomposition.

In addition, the *p*-benzyloxybenzylamine (BOBA) resin (**7**) developed by Kobayashi and Aoki [32, 504] shows interesting possibilities, such as reductive amination and acylation. Furthermore, it can be cleaved under oxidative conditions to yield the amidic structures.

Recently, a new backbone amide linker has been devised using indole chemistry [85] (Scheme 4.105). In addition, various other related benzyl-type linkers [503, 505, 506] have been used for the synthesis of secondary amides. The Rink linker (**28**) [507], the Merrifield α-methoxybenzyl (MAMP) linker (**24**) [61], methylbenzylhydryl amine (MBHA) resin (**23**) (cleavable with HF/anisole) [59], or AHB (4-alkoxy-2-hydroxybenzaldehyde) linker (**12**) (R = H) [45], a reductively aminated acetophenone linker [54], and SASRIN-related systems [44, 51] are also suitable.

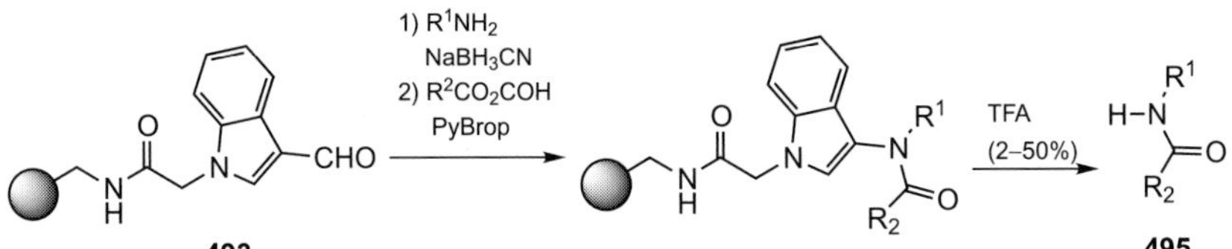

Scheme 4.105. The indole linker according to Estep et al. [85].

The T2 linker has recently been shown to be a versatile backbone amide anchor [194]. Immobilized disubstituted triazenes were acylated with carboxylic acid anhydrides or chlorides to give amidic structures (**482**) [194] (Scheme 4.106).

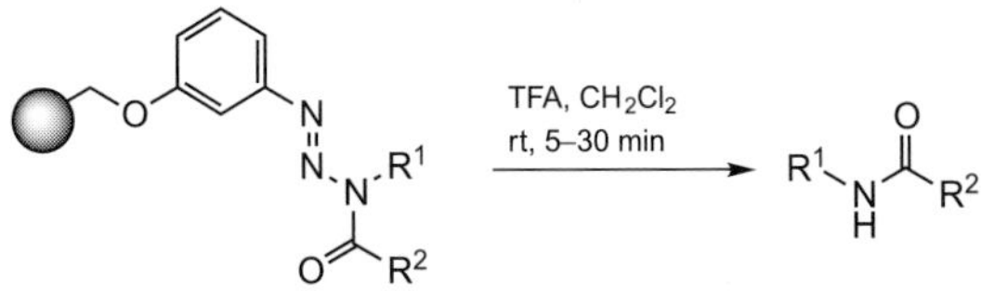

Scheme 4.106. Synthesis of amides using the triazene T2 linker [194].

A new backbone linker for β-lactams and secondary amides was presented using a benzyloxyaniline linker, which is cleavable by ceric ammonium nitrate [508].

Secondary amides and ureas can be synthesized using supported hydroxylamine, which can be cleaved by samarium(II) iodide [509].

Linkers for tertiary amides   Linkers for tertiary amides are the ester anchors cleavable with secondary amines. In addition, the triazene T2 linker is also a precursor for amides when cleaved by acid chlorides [193, 510].

Linkers for carbamates   Both the SASRIN and the indole linker are suitable for the synthesis of carbamates, however the SASRIN linker can be cleaved with 5% TFA in dichloromethane [41], whereas 50% TFA in dichloromethane is required for the indole linker [85].

Linkers for ureas   Urea derivatives, which are important biologically active compounds and building blocks for organic syntheses, have been synthesized on solid support using various strategies.

A useful linker for ureas is the phoxime resin (**498**) (phosgenated oxime resin) [511, 512], which is cleavable at elevated temperature with amines to yield ureas (Scheme 4.107). Other linkers for ureas are the SASRIN linker [41, 44] and other systems [422, 513].

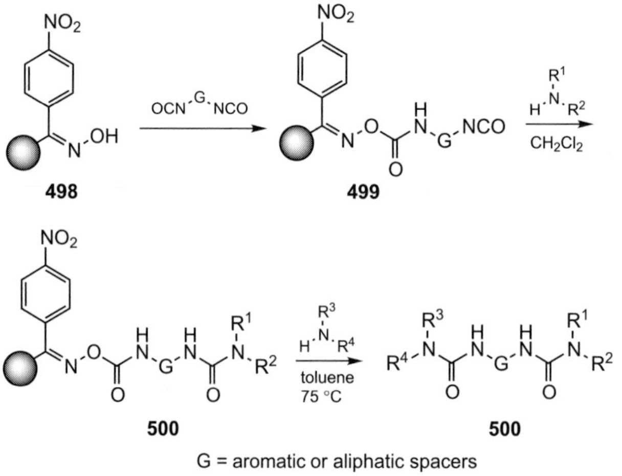

G = aromatic or aliphatic spacers

Scheme 4.107.   The phoxime resin for the synthesis of urea derivatives [511].

The T2 linker has also being used for the attachment and modification of primary amines, yielding urea derivatives after mild cleavage (Scheme 4.108).

Aminosulfonylureas have been released from carbamate linkers using primary and secondary amines (multifunctional: Sect. 4.5.6) [420]. The samarium(II) iodide-promoted cleavage of acylated polymer-bound hydroxylamines gives general access to urea derivatives [509].

Linkers for thioureas   Thioureas have been synthesized using the T2 linker [194] (Scheme 4.109).

Linkers for isothioureas   Isothioureas have been detached from solid support using the T2 linker under mild conditions [514] (Scheme 4.110).

Linkers for guanidines   Guanidines are basic molecules with the capacity of forming H-bonding interactions. They are therefore a promising class of potentially use-

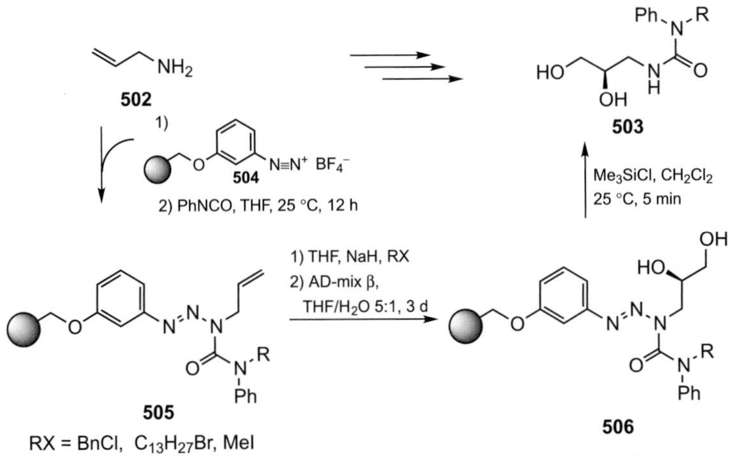

Scheme 4.108. The backbone amide-anchoring mode of the T2 linker [194].

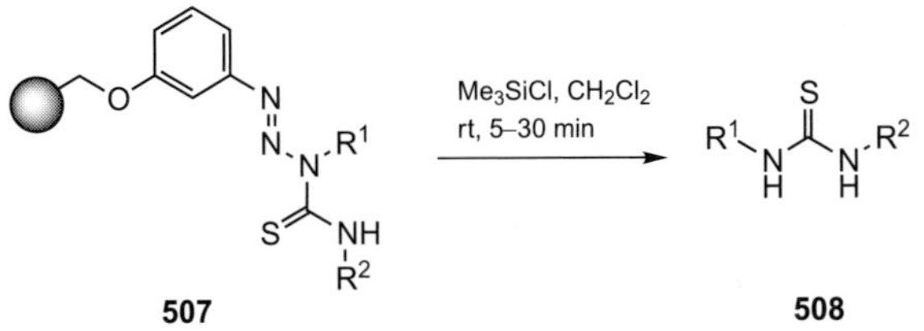

Scheme 4.109. Synthesis of thioureas using the T2 linker [194].

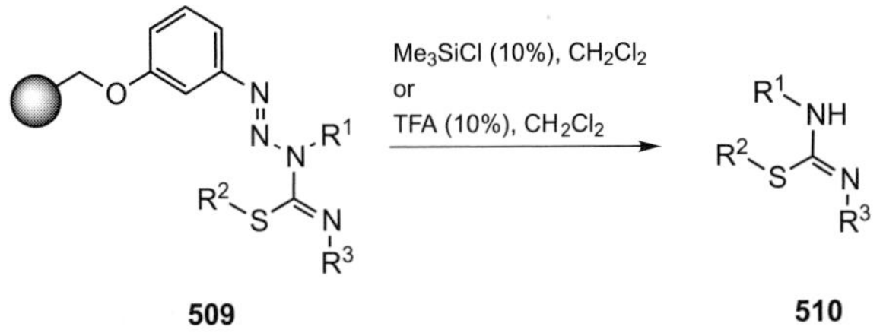

Scheme 4.110. Synthesis of isothioureas using the T2 linker [514].

ful pharmacologically active compounds and the synthesis of guanidines in liquid phase has found widespread application in organic chemistry [515].

The T2 linker [193, 194] and the improved T2*-linker [183, 198] offer a unique possibility to immobilize and modify amine derivatives on solid support. Starting from immobilized amines, a three-step sequence based on coupling with isothiocyanates, conversion to guanidines with amines, and subsequent cleavage presents an approach to the formation of guanidines in which all three substituents can be varied to a wide extent (Scheme 4.111) [195]. In addition, various other linkers for guanidines are available [85, 427, 516–524].

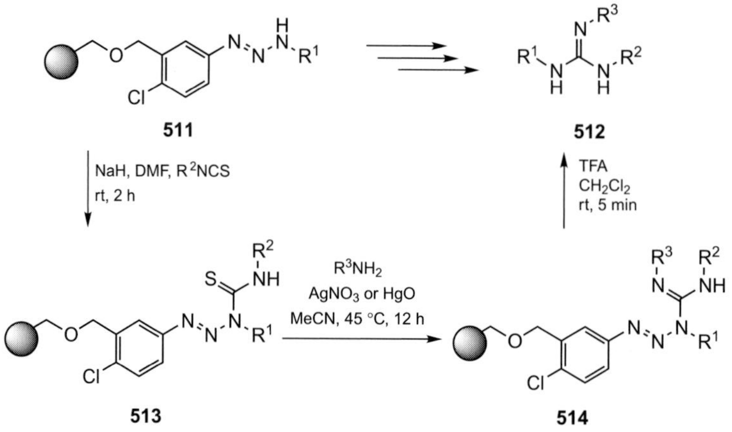

Scheme 4.111. Synthesis of functionalized guanidines [195].

Linkers for amidines Several techniques enable the synthesis of amidines on solid support. In most cases, benzyl-type linkers such as the indole linker [85], the Wang linker [525], or a carbamate system on the Wang linker [525] have been used.

Linkers for hydroxamic acids Hydroxamic acids are important building blocks in metalloproteinase inhibitors. Therefore, various linkers have been developed to satisfy these requirements.

The Wang linker is, for example, suitable for the detachment of hydroxamic acids as demonstrated in a cascade carbopalladation reaction [526] (Scheme 4.112).

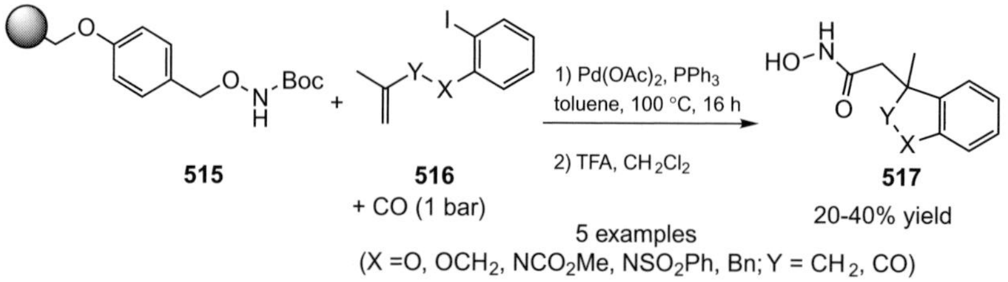

Scheme 4.112. Carbonylation cascade on solid support toward hydroxamic acids according to Grigg et al. [526].

Alternatively, the THP, the Wang resin [527, 528], the Rink linker [529], trityl resin [530], PAL resin [528, 531], oxime resins [532], and others [285, 437, 533–537] are suitable linkers for hydroxamic acids.

4.6.2.5 Linkers for Hydrazides and Semicarbazones

Only a few linkers have been used for the synthesis of hydrazides and semicarbazones, including a phthalamide linker cleavable with hydrazines [538] and the

T2 linker (Sect. 4.14.3.8) [466] (Scheme 4.113). Hydrazides can be cleaved from solid support using the trityl linkers, for example.

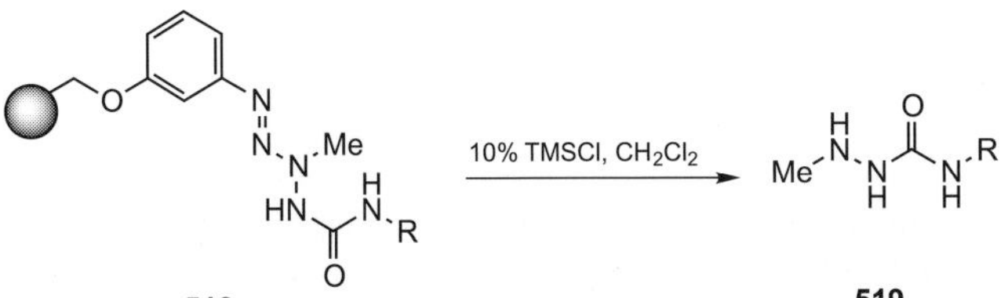

Scheme 4.113. Cleavage of semicarbazones from triazene resins [466].

4.6.2.6 Linkers for Cyclic Amides and Related Structures

Cyclic amide structures have found widespread application in the synthesis of biologically active compounds. Most of the linkers mentioned above for tertiary and secondary amide structures, ureas, and so forth, are suitable for the synthesis of cyclic amidic moieties; however, the most widespread applications are dedicated to the cyclative cleavage from the bead (Sect. 4.5.2).

Linkers for lactams   The synthesis of β-lactams has been achieved using the hydroxyaniline linker [508] (Scheme 4.114).

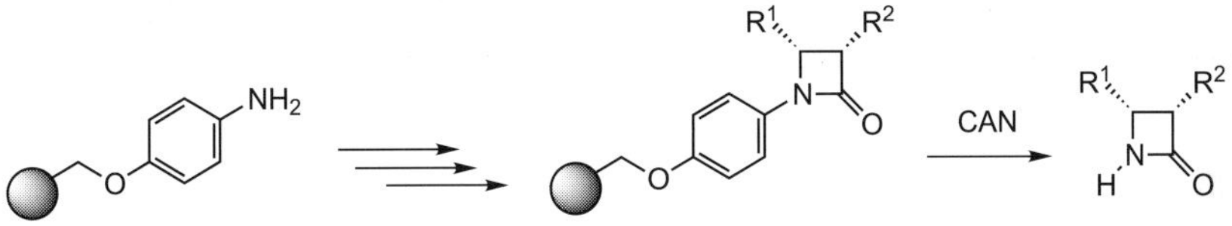

Scheme 4.114.   Linker for β-lactams and secondary amides [508].

Linkers for other heterocycle-containing amidic structures   Various techniques have been used for the synthesis of hydantoins [307, 308–313, 319], thiohydantoins [315], oxazolidinones [320, 321], diketopiperazines, frequently the "byproducts" of peptide synthesis [322–331], benzodiazepines and benzodiazepinones [306, 318, 332, 333], pyrazolones [335, 336], diketomorpholines [323], tetramic acids [337–340], quinazolinediones [341], dihydropyrimidine-2,4-diones [342], quinolinones [343], tetrahydrocarbolines [326], thiazoles [317], and perhydrodiazepinones [327]. Sulfahydantoins (1,2,5-thiadiazolidin-3-one 1,1-dioxides) were prepared from ester-bound amino acids, which were first reductively alkylated, then reacted with sulfamoyl chloride, and finally cleaved from the resin using 1,8-diazabicyclo[5.4.0]-undecene-7 (DBU) [344].

4.6.3

Linkers for Ketones and Aldehydes

The anchoring of carbonyl compounds is generally based on established protecting groups. For example, alkyl hydrazones (Scheme 4.74) [449], sulfonyl hydrazones [539], and semicarbazones [540] can be used as linkers for aldehydes and ketones. Cleavage is conducted with acids, preferably in the presence of formaldehyde. In addition, aldehydes and ketones can be synthesized using photo-labile linkers, reduction of or nucleophilic additions to Weinreb amides [541–543], nucleophilic addition to thioesters [115, 544], oxidative cleavage (ozonolysis) of alkenes [545, 546], hydrolysis of enamines [419], and using various other methods [547].

Since their introduction some 20 years ago by Leznoff and coworkers [101, 102], acetal linkers have been used in the solid-phase synthesis of ketones and aldehydes [104, 548] (Scheme 4.115). Recently, it was demonstrated that even sterically hindered ketones could be attached to this support using scandium triflate and trimethylorthoformate by acetal exchange reaction [104]. Similarly, thioketal structures are also suitable linkers for ketones [105] (Sect. 4.3.7.1).

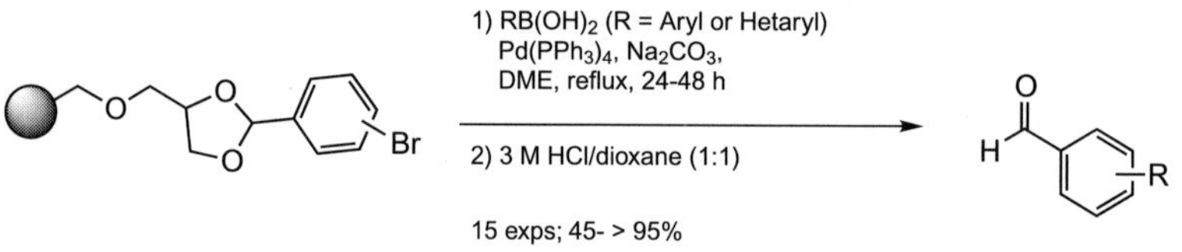

1) RB(OH)$_2$ (R = Aryl or Hetaryl)
Pd(PPh$_3$)$_4$, Na$_2$CO$_3$,
DME, reflux, 24-48 h

2) 3 M HCl/dioxane (1:1)

15 exps; 45- > 95%

523   **524**

Scheme 4.115. Cleavage of an acetal resin by Snieckus and coworkers [548].

4.6.4

Linkers for Alcohols, Phenols, Ethers, and Ketals

4.6.4.1 **Linkers for Alcohols**

Various linkers are suitable for the anchoring and release of alcohols; in particular, silyl ethers have been frequently used [20]. An early example of this type was provided by Farrall and Frechet [549]. Furthermore, these linkers are suitable for glycopeptides [550], oligosaccharides [20], and prostaglandins (**511**) [123] (Scheme 4.116).

A direct loading of alcohols onto a silyl resin is possible using a hydridosilane [133], which might be better than the procedure described using silyl chloride [135]. A SEM (2-trimethylsilylethoxymethyl) linker is also suitable for the attachment and detachment of sterically encumbered alcohols. In this case, cleavage is conducted by tetrabutylammonium fluoride [21]. In addition to the silyl linkers, various ketal-based methods have been reported recently [23, 49, 69, 543, 551–553] (for a review, see [554]).

The method of choice for various applications are the THP-type linkers [99,

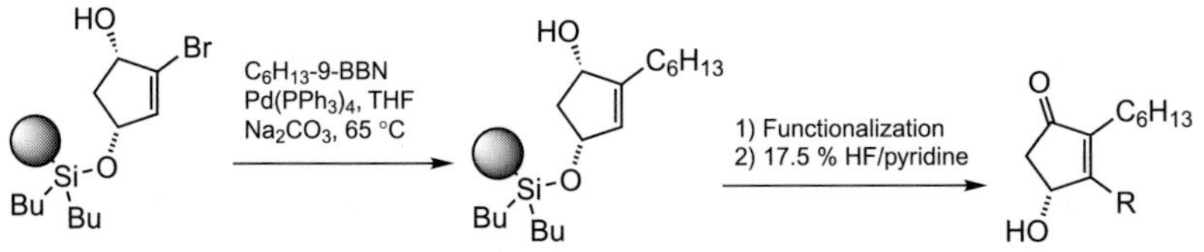

Scheme 4.116. Synthesis of prostaglandin by Ellman and coworkers [123].

108–114] (Sect. 4.3.3), as demonstrated in the synthesis of indolactams on solid support [555] (Scheme 4.117). However, other acetal linkers [100, 106] (Sect. 4.3.3) (Scheme 4.118) have also been used in this context. In addition, alcohols can be attached to a Wang or trityl resin, and can be cleaved in turn by the action of mild acids [533].

Alternatively, the benzyl group can be attached to the solid support, and hydrogenolytic cleavage has been used to detach the molecules which are then usu-

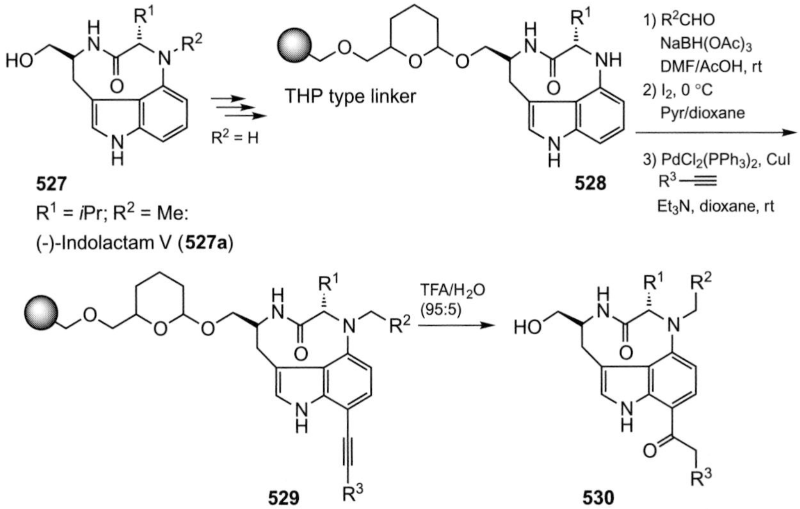

Scheme 4.117. Synthesis of indolactam analogs according to Waldmann and coworkers [555].

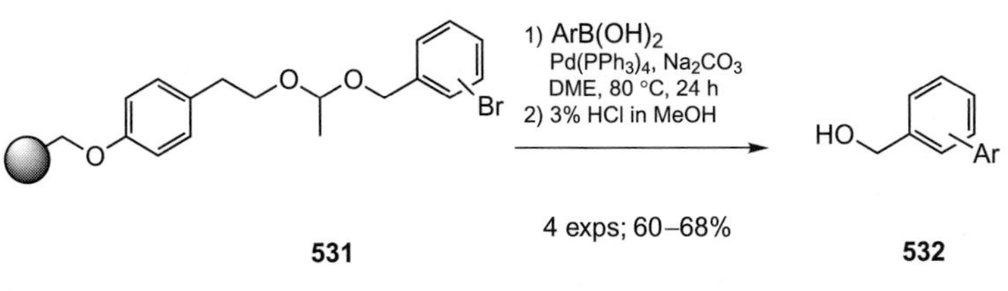

4 exps; 60–68%

Ar = Ph, *p*-MeC$_6$H$_4$

Scheme 4.118. Synthesis of benzyl alcohols on solid support [100].

ally left with an oxygen or nitrogen functionality (cleavage of C–O Bn and C–N bond respectively) [29, 447]. The polymers in these cases are formally immobilized Z or Cbz groups. Interestingly, TentaGel and polystyrene give similar yields under identical conditions. Benzylic linkers can also be used advantageously in the presence of other benzylic protecting groups, since they can be removed in the same step [29] (Scheme 4.119).

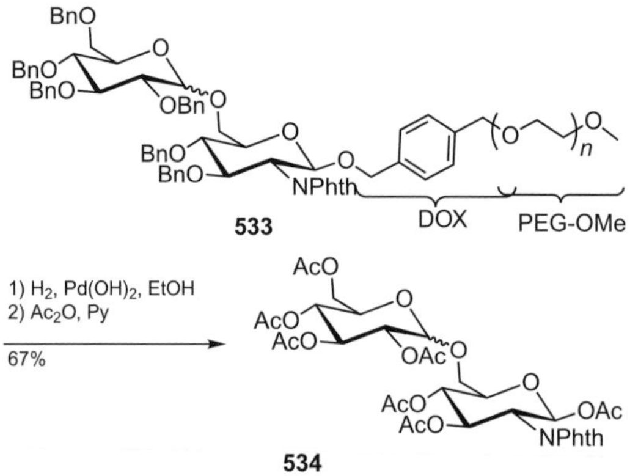

Scheme 4.119. Detachment of saccharides from polymeric benzyl-type protecting groups [29].

Finally, alcohols can be released from ester [117, 556], thioester [115], and amide resins using reductive methods (Sect. 4.4.2, Table 4.13). Furthermore, the reaction of Grignard reagents with thioesters has been reported to give tertiary alcohols [115]. Other anchoring groups are enzyme-labile linkers [285] and fluoride-sensitive linkers [557]. Cyclic ethers (tetrahydrofurans) are accessible via an iodolactonization approach [558, 559].

4.6.4.2 Linkers for Phenols
Similar to alcohols, phenols have been linked to anchors such as Wang [560], Rink [82], trityl [561], 2-chlorotrityl [562], and modified trityl [563] linkers. However, since these linkers are more acid stable, simple hydroxymethyl polystyrene can serve as a linker when cleaved by triflic acid [549, 557, 564, 565] (Scheme 4.120).

A direct method for the synthesis of phenol acetates has been demonstrated with the triazene T1 linker. An *in situ* cleavage and acylation proceeds with good yields when cleaved by acetic acid/acetic anhydride (Scheme 4.121) [186].

4.6.5
Linkers for Sulfur Compounds

4.6.5.1 Linkers for Thiols and Thioethers
Thiols can be synthesized using disulfides as remarkably stable linkers. The detachment proceeds in the presence of certain phosphines [232–235] or by exchange

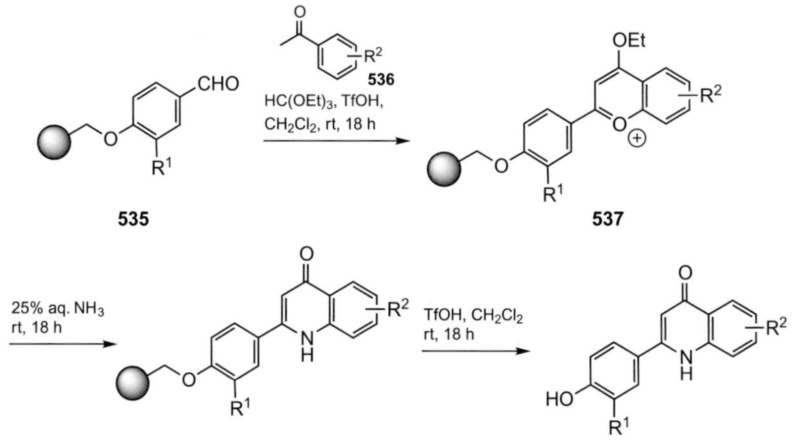

Scheme 4.120. Synthesis of quinolones from flavilylium salts according to Sato et al. [564].

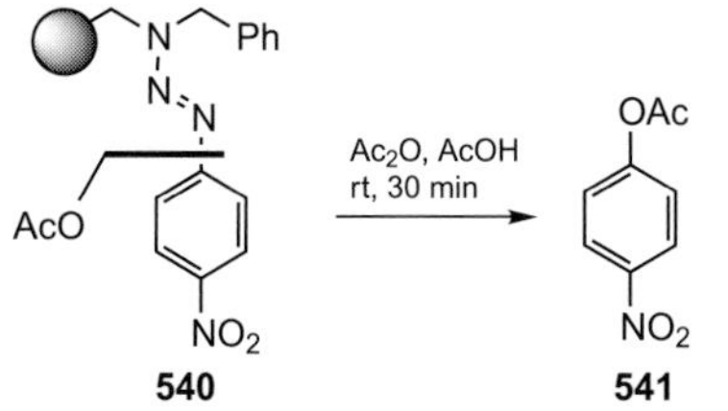

Scheme 4.121. Synthesis of phenol acetates (**541**) [186].

with thiols [566]. However, other suitable linkers are the Wang linker, cleavable by HF [33], Rink-type linkers [82, 567], thiocarbamates, cleavable by bases [568], dinitroaryl linker, cleavable by nucleophiles [229]; and others [569, 570]. Because thiols are prone to oxidation to give symmetrical disulfides, the latter were frequently found during cleavage if oxygen had not been strictly excluded.

4.6.5.2 Linkers for Sulfonamides
The synthesis of sulfonamides is similar to the strategies used for carboxamides. Therefore, primary sulfonamides were synthesized using, for example, the Rink resin (Scheme 4.122) [67].

The indole resin [85], the SASRIN linker [41, 44], and a reductively aminated acetophenone linker [54] are also suitable for the detachment for sulfonamides [24, 41, 67, 415, 571–574].

4.6.5.3 Linkers for Sulfonic Acids
Aryl and alkyl sulfonic acids have been detached from both Wang resin and SASRIN-type resins under mild conditions (20% TFA in CH_2Cl_2) after being at-

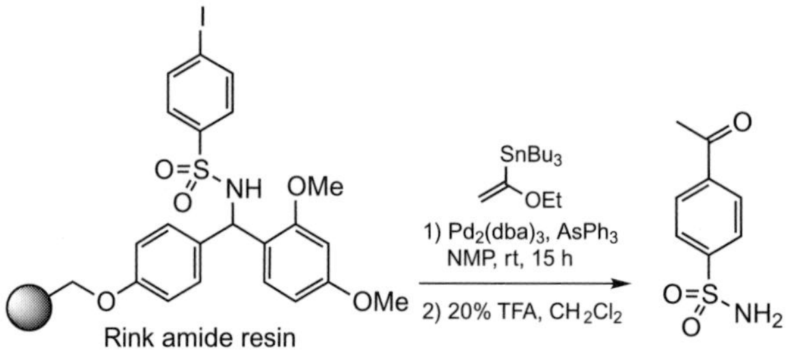

Scheme 4.122. Detachment of sulfonamides from the Rink resin [67].

tached to a solid support using the corresponding benzyl alcohol resins and sulfonyl chlorides [575].

4.6.5.4 Linkers for Sulfones and Sulfoxides

Only one example has been presented so far for the synthesis of sulfones [405]. Starting from Merrifield resin, attachment of enantiopure sulfoximines and subsequent aldol-type coupling gives access to highly substituted sulfoximines. Cleavage proceeds under oxidative conditions using *meta*-chloroperbenzoic acid (*m*CPBA) to give sulfones (Scheme 4.123) [405].

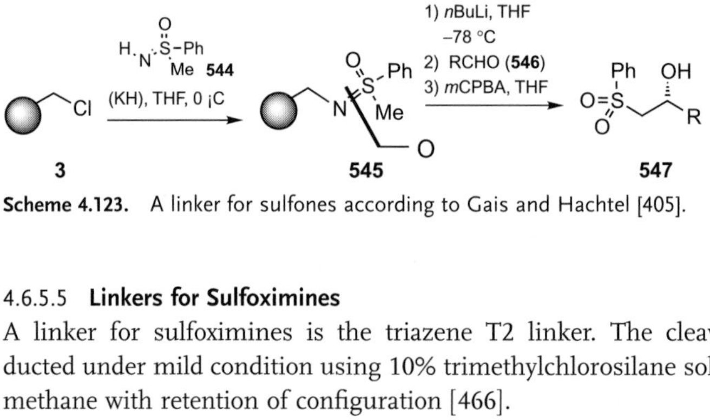

Scheme 4.123. A linker for sulfones according to Gais and Hachtel [405].

4.6.5.5 Linkers for Sulfoximines

A linker for sulfoximines is the triazene T2 linker. The cleavage can be conducted under mild condition using 10% trimethylchlorosilane solution in dichloromethane with retention of configuration [466].

4.6.6
Linkers for Hydrocarbons

Linkers for hydrocarbons are important tools in combinatorial chemistry for the synthesis of the lipophilic compounds required for modern drug research. As described above, access to hydrocarbons can proceed via the formation of either a C–C bond or a C–H bond. The latter strategy has been discussed in detail in Sect. 4.5.5 (traceless linkers), since the introduction of a hydrogen atom clearly is the prototype for this kind of linker.

4.6.6.1 Linkers for Alkanes

Alkanes have been synthesized on solid support, mostly by reduction of a C–X bond to a C–H functionality. This concept has been demonstrated using the selenium [165], sulfur [120], and stannane linkers. The formation of C–C bonds for the synthesis of alkanes has been described by Schiemann and Showalter [576]. Beginning with an aryl fragment attached to an immobilized benzotriazole, cleavage and subsequent C–C bond formation are achieved using organomagnesium compounds.

4.6.6.2 Linkers for Arenes and Heteroarenes

Alternatively, C–aryl bonds were formed using various cross-coupling methods, including stannanes [163, 169, 577], triazenes [187, 188] (Scheme 4.34), and boronates [148] as precursors.

Naphthalene derivatives are accessible via an electrocyclic ring opening of benzocyclobutane derivatives and a subsequent Diels–Alder reaction with dienophiles [578]. Hydrogenolytic removal of substrates from the solid support is important as it cleaves the substrate to form a C–H bond at the former binding site of the polymer. These types of linkers are also called traceless linkers (Sect. 4.5.5) [183].

The detachment of substituted arylsulfonates in the presence of a reducing agent such as formic acid provides a traceless cleavage. In this case, it is important that the arene core is substituted with electron-withdrawing substituents to enhance the yields [152]. This approach has been described previously (without experimental details) quite early in a patent and includes the possible derivatization of the intermediate σ-aryl palladium aryl complex [579].

4.6.6.3 Linkers for Alkenes

Most of the linkers for alkenes are traceless linkers, such as those described in Sect. 4.5.5. Besides these, classical double-bond-forming reactions, such as the Wittig–Horner–Emmons [304, 305] or the Wittig reaction, can be used for the formation of C=C bonds [364]. Syntheses via metathesis (Sect. 4.4.4.2), for example the ring-closing metathesis of olefins [277, 580], have been used for the preparation of alkenes on solid support. In addition, multifunctional cleavage (Sect. 4.5.6) can be achieved using cross-metathesis.

Allylic groups can be attached to solid support via a sulfone, which is prepared by lithiation of polystyrene and subsequent treatment with sulfur dioxides and an allylbromide. After modification on the bead, cleavage proceeds with the action of a Grignard reagent in the presence of copper iodide. This overall S_N2' alkylation provides a route to substituted alkenes [581] (Scheme 4.124).

The β-elimination generating alkenes has been used in the chemistry of the sulfone and selenium linkers [164, 582, 583] (Scheme 4.22). Similarly, polymer-bound 1-alkenylcyclobutylsulfones [274] (Scheme 4.33) or pentadienol carboxylates [273] (Scheme 4.32) were cleaved from a resin in the presence of suitable nucleophiles and palladium catalysts to give substituted cyclobutylidene derivatives or dienes, respectively.

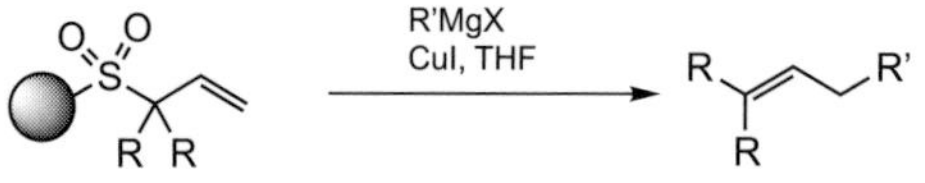

548   **549**

Scheme 4.124. Nucleophilic substitution of an allylic sulfone [581].

4.6.6.4 Linkers for Alkynes

Two linkers have been used for the detachment of alkynes from solid support. Gibson and coworkers [355] have described the immobilization of alkynes onto polymer-bound triphenyl phosphine via a dicobaltoctacarbonyl arm. The detachment was conducted using air as the final oxidant.

Alkynes were obtained by the cleavage cross-coupling strategy of the T1 triazene resin (Scheme 4.26). In contrast to the Heck cleavage, these cleavage conditions give rise to di- and trimerization, thus making a chromatographic separation necessary [187] (Scheme 4.38).

4.6.7
Linkers for Aryl and Alkyl Halides

Aryl iodides have been synthesized by Moore et al. [179], starting from triazene resin by the action of methyl iodide (Sect. 4.14.3.8) (Scheme 4.125). Aryl iodides, bromides, and chlorides are also accessible from the triazene T1 linker using the corresponding trimethylsilyl halide (Scheme 4.126) [190].

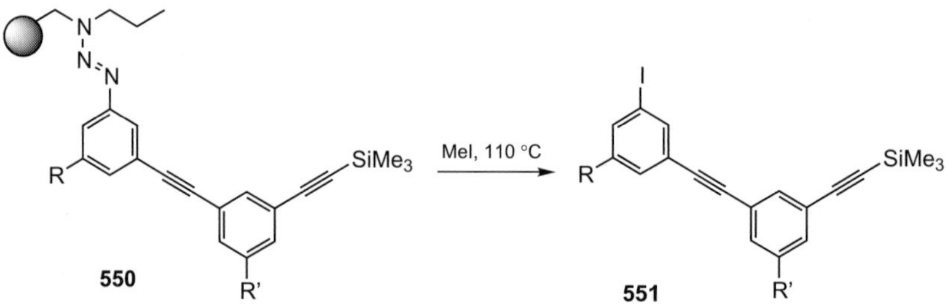

550   **551**

Scheme 4.125. The use of triazene anchoring groups in the synthesis of iodo arenes by Moore and coworkers [179].

The cleavage of triazene T2-linked primary amines with trimethylsilyl chloride, bromide, or iodide proceeds smoothly to give alkyl halides. This reaction proceeds presumably via the aliphatic diazonium ion. In some cases, a rearrangement was observed (Scheme 4.127).

Arylbromides and -iodides are accessible from silicon or germanium-linked arene fragments (Sect. 4.3.5). The released can be conducted with either bromine/pyridine [126, 131] or iodochloride [126, 138].

Scheme 4.126. Synthesis of aryl halides via the T1 triazene resin [190].

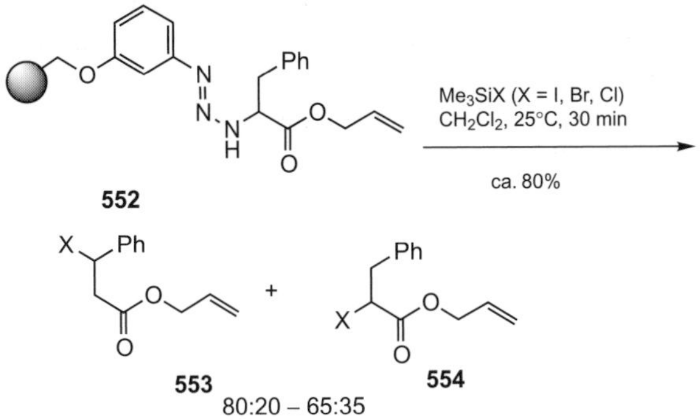

Scheme 4.127. Synthesis of alkyl halides via the triazene T2 linker [183, 196, 197].

The nucleophilic substitution of alkylsulfonates was used for the synthesis of alkyl iodides. Starting from the corresponding alcohols, attachment to a sulfonyl chloride and subsequent release from the bead was performed using sodium iodide [382, 465]. Allyl bromides can be released from a trityl linker if cleaved with hydrobromic acid in acetic acid [74]. Other methods start from with trityl linkers [74, 163, 382, 559].

4.6.8
Linkers for Heterocycles

Various methods are applicable in the synthesis of heterocycles [361, 362, 384, 584]. The cyclofragmentation of a certain class of sulfones leads to 3-arylbenzo-furans [160].

4.6.9
Linkers for Reactive Intermediates

Reactive intermediates cleaved from a solid support can be used for a subsequent functionalization. Thus, radicals, carbanions, and carbocations might then react with additional building blocks. This multifunctional cleavage mode (Sect. 4.14.5.6) has, for example, been used with the stannane, selenium (Sect. 4.14.3.7), and tri-azene linkers (Sect. 4.14.3.8).

4.6.10
Linkers for Other Functional Groups

4.6.10.1 Linkers for Phosphonates

Phosphonates can be released from the resin using trimethylsilyl iodide (Scheme 4.128) [585]. They have been attached to a solid support using the Wang linker [586, 587, 588].

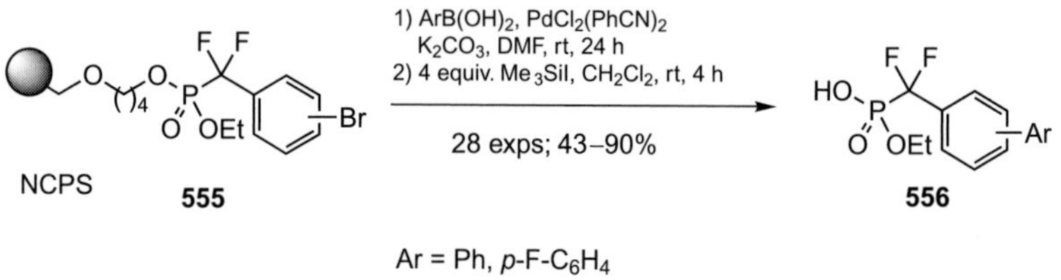

Scheme 4.128. Synthesis of phosphonates using noncrosslinked polystyrene (NCPS) as support [585].

4.6.10.2 Linkers for Boronates

Diols are suitable anchors for boronic acids, as shown in the synthesis of hepatitis C virus proteinases [589].

4.6.10.3 Linkers for Silanes and Silanols

Silanols are accessible from silyl ether linker [127]. The silicon–oxygen bond can be cleaved with TFA via a protio-*ipso*desilylation.

4.7
Overview for Linkers for Functional Groups

Table 4.20 gives a short overview of the different linker families, as described in Chapter X.

4.8
Conclusion, Summary and Outlook

In recent years, various new types of linkers have emerged. The design of a new anchoring group can be essential for the success of a synthesis, especially for small molecules on a solid support. Linker, cleavage conditions, and functional groups are associated with each other. Therefore, the decision to use one specific linker type has to be balanced with the requirements of the library to be synthesized.

Although the "perfect" or "universal" linker has not yet been developed, and will prove unattainable, interesting new developments increase the flexibility of solid-phase synthesis by traceless (Sect. 4.5.5) and multifunctional cleavage (Sect.

Tab. 4.20. Short overview for various linker types.

Functional group	Benzyl-type linkers	Ketal/acetal linkers	Esters/amide linkers	Silane linkers	Triazene linkers	Selenium/sulfur/stannyl linkers
R_3N	✓	✓	✓		✓	✓
ROH	✓	✓	✓	✓		
R_2NCOR	✓		✓		✓	
RH (traceless) (Sect. 4.5.5)	✓		✓	✓	✓	✓
RCO_2H	✓		✓			
Heterocycles	✓	✓			✓	
BAL	✓				✓	
RX				✓	✓	
Safety-catch option	✓		✓			✓
Multifunctional cleavage				✓	✓	✓
Photo cleavage	✓					

4.5.6). While traceless linkers provide access to unsubstituted compounds with "no memory" of solid-phase synthesis, multifunctional cleavage allows the introduction of various new functionalities during cleavage from the resin. Backbone amide linkers present new opportunities for solid-phase synthesis of small amidic structures, and cyclization-release strategies provide an opportunity to create novel carbo- and heterocyclic structures upon cleavage.

An anchor for traceless linking can also be a safety-catch linker (Sect. 4.5.1), or it can be suitable for multifunctional cleavage. Linker systems allow the introduction of certain atoms or molecule fragments and will play an important role in the development of diverse organic substance libraries. It is important to point out that the final diversification is achieved in the cleavage step and not in an additional solution-phase reaction step after the cleavage. However, only a few linker systems that are applicable to a wider range of substrates have been developed so far. As these linker systems offer the widest possibilities for the final diversification of a synthesized library, they will be the subject of increasing attention in the future.

References

1 E. M. GORDON, R. W. BARRETT, W. J. DOWER, S. P. A. FODOR, M. A. GALLOP, *J. Med. Chem.* **1994**, *37*, 1385–1401.

2 J. A. ELLMAN, *Acc. Chem. Res.* **1996**, *29*, 132–143.

3 R. W. ARMSTRONG, A. P. COMBS, P. A. TEMPEST, S. D. BROWN, T. A. KEATING, *Acc. Chem. Res.* **1996**, *29*, 123–131.

4 F. BALKENHOHL, C. VON DEM BUSSCHE-HÜNNEFELD, A. LANSKY, C. ZECHEL, *Angew. Chem. Int. Ed.* **1996**, *35*, 2289–2337; *Angew. Chem.* **1996**, *108*, 2436–2488.

5 J. S. FRÜCHTEL, G. JUNG in: Combi-

natorial Peptide and Nonpeptide Libraries: A Handbook. Jung, G. (ed.), VCH, Weinheim 1996, pp. 19–78.

6 P. H. H. Hermkens, H. C. J. Ottenheijm, D. Rees, *Tetrahedron* **1996**, *52*, 4527–4554.

7 I. W. James, *Tetrahedron* **1999**, *55*, 4855–4946.

8 F. Guillier, D. Orain, M. Bradley, *Chem. Rev.* **2000**, *100*, 2091–2157.

9 A. C. Comely, S. E. Gibson, *Angew. Chem. Int. Ed.* **2001**, *40*, 1012–1032; *Angew. Chem.* **2001**, *113*, 1043–1063.

10 F. Zaragoza Dörwald, *Organic Synthesis on Solid-phase: Supports, Linkers, Reactions*, Wiley-VCH, Weinheim 2000.

11 B. Carboni, F. Carreaux, J. F. Pilard, *Actual. Chimique* **2000**, 9–13.

12 D. Maclean, J. J. Baldwin, V. T. Ivanov, Y. Kato, A. Shaw, P. Schneider, E. M. Gordon, *Pure Appl. Chem.* **1999**, *71*, 2349–2365.

13 B. J. Backes, J. Ellman, *Curr. Opin. Chem. Biol.* **1997**, *1*, 86–93.

14 C. T. Bui, F. A. Rasoul, F. Ercole, Y. Pham, N. J. Maeji, *Tetrahedron Lett.* **1998**, *39*, 9279–9282.

15 W. Rapp, in: Combinatorial Peptide and Nonpeptide Libraries: A Handbook. Jung, G. (ed.), VCH, Weinheim 1996, pp. 425–464.

16 A. Svensson, T. Fex, J. Kihlberg, *J. Comb. Chem.* **2000**, *2*, 736–748.

17 J. R. Hauske, P. Dorff, *Tetrahedron Lett.* **1995**, *36*, 1589–1592.

18 K. Akaji, Y. Kiso, L. A. Carpino, *J. Chem. Soc., Chem. Commun.* **1990**, 584–586.

19 T. H. Chan, W. Q. Huang, *J. Chem. Soc., Chem. Commun.* **1985**, 909–911.

20 J. T. Randolph, K. F. McClure, S. J. Danishefsky, *J. Am. Chem. Soc.* **1995**, *117*, 5712–5719.

21 W. J. Koot, *J. Comb. Chem.* **1999**, *1*, 467–473.

22 R. B. Merrifield, *J. Am. Chem. Soc.* **1963**, *85*, 2149–2154.

23 S.-S. Wang, *J. Am. Chem. Soc.* **1973**, *95*, 1328–1333.

24 B. Yan, N. Nguyen, L. Liu, G. Holland, B. Raju, *J. Comb. Chem.* **2000**, *2*, 66–74.

25 M. Mergler, R. Tanner, J. Gosteli, P. Grogg, *Tetrahedron Lett.* **1988**, *22*, 4005–4008.

26 M. Mergler, J. Gosteli, P. Grogg, P. Nyfeler, R. Tanner, *Chimia* **1999**, *53*, 29–34.

27 F. Albericio, G. Barany, *Tetrahedron Lett.* **1991**, *32*, 1015–1018.

28 E. Atherton, C. J. Logan, R. C. Sheppard, *J. Chem. Soc., Perkin Trans. 1* **1981**, 538–546.

29 S. Manabe, Y. Ito, T. Ogawa, *Synlett* **1998**, 628–630.

30 G.-s. Lu, S. Mojsov, J. P. Tam, R. B. Merrifield, *J. Org. Chem.* **1981**, *46*, 3433–3436.

31 T. L. Deegan, O. W. Gooding, S. Baudart, J. A. Porco, *Abstr., Pap. Am. Chem Soc.* **1997**, *214*, 238–ORGN.

32 S. Kobayashi, Y. Aoki, *Tetrahedron Lett.* **1998**, *39*, 7345–7348.

33 D. R. Englebretsen, B. G. Garnham, D. A. Bergman, P. F. Alewood, *Tetrahedron Lett.* **1995**, *36*, 8871–8874.

34 G. Barany, R. B. Merrifield in: The Peptides. Gross, E., Meienhofer, J. (eds), Academic Press, New York **1979**.

35 A. R. Mitchell, B. W. Erickson, M. N. Ryabtsev, R. S. Hidges, R. B. Merrifield, *J. Am. Chem. Soc.* **1976**, *98*, 7357–7362.

36 D. Seebach, A. Thaler, D. Blaser, S. Ko, *Helv. Chim. Acta* **1991**, *74*, 1102–1119.

37 J. V. Aldrich, S. C. Story, *Int. J. Pept. Protein Res.* **1992**, *39*, 87–92.

38 R. C. Sheppard, B. Williams, *Int. J. Pept. Protein Res.* **1982**, *20*, 451–454.

39 R. M. Valerio, A. M. Bray, N. J. Maeji, *Int. J. Pept. Protein Res.* **1994**, *44*, 158–165.

40 F. Albericio, G. Barany, *Int. J. Pept. Protein Res.* **1985**, *26*, 92–97.

41 A. M. Fivush, T. M. Willson, *Tetrahedron Lett.* **1997**, *38*, 7151–7154.

42 X. Ouyang, N. Tamayo, A. S. Kiselyov, *Tetrahedron* **1999**, *55*, 2827–2834.

43 D. Sarantakis, J. J. Bicksler, *Tetrahedron Lett.* **1997**, *38*, 7325–7328.

44 E. E. Swayze, *Tetrahedron Lett.* **1997**, *38*, 8465–8468.

45 T. Okayama, A. Burritt, V. J. Hruby, *Org. Lett.* **2000**, *2*, 1787–1790.

46 F. Albericio, N. Kneib-Cordonier, S. Biancalana, L. Gera, R. I. Masada, D. Huson, G. Barany, *J. Org. Chem.* **1990**, *55*, 3730–3743.

47 C. G. Boojamra, K. M. Burow, J. A. Ellman, *J. Org. Chem.* **1995**, *60*, 5742–5743.

48 K. Barlos, O. Chatzi, D. Gratos, G. Stavropulos, *Int. J. Pept. Protein Res.* **1991**, *37*, 513–520.

49 V. Krchňák, L. Szabo, J. Vagner, *Tetrahedron Lett.* **2000**, *41*, 2835–2848.

50 K. J. Jensen, J. Alsina, M. F. Songster, J. Vagner, F. Albericio, G. Barany, *J. Am. Chem. Soc.* **1998**, *120*, 5441–5452.

51 L. S. Harikrishnan, H. D. H. Showalter, *Synlett* **2000**, 1339–1341.

52 R. Ramage, C. A. Barron, S. Bidecki, D. W. Thomas, *Tetrahedron Lett.* **1987**, *28*, 4105–4108.

53 R. Ramage, C. A. Barron, S. Bielecki, R. Holden, D. W. Thomas, *Tetrahedron* **1992**, *48*, 499–514.

54 C. T. Bui, A. M. Bray, F. Ercole, Y. Pham, F. A. Rasoul, N. J. Maeji, *Tetrahedron Lett.* **1999**, *40*, 3471–3474.

55 H. Chao, M. S. Bernatowicz, G. R. Matsueda, *J. Org. Chem.* **1993**, *58*, 2640–2644.

56 H. G. Chao, M. S. Bernatowicz, P. D. Reiss, C. E. Klimas, G. R. Matsueda, *J. Am. Chem. Soc.* **1994**, *116*, 1746–1752.

57 P. G. Pietta, G. R. Marshall, *J. Chem. Soc. D* **1970**, 650–651.

58 G. R. Matsueda, J. M. Stewart, *Peptides* **1981**, *22*, 45–50.

59 M. E. Theoclitou, J. M. Ostresh, V. Hamashin, R. A. Houghten, *Tetrahedron Lett.* **2000**, *41*, 2051–2054.

60 R. C. Orlowski, R. Walter, D. Winkler, *J. Org. Chem.* **1976**, *41*, 3701–3705.

61 D. S. Brown, J. M. Revill, R. E. Shute, *Tetrahedron Lett.* **1998**, *39*, 8533–8536.

62 J. P. Tam, *J. Org. Chem.* **1985**, *50*, 5291–5298.

63 J. P. Tam, R. D. DiMarchi, R. B. Merrifield, *Tetrahedron Lett.* **1981**, *22*, 2851–2854.

64 M. Patek, M. Lebl, *Tetrahedron Lett.* **1991**, *32*, 3891–3894.

65 Y. Kiso, T. Fukui, S. Tanaka, T. Kimura, K. Akaj, *Tetrahedron Lett.* **1994**, *35*, 3571–3574.

66 H. Rink, *Tetrahedron Lett.* **1987**, *28*, 3787–3790.

67 K. A. Beaver, A. C. Siegmund, K. L. Spear, *Tetrahedron Lett.* **1996**, *37*, 1145–1148.

68 A. L. Marzinzik, E. R. Felder, *Tetrahedron Lett.* **1996**, *37*, 1003–1006.

69 A. Routledge, H. T. Stock, S. L. Flitsch, N. J. Turner, *Tetrahedron Lett.* **1997**, *38*, 8287–8290.

70 P. Sieber, *Tetrahedron Lett.* **1987**, *28*, 2107–2110.

71 Y. X. Han, S. L. Bontems, P. Hegyes, M. C. Munson, C. A. Minor, S. A. Kates, F. Albericio, G. Barany, *J. Org. Chem.* **1996**, *61*, 6326–6339.

72 W. C. Chan, S. L. Mellor, *J. Chem. Soc., Chem. Commun.* **1995**, 1475–1477.

73 M. Noda, M. Yamaguchi, E. Ando, K. Takeda, K. Nokihara, *J. Org. Chem.* **1994**, *59*, 7968–7975.

74 J. M. J. Frechet, L. Nuyens, *Can. J. Chem.* **1976**, *54*, 926–934.

75 T. M. Fyles, C. C. Leznoff, *Can. J. Chem.* **1976**, *54*, 935–942.

76 K. Barlos, D. Gatos, J. Kallitsis, D. Papaioannou, P. Sotiriu, *Liebigs Ann. Chem.* **1988**, 1079–1081.

77 K. Barlos, D. Gatos, J. Kallitsis, G. Papaphotiu, W. Yao, W. Schäfer, *Tetrahedron Lett.* **1989**, *30*, 3943–3946.

78 S. Eleftheriou, D. Gatos, A. Panagopoulos, S. Stathopoulos, K. Barlos, *Tetrahedron Lett.* **1999**, *40*, 2825–2828.

79 C. C. Zikos, N. G. Ferdierigos, *Tetrahedron Lett.* **1994**, *35*, 1767–1768.

80 L. Leondiadis, I. Vassiliadou, C. Zikos, N. Ferderigos, E. Livaniou, D. S. Ithakissios, G. P. Evangelatos, *J. Chem. Soc., Perkin Trans I* **1996**, *10*, 971–975.

81 T. Wieland, C. Birr, P. Fleckenstein, *Liebigs Ann. Chem.* **1972**, *756*, 14–19.

82 R. S. Garigipati, *Tetrahedron Lett.* **1997**, *38*, 6807–6810.

83 J. Alsina, T. S. Yokum, F. Albericio,

G. Barany, *J. Org. Chem.* **1999**, *64*, 8761–8769.

84 N. Thieriet, F. Guibe, F. Albericio, *Org. Lett.* **2000**, *2*, 1815–1817.

85 K. G. Estep, C. E. Neipp, L. M. S. Stramiello, M. D. Adam, M. P. Allen, S. Robinson, E. J. Roskamp, *J. Org. Chem.* **1998**, *63*, 5300–5301.

86 H. Kunz, B. Dombo, *Angew. Chem. Int. Ed. Engl.* **1988**, *12*, 711–712; *Angew. Chem.* **1988**, *100*, 732–734.

87 H. Kunz, W. Kosch, J. März (Orpegen GmbH), Patent No. US5214195, **1990**.

88 C. Schumann, L. Seyfarth, G. Greiner, S. Reissmann, *J. Pept. Res.* **2000**, *55*, 428–435.

89 T. Johnson, R. C. Sheppard, *J. Chem. Soc., Chem. Commun.* **1991**, 1653–1655.

90 O. Seitz, H. Kunz, *J. Org. Chem.* **1997**, *62*, 813–826.

91 O. Seitz, *Tetrahedron Lett.* **1999**, *40*, 4161–4164.

92 O. Seitz, C. H. Wong, *J. Am. Chem. Soc.* **1997**, *119*, 8766–8776.

93 B. Blankemeyer-Menge, R. Frank, *Tetrahedron Lett.* **1988**, *29*, 5871–5874.

94 K. Kaljuste, A. Unden, *Tetrahedron Lett.* **1996**, *37*, 3031–3034.

95 F. Guibé, O. Dangles, G. Balavoine, A. Loffet, *Tetrahedron Lett.* **1989**, *30*, 2641–2644.

96 X. H. Zhang, R. A. Jones, *Tetrahedron Lett.* **1996**, *37*, 3789–3790.

97 K. C. Nicolaou, N. Winssinger, R. Hughes, C. Smethurst, S. Y. Cho, *Angew. Chem. Int. Ed.* **2000**, *39*, 1084–1089; *Angew. Chem.* **2000**, *112*, 1126–1130.

98 O. Seitz, H. Kunz, *Angew. Chem. Int. Ed.* **1995**, *34*, 803–805; *Angew. Chem.* **1995**, *107*, 901–904.

99 L. A. Thompson, J. A. Ellman, *Tetrahedron Lett.* **1994**, *35*, 9333–9336.

100 S.-e. Yoo, Y.-D. Gong, M.-Y. Choi, J.-s. Seo, K. Y. Yi, *Tetrahedron Lett.* **2000**, *41*, 6415–6418.

101 C. C. Leznoff, J. Y. Wong, *Can. J. Chem.* **1973**, *51*, 3756–3764.

102 C. C. Leznoff, S. Greenberg, *Can. J. Chem.* **1976**, *54*, 3824–3829.

103 Z. H. Xu, C. R. McArthur, C. C. Leznoff, *Can. J. Chem.* **1983**, *61*, 1405–1409.

104 R. Maltais, M. Berube, O. Marion, R. Labrecque, D. Poirier, *Tetrahedron Lett.* **2000**, *41*, 1691–1694.

105 C. M. Huwe, H. Künzer, *Tetrahedron Lett.* **1999**, *40*, 683–686.

106 G. T. Wang, S. Li, N. Wideburg, G. A. Krafft, D. J. Kempf, *J. Med. Chem.* **1995**, *38*, 2995–3002.

107 S. Q. Chen, K. D. Janda, *Tetrahedron Lett.* **1998**, *39*, 3943–3946.

108 J. S. Koh, J. A. Ellman, *J. Org. Chem.* **1996**, *61*, 4494–4495.

109 E. K. Kick, J. A. Ellman, *J. Med. Chem.* **1995**, *38*, 1427–1430.

110 G. Wess, K. Bock, H. Kleine, M. Kurz, W. Guba, H. Hemmerle, E. Lopez-Calle, K. H. Baringhaus, H. Glombik, A. Enhsen, W. Kramer, *Angew. Chem. Int. Ed.* **1996**, *35*, 2222–2224; *Angew. Chem.* **1996**, *108*, 2363–2366.

111 W. H. Pearson, R. B. Clark, *Tetrahedron Lett.* **1997**, *38*, 7669–7672.

112 S. E. Yoo, J. S. Seo, K. Y. Yi, Y. D. Gong, *Tetrahedron Lett.* **1997**, *38*, 1203–1206.

113 J. H. Ryu, J. H. Jeong, *Arch. Pharm. Res.* **1999**, *22*, 585–591.

114 M. Ramaseshan, J. W. Ellingboe, Y. L. Dory, P. Deslongchamps, *Tetrahedron Lett.* **2000**, *41*, 4743–4749.

115 P. J. May, M. Bradley, D. C. Harrowven, D. Pallin, *Tetrahedron Lett.* **2000**, *41*, 1627–1630.

116 Y. Kondo, T. Komine, M. Fujinami, M. Uchiyama, T. Sakamoto, *J. Comb. Chem.* **1999**, *1*, 123–126.

117 L. F. Tietze, A. Steinmetz, *Angew. Chem. Int. Ed.* **1996**, *35*, 651–652; *Angew. Chem.* **1996**, *108*, 682–683.

118 U. Grether, H. Waldmann, *Angew. Chem. Int. Ed.* **2000**, *39*, 1629–1632; *Angew. Chem.* **2000**, *112*, 1688–1691.

119 C. R. Millington, R. Quarrel, G. Lowe, *Tetrahedron Lett.* **1998**, *39*, 7201–7204.

120 J. R. Horton, L. M. Stamp, A. Routledge, *Tetrahedron Lett.* **2000**, *41*, 9181–9184.

121 P. Soucy, Y. L. Dory, P. Deslongchamps, *Synlett* **2000**, 1123–1126.

122 D. G. Mullen, G. Barany, *J. Org. Chem.* **1988**, *53*, 5240–5248.

123 L. A. Thompson, F. L. Moore, Y. C. Moon, J. A. Ellman, *J. Org. Chem.* **1998**, *63*, 2066–2067.

124 B. Chenera, J. A. Finkelstein, D. F. Veber, *J. Am. Chem. Soc.* **1995**, *117*, 11999–12000.

125 M. J. Plunkett, J. A. Ellman, *J. Org. Chem.* **1995**, *60*, 6006–6007.

126 Y. Han, S. D. Walker, R. N. Young, *Tetrahedron Lett.* **1996**, *37*, 2703–2706.

127 T. L. Boehm, H. D. H. Showalter, *J. Org. Chem.* **1996**, *61*, 6498–6499.

128 C. A. Briehn, T. Kirschbaum, P. Bäuerle, *J. Org. Chem.* **2000**, *65*, 352–359.

129 T. Kirschbaum, C. A. Briehn, P. Bäuerle, *Perkin 1* **2000**, 1211–1216.

130 F. X. Woolard, J. Paetsch, J. A. Ellman, *J. Org. Chem.* **1997**, *62*, 6102–6103.

131 M. J. Plunkett, J. A. Ellman, *J. Org. Chem.* **1997**, *62*, 2885–2893.

132 NovaBiochem, *Catalog and Peptide Synthesis Handbook* **2000**.

133 Y. Hu, J. J. A. Porco, *Tetrahedron Lett.* **1998**, *39*, 2711–2714.

134 K. A. Newlander, B. Chenera, D. F. Veber, N. C. F. Yim, M. L. Moore, *J. Org. Chem.* **1997**, *62*, 6726–6732.

135 Y. H. Hu, J. A. Porco, J. W. Labadie, O. W. Gooding, B. M. Trost, *J. Org. Chem.* **1998**, *63*, 4518–4521.

136 N. D. Hone, S. G. Davies, N. J. Devereux, S. L. Taylor, A. D. Baxter, *Tetrahedron Lett.* **1998**, *39*, 897–900.

137 R. Maltais, M. R. Tremblay, D. Poirier, *J. Comb. Chem.* **2000**, *2*, 604–614.

138 S. D. Brown, R. W. Armstrong, *J. Org. Chem.* **1997**, *62*, 7076–7077.

139 R. H. Crabtree, *Chem. Commun.* **1999**, 1611–1616.

140 Y. Lee, R. B. Silverman, *J. Am. Chem. Soc.* **1999**, *121*, 8407–8408.

141 S. Curtet, M. Langlois, *Tetrahedron Lett.* **1999**, *40*, 8563–8566.

142 A. Studer, S. Hadida, R. Ferritto, S. Y. Kim, P. Jeger, P. Wipf, D. P. Curran, *Science* **1997**, *275*, 823–826.

143 A. Studer, P. Jeger, P. Wipf, D. P.

Curran, *J. Org. Chem.* **1997**, *62*, 2917–2924.

144 M. Schuster, N. Lucas, S. Blechert, *Chem. Commun.* **1997**, 823–824.

145 L. S. Harikrishnan, H. D. H. Showalter, *Tetrahedron* **2000**, *56*, 515–519.

146 J. M. J. Frechet, L. J. Nuyens, E. Seymour, *J. Am. Chem. Soc.* **1979**, *101*, 432–436.

147 C. Pourbaix, F. Carreaux, B. Carboni, H. Deleuze, *Chem. Commun.* **2000**, 1275–1276.

148 W. Li, K. Burgess, *Tetrahedron Lett.* **1999**, *40*, 6527–6530.

149 H. S. Overkleeft, P. R. Bos, B. G. Hekking, E. J. Gordon, H. L. Ploegh, B. M. Kessler, *Tetrahedron Lett.* **2000**, *41*, 6005–6009.

150 D. L. Marshall, I. E. Liener, *J. Org. Chem.* **1970**, *35*, 867–868.

151 Z. Timar, T. Gallagher, *Tetrahedron Lett.* **2000**, *41*, 3173–3176.

152 S. J. Jin, D. P. Holub, D. J. Wustrow, *Tetrahedron Lett.* **1998**, *39*, 3651–3654.

153 K. C. Nicolaou, P. S. Baran, Y. L. Zhong, *J. Am. Chem. Soc.* **2000**, *122*, 10246–10248.

154 C. Vanier, F. Lorgé, A. Wagner, C. Mioskowski, *Angew. Chem. Int. Ed.* **2000**, *39*, 1679–1683; *Angew. Chem.* **2000**, *112*, 1745–1749.

155 K. W. Jung, X. Y. Zhao, K. D. Janda, *Tetrahedron* **1997**, *53*, 6645–6652.

156 K. W. Jung, X. Y. Zhao, K. D. Janda, *Tetrahedron Lett.* **1996**, *37*, 6491–6494.

157 X. Zhao, K. D. Janda, *Bioorg. Med. Chem. Lett.* **1998**, *8*, 2439–2442.

158 X. Y. Zhao, K. D. Janda, *Tetrahedron Lett.* **1997**, *38*, 5437–5440.

159 H. C. Zhang, K. K. Brumfield, B. E. Maryanoff, *Tetrahedron Lett.* **1997**, *38*, 2439–2442.

160 K. C. Nicolaou, S. A. Snyder, A. Bigot, J. A. Pfefferkorn, *Angew. Chem. Int. Ed.* **2000**, *39*, 1093–1096; *Angew. Chem.* **2000**, *112*, 1135–1138.

161 I. Sucholeiki, *Tetrahedron Lett.* **1994**, *35*, 7307–7310.

162 F. W. Forman, I. Sucholeiki, *J. Org. Chem.* **1995**, *60*, 523–528.

163 K. C. Nicolaou, N. Winssinger, J. Pastor, F. Murphy, *Angew. Chem.*

Int. Ed. **1998**, *37*, 2534–2537; *Angew. Chem.* **1998**, *110*, 2677–2680.

164 K. C. NICOLAOU, J. PASTOR, S. BARLUENGA, N. WINSSINGER, *Chem. Commun.* **1998**, 1947–1948.

165 T. RUHLAND, K. ANDERSEN, H. PEDERSEN, *J. Org. Chem.* **1998**, *63*, 9204–9211.

166 B. CHENERA (Smithkline Beecham Corporation), Patent No. WO PCT. 98/17695, **1995**.

167 G. MARCHAND, J. F. PILARD, J. SIMONET, *Tetrahedron Lett.* **2000**, *41*, 883–885.

168 C. GARCIA-ECHEVERRIA, *Tetrahedron Lett.* **1997**, *38*, 8933–8934.

169 H. KUHN, W. P. NEUMANN, *Synlett* **1994**, 123–124.

170 R. MICHELS, M. KATO, W. HEITZ, *Makromol. Chem.* **1976**, *177*, 2311–2320.

171 H. E. RUSSELL, R. W. A. LUKE, M. BRADLEY, *Tetrahedron Lett.* **2000**, *41*, 5287–5290.

172 K. C. NICOLAOU, J. A. PFEFFERKORN, A. J. ROECKER, G. Q. CAO, S. BARLUENGA, H. J. MITCHELL, *J. Am. Chem. Soc.* **2000**, *122*, 9939–9953.

173 K. C. NICOLAOU, J. A. PFEFFERKORN, H. J. MITCHELL, A. J. ROECKER, S. BARLUENGA, G. Q. CAO, R. L. AFFLECK, J. E. LILLIG, *J. Am. Chem. Soc.* **2000**, *122*, 9954–9967.

174 K. C. NICOLAOU, J. A. PFEFFERKORN, G. Q. CAO, *Angew. Chem. Int. Ed.* **2000**, *39*, 734–739; *Angew. Chem.* **2000**, *112*, 750–755.

175 K. C. NICOLAOU, G. Q. CAO, J. A. PFEFFERKORN, *Angew. Chem. Int. Ed.* **2000**, *39*, 739–743; *Angew. Chem.* **2000**, *112*, 755–759.

176 K. C. NICOLAOU, A. J. ROECKER, J. A. PFEFFERKORN, G. Q. CAO, *J. Am. Chem. Soc.* **2000**, *122*, 2966–2967.

177 K. C. NICOLAOU, H. J. MITCHELL, K. C. FYLAKTAKIDOU, H. SUZUKI, R. M. RODRÍGUEZ, *Angew. Chem. Int. Ed.* **2000**, *39*, 1089–1093; *Angew. Chem.* **2000**, *112*, 1131–1135.

178 K. C. NICOLAOU, C. N. C. BODDY, S. BRÄSE, N. WINSSINGER, *Angew. Chem. Int. Ed.* **1999**, *38*, 2096–2152; *Angew. Chem.* **1999**, *111*, 2230–2287.

179 J. C. NELSON, J. K. YOUNG, J. S.

MOORE, *J. Org. Chem.* **1996**, *61*, 8160–8168.

180 J. K. YOUNG, J. C. NELSON, J. S. MOORE, *J. Am. Chem. Soc.* **1994**, *116*, 10841–10842.

181 L. JONES, J. S. SCHUMM, J. M. TOUR, *J. Org. Chem.* **1997**, *62*, 1388–1410.

182 S. BRÄSE, D. ENDERS, J. KÖBBERLING, F. AVEMARIA, *Angew. Chem. Int. Ed.* **1998**, *37*, 3413–3415; *Angew. Chem.* **1998**, *110*, 3614–3616.

183 S. BRÄSE, S. DAHMEN, *Chem. Eur. J.* **2000**, *6*, 1899–1905.

184 M. LORMANN, S. DAHMEN, S. BRÄSE, *Tetrahedron Lett.* **2000**, *41*, 3813–3816.

185 S. SCHUNK, D. ENDERS, *Org. Lett.* **2000**, *2*, 907–910.

186 M. LORMANN, S. BRÄSE, unpublished.

187 S. BRÄSE, M. SCHROEN, *Angew. Chem. Int. Ed.* **1999**, *38*, 1071–1073; *Angew. Chem.* **1999**, *111*, 1139–1142.

188 A. DE MEIJERE, H. NÜSKE, M. ES-SAYED, T. LABAHN, M. SCHROEN, S. BRÄSE, *Angew. Chem. Int. Ed.* **1999**, *38*, 3669–3672; *Angew. Chem.* **1999**, *111*, 3881–3884.

189 F. AVEMARIA, V. ZIMMERMANN, S. BRÄSE, *Org. Lett.* **2001**, submitted.

190 S. BRÄSE, M. LORMANN, J. HEUTS, *Chem. Eur. J.* **2001**, in preparation.

191 S. BRÄSE, S. DAHMEN, J. HEUTS, *Tetrahedron Lett.* **1999**, *40*, 6201–6203.

192 M. E. P. LORMANN, C. H. WALKER, S. BRÄSE, *Chem. Commun.* **2001**, submitted.

193 S. BRÄSE, J. KÖBBERLING, D. ENDERS, M. WANG, R. LAZNY, S. BRANDTNER, *Tetrahedron Lett.* **1999**, *40*, 2105–2108.

194 S. BRÄSE, S. DAHMEN, M. PFEFFER-KORN, *J. Comb. Chem.* **2000**, *2*, 710–717.

195 S. DAHMEN, S. BRÄSE, *Org. Lett.* **2000**, *2*, 3563–3565.

196 a) C. PILOT, MAÍTRISE DE CHIMIE **1999**, RWTH Aachen/Université Strasbourg. b) C. PILOT, S. DAHMEN, F. LAUTERWASSER, S. BRÄSE, *Tetrahedron Lett.* **2001**, *42*, 9179–9181.

197 S. BRÄSE, *Chimica Oggi* **2000**, *18(9)*, 14–18.

198 S. DAHMEN, S. BRÄSE, *Angew. Chem.*

Int. Ed. **2000**, *39*, 3681–3683; *Angew. Chem.* **2000**, *112*, 3827–3830.

199 J. Heuts, S. Bräse, unpublished.

200 J. Rademann, J. Smerdka, G. Jung, P. Grosche, D. Schmid, *Angew. Chem. Int. Ed.* **2001**, *40*, 381–385.

201 S. Bräse, S. Dahmen, M. Schroen, unpublished.

202 S. Bräse, S. Dahmen, C. Popescu, M. Schroen, F.-J. Wortmann, *Polym. Degr. Stab.* in press.

203 W. D. F. Meutermans, P. F. Alewood, *Tetrahedron Lett.* **1995**, *36*, 7709–7712.

204 R. B. Merrifield, L. D. Vizioli, H. G. Boman, *Biochemistry* **1982**, *21*, 5020–5031.

205 J. P. Tam, W. F. Heath, R. B. Merrifield, *J. Am. Chem. Soc.* **1983**, *105*, 6442–6455.

206 H. Yaijima, N. Fujii, H. Ogawa, H. Kawatani, *J. Chem. Soc., Chem. Commun.* **1974**, 107–108.

207 A. R. Mitchell, S. B. H. Kent, M. Engelhard, R. B. Merrifield, *J. Org. Chem.* **1978**, *43*, 2845–2852.

208 K. C. Nicolaou, N. Winssinger, J. Pastor, F. DeRoose, *J. Am. Chem. Soc.* **1997**, *119*, 449–450.

209 A. Mazurov, *Tetrahedron Lett.* **2000**, *41*, 7–10.

210 D. R. Englebretsen, C. T. Choma, G. T. Robillard, *Tetrahedron Lett.* **1998**, *39*, 4929–4932.

211 P. R. Hansen, C. E. Olsen, A. Holm, *Bioconj. Chem.* **1998**, *9*, 126–131.

212 Y. X. Han, G. Barany, *J. Org. Chem.* **1997**, *62*, 3841–3848.

213 M. C. Munson, G. Barany, *J. Am. Chem. Soc.* **1993**, *115*, 10203–10210.

214 G. Mezo, N. Mihala, G. Koczan, F. Hudecz, *Tetrahedron* **1998**, *54*, 6757–6766.

215 D. Limal, J. P. Briand, P. Dalbon, M. Jolivet, *J. Pept. Res.* **1998**, *52*, 121–129.

216 B. Yan, H. Gstach, *Tetrahedron Lett.* **1996**, *37*, 8325–8328.

217 J. Blake, C. H. Li, *J. Am. Chem. Soc.* **1968**, *90*, 5882–5884.

218 E. G. Mata, *Tetrahedron Lett.* **1997**, *38*, 6335–6338.

219 J. D. Winkler, W. McCoull, *Tetrahedron Lett.* **1998**, *39*, 4935–4936.

220 M. J. Plater, A. M. Murdoch, J. R. Morphy, Z. Rankovic, D. C. Rees, *J. Comb. Chem.* **2000**, *2*, 508–512.

221 J. R. Morphy, Z. Rankovic, D. C. Rees, *Tetrahedron Lett.* **1996**, *37*, 3209–3212.

222 P. H. Toy, T. S. Reger, K. D. Janda, *J. Comb. Chem.* **2000**, *2*, 2205–2207.

223 A. R. Brown, D. C. Rees, Z. Rankovic, J. R. Morphy, *J. Am. Chem. Soc.* **1997**, *119*, 3288–3295.

224 X. H. Ouyang, R. W. Armstrong, M. M. Murphy, *J. Org. Chem.* **1998**, *63*, 1027–1032.

225 X. H. Ouyang, A. S. Kiselyov, *Tetrahedron Lett.* **1999**, *40*, 5827–5830.

226 W. Bannwarth, J. Huebscher, R. Barner, *Bioorg. Med. Chem. Lett.* **1996**, *6*, 1525–1528.

227 S. R. Chhabra, H. Parekh, A. N. Khan, B. W. Bycroft, B. Kellam, *Tetrahedron Lett.* **2001**, *42*, 2189–2192.

228 S. R. Chhabra, A. N. Khan, B. W. Bycroft, *Tetrahedron Lett.* **2000**, *41*, 1099–1102.

229 J. D. Glass, A. Talansky, Z. Gronka, I. L. Schwartz, R. Walter, *J. Am. Chem. Soc.* **1974**, *96*, 6476–6480.

230 A. M. Bray, N. J. Maeji, A. G. Jhingran, R. M. Valerio, *Tetrahedron Lett.* **1991**, *32*, 6163–6166.

231 G. Faita, A. Paio, P. Quadrelli, F. Rancati, P. Seneci, *Tetrahedron Lett.* **2000**, *41*, 1265–1269.

232 A. J. Souers, A. A. Virgilio, S. S. Schürer, J. A. Ellman, T. P. Kogan, H. E. West, W. Ankener, P. Vanderslice, *Bioorg. Med. Chem. Lett.* **1998**, *8*, 2297–2302.

233 K. Kurokawa, H. Kumihara, H. Kondo, *Bioorg. Med. Chem. Lett.* **2000**, *10*, 1827–1830.

234 A. A. Virgilio, S. C. Schürer, J. A. Ellman, *Tetrahedron Lett.* **1996**, *37*, 6961–6964.

235 A. J. Souers, A. A. Virgilio, A. Rosenquist, W. Fenuik, J. A. Ellman, *J. Am. Chem. Soc.* **1999**, *121*, 1817–1825.

236 F. Berst, A. B. Holmes, M. Ladlow, P. J. Murray, *Tetrahedron Lett.* **2000**, *41*, 6649–6653.

237 F. Stieber, U. Grether, H.

WALDMANN, *Angew. Chem. Int. Ed.* **1999**, *38*, 1073–1077; *Angew. Chem.* **1999**, *111*, 1142–1145.

238 D. H. RICH, S. K. GURWARA, *J. Am. Chem. Soc.* **1975**, *97*, 1575–1579.

239 S. J. TEAGUE, *Tetrahedron Lett.* **1996**, *37*, 5751–5754.

240 R. P. HAMMER, F. ALBERICIO, E. GIRALT, G. BARANY, *Int. J. Pept. Protein Res.* **1990**, *28*, 31–45.

241 C. P. HOLMES, D. G. JONES, *J. Org. Chem.* **1995**, *60*, 2318–2319.

242 S. PEUKERT, B. GIESE, *J. Org. Chem.* **1998**, *63*, 9045–9051.

243 R. GLATTHAR, B. GIESE, *Org. Lett.* **2000**, *2*, 2315–2317.

244 H. B. LEE, S. BALASUBRAMANIAN, *J. Org. Chem.* **1999**, *64*, 3454–3460.

245 S.-S. WANG, *J. Org. Chem.* **1976**, *41*, 3258–3261.

246 J. P. TAM, R. D. DIMARCHI, R. B. MERRIFIELD, *Int. J. Pept. Protein Res.* **1980**, *16*, 412–425.

247 D. H. RICH, S. K. GURWARA, *J. Chem. Soc., Chem. Commun.* **1973**, 610–611.

248 A. AIYAGHOSH, V. N. R. PILLAI, *Tetrahedron* **1988**, *44*, 6661–6666.

249 U. ZEHAVI, A. PATCHORNIK, *J. Am. Chem. Soc.* **1973**, *95*, 5673–5677.

250 D. L. WHITEHOUSE, S. N. SAVINOV, D. J. AUSTIN, *Tetrahedron Lett.* **1997**, *38*, 7851–7852.

251 D. J. YOO, M. M. GREENBERG, *J. Org. Chem.* **1995**, *60*, 3358–3364.

252 M. RINNOVA, M. NOVAKOVA, V. KASICKA, J. JIRACEK, *J. Pept. Sci.* **2000**, *6*, 355–365.

253 A. AJAYAGOSH, V. N. R. PILLAI, *J. Org. Chem.* **1990**, *55*, 2826–2829.

254 C. P. HOLMES, *J. Org. Chem.* **1997**, *62*, 2370–2380.

255 M. SMET, L. X. LIAO, W. DEHAEN, D. V. MCGRATH, *Org. Lett.* **2000**, *2*, 511–513.

256 B. B. BROWN, D. S. WAGNER, H. M. GEYSEN, *Mol. Diversity* **1995**, *1*, 4–12.

257 R. RODEBAUGH, B. FRASER-REID, H. M. GEYSEN, *Tetrahedron Lett.* **1997**, *38*, 7653–7656.

258 S. M. STERNSON, S. L. SCHREIBER, *Tetrahedron Lett.* **1998**, *39*, 7451–7454.

259 J. J. BALDWIN, J. J. BURBAUM, I. HENDERSON, M. H. J. OHLMEYER, *J. Am. Chem. Soc.* **1995**, *117*, 5588–5589.

260 A. AJAYAGOSH, V. N. R. PILLAI, *J. Org. Chem.* **1987**, *52*, 5714–5717.

261 P. LLOYD-WILLIAMS, M. GAIRI, F. ALBERICIO, E. GIRALT, *Tetrahedron* **1993**, *49*, 10069–10078.

262 H. VENKATESAN, M. M. GREENBERG, *J. Org. Chem.* **1996**, *61*, 525–529.

263 P. LLOYD-WILLIAMS, M. GAIRI, F. ALBERICIO, E. GIRALT, *Tetrahedron* **1991**, *49*, 9867–9880.

264 R. S. ROCK, S. I. CHAN, *J. Org. Chem.* **1996**, *61*, 1526–1529.

265 J. C. SHEEHAN, K. UMEZAWA, *J. Org. Chem.* **1973**, *38*, 3771–3774.

266 A. ROUTLEDGE, C. ABELL, S. BALASUBRAMANIAN, *Tetrahedron Lett.* **1997**, *38*, 1227–1230.

267 T. D. RYBA, P. G. HARRAN, *Org. Lett.* **2000**, *2*, 851–853.

268 S. C. MCKEOWN, S. P. WATSON, R. A. E. CARR, P. MARSHALL, *Tetrahedron Lett.* **1999**, *40*, 2407–2410.

269 E. B. AKERBLOM, *Mol. Diversity* **1999**, *4*, 53–69.

270 C. DELLAQUILA, J. L. IMBACH, B. RAYNER, *Tetrahedron Lett.* **1997**, *38*, 5289–5292.

271 E. B. AKERBLOM, A. S. NYGREN, K. H. AGBACK, *Mol. Diversity* **1998**, *3*, 137–148.

272 M. R. TREMBLAY, D. POIRIER, *Tetrahedron Lett.* **1999**, *40*, 1277–1280.

273 S. C. SCHÜRER, S. BLECHERT, *Synlett* **1998**, 166–168.

274 W. C. CHENG, C. HALM, J. B. EVARTS, M. M. OLMSTEAD, M. J. KURTH, *J. Org. Chem.* **1999**, *64*, 8557–8562.

275 J. U. PETERS, S. BLECHERT, *Synlett* **1997**, 348–350.

276 J. PERNERSTORFER, M. SCHUSTER, S. BLECHERT, *Chem. Commun.* **1997**, 1949–1950.

277 K. C. NICOLAOU, N. WINSSINGER, J. PASTOR, S. NINKOVIC, F. SARABIA, Y. HE, D. VOURLOUMIS, Z. YANG, T. LI, P. GIANNAKAKOU, E. HAMEL, *Nature* **1997**, *387*, 268–272.

278 A. D. PISCOPIO, J. F. MILLER, K. KOCH, *Tetrahedron Lett.* **1997**, *38*, 7143–7146.

279 A. D. PISCOPIO, J. F. MILLER, K. KOCH, *Tetrahedron Lett.* **1998**, *39*, 2667–2670.

280 A. D. PISCOPIO, J. F. MILLER, K.

KOCH, *Tetrahedron* **1999**, *55*, 8189–8198.

281 J. H. VAN MAARSEVEEN, J. A. J. DEN HARTOG, V. ENGELEN, E. FINNER, G. VISSER, C. G. KRUSE, *Tetrahedron Lett.* **1996**, *37*, 8249–8252.

282 J. J. N. VEERMAN, J. H. VAN MAARSEVEEN, G. M. VISSER, C. G. KRUSE, H. E. SCHOEMAKER, H. HIEMSTRA, F. P. J. T. RUTJES, *Eur. J. Org. Chem.* **1998**, 2583–2589.

283 L. G. MELEAN, W.-C. HAASE, P. H. SEEBERGER, *Tetrahedron Lett.* **2000**, *41*, 4329–4333.

284 B. SAUERBREI, V. JUNGMANN, H. WALDMANN, *Angew. Chem. Int. Ed.* **1998**, *37*, 1143–1146; *Angew. Chem.* **1998**, *110*, 1187–1190.

285 G. BÖHM, J. DOWDEN, D. C. RICE, I. BURGESS, J. F. PILARD, B. GUILBERT, A. HAXTON, R. C. HUNTER, N. J. TURNER, S. L. FLITSCH, *Tetrahedron Lett.* **1998**, *39*, 3819–3822.

286 B. J. BACKES, J. A. ELLMAN, *J. Am. Chem. Soc.* **1994**, *116*, 11171–11172.

287 B. J. BACKES, J. A. ELLMAN, *J. Org. Chem.* **1999**, *64*, 2322–2330.

288 B. J. BACKES, A. A. VIRGILIO, J. A. ELLMAN, *J. Am. Chem. Soc.* **1996**, *118*, 3055–3056.

289 G. W. KENNER, J. R. McDERMOTT, R. C. SHEPPARD, *J. Chem. Soc., Chem. Commun.* **1971**, 636–637.

290 C. HULME, J. PENG, G. MORTON, J. M. SALVINO, T. HERPIN, R. LABAUDINIERE, *Tetrahedron Lett.* **1998**, *39*, 7227–7230.

291 R. SOLA, R. SAGUER, M. L. DAVID, R. PASCAL, *J. Chem. Soc., Chem. Commun.* **1995**, 1786–1788.

292 R. SOLA, J. MERY, R. PASCAL, *Tetrahedron Lett.* **1996**, *37*, 9195–9198.

293 M. H. TODD, S. F. OLIVER, C. ABELL, *Org. Lett.* **1999**, *1*, 1149–1151.

294 A. LINK, S. VANCALENBERGH, P. HERDEWIJN, *Tetrahedron Lett.* **1998**, *39*, 5175–5176.

295 A. GOLISADE, J. C. BRESSI, S. VAN CALENBERGH, M. H. GELB, A. LINK, *J. Comb. Chem.* **2000**, *2*, 537–544.

296 M. H. LYTTLE, D. HUDSON, R. M. COOK, *Nucl. Acids Res.* **1996**, *24*, 2793–2798.

297 C. L. BEECH, J. F. COOPE, G. FAIRLEY, P. S. GILBERT, B. G. MAIN, K. PLÉ, *J. Org. Chem.* **2001**, *66*, 2240–2245.

298 S. HOFFMANN, R. FRANK, *Tetrahedron Lett.* **1994**, *35*, 7763–7766.

299 O. LORTHIOIR, S. C. McKEOWN, N. J. PARR, M. WASHINGTON, S. P. WATSON, *Tetrahedron Lett.* **2000**, *41*, 8609–8613.

300 T. MASQUELIN, N. MEUNIER, F. GERBER, G. ROSSE, *Hetrocycles* **1998**, *48*, 2489–2505.

301 L. YANG, *Tetrahedron Lett.* **2000**, *41*, 6981–6984.

302 H. ZHANG, H. YE, A. MORETTO, K. BRUMFIELD, B. MARYANOFF, *Org. Lett.* **2000**, *2*, 89–92.

303 C. ATRASH, M. BRADLEY, *Chem. Commun.* **1997**, 1397–1398.

304 K. C. NICOLAOU, J. PASTOR, N. WINSSINGER, F. MURPHY, *J. Am. Chem. Soc.* **1998**, *120*, 5132–5133.

305 C. R. JOHNSON, B. R. ZHANG, *Tetrahedron Lett.* **1995**, *36*, 9253–9256.

306 F. CAMPS, J. CASTELLS, J. PI, *Ann. Quim.* **1974**, *70*, 848–849.

307 Y. D. GONG, S. NAJDI, M. M. OLMSTEAD, M. J. KURTH, *J. Org. Chem.* **1998**, *63*, 3081–3086.

308 S. H. LEE, S. H. CHUNG, Y. S. LEE, *Tetrahedron Lett.* **1998**, *39*, 9469–9472.

309 K. H. PARK, E. ABBATE, S. NAJDI, M. M. OLMSTEAD, M. J. KURTH, *Chem. Commun.* **1998**, 1679–1680.

310 K. H. PARK, M. M. OLMSTEAD, M. J. KURTH, *J. Org. Chem.* **1998**, *63*, 6579–6585.

311 S. HANESSIAN, R. Y. YANG, *Tetrahedron Lett.* **1996**, *37*, 5835–5838.

312 L. J. WILSON, M. LI, D. E. PORTLOCK, *Tetrahedron Lett.* **1998**, *39*, 5135–5138.

313 S. W. KIM, J. S. KOH, E. J. LEE, S. RO, *Mol. Diversity* **1998**, *3*, 129–132.

314 S. W. KIM, S. Y. AHN, J. S. KOH, J. H. LEE, S. RO, H. Y. CHO, *Tetrahedron Lett.* **1997**, *38*, 4603–4606.

315 J. MATTHEWS, R. A. RIVERO, *J. Org. Chem.* **1997**, *62*, 6090–6092.

316 B. A. DRESSMAN, L. A. SPANGLE, S. W. KALDOR, *Tetrahedron Lett.* **1996**, *37*, 937–940.

317 J. STADLWIESER, E. P. ELLMERER-MÜLLER, A. TAKO, N. MASLOUH, W. BANNWARTH, *Angew. Chem. Int. Ed.*

1998, *37*, 1402–1404; *Angew. Chem.* 1998, *110*, 1487–1489.

318 S. H. DeWitt, J. K. Kiely, C. J. Stankovic, M. C. Schroeder, D. M. R. Cody, M. R. Pavia, *Proc. Natl. Acad. Sci. USA* 1993, *90*, 6909–6913.

319 A. Boeijen, J. A. W. Kruijtzer, R. M. J. Liskamp, *Bioorg. Med. Chem. Lett.* 1998, *8*, 2375–2380.

320 H. P. Buchstaller, *Tetrahedron* 1998, *54*, 3465–3470.

321 P. ten Holte, L. Thijs, B. Zwanen- burg, *Tetrahedron Lett.* 1998, *39*, 7407–7410.

322 V. S. Goodfellow, C. P. Laudeman, J. I. Gerrity, M. Burkard, E. Strobel, J. S. Zuzack, D. A. McLeod, *Mol. Diversity* 1996, *2*, 97–102.

323 A. K. Szardenings, T. S. Burkoth, H. H. Lu, D. W. Tien, D. A. Campbell, *Tetrahedron* 1997, *53*, 6573–6593.

324 J. Kowalski, M. A. Lipton, *Tetra- hedron Lett.* 1996, *37*, 5839–5840.

325 A. Van Loevezijn, J. H. Van Maarseveen, K. Stegman, G. M. Visser, G. J. Koomen, *Tetrahedron Lett.* 1998, *39*, 4737–4740.

326 P. P. Fantauzzi, K. M. Yager, *Tetrahedron Lett.* 1998, *39*, 1291–1294.

327 R. A. Smith, M. A. Bobko, W. Lee, *Bioorg. Med. Chem. Lett.* 1998, *8*, 2369–2374.

328 W. R. Li, S. Z. Peng, *Tetrahedron Lett.* 1998, *39*, 7373–7376.

329 A. K. Szardenings, D. Harris, S. Lam, L. H. Shi, D. Tien, Y. W. Wang, D. V. Patel, M. Navre, D. A. Campbell, *J. Med. Chem.* 1998, *41*, 2194–2200.

330 B. O. Scott, A. C. Siegmund, C. K. Marlowe, Y. Pei, K. L. Spear, *Mol. Diversity* 1996, *1*, 125–134.

331 A. Golebiowski, S. R. Klopfenstein, J. J. Chen, X. Shao, *Tetrahedron Lett.* 2000, *41*, 4841–4844.

332 D. A. Goff, R. N. Zuckermann, *J. Org. Chem.* 1995, *60*, 5744–5745.

333 J. P. Mayer, J. W. Zhang, K. Bjergarde, D. M. Lenz, J. J. Gaudino, *Tetrahedron Lett.* 1996, *37*, 8081–8084.

334 L. Moroder, J. Lutz, F. Grams, S. Rudolph-Böhner, G. Ösapay, M.

Goodman, W. Kolbeck, *Biopolymers* 1996, *38*, 295–300.

335 L. F. Tietze, A. Steinmetz, *Synlett* 1996, 667–668.

336 L. F. Tietze, A. Steinmetz, F. Balkenhohl, *Bioorg. Med. Chem. Lett.* 1997, *7*, 1303–1306.

337 T. T. Romoff, L. Ma, Y. W. Wang, D. A. Campbell, *Synlett* 1998, 1341– 1342.

338 L. Weber, P. Iaiza, G. Biringer, P. Barbier, *Synlett* 1998, 1156–1158.

339 J. Matthews, R. A. Rivero, *J. Org. Chem.* 1998, *63*, 4808–4810.

340 B. A. Kulkarni, A. Ganesan, *Tetrahedron Lett.* 1998, *39*, 4369–4373.

341 A. L. Smith, C. G. Thomson, P. D. Leeson, *Bioorg. Med. Chem. Lett.* 1996, *6*, 1483–1486.

342 S. A. Kolodziej, B. C. Hamper, *Tetrahedron Lett.* 1996, *37*, 5277–5280.

343 M. M. Sim, C. L. Lee, A. Ganesan, *Tetrahedron Lett.* 1998, *39*, 6399–6402.

344 F. Albericio, J. Garcia, E. L. Michelotti, E. Nicolas, C. M. Tice, *Tetrahedron Lett.* 2000, *41*, 3161–3163.

345 W. L. Huang, R. M. Scarborough, *Tetrahedron Lett.* 1999, *40*, 2665– 2668.

346 C. Le Hetet, M. David, F. Carreaux, B. Carboni, A. Sauleau, *Tetrahedron Lett.* 1997, *38*, 5153–5156.

347 J. F. Pons, Q. Mishir, A. Nouvet, F. Brookfield, *Tetrahedron Lett.* 2000, *41*, 4965–4968.

348 E. J. Guthrie, J. Macritchie, R. C. Hartley, *Tetrahedron Lett.* 2000, *41*, 4987–4990.

349 A. Mazurov, *Bioorg. Med. Chem. Lett.* 2000, *10*, 67–70.

350 S. Bräse, R. Lazny, unpublished.

351 B. K. Lorsbach, R. B. Miller, M. J. Kurth, *J. Org. Chem.* 1996, *61*, 8716– 8717.

352 B. A. Lorsbach, J. T. Bagdanoff, R. B. Miller, M. J. Kurth, *J. Org. Chem.* 1998, *63*, 2244–2250.

353 S. E. Gibson, N. J. Hales, M. A. Peplow, *Tetrahedron Lett.* 1999, *40*, 1417–1418.

354 M. F. Semmelhack, G. Hilt, J. H. Colley, *Tetrahedron Lett.* 1998, *39*, 7683–7686.

355 A. C. Comely, S. E. Gibson, N. J.

HALES, *Chem. Commun.* **1999**, 2075–2076.

356 M. A. GALLOP (Glaxo Wellcome Inc., USA), US Patent No. 6057465 A, **2000**.

357 A. C. COMELY, S. E. GIBSON, N. J. HALES, M. A. PEPLOW, PCT Int. Appl. WO 0007966, **2000**.

358 K. I. WASHIZUKA, K. NAGAI, S. MINAKATA, I. RYU, M. KOMATSU, *Tetrahedron Lett.* **2000**, *41*, 691–695.

359 L. BLANCO, R. BLOCH, E. BUGNET, S. DELOISY, *Tetrahedron Lett.* **2000**, *41*, 7875–7878.

360 D. L. WHITEHOUSE, K. H. NELSON, S. N. SAVINOV, R. S. LOWE, D. J. AUSTIN, *Bioorg. Med. Chem. Lett.* **1998**, *6*, 1273–1282.

361 M. R. GOWRAVARAM, M. A. GALLOP, *Tetrahedron Lett.* **1997**, *38*, 6973–6976.

362 D. L. WHITEHOUSE, K. H. J. NELSON, S. N. SAVINOV, D. J. AUSTIN, *Tetrahedron Lett.* **1997**, *38*, 7139–7142.

363 A. C. SPIVEY, C. M. DIAPER, H. ADAMS, A. J. RUDGE, *J. Org. Chem.* **2000**, *65*, 5253–5263.

364 I. HUGHES, *Tetrahedron Lett.* **1996**, *37*, 7595–7598.

365 R. M. SLADE, M. A. PHILLIPS, J. G. BERGER, *Mol. Diversity* **1998**, *4*, 215–219.

366 I. SUCHOLEIKI (Solid-phase Sciences Corporation), Patent No. US5684130, **1997**.

367 H. KAMOGAWA, A. KANZAWA, M. KADOYA, T. NAITO, M. M. NANASAWA, *Bull. Chem. Soc. Jpn.* **1983**, *56*, 762–765.

368 P. GARIBAY, J. NIELSEN, T. HOEG-JENSEN, *Tetrahedron Lett.* **1998**, *39*, 2207–2210.

369 Commercially available from CalBioChem-NovaBioChem (Swiss).

370 Commercially available from Advanced Chem Tech.

371 J. M. COBB, M. T. FIORINI, C. R. GODDARD, M.-E. THEOCLITOU, C. ABELL, *Tetrahedron Lett.* **1999**, *40*, 1045–1048.

372 M. M. SIM, C. L. LEE, A. GANESAN, *Tetrahedron Lett.* **1998**, *39*, 2195–2198.

373 F. ZARAGOZA, *Tetrahedron Lett.* **1997**, *38*, 7291–7294.

374 A. PATCHORNIK, M. A. KRAUS, *J. Am. Chem. Soc.* **1970**, *92*, 7287–7289.

375 B. C. HAMPER, K. Z. GAN, T. J. OWEN, *Tetrahedron Lett.* **1999**, *40*, 4973–4976.

376 S. P. DOUGLAS, D. M. WHITFIELD, J. J. KREPINSKY, *J. Am. Chem. Soc.* **1995**, *117*, 2116–2117.

377 M. R. PAVIA, G. WHITESIDES, D. G. HANGAUER, M. E. HEDIGER (Sphinx Pharmaceuticals Corporation), Patent No. WO PCT. 95/04277, **1995**.

378 A. B. REITZ, *Curr. Opin. Drug. Disc. Develop.* **1999**, *2*, 358–364.

379 D. OBRECHT, J. M. VILLALGORDO in: Solid-Supported Combinatorial and Parallel Synthesis of Small-Molecular-Weight Compound Libraries, Elsevier, Oxford, 1998, p. 98.

380 F. ZARAGOZA, *Angew. Chem. Int. Ed.* **2000**, *39*, 2077–2079; *Angew. Chem.* **2000**, *112*, 2158–2159.

381 J. K. RUETER, S. O. NORTEY, E. W. BAXTER, G. C. LEO, A. B. REITZ, *Tetrahedron Lett.* **1998**, *39*, 975–978.

382 J. A. HUNT, W. R. ROUSH, *J. Am. Chem. Soc.* **1996**, *118*, 9998–9999.

383 R. MOHAN, Y.-L. CHOU, M. M. MORRISSEY, *Tetrahedron Lett.* **1996**, *37*, 3963–3966.

384 L. M. GAYO, M. J. SUTO, *Tetrahedron Lett.* **1997**, *38*, 211–214.

385 M. SCHUSTER, J. PERNERSTORFER, S. BLECHERT, *Angew. Chem. Int. Ed.* **1996**, *35*, 1979–1980; *Angew. Chem.* **1996**, *108*, 2111–2112.

386 S. C. STINSON, *Chem. Ind. News* **1999**, *77(41)*, 101–120.

387 S. C. STINSON, *Chem. Ind. News* **1999**, *78(43)*, 55–78.

388 J. H. KIRCHHOFF, M. E. P. LORMANN, S. BRÄSE, *Chimica Oggi* **2001**, in press.

389 M. KAWANA, S. EMOTO, *Tetrahedron Lett.* **1972**, *13*, 4855–4858.

390 M. KAWANA, S. EMOTO, *Bull. Chem. Soc. Jpn.* **1974**, *47*, 160–165.

391 P. M. WORSTER, C. R. MCARTHUR, C. C. LEZNOFF, *Angew. Chem. Int. Ed. Engl.* **1979**, *18*, 221; *Angew. Chem.* **1979**, *91*, 255.

392 C. R. MCARTHUR, P. M. WORSTER, J.-L. JIANG, C. C. LEZNOFF, *Can. J. Chem.* **1982**, *60*, 1836–1841.

393 S. M. ALLIN, S. J. SHUTTLEWORTH,

Tetrahedron Lett. **1996**, *37*, 8023–8026.

394 C. W. PHOON, C. ABELL, *Tetrahedron Lett.* **1998**, *39*, 2655–2658.

395 V. PURANDARE, S. NATARAJAN, *Tetrahedron Lett.* **1997**, *38*, 8777–8780.

396 K. BURGESS, D. LIM, *Chem. Commun.* **1997**, 785–786.

397 S. P. BEW, S. D. BULL, S. G. DAVIES, *Tetrahedron Lett.* **2000**, *41*, 7577–7581.

398 A. VOLONTERIO, P. BRAVO, N. MOUSSIER, M. ZANDA, *Tetrahedron Lett.* **2000**, *41*, 6517–6521.

399 A. R. COLWELL, L. R. DUCKWALL, R. BROOKS, S. P. MCMANUS, *J. Org. Chem.* **1981**, *46*, 3097–3102.

400 H. S. MOON, N. E. SCHORE, M. J. KURTH, *J. Org. Chem.* **1992**, *57*, 6088–6089.

401 H. S. MOON, N. E. SCHORE, M. J. KURTH, *Tetrahedron Lett.* **1994**, *35*, 8915–8918.

402 J. H. KIRCHHOFF, J. KÖBBERLING, D. ENDERS, S. BRÄSE, Deutsche Patentanmeldung 100 07 704.8, **2000**.

403 D. ENDERS, J. H. KIRCHHOFF, J. KÖBBERLING, T. H. PEIFFER, *Org. Lett.* **2001**, *3*, 1241–1244.

404 B. J. BACKES, D. R. DRAGOLI, J. A. ELLMAN, *J. Org. Chem.* **1999**, *64*, 5472–5478.

405 J. HACHTEL, H. J. GAIS, *Eur. J. Org. Chem.* **2000**, 1457–1465.

406 S. ITSUNO, A. A. EL-SHEHAWY, M. Y. ABDELAAL, K. ITO, *New J. Chem.* **1998**, *22*, 775–777.

407 K. OERTEL, G. ZECH, H. KUNZ, *Angew. Chem. Int. Ed.* **2000**, *39*, 1431–1433; *Angew. Chem.* **2000**, *112*, 1489–1491.

408 T. HENKEL, R. M. BRUNNE, H. MÜLLER, F. REICHEL, *Angew. Chem. Int. Ed.* **1999**, *38*, 643–647.

409 J. G. BREITENBUCHER, G. FIGLIOZZI, *Tetrahedron Lett.* **2000**, *41*, 4311–4315.

410 M. MEISENBACH, W. VOELTER, *Chem. Lett.* **1997**, 1265–1266.

411 J. ALSINA, F. RABANAL, E. GIRALT, F. ALBERICIO, *Tetrahedron Lett.* **1994**, *35*, 9633–9636.

412 L. GOUILLEUX, J. A. FEHRENTZ, F. WINTERNITZ, J. MARTINEZ, *Tetrahedron Lett.* **1996**, *37*, 7031–7034.

413 C. G. BOOJAMRA, K. M. BUROW, L. A. THOMPSON, J. A. ELLMAN, *J. Org. Chem.* **1997**, *62*, 1240–1256.

414 A. S. HERNANDEZ, J. C. HODGES, *J. Org. Chem.* **1997**, *62*, 4861–4864.

415 K. NGU, D. V. PATEL, *Tetrahedron Lett.* **1997**, *38*, 973–976.

416 P. S. FURTH, M. S. REITMAN, A. F. COOK, *Tetrahedron Lett.* **1997**, *38*, 5403–5406.

417 P. CONTI, D. DEMONT, J. CALS, H. C. J. OTTENHEIJM, D. LEYSEN, *Tetrahedron Lett.* **1997**, *38*, 2915–2918.

418 C. KAY, P. J. MURRAY, L. SANDOW, A. B. HOLMES, *Tetrahedron Lett.* **1997**, *38*, 6941–6944.

419 N. W. HIRD, K. IRIE, K. NAGAI, *Tetrahedron Lett.* **1997**, *38*, 7111–7114.

420 L. J. FITZPATRICK, R. A. RIVERO, *Tetrahedron Lett.* **1997**, *38*, 7479–7482.

421 H.-P. HSIEH, Y.-T. WU, S.-T. CHEN, *Chem. Commun.* **1998**, 649–650.

422 B. A. DRESSMANN, U. SINGH, S. W. KALDOR, *Tetrahedron Lett.* **1998**, *39*, 3631–3634.

423 J. G. BREITENBUCHER, C. R. JOHNSON, M. HAIGHT, J. C. PHELAN, *Tetrahedron Lett.* **1998**, *39*, 1295–1298.

424 M. DEL FRESNO, J. ALSINA, M. ROYO, G. BARANY, F. ALBERICIO, *Tetrahedron Lett.* **1998**, *39*, 2639–2642.

425 S. R. CHHABRA, A. N. KHAN, B. W. BYCROFT, *Tetrahedron Lett.* **1998**, *39*, 3585–3588.

426 A. NEFZI, M. A. GIULIANOTTI, N. A. ONG, R. A. HOUGHTEN, *Org. Lett.* **2000**, *2*, 3349–3350.

427 J. M. OSTRESH, C. C. SCHONER, V. T. HAMASHIN, A. NEFZI, J. P. MEYER, R. A. HOUGHTEN, *J. Org. Chem.* **1998**, *63*, 8622–8623.

428 R. A. TOMMASI, P. G. NANTERMET, M. J. SHAPIRO, J. CHIN, W. K. D. BRILL, K. ANG, *Tetrahedron Lett.* **1998**, *39*, 5477–5480.

429 A. V. PURANDARE, M. A. POSS, *Tetrahedron Lett.* **1998**, *39*, 935–938.

430 L. GAUZY, Y. LE MERRER, J. C. DEPEZAY, F. CLERC, S. MIGNANI, *Tetrahedron Lett.* **1999**, *40*, 6005–6008.

431 A. R. KATRITZKY, L. XIE, G. ZHANG, *Tetrahedron Lett.* **1997**, *38*, 7011–7014.

432 E. A. BOYD, W. C. CHAN, V. M. LOH, *Tetrahedron Lett.* **1996**, *37*, 1647–1650.

433 M. A. Youngman, S. L. Dax, *Tetrahedron Lett.* **1997**, *38*, 6347–6350.

434 J. J. McNally, M. A. Youngman, S. L. Dax, *Tetrahedron Lett.* **1998**, *39*, 967–970.

435 W. J. Hoekstra, B. E. Maryanoff, P. Andrade-Gordon, J. H. Cohen, M. J. Costanzo, B. P. Damiano, B. J. Haertlein, B. D. Harris, J. A. Kauffman, P. M. Keane, D. F. McComsey, F. J. Villani, S. C. Yabut, *Bioorg. Med. Chem. Lett.* **1996**, *6*, 2371–2376.

436 W. J. Hoekstra, M. N. Greco, S. C. Yabut, B. L. Hulshizer, B. E. Maryanoff, *Tetrahedron Lett.* **1997**, *38*, 2629–2632.

437 K. H. Bleicher, J. R. Wareing, *Tetrahedron Lett.* **1998**, *39*, 4591–4594.

438 Y. Hidai, T. Kan, T. Fukuyama, *Tetrahedron Lett.* **1999**, *40*, 4711–4714.

439 K. Yamada, S. I. Nishimura, *Tetrahedron Lett.* **1995**, *36*, 9493–9496.

440 H. Stephensen, F. Zaragoza, *J. Org. Chem.* **1997**, *62*, 6096–6097.

441 J. F. W. Keana, M. Shimizu, K. K. Jernsted, *J. Org. Chem.* **1986**, *51*, 1641–1644.

442 M. W. Wilson, A. S. Hernández, A. P. Calvet, J. C. Hodges, *Mol. Diversity* **1998**, *3*, 95–112.

443 R. Léger, R. Yen, M. W. She, V. J. Lee, S. J. Hecker, *Tetrahedron Lett.* **1998**, *39*, 4171–4174.

444 D. L. McMinn, M. M. Greenberg, *Tetrahedron* **1996**, *52*, 3827–3840.

445 L. E. Canne, R. L. Winston, S. B. H. Kent, *Tetrahedron Lett.* **1997**, *38*, 3361–3364.

446 F. J. Wang, J. R. Hauske, *Tetrahedron Lett.* **1997**, *38*, 6529–6532.

447 M. F. Gordeev, G. W. Luehr, H. C. Hui, E. M. Gordon, D. V. Patel, *Tetrahedron* **1998**, *54*, 15879–15890.

448 K. H. Bleicher, J. R. Wareing, *Tetrahedron Lett.* **1998**, *39*, 4587–4590.

449 J. H. Kirchhoff, S. Bräse, D. Enders, *J. Comb. Chem.* **2001**, *3*, 71–77.

450 P. S. Furth, M. S. Reitman, R. Gentles, A. F. Cook, *Tetrahedron Lett.* **1997**, *38*, 6643–6646.

451 X. Chen, I. A. McDonald, B. Munoz, *Tetrahedron Lett.* **1998**, *39*, 217–220.

452 C. X. Chen, B. Munoz, *Tetrahedron Lett.* **1998**, *39*, 3401–3404.

453 M. F. Gordeev, D. V. Patel, E. M. Gordon, *J. Org. Chem.* **1996**, *61*, 924–928.

454 J. F. Tolborg, K. J. Jensen, *Chem. Commun.* **2000**, 147–148.

455 J. M. Smith, V. Krchnák, *Tetrahedron Lett.* **1999**, *40*, 7633–7636.

456 J. Kraxner, M. Arlt, P. Gmeiner, *Synlett* **2000**, 125–127.

457 A. L. Smith, G. I. Stevenson, C. J. Swain, J. L. Castro, *Tetrahedron Lett.* **1998**, *39*, 8317–8320.

458 F. E. K. Kroll, R. Morphy, D. Rees, D. Gani, *Tetrahedron Lett.* **1997**, *38*, 8573–8576.

459 M. Wang, S. Bräse, unpublished.

460 P. Blaney, R. Grigg, Z. Rankovic, M. Thoroughgood, *Tetrahedron Lett.* **2000**, *41*, 6635–6638.

461 P. Blaney, R. Grigg, Z. Rankovic, M. Thoroughgood, *Tetrahedron Lett.* **2000**, *41*, 6639–6642.

462 C. Y. Ho, M. J. Kukla, *Tetrahedron Lett.* **1997**, *38*, 2799–2802.

463 J. J. N. Veerman, F. P. J. T. Rutjes, J. H. van Maarseveen, H. Hiemstra, *Tetrahedron Lett.* **1999**, *40*, 6079–6082.

464 E. W. Baxter, J. K. Rueter, S. O. Nortey, A. B. Reitz, *Tetrahedron Lett.* **1998**, *39*, 979–982.

465 T. Takahashi, S. Tomida, H. Inoue, T. Doi, *Synlett* **1998**, 1261–1263.

466 T. Dahmen, S. Bräse, unpublished.

467 M. Lormann, S. Bräse, H. Vogt in: Proceedings of ECSOC-4, The Third International Electronic Conference on Synthetic Organic Chemistry, http://www.mdpi.org/ecsoc-4.htm, Scope and Limitation of a Tin Promoted Amidation on Solid-phase: A New Monitoring for the T1 Triazene Linker, September 1–30, 2000.

468 D. Obrecht, C. Abrecht, A. Grieder, J. M. Villalgordo, *Helv. Chim. Acta* **1997**, *80*, 65–72.

469 P. J. Das, S. Khound, J. Dutta, *Indian J. Chem. Sect. B* **1995**, *34*, 161–163.

470 S. Bräse, C. Gil, K. Knepper, *Bioorg. Med. Chem.* **2001**, submitted.

471 D. TUMELTY, K. CAO, C. P. HOLMES, *Org. Lett.* **2001**, *3*, 83–86.

472 S. KOBAYASHI, R. AKIYAMA, *Tetrahedron Lett.* **1998**, *39*, 9211–9214.

473 S. D. LEPORE, M. R. WILEY, *J. Org. Chem.* **1999**, *64*, 4547–4550.

474 Y. H. HU, S. BAUDART, J. A. PORCO, *J. Org. Chem.* **1999**, *64*, 1049–1051.

475 C. T. BUI, A. M. BRAY, T. NGUYEN, F. ERCOLE, F. RASOUL, W. SAMPSON, N. J. MAEJI, *J. Pept. Sci.* **2000**, *6*, 49–56.

476 S. R. KLOPFENSTEIN, J. J. CHEN, A. GOLEBIOWSKI, M. LI, S. X. PENG, X. SHAO, *Tetrahedron Lett.* **2000**, *41*, 4835–4839.

477 S. B. KATTI, *J. Chem. Soc., Chem. Commun.* **1992**, 843–844.

478 W. F. DEGRADO, E. T. KAISER, *J. Org. Chem.* **1980**, *45*, 1295–1300.

479 F. S. TJOENG, G. A. HEAVNER, *J. Org. Chem.* **1983**, *48*, 355–359.

480 F. ALBERICIO, E. GIRALT, R. ERITJA, *Tetrahedron Lett.* **1991**, *32*, 1515–1518.

481 C. BLACKBURN, A. PINGALI, T. KEHOE, L. W. HERMAN, H. Q. WANG, S. A. KATES, *Bioorg. Med. Chem. Lett.* **1997**, *7*, 823–826.

482 W. R. LI, N. M. HSU, H. H. CHOU, S. T. LIN, Y. S. LIN, *Chem. Commun.* **2000**, 401–402.

483 C. ALONSO, M. H. NANTZ, M. J. KURTH, *Tetrahedron Lett.* **2000**, *41*, 5617–5622.

484 A. N. SEMENOV, K. Y. GORDEEV, *Int. J. Pept. Protein Res.* **1995**, *45*, 303–304.

485 M. ROTTLÄNDER, P. KNOCHEL, *Synlett* **1997**, 1084–1086.

486 R. F. W. JACKSON, L. J. OATES, M. H. BLOCK, *Chem. Commun.* **2000**, 1401–1402.

487 Y. HAN, A. ROY, A. GIROUX, *Tetrahedron Lett.* **2000**, *41*, 5447–5451.

488 F. HOMSI, K. NOZAKI, T. HIYAMA, *Tetrahedron Lett.* **2000**, *41*, 5869–5872.

489 S. MA, D. DUAN, Z. SHI, *Org. Lett.* **2000**, *2*, 1419–1422.

490 D. H. KO, D. J. KIM, C. S. LYU, I. K. MIN, H. S. MOON, *Tetrahedron Lett.* **1998**, *39*, 297–300.

491 L. E. CANNE, S. M. WALKER, S. B. H. KENT, *Tetrahedron Lett.* **1995**, *36*, 1217–1220.

492 M. A. FINDEIS, E. T. KAISER, *Tetrahedron Lett.* **1993**, *34*, 1269–1270.

493 D. R. BARN, M. J. R., D. C. REES, *Tetrahedron Lett.* **1996**, *37*, 3213–3216.

494 Y. WANG, T.-N. HUANG, *Tetrahedron Lett.* **1999**, *40*, 5837–5840.

495 D. M. KETCHA, L. J. WILSON, D. E. PORTLOCK, *Tetrahedron Lett.* **2000**, *41*, 6253–6257.

496 E. NICOLAS, M. PUJADES, J. BACARDIT, E. GIRALT, F. ALBERICIO, *Tetrahedron Lett.* **1997**, *38*, 2317–2320.

497 J. D. WADE, M. N. MATHIEU, M. MACRIS, G. W. TREGEAR, *Lett. Pept. Sci.* **2000**, *7*, 107–112.

498 J. HOWE, M. QUIBELL, T. JOHNSON, *Tetrahedron Lett.* **2000**, *41*, 3997–4001.

499 J. ALSINA, K. J. JENSEN, F. ALBERICIO, G. BARANY, *Chem. Eur. J.* **1999**, *5*, 2787–2795.

500 J. ALSINA, T. S. YOKUM, F. ALBERICIO, G. BARANY, *Tetrahedron Lett.* **2000**, *41*, 7277–7280.

501 F. GUILLAUMIE, J. C. KAPPEL, N. M. KELLY, G. BARANY, K. J. JENSEN, *Tetrahedron Lett.* **2000**, *41*, 6131–6135.

502 J. A. OLSEN, K. J. JENSEN, J. NIELSEN, *J. Comb. Chem.* **2000**, *2*, 143–150.

503 G. T. BOURNE, W. D. F. MEUTERMANS, M. L. SMYTHE, *Tetrahedron Lett.* **1999**, *40*, 7271–7274.

504 Y. AOKI, S. KOBAYASHI, *J. Comb. Chem.* **1999**, *1*, 371–372.

505 G. T. BOURNE, W. D. F. MEUTERMANS, P. F. ALEWOOD, R. P. McGEARY, M. SCANLON, A. A. WATSON, M. L. SMYTHE, *J. Org. Chem.* **1999**, *64*, 3095–3101.

506 W. D. F. MEUTERMANS, S. W. GOLDING, G. T. BOURNE, L. P. MIRANDA, M. J. DOOLEY, P. F. ALEWOOD, M. L. SMYTHE, *J. Am. Chem. Soc.* **1999**, *121*, 9790–9786.

507 E. G. BROWN, J. M. NUSS, *Tetrahedron Lett.* **1997**, *38*, 8457–8460.

508 K. H. GORDON, S. BALASUBRAMANIAN, *Org. Lett.* **2001**, *3*, 53–56.

509 R. M. MYERS, S. P. LANGSTON, S. P. CONWAY, C. ABELL, *Org. Lett.* **2000**, *2*, 1349–1352.

510 J. KÖBBERLING, Dissertation, RWTH Aachen **2001**.

511 M. A. SCIALDONE, *Tetrahedron Lett.* **1996**, *37*, 8141–8144.

512 M. A. SCIALDONE, S. W. SHUEY, P. SOPER, Y. HAMURO, D. M. BURNS, *J. Org. Chem.* **1998**, *63*, 4802–4807.

513 E. E. SWAYZE, *Tetrahedron Lett.* **1997**, *38*, 8643–8646.

514 S. DAHMEN, S. BRÄSE, unpublished.

515 K. BURGESS, J. CHEN in: Solid-Phase Organic Synthesis. BURGESS, K. (ed.), John Wiley & Sons, New York 2000, S. 1–23.

516 J. CHEN, M. PATTARAWARAPAN, A. J. ZHANG, K. BURGESS, *J. Comb. Chem.* **2000**, *2*, 276–281.

517 D. S. DODD, O. B. WALLACE, *Tetrahedron Lett.* **1998**, *39*, 5701–5704.

518 L. J. WILSON, S. R. KLOPFENSTEIN, M. LI, *Tetrahedron Lett.* **1999**, *40*, 3999–4002.

519 M. PÁTEK, M. SMRČINA, E. NAKANISHI, H. IZAWA, *J. Comb. Chem.* **2000**, *2*, 370–377.

520 J. A. JOSEY, C. A. TARLTON, C. E. PAYNE, *Tetrahedron Lett.* **1998**, *39*, 5899–5902.

521 M. BONNAT, M. BRADLEY, J. D. KILBURN, *Tetrahedron Lett.* **1996**, *37*, 5409–5412.

522 H. M. ZHONG, M. N. GRECO, B. E. MARYANOFF, *J. Org. Chem.* **1997**, *62*, 9326–9330.

523 L. GOMEZ, F. GELLIBERT, A. WAGNER, C. MIOSKOWSKI, *Chem. Eur. J.* **2000**, *6*, 4016–4020.

524 D. P. ARYA, T. C. BRUICE, *J. Am. Chem. Soc.* **1998**, *120*, 6619–6620.

525 P. ROUSSEL, M. BRADLEY, I. MATTHEWS, P. KANE, *Tetrahedron Lett.* **1997**, *38*, 4861–4864.

526 R. GRIGG, J. P. MAJOR, F. M. MARTIN, M. WHITTAKER, *Tetrahedron Lett.* **1999**, *40*, 7709–7711.

527 L. S. RICHTER, M. C. DESAI, *Tetrahedron Lett.* **1997**, *38*, 321–322.

528 C. D. FLOYD, C. N. LEWIS, S. R. PATEL, M. WHITTAKER, *Tetrahedron Lett.* **1996**, *37*, 8045–8048.

529 S. L. MELLOR, W. C. CHAN, *Chem. Commun.* **1998**, 2005–2006.

530 S. L. M. C. MELLOR, W. C. CHAN, *Tetrahedron Lett.* **1997**, *38*, 3311–3314.

531 K. NGU, D. PATEL, *J. Org. Chem.* **1997**, *62*, 7088–7089.

532 A. GOLEBIOWSKI, S. KLOPFENSTEIN, *Tetrahedron Lett.* **1998**, *39*, 3397–3400.

533 S. HANESSIAN, F. XIE, *Tetrahedron Lett.* **1998**, *39*, 737–740.

534 U. BAUER, W. B. HO, A. M. P. KOSKINEN, *Tetrahedron Lett.* **1997**, *38*, 7233–7236.

535 G. E. ATKINSON, P. M. FISCHER, W. C. CHAN, *J. Org. Chem.* **2000**, *65*, 5048–5056.

536 M. F. GORDEEV, H. C. HUI, E. M. GORDON, D. V. PATEL, *Tetrahedron Lett.* **1997**, *38*, 1729–1732.

537 S. M. DANKWARDT, *Synlett* **1998**, 761.

538 J. NIELSEN, P. H. RASMUSSEN, *Tetrahedron Lett.* **1996**, *37*, 3351–3354.

539 D. W. EMERSON, R. R. EMERSON, S. C. JOSHI, E. M. SORENSEN, J. E. TUREK, *J. Org. Chem.* **1979**, *44*, 4634–4640.

540 A. M. MURPHY, R. DAGNINO, JR, P. L. VALLAR, A. J. TRIPPE, S. L. SHERMAN, R. H. LUMPKIN, S. Y. TAMURA, T. R. WEBB, *J. Am. Chem. Soc.* **1992**, *114*, 3156–3157.

541 T. Q. DINH, R. W. ARMSTRONG, *Tetrahedron Lett.* **1996**, *37*, 1161–1164.

542 J. A. FEHRENTZ, M. PARIS, A. HEITZ, J. VELEK, C. F. LIU, F. WINTERNITZ, J. MARTINEZ, *Tetrahedron Lett.* **1995**, *36*, 7871–7874.

543 S. KOBAYASHI, I. HACHIYA, M. YASUDA, *Tetrahedron Lett.* **1996**, *37*, 5569–5572.

544 I. VLATTAS, J. DELLUREFICIO, R. DUNN, I. I. SYTWU, J. STANTON, *Tetrahedron Lett.* **1997**, *38*, 7321–7324.

545 B. J. HALL, J. D. SUTHERLAND, *Tetrahedron Lett.* **1998**, *39*, 6593–6596.

546 C. POTHION, M. PARIS, A. HEITZ, L. ROCHEBLAVE, F. ROUCH, J. A. FEHRENTZ, J. MARTINEZ, *Tetrahedron Lett.* **1997**, *38*, 7749–7752.

547 D. R. CODY, S. H. DEWITT, J. C. HODGES, B. D. ROTH, M. C. SCHROEDER, C. J. STANKOVIC, W. H. MOOS, M. R. PAVIA, J. S. KIELY (Warner-Lambert Co.), Patent Nos WO PCT 9408711; US 958,383; CAN 122:106536, **1994**.

548 S. CHAMOIN, S. HOULDSWORTH, C. G.

KRUSE, W. I. BAKKER, V. SNIECKUS, *Tetrahedron Lett.* **1998**, *39*, 4179–4182.

549 M. J. FARRALL, J. M. J. FRECHET, *J. Org. Chem.* **1976**, *41*, 3877–3882.

550 A. ISHII, H. HOJO, A. KOBAYASHI, K. NAKAMURA, Y. NAKAHARA, Y. ITO, Y. NAKAHARA, *Tetrahedron* **2000**, *56*, 6235–6243.

551 J. SWISTOK, J. W. TILLEY, W. DANHOHO, R. WAGNER, K. MULKERINS, *Tetrahedron Lett.* **1989**, *30*, 5045–5048.

552 W. NEUGEBAUER, E. ESCHER, *Helv. Chim. Acta* **1989**, *72*, 1319–1323.

553 J. W. APSIMON, D. M. DIXIT, *Synth. Commun.* **1982**, *12*, 113–116.

554 J. M. J. FRÉCHET, *Tetrahedron* **1981**, 663–683.

555 B. MESEGUER, D. ALONSO-DIAZ, N. GRIEBENOW, T. HERGET, H. WALDMANN, *Angew. Chem. Int. Ed.* **1999**, *38*, 2902–2906; *Angew. Chem.* **1999**, *111*, 3083–3087.

556 L. F. TIETZE, T. HIPPE, A. STEINMETZ, *Chem. Commun.* **1998**, 793–794.

557 Y. L. CHOU, M. M. MORRISSEY, R. MOHAN, *Tetrahedron Lett.* **1998**, *39*, 757–760.

558 X. BEEBE, N. E. SCHORE, M. J. KURTH, *J. Am. Chem. Soc.* **1992**, *114*, 10061–10062.

559 X. BEEBE, N. E. SCHORE, M. J. KURTH, *J. Org. Chem.* **1995**, *60*, 4196–4203.

560 B. C. HAMPER, D. R. DUKESHERER, M. S. SOUTH, *Tetrahedron Lett.* **1996**, *37*, 3671–3674.

561 B. B. SHANKAR, D. Y. YANG, S. GIRTON, A. K. GANGULY, *Tetrahedron Lett.* **1998**, *39*, 2447–2448.

562 K. J. ELGIE, M. SCOBIE, R. W. BOYLE, *Tetrahedron Lett.* **2000**, *41*, 2753–2757.

563 T. TAKAHASHI, S. EBATA, T. DOI, *Tetrahedron Lett.* **1998**, *39*, 1369–1372.

564 S. SATO, Y. KUBOTA, H. KUMAGAI, T. KUMAZAWA, S. MATSUBA, J. ONODERA, M. SUZUKI, *Heterocycles* **2000**, *53*, 1523–1532.

565 H. V. MEYERS, G. J. DILLEY, T. L. DURGIN, T. S. POWERS, N. A. WINSSINGER, H. ZHU, M. R. PAVIA, *Mol. Diversity* **1995**, *1*, 13–20.

566 T. R. LEE, D. S. LAWRENCE, *J. Med. Chem.* **1999**, *42*, 784–787.

567 W. K. D. BRILL, E. SCHMIDT, R. A. TOMMASI, *Synlett* **1998**, 906–908.

568 G. L. STAHL, R. WALTER, C. W. SMITH, *J. Am. Chem. Soc.* **1979**, *101*, 5383–5394.

569 J. M. CAMPAGNE, J. COSTE, P. JOUIN, *Tetrahedron Lett.* **1995**, *36*, 2079–2082.

570 J. MERY, C. GRANIER, M. JUIN, J. BRUGIDUO, *Int. J. Pept. Protein Res.* **1995**, *42*, 44–52.

571 M. F. GORDEEV, E. M. GORDON, D. V. PATEL, *J. Org. Chem.* **1997**, *62*, 8177–8181.

572 S. M. DANKWARDT, D. B. SMITH, J. A. PORCO, C. H. NGUYEN, *Synlett* **1997**, 854–856.

573 B. RAJU, T. P. KOGAN, *Tetrahedron Lett.* **1997**, *38*, 4965–4968.

574 B. RAJU, T. P. KOGAN, *Tetrahedron Lett.* **1997**, *38*, 3373–3376.

575 A. HARI, B. L. MILLER, *Org. Lett.* **1999**, *1*, 2109–2111.

576 K. SCHIEMANN, H. D. H. SHOWALTER, *J. Org. Chem.* **1999**, *64*, 4972–4975.

577 D. P. CURRAN, M. HOSHINO, *J. Org. Chem.* **1996**, *61*, 6480–6481.

578 D. CRAIG, M. J. ROBSON, S. J. SHAW, *Synlett* **1998**, 1381–1383.

579 B. CHENERA (Smithkline Beecham Corporation), Patent No. WO PCT. 98/17695, **1995**.

580 K. C. NICOLAOU, D. VOURLOUMIS, T. H. LI, J. PASTOR, N. WINSSINGER, Y. HE, S. NINKOVIC, F. SARABIA, H. VALLBERG, F. ROSCHANGAR, N. P. KING, M. R. V. FINLAY, P. GIANNAKAKOU, P. VERDIER-PINARD, E. HAMEL, *Angew. Chem. Int. Ed.* **1997**, *36*, 2097–2103.

581 C. E. J. HALM, M. J. KURTH, *Tetrahedron Lett.* **1997**, *38*, 7709–7712.

582 M. YAMADA, T. MIYAJIMA, H. HORIKAWA, *Tetrahedron Lett.* **1998**, *39*, 289–292.

583 A. BARCO, S. BENETTI, C. DERISI, P. MARCHETTI, G. P. POLLINI, V. ZANIRATO, *Tetrahedron Lett.* **1998**, *39*, 7591–7594.

584 D. A. NUGIEL, L. A. M. CORNELIUS, J. W. CORBETT, *J. Org. Chem.* **1997**, *62*, 201–203.

585 G. HUM, J. GRZYB, S. D. TAYLOR, *J. Comb. Chem.* **2000**, *2*, 234–242.

586 X. D. Cao, A. M. M. Mjalli, *Tetrahedron Lett.* **1996**, *37*, 6073–6076.

587 C. Z. Zhang, A. M. M. Mjalli, *Tetrahedron Lett.* **1996**, *37*, 5457–5460.

588 C. A. Metcalf, C. B. Vu, R. Sundaramoorthi, V. A. Jacobsen, E. A. Laborde, J. Green, Y. Green, K. J. Macek, T. J. Merry, S. G. Pradeepan, M. Uesugi, V. M. Varkhedkar, D. A. Holt, *Tetrahedron Lett.* **1998**, *39*, 3435–3438.

589 R. M. Dunsdon, J. R. Greening, P. S. Jones, S. Jordan, F. X. Wilson, *Bioorg. Med. Chem. Lett.* **2000**, *10*, 1577–1579.

5
Encoding Technologies

Thomas Krämer, Valery V. Antonenko, Reza Mortezaei,
Nicolay V. Kulikov

5.1
Introduction

With the application of combinatorial chemistry methods large collections of individual chemical entities (chemical libraries) have been synthesized that require some sort of organization to be used and handled efficiently. Conventional and convenient tracking methods such as writing codes on reaction flasks and vials with waterproof pens were no longer appropriate. Assuming that the solid- and solution-phase reactions used to synthesize a combinatorial library member are successful, then knowledge of the specific reaction sequence is equivalent to knowing the member's chemical identity. Because the determination of chemical identity is typically not automatable and requires a substantial amount of material, schemes that encode a member's reaction history onto the synthesis platform are of value [1, 2].

One of the most obvious methods for the encoding of chemical compounds is spatial (or positional) encoding. The structure of a compound or its chemical history is encoded by the position of the corresponding reaction vessel in a spatially fixed, two-dimensional matrix. However, there are a number of options for encoding a chemical structure: graphical encoding methods; the use of chemical tags consisting of peptides, oligonucleotides or aromatic compounds; spectrometric encoding; or radiofrequency encoding, which is the most advanced technique.

Encoding is a prerequisite for the efficient handling of compound libraries and offers two major advantages: establishing a relational nomenclature and providing automated compound and data handling (if the tags were machine-readable). Readability of tags by technical devices is of major importance since manual handling of a large number of chemical entities is either very tedious or simply not possible.

Post-synthesis encoding is successfully being applied in the compound repositories of major pharmaceutical companies where hundreds of thousands of archived test samples are stored in bar-coded vials. In this case the bar code maintains the relation between the vial and the corresponding compound data [structure, molecular weight (MW), amount, etc.] and offers robotic vial handling. However, for combinatorial library syntheses before or during synthesis encoding (i.e. encoding compound libraries at the time of their synthesis, rather than afterwards) is far more efficient. Again, it is the synthetic history of a compound that is encoded rather than its chemical identity.

Furthermore, a major requirement of encoding during synthesis is that the tag must invariably be connected to the compound itself (like a protecting group) or to the polymer on which the compound is being synthesized. Consequently, the tags need to be chemically inert under the reaction conditions used to synthesize the library compounds, i.e. tag chemistry and library chemistry have to be orthogonal. Since some compounds are screened for biological activity in an on-bead screening assay (with the tag still attached), the tags should also not interfere with the biological properties of the library compound. Therefore, it is essential that tags have to be chemically and biologically inert.

In the following sections a review of the existing methods for encoding combinatorial libraries will be given that divide encoding strategies into two categories: chemical encoding and non-chemical (physical) encoding.

5.2
Chemical Encoding Methods

One of the solutions to the structure elucidation problem utilizes a number of chemical tags that can unambiguously identify the chemical entities with which they are associated. The tags should be incorporated on the same bead on which the compound they encode is synthesized. Therefore, the tagging process should not interfere with the synthesis. The tags should not occupy much of the bead capacity. It should be possible to cleave the synthesized compounds from the bead selectively in the presence of the encoding elements. The decoding process should be quick and reliable, while the chemical nature of the tags should permit their rapid determination in small quantities using conventional analytical technologies. Time-consuming methods may defeat the purpose of the encoding.

5.2.1
Oligonucleotide Tags

In 1992, Brenner and Lerner [3] were the first to suggest in the literature the concept of chemical encoding when they proposed a method for producing an oligonucleotide-encoded peptide library. Addition of each amino acid to the polymeric bead is followed by the attachment of two preselected oligonucleotides to a different "tag site" on the same bead. Each base pair encodes one, and only one, amino acid. In the "split-and-pool" strategy, the oligonucleotide chain grows in parallel with the peptide chain; thus, each unique peptide sequence is encoded by a unique oligonucleotide. The decoding process starts by amplifying the encoding oligonucleotide using the polymerase chain reaction (PCR). To reveal the primary structure, the encoding oligonucleotide is sequenced, unambiguously unveiling the amino acid sequence of the satellite peptide.

Oligonucleotide encoding was reported by Nielsen and colleagues [4], who used the construct presented in Scheme 5.1. α-Amino functionality of the serine residue was used as an attachment point for the growing peptide chain, while the β-hydroxyl of the side-chain was used for the encoding oligonucleotide assembly. Therefore, the tag-to-peptide ratio used in this work was 1:1. Synthesis was carried out on controlled porous glass (CPG) beads.

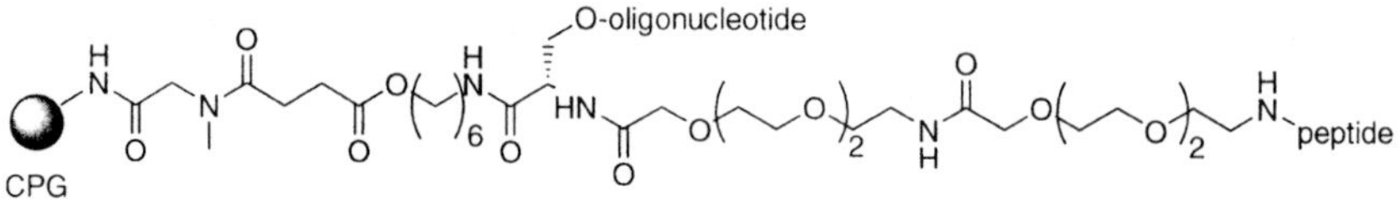

Scheme 5.1. Nielsen construct for oligonucleotide encoding [4].

Gallop and coworkers from the Affymax Research Institute used the oligo-nucleotide tagging strategy for the encoding of a library containing 823,543 hepta-peptides [5]. The synthesis was carried out on 10-μm-diameter monodisperse beads. The beads were used in sufficient number to synthesize 200 copies of each sequence in the library. The beads were made of polystyrene crosslinked with di-vinylbenzene and derivated with a 1,12-diaminododecane linker. The capacity of the resin was 10 μmol g^{-1}, corresponding to 10 fmol per bead. The resin was fur-ther functionalized with two chemical linkers: a mixture of 9-fluorenylmethoxy-carbonyl (Fmoc)–Thr(t-Bu)–OBt and succinimidyl activated ester of 4-O-(dimeth-oxytrityl) oxobutyrate. The former is introduced to support the growing peptide chain; the latter to support the synthesis of the encoding oligonucleotide. The re-sulting construct contained a 20:1 ratio of peptides to oligonucleotides on the bead. The structure of the construct is shown in Scheme 5.2.

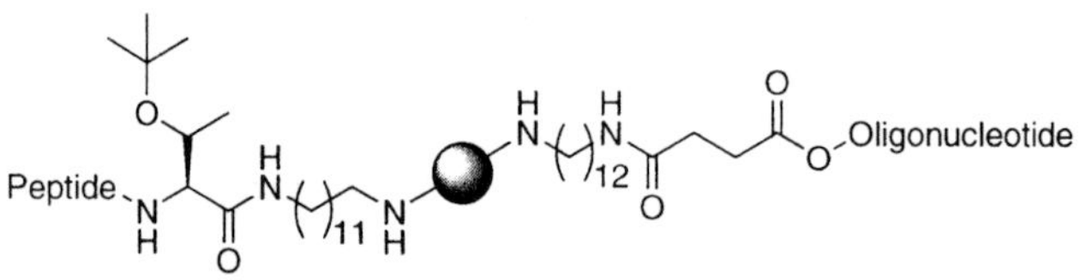

Scheme 5.2. Affymax's construct for oligonucleotide encoding [5].

The peptides were synthesized using Fmoc-protecting groups for α-amino func-tionalities and t-butyl-type protection for the side-chains of amino acids. Oligo-nucleotides were assembled using dimethoxytrityl-protected 3′-O-methyl-N,N-diisopropyl phosphorimidates. To protect the encoded oligonucleotides from the depurination side reaction upon trifluoracetic acid (TFA)-mediated side-chain de-protection of t-butyl-type protecting groups, the susceptible 2′-deoxyguanosine (dG) unit was not used. 2′-Deoxyadenosine (dA) was substituted with 7-deaza-2 deoxyadenosine (c7dA), which is stable to TFA treatment. To make the encoding compatible with the peptide synthesis, phosphorimidites carried 3'-O-methyl protection, which is, contrary to the β-cyanomethyl protective group, stable to piperidine used for Fmoc-deprotection.

The library was built to construct 7^7 analogs of the C-terminal heptapeptide RQFKVVT of opioid peptide dynorphin B. Amino acids Arg, Gln, Phe, Lys, Val, Thr and D-Val were used in a seven-step "split-and-pool" synthesis. Each amino acid was encoded by a unique dinucleotide. Thus, each heptapeptide was synthesized with the corresponding sequence of 14 nucleotide bases. At the end of the synthesis, the encoding sequences were framed by the degenerate DNA PCR primer sequence (55 bases). The library was screened against the fluorescently labeled anti-dynorphin B antibody, D32.39, with the fully deprotected peptides attached to the beads. A sample containing a sufficient number of beads to represent all synthesized sequences with high statistical certainty was analyzed. The beads containing binding sequences became fluorescent. The top 0.17% of the beads with the most intense fluorescence was collected using a fluorescence-activated cell-sorting (FACS) instrument. The collected beads were subjected to PCR amplification and subsequent sequencing of the encoding oligonucleotides. The sequences of the binding peptides were revealed from their primary structures.

Oligonucleotide tagging was developed and used successfully for the synthesis of peptides; many methodological issues, related mostly to the chemical compatibility of the approach, were solved. However, the compatibility of the encoding strategy with the synthesis of other classes of organic molecules remains limited.

5.2.2
Peptide Tags

The technology for automated Edman degradation of peptides to determine their sequence is well developed. Therefore, peptides can be used similarly to oligonucleotides as encoding molecules. Several groups have reported using peptide encoding [6, 7]. Unfortunately, applications of the peptide encoding are very limited because of severe restrictions imposed on the scope of chemical methods that are compatible with this approach. With the development of alternative encoding strategies, peptide encoding has very limited applications in the modern approaches to the synthesis of small organic molecules.

5.2.3
Haloaromatic Binary Coding

Several laboratories have worked on the development of robust encoding strategies that can be used for combinatorial synthesis of pharmacologically appealing chemical libraries. In 1993, Still and colleagues were the first to report binary encoding [8] using chromatographically resolvable haloaromatic reagents as tagging molecules. The tags are attached to the beads during library synthesis in a binary coding strategy intended to keep a molecular record of all chemical transformations to which the beads are subjected. The tags are incorporated via amide bond forma-

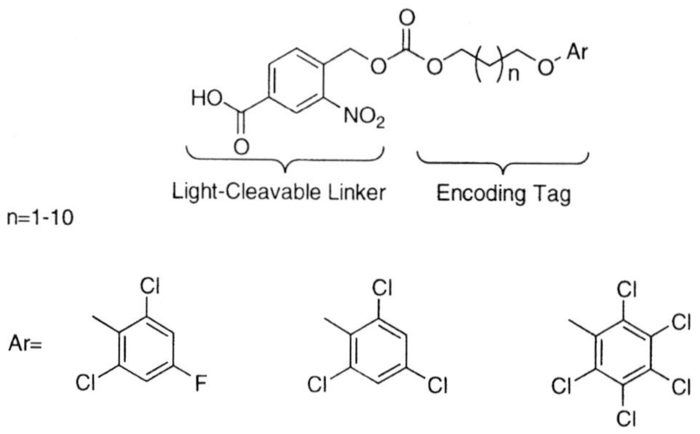

Scheme 5.3. Haloaromatic tags on a photolinker [8].

tion as part of a construct with a photolabile linker (Scheme 5.3) at the expense of the ligand synthesis sites. Each compound in the library is encoded by a limited set of tags. The presence, as well as the absence, of each member of the set carries information about the specific encoded structure. In a binary code, a set of n tags can encode $2^n - 1$ different structures. For example, 20 tags can encode $2^{20} - 1 = 1,048,575$ different library members. After being released from the beads by photolysis, the tags are detected by capillary gas chromatography using electron capture detection (ECGC), a detection method that is particularly sensitive to heavily chlorinated aromatics. The haloaromatic compounds were selected to ensure reliable and reproducible separation by ECGC, which is capable of detecting sub-picomolar amounts of the tags. Consequently, the beads can be tagged at only 0.5–1% of the resin loading (0.5–1 pmol per bead) without detectable interference with the library synthesis.

Later, Still and coworkers modified the original strategy by developing a new type of tagging reagent TnC [9], presented in Scheme 5.4. A derivative of vanillic acid (3-methoxy-4-hydroxybenzoic acid) was chosen as a linker. Synthesis of the tagging reagent TnC begins with a Mitsunobu reaction of a tag alcohol Tn with methyl vanillate, followed by LiOH hydrolysis of the methyl ester. This produces free acid TnA, which is converted to acid chloride TnB. Excess of diazomethane converts acid chloride TnB into the tagging diazoketone TnC. In the presence of rhodium reagents $Rh_2(OAc)_4$ or $Rh_2(O_2CCF_3)_4$ TnC forms an acylcarbene, which rapidly and cleanly reacts with benzene, forming a derivative of cycloheptatriene Tn1. Benzene was used as a soluble analog of polystyrene resin. Derivatives of TnC containing different numbers of methylene groups n have been prepared. The researchers found that diazoketones TnC are stable solids and can be stored at room temperature for months. They react easily with polystyrene resin, providing a means for binary encoding. At the decoding stage, the tags are oxidatively cleaved by ceric ammonium nitrate and analyzed by ECGC.

The acylcarbene tagging strategy does not require any specific functional group for tag attachment, and the tags and linkers are generally compatible with a wide

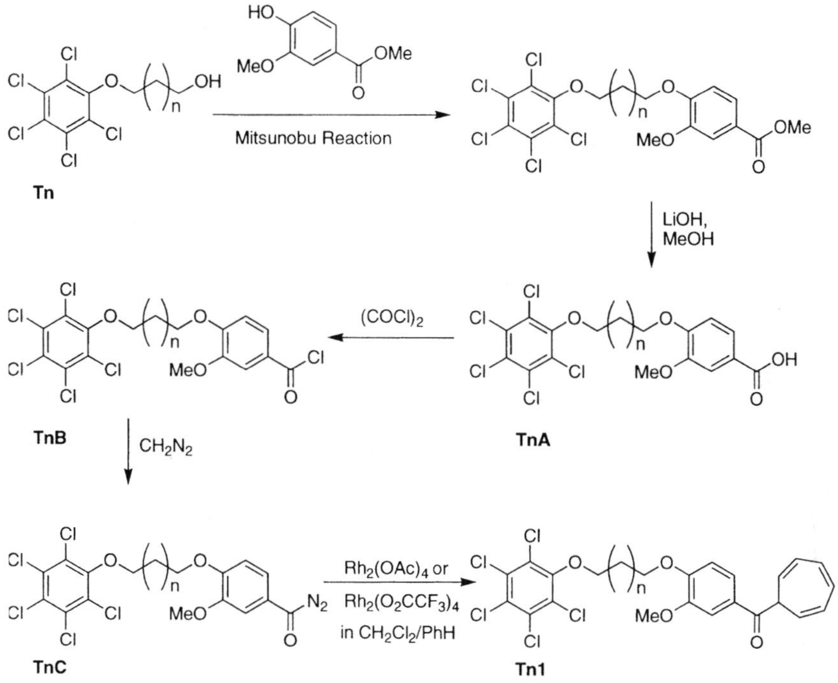

Scheme 5.4. Acylcarbene-generating tagging reagents [9].

range of chemical reactions. The acylcarbenes can unselectively add to some of the synthesized compounds. However, the bulk of the library is represented by the polymeric support, which accepts the major portion of the tag molecules. Because tags are added at molar levels corresponding to 1% or less of the library members, interference with the synthesis is minimal [10].

5.2.4
Secondary Amine Binary Coding

In 1996, Gallop and coworkers developed another robust encoding strategy [11, 12], based on secondary amine tags, which are incorporated into a polyamide backbone. The secondary amine binary coding scheme utilizes an amine-based polymeric resin that is differentially functionalized with sites for both ligand synthesis and tag addition (Scheme 5.5). The ligand synthesis site is derivatized with a N-Fmoc-protected photocleavable linker group [13]. This linker allows for the release of the ligand from the resin by exposure to ultraviolet (UV) light. The amino group of the tag site is protected with orthogonal to the Fmoc group functionalities (e.g. Boc or Alloc). The tag site occupies only 10% of the total number of amino groups on the resin. Each 130-μm-diameter bead of the TentaGel S resin, which is recommended by the authors of this methodology, contains about 300 pmol of amino groups. Therefore, the theoretical yield of the ligand is 270 pmol per bead. Such

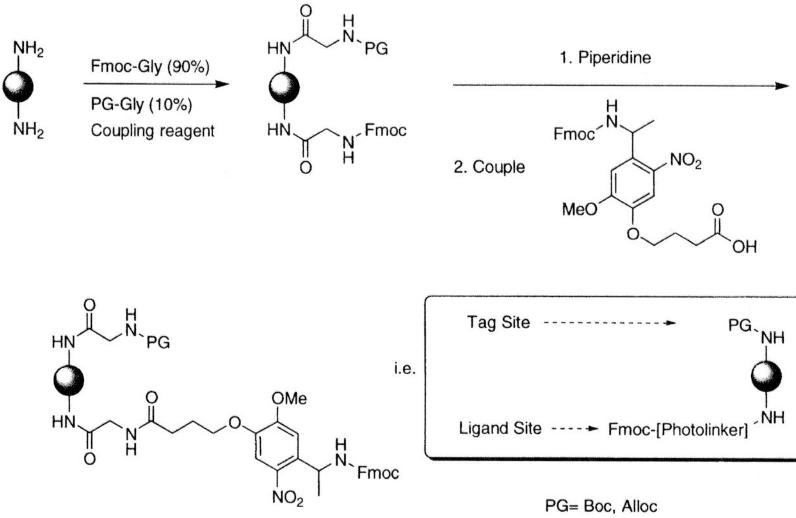

Scheme 5.5. Resin construct for secondary amine binary coding [11, 12].

quantities being photo-released in 100 μl of a solvent would be adequate for the concentration of the ligand in most biological assays.

The tags are a set of relatively hydrophobic amines, such as HN(Et)(Bu), HNMe(C$_6$H$_{13}$), HNBu$_2$, HNMe(C$_7$H$_{15}$), and HN[CH$_2$CH(Et)C$_4$H$_9$]$_2$. The set is selected to ensure reliable separation of the dansyl derivatives of the amines by reversed-phase high-performance liquid chromatography (HPLC). The tagging monomer units are synthesized by reaction of an N-protected iminodiacetic anhydride with a secondary amine from the set (Scheme 5.6).

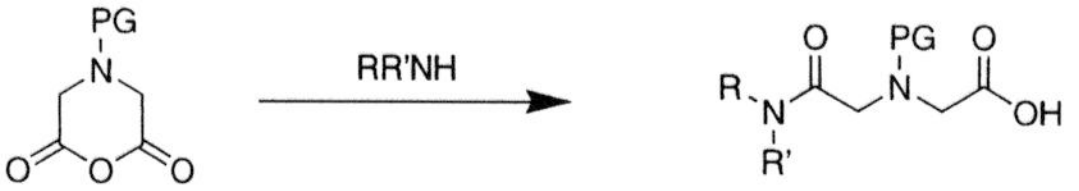

Scheme 5.6. Preparation of tag monomers for secondary amine coding.

The resulting N-protected N-[(dialkylcarbamoyl)methyl]glycines are assembled into binary mixtures, which are incorporated into the tag sites of the resin beads by using HATU or other peptide-coupling reagents (Scheme 5.7). Addition of each new building block at the ligand synthesis site in the course of a "split-and-pool" combinatorial synthesis is accompanied by the incorporation of the preselected mixture of the monomer units at the tag addition site. Selection of the protecting groups allows for the addition of the tags either before or after the addition of the building block to the ligand. The ability to choose different protecting groups for the N-protected tag monomers helps to resolve potential chemical compatibility issues.

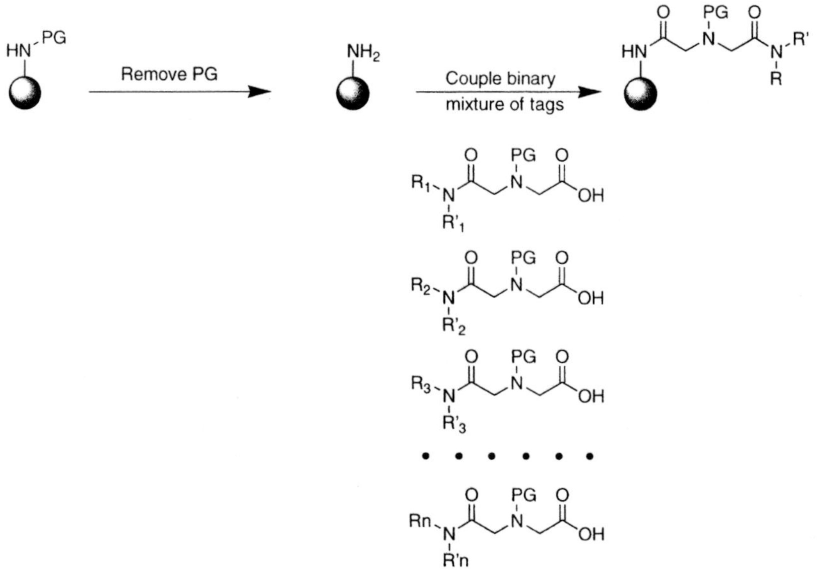

Scheme 5.7. Coupling of secondary amine tag constructs to solid support.

Upon completion of the library synthesis, each bead is distributed into a separate well of a microtiter plate. The ligands are released from the beads into the assay medium by exposure to UV light at 365 nm. The tag residues remain covalently attached to the beads. After screening, beads from the wells containing active compounds are collected for decoding. The decoding process is shown schematically in Scheme 5.8, and begins with the acid hydrolysis of the beads in 6 N hydrochloric acid. Under these conditions all amide bonds are hydrolyzed, releasing free secondary amine tags into the solution. After evaporation of the HCl, the amines are converted into the corresponding dansyl derivatives by treatment with dansyl chloride. Analysis of the resulting mixture of dansylated tags is carried out

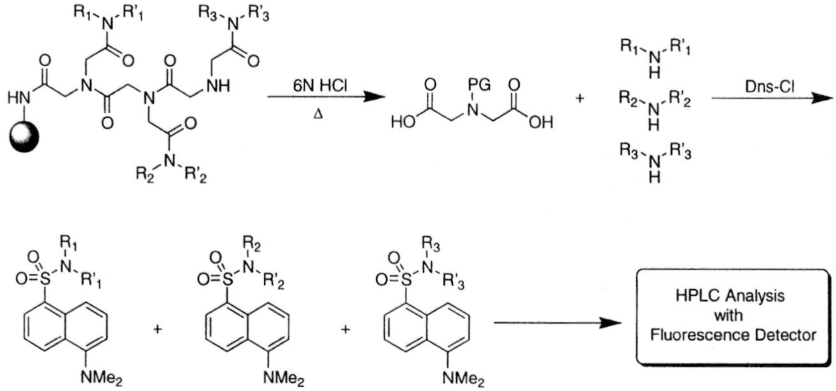

Scheme 5.8. The decoding process for secondary amine binary coding.

by reversed-phase HPLC on a microbore column. Fluorescence detection allows for the reliable analysis of 20–30 fmol of a dansylated tag. Only 2–5% of the entire dansylated hydrolyzate from a single bead is sufficient to obtain unambiguous results.

The secondary amine encoding method was used in the synthesis of a library of pyrrolidine-based inhibitors of angiotensin converting enzyme (ACE) via [2 + 3] cycloaddition [14].

5.2.5
Mass Encoding

As mentioned previously, conventional methods of characterization of chemical compounds cannot be applied to most combinatorial technologies owing to the insufficient quantities of the analytes. One fortunate exception is mass spectroscopy. Matrix-assisted laser desorption and electrospray ionization mass spectroscopy (MS) have been used [15, 16] to identify compounds synthesized on a single bead. For beads with diameters larger than 100 μm, a mass spectrum can be acquired in less than 1 min. In 1996, Geysen and colleagues from Glaxo-Wellcome proposed a novel encoding method [17], which takes advantage of modern MS techniques. In general, when recording the chemical history of a compound synthesized on a bead mass encoding incorporates stable isotopes that give distinct isotopic patterns in mass spectra.

In one of the proposed strategies, the resin is derivatized with a linker (Scheme 5.9) to which an MS code is attached. The mass of the coding block is designed to appear in a convenient region of the mass spectrum. As an example, the code can be a dipeptide built from the combinations of the natural amino acids glycine (Gly) and alanine (Ala), and their ^{13}C-labeled derivatives: [^{13}C]Gly, [^{13}C]$_2$Gly, [^{13}C]Ala, and [^{13}C]$_4$Ala. As an example, the authors present mass spectra recorded in the 295–330 MW range for the following ten mass codes: Gly–Gly, Gly–[^{13}C]Gly, [^{13}C]Gly–[^{13}C]Gly, Gly–[^{13}C]$_2$Gly, Gly–Ala, Gly–[^{13}C]Ala, [^{13}C]$_2$Gly–Ala, [^{13}C]$_2$Gly–[^{13}C]Ala, and[^{13}C]$_2$Gly–[^{13}C]$_4$Ala. After the code (Scheme 5.10), an additional linker is introduced; this should be orthogonal to the first one and is used for the release of the synthesized compound ABC. In a "split-and-pool" synthesis, the code defines the identity of the first building block A. The third building block C can be known from the final pool in which its addition was carried out. The identity of the building block B can be calculated from the molecular weight of the ligand and the molecular weight of monomers A and C.

For unambiguous results, the set of building blocks used at the second step of the library synthesis should not contain any compounds with the same mass as building block B.

Scheme 5.9. Resin construct for mass encoding.

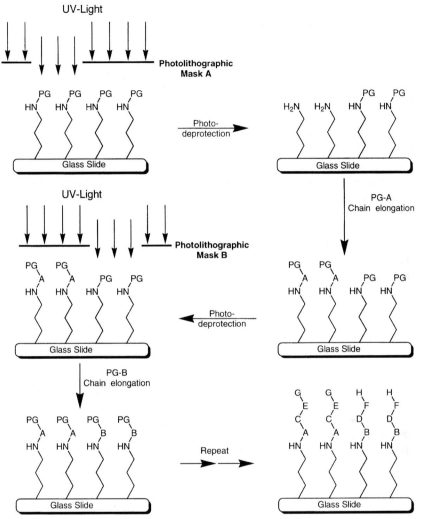

Scheme 5.10. Principle of light-directed, spatially addressable synthesis.

The code block can also consist of equimolar mixtures built from a set of isotopically labeled dipeptides: Gly–Gly, Gly–[^{13}C]Gly, Gly–[^{13}C]$_2$Gly, [^{13}C]Gly–[^{13}C]Gly, [^{13}C]Gly–[^{13}C]$_2$Gly, and [^{13}C]Gly–[^{13}C]$_2$Gly. These dipeptides incorporate equal to their content in the mixture, because isotopic labeling does not affect chemical reactivity. The resulting mass spectra of these mixture codes serve as "bar codes".

In the ratio-encoding strategy, the encoded information is derived from just two peaks in the mass spectrum. A mixture of a reagent, common to all members of the library, is prepared from different ratios of the isotopic isomers of the reagent. The resulting mass spectrum will reveal two distinct peaks corresponding to each of the isomers. The relative heights of the peaks provides the encoded information.

The mass encoding strategy was used by Wagner and coworkers [18] for the synthesis of a peptoid library. A set of 20 ^{14}N:^{15}N ratio-encoded imidazoles was synthesized "to investigate pharmacokinetic applications of isotopic labeling".

5.3
Non-Chemical Encoding Methods

Chemical methods of encoding combinatorial libraries can be used for synthesis on very small elements of solid supports (i.e. 10-μm-diameter beads). However, none of the procedures developed so far is totally compatible with all chemical transformations that might be necessary for the implementation of successful drug discovery programs. In principle, physical methods of encoding are fully compatible. Unfortunately, none of the conventionally used labeling techniques (bar codes, labels, etc.) is small enough to be used with resin beads. It is often necessary to obtain compounds via combinatorial chemistry approaches in substantially larger quantities than the single-bead approach is capable of. As an alternative, macroscopic pieces of a solid support can be used, the larger physical dimensions allowing for the application of more space-demanding encoding strategies.

5.3.1
Positional Encoding

Positional, or spatial, encoding directly links the reaction history of a compound to the particular location of the corresponding synthesis compartment (reactor) in a spatially arrayed, fixed matrix of reactors, i.e. the 8×12 two-dimensional (2D) matrix in a 96-well plate or special chip formats.

5.3.1.1 Light-directed Synthesis
In 1991, Fodor and colleagues combined photolithography, photochemistry, and solid-phase synthesis in a new technology [19–22] called light-directed, spatially addressable parallel synthesis or VLSIPS (very-large-scale immobilized polymer synthesis). The principal points of the technology are illustrated in Scheme 5.10.

The synthesis occurs on a flat glass surface modified with an appropriate linker (e.g. 3-aminopropyl-triethoxysilane) to allow for the covalent attachment of protected amino acids. The entire synthesis area of the slide is derivatized with a photolabile protecting group (PG). At the first step of the synthesis, selected sites of the synthesis area (typically three squares per slide, 1.28 cm × 1.28 cm each) are exposed to UV light through photolithographic mask A. The variety of patterns available for photolithography is essentially unlimited. The exposure to light causes removal of the photolabile groups, thus elaborating amino functionality. At the next step, the synthesis area is treated with the reagents necessary for the elongation of the peptide chain. Only the sites that were previously photodeprotected will participate in a coupling reaction; the rest of the synthesis area remains protected and intact. Synthesis continues by illuminating another part of the surface through photolithographic mask B, followed by the next chain elongation reaction. By repeating the photodeprotection and coupling steps, highly dense arrays, each consisting of thousands of peptides, can by synthesized. Importantly, the primary structure of each peptide in the array is sufficiently defined by the sequence of coupling and photolysis steps, and by photolithography mask patterns. Therefore, the structure can be easily deduced from the (x, y) coordinates of the peptide on the slide. This eliminates the need for encoding–decoding procedures required by some other combinatorial technologies. After completion of the synthesis, the synthesis area is exposed to reagents necessary for the elimination of side-chain protecting groups. To assess the binding properties of all synthesized peptides, the entire array is incubated with a fluorescently labeled target molecule and scanned using a stage-scanning confocal fluorescence microscope. Sites, containing peptides that bind to the target, become fluorescent. Affinity data on all peptides in the entire array are obtained in one step.

The consumption of chemical reagents required for the synthesis of thousands of peptides composing the array, together with the biological reagents necessary for bioassay, is very small, because the capacity of the flat glass surface is only 5–20 pmol cm^{-2}. Biological reagents used in this technology are recoverable and can be reused. Moreover, after performing an assay with one target molecule, the bound target can be easily dissociated from the array (e.g. by treating it with 6 M guanidine hydrochloride), making the array available for subsequent screening with other targets. These arrays are reusable for at least 6 months.

With special (orthogonal) masking strategies the number of synthetic regions on the glass surface can be increased until the limit of photolithographic resolution is achieved (10–20 μm). With this resolution, 250,000–1,000,000 compounds can be synthesized in 1 cm^2. Routinely, 50-μm resolution is practiced and allows for the production of 40,000 compounds in the same area.

Light-directed, spatially addressable synthesis is a powerful technology for generating chemical diversity. Unfortunately, the technique is limited to peptides, oligonucleotides and other linear oligomeric structures.

5.3.1.2 Microtiter Plate-based Positional Encoding

The 96-well microtiter format has been a common platform for high throughput screening (HTS) for many years, and also offers many advantages as a platform for parallel synthesis. There are multiple liquid-handling devices (multichannel pipettes, robotic systems, etc.) that significantly simplify and accelerate the process of the reagent delivery to the wells on the plate. Moreover, all 96 compounds can be handled as one synthesis entity. Their structures are spatially encoded by their location on the plate. Contemporary analytical instrumentation, such as mass spectrometers and HPLC systems, can also easily handle compounds in the 96-well microtiter format. Combined with the commonly used 96-well-based systems for screening, this eliminates the need for time-consuming redistributing and relabeling of a large number of synthesized compounds for analytical and biological characterization.

Synthesis on multipins In 1984, Geysen and coworkers [23] introduced an alternative to the synthesis on polymeric beads. As a solid support, they proposed the use of "pins"; these are reusable polyethylene rods, which have a diameter of 4 mm and a length of 40 mm, and are grafted with polyacrylic acid. The pins are attached to a supporting block arranged in an 8×12 array to fit into a 96-well microtiter plate. The 96-well plate is used as 96 separate reaction vessels. Each well can be used for the attachment of a different building block. Washing and deprotection steps can be carried out in a separate common reaction vessel. After the synthesis is completed and side-chain protecting groups are removed, peptides or small organic molecules can be assayed while attached to the pins [24, 25]. If a cleavable linker was used, individual peptides are separated from the solid support in a 96-well microtiter plate containing appropriate reagents [26], or by exposing the array to vapor-phase ammonolysis [27, 28]. Originally, the capacity of each pin was 10–100 nmol. Later, a much greater variety of multipins of different capacities with a wide selection of linkers became commercially available from Chiron Technologies Pty. Ltd. In the latest development, the pins can be attached to the radio-frequency tags to allow for encoding in the "split-and-pool" mode.

HiTOPS system Over time, original approaches have been developed by many companies to perform the synthesis of organic compounds on the footprint of a 96-well plate. The HiTOPS (high-throughput organic parallel synthesis) system [29] (Scheme 5.11) utilizes a variety of 96-deep-well filtration microtiter plates available from Polyfiltronics/Whatman.

The plates are made of polypropylene and other polymers, and are available with a selection of different filters. The volume of each well is 2 mL and allows the use of up to 50 mg of resin. For larger scale syntheses, several wells or even the entire row or the entire column can be used for the preparation of the same compound. Reactants are retained in wells by the positive pressure of an inert gas.

Scheme 5.11. HiTOPS system. Synthesis device (left) and cleavage device (right).

5.3.2
Non-Positional Encoding

5.3.2.1 **Tea-Bag Approach**
In 1985, Houghten [30] reported on peptide synthesis carried out on a resin sealed
in porous polypropylene packets. The pore size of the polypropylene mesh (74 μm)
allows the free access of chemicals to and from the contained resin. Each packet
or "tea-bag" can be individually labeled to identify the peptide synthesized on
the entrapped polymer. Many "tea-bags" can be combined in the same reaction to
carry out common synthetic steps, such as washing and deprotection. The packets
are sorted in separate reaction vessels according to the specific amino acid that will
be coupled next. Cleavage, depending on the amount of used resin, can be carried
out in separate vessels or in a 96-well microtiter plate with a 2-mL well volume. In
the original paper, 248 different tridecapeptides were synthesized in 10- to 20-mg
quantities and characterized in less than 4 weeks.

The "tea-bag" method is very practical as it does not require any special tools,
except a sealing device to make the "tea-bags". A number of mesh materials of
different porosity are available form Spectrum Medical Industries. Almost all
commercially available resins can be used, as long as the size of the beads is larger
than the mesh size; in addition, the mesh should be stable to reaction conditions.
The scale of the synthesis is easy to control by the size of the packet chosen
to contain the desired amount of a polymer support. Multiple synthesis can be
carried out manually or on a synthesizer. The "tea-bags" can be mechanically
labeled, or radiofrequency tags can be used for this purpose.

5.3.2.2 **Cellulose and Laminar Supports**
Similarly to the "tea-bag" approach, pieces of any solid-phase support suitable
for solid-phase synthesis can be used for parallel peptide preparation. One of the
first materials used in this manner was paper. Frank used paper disks [31, 32]
(Whatman 3MM, 1.5 cm diameter) packed into columns of a multicolumn continuous flow synthesizer. Prior to the synthesis, the paper was derivatized with *p*-
alkoxybenzyl linkage to allow for the cleavage with trifluoroacetic acid. The disks
can easily be labeled with a pencil, sorted, and combined depending on the common amino acid to be coupled at the next step.

Cotton has mechanical properties superior to those of paper, and was used in
a similar way by Lebl and colleagues [33, 34]. Researchers from the Rockefeller
University used polystyrene-grafted polyethylene film [35, 36], whereas scientists
from Pfizer used material prepared from two sheets of polypropylene mesh with
resin sandwiched in between [37]. The "sandwich" was fused together with a
polymer with a low melting point.

5.3.2.3 **Radiofrequency Tags**
One of the most successful automated solutions for the synthesis of encoded
chemical libraries is utilizing electronic identification devices. Scheme 5.12 illustrates one such device – the radiofrequency (RF) identification (ID) tag. Similar

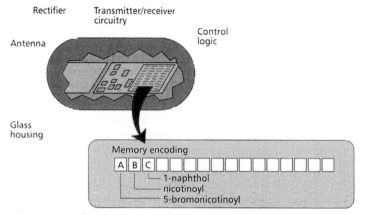

Scheme 5.12. Schematic of an RF memory chip useful for encoded synthesis.

devices have been used for years to tag laboratory mice subcutaneously, and more recently these have been used to provide security for automobile ignition keys, to secure building entrance identification, and for a variety of other functions. The application of RF tags to combinatorial syntheses was reported in 1995 by two groups, one working at IRORI [38] and the other at Ontogen [39]. The RF ID tag is about the size of a "flea" stirbar and is encased in a thick-walled glass shell. As shown in Scheme 5.12, the essential components are an antenna (the largest component) and a microelectronic chip. Each chip has a unique, nonvolatile 40-bit ID code laser-etched into it. With 40 bits available, a total of 2^{40} (over 1 trillion) unique ID codes is possible. The virtually inexhaustible range allows one to guarantee the uniqueness of all present and future RF tag ID codes. Additional bits are used to perform extensive digital error detection, which prevents incorrect reporting. (Bar codes are a more common graphical embodiment of encoding, but of course one that is substantially less information-rich.)

A transceiver controlled by a computer is used to interrogate and receive the ID code of each RF tag. The transceiver antenna transmits a specially modulated, 125-kHz electromagnetic field. This field is of very low energy and is not harmful. When an RF tag is held within about 1 cm from the transceiver's antenna, energy is picked up by the RF tag's antenna. A rectifier in the chip converts this energy to microwatt levels of direct current (DC) power, which is enough to power-up the logic circuitry on the chip. In a very real sense, the RF tag is similar to a crystal radio (which does not require an external power source), except that the device serves as both receiver and transmitter. It is "self-contained", in that the chip used no internal batteries and has no external metallic connections.

A synchronization signal modulated onto the transceiver's signal allows the chip to respond with its ID code (a serial sequence of ID bits) and error-checking bits. The time elapsed between placing the chip on the transceiver and seeing the ID code on the interfaced computer screen is about 0.5 s.

If one could physically associate this chip with a compound undergoing synthetic transformations, then the ID code on the chip would permit one to pick up a sample at any point of the combinatorial synthesis and know which reaction(s) it had already been through. Of course, just knowing the reaction history of a sample is not equivalent to knowing the chemical concept of a sample, but it is much better than nothing. If each reactor ("compound-carrying unit") must be present in a series of reaction flasks containing lots of other reactors, then having them tagged makes it possible to put them into the right flask for each step of the synthesis. Moreover, if that tag can be read easily, then that movement of reactors can be automated. Electronic tagging permits all of these benefits.

How this works in practice is detailed as follows. After a compound has been identified for which several hundred to several thousand derivatives would be of value, a synthetic route is chosen that: (i) permits linkage to a solid-phase support; (ii) utilizes reaction steps that appear possible to optimize to ≥90% yield; and (iii) affords reagents in each step for which desirable variants can be purchased (or, less optimally, can be made trivially). In the synthesis itself, one of the significant advantages of the "microreactor" approach becomes evident: one can use standard laboratory glassware and equipment to accomplish the library synthesis. There is no need for the automation of liquid-handling steps, and indeed there is no need for automation at all until rather large libraries are desired.

IRORI offers several miniature reactors – MicroKans® and MicroTubes® (Scheme 5.13). MicroKans® is a small cylindrical container with mesh walls; the internal volume of the container is 330 μl. In addition to the RF tag, the container hold up to 30 mg of any commercially available resin.

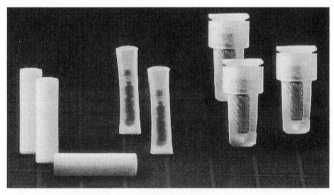

Scheme 5.13. MicroKans® and MicroTubes®.

Manual sorting of microreactors can become tedious when a large number (more than 1000–2000) of them is used. IRORI designed the AutoSort™ – 10-K microreactor sorting system (Scheme 5.14) to solve this problem. The system accommodates up to 10,000 microreactors that can be distributed between chemical steps into 48 different containers. The distribution rate is 1000 microreactors per hour. The device is also very useful for sorting MicroKrans® and MicroTubes® into microreactor carriers for further cleavage, which takes place on the Accu-Cleave-96

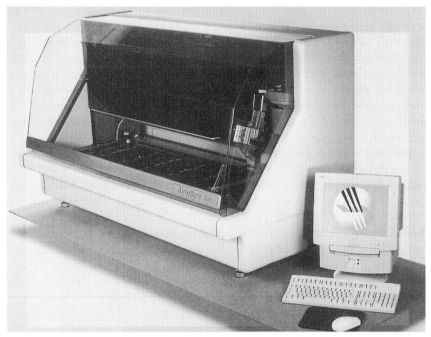

Scheme 5.14. IRORI AutoSort™ – 10-K microreactor sorting system.

cleavage station. Twelve 8×12 microreactor carriers can be used simultaneously on the sorting system.

One of the first applications of the RF tagging technology was demonstrated by the synthesis of a 432-member library ($18 \times 8 \times 3$ array) of tyrphostins (Scheme 5.15) [40].

5.3.2.4 Laser Encoding

Another example of encoding that does not interfere with chemistry was suggested in 1997 by Xiao et al. [41]. The coding structure is an inert ceramic plate with a two-dimensional, laser-etched bar code. The encoded plate, which is a 3 mm $\times$ 3 mm square, is placed in the center of a laser optical synthesis chip (LOSC). The chip is a 1 cm $\times$ 1 cm square made of polypropylene grafted with polystyrene. The smallest possible size of the encoding ceramic plate is 0.5 mm $\times$ 0.5 mm. Unfortunately, the bar code cannot be modified during the course of the synthesis.

5.4
Conclusion

The development of various encoding methods for combinatorial chemistry played a crucial role in the chemist's ability to synthesize large numbers of organic mol-

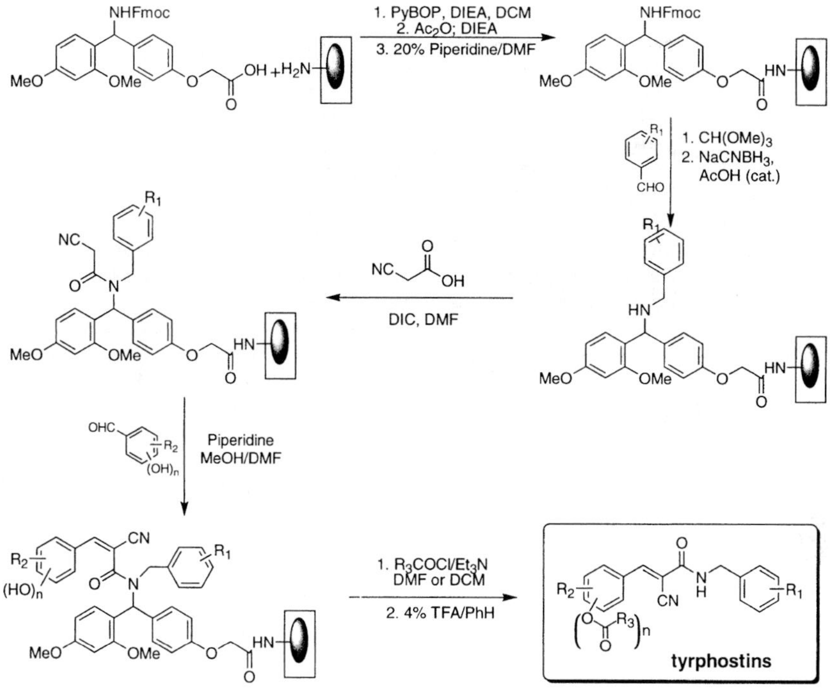

Scheme 5.15. A combinatorial synthesis of 432 tyrphostins (prepared using 29 reaction vessels) [40].

ecules rapidly in distinct formats. If the focus in library synthesis is on single compounds, milligram quantities, automated handling, flexibility and efficiency radiofrequency encoding being broadly applied in pharmaceutical companies is considered to be the most promising technology.

References

1 a) V. V. ANTONENKO, N. V. KULIKOV, R. MORTEZAEI, F. GUALTIERI (eds) *Methods and Priniples in Medicinal Chemistry: New Trends in Medicinal Chemistry*, Wiley-VCH, Weinheim 2000, pp. 39–80; b) A. W. CZARNIK, F. GUALTIERI (eds) *Methods and Priniples in Medicinal Chemistry: New Trends in Medicinal Chemistry*, Wiley-VCH, Weinheim 2000, pp. 81–96.

2 A. W. CZARNIK, *Curr. Opin. Chem. Biol.* **1997**, *1*, 60–66.

3 S. BRENNER, R. A. LERNER, *Proc. Natl. Acad. Sci USA* **1992**, *89*, 5381–5383.

4 J. NIELSEN, S. BRENNER, K. D. JANDA, *J. Am. Chem. Soc.* **1993**, *115*, 9812–9813.

5 M. C. NEEDELS, D. G. JONES, E. H. TATE, G. L. HEINKEL, L. M. KOCHERSPERGER, W. J. DOWER, R. W. BARRETT, M. A. GALLOP, *Proc. Natl. Acad. Sci. USA* **1993**, *90*, 10700–10704.

6 V. NIKOLAEV, A. STIERANDOVA, V. KRCHNAK, B. SELIGMANN, K. S. LAM,

S. E. Salmon, M. Lebl, *Pept. Res.* **1993**, *6*, 161–170.

7 J. M. Kerr, S. C. Banville, R. N. Zuckermann, *J. Am. Chem. Soc.* **1993**, *115*, 2529–2531.

8 M. H. J. Ohlmeyer, R. N. Swanson, L. W. Dillard, J. C. Reader, G. Asouline, R. Kobayashi, M. Wigler, W. C. Still, *Proc. Natl. Sci. USA* **1993**, *90*, 10922–10926.

9 H. P. Nestler, P. A. Bartlett, W. C. Still, *J. Org. Chem.* **1994**, *59*, 4723–4724.

10 W. C. Still, *Acc. Chem. Res.* **1996**, *29*, 155–163.

11 Z.-J. Ni, D. Maclean, C. P. Holmes, M. M. Murphy, B. Ruhland, J. W. Jacobs, E. M. Gordon, M. M. Gallop, *J. Med. Chem.* **1996**, *39*, 1601–1608.

12 Z.-J. Ni, D. Maclean, C. P. Holmes, M. M. Gallop in: Methods in Enzymology, vol. 267, Combinatorial Chemistry. Abelson, J. N. (ed.). Academic Press, San Diego 1996, pp. 261–272.

13 C. P. Holmes, D. G. Jones, *J. Org. Chem.* **1995**, *60*, 2318–2319.

14 D. Maclean, J. R. Schullek, M. M. Murphy, Z.-J. Ni, E. M. Gordon, M. A. Gallop, *Proc. Natl. Acad. Sci. USA* **1997**, *94*, 2805–2810.

15 M. Stankova, O. Issakova, N. F. Sepetov, V. Krchnak, K. S. Lam, M. Lebl, *Drug. Dev. Res.* **1994**, *33*, 146–156.

16 R. A. Zambias, D. A. Boulton, P. R. Griffin, *Tetrahedron Lett.* **1994**, *35*, 4283–4286.

17 H. M. Geysen, C. D. Wagner, W. M. Bodnar, C. J. Markworth, G. J. Parke, F. J. Schoenen, D. S. Wagner, D. S. Kinder, *Chem. Biol.* **1996**, *3*, 679–688.

18 D. S. Wagner, C. J. Markworth, C. D. Wagner, F. J. Schoenen, C. E. Rewerts, B. K. Kay, H. M. Geysen, *Comb. Chem. High Throughput Screening* **1998**, *1*, 143–153.

19 S. P. A. Fodor, L. J. Read, M. C. Pirrung, L. Stryer, A. T. Lu, D. Solas, *Science* **1991**, *251*, 767–773.

20 M. C. Pirrung, J. L. Read, S. P. A.

Fodor, L. Stryer, US Patent No. 5,143,854, **1992**.

21 J. W. Jacobs, S. P. A. Fodor, *Trends. Biotechnol.* **1994**, *12*, 19–26.

22 M. C. Pirrung, *Chem. Rev.* **1997**, *97*, 473–488.

23 H. M. Geysen, R. H. Meloen, S. J. Barteling, *Proc. Natl. Acad. Sci. USA* **1984**, *81*, 3998–4002.

24 H. M. Geysen, S. J. Rodda, T. J. Mason, *Mol. Immunol.* **1986**, *23*, 709–715.

25 H. M. Geysen, S. J. Rodda, T. J. Mason, G. Tribbick, P. G. Scoofs, *J. Immunol. Methods* **1987**, *102*, 259–274.

26 A. M. Bray, N. J. Maeji, R. M. Valerio, R. A. Campbell, H. M. Geysen, *J. Org. Chem.* **1991**, *56*, 6659–6666.

27 A. M. Bray, N. J. Maeji, A. G. Jhingran, R. M. Valerio, *Tetrahedron Lett.* **1991**, *32*, 6163–6166.

28 A. M. Bray, A. G. Jhingran, R. M. Valerio, N. J. Maeji, *J. Org. Chem.* **1994**, *59*, 2197–2203.

29 V. V. Antonenko in: Combinatorial Chemistry and Combinatorial Technologies Methods and Applications. Miertus, S., Fassina, G. (eds), Marcel Dekker 1998, pp. 205–232.

30 R. A. Houghten, *Proc. Natl. Acad. Sci. USA* **1985**, *82*, 5131–5135.

31 R. Frank, R. Doring, *Tetrahedron* **1988**, *44*, 6031–6040.

32 B. Blankenmayer-Menge, R. Frank, *Tetrahedron Lett,* **1988**, *29*, 5871–5874.

33 M. Lebl, V. Gut, J. Eichler in: Peptides 1990. Giralt, E., Andreu, D. (eds), ESCOM, Leiden 1991, pp. 1059–1060.

34 M. Lebl, J. Eichler, *Peptide Res.* **1989**, *2*, 297–300.

35 R. H. Berg, K. Almdal, W. Watsberg Pedersen, A. Holm, J. P. Tam, R. B. Merrifield, *J. Am. Chem. Soc.* **1989**, *111*, 8024–8026.

36 R. H. Berg, K. Almdal, W. Watsberg Pedersen, A. Holm, J. P. Tam, R. B. Merrifield in: Innovation and Perspectives in Solid Phase Synthesis:

Peptides, Polypeptides and Oligonucleotides, Macro-Organic Reagents
and Catalysts-1990. EPTON, R. (ed.),
SPCC, Birmingham 1990, pp. 453–
459.

37 N. K. TERRETT, M. GARDNER,
D. W. GORDON, R. J. KOBYLECKI,
J. STEELE, *Chem.-Eur. J.* **1997**, *3*,
1917–1920.

38 K. C. NICOLAOU, X. Y. XIAO, Z.
PARANDOOSH, A. SENYEI, M. NOVA,
Angew. Chem. Int. Ed. **1995**, *3*, 2289–
2291.

39 E. J. MORAN, S. SARSHAR, J. F.
CARGILL, M. M. SHABAZ, A. LIO, A. M.
M. MJALLI, R. W. ARMSTRONG, *J. Am.
Chem. Soc.* **1995**, *117*, 10787–10788.

40 T. CZARNIK, M. NOVA, *Chem. Britain*
1997, *33*, 39–41.

41 X. Y. XIAO, C. ZHAO, H. POTASH, M.
P. NOVA, *Angew. Chem. Int. Ed. Engl.*
1997, *36*, 780–782.

6
Instrumentation for Combinatorial Chemistry

Marcus Bauser and Hubertus Stakemeier

6.1
Automation in Combinatorial Synthesis

6.1.1
General Remarks

Economic pressure to speed up the drug discovery process has had a huge impact on all fields of medicinal chemistry [1], therefore automation has increasingly become one of the main strategies to fulfill this demand [2]. While automation was successful in high-throughput screening (HTS) [3] and peptide synthesis, it played a minor role in mainstream organic synthesis. Automated systems are recommended for procedures that are highly predictable and repetitive [4]. However, organic chemistry is seldom defined in this way. The successful synthesis of organic molecules depends strongly on the chemical properties of the reagents and reactants. Within the library production process solubility and reactivity of synthons can be highly different, therefore it is very difficult to find one protocol that works for every building block. Since multiple parallel synthesis began, there has been a wide range of different approaches and concepts for the design of automated systems to overcome these problems [5–7]. The first attempts at automation were the simple parallelization of commercially available reaction vessels. Secondly, reaction blocks were designed and used in combination with existing liquid-handling systems. Another approach uses stand alone systems that mimic the typical action of a chemist. Today, modern automated systems are modular workstation approaches. An overview is given in Table 6.1.

6.1.2
Fully Automated Systems for Solid- and Solution-phase Synthesis

6.1.2.1 Robot-arm-based Systems
The main advantage of fully automated robot-arm-based systems is that they offer customized solutions similar to HTS equipment [8]. The individual user needs to define the architecture, the layout, and the processes of the robotic system. Within

Tab. 6.1. Automated systems for solid- and solution-phase chemistry.

	Accelab Acrosyn98	ACT BenchMark Omega Series	ACT Vantage	ACT Venture
Automation				
Type	Unit	Unit	Unit	Unit
Arm	Cylindric arm	Two multiprobe XYZ arms	Two multiprobe XYZ arms	Two multiprobe XYZ arms
Chemistry	Solid phase Solution phase	Solid phase Solution phase	Solid phase Solution phase	Solid phase Solution phase
Reaction block (RB)				
Array	20 (4 × 5)	Monomer rack 8-, 16-, 40-, 96-well reactor	Ares reactor 40, 96, 384	40, 96, 384
Number of possible RBs	Up to 5 RBs	Up to 4 RBs (Omega 384)	1	Multiple
Type	Fixed during synthesis	Fixed during synthesis	Fixed during synthesis	Fixed during synthesis
Reaction vessels				
Number/volume	20/5 ml	Solid phase: 96/3.5 ml; 40/9 ml; 16/15 ml; 8/30 ml; Solution phase: 96/6 ml; 40/14.5 ml	384/0.5 ml; 96/2 ml; 40/6 ml	384/0.5 ml; 96/2 ml; 40/6 ml
Material	Polypropylene or glass syringes with frit or customized blocks	Multiple-well teflon reactor block with frits for solid phase	Teflon block with fritted wells	Teflon block with fritted wells
Reaction procedures				
Reflux	+	+ (condenser module)	Completely sealed reaction block	Completely sealed reaction block
Filtration	Bottom	Bottom	Bottom	–
Evaporation in rv	–	–	–	
Agitation	Magnetic levitation stirring	Orbital shaking	Orbital shaking	–
Temperature range	–80 to 160 °C	–70 to 150 °C	–70 to 150 °C	
Pressure	–	–	Up to 150 psi	Up to 150 psi
Inert atmosphere	In rv (rv = reaction vessel)	In rv	In rv	In rv

Tab. 6.1. (continued)

	Accelab Acrosyn98	ACT BenchMark Omega Series	ACT Vantage	ACT Venture
Work-up				
Liquid–liquid extraction	+	–	–	+
Solid-phase extraction	+	–	–	–
Analysis	HPLC upgradable			–
Special features	• Fully enclosed robot-arm-based system (SCARA) with balance, vortexer, vacuum centrifuge • Flexible control software	• Automatic on-board cleavage of all products • Multiple segregation of reactor waste for hazard classification or reagent recovery • Fully enclosed	• Heating above boiling point is possible • Multichannel solvent/reagent delivery • Fully enclosed	• Conductometric detection of phase boundry • Up to 10,000 parallel reactions • Fully enclosed

	Argonaut Trident	Charybdis Technologies Illiad PS²	Chemspeed ASW200	Irori
Automation				
Type	Modular	Unit	Unit	Modular
Arm	XYZ arm	Two XYZ arms	XYZ arm (Gilson 222)	
Chemistry	Solid phase Solution phase	Solid phase Solution phase	Solid phase Solution phase	Solid phase
Reaction block (RB)				
Array (wells/vials)	Trident Reaction Cassette™ 48 (6 × 8)	Calypso Reaction Block Microplate footprint e.g. 96 (8 × 12)	16 (8 × 2)	–
Number of possible RBs	Up to 4 RBs	Up to 4 RBs	7	–
Type	Flexible during synthesis	Fixed during synthesis	Fixed during synthesis	–

				KAN™
Reaction vessels				
Number/volume	48/5 m, 24/14 ml	6/50 ml; 12/25 ml; 24/10 ml; 48/5 ml; 96/2 ml	112/13 ml; 64/27 ml; 32/75 ml; 16/100 ml	Kans for 30 mg, 60 mg and 300 mg resin capacity are available
Material	Glass vessels	Multiple-well teflon reactor block with frits	Individual glass vials	Polypropylene
Reaction procedures				
Reflux	– (Sealed vessels)	– (Sealed vessels)	Reflux condensers for each vial	+ (standard glassware)
Filtration	Top	Bottom	Parallel vessel to vessel	+
Evaporation in rv (online)	–	–	+	+
Agitation	Orbital shaking	Orbital shaking	Orbital shaking	+
Temperature range	–40 to 150 °C		–70 to 150 °C	–60 to 120 °C
Pressure	–	–	+ (upgradable up to 10 bar)	+
Inert atmosphere	In rv	In rv	In rv	In rv
Work-up				
Liquid–liquid extraction	+ (Trident Processing Station)	–	+ (by volume)	–
Solid-phase extraction	+ (Trident Processing Station)	–	+	–
Analysis	HPLC upgradable	–	Online TLC; online HPLC possible valve preinstalled	–
Special features	• Fully enclosed • Temperature and agitation speed flexible for each reaction cassette • Rvs sealed with rotatable Teflon valve	• Modular bench layout and format	• Reagent additions while shaking or stirring and heating or cooling • Closure of the reactors by a ceramic valve allowing for efficient evaporation (DMSO, DMF) • Fully enclosed • Handling of slurries and suspensions	• Modular, expandable system • AutoSortTM10K automatically sorts microreactors between reaction steps • Sort 1000 microreactors per hour

Tab. 6.1. *(continued)*

	ISRA	Mettler Toledo Myriad Core system	MultiSyn Tech Syro II	Perkin-Elmer Solaris 530
Automation				
Type	Unit	Modular	Unit	Unit
Arm	Articulated arm	Transfer of RBs by a conveyor belt	Two XYZ arms	XYZ arm (TECAN)
Chemistry	Solution phase	Solid phase Solution phase	Solid phase	Solid phase Solution phase
Reaction block (RB)				
Array (wells/vials)	–	Minitray 12 (2 × 6)	96; 60; 40	48 (6 × 8)
Number of possible RBs	–	Multiple	1	Up to 4 RBs
Type	–	Flexible during synthesis	Fixed during synthesis	Fixed during synthesis
Reaction vessels				
Number/volume	50/35 ml	12/10 ml	96/2 ml; 60/5 ml, 40/10 ml	48/10 ml
Material	Glass vessels	Glass vessels	Glass or polypropylene syringes	Single glass reactors
Reaction procedures				
Reflux	Closed reactors	Closed rvs with overpressure value	+	+ (reflux channel)
Filtration	+	Bottom	Bottom	Top
Evaporation in rv (online)	+	–	–	–
Agitation	Magnetic stirring	Magnetic stirring, gas bubbling	Magnetic levitation stirring	Orbital shaking
Temperature range	–40 to 150 °C	–60 to 150 °C	–60 to 150 °C	–30 to 150 °C
Pressure	–	–	–	–
Inert atmosphere	In rv	In rv	In rv	In rv

	Zenyx Magellan	Zymark	Zinsser Sophas	
Work-up				
Liquid–liquid extraction	+	+ (Allex™)	–	+
Solid-phase extraction	+	–	–	–
Analysis	Online HPLC	–	–	–
Special features	• Detection of phase boundary via a camera system • Online error detection	• Modular, expandable system • Automated scheduling and simultaneous module operation • Compatibility with the Myriad Discoverer series • Septumless rvs with twist cap	• Fully enclosed • Reagent and solvent addition under agitation possible	• Offline incubator available • Seven sensors to monitor critical instrument functions • Fully enclosed
Automation				
Type	Unit	Unit	Unit	
Arm	XYZ arm	Cylindrical arm	XYZ arm	
Chemistry	Solid phase (Solution phase)	Solid phase / Solution phase	Solid phase	
Reaction block (RB)	Stem Block			
Array (wells/vials)	98	Customized		
Number of possible RBs	1	Customized		
Type	Fixed during synthesis	Fixed during synthesis	Fixed during synthesis	
Reaction vessels				
Number/volume	96/10 ml	Customized	96/2 ml; 60/5 ml, 40/10 ml	
Material	Glass vessels	Glass vessels	Teflon, glass vessels	

Tab. 6.1. *(continued)*

	Zenyx Magellan	Zymark	Zinsser Sophas
Reaction procedures			
Reflux	Closed reactors	Closed reactors	Optional
Filtration	Top	+	Top
Evaporation in rv (online)	–	Optional	–
Agitation	Orbital shaking	Orbital shaking	Orbital shaking
Temperature range	Ambient to 150 °C	Ambient to 150 °C	–80 to 150 °C
Pressure	–	–	–
Inert atmosphere	–	In rv	In rv
Work-up			
Liquid–liquid extraction	+	+	–
Solid-phase extraction	+	+	+
Analysis	–	Optional	+
Special features	• Offline incubation possible • 96-channel filtration head	• Modular, expandable system • Fully enclosed	• Offline incubation possible • Reaction vessels are moved between functional unit on workbench

the typical workflow of an organic synthesis every step can be fully automated via specialized devices that work totally independently but are at the same time connected and controlled through an intelligent and easy-to-use software program. Therefore, robotic systems are one of the most flexible solutions to laboratory automation, and established systems can be further developed or redesigned if necessary. Once a robotic system is established it is subject to constant optimization, which is in contrast to workstation approaches where fixed hardware and software subroutines are usually used. In customized robotic systems, the control software is used with a wide range of devices from different suppliers, and subroutines for the specialized stations can be easily redefined. With flexible control software, automated error identification mechanisms and correction tasks are possible that allow an unattended "round-the-clock" operation. But maximum versatility comes at a price, which is one of the main disadvantages of customized robotic systems. The size of the instrument requires custom-built ventilation cabinets and safety installations such as a fire extinguisher system, therefore the timeline for the successful implementation of an automated "robotlab" is often a critical point for pharmaceutical companies. The reliability of the equipment is difficult to predict and can only be tested during installation, which can be a time-intensive process.

The key components of robot-arm systems are industrial robots that are used extensively in many areas of industry, such as the car industry, for a wide range of different tasks. Some of the main suppliers of robots are CRS [9], Beckman Coulter [10], Mitsubishi [11], and Zymark [12] (robot-arm features are shown in Table 6.2).

Isra system A one-arm-based system for solution-phase synthesis was manufactured by Isra (Darmstadt, Germany). This system was developed in cooperation with different companies, among them Bayer [13], at the beginning of 1997. The system was designed for the synthesis, work-up, and analysis of arrays of 50 compounds per run in amounts of up to 2 g. In this system, the CRS robot arm is used for transport functions. It can be easily programmed via access tasks by using get-and-put commands. Synthesis planning and navigation of the robotic system as

Tab. 6.2. Robot arm features.

Physical characteristics	
Travel length	Up to 5 m
Height to enclosure	138.7 mm
Height to saddle mounting surface	201 mm
Weight approximately	50 kg/m
Performance specifications	
Speeds	0.01–0.9 m s^{-1}
Acceleration	3 m s^{-2}
Repeatability	0.08 mm

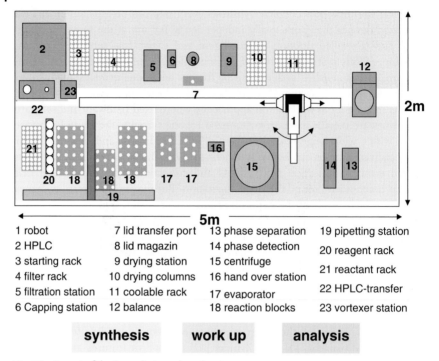

1 robot
2 HPLC
3 starting rack
4 filter rack
5 filtration station
6 Capping station

7 lid transfer port
8 lid magazin
9 drying station
10 drying columns
11 coolable rack
12 balance

13 phase separation
14 phase detection
15 centrifuge
16 hand over station
17 evaporator
18 reaction blocks

19 pipetting station
20 reagent rack
21 reactant rack
22 HPLC-transfer
23 vortexer station

synthesis   **work up**   **analysis**

Fig. 6.1. Layout of the Isra robot-arm-based system.
Implemented at Bayer's central research center.

well as documentation of the chemistry undertaken are also very important features of the control software. Therefore, a software concept was developed that offers complete and legally admissible documentation of all substances which are to be prepared by the robotic system. The data generated can easily be transferred into widely used electronic laboratory journals. The whole system has a footprint of 5 m long by 2 m wide and is divided into three functional parts for the synthesis, work-up, and analysis; the layout is outlined in Fig. 6.1.

The individual components for the synthesis are a needle XYZ-pipetting robot, heating and cooling reaction blocks with magnetic stirrers, a rack for reagents, and a rack for starting materials. These devices have been successfully used in many different syntheses under various conditions. The largest part of the machine is used for storage and work-up procedures. The main devices for work-up are a filter station, a drying station, a balance, a centrifuge, a solvent evaporator, and a phase separation and a phase boundary recognition station. The analytic part of the robot is represented by a Shimadzu high-performance liquid chromatography/ultraviolet (HPLC-UV) system. Via an interface it can be used for online analytical characterization of the synthesized products. The main features of the system are:

• liquid capability per vial: 35 ml;
• all manipulations under argon;

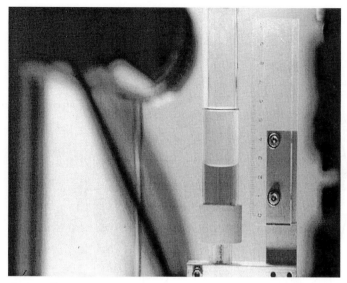

Fig. 6.2. Phase separation.

- classical laboratory procedures:
  - mixing, vortexing, and filtration,
  - liquid–liquid extraction, centrifugation,
  - drying, evaporation;
- online analysis of purity and yield.

The system contains a specially designed phase-separation station. An integrated camera (Fig. 6.2) system is able to measure the z-height; the registered value is used as the input command for the phase-separation device.

Zymark   One of the most widely used robot arms in fully automated systems for process development and organic synthesis is the Zymate XP arm [14], which has been commercially available since the early 1980s. The example shown uses a cylindrical Zymark robot with interchangeable effectors for performing different manipulations (Fig. 6.3).

This system was developed as a result of collaboration between Zeneca and Zymark (Runcorn, UK) [15]. Depending on the individual components, the main features of the system are:

- synthesis of 40–50 reactions at a time in gram scale;
- classical laboratory procedures:
  - heating, cooling, mixing, vortexing,
  - filtration, liquid–liquid extraction;
- sample preparation for analytical purposes [16].

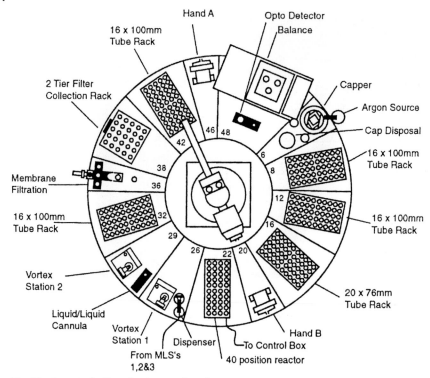

Fig. 6.3. Layout of a Zymark robot-arm-based system.

The customized synthesizer has an open architecture with considerable flexibility. Based on a Zymark synthesizer, SmithKline Beecham developed a system for synthesis and online purification via column chromatography [17].

Accelab Arcosyn98 The Arcosyn98 manufactured by Accelab [18] is a fully automated system for solid- and solution-phase synthesis. A fixed Scara [19] robot arm is equipped with a gripper change system, enabling it to change its effector tool. Based on this technology, the Scara arm is in charge of all vessel transfers. In addition, other functions such as pipetting and vial opening/closing mechanisms can be performed without an in-built peripheral station (Fig. 6.4).

To achieve high throughput, the system includes devices for heating and cooling, evaporation, dissolution, liquid–liquid extractions, yield determination, and sample preparation for analytical purposes, as shown. The Arcosyn98 can be used with five reaction blocks at a time, therefore 100 parallel reactions are possible in one run (Fig. 6.5).

The whole system can be controlled and programmed with a software package

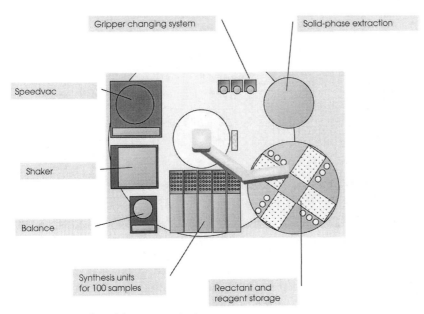

Fig. 6.4. Layout of Accelab's Arcosyn98 robot system.

using one master computer. The multifunctional software can also be used for planning, scheduling, and administration of data entries and results.

6.1.2.2 Fully Automated Workstation Systems

Chemspeed ASW2000 system Chemspeed [20] was founded in 1997 by a team of chemists who had previously worked in a medicinal chemistry laboratory at Roche.

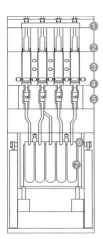

Fig. 6.5. Reaction block Arcosysn98.

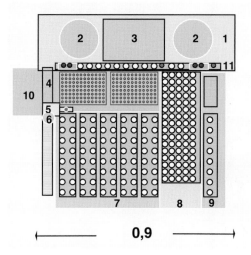

1   Gilson 222
2   reservoir
3   syringe
4   xyz-roboticarm
5   arm with needle
6   valve
7   reactionzone with glassware
8   variable reagents/TLC/SPE
9   central reagents
10   microplates
11   bar (Waste/HPLC/transfer ports)

Fig. 6.6. ASW2000 layout.

Their main objective was to develop an unattended automated system that allowed parallel preparation of compounds in glassware reactors with online purification and analysis in a compact workstation approach (Fig. 6.6).

Chemspeed delivers the system on a trolley (footprint 1.4 m × 0.8 m) covered with a hood. Because of this compact design, the very flexible system can be used in nearly every laboratory. Within one run, 80 parallel reactions (with an option of 112) can be performed that have the following features:

- liquid capability of 13, 27, 75 or 100 ml per vial;
- fully inert environment inside and outside the reactors;
- reagent additions while shaking or stirring and while heating or cooling;
- classical laboratory procedures:
  - mixing, vortexing, filtration,
  - liquid–liquid extraction,
  - drying, evaporation, cold finger refluxing;

- solid-phase extraction (SPE);
- temperature range from −70 to +150 °C, measured and controlled in the reaction mixture;
- online analysis of purity and yield by thin layer chromatography (TLC) or analytical HPLC (optional);
- rheodyne valves for preparative HPLC.

The whole system can be upgraded into the ASW2000P workstation, which allows pressurized reactions that integrate work-up and analysis procedures during the synthesis. The pressure capabilities of the system are:

- up to five reactor blocks (80 parallel reactions at maximum);
- up to 12 bar;
- pressure sensor;
- autosecurity system to provide optimum seal performance.

Advanced ChemTech Since 1985, Advanced ChemTech (ACT) [21] has supplied systems for organic synthesis, which in the beginning were especially for peptide synthesis. Today ACT offers a broad "family" of manual, semiautomated, and automated synthesizers for solution- and solid-phase synthesis (Table 6.1).

The Venture 576 (launched in May 1999) is designed to be a platform for fully automated high-throughput synthesis (Fig. 6.7).

The heart of the Venture 576 is a special reaction block combined with two multiprobe XYZ arms. The microplate reactor block contains 96 wells and is constructed of glass-impregnated polytetrafluoroethylene (PTFE) constructions. The usable reaction well volume is approximately 3 ml compared with a total volume of 6 ml and allows parallel reactions in milligram scale even under a reactive gas atmosphere because reaction vessels can be pressurized up to 150 psi. The system is capable of up to 10,000 simultaneous reactions that are determined by the customer. The following description summarizes the main features of the system:

- inert atmosphere for reactor, reactants, and the whole platform;
- classical laboratory procedures:
  - vortexing, filtration, liquid–liquid extraction;
- reactor block with integrated condenser module;
- temperature range from −70 to +150 °C, electric resistive heating, cooling by nitrogen gas generated from liquid nitrogen.

The floor-standing cabinet is equipped with connections for a ventilation system. Because the system is capable of performing up to 10,000 reactions per run, it is recommended for combinatorial laboratories producing large numbers libraries.

MultiSyn Tech Syro II MultiSyn Tech [22] offers manual, semiautomated, and fully automated systems for combinatorial chemistry. The fully automated Syro system

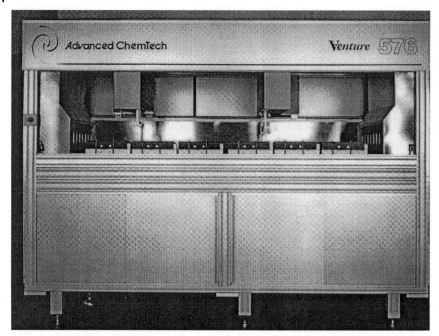

Fig. 6.7. ACT Venture 576.

(Fig. 6.8) consists of two independent XYZ robot arms and a specially designed reaction block. The system can be equipped with different reaction vessels. For solid-phase synthesis the removable reaction vessels (number/volume: 96/2 ml; 60/5 ml; 40/10 ml) in glass or polypropylene with glass or PTFE frits are rec-

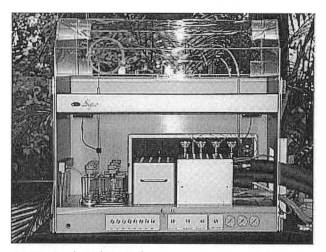

Fig. 6.8. Syro by MultiSyn Tech.

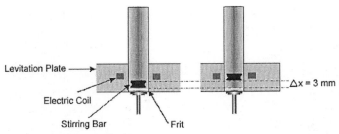

Fig. 6.9. Levitation mechanism.

ommended. Owing to the reaction block design all manipulations can be performed under inert gas atmosphere using reaction temperatures between −60 and +150 °C.

The system has a specially designed agitation mechanism. Each reaction tube is circumvented by electric coils that are used to generate a magnetic field. The coils are placed in a movable levitation plate, the center of the magnetic field is about 6 mm above the frit of the reaction tubes. This special set-up ensures that, during agitation, the resin is not crushed between the stirring bar and the frit (Fig. 6.9).

Zenyx Magellan synthesizer In 1996, Zeneca Pharmaceuticals entered into a collaboration with Zenyx Scientific to design and build an instrument for multiple parallel synthesis [23]. The new system was intended to fulfill the following main objectives:

- access to libraries containing up to 1000 single compounds;
- scale: up to 30 mg for primary screening and repository;
- microplates should be used as transfer racks.

The Magellan system is a fully automated synthesizer controlled by a computer. An Excel interface within the control software allows substance data to be imported for the reprogramming of standard protocols. With this system 96 reactions can be performed within one run, using a Stem [24] reaction block. The set-up is very flexible because the layout of the workstation can easily be changed. The XYZ robot arm is in charge of all pipetting jobs. After cleavage from the resin, the compounds can be delivered directly into disposable vials using a special 96-filter station with 96 individual needles with a filter frit at the end.

Perkin-Elmer Solaris 530 The Solaris™ 530 [25] organic synthesizer for automated combinatorial chemistry was introduced in late 1998 by PE Biosystems, a division of the PE Corporation (Fig. 6.10).

The system is able to synthesize 48 discrete molecules per run in parallel. The portable synthesis module contains an array of 8 × 6 reaction vessels with a volume of 10 ml per vial (Fig. 6.11). A reflux channel is built in the chemically inert module to provide reflux conditions. The dual septa secures an inert atmosphere.

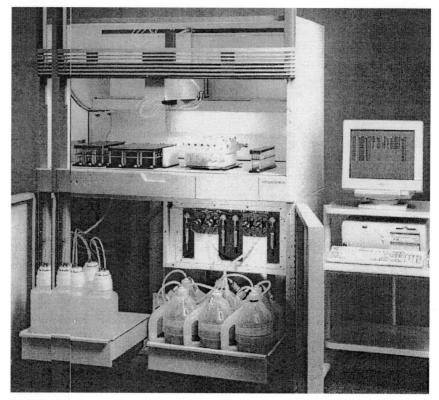

Fig. 6.10. Solaris.

Key features of the system include:

- eight liquid-handling tips;
- seven sensors to monitor critical instrument functions;
- completely enclosed system with flow-through ventilation;
- multiple instrument sensors;
- offline workstation for heating, mixing, and cooling.

An offline incubation workstation enables multiuser access and staggered synthesis runs, which increases throughput. The system is designed for the automated lead optimization using solid-phase synthesis techniques.

Zinsser Sophas   In 1998 Zinsser [26] launched the Sophas system, which is specially designed for solid-phase synthesis. The synthesizer uses a robotic XYZ arm with four independent probes that are manufactured by Rosys [27]. All liquid-handling tasks are controlled via a personal computer. The easy-to-use software is very flexible, can import data from any database, and allows customized layouts to

Fig. 6.11. Glass vials.

be defined. The automated workbench offers the opportunity to choose from a set of movable reactors that range from 96-well plates to 25-ml reaction vessels.

Charybdis Technologies Illiad PS² Charybdis Technologies [28] was founded in 1996, and provides solutions in the areas of high-throughput organic synthesis, laboratory automation, and chemical discovery. The Iliad PS² Series are multi-tasking robotic workstations for solid- and solution-phase chemistry. The two independent XYZ robotic arms work in parallel and are controlled via the OASys software. The heart of all Charybdis synthesizers is the Calypso reaction block system, which consists of a top frame with a specially designed top cover plate and a base frame with a base cover plate. The reaction block accepts any array with the standard footprint of a microplate with well volumes of 50, 25, 10, 5, and 2 ml. The Iliad systems can be upgraded to provide online filtration and agitation.

6.1.2.3 Modular Systems

Argonaut Trident Argonaut Technologies [29] was founded in 1994, and provides systems for parallel solution- and solid-phase chemistry. The Trident [30] family consists of a reaction cassette, a library synthesizer, a workstation, and a processing station. The core of the system is the Trident Reaction Cassette™ that contains 48 5-ml vessels or 24 14-ml vessels which are made from glass. The reaction cassette fits into every member of the Trident family and can be easily removed during synthesis for spot checks on the reactions.

With one run, the fully automated Trident library synthesizer can prepare up to 192 reactions in parallel. The whole system is controlled via the Trident software, which controls, for example, the temperature, solvent deliveries, or product collection. The synthesizer can handle up to four different reaction cassettes, and each cassette position can be set to a different temperature and agitation speed.

The Trident workstation is a straightforward supplement to the synthesizer, because it can be used for the manual delivery of solvent and reagent, so it performs parallel resin washing and sample collection in a very effective way. The workstation is ideal for development chemistry, because the methods are developed and adopted in the same cassette as that used for library production. The Trident Processing Station completes the family. It is a multipurpose liquid handler with a special interface to the reaction cassette. The key features of the station are:

- liquid–liquid extraction;
- solid-phase extraction;
- clean up with scavenger resins;
- reverse filtration;
- dry solid loading to open vessels;
- reformatting (e.g. from tubes to microplates).

Mettler Toledo Myriad Core System The Myrid™ Core System (MCS) [31] was developed by a consortium of pharmaceutical companies (SmithKline Beecham, Pfizer, BASF, Novartis, Merck, Takeda, and Chiroscience) and The Technology Partnership [32]. In December 1998 the MCS was sold to Mettler Toledo [33].

The MCS was created as a synthesis system based on a series of robotic processing modules combined with a unique reaction vessel design, fulfilling the following objectives:

- solid- and solution-phase chemistry should be possible;
- synthesis of large numbers of compounds for high-throughput screening;
- multistep directed synthesis of pure compounds;
- method development should be possible;
- modular, expandable design;
- automatic transfer of vessels between processing modules;
- automated scheduling and simultaneous module operation.

The reaction block consists of an array of 12 reaction vessels with twist caps. The whole blocks can be transferred to the different units (incubator and processing module) by a conveyor belt. The system is controlled by the chemist using software that allows the processing of four completely independent batches (48 reactions).

Irori In 1996 Irori [34] introduced the AccuTag™-100 Combinatorial Chemistry System. With this system, large numbers of discrete compounds can be produced using the "directed sorting" split-and-pool technique [35]. The reactions are performed in single microreactors which are identified using miniature electronic tags. The Irori Kan™ reactor family is specially designed for solid-phase synthesis. Three different Kans – the MicroKan™, the MiniKan™, and the MacroKan™ – with a resin capacity of up to 30, 60, and 300 mg are available, therefore compounds can be synthesized in approximately 10-, 20- and 100-mg batches (Fig.

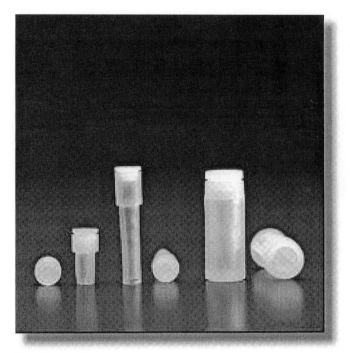

Fig. 6.12. Irori Kans.

6.12). Another big advantage of this approach is that standard glassware can be used for performing the library synthesis.

To achieve high throughput, the Kans are used together with the AutoSort-10K. This workstation has been designed to extend the directed sorting technique for use with libraries in the range of 1000 to 10,000 compounds. The AutoSort™-10K Microreactor Sorting System (Fig. 6.13) automatically sorts microreactors between reaction steps.

The key features of the system are:

• accommodates up to 10,000 microreactors;
• sorts 1000 microreactors per hour;
• automatically sorts for cleavage into the AccuCleave-96;
• standard glassware applicable.

6.2
Purification of Combinatorial Libraries

In the early days of combinatorial chemistry [36], mixtures derived from solid-phase split-and-mix libraries [37] as well as nonpurified compounds from solution-

Fig. 6.13. Autosort.

phase synthesis played an important role as test compounds for biological screening, basically because it was a very easy way to produce the promised numbers of compounds. It was soon recognized that those compounds often led to false-positive test results, and that the deconvolution of mixtures and extraction of biologically active molecule in an HTS mode are difficult tasks [38]. This problem resulted in the synthesis of single compounds fulfilling properties such as diversity, drug likeness, and a high degree of purity [39]. Classical purification procedures such as liquid–liquid extraction and chromatography were automated, solid-phase extractions with ion exchangers were adopted, and scavenger reagents for trapping excess starting material or reagents were developed. Automated preparative reversed phase HPLC systems were set up to address high-throughput purification issues.

6.2.1
Automated Liquid–Liquid Extraction

Liquid–liquid extraction is one of the standard methods for removing hydrophilic byproducts, reactants, or reagents. Several examples of the parallel usage of this technique for purification of combinatorial libraries have been described [40]. This method is only appropriate for very simple separation problems and can be performed in an automated fashion by using a liquid handler. The method totally fails when polar compounds have to be purified because emulsions can appear or products remain in the aqueous phase. Coupled to a solid-phase extraction system, automated liquid–liquid extraction can be performed together with a 96-needle pipetting system such as Quadra 96 [41]. Separation problems, for example removal of amines from a combinatorial library, were solved by applying several extraction cycles. Dichloromethane and dilute HCl were mixed and separated through filtration using solid-phase diatomaceous earth. The phase separation was achieved via vacuum filtration through a 96-well filterplate carrying a hydrophobic membrane which held the aqueous phase back. Liquid–liquid extraction can be easily automated by using pipetting workstations. Reaction mixtures are mixed with buffers and the upper or lower phase, depending on the density, is removed. Efficient mixing can be achieved through redispensing, which can be repeated several times. The use of a capacity sensor represents a more sophisticated approach. This sensor type is available for several liquid handlers (Zinsser, Tecan). A drawback of this method is when phase separation is not complete or undetectable, erratic results are obtained.

6.2.2
Solid-phase Extraction

Solid-phase extraction is widely used in bio- or environmental analyses in order to enrich or discover organic molecules from complex aqueous matrices. It can also be very useful for the purification of combinatorial libraries. There are rather simple desktop instruments available which are designed for this application – Zymark Benchmate™, Gilson Aspec™, and Hamilton Microlab™ – in addition to integrated devices in automated systems (e.g. Chemspeed); simple self-constructed systems have also been reported [42]. One of the big advantages of solid-phase extraction is the availability of various solid phases for the separation of diverse molecules, which range from hard and soft anion or cation exchangers to reversed phase silica columns which are all available in prepacked formats. One disadvantage of the system is that the separation problem has to be evaluated very carefully to achieve optimal results. This becomes a problem when the library is diverse in terms of biophysical behavior and limits the universality of this methodology. An example of the successful application of solid-phase extraction is the synthesis of 225 basic amines which were adsorbed on an acidic ion exchanger. Neutral side products were eluted, and after several washing steps the products were cleaved using ammonia in methanol [43]. Another example is the separation of a library containing neutral compounds from basic and acidic side products via inline solid-phase extraction using ion exchangers. The neutral products were not affected by the solid phases and only side products were absorbed. Excess isocyanates were also successfully removed by trapping them as ureas using an excess of a basic amine. This side product was then efficiently absorbed to an acidic solid-phase absorber [44]. An enormous increase in the versatility of this methodology can be achieved with scavenging reagents. Scavenging reagents offer the opportunity to use chemical reactivity as a mode of irreversible absorption to a solid support. This allows the selective and efficient removal of many excess reagents. Examples are the removal of excess electrophiles such as acid chlorides, sulfonic chlorides, and isocyanates as well as the trapping of excess nucleophiles such as amines [45]. Despite these simple separation problems rather complex multistep transformations have been performed using this separation technique [46]. Several of these polymer-supported reagents are commercially available [47]. Furthermore, solid-phase reagents prepacked in columns or microplates would be a great advantage in performing parallel reactions.

6.2.3
Normal Phase Chromatography

Together with crystallization and liquid–liquid extraction, normal phase chromatography is one of the most important purification operations in organic synthesis. For automation and parallelization normal phase chromatography is also appropriate and several systems are commercially available, e.g. CombiFlash™ from Isco [48], Quad3™ from Biotage [49], and FlashMaster™ from Jones Chromatography [50]. Parallel flash chromatography ranging from 4 to 12 single-use columns is widely used. The Isco and Biotage systems provide time-triggered fractionation; the system from Jones chromatography collects fractions via UV detection, which minimizes the number of fractions. Prepacked columns are available and have to be exchanged after each run, therefore only batchwise working is possible. If salts and other polar byproducts are carefully removed before chromatography, the columns can be used several times. The commercially available columns can also be self-packed with very simple equipment. The loading of these columns is usually very high: a 20-g cartridge can be used for the purification of up to gram quantities of product.

6.2.3.1 **CombiFlash™ from Isco [48]**
This instrument is designed for parallel separation with ten columns and is equipped with an injection system in which dissolved samples or samples absorbed on silica gel can be loaded. Besides the commercially available Isco Red-Sep™ columns all other columns with luer fittings can be used. Fractionation is time guided, and a maximum of 40 fractions per run can be collected. An advantage of this system is the very small hood space of 50 cm which is required.

6.2.3.2 **Quad3™ from Biotage [49]**
With the Quad3™ 12 columns can be used in parallel together with a fraction collector which is able to handle different vial types as well as microplates. The microplates simplify the evaporation process because vacuum centrifuges can be used. A step gradient can be applied, and an injection system for the delivery of

Tab. 6.3. Common stationary phases for reversed phase preparative HPLC.

Material	Application
– C18 endcapped	Drug-like compounds
– C18 hydropylic endcapped	Drug-like compounds especially for polar compounds injected in DMSO
– C8	Drug-like compounds, amines, lipids
– CN	Polar compounds

dissolved samples or samples absorbed on silica gel is available. In addition to this system a simple desktop version, the MultiElute™ system for 12 columns with fraction collector, is also available.

6.2.3.3 FlashMaster™ from Jones Chromatography [50]

FlashMaster™ is a computer-controlled desktop system for automated flash chromatography of ten columns, which allows gradient elution. With an integrated UV detector, UV-triggered fractionation minimizes identification logistics. Together with postrun reporting of chromatograms, product identification can be easily realized. In addition to this system, Jones Chromatography also offers simple solutions for parallel flash chromatography.

6.2.4
Reversed Phase Chromatography

Reversed phase chromatography (RP-HPLC) is widely used in analyses and purification of combinatorial libraries (Tables 6.3 and 6.4). The method is very universal in terms of polarity of the analytes and a wide range of compound types can be purified with RP-HPLC. This makes RP-HPLC a favorable method for purifying combinatorial libraries, mostly because no method development is necessary. The available equipment is very rugged and has been developed for high throughput. Columns can be used up to 1000 times and continuous working is applicable, enabling fairly high throughput which compensates for the serial nature of HPLC. In theory, HPLC can be used 24 h per day, but normally this does not happen because of solvent delivery and removal, sample supply, and fraction removal. Limited throughput because of the serial nature of the system is compensated by very

Tab. 6.4. HPLC columns, dimensions, and applicable loading.

Dimension (mm)/material/particle size (μm)	Loading (μmol)	Flow rate (ml min^{-1})
8 × 75/C18/5	10	8–14
20 × 50/C18/5	50	25–40
30 × 125/C18/5	75–100	50–70

fast chromatography, which allows the semipreparative separation of one sample within minutes and a daily throughput of approximately 200 samples is possible. Fractions are usually triggered via UV and only compound-containing fractions are collected. Difficulties can be encountered in the identification of the target product-containing fraction, which is usually done offline with FIA-MS (FIA = flow injection analysis) or HPLC-MS. Column switching is desirable for speeding up cycle times. One column is used for the chromatography while the other column is equilibrated to starting conditions, saving about 10–20% of the cycle time. There are systems on the market which are able to run two columns. This doubles throughput and also the price of the system because further pumps, a detector, and fraction collector must be used. The advantage of running two columns is small compared with running two individual systems.

Because of the complexity of the whole set-up – consisting of HPLC plus downstream processing – large efforts are required for the implementation of this technology in a laboratory workflow in order to run the system in an optimal fashion. Depending on the separation problem, and on the number of byproducts, up to ten fractions can easily be obtained per injection. For identification of the target product, each fraction must be analyzed with FIA-MS. This issue consequently demands very close interaction with the analytical group performing these experiments. RP-HPLC is complementary to RP-HPLC-MS and the information from analytical HPLC-MS can be directly used for identification of the desired product fraction without any FIA-MS [51]. For every kind of set-up the logistics are complex, and adoption or development of useful software solutions are desirable for tracking the large amounts of data and to prevent bottlenecks. A reduction in the amount of solvent used may be achieved through supercritical fluid chromatography (SFC). This method offers the opportunity to use a supercritical gas, for example CO_2, as a mobile phase. CO_2 is cheap, it has high flow rates because of low backpressure, and it "evaporates" when external pressure is removed [52].

6.2.4.1 Biotage [49]
The Parallex™ System from Biotage is a four times parallel HPLC-UV workstation. The system has four channels with four individual fraction collectors, detectors, and pump heads. Throughput lies in the range of 40 injections per hour and the injection system can handle up to 768 samples per run. Loading can be from 1 mg up to 200 mg depending on the column used. The system is fully software controlled and a detailed reporting software package is available.

6.2.4.2 Gilson [53]
Gilson offers a very rugged preparative HPLC system named Nebula Series™. The pumps especially are outstanding. Flow rates of 200 ml min^{-1} and pressure limits of 400 bar can be achieved. The fraction collector 215™ has a capacity of up to 12 microplates and completes the high-throughput system. The system is totally software controlled and detailed reporting allowing fraction tracking is also included.

6.2.4.3 Merck [54]

Merck offers an automated preparative HPLC for parallel chromatography of two columns named the High Throughput Purifier™ (HTP). Four columns are integrated: two are conditioned while the other pair is used for chromatography. The optionally available Merck-Hitachi Ion Trap MS offers online or offline product identification. The MS is fully integrated into the workflow and offers very simple reporting to Excel, which is essential for integration into a laboratory workflow.

6.2.4.4 Varian [55]

Varian distributes an HPLC equipped with Dynamax HPLC Pumps. Flow rates of up to 200 ml min^{-1} with a pressure limit of 410 bar are applicable. High throughput is not possible because the system lacks a useful fraction collector that can handle microplates. In addition to this, the Apple software is hard to integrate with Microsoft-based laboratory environment.

6.2.4.5 Shimadzu [56]

Shimadzu is one of the leading companies in preparative HPLC. The systems offered are very robust and the equipment, such as the UV detector and pumps, can be very easily controlled. A disadvantage is the fraction collector, which cannot handle microplates – this feature minimizes throughput and capacity.

6.2.5
Preparative HPLC-MS

The search for more powerful techniques for the purification of combinatorial libraries using HPLC led to collaboration between researchers from the pharmaceutical and analytical industries, namely PE Sciex and Micromass, with the goal of developing preparative HPLC-MS [57, 58]. The advantage of this technique is that the molecular mass of the target molecule triggers the fraction collector. This allows the collection of the desired compounds and the online identification of the target molecules. Furthermore, this method offers logistical benefits. If just the target molecules are collected, fraction collector capacity is not critical and downstream processing can be performed very efficiently. This enables the technique to work very efficiently in terms of throughput. For setting up such a system integration of the mass spectrometer and a reliable software platform which ensures fraction collecting and sample tracking are necessary. This application is now offered from more vendors than the pioneers Micromass and PE Sciex, e.g. Merck-Hitachi, Gilson in collaboration with Thermoquest and Shimadzu.

6.2.5.1 PE Sciex [59]

PE Sciex, the first vendor with this application type on the market, offers a preparative HPLC-MS/UV system with Shimadzu pumps and a Gilson 215 as a fraction collector. The hardware components, especially the API 150EX mass spectrometer, are very rugged, and from this point of view the system is very useful for high throughput. One of the main disadvantages is the Apple-based software and the unstable fraction collector script used for tracking fractions. Besides lacking software stability, it is also not easy to integrate an Apple-based system into a usually Microsoft-based laboratory environment.

6.2.5.2 Waters Micromass [60]

Micromass merged with Waters and was the second vendor offering this application. As HPLC components, Waters devices can be integrated with equipment from other HPLC providers. The Gilson 215 is used as a fraction collector device that is fully integrated in the software platform. The Masslynx™ software is fairly well adapted to high throughput, making it easy to integrate these systems into laboratory workflow.

6.2.5.3 Merck-Hitachi

As already mentioned above, Merck-Hitachi offers a system that is designed for this application. By using a special valve as an interface, it is possible to switch between two flows running into the MS inlet. Therefore two preparative separations at a time can be performed and analyzed, thus doubling the throughput and reducing hardware costs. Merck-Hitachi devices together with Gilson fraction collectors are compatible HPLC components.

6.2.5.4 Shimadzu

Shimadzu offers a preparative HPLC-MS/UV system with a Gilson 215 fraction collector. Besides being good HPLC equipment, a reliable and flexible software package is available. In addition to the software package operating the instrument, a Shimadzu LIMS (LIMS = laboratory information management system) system is available which can also be used for documentation of data other than those from the HPLC-MS.

6.2.5.5 Gilson ThermoQuest

Gilson, one of the leading experts in HPLC, has established a preparative HPLC-MS system in collaboration with Thermoquest. The HPLC part of the system, supplied by Gilson, is controlled by Uni Point software (Gilson); the ThermoQuest MS however is controlled by Excalibur software (Thermoquest). Both software applications are controlled via special interface software, which has the consequence that the MS detector is only marginally controlled by the HPLC part of the equipment, by pulsing externally, and the two independent software packages always have to run on a single PC.

6.3
Analysis of Combinatorial Libraries

Combinatorial syntheses need fast and reliable analyses in order to determine the identity of products and intermediates in each step of a parallel reaction sequence.

Tab. 6.5. Analytical methods and their applications in combinatorial chemistry.

Method	Application
FT-IR	Characterization of resin-bound intermediates in reflection (ATR (ATR = attenuated total reflectance) DRIFTS (DRIFTS = diffuse reflectance infrared fourier transform spectroscopy))
NMR	Characterization and quantification of combinatorial libraries with fast FIA or sample exchanger devices
ESI-MS	Characterization of libraries in coupling to HPLC or CE. Identification of fractions from preparative HPLC in FIA mode

Consequently, analyses were adopted that operate at the same speed as combinatorial methodologies when these methods became popular for synthesis. The analysis of the final products from solid- and solution-phase synthesis – usually compounds for biological testing – are the same because similar purity criteria are applicable. Because of the nature of solid-phase synthesis, the analysis of polymer-bound intermediates is rather difficult. Special nuclear magnetic resonance (NMR) and FT-IR methods (Table 6.5) have been developed for the purpose of characterizing polymer-bound intermediates. In the case of FT-IR techniques, reflection measurements are usually used, e.g. ATR and DRIFTS [61]. These methods can be performed directly on bead and no KBr solids have to be made, as is the case in transmission experiments. In the case of NMR, magic angle spinning can be used to suppress signals from the polymer matrix [62]. All these special analytical techniques are usually used for developing the method owing to the fact that they are very time-consuming and therefore have a fairly low throughput.

The minimal requirements for combinatorial libraries are usually purity, identity, and quantity.

6.3.1
Purity of Combinatorial Libraries

Purity assessment is historically carried out using chromatographic methodologies. Thin layer chromatography (TLC), reversed phase (RP)-TLC, and HPLC are therefore widely used; these methods may also be used for spot tests in parallel synthesis. For a purity check of a whole library, methods with a higher throughput such as automated HPLC, supercritical fluid chromatography (SFC) [63], and capillary electrophoresis (CE) are applicable. In addition to chromatographic separation, detection plays an important role. Standard detection is UV absorption at different wavelengths. In addition, another optical detector – the light-scattering detector (ELSD) – can be used [64]. Light-scattering detectors have a more linear signal to mass response and are more likely to be quantitative. For use in quantitative analysis of combinatorial libraries, a nitrogen chemoluminescence detector is even more appropriate because the nitrogen signal is proportional to the amount of nitrogen present in the sample [65]. HPLC-MS coupling is the most widely used method of purity assessment in parallel synthesis, mainly because separation and identification of the target molecules is carried out in a parallel fashion. Mild ionization techniques such as ESI and APCI are used. They almost always show the molecular ion peak, which can be directly taken from the synthesis protocol. Cycle times are highly optimized. It is possible to perform 400 HPLC-MS runs per day [66]. The resulting data have to be stored in an appropriate LIMS system. Tools are available from hardware suppliers for the semiautomatic processing of these data. Using these tools, further steps such as adaptation to the laboratory workflow are additional issues, and efforts to optimize these interfaces are necessary [67]. The widespread use of HPLC-MS for purity evaluation gives a limited insight into the actual purity of a product because only UV-absorbing side products and those on RP-absorbable compounds can be detected. Salts and polymer fragments are usually not detected. One should take this feature into account when assessing the purity of combinatorial libraries if they are not purified chromatographically before biological testing. Many suppliers sell fast HPLC-MS applications especially designed for library purity checks, and all have semiautomated software packages available for data processing.

6.3.2
Identity of Combinatorial Libraries

Besides purity, determination of the identity of the individual library member is one of the important analytical tasks in combinatorial chemistry. Usually, classical methods such as NMR and mass spectrometry are used. NMR has a disadvantage in that it is possible to measure hundreds of NMR spectra daily but the automatic processing of data is still a bottleneck. Therefore, NMR is mostly used during method development, especially when stereochemical aspects have to be addressed. As already mentioned ESI (ESI = electro spray ionisation) and APCI (APCI = atmospheric pressure chemical ionisation) mass spectrometry are widely used ionization methods. The advantages of these ionization techniques are that detection from a liquid spray is possible and that an FIA system or an HPLC system can be directly coupled. Additionally almost always a molecular peak is obtained which makes the semiautomatic processing of the data very easy. Classical ionization methods such as EI (EI = electron ionisation) and CI (CI = chemical ionisation) are not applicable for polar molecules in the mass range of 300 to 500. MALDI is also very popular, especially for peptides and proteins in bioanalyses [68]. Critical sample preparation procedures and the impossibility of coupling to an HPLC system limit the methodology. For mass analyses, the highly robust quadruple technique is widely used, as are TOF [69] and FT-ICR [70]. The advantage of TOF and especially FT-ICR is the very good mass resolution of 5 ppm in the case of TOF (TOF = time of flight) and better than 1 ppm in the case of FT-ICR (FT-ICR = fourier transform ion cyclotron resonance spectroscopy). Owing to this high resolution, the molecular composition can be obtained directly from the mass signal.

For fast analysis of combinatorial libraries and for identification of products in fractions coming from preparative HPLC, FIA-ESI-MS is the method of choice [71,

72]. Cycle times below 10 s can be achieved even with standard equipment. Problems can arise because of the vast amounts of data, but software packages are available from all suppliers of ESI mass spectrometers for automated processing. The implementation of these data into the workflow of combinatorial laboratories is of course another issue and is only achievable with programming resources which can be limited to automated reformatting of Excel tables. FIA-ESI-MS does not give any results concerning purity, and if mixtures are obtained ionization yields can vary dramatically and lead to false results if the product is suppressed, e.g. by basic impurities. In addition to FIA-ESI-MS, MALDI-MS (MALDI = matrix aided laser desorption ionisation) in an offline FIA mode can also be performed. Through an automated process MALDI targets have been coated and prepared for measurement [73]. This application is rather complex because it is impossible to work directly from solution. The advantage however is that it is the only method which gives a molecular peak from larger molecules such as proteins.

6.3.3
Quantification of Combinatorial Libraries

In addition to purity and identity, quantification of combinatorial libraries is one of the major issues in combinatorial analyses because it is essential to know the amount of compound in order to quantify the biological effects. Unfortunately, there is no simple way to automate quantification. The best way to quantify the system, but also the worst way to automate it, is by weighing. Besides weighing, it is also possible to use a nitrogen detector, but accuracy can be very low for diverse sets of compounds. Therefore weighing as the method of choice must be automated. This can be achieved with liquid handlers carrying a small robotic arm. These systems, which are commercially available from Zinsser Calli™ and Mettler Toledo Bodan™, can handle single vials. The bar-coded vials are weighed empty and the weight is recorded in a database. After weighing all the vials, the dispenser transfers solutions of the synthesized compounds into the vials. After evaporation of the solvent in a vacuum centrifuge, the robot arm again weighs the vials and the quantity of compound is determined.

6.4
Automated Sample Processing

6.4.1
Sample Logistics

Parallel high-throughput synthesis is a mass production process. The process consists of several stages that must operate in an interlocking way. It is very important to identify and remove bottlenecks in order to obtain optimal throughput. A very important role that is often regarded as a bottleneck is played by the sample logistics (Fig. 6.14).

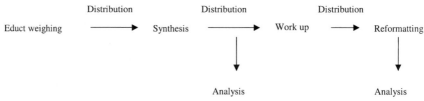

Fig. 6.14. Flow chart of an automated processes in parallel synthesis.

After evaluation of a synthesis that is considered useful for library production, the production process begins with weighing of the starting materials. The weighing process cannot be performed in an automated fashion because of the diverse physical nature of common starting materials, which can be liquids, solids, or oils. After weighing the starting materials, they can be dissolved in an appropriate solvent and then transferred very efficiently to the reaction vessels by using liquid handlers; some four- or eight-needle systems are described in Table 6.6. When the reaction is complete, the work-up protocols described in Chapter 6.2 are used for purification of the library. For liquid–liquid extractions or solid-phase extractions, the samples are produced in ratios of 1:1, which simplifies the sample logistics for reformatting and analysis. Samples can be taken directly with pipetting robots and then diluted or filtered for analysis. After analysis, those samples that are to be processed further to determine their biological activity are selected. This decision is based on the analytic results, usually those from HPLC-ESI-MS. Qualitatively acceptable compounds are taken through a quantification and reformatting step. Sometimes further analytical and quality control steps are necessary. When HPLC or HPLC-MS are used for purification of the library, the logistics are much more difficult than for the 1:1 purification strategies. The raw products from synthesis have to be filtered and dissolved in an HPLC-compatible solvent. Additionally, all of them have to be reprocessed into an HPLC-suitable format, usually a microplate. After chromatographic purification a large number of fractions have to be identified and evaporated. A pooling step is necessary for those fractions that give

Tab. 6.6. Liquid handlers suitable for sample processing in parallel synthesis.

Liquid handler	Application
Tecan	Four- and eight-needle systems suitable for reformatting and synthesis in MTPs. A robot arm can also be adopted for use in vial handling, weighing, etc. Further equipment for evaporating of filter plates, etc. available
Myriad Mettler Toledo	Four-needle system with robot arm specially designed for weighing, SPE
Zinsser Lissy	Four-needle system for reformatting and synthesis. Robot arm and further equipment for weighing, SPE and synthesis available
Packard	Eight-needle system, second arm with disposable tips, evaporation manifold available

MTP, microtiter plate; SPE, solid-phase extraction.

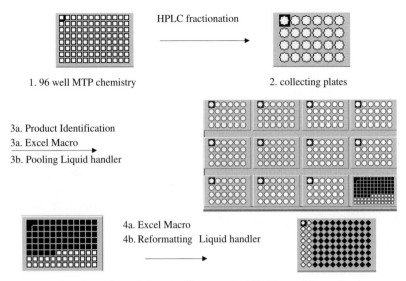

1. 96 well MTP chemistry   HPLC fractionation   2. collecting plates

3a. Product Identification
3a. Excel Macro
3b. Pooling Liquid handler

4a. Excel Macro
4b. Reformatting Liquid handler

Fig. 6.15. Flow chart of sample logistics if preparative HPLC is used for purification.

a positive result. This step can be performed before or after evaporation of the fractions. Reformatting before evaporation has the advantage that only the product-containing fractions are evaporated, thus saving capacity in the vacuum centrifuges. After evaporation, the fractions are pooled and analyzed and normally redissolved and processed into vials or microplates (MTPs), from where biological testing can be performed. The whole process is described in Fig. 6.15. The HPLC system distributes the fractions from one injection into different cavities of the collecting MTP. After identification of the products a software tool, usually an Excel macro, produces a run file for the liquid handler, which pools the fractions back into a 96-well MTP. From there, further analysis and reformatting steps are performed after the Excel macro has produced the second run file pooling the quality-controlled samples into the desired final formats.

6.4.2
Evaporation

Solvent removal is one of the most important downstream processes in parallel synthesis. Several attempts to address this issue have been made, and vacuum centrifuges and freeze-drying for water-containing mixtures are the most appropriate methods. Besides these techniques, shakers using infrared (IR) radiation [74] or stirrers in combination with air or nitrogen streams [75] are also in use.

The main advantages of freeze-drying are that no thermal stress is applied to the sample and the loading of liquid can be fairly high. Also, carrier formats are not limited and almost every type of carrier, for example microplates, racks, and single vials, can be loaded into the freeze-dryer. The disadvantages are that the evaporation time can be long and that the methodology is limited to aqueous solutions.

Tab. 6.7. Vacuum centrifuges commonly used in parallel synthesis.

Centrifuge	Advantages/disadvantages
Genevac	Robust in terms of imbalance, high capacity, fast evaporation/price, stability
Christ	Price, fast evaporation/capacity, unstable in case of imbalance
Savant	Vapor stability/capacity

Vacuum centrifuges have the advantage that almost every solvent can be evaporated, even those with very high boiling points such as dimethylsulfoxide (DMSO). Evaporation times are usually fast and the samples are concentrated and centrifuged at once, which results in all the material ending up at the bottom of the vial. Disadvantages are the thermal stress, which can lead to degradation of the products, and the possibility of imbalance due to the different concentrations of solvent mixtures.

Several suppliers sell vacuum centrifuges for evaporation of HPLC fractions or cleaving solutions from solid-phase chemistry. Usually, carrier formats are fixed or limited. The microplate footprint is the most common, which makes it necessary to optimize the whole workflow to this footprint. Advantages and disadvantages of several commercially available instruments are shown in Table 6.7.

References

1 a) W. HARRISON, *Drug Discovery Today* **1998**, *3*, 343–349; b) J. DREWS, S. RYSER, *Drug Discovery Today* **1997**, *2*, 365–372.
2 a) A. T. MERRIT, *Drug Discovery Today* **1998**, *3*, 505–510; b) J. S. LINDSEY, *Lab. Information Manage.* **1992**, *17*, 15–45; c) J. F. CARGILL, M. LEBL, *Curr. Opin. Chem. Biol.* **1997**, *1*, 67–71.
3 a) D. HARDING, M. BANKS, S. FOGARTY, A. BINNIE, *Drug Discovery Today* **1997**, *9*, 385–390; b) J. R. BROACH, J. THORNER, *Nature* **1996**, *7*, 14–16.
4 N. W. HIRD, *Drug Discovery Today* **1999**, *4*, 265–274.
5 S. H. DeWITT, A. W. CZARNIK, *Curr. Opin. Biotechnol.* **1995**, *6*, 640–645.
6 J. H. HARDIN, F. R. SMIETANA in: High Throughput Screening. DEVLI, J. P. (ed.), New York, Basel, Decker 1997, pp. 251–261.
7 A. M. M. MJALI, B. E. TOYONAGA in: High Throughput Screening. DEVLI,
J. P. (ed.), New York, Basel, Decker 1997, pp. 209–221.
8 http://www.robocon.com
9 http://www.crsrobotics.com
10 http://www.beckmancoulter.com
11 http://www.mitsubishi-automation.de
12 http://www.zymark.com
13 H. BRÜMMER, R. L. M. MARKERT, C. SCHWEMLER, *GIT Laborfachzeitschrift,* **1999**, *43*, 598–601.
14 a) T. E. WEGLARZ, S. C. ATKIN in: Advances in Laboratory Automation Robotics, vol. 6. STRIMAITIS, J. R., HELFRICH, J. P. (eds), Zymark, Hopkinition 1990, pp. 435–461; b) R. MATSUDA, M. ISHIBASHI, Y. TAKEDA, *Chem. Pharm. Bull.* **1988**, *36*, 3512–3518.
15 B. G. MAIN, D. A. RUDGE in: ISLAR'93 Proceedings, http://www.islar.com
16 E. WEGRZYNIAK et al. in: ISLAR'98 Proceedings, http://www.islar.com
17 D. A. RUDGE, M. L. CROWTHER in:

ISLAR'97 Proceedings, http://www.islar.com

18 http://www.accelab.de

19 http://www.Sankyo.com

20 http://www.chemspeed.com

21 http://www.peptide.com

22 http://www.multisyntech.com

23 J. COOPE, Zeneca Pharmaceuticals in: High Throughput Multiple Parallel Synthesis using the Zenyx Magellan Synthesizer, on MipTec-ICAR'99, 1999.

24 http://www.prosence.net

25 a) http://www.appliedbiosystems.com/pa/solaris/index_.html; b) http://www.appliedbiosystems.com/molecularbiology/press_releases/solaris/081098.html

26 http://www.zinsser-analytic.com

27 http://www.qiageninstruments.com/automation

28 http://www.charybdis.com

29 http://www.argotech.com

30 J. LABADIE, Argonaut Technologies Inc. in: The Trident System: A Flexible Approach to Automated Synthesis, on MipTec-ICAR'99, 1999.

31 a) http://www.techprt.co.uk/ttp/m-t_myriad.htm; b) B. MACLACHLAN, SmithKline Beecham in: Automated Chemical Synthesis using Myriad System, on MipTec-ICAR'99, 1999.

32 http://www.techprt.co.uk/ttp/index.html

33 http://www.mt.com/home/mettlertoledo.asp

34 a) http://www.irori.com; b) X. XIAO, C. ZHAO, H. POTASH, M. P. NOVA, *Angew. Chem.* **1997**, *109*, 799–801.

35 N. K. TERRET, M. GARDNER, D. W. GORDON, R. J. KOBYLECKI, J. STEEL, *Tetrahedron* **1995**, *51*, 8135–8173.

36 M. A. GALLOP, R. W. BARETT, W. J. DOWER, S. P. A. FODOR, E. M. GORDON, *J. Med. Chem.* **1994**, *37*, 1233–1236.

37 J. S. FRÜCHTEL, G. JUNG, *Angew. Chem.* **1996**, *108*, 19–46.

38 R. H. GRIFFEY, A. HAOYUN, L. CUMMINS, H.-J. GAUS, B. HALY, R. HERRMANN, P.-D. COOK, *Tetahedron* **1998**, *54*, 4067–4076

39 H. N. WELLER, *Mol. Diversity* **1999**, *4*, 47–52.

40 S. CHENG, D. D. CORNER, J. P. WILLIAMS, P. L. MEYERS, D. L. BOGER, *J. Am. Chem. Soc.* **1996**, *118*, 2567–2573

41 S. X. PENG, C. HENSON, M. J. STROJNOWSKY, A. GOLEBIOWSKI, S. R. KLOPFSTEIN, *Anal. Chem.* **2000**, *72*, 261–266.

42 B. KAYE, W. J. HERRON, P. V. MACRAE, S. ROBINSON, D. S. STOPHER, R. F. VENN, W. WILD, *Anal. Chem.* **1996**, *68*, 1658–1660.

43 R. M. LAWRENCE, S. A. BILLER, O. M. FRYSZMAN, M. A. POSS, *Synthesis* **1997**, 553–558.

44 M. G. SIEGEL, P. J. HAHN, B. A. DRESSMAN, J. E. FRITZ, J. R. GRUNWELL, S. W. KALDOR, *Tetrahedron Lett.* **1997**, *38*, 3357–3360.

45 R. J. BOOTH, J. C. HODGES, *J. Am. Chem. Soc.* **1997**, *119*, 4882.

46 S. V. LEY, I. R. BAXENDALE, R. N. BREAM, P. S. JACKSON, A. G. LEACH, D. A. LONGBOTTOM, M. NESI, J. S. SCOTT, R. I. STORER, S. J. TAYLOR, *J. Chem. Soc. Perkin Trans. 1* **2000**, 3815–4195.

47 Calbiochem-Novabiochem AG, Weidenmattweg 4, CH-4448 Läuflingen.

48 http://www.isco.com

49 http//www.biotage.com

50 http://www.joneschrom.com

51 I. HUGHES, *J. Assoc. Lab. Autom.* **2000**, *5*, 69–71.

52 W. C. RIPKA, G. BARKER, J. KRAKOVER, *Drug Discovery Today* **2001**, *6*, 471–477.

53 http//www.gilson.com

54 http//www.hii-hitachi.com

55 http://www.varianinc.com

56 http://www.shimadzu.com

57 L. ZENG, L. BURTON, K. YUNG, B. SHUSHAN, D. B. KASSEL, *J. Chromatogr. A* **1998**, *794*, 3–13.

58 J. P. KIPLINGER, R. S. WARE, E. ROSKAMP, P. DORF, R. G. COLE, S. ROBINSON, A. BRAILSFORD, R. RASO, C. FREY in: Pittcon'97 Conference, Atlanta 1997.

59 http://www.appliedbiosystems.com

60 http://www.waters.com

61 T. Y. Chan, R. Chen, M. J. Sofia, B. C. Smith, D. Glennon, *Tetrahedron Lett.* **1997**, *38*, 2821–2824.

62 R. Warras, J.-M. Wieruszeski, C. Boutillon, G. J. Lippens, *J. Am. Chem. Soc.* **2000**, *122*, 1789–1795.

63 M. C. Ventura, W. P. Farrell, C. M. Aurigemma, M. J. Greig, *Anal. Chem.* **1999**, *71*, 2410–2416.

64 B. H. Hsu, E. Orton, S.-Y. Tang, R. A. Carlton, *J. Chrom. B* **1999**, *725*, 103–112.

65 E. W. Taylor, M. K. Qian, G. D. Dollinger, *Anal. Chem.* **1998**, *70*, 3339–3347.

66 W. K. Götzinger, J. N. Kyranos, *Am. Lab.* **1998**, *30*, 27.

67 B. D. Duléry, J. Verne-Mismer, E. Wolf, C. Kugel, L. Van Hijfte, *J. Chrom. B* **1999**, *7*, 25, 39–47.

68 R. J. Cotter, *Anal. Chem.* **1999**, *71*, 445A–451A.

69 U. Selditz, S. Nilsson, D. Barnidge, K. E. Markides, *Chimia* **1999**, *53*, 506–510.

70 M. Wigger, J. P. Nawrocki, C. H. Watson, J. R. Eyler, S. A. Benner, *Rapid Commun. Mass Spectrom.* **1997**, *11*, 1749–1752.

71 R. Richmond, E. Görlach, *Anal. Chimica Acta* **1999**, *394*, 33–42.

72 R. Richmond, E. Görlach, J.-M. Seifert, *J. Chrom. A* **1999**, *835*, 29–39.

73 D. A. Lake, V. Johnson, C. N. McEwen, B. S. Larsen, *Rapid Commun. Mass Spectrom.* **2000**, *14*, 1008–1013.

74 http://www.hettlab.ch/IRclancer.htm

75 http://www.comgenex.hu

Part II
Synthetic Chemistry

7

Radical Reactions in Combinatorial Chemistry

A. Ganesan and Mukund P. Sibi

7.1
Introduction

Organic synthesis is dominated by polar transformations, in which an electron-rich center reacts with an electron-deficient center. By contrast, homolytic processes involving organic radicals were largely unexploited for many years. The general belief was that such "free radicals" would be undisciplined in their reactions, and prone to undesirable pathways such as premature radical–radical recombination or hydrogen atom abstraction from the solvent. Even under successful chain-propagating conditions, the chemistry seemed best suited to polymerization. More recently, a deeper understanding of the kinetics of radical reactions has enabled [1] the orchestration of a complicated series of elementary steps, and there is now a vast number of synthetically useful radical reactions which are complementary to traditional polar processes. For example, the reaction conditions are relatively mild, avoiding strongly acidic or basic reagents, and many functional groups are tolerated without requiring protection. Additionally, the product of a radical abstraction reaction or a radical addition to an unsaturated center is a new radical that can participate in further tandem reactions. Such reaction cascades can represent strategically powerful disconnections, as a relatively complex product is derived from a much simpler starting material in one step.

Combinatorial chemistry is also heavily dominated by polar reactions. Early efforts were predominantly based around peptides and nucleotides, as their synthesis was adaptable to the preparation of large libraries and compatible with existing automated equipment. Solid-phase synthesis techniques for such oligomers were already highly advanced [2], with nearly quantitative coupling yields at each cycle. However, the emphasis in combinatorial libraries rapidly shifted toward drug-like "small molecules". The resultant need to assemble diverse nonoligomeric carbocyclic and heterocyclic scaffolds has led to an intense effort to devise combinatorial versions of more sophisticated organic reactions, including those in which carbon–carbon bonds [3] are made. In recent years, the potential of radical reactions for combinatorial purposes has become apparent.

The majority of radical reactions explored for combinatorial applications have been reported on solid phase. While this does not change the actual chemistry compared with traditional solution-phase radical reactions, there are contexts in which the solid-phase environment may be advantageous. Even in the swollen gel phase, reactions are kinetically slower than in homogeneous solution-phase conditions, which may be helpful in determining the relative partitioning of a radical intermediate into various pathways. A second unique feature is the loading of polymeric resins, typically ≤ 1 mmol g^{-1} of beads, thus effectively placing an upper limit on the maximum concentration attainable with solid-phase reactions. This enforced dilution can be useful for radical reactions, although gel-phase polymer chains do exhibit significant freedom of motion and certainly do not approach "infinite dilution". A further variable is the spacer length between the substrate and the matrix. Many reactions, including Merrifield's original peptide synthesis, were carried out with the substrate attached directly to the polystyrene matrix. Today, it is more common to include a "linker" [4] between the polymer and the substrate. The linker can profoundly influence chain mobility as well as the polymer microenvironment where the reaction is taking place. Linkers are often used for solid-phase radical reactions, although the reasons for selecting a particular linker are seldom described. Finally, the ability to filter off reagents and byproducts can certainly be a bonus for solid-phase radical reactions, especially those mediated by tin reagents, whose removal is not always trivial in solution phase (see below).

7.2
Intramolecular Radical Additions to sp^2 and sp Carbon

The first carbon–carbon bond-forming radical reactions reported on solid phase were intramolecular aryl radical 5-*exo* cyclizations, which have ample solution-phase precedents [5]. Routledge et al. [6] investigated the formation of dihydrobenzofuran (**2**) from an aryl halide precursor (Scheme 7.1). The efficiency of the reaction was found to depend on the resin: with polystyrene, more than 1 equivalent of AIBN was required as radical initiator, whereas the reaction was complete within 20 h using 6 mol% of AIBN on TentaGel resin. The intermediate alkyl radical underwent two different types of reactions: a β-elimination or a H-atom abstraction from tributyltinhydride. The β-elimination process could be suppressed

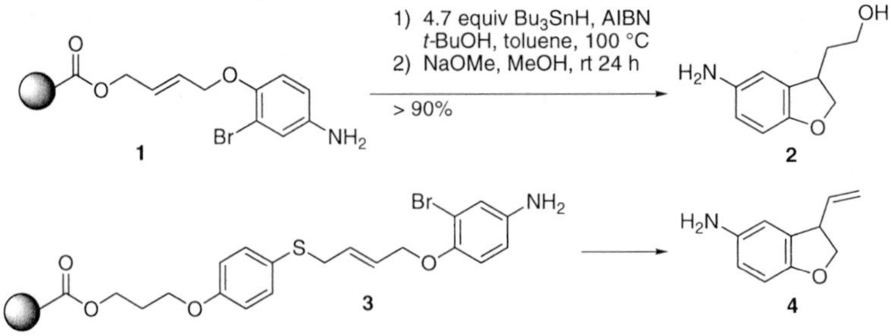

Scheme 7.1. Aryl halide cyclizations by Routledge et al. [6].

by the addition of *t*-butanol. An attempt to make the *β*-elimination the major pathway by switching to thiyl linker (**3**) only yielded a small amount of **4**. 5-*Exo* cyclizations of an alkyl radical onto an acetylene group leading to exomethylene furans were also reported in this study.

Du and Armstrong [7] reported similar cyclizations of aryl iodides attached to polystyrene–Rink resin using SmI₂ for radical generation (Scheme 7.2). The reactions can be carried out under mild conditions without the solvent degassing. However, a large excess of HMPA was found to be necessary for efficient reaction. Use of a TentaGel-type resin allowed polymer swelling in aqueous solvents, enabling Sm(III) impurities in the beads to be removed by saturated NaHCO₃ prior to resin cleavage. The feasibility of radical–polar crossover reactions by anionic capture of the intermediary Sm(III) species by electrophiles was attempted [8]. The reaction was unsuccessful when the carboxylic acid was immobilized on polystyrene–Rink resin as an amide, possibly due to quenching of the anion by the amide proton. On the other hand, when substrate **7** on TentaGel–Wang resin was treated with HMPA, 3-pentanone, and SmI₂, the crossover product **8** was obtained in moderate yield after resin cleavage. Reaction efficiency with the TentaGel–Wang-supported substrate (33% yield) was similar to that in solution phase (40% yield), although perhaps still too low for reliable generation of libraries.

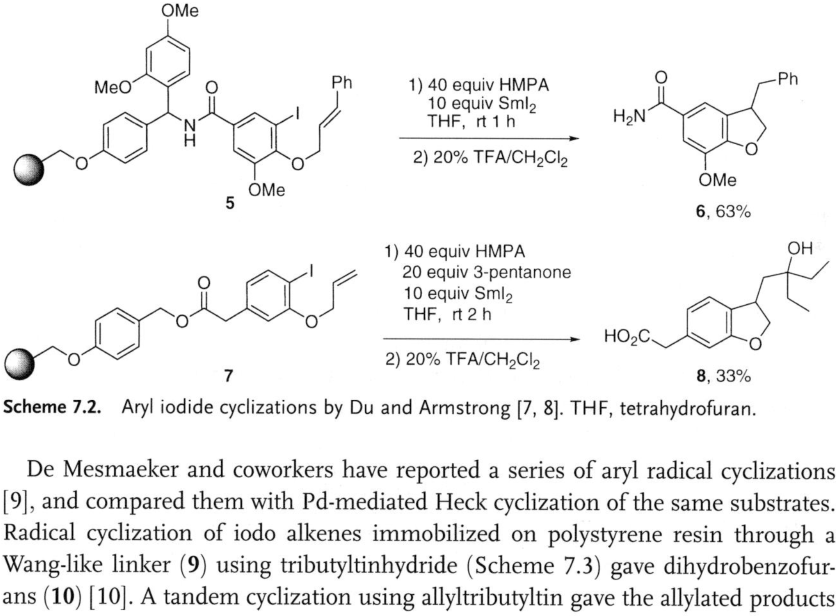

Scheme 7.2. Aryl iodide cyclizations by Du and Armstrong [7, 8]. THF, tetrahydrofuran.

De Mesmaeker and coworkers have reported a series of aryl radical cyclizations [9], and compared them with Pd-mediated Heck cyclization of the same substrates. Radical cyclization of iodo alkenes immobilized on polystyrene resin through a Wang-like linker (**9**) using tributyltinhydride (Scheme 7.3) gave dihydrobenzofurans (**10**) [10]. A tandem cyclization using allyltributyltin gave the allylated products (**11**) in low to moderate yields. Formation of isomeric dihydrobenzofurans (**13**) could be accomplished by radical cyclization onto *β*-alkoxy esters [11]. For best results, the tributyltinhydride and AIBN were added portionwise every 5–8 h. The impressive 95% yield was in fact higher than that for the solid-phase Heck cyclization of **12**.

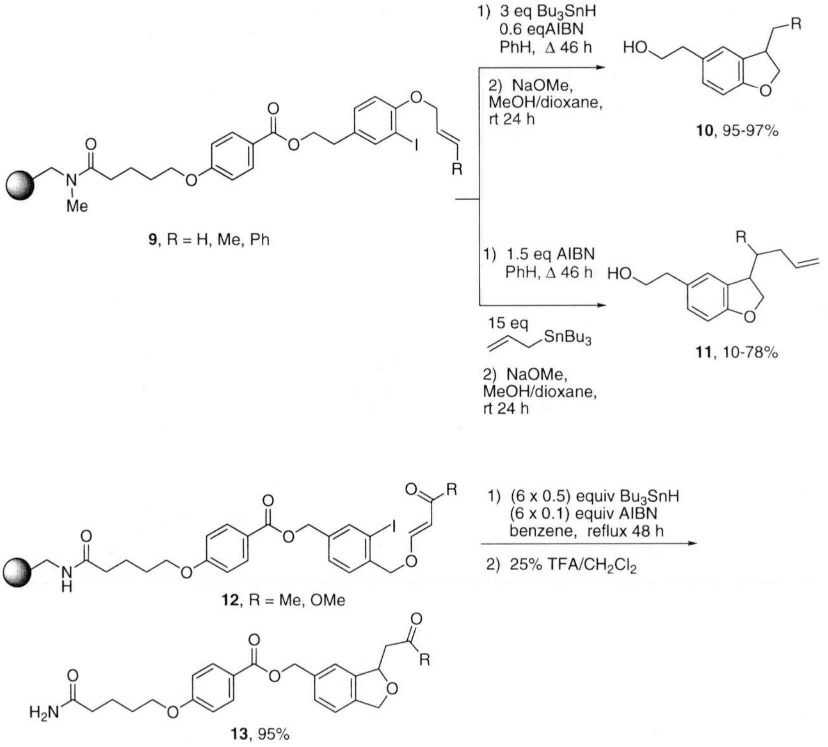

Scheme 7.3. Aryl iodide cyclizations by the Novartis group [10, 11].

The Novartis group has also studied [12] the radical cyclization of cyclohexene-diols, immobilized by a ketal linkage on polystyrene (Scheme 7.4). The reaction of **14** gave the desired dihydrobenzofuran (**15**) and the uncyclized product of direct reduction (**16**). Jia et al. [13] have reported a related cyclization with allylamine (**17**) immobilized on polystyrene–Wang resin. The reaction was monitored by acetylation and cleavage to yield **18**, as a mixture of free and Boc-protected amines. This solid-phase synthesis of *seco*-CBI (**18**, R = H), related to the pharmacophore of the CC-1065 and duocarmycin class of cyclopropylindole antitumor antibiotics, has potential for the preparation of analog libraries, and an example of further conversion of resin-bound **18** to a polyamide has been presented.

A series of bromoacetals (**19**) (Scheme 7.5) linked to polystyrene was cyclized to the corresponding acetals by Watanabe et al. [14]. Oxidative cleavage of the resin using Jones reagent gave easy access to γ-butyrolactones (**20**).

7.3
Intermolecular Radical Additions

The carbon–carbon bond-forming steps of intramolecular reactions are facilitated by entropic acceleration. *Intermolecular* reactions offer a more stringent test of the

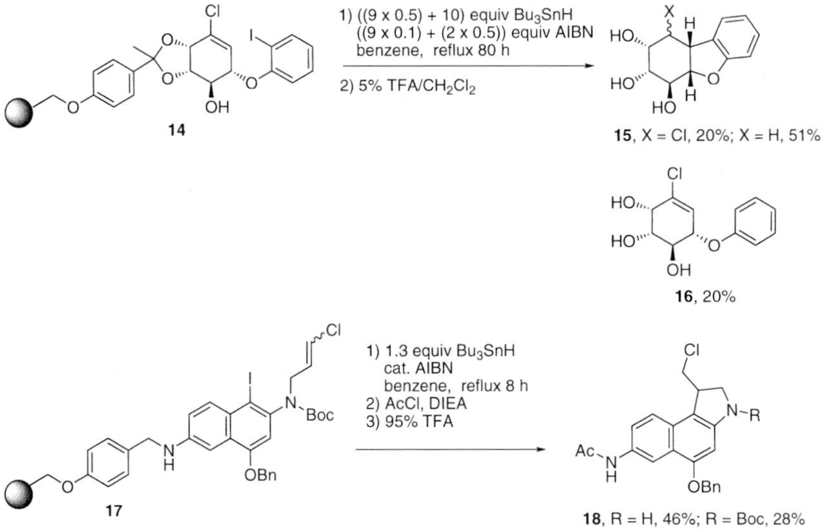

Scheme 7.4. Aryl and vinyl iodide cyclizations by Berteina et al. and Jia et al. [12, 13].

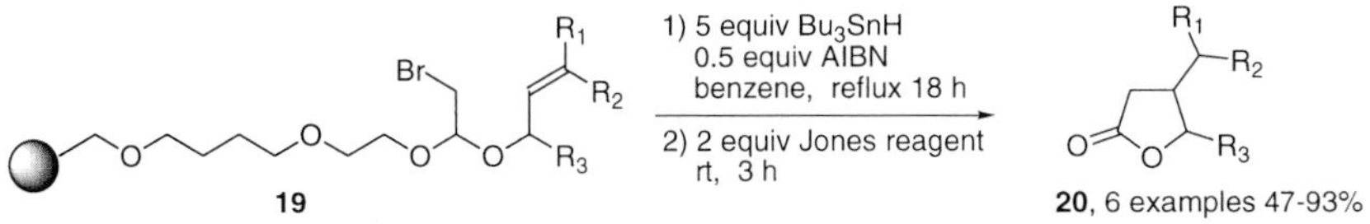

Scheme 7.5. Intramolecular cyclizations by Watanabe et al. [14].

feasibility of solid-phase radical reactions. Sibi and Chandramouli [15] reported the first examples of intermolecular radical allylations. The polymer-bound electrophilic C-radicals generated from α-bromo esters (**21**) gave γ,δ-unsaturated acids (**22**) (Scheme 7.6). Large excesses of allylstannane and AIBN were required for good yields. Radical initiation at room temperature using Et_3B/O_2 was less efficient. The yields were similar to those for solution-phase reactions, while reduction of **21** with tributyltin deuteride gave ~93% deuterium incorporation, implying <7% hydrogen atom transfer from the polymer matrix. Electron-withdrawing

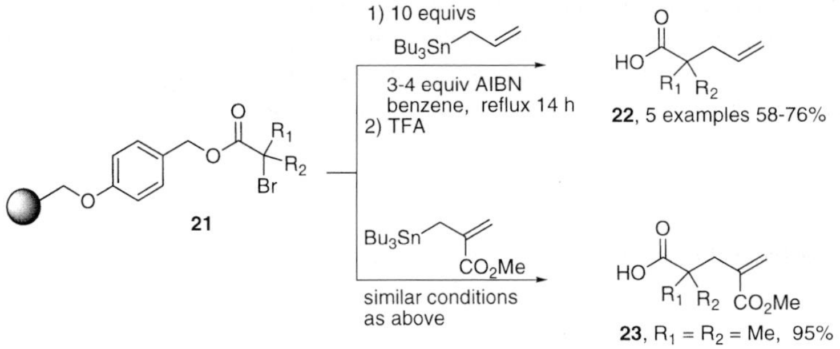

Scheme 7.6. Intermolecular allylations by Sibi and Chandramouli [15].

groups at the 2-position of the allylstannane were also found to improve the reaction yield, as in the formation of **23**.

Enholm et al. have reported the first example of a diastereoselective radical reaction on a polymer (Scheme 7.7) [16]. A sugar-based auxiliary attached to a non-crosslinked soluble polystyrene polymer (**24**) as the support was used in the experiments. Allylation under standard conditions gave **25** in high yield and diastereoselectivity (97%). Use of Lewis acids for the allylation was detrimental in that chiral auxiliary cleavage was observed.

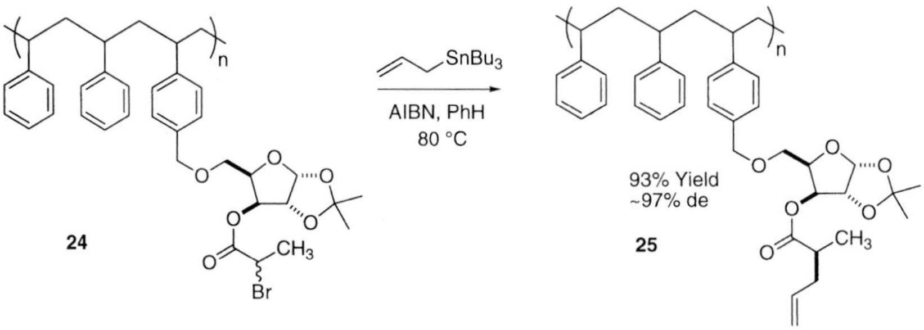

Scheme 7.7. Polymer-supported diastereoselective allylation [16].

Although organotin reagents are the most popular reagents for generating radicals in solid-phase reactions, other methods are also being explored. Zhu and Ganesan have investigated [17] the conjugate addition of radicals generated from Barton esters. Acrylic acid was immobilized as the ester (**26**) (Scheme 7.8) using the Wang linker on both polystyrene and TentaGel resins. Conjugate radical addition and cleavage gave acids (**27**) in high yields. Also noteworthy, as in the tin-mediated reactions, is the ability to use large excesses of reagent without problems in product isolation. The yields with polystyrene resin were similar to solution-phase results with methyl acrylate and appear to be relatively insensitive to the nature of the radical: primary, secondary, or tertiary. Benzyl radical gave **27** in only 32% yield, presumably due to its highly stabilized nature. The TentaGel resin gave consistently lower yields than polystyrene, suggesting interference from the polyether spacer. When acrylic acid was immobilized as the amide on polystyrene–Rink resin, yields were also lower, which is consistent with the lower solution-phase yields with acrylamide than with methyl acrylate.

An interesting cascade reaction was observed with the Barton ester (**28**) of cyclopent-2-enylacetic acid. Radical generation and addition to the acrylate resin results in electrophilic acyl-substituted radical **29**, which can undergo chain transfer to produce **30** (after cleavage and esterification). The major pathway is intramolecular 5-*exo* cyclization to provide radical **29a**. The intermediate nucleophilic radical **29a** gave two different products depending on the termination step. Chain transfer gave **31** as a minor product. Alternatively, addition of a second acrylate followed by thiyl transfer gave **32** as the major product. Formation of **32** requires the crosslinking of two polymer chains and shows that site isolation is not significant.

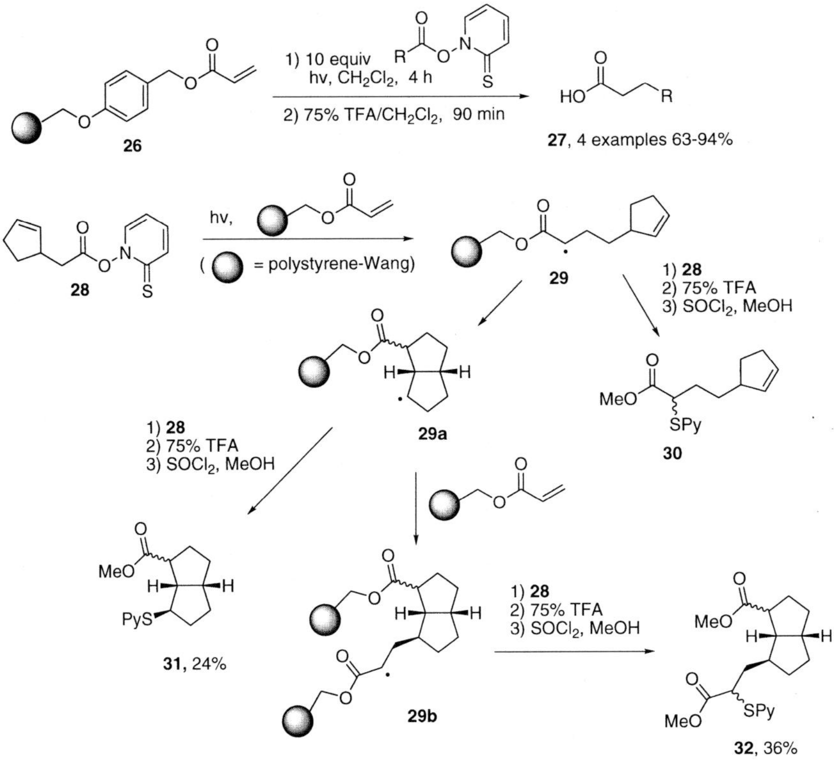

Scheme 7.8. Intermolecular conjugate additions by Zhu and Ganesan [17].

Barton ester chemistry also featured in the work by Attardi and Taddei [18]. Irradiation of a solid-phase-bound Barton ester derived from aspartic acid (Scheme 7.9) gave the corresponding alkyl radical, which could be trapped by CBrCl₃ provided it was in very large excess, to give the alkyl bromide (**34**) after cleavage. Also presented were examples in which the alkyl bromide was displaced by an amine, providing a route to the solid-phase synthesis of unnatural amino acids. Trapping the alkyl radical by intermolecular conjugate addition to methyl acrylate or a nitroolefin was also attempted, although desired products such as **35** were contaminated by 10–20% of **36**.

A route to α-amino acids by conjugate radical addition to the dehydroalanine derivative (**37**) on solid phase has been reported by Yim et al. [19]. Organomercurials as radical precursors (Giese method) gave better yields of **38** in contrast to the standard conditions of reaction between alkyl halides with tributyltinhydride (Scheme 7.10). The intermolecular addition of tosyl radicals to an unactivated alkene and alkyne (**39**) has also been reported [20], and the reaction has been found to be quite sensitive to solvent – toluene giving optimum results. The related free radical addition of thiols to unactivated alkenes has been described by Plourde et al. [21]. An inositol derivative was directly immobilized on carboxypolystyrene

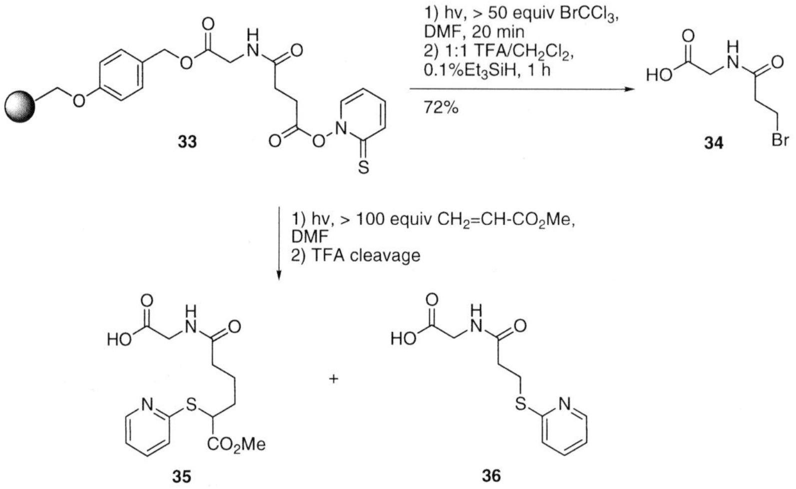

Scheme 7.9. Intermolecular radical additions by Attardi and Taddei [18].

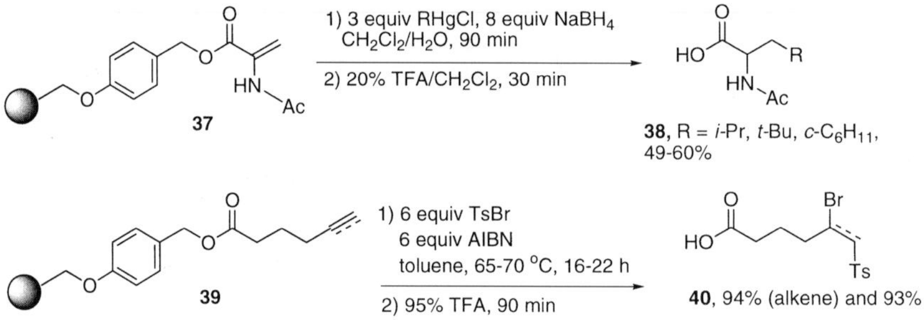

Scheme 7.10. Intermolecular alkene additions [19, 20].

resin and reacted with aliphatic thiyl radicals generated by AIBN, yielding **42** (Scheme 7.11) after cleavage. The reaction was not observed with arylthiyl radicals, possibly because of their increased stability. Similarly, functionalization of an aminocyclitol linked to the Wang resin to give **44** was accomplished.

Amines can be readily prepared by intermolecular radical additions to oxime ethers. In a nice extension of its work in solution-phase chemistry, Naito's group has published several papers on amine synthesis on solid support. Immobilized glyoxylic oxime ether (**45**) underwent radical addition to give amino acid derivatives (**46**) in moderate to good yields (Scheme 7.12) [22]. Preparation of enantioenriched amino acids by this strategy has also been reported. Chirality was incorporated in the form of Oppolzer's camphorsultam and the solid support was attached on the oxime hydroxyl (**47**). Ethyl radical addition to **47** using either triethylborane or diethylzinc as radical initiators gave product (**48a**) with >95% diastereoselectivity [23]. This was better than the analogous solution-phase reaction, suggesting that the immobilized oxime is less reactive. The addition of isopropyl

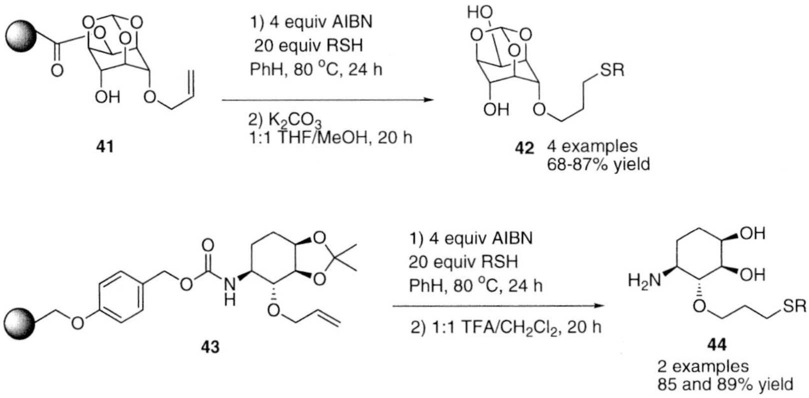

Scheme 7.11. Intermolecular alkene additions [21].

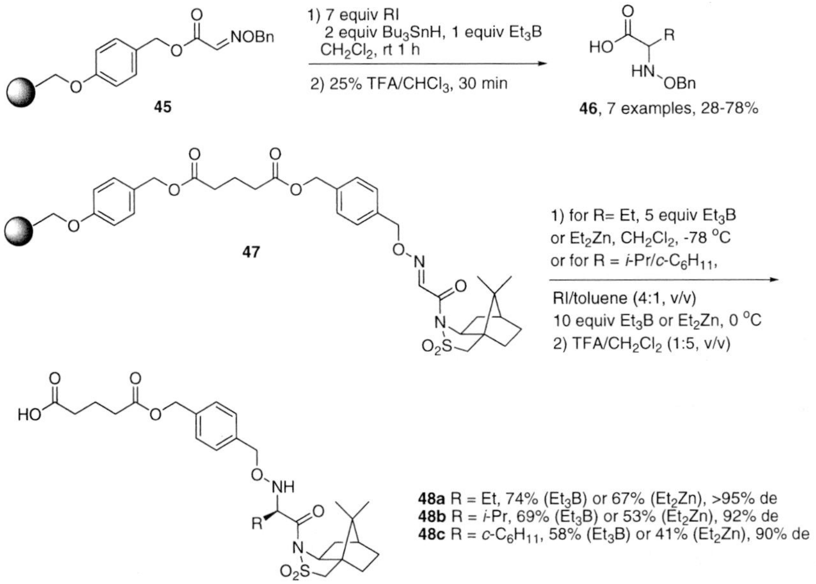

Scheme 7.12. Intermolecular oxime ether additions by Miyabe et al. [22, 23].

and cyclohexyl radicals by atom transfer from the corresponding iodide, to furnish **48b,c**, has also been demonstrated. In these cases, the product was contaminated by the competing addition of initiating ethyl radical, and this was avoided by using a large excess of the alkyl iodide.

Pyrrolidines have been synthesized on solid phase by a combination of an intermolecular radical addition followed by an intramolecular oxime ether cyclization, as exemplified in the preparation of **50** [24] and **52** [25] (Scheme 7.13). The solid-phase reactions were sluggish with triethylborane as an initiator at room temperature, while the analogous solution-phase process was kinetically much faster. Rad-

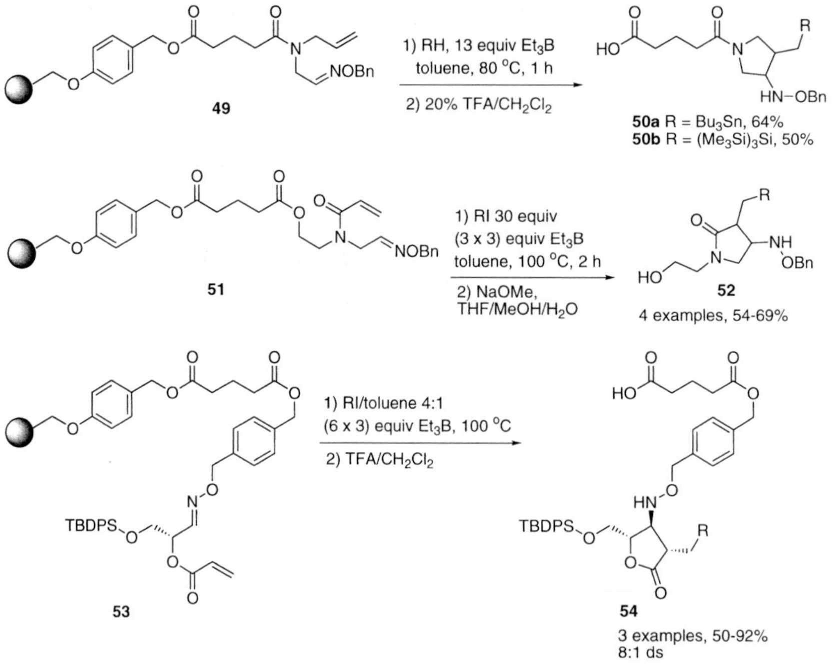

Scheme 7.13. Intermolecular oxime ether additions by Miyabe et al. [24, 25].

ical addition to chiral substrate **53** using triethylborane as an initiator gave mostly the ethyl addition product **54** (R = Et). Successful incorporation of the alkyl radical generated from the alkyl iodide required the reaction to be performed almost neat in the latter, and proceeded with decent diastereoselectivity.

Radical additions to the phenylsulfonyl oxime ether **55** (Scheme 7.14) have been reported by Jeon et al. [26]. Yields were better with primary and secondary alkyl

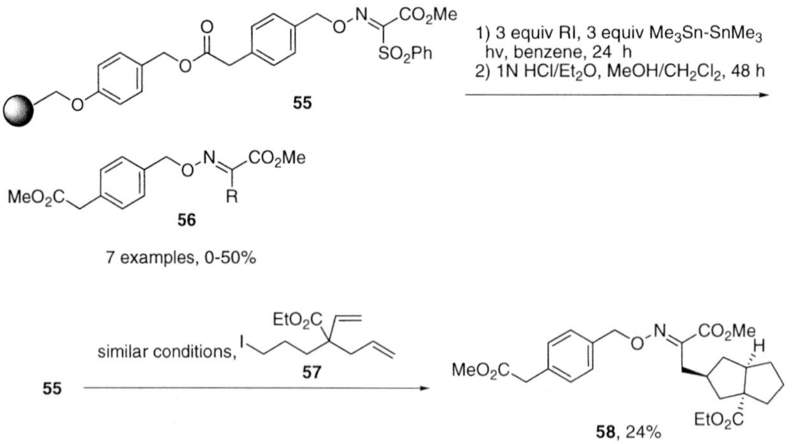

Scheme 7.14. Intermolecular oxime ether additions by Jeon et al. [26].

iodides, and the tandem cyclization sequence with iodide (**57**) to afford bicyclic **58** has been accomplished, albeit in modest yield.

7.4
Functional Group Removal

Solid-phase synthesis necessarily requires attachment to the polymer matrix by a functional group, which will be unmasked at the end of the synthetic sequence upon cleavage from the resin. In peptide and nucleotide synthesis, this respectively reveals the C-terminus carboxylic acid and the 5′ alcohol. As these groups are inherently part of the final biopolymer, this does not pose a problem. For small-molecule synthesis, however, it is not always desirable to have a dangling functional group whose sole purpose had been to enable immobilization. One possible solution is to devise "traceless" linkers [27] that produce C–H bonds upon cleavage, as realized by Plunkett and Ellman [28] with an arylsilane cleaved by protodesilylation.

Homolytic carbon–heteroatom bond fission followed by quenching of the organic radical represents an alternative means of achieving a "traceless" solid-phase synthesis. A pioneering effort by Sucholeiki [29] involved the photochemical irradiation of resin-bound thioethers (Scheme 7.15) attached via two different linkers to TentaGel resin. The isolated yields of biphenyl **60** were reportedly low owing to product volatility, and degassed solvent was crucial to minimize the formation of byproduct aldehyde **61**.

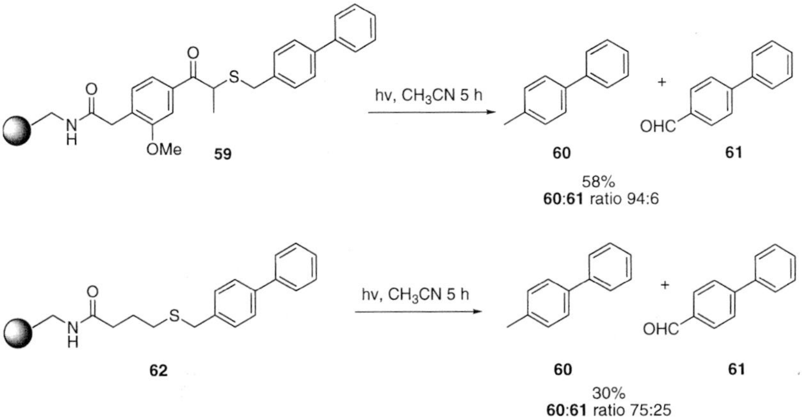

Scheme 7.15. Thioether cleavage by Sucholeiki [29].

More recently, several groups have explored the homolytic cleavage of resin-bound selenides as a route to "traceless" synthesis. Nicolaou et al. [30] started with lithiated polystyrene, which was trapped by dimethyl diselenide to furnish the alkylselenide resin. Reaction with bromine generated selenylbromide resin (**63**) (Scheme 7.16), which could be reduced to the selenide anion (**64**). This resin can

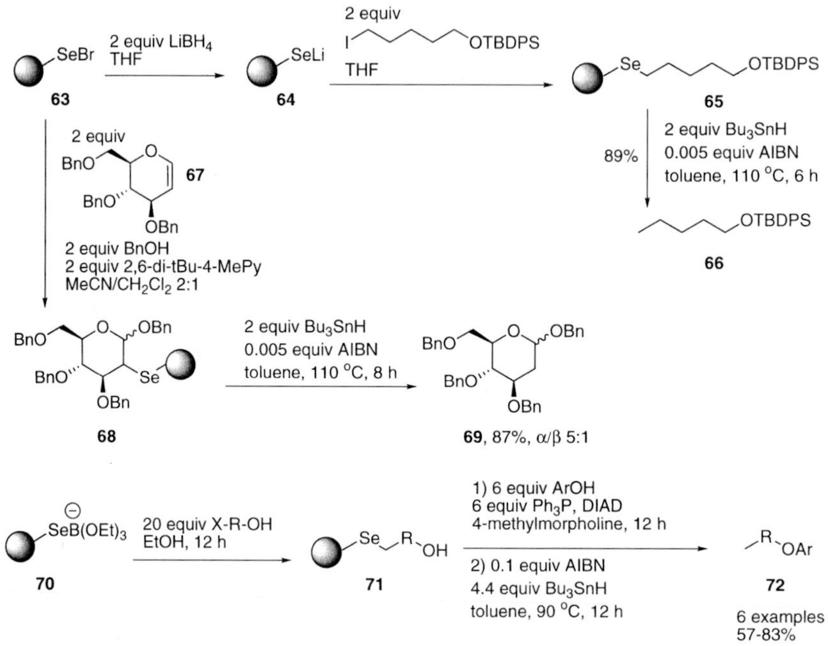

Scheme 7.16. Traceless selenide cleavage by Nicolaou et al. and Ruhland et al. [30, 31].

then be reacted with alkyl halides, followed by "traceless" reductive cleavage with tributyltinhydride. A similar dealkylation was simultaneously reported by Ruhland et al. [31], in which a chloro- and a bromoalcohol were immobilized by the selenide resin (**70**). The free alcohol was then functionalized by a Mitsunobu reaction prior to "traceless" cleavage. Nicolaou also described the use of resin **63** in promoting the glycosidations of sugar glycals, resulting in a polymer-bound selenide whose radical cleavage yields 2-deoxyglycosides. The Nicolaou group has since reported a number of further applications with these selenide resins.

In the previous sections, many solid-phase radical reactions were described that had an extended linker between the substrate and the polymer matrix, unlike the Nicolaou and Ruhland examples. Fujita et al. were the first to report [32] a "traceless" selenide cleavage with a linker. However, the yield from **73** (Scheme 7.17) was poor, and the authors suggest that the tinhydride reagent is reacting with the linker itself. A more robust ether-based selenide linker (**76**) has recently been described by Li et al. [33], and an example of homolytic cleavage proceeded with good yields.

7.5
Polymer-supported Reagents for Radical Chemistry

As was illustrated in the above sections, radical reactions are feasible on solid support and are relatively new entrants to the field. In contrast, the use of polymer-

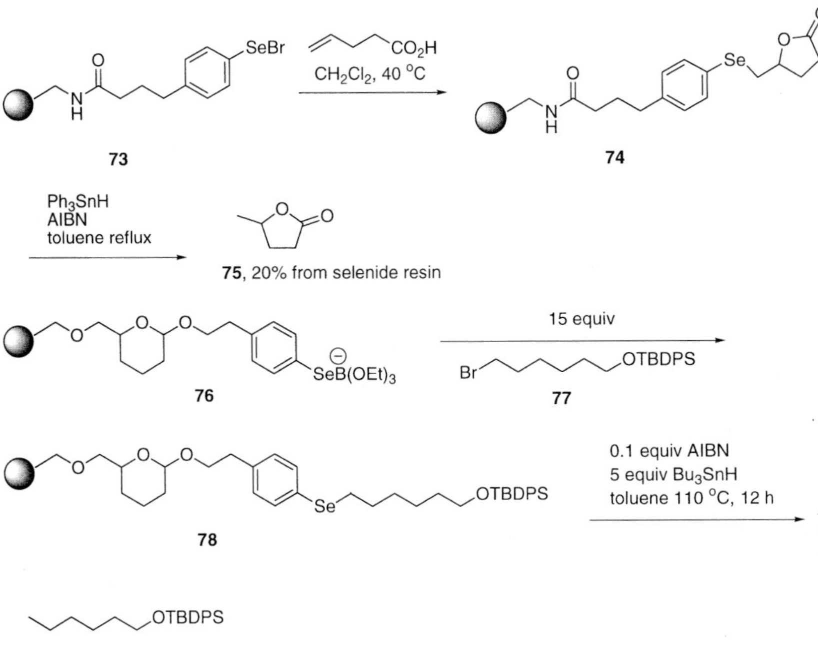

Scheme 7.17. Traceless selenide cleavage by Fujita et al. and Li et al. [32, 33].

supported organic reagents has a much longer history in radical chemistry. Developments in this area can be attributed to the difficulty in removal of organotin by-products, since tin-derived compounds are used extensively for radical chemistry. Use of polymer-supported reagents alleviates most of the difficulties associated with product purification. The tin-containing polymer can be easily removed by simple washing.

7.5.1
Polymer-supported Tinhydrides

Most of the activity in polymer-supported reagents has been in the preparation of tinhydride derivatives. Early work in this area was carried out by Neumann and coworkers [34]. A polystyrene-derived reagent (**80**), which mimics tributyltinhydride, was prepared and evaluated as a reducing and chain transfer reagent. Representative examples are shown in Scheme 7.18. Similar chemical efficiency in the cyclization of **81** and in the reductions of **82–84** demonstrated that the polymeric tinhydride has similar reaction characteristics to tributyltinhydride.

Dumartin and coworkers have evaluated several polystyrene-supported tinhydrides prepared from Amberlite XE305 in reductions [35]. These reagents (**85–86**) contain different spacers between the tin atom and the phenyl group. The amount of residual tin was determined by ICP-MS for the reduction of **87** (Scheme 7.19). The polymer hydride produced 45 ppm of residual tin compared with 7000 ppm

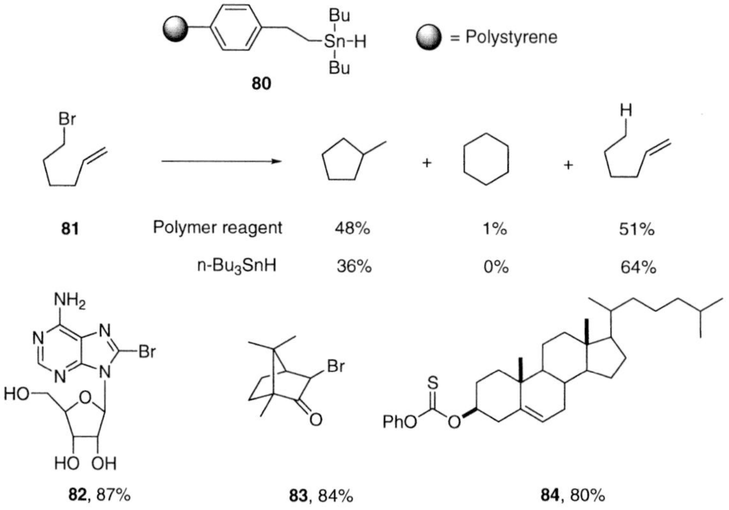

Scheme 7.18. Reductions using polystyrene-supported tinhydride [34].

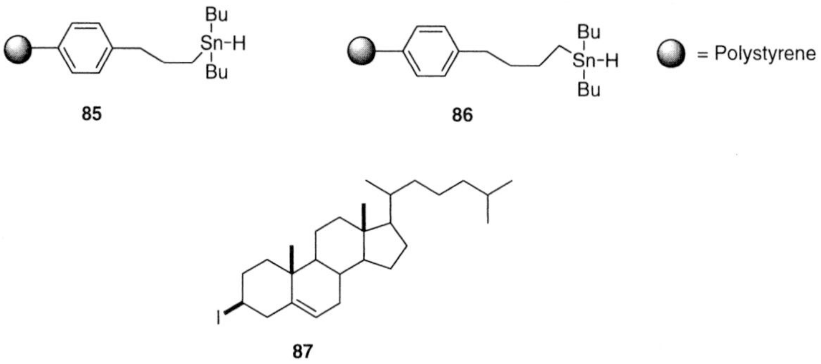

Scheme 7.19. Reductions using polystyrene-supported tinhydride [35].

from tributyltinhydride. *In situ* generation of the polymer-supported reagent from the corresponding halide and sodium borohydride and reactions without mechanical stirring significantly reduced the tin pollution level. The Dumartin group has also reported Barton–McCombie deoxygenation of secondary alcohols using catalytic amounts of polymer-supported tinhydride [36].

Deluze and coworkers [37] have prepared a new type of macroporous polymer-supported tinhydride using suspension copolymerization (Scheme 7.20). The authors have determined the swelling characteristics of the resin in different solvents as well as the specific surface areas. The organotin chloride-functionalized beads showed good stability and reactivity in reductions using sodium borohydride as the co-reductant.

Recently, Enholm and Schulte [38] prepared a soluble tinhydride on a non-crosslinked polystyrene support (**89**) and demonstrated its utility in reductions

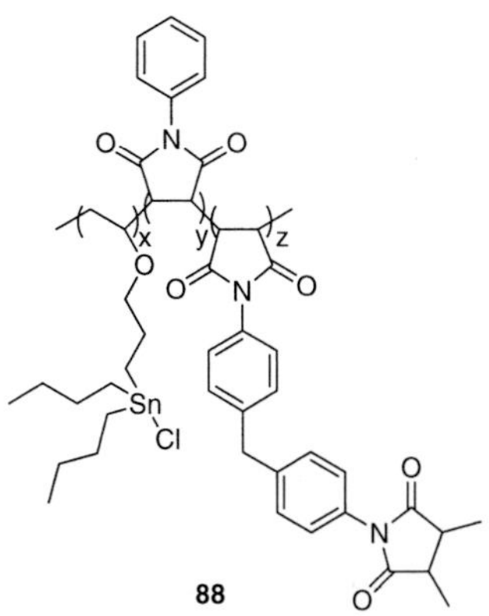

Scheme 7.20. A macroporous polymer-supported tinhydride precursor [37].

(Scheme 7.21). The reductions use only catalytic amounts of the polymeric tinhydride and sodium borohydride as the co-reductant. Several examples (**90–94**) of chemoselective reductions have been reported. The tin contents in the products were also determined and shown to be in the ppm range.

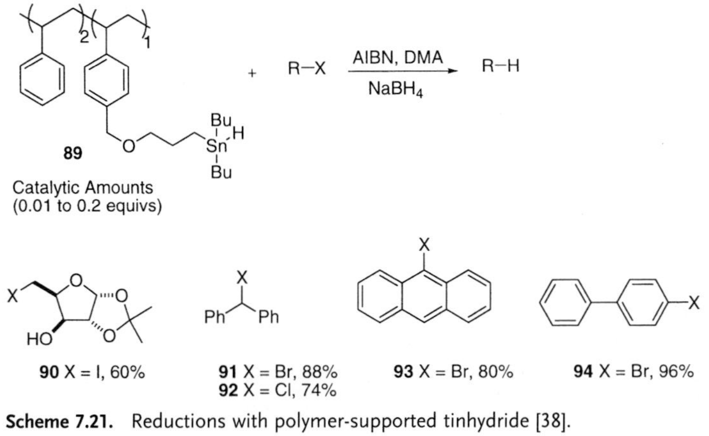

Scheme 7.21. Reductions with polymer-supported tinhydride [38].

Curran and coworkers [39] have developed novel alternatives for polymer-supported reagents. They have successfully demonstrated the use of very versatile fluorous reagents (**95**) for free radical chemistry [39]. The fluorous tinhydrides (Scheme 7.22) are commercially available and have the advantages of ease of product purification by simple separation and high reaction rates since the reactions are carried out in the solution phase.

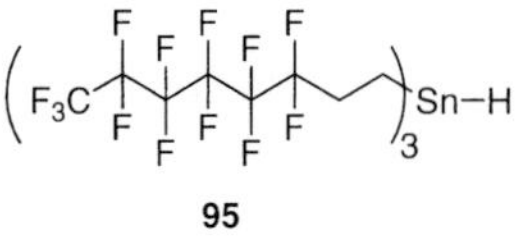

95

Scheme 7.22. Fluorous tinhydride [39].

7.5.2
Polymer-supported Allyl Stannane

Enholm and coworkers [40] have prepared a soluble noncrosslinked polystyrene-supported allyltin reagent (**96**) (Scheme 7.23). The reagent is quite successful in transferring an allyl group to a variety of substrates in moderate to good yields. The products (**98–101**) obtained from allylation show reasonable structural diversity. The selective allyl transfer to the more electrophilic site in **101** is noteworthy. Ryu, Curran, and coworkers [41] have demonstrated the utility of fluorous allyltin reagents in organic synthesis.

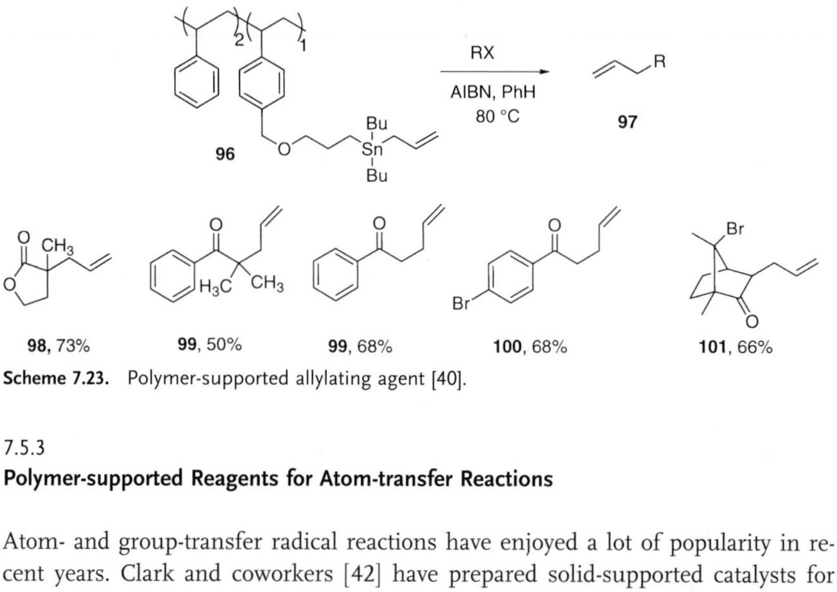

Scheme 7.23. Polymer-supported allylating agent [40].

7.5.3
Polymer-supported Reagents for Atom-transfer Reactions

Atom- and group-transfer radical reactions have enjoyed a lot of popularity in recent years. Clark and coworkers [42] have prepared solid-supported catalysts for atom-transfer radical cyclizations (Scheme 7.24). Functionalized aminopropyl silicagel was coupled to pyridine 2-carboxaldehye to provide an orange solid (**102**). The catalyst (**103**) was prepared by stirring a mixture of **102** and copper halide in acetonitrile. Atom transfer cyclization of several *N*-allyl amides using 30 mol% of the catalyst gave the product in high yields. The catalyst could be recovered by simple filtration. Recently, a soluble copper catalyst supported on poly(ethylene)-block-poly(ethylene glycol)-polymer (**108**) has been effectively used for atom-transfer radical polymerizations [43]. Marsh and coworkers [44] have reported Cu(I)-mediated radical polymerization of nucleoside monomers on silica support.

The preparation of polystyrene-block-polymethylmethacrylate films by a sequential carbocationic polymerization of styrene followed by radical polymerization of methylmethacrylate has been reported previously [45].

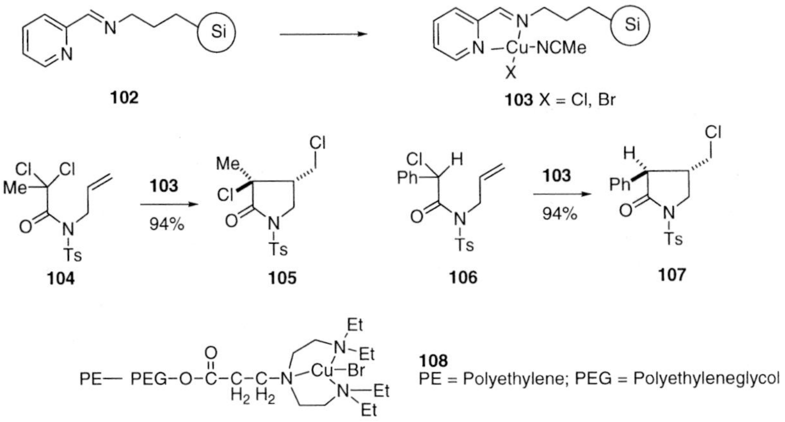

Scheme 7.24. Atom-transfer radical cyclizations and polymerizations with polymer-supported catalysts [42, 43].

7.5.4
Photochemical Generation of Radicals

O-Acyl and *O*-alkylthiohydroxamates are convenient precursors for C- and O-centered radicals. Product purification at times is difficult because of the sulfur byproducts. De Luca and coworkers [46] have reported a solution to this problem. The polymer support with appropriate functionality was prepared from *N*-hydroxy-thiazole 2(3)-thione and functionalized Wang resin (**109**). A Hunsdicker reaction of **110** furnished the corresponding bromide (**111**) cleanly. Similarly, the cyclization of the *O*-centered radical under reductive conditions furnished ether **113**. In both cases, product purification was simple (Scheme 7.25).

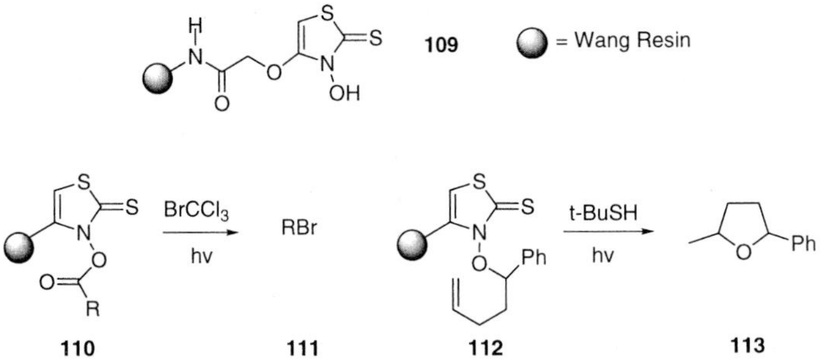

Scheme 7.25. Photochemical generation of radicals [46].

Tab. 7.1. Methods used in solid-phase radical reactions.

Radical precursor	Method of radical generation
Alkyl/aryl halide RBr or RI	AIBN, Bu_3SnH
	$Me_3SnSnMe_3$, $h\nu$
	SmI_2
	Et_3B, O_2
	Et_2Zn, O_2
Organomercurial RHgCl	$NaBH_4$
Thioether RSR	$h\nu$
Thiol RSH	AIBN
Sulfonyl bromide $ArSO_2Br$	AIBN
Alkyl selenide RSeAr	AIBN, Bu_3SnH
Barton ester RCO_2–NAr	$h\nu$

7.6
Summary

Radical reactions have yet to be applied to the synthesis of compound libraries, whether in combinatorial fashion or as parallel arrays. This is perhaps not surprising, as it is only in the last few years that the feasibility of carrying out radical reactions at all on solid phase has been demonstrated. The examples in this chapter convincingly illustrate that there is no intrinsic limitation to such processes. Now that the gestation period is over, one can predict that radical reactions will be part and parcel of the combinatorial chemist's toolkit of organic reactions. The various precursors to radicals used in solid-phase reactions, and the methods of generation, are summarized in Table 7.1.

The early preponderance of solid-phase reactions in combinatorial chemistry has also evolved to a more equal weighting with solution-phase synthesis. Here also, the field of radical reactions has not been left out. A number of polymer-supported reagents and scavenger resins are now available that facilitate radical reactions in parallel, with the potential for considerably simplified work-up and purification compared with the original procedures. Further advances in combinatorial radical chemistry, both solution and solid phase, can certainly be anticipated.

References

1 P. Renaud, M. P. Sibi (eds) in: Radicals in Organic Synthesis, Wiley-VCH, Weinheim, 2001.
2 For an account of the early history of solid-phase synthesis, see: a) D. Hudson, *J. Comb. Chem.* 1999, *1*, 333–360. b) D. Hudson, *J. Comb. Chem.* 1999, *1*, 403–457.
3 For reviews, see: a) B. A. Lorsbach, M. J. Kurth, *Chem. Rev.* 1999, *99*, 1549–1581; b) R. E. Sammelson, M. J. Kurth, *Chem. Rev.* 2001, *101*, 137–202.
4 For reviews, see: a) I. W. James, *Tetrahedron* 1999, *55*, 4855–4946; b) F. Guillier, D. Orain, M. Bradley, *Chem. Rev.* 2000, *100*, 2091–2157.

5 For a review, see: B. K. Banik, *Curr. Org. Chem.* **1999**, *3*, 469–496.

6 A. Routledge, C. Abell, S. Balasubramanian, *Synlett* **1997**, 61–62.

7 X. Du, R. W. Armstrong, *J. Org. Chem.* **1997**, *62*, 5678–5679.

8 X. Du, R. W. Armstrong, *Tetrahedron Lett.* **1998**, *39*, 2281–2284.

9 For a review, see: S. Wendeborn, A. De Mesmaeker, W. K. D. Brill, S. Berteina, *Acc. Chem. Res.* **2000**, *33*, 215–224.

10 S. Berteina, A. De Mesmaeker, *Tetrahedron Lett.* **1998**, *39*, 5759–5762.

11 S. Berteina, S. Wendeborn, A. De Mesmaeker, *Synlett* **1998**, 1231–1233.

12 S. Berteina, A. De Mesmaeker, S. Wendeborn, *Synlett* **1999**, 1121–1123.

13 G. Jia, H. Iida, J. W. Lown, *Synlett* **2000**, 603–606.

14 Y. Watanabe, S. Ishikawa, G. Takao, T. Toru, *Tetrahedron Lett.* **1999**, *40*, 3411–3414.

15 M. P. Sibi, S. V. Chandramouli, *Tetrahedron Lett.* **1997**, *38*, 8929–8932.

16 E. J. Enholm, M. E. Gallagher, S. Jiang, W. A. Batson, *Org. Lett.* **2000**, *2*, 3355–3357.

17 X. Zhu, A. Ganesan, *J. Comb. Chem.* **1999**, *1*, 157–162.

18 M. E. Attardi, M. Taddei, *Tetrahedron Lett.* **2001**, *42*, 3519–3522.

19 A.-M. Yim, Y. Vidal, P. Viallefont, J. Martinez, *Tetrahedron Lett.* **1999**, *40*, 4535–4538.

20 S. Caddick, D. Hamza, S. N. Wadman, *Tetrahedron Lett.* **1999**, *40*, 7285–7288.

21 R. Plourde Jr, L. L. Johnson Jr, R. K. Longo, *Synlett* **2001**, 439–441.

22 H. Miyabe, Y. Fujishima, T. Naito, *J. Org. Chem.* **1999**, *64*, 2174–2175.

23 H. Miyabe, C. Konishi, T. Naito, *Org. Lett.* **2000**, *2*, 1443–1445.

24 H. Miyabe, H. Tanaka, T. Naito, *Tetrahedron Lett.* **1999**, *40*, 8387–8390.

25 H. Miyabe, K. Fujii, H. Tanaka, T. Naito, *Chem. Commun.* **2001**, 831–832.

26 G.-H. Jeon, J.-Y. Yoon, S. Kim, S. S. Kim, *Synlett* **2000**, 128–130.

27 For a review, see: S. Bräse, S. Dahmen, *Chem. Eur. J.* **2000**, *6*, 1899–1905.

28 M. J. Plunkett, J. A. Ellman, *J. Org. Chem.* **1995**, *60*, 6006–6007.

29 I. Sucholeiki, *Tetrahedron Lett.* **1994**, *35*, 7307–7310.

30 K. C. Nicolaou, J. Pastor, S. Barluenga, N. Winssinger, *Chem. Commun.* **1998**, 1947–1948.

31 T. Ruhland, K. Anderson, H. Pedersen, *J. Org. Chem.* **1998**, *63*, 9204–9211.

32 K. Fujita, K. Watanabe, A. Oishi, Y. Ikeda, Y. Taguchi, Y. *Synlett* **1999**, 1760–1762.

33 Z. Li, B. A. Kulkarni, A. Ganesan, *Biotechnol. Bioeng.* **2001**, *71*, 104–106.

34 a) U. Gerigk, M. Gerlach, W. P. Neumann, R. Vieler, V. Weintritt, *Synthesis* **1990**, 458–452; b) M. Gerlach, F. Jordens, H. Kuhn, W. P. Neumann, M. Peerseim, *J. Org. Chem.* **1991**, *56*, 5971–5972; c) C. Bokelmann, W. P. Neumann, M. Peterseim, *J. Chem. Soc., Perkin Trans. 1* **1992**, 3165.

35 a) G. Dumartin, M. Pourcel, B. Delmond, O. Donard, M. Pereyre, *Tetrahedron Lett.* **1998**, *39*, 4663–4666; b) G. Dumartin, G. Ruel, J. Kharboutli, B. Delmond, M.-F. Connil, B. Jousseaume, M. Pereyre, *Synlett* **1994**, 952–954; c) G. Ruel, N. K. The, G. Dumartin, B. Delmond, M. Pereyre, *J. Organomet. Chem.* **1993**, *444*, C18–C20.

36 P. Boussaguet, B. Delmond, G. Dumartin, M. Pereyre, *Tetrahedron Lett.* **2000**, *41*, 3377–3380.

37 a) A. Chemin, H. Deleuze, B. Maillard, *Eur. Polym. J.* **1998**, *34*, 1395; b) A. Chemin, H. Deleuze, B. Mailard, *J. Chem. Soc., Perkin Trans. 1*, **1999**, 137–142.

38 E. J. Enholm, J. P. Schulte, II, *Org. Lett.* **1999**, *1*, 1275–1277.

39 a) D. P. Curran, S. Hadida, S.-Y. Kim, Z. Luo, *J. Am. Chem. Soc.* **1999**, *121*, 6607–6615; b) D. P. Curran, S. Hadida, *J. Am. Chem. Soc.* **1996**, *118*, 2531–2532; c) I. Ryu, T. Niguma, S. Minakata, M. Komatsu, S. Hadida, D. P. Curran, *Tetrahedron Lett.* **1997**, *38*, 7883–7886.

40 E. J. Enholm, M. E. Gallagher, K. M. Moran, J. S. Lombardi, J. P. Schulte, II, *Org. Lett.* **1999**, *1*, 689–691.

41 a) I. Ryu, T. Niguma, S. Minakata, M. Komatsu, Z. Luo, D. P. Curran, *Tetrahedron Lett.* **1999**, *40*, 2367–2370; b) see also: D. P. Curran, S. Hadida, M. He, *J. Am. Chem. Soc.* **1997**, *119*, 6714–6715.

42 A. J. Clark, R. P. Filik, D. M. Haddleton, A. Radigue, C. J. Sanders, G. H. Thomas, M. E. Smith, *J. Org. Chem.* **1999**, *64*, 8954–8957.

43 Y. Shen, S. Zhu, R. Pelton, *Macromolecules* **2001**, *34*, 3182–3185.

44 A. Marsh, A. Khan, M. Garcia, D. M. Haddleton, *Chem. Commun.* **2000**, 2083–2084.

45 B. Zhao, W. J. Brittain, *J. Am. Chem. Soc.* **1999**, *121*, 3557–3558.

46 L. De Luca, G. Giacomelli, G. Porcu, M. Taddei, *Org. Lett.* **2001**, *3*, 855–857.

8

Nucleophilic Substitution in Combinatorial and Solid-phase Synthesis

Jan-Gerd Hansel and Stephan Jordan

8.1
Introduction

Nucleophilic substitution (S_N) reactions are widely used in combinatorial chemistry. Two reasons account for this prevalence: first the potential to combine building blocks by S_N and second the availability of building blocks that can participate in S_N reactions. In addition, S_N reactions are typically very reliable transformations, robust, and thus ideally suited for high-throughput synthesis of organic compounds.

In this chapter the application of S_N reactions to combinatorial chemistry using both solution- and solid-phase formats will be discussed. The sections on aliphatic and aromatic substrates are subdivided by the classes of nucleophiles that participate in the reaction. Reactions used to form medium-sized or large rings are treated separately owing to their significance in macrocyclic ring chemistry. This chapter will not cover nucleophilic substitution reactions at acyl carbons, reactions of organometallic reagents, reactions of carbon nucleophiles, and transition metal-catalyzed substitution reactions of nucleophiles (see other Chapters).

8.2
Nucleophilic Substitution at Aliphatic Carbons

8.2.1
General Remarks

Nucleophilic substitutions (S_N) at aliphatic carbons play an important role in the combinatorial synthesis of diverse classes of compounds. The reactions have had great impact right from the beginning of combinatorial chemistry. R. B. Merrifield used S_N of chloromethyl-polystyrene with cesium carboxylates as a means to attach amino acids to a solid support (see Scheme 8.1) [1].

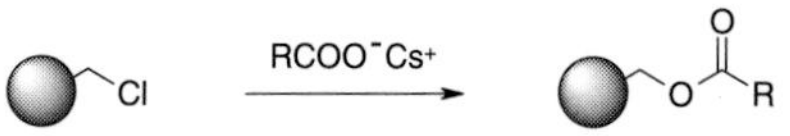

Scheme 8.1. Immobilization of amino acids by Merrifield.

A broad range of electrophilic substrates can be used in S_N reactions. Alkyl chlorides and bromides are fairly common whereas alkyl iodides are used less often [2]. Alkyl chlorides are usually activated with catalytic amounts of iodide salts such as potassium iodide or tetrabutylammonium iodide (TBAI). Alkyl esters of sulfonic acids such as alkyl tosylates or alkyl mesylates are also common electrophiles. They are prepared from a sulfonyl halide and the corresponding alcohol. The alcohol itself can be employed as an electrophilic substrate when activated *in situ*. The Mitsunobu reaction [3] and related transformations are frequently utilized for this purpose. Recently, the activation of alcohols as trichloroacetimidates has been applied to parallel synthesis [4].

Epoxides and cyclic sulfates [5] are synthetically valuable substrates for library generation since the reaction with a nucleophilic building block liberates another hydroxyl function ready for further derivatization.

8.2.2
Halogen Nucleophiles

Benzylic alcohols are converted into the corresponding benzyl chlorides or bromides by reaction with phosphorus trihalides or under milder conditions with triphenylphosphine (TPP) and tetrahalogenomethanes. The reaction has found widespread application in the activation of benzyl alcohol-type linkers for subsequent attachment of nucleophilic substrates to a solid support [6].

The reaction was also applied to one of the masterpieces of multistep solution-phase synthesis involving polymer-supported reagents and scavengers. During Ley's pyrrole synthesis, all six reaction steps to the highly diverse pyrroles used these reagents. In those cases where the benzylic halides are not commercially available, the alkyl halides used for N-alkylation are prepared from the corresponding alcohols. In the bromination step, a reagent combination of polymer-supported TPP and tetrabromomethane is used (see Scheme 8.2) [7].

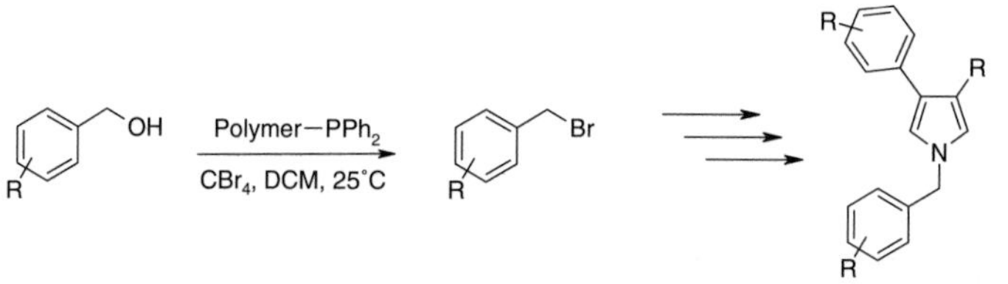

Scheme 8.2. Solution-phase benzyl bromide synthesis.

8.2.3
Oxygen Nucleophiles

Carboxylate alkylation by alkyl halides can be used both on solid phase (see above) and in solution [8] to give alkyl carboxylic acid esters. For optimum results, cesium salts are employed. Alternatively, carboxylate alkylation can be performed under Mitsunobu conditions using alcohols as electrophiles. For solid-phase applications the common reagents TPP and diethyl azodicarboxylate (DEAD) can be used. In solution-phase parallel synthesis, it is advantageous to use reagents that do not require a chromatographic product isolation. A combination of polymer-bound TPP [9] and di-*tert*-butyl azodicarboxylate has been found to be particularly useful (see Scheme 8.3) [10]. While the former reagent is filtered off easily, the latter (and its byproduct di-*tert*-butylhydrazodicarboxylate) is readily converted into volatile compounds by treatment with trifluoroacetic acid (TFA).

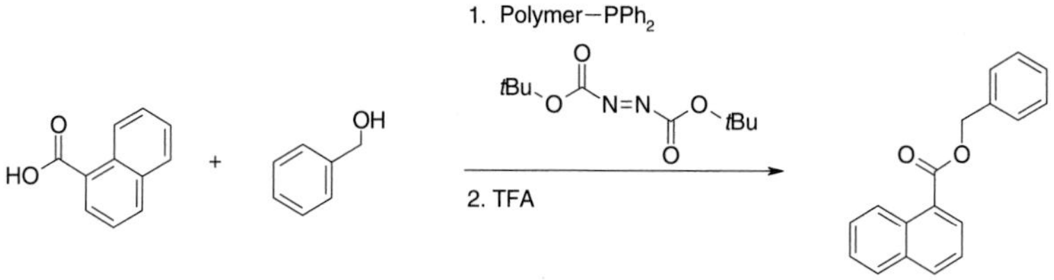

Scheme 8.3. Solution-phase Mitsunobu carboxylate alkylation.

Cesium carbonate in the presence of an alcohol and carbon dioxide can be alkylated by Merrifield resin to give solid-supported carbonates (see Scheme 8.4). Adding an amine instead of an alcohol gives rise to carbamate formation [11].

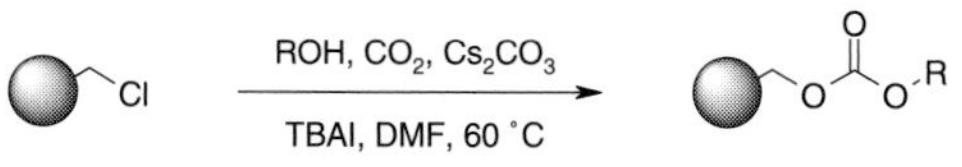

Scheme 8.4. Solid-phase carbonate formation.

Alcohols are transformed into ethers by reaction with an alkylating agent in the presence of a very strong base such as sodium hydride. The reaction, known as the Williamson ether synthesis, is frequently used for attaching alcoholic substrates to Merrifield-type resins (see Scheme 8.5) [12].

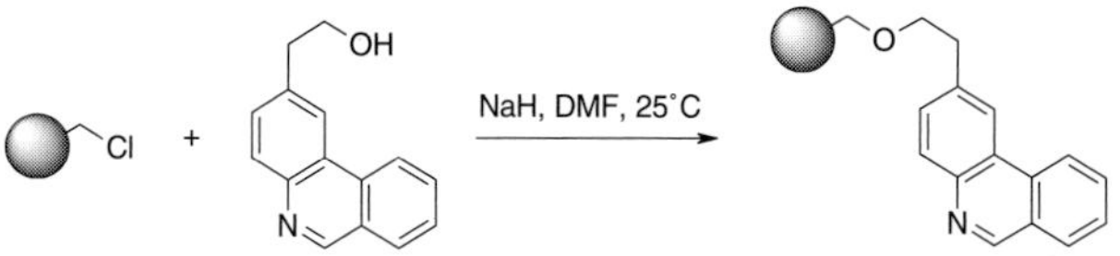

Scheme 8.5. Solid-phase Williamson ether synthesis.

In a similar reaction, Wang resin is converted into a trichloroacetimidate derivative and used as a polymer-bound benzylating agent. Attachment of alcohols to this resin is achieved under acid catalysis [4].

Ether-forming reactions involving phenolate nucleophiles proceed much more readily. All types of electrophiles have been used in both solution- and solid-phase chemistry. Usually, the phenolates are generated *in situ* from phenols and bases such as 1,8-diazabicyclo[5.4.0]undecene-7 (DBU) or potassium carbonate (see Scheme 8.6) [13].

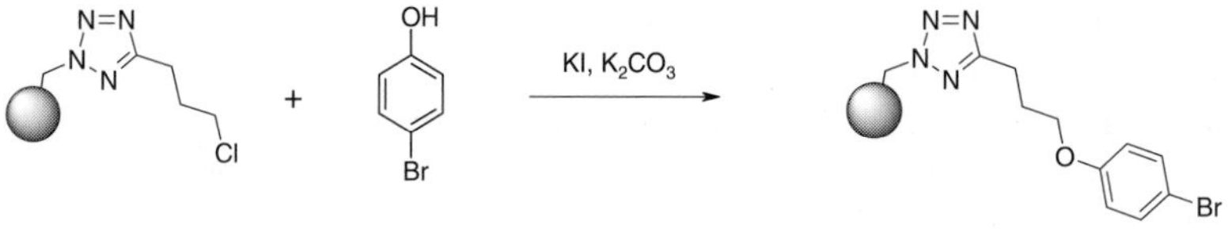

Scheme 8.6. Solid-phase phenolate alkylation.

The solution-phase synthesis of alkyl aryl ethers is simplified by using ion-exchange resins. The reaction of various phenolates loaded on Amberlite IRA-900 with alkyl bromides gives alkyl aryl ethers in solution. Thereby the ammonium bromide byproduct remains on the Amberlite resin [14].

Aryl alkyl ethers are also obtained by Mitsunobu reaction of phenols and alcohols. For solid-phase applications, the standard reagents TPP and DEAD have been used successfully. In some cases, Castro's sulfonamide betaine 1 (see Scheme 8.7) [15] offers cleaner reactions and is easier to handle especially with automated equipment. Since the betaine decomposes upon exposure to air into inactive components without a change of appearance, care should be taken to ensure pure material is used.

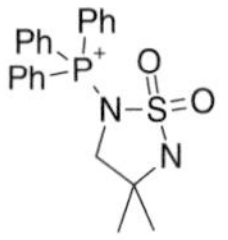

Scheme 8.7. Castro and Matassa's sulfonamide betaine (**1**) [15].

The solid-phase Mitsunobu reaction has been employed for linker attachment [16] as well as diversity generation [17]. In solution-phase chemistry, specially designed reagents are used to facilitate product separation (see above). Thus, polymer-bound TPP and DEAD derivatives that are easily filtered off are commercially available. Also, ionophoric side-chains have been attached to both TPP and DEAD to make them (and the corresponding byproducts) easily separable using ion-exchange resins [18]. This technique is an example of a general concept called phase tagging or phase labeling [19].

The solution-phase reaction of phenols with epoxides is catalyzed by polymer-supported Co(salen) complexes [20]. Using chiral catalysts the corresponding 1-aryloxy-2-alcohols are obtained in high yields, purities, and enantiomeric excesses.

8.2.4
Sulfur Nucleophiles

The reactions of sulfur nucleophiles show similarities to those of oxygen nucleophiles. The greater nucleophilicity of sulfur compounds, however, makes a range
of unique additional transformations possible.

A convenient way to prepare the thiol function involves thioacetate alkylation.
This method has been adapted to polymer-supported solution-phase synthesis.
Displacement of halides or tosylates with a polymer-supported thioacetate reagent
affords an intermediate thioester, which can be reduced to the thiol by the addition
of borohydride exchange resin [21].

As expected, alkane thiols [22] and thiophenols [23] are readily alkylated by alkyl
halides. The substitution proceeds at room temperature in the presence of a base
(DIEA for aromatic and DBU for aliphatic thiols).

Thioether formation can also be used for synthesis of heterocycles, as shown in
the preparation of 2,4,5-trisubstituted thiomorpholin-3-ones [24].

Dithiocarbamates, which are formed *in situ* from the reaction of carbon disulfide
with amines, are alkylated by Merrifield resin to obtain resin-bound dithiocarbamates (see Scheme 8.8). These compounds are intermediates in a reaction sequence leading to guanidines carrying three different side-chains [25].

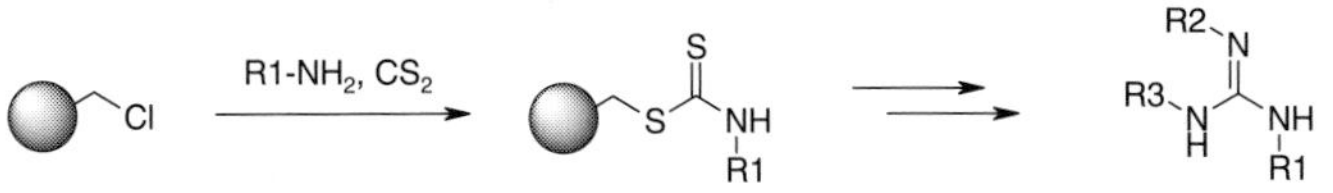

Scheme 8.8. Solid-phase dithiocarbamate alkylation.

Similarly, thiourea is attached to Merrifield resin by *S*-alkylation to give a *S*-alkyl-
isothiouronium salt, which is further transformed into guanidine derivatives [26]
and pyrimidines (see Sect. 8.3.2).

Thiourea alkylation is also involved in the classical Hantzsch thiazole synthesis.
N-Substituted thioureas are alkylated with α-bromoketones, and after immediate
cyclization aminothiazoles are obtained (see Scheme 8.9). The reaction has been
adapted to a solution-phase library synthesis by simply mixing the components as
dimethylformamide (DMF) solutions and heating the mixtures to 70 °C [27].

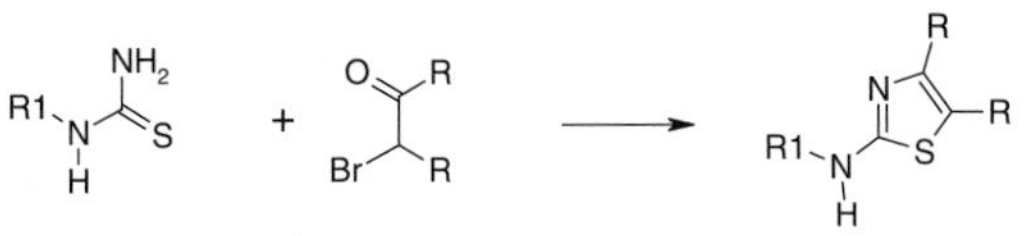

Scheme 8.9. Solution-phase Hantzsch thiazole synthesis.

8.2.5
Nitrogen Nucleophiles

The direct *N*-alkylation of primary amines is in general not the method of choice
for the synthesis of secondary amines because of potential of overalkylation. There-

by a reductive alkylation with a carbonyl compound is to be preferred. Most often the primary amine is reacted with an aldehyde in the presence of a water-removing agent such as trimethyl orthoformate (TMOF) or molecular sieves to form an imine intermediate. This is reduced to the secondary amine with a suitable boro-hydride agent, preferably tris(acetoxy)sodiumborohydride.

A different situation occurs in solid-phase synthesis when the alkylating agent is attached to the solid support. Upon reaction with a primary amine – usually applied in great excess – a clean conversion to the immobilized secondary amine is observed. The reaction can be used for the attachment of amines to the resin [28]. The most important application of this reaction is in the solid-phase synthesis of N-substituted glycine (NSG) oligomers, so-called peptoids [29]. These oligomers resemble peptides. Their side-chains are not attached at the α-carbon as in peptides but rather at the nitrogen atom. Peptoids are synthesized by a repeated sequence of acylation with bromoacetic acid followed by S_N reaction with a primary amine (see Scheme 8.10). Iodoacetic acid can also be used, but chloroacetic acid gives inferior results.

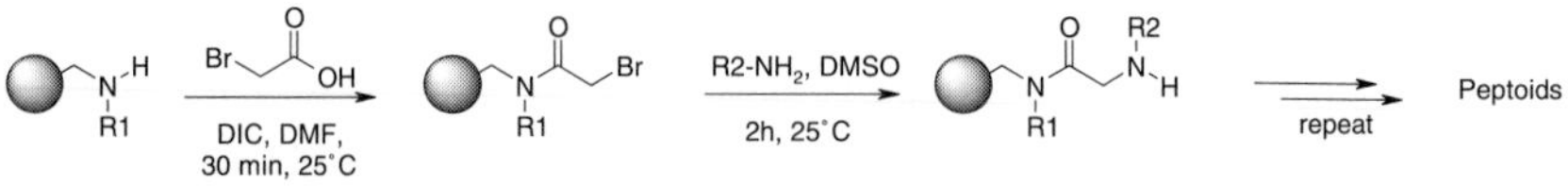

Scheme 8.10. Solid-phase peptoid synthesis.

The straightforward synthesis has mostly been developed by Chiron researchers. Exploiting the great number of available amines, they were able to identify rapidly nanomolar ligands for α_1-adrenergic and opiate receptors out of huge peptoid libraries.

Peptoids offer a good example of the evolution of methods that mimic natural oligomers with easy-to-make unnatural compounds. A recent similar development has been peptide nucleic acids (PNAs). PNAs resemble DNA with the phospho-diester backbone of the DNA being replaced by an oligo-[N-(2-aminoethyl)glycine] motif. They have been investigated for diagnostic and antisense purposes [30].

The S_N reaction of alkylating agents with secondary amines can be controlled to give tertiary amines selectively. Thus, reaction of polymer-bound alkyl sulfonates with secondary amines gives immobilized tertiary amines [31]. In solution-phase synthesis, selective monoalkylation of secondary amines has been achieved mostly with monosubstituted piperazine substrates (see Scheme 8.11) [32]. The reaction can be performed with an excess of piperazine or with an additional base. To facilitate product isolation in solution-phase chemistry, either a water-soluble base in combination with an aqueous work-up [33] or a polymer-bound base [34] can be employed.

Scheme 8.11. Solution-phase piperazine alkylation.

Intramolecular amine alkylation does not usually bear the risk of overalkylation and is an excellent way to close heterocyclic rings. An Epibatidine synthesis using polymer-supported reagents provides an example that involves a mesylate substitution [35]. Even strained rings such as aziridines can be formed (Gabriel–Cromwell reaction) [36].

Epoxide opening with amine nucleophiles is frequently used in combinatorial chemistry since it leads to the attractive aminoalcohol substructure. When catalyzed with Lewis acids, for example lithium perchlorate, the reaction proceeds smoothly with a range of alkyl amines, anilines, and heteroaromatic amines. The reaction is useful in solution-phase synthesis (see Scheme 8.12) [37] as well as in solid-phase synthesis [38].

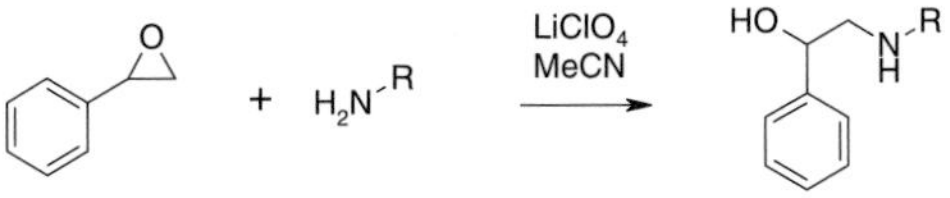

Scheme 8.12. Solution-phase epoxide opening by amines.

The nucleophilic attack of an epoxide by amines can be followed by reactions involving the newly generated hydroxyl function. In Scheme 8.13 a carbamate group at a suitable distance is attacked by the hydroxyl group that is exposed during epoxide opening. The resulting oxazoline formation occurs with concomitant cleavage off the resin [39].

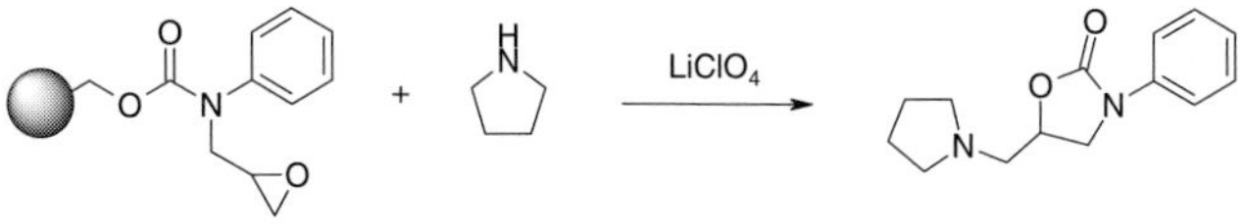

Scheme 8.13. Solid-phase oxazolidinone synthesis.

Other nucleophiles that are readily alkylated on the solid phase by alkyl halides or sulfonates include hydroxyl amine [40], O-benzyl hydroxylamine [41], hydrazine derivatives [42], azide ion [43], and sulfonamides [44].

Nitrogen compounds bearing hydrogens of sufficient acidity can also be alkylated under Mitsunobu conditions (see above). Thus, a polymer-bound imidodicarbonate can be alkylated with primary and secondary alcohols using the standard reagents TPP and DEAD. After cleavage off the resin, primary amines are obtained (see Scheme 8.14) [45].

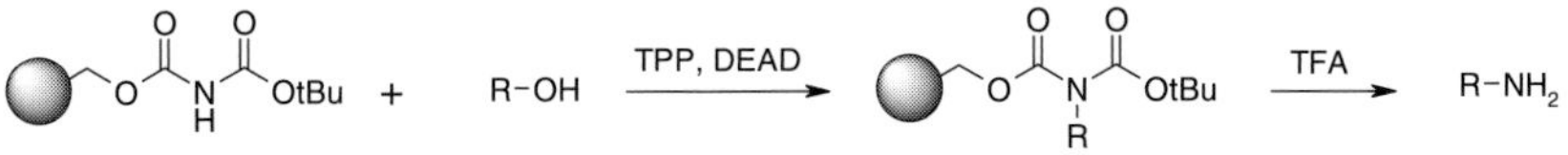

Scheme 8.14. Solid-phase Mitsunobu reaction of imidodicarbonate.

8.2.6
Ring-closing Reactions

The formation of medium-sized or large rings in combinatorial synthesis is frequently accomplished with S_N reactions. Especially *S*-alkylation reactions are used in the ring-closing step. Representative examples involve the syntheses of *β*-turn mimetics containing nine- or ten-member rings [46] and a cyclic oligocarbamate consisting of a 26-membered ring [47].

An unusual case of macrocyclization is observed in solid-phase synthesis of [Arg-8]-vasopressin. The thiol groups of two cysteines of the peptide can be linked by a methylene unit to form a macrocyclic methylenedithioether [48]. The transformation is achieved by simply treating the resin with tetrabutylammonium fluoride in dichloromethane.

8.3
Nucleophilic Substitution at Aromatic Carbons

8.3.1
General Remarks

Nucleophilic aromatic substitution (S_NAr) is an attractive approach to the functionalization of electron-deficient aromatic systems, in solution phase as well as on solid support. It has become an invaluable part of combinatorial transformations, particularly for the installation of nitrogen- or oxygen-linked substituents. A wide range of readily accessible nucleophiles can be introduced, making S_NAr reactions as popular as amide formations and outstanding for the synthesis of combinatorial libraries.

The large majority of these reactions is based on two classes of reactive scaffolds. Di- and trihalogenated heterocycles with two or more heteroatoms in the heterocyclic ring (mostly nitrogen atoms) have been used extensively. In these molecules the halogen atoms can be replaced selectively and sequentially with various nucleophiles, making the reaction ideal for combinatorial library production. On the other hand, there is a large number of examples for attractive and practical methods to cleanly and efficiently prepare novel libraries using substituted halogeno-nitrobenzenes as templates. First, the nitro group activates the aromatic system for nucleophilic attack. After halogen substitution the nitro group can be reduced easily and used in an excellent way for further diversification of the libraries. If the chloro derivatives prove not to be reactive enough in the S_NAr reactions, the chloro substrates can be transformed into the fluoro analogs by halex reaction (which is a S_NAr reaction in itself). Fluoroaromatics are significantly more reactive toward nucleophiles [49].

8.3.2
Nitrogen Nucleophiles

From a historical point of view, the first template for S_NAr reactions was cyanuric chloride (2,4,6-trichloro-1,3,5-triazine), which is commercially available and inexpensive [50]. Compounds containing this core element have shown biological activity, e.g. as herbicides (Atrazin).

A research group at ArQule showed that the selective and sequential derivatization of cyanuric chloride could be achieved by simply controlling the reaction temperature. The generality of the research group's method has been proven by the solution-phase synthesis of a large combinatorial library of over 40,000 individual compounds (see Scheme 8.15) [51]. The first chloride substitution proceeds at −20 °C using N,N-diisopropylethylamine (DIPEA) as the base and acetonitrile as the solvent. Even anilines react with the very reactive cyanuric chloride in the proposed way. Usually, the arylation of the primary amine reduces its nucleophilicity strongly. Therefore, no bis-arylation occurs at this position and the second chlorine atom can be substituted at room temperature. Owing to the relatively low reactivity of the third position, heating is required for the last exchange and only strong nucleophiles such as secondary amines give pure products and high conversions. Besides amines and anilines, also carbohydrates, dipeptides, amidines, and α-ketoamides can be incorporated, giving access to more complex structures. Upon screening of these compounds, a series of hits in the cardiovascular area has emerged.

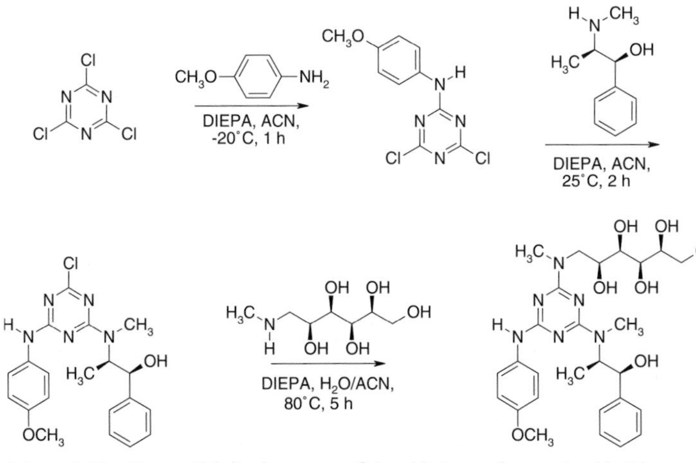

Scheme 8.15. Sequential displacement of the chlorines of cyanuric chloride.

Using the cyanuric chloride template for solid-phase synthesis has become a very efficient method for the production of large combinatorial libraries. This was also demonstrated by the synthesis of a 12,000-compound library by Stanková and Lebl [52]. The first chlorine atom was selectively substituted by coupling the template to a resin loaded with amino acids (see Scheme 8.16). Taking advantage of the decreasing reactivity of tri-, di-, and monochlorotriazines, various nucleophiles (amines, anilines, hydrazines) were introduced at different temperatures [53]. Likewise, an 8000-membered library on a cellulose-based polymeric membrane has been synthesized [54].

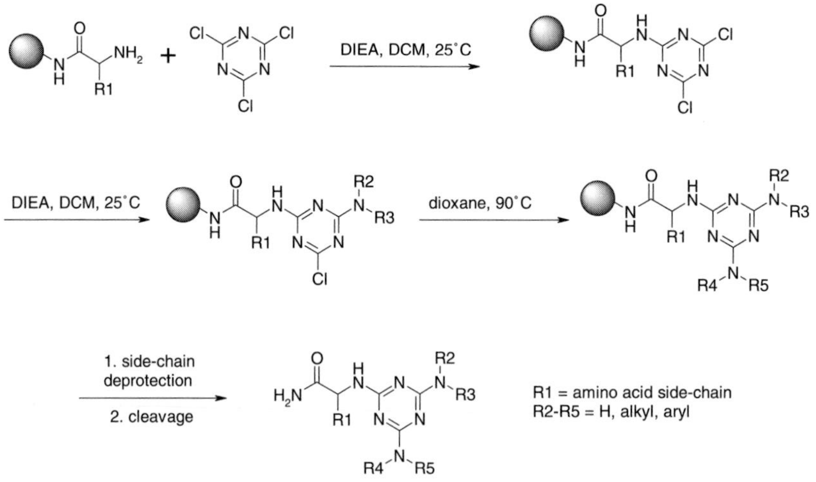

Scheme 8.16. Solid-phase library based on the triazine template.

The reaction sequence can be extended to related starting materials such as 2,6-dichloropurines, although the reaction conditions need to be harsher [55]. In a representative example, the dichloropurine was treated with a primary amine at elevated temperature. For the second substitution, reflux conditions and five equivalents of amine were necessary. Excess amine was removed by the use of formyl-polystyrene beads. The compounds could be benzylated at the N-9 position by an alkylation protocol or by using the Mitsunobu reaction (see Scheme 8.17) [56].

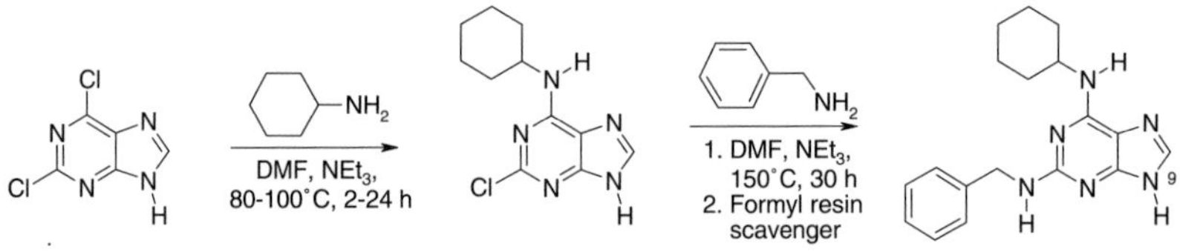

Scheme 8.17. Liquid-phase synthesis of a purine-based library.

S$_N$Ar reactions on perhalogenated heterocyclic systems are well established in solution phase and these reactions have been well adapted to solid-phase synthesis. They have been shown to be useful – indeed, more advantageous in many cases – than their solution-phase counterparts. Special attention in this area should be given to purine templates as purines are involved in signal pathways and metabolic processes in all living organisms (see Scheme 8.18). For example, the discov-

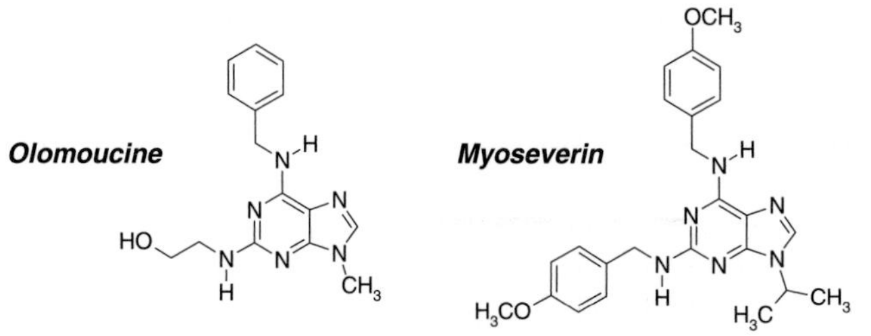

Scheme 8.18. Natural products containing the purine core.

ery of the biologically active natural product olomoucine stimulated attempts to generate diverse analogs based on the adenosine template on solid support [57].

First an aldehyde-functionalized resin preloaded with benzylamines was reacted with the purine scaffold. The authors chose N-9-SEM (SEM: trimethylsilylethoxy-methylene)-protected 2-fluoro-6-chloropurine as the starting material because of the activating potential of the SEM group in the amination reaction. After removal of this group using tetrabutylammonium fluoride in tetrahydrofuran (THF), a Mitsunobu reaction introduces the isopropyl moiety. A second nucleophilic substitution and final cleavage from the resin led to the desired purine analogs (see Scheme 8.19). Since the molecules were attached to the resin via the amino group at position 6, only primary amines could be introduced into the reaction sequence [58]. As a result, myoseverin, a novel microtubule-binding molecule, was identified upon screening these libraries [59].

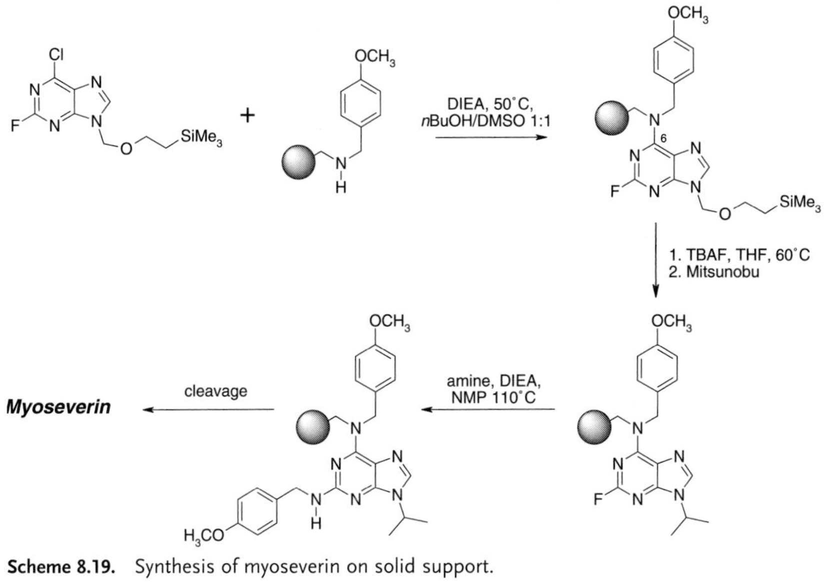

Scheme 8.19. Synthesis of myoseverin on solid support.

The production of new compound libraries from polyhalogenated heterocycles is very common in combinatorial chemistry – for example reactions involving nucleophilic amines. Besides cyanuric chloride and chloropurines, many other templates have been used as starting materials (see Scheme 8.20) [50b] [60]. For example, for the synthesis of a library of 160 pyrimidine carboxamides, Suto et al. [61b] took advantage of the difference in reactivity between the two reactive sites of the substituted 2-chloropyrimidine-5-carboxylic acid chloride core (see Scheme 8.21). After amidation of the carboxylic function, some of the products were treated with amines to increase the polarity [61].

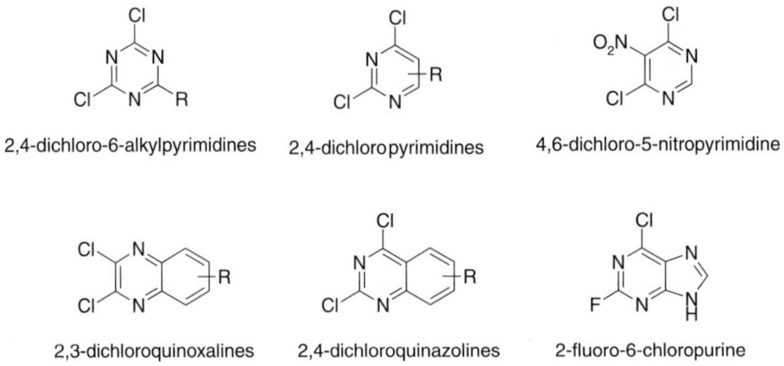

2,4-dichloro-6-alkylpyrimidines   2,4-dichloropyrimidines   4,6-dichloro-5-nitropyrimidine

2,3-dichloroquinoxalines   2,4-dichloroquinazolines   2-fluoro-6-chloropurine

Scheme 8.20. Useful templates for S_NAr reactions.

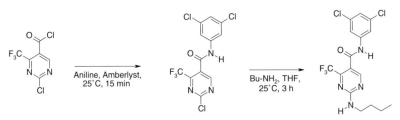

Scheme 8.21. Liquid-phase synthesis of a pyrimidine-based library.

Alternatively, structurally diverse pyrimidines can be obtained by a *de novo* synthesis. The synthesis commencing with isothiouronium salts (R1 = Ar-CH$_2$– or resin-CH$_2$–, see Section 2.3) is amenable to solution as well as to solid-phase applications. When condensing the isothiouronium salts with ketene derivatives, a pyrimidine skeleton with versatile functional groups is obtained. Oxidation of the alkylthio-linkage with *m*-chloroperbenzoic acid (*m*-CPBA) activates the molecule for S_NAr derivatization (see Scheme 8.22). The corresponding sulfinyl or sulfonyl compounds are then easily substituted with various amines [62].

In a very similar approach, the solution- and solid-phase synthesis of libraries of trisubstituted 1,3,5-triazines has been described previously [63].

Arylpiperazines have been identified as a privileged structural element in various biologically active compounds. Besides strategies involving the cyclization of a

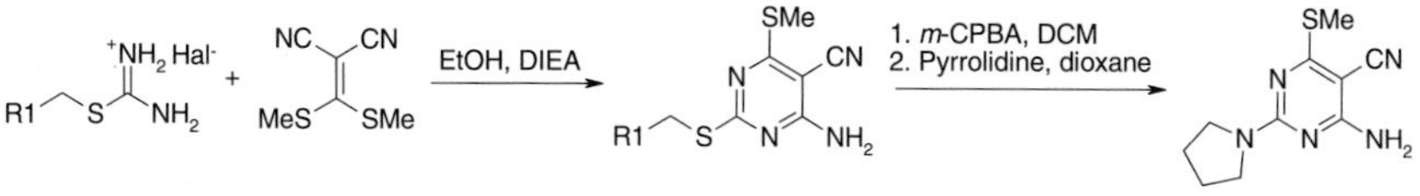

Scheme 8.22. *De novo* synthesis of pyrimidines.

substituted aniline with bis-(2-chloroethyl)amine, a synthetic route based on nu-cleophilic aromatic substitutions was also required. Different fluorobenzenes with a nitro group either at the ortho or para position underwent S_NAr reactions with N-Boc-piperazine (see Scheme 8.23). After removal of the protection group, acyla-tion under Schotten–Baumann conditions with a set of eight carboxylic acid chlo-rides gave 48 N-alkyl-N'-acylpiperazines [64]. This methodology has been well adapted to solid-phase chemistry, as reflected in recent reviews.

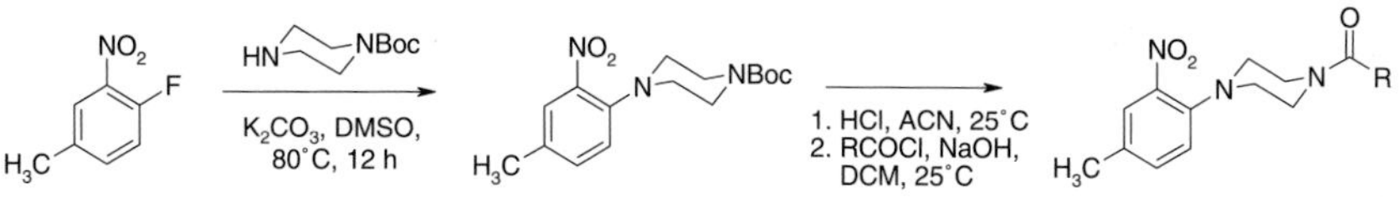

Scheme 8.23. Synthesis of an arylpiperazine library.

Resin-bound 4-fluoro-3-nitrobenzoic acid is also an outstanding template for nucleophilic aromatic substitution reactions with nitrogen nucleophiles. An enor-mous number of publications report the syntheses of benzodiazepin-2-ones, ben-zimidazoles, and related structures. Not surprisingly, only a short selection of ex-amples can be described here.

Owing to the importance of benzodiazepines in many therapeutic areas, funda-mental work in the area of solid-phase synthesis has been carried out using this structural element [65].

The 4-fluoro-3-nitrobenzoic acid has been immobilized via the acid group and reacted with a variety of α- and/or β-substituted β-amino esters in DMF in the presence of DIEA. The reduction of the arylic nitro compound to the aniline was carried out using standard conditions [2 M SnCl$_2$·H$_2$O in DMF, room temperature (rt)] (see Scheme 8.24). After hydrolysis of the ester with a mixture of 1 N NaOH/

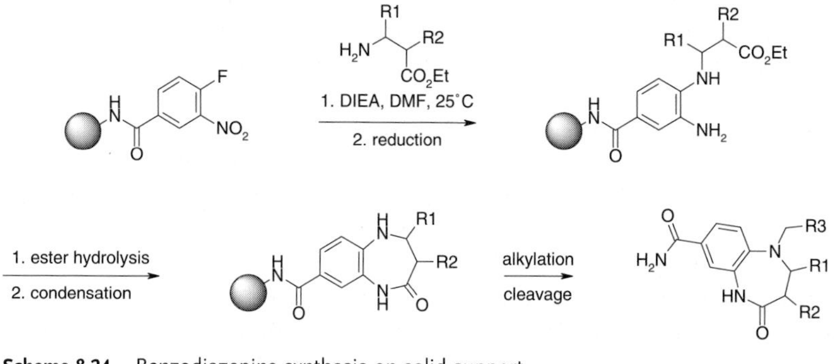

Scheme 8.24. Benzodiazepine synthesis on solid support.

THF, the resulting compound was cyclized with DIC and HOBt and the 1,5-benzodiazepin-2-one was obtained. Alternatively, selective alkylation at the N-5 position adds further diversity to the library [66].

Using α-amino acids in the place of the β-amino acids, the [6,6]-fused ring system of quinoxalin-2-ones has been accessed [67]. The synthetic strategy is an adaptation of TenBrinks and coworkers' solution-phase synthesis on solid phase [68]. Variably substituted tetrahydroquinoxalin-2-ones can also be prepared based on 4-fluoro-3-nitrobenzoic acid. After substitution of the fluorine with primary aliphatic amines at room temperature and reduction of the nitro group, double acylation with chloroacetic anhydride has been shown to be the key step in the synthesis [69].

Further ring contraction to [6,5]-fused systems such as benzimidazoles has been the aim of other synthetic efforts – in solution and in solid-phase synthesis. Again, 4-fluoro-3-nitroarenes were linked to a solid support via an ether linkage [70] or via a carboxylic acid [71]. Commonly, both strategies use a S_NAr displacement reaction of the fluorine atom by an amine with subsequent reduction of the nitro group. Whereas Phillips and Wie [70] achieved immediate cyclization by condensation with benzimidates, a research group at Affymax acylated the intermediate with an activated bromoacetic acetic acid first (see Scheme 8.25). After displacement of the bromide groups by nucleophiles, cyclization occurred upon cleavage with a concomitant dehydration [72].

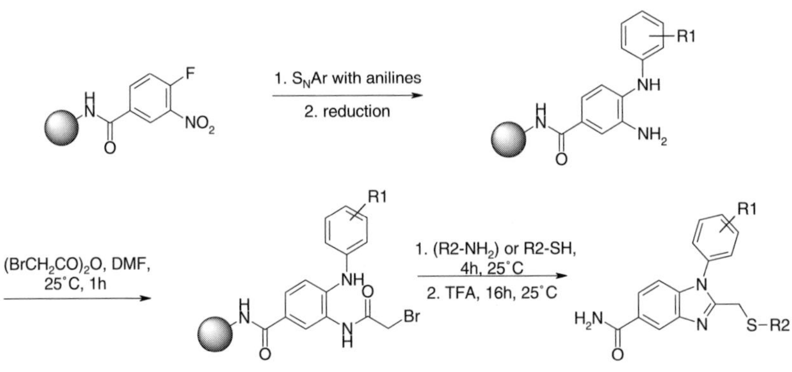

Scheme 8.25. Solid-phase synthesis of benzimidazoles.

Finally, benzimidazolones can also be prepared from solid-supported 4-fluoro-3-nitrobenzoic acid. The key step in the synthesis was again the displacement of fluorine by a nitrogen nucleophile. The nitro group was reduced as described above and the resulting molecules underwent cyclization with the phosgene equivalent disuccinimidocarbonate [73].

An efficient liquid-phase synthesis of substituted benzimidazolones has also been described using a soluble polymer support [MeO-PEG, molecular weight (MW) 5000] [74]. This polymer support dissolves in many organic solvents (e.g.

DMF, THF) and precipitates in particular solvents (e.g. diethyl ether) (see Scheme 8.26). Again, 4-fluoro-3-nitrobenzoic acid was loaded onto the support and was then allowed to react with a variety of amines. After reduction of the nitro group, cyclization was achieved with trichlorophosgene.

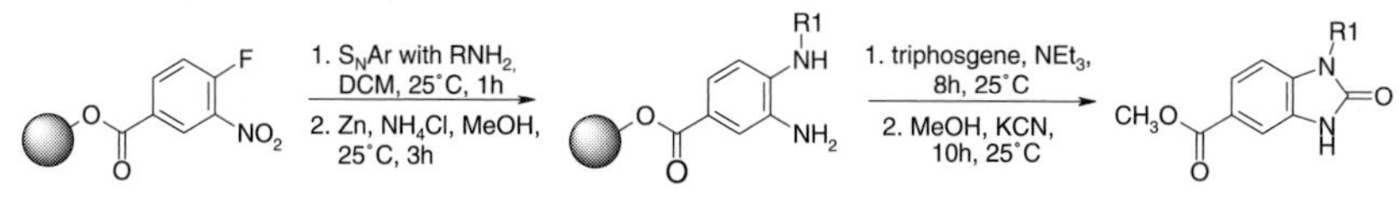

soluble polymer support

Scheme 8.26. Benzimidazolones via solid-phase chemistry.

Recently, a synthetic route to substituted 7-azabenzimidazoles was published. As a key template the highly reactive 6-chloro-5-nitro-nicotinyl chloride was used. The sequential alkylation with different amines by replacement of the strongly activated chloro atoms proceeds easily at room temperature [75].

Another scaffold well suited to the generation of huge libraries by combinatorial methods is 1,5-difluoro-2,4-dinitrobenzene. The two fluoro groups in the ortho positions of two aromatic nitro groups can be sequentially substituted with two amines. As the scaffold is planar and symmetrical, no problems with regioselectivity occur. Lam and coworkers [76] demonstrated the viability of this concept by the production of a 2485-membered library designed for screening for antibacterial activity (see Scheme 8.27) – whereas the first substitution takes place within hours, the second runs overnight [76]. The authors then used the same concept for a solid-phase synthesis on 2-chloro-trityl resin, demonstrating that the two methods are complementary.

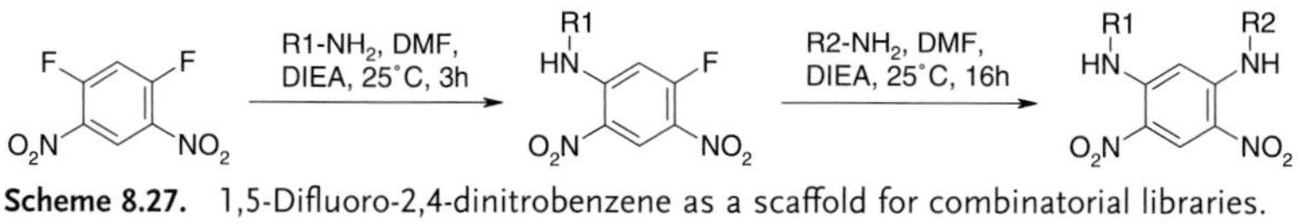

Scheme 8.27. 1,5-Difluoro-2,4-dinitrobenzene as a scaffold for combinatorial libraries.

The displacement of activated halides with nucleophilic amines such as piperazines is also a key step in the synthesis of antiviral quinolones and other pharmaceutically relevant compounds [77].

The starting materials, suitably substituted 3-oxo-3-phenyl-propanoates, are converted into enamines and cyclized to the quinolone core via an intramolecular $S_N Ar$ reaction (see Scheme 8.28). After ester hydrolysis, a wide range of amines can be introduced by displacement of an activated halogen, e.g. a fluorine atom. A library of related compounds was synthesized and screened for human immunodeficiency virus (HIV) suppression [78].

A solid-phase approach to quinolones was published using the same cycloarylation procedure (see Scheme 8.29). A resin-bound β-keto ester was transferred

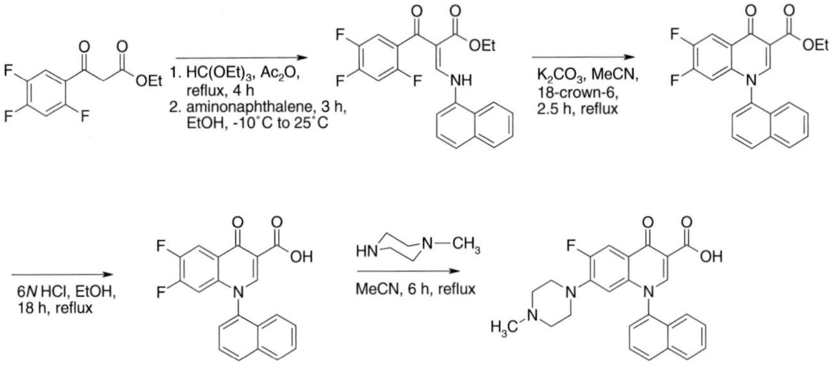

Scheme 8.28. Synthesis of quinolones.

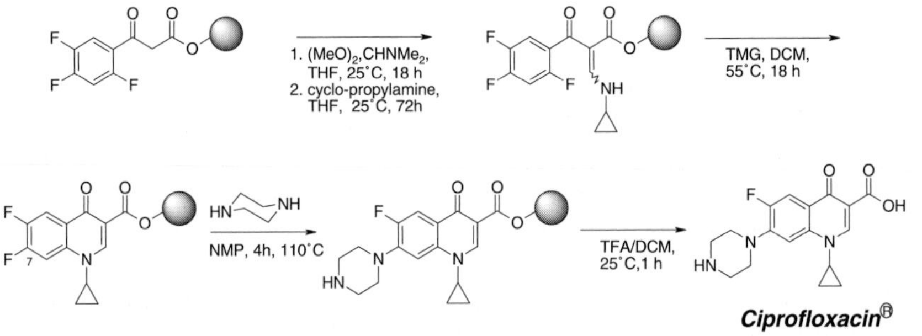

Scheme 8.29. Solid-phase approach to Ciprofloxacin.

into the enamine and cyclized using tetramethylguanidine (TMG). Many amines can be incorporated by displacement of the fluorine atom at C-7 before the products are cleaved off the resin [79].

8.3.3
Oxygen Nucleophiles

Reactive arenes such as fluoro-nitroarenes, halopyridines, halopyrimidines, and halotriazines are preferred for reactions with oxygen nucleophiles [80]. Diary-lethers can be prepared by simply reacting the well-known 4-fluoro-3-nitrobenzoic acid template with a wide range of functionalized phenols in a solid-phase reaction (see Scheme 8.30) [81].

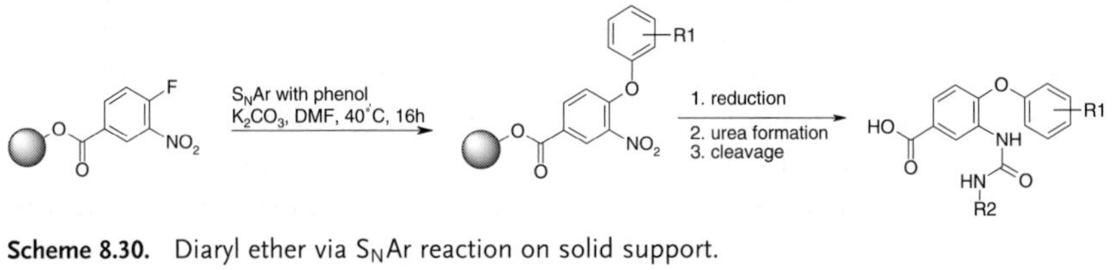

Scheme 8.30. Diaryl ether via S_NAr reaction on solid support.

In solution-phase chemistry, diarylethers can be produced utilizing a polymer-supported guanidine base. The reaction requires an excess of phenol to achieve complete conversion. The polymer-supported base deprotonates the phenol and also traps unreacted starting material (see Scheme 8.31). Since numerous phenols are commercially available, the method is well suited to library synthesis in an automated and parallel manner [82].

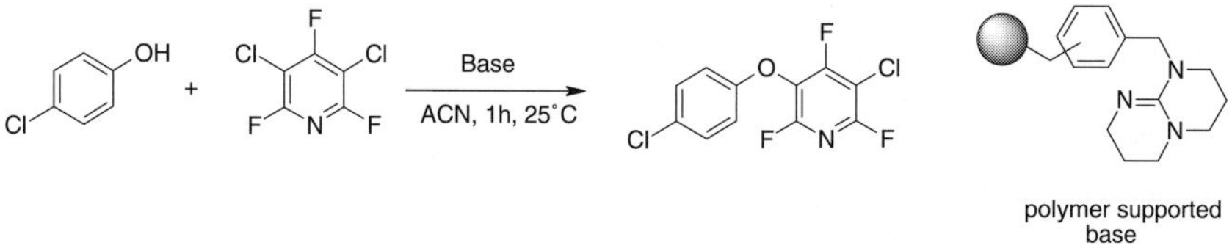

polymer supported base

Scheme 8.31. Diaryl ether synthesis using polymer-supported reagents.

The seven- and eight-member ring systems of dibenzoxazepins and dibenzoxazocines, respectively, have been well investigated in combinatorial chemistry using solution- and solid-phase synthesis [83]. The target heterocycles are efficiently assembled via intramolecular aromatic substitution of the fluorine in 2-fluoro-5-nitroarenes with the OH function of various phenols (see Scheme 8.32). For the solid-phase approach, the cyclization step was achieved using a 5% solution of DBU in DMF. DBU was found to give superior results when compared with TMG or *N*-methylmorpholine. The authors preferred a solid-phase approach rather than a solution-phase approach because yields were generally better while purities were identical.

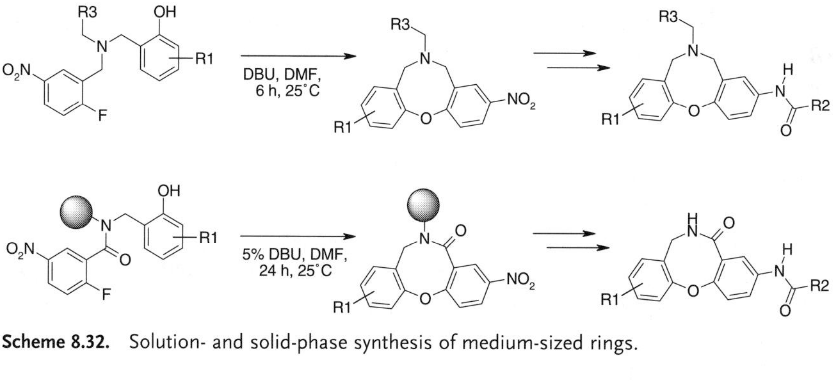

Scheme 8.32. Solution- and solid-phase synthesis of medium-sized rings.

The value of trichlorotriazine as a template for library synthesis by sequential substitution of the chloro atoms is discussed in Section 8.3.2. Besides this, trichlorotriazine can react with a soluble polymer support (MeO-PEG-OH, MW 5000) to give PEG-bound dichlorotriazine, a new soluble electrophilic scavenger (see Scheme 8.33). Because of the high reactivity of the scavenger toward nucleophiles, it was used to remove alcohols at the end of ester or silyl ether-forming reactions [84].

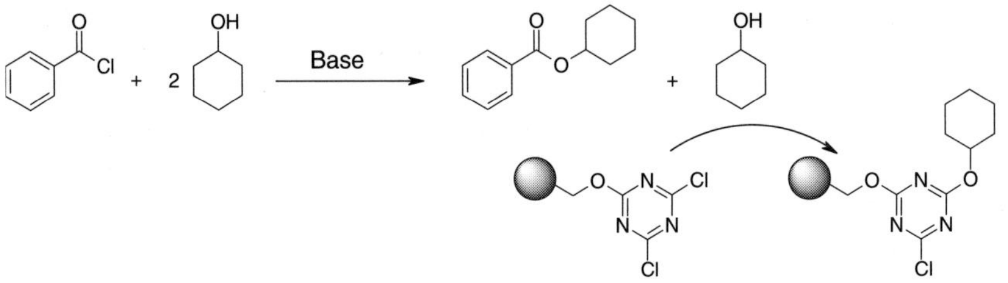

Scheme 8.33. Ester synthesis using dichlorotriazine scavenger.

8.3.4
Sulfur Nucleophiles

The fluoro-nitroarene motif is also the most preferred template for $S_N Ar$ for sulfur nucleophiles. In many applications, a suitably protected form of cysteine as a β-mercapto acid is reacted with 4-fluoro-3-nitrobenzoic acid to form 1,5-benzothiazepin-4-ones, an important class of drugs in the treatment of cardiovascular disorders. For example, 4-fluoro-3-nitrobenzoic acid was treated with 1.5 equivalents of 9-fluorenylmethoxycarbonyl (Fmoc)-l-cysteine in DMF to be converted to the 2-nitro-thioether (see Scheme 8.34) [85]. After the reduction of the nitro group and a subsequent reductive alkylation (difficult because of the poor nucleophilicity of some anilines), the resulting secondary anilines were cyclized to form the seven-member thiazepine ring.

Scheme 8.34. Solid-phase synthesis of 1,5-benzothiazepin-4-ones.

Fmoc-homocysteine reacts with similar effectiveness to a nucleophile in the same reaction sequence. Shortening the side-chain from β-mercapto acid to α-mercapto acids, the $S_N Ar$ reaction is also reliable in this case. However, the synthetic utility is limited owing to the fact that only a very small number of α-mercapto acids are commercially available.

Using a reversed synthetic strategy, a great variety of other fused heterocycles is accessible [86]. Wang resin loaded with cysteine via a carbamate linker can be reacted with numerous halo-nitroarenes bearing diverse functional groups.

Recently, immobilized 4-fluoro-3-nitrobenzoic acid was transformed into 2-amino-4-carboxythiophenol as an intermediate for the synthesis of several classes of heterocyclic compounds such as benzothiazoles or 3,4-dihydro-1,4-benzothiazines (see Scheme 8.35) [87]. The resin-bound 2-amino-4-carboxythiophenol was prepared by displacement of the fluorine atom with triphenylmethylmercaptan, reduction of the nitro group, and removal of the sulfur-protecting group.

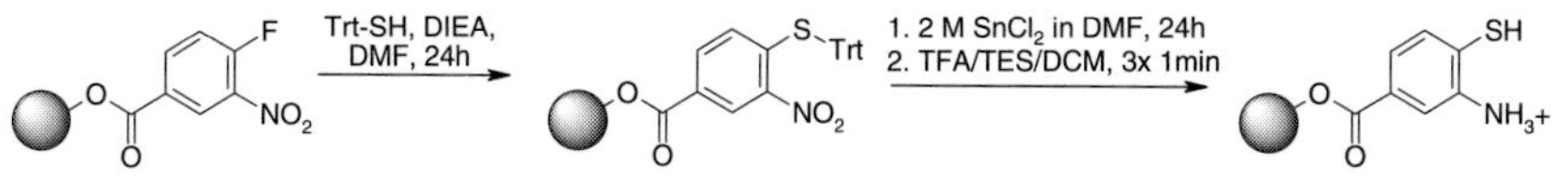

Scheme 8.35. Solid-phase synthesis of 2-amino-4-carboxythiophenol.

8.3.5
Macrocyclization Reactions

In this section the reaction for closing large rings by nucleophilic substitution of activated arenes will be discussed. Cyclorelease reactions such as those used in the well-established syntheses of hydantoins [88] or diketopiperazines [89] will not be covered.

At first sight, the closure of medium-sized or large rings by nucleophilic displacement does not meet the demands of combinatorial synthesis. No new substituents are incorporated and therefore no diversification is achieved. On closer inspection, the great importance of this reaction becomes obvious in view of the immense difference between the three-dimensional structure of a linear oligopeptide and the corresponding cyclic analog. So S_NAr macrocyclizations have become an important part of solid-phase organic synthesis, especially for the preparation of libraries of β-turn mimics.

Based on the experiences presented earlier in this chapter, suitably substituted fluoro-nitrobenzoic acids are the substrates of choice for intramolecular S_NAr reactions [90].

Small libraries of 14-membered macrocyclic diaryl ethers and thioethers can be produced using a very similar procedure. Precursors are synthesized by the acylation reaction of solid-supported peptides with 3-fluoro-4-nitro benzoic acid. The substrates undergo cyclization by displacement of the fluorine with the phenolic oxygen of a tyrosine derivative [91] or with the thiol group of cysteines [92] under exceptionally mild conditions (see Scheme 8.36). Diversity could be increased via postmodification reactions of the nitro group.

Recently, Burgess and coworkers reported in detail on their research on libraries of peptide turn mimetics [93]. The effects on nucleophilicity, product ring size, resins, and other reaction conditions were examined. Optimized procedures to produce 13- to 16-membered ring systems are described [93].

Larger rings are synthesized by the replacement of the fluorine atom in three regioisomeric fluoro-nitrobenzoic acids to prepare analogs of tocinoic acid (see Scheme 8.37) [94]. In this synthesis the amino group of lysine serves as an internal nucleophile for the closure of the macrocycle.

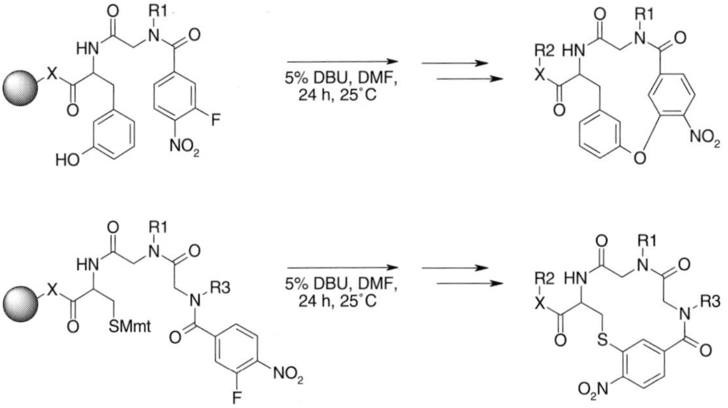
5% DBU, DMF,
24 h, 25°C
5% DBU, DMF,
24 h, 25°C

Scheme 8.36. Macrocyclizations using S_NAr reactions.

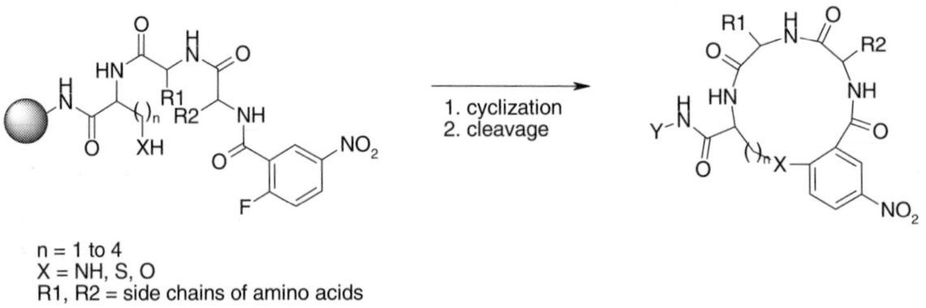
1. cyclization
2. cleavage
n = 1 to 4
X = NH, S, O
R1, R2 = side chains of amino acids

Scheme 8.37. Synthesis of large-ring systems.

References

1 R. B. Merrifield, *J. Am. Chem. Soc.* **1963**, *85*, 2149–2154.

2 P. Karoyan, A. Triolo, R. Nannicini, D. Giannotti, M. Altamura, G. Chaissaing, E. Perrotta, *Tetrahedron Lett.* **1999**, *40*, 71–74.

3 O. Mitsunobu, *Synthesis* **1981**, 1–28.

4 S. Hanessian, F. Xie, *Tetrahedron Lett.* **1998**, *39*, 733–736.

5 N. Bouloc, N. Walshe, M. Bradley, *J. Comb. Chem.* **2001**, *3*, 6–8.

6 G. A. Morales, J. W. Corbett, W. F. DeGrado, *J. Org. Chem.* **1998**, *63*, 2583–2589.

7 M. Caldarelli, J. Habermann, S. V. Ley, *J. Chem. Soc., Perkin Trans. 1* **1999**, 107–110.

8 H. Perrier, M. Labelle, *J. Org. Chem.* **1999**, *64*, 2110–2113.

9 M. Bernard, W. T. Ford, *J. Org. Chem.* **1983**, *48*, 326–332.

10 J. C. Pelletier, S. Kincaid, *Tetrahedron Lett.* **2000**, *41*, 797–800.

11 R. N. Salvatore, V. L. Flanders, D. Ha, K. W. Jung, *Org. Lett.* **2000**, *2*, 2797–2800.

12 See for example W.-R. Li, N.-M. Hsu, H.-H. Chou, S.-T. Lin, Y.-S. Lin, *Chem. Commun.* **2000**, 401–402.

13 D. P. Matthews, J. E. Green, A. J. Shuker, *J. Comb. Chem.* **2000**, *2*, 19–23.

14 J. J. Parlow, *Tetrahedron Lett.* **1996**, *37*, 5257–5260.

15 J. L. Castro, V. G. Matassa, *J. Org. Chem.* **1994**, *59*, 2289–2291.

16 E. E. Swayze, *Tetrahedron Lett.* **1997**, *38*, 8465–8468.

17 M. G. Johnson, D. D. Bronson, J. E.

GILLESPIE, D. S. GIFFORD-MOORE, K. KALTER, M. P. LYNCH, J. R. McCOWAN, C. C. REDICK, D. J. SALL, G. F. SMITH, R. F. FOGLESONG, *Tetrahedron* **1999**, *55*, 11641–11652.

18 G. W. STARKEY, J. J. PARLOW, D. L. FLYNN, *Bioorg. Med. Chem. Lett.* **1998**, *8*, 2385–2390.

19 D. P. CURRAN, *Angew. Chem., Int. Ed.* **1998**, *37*, 1175–1196.

20 S. PEUKERT, E. N. JACOBSEN, *Org. Lett.* **1999**, *1*, 1245–1248.

21 J. CHOI, N. M. YOON, *Synth. Commun.* **1995**, *25*, 2655–2663.

22 J. RADEMANN, R. R. SCHMIDT, *J. Org. Chem.* **1997**, *62*, 3650–3653.

23 A. BHANDARI, D. G. JONES, J. R. SCHULLEK, K. VO, C. A. SCHUNK, L. L. TAMANAHA, D. CHEN, Z. Y. YUAN, M. C. NEEDELS, M. A. GALLOP, *Bioorg. Med. Chem. Lett.* **1998**, *8*, 2303–2308.

24 A. NEFZI, M. GIULIANOTTI, R. A. HOUGHTEN, *Tetrahedron Lett.* **1998**, *39*, 3671–3674.

25 L. GOMEZ, F. GELLIBERT, A. WAGNER, C. MIOSKOWSKI, *Chem. Eur. J.* **2000**, *6*, 4016–4020.

26 D. S. DODD, O. B. WALLACE, *Tetrahedron Lett.* **1998**, *39*, 5701–5704.

27 N. BAILEY, A. W. DEAN, D. B. JUDD, D. MIDDLEMISS, R. STORER, S. P. WATSON, *Bioorg. Med. Chem. Lett.* **1996**, *6*, 1409–1414.

28 S. BRÄSE, D. ENDERS, J. KÖBBERLING, F. AVEMARIA, *Angew. Chem., Int. Ed.* **1998**, *37*, 3413–3415.

29 a) R. N. ZUCKERMANN, J. M. KERR, S. B. H. KENT, W. H. MOOS, *J. Am. Chem. Soc.* **1992**, *114*, 10646–10647; b) R. N. ZUCKERMANN, E. J. MARTIN, D. C. SPELLMEYER, G. B. STAUBER, K. R. SHOEMAKER, J. M. KERR, *J. Med. Chem.* **1994**, *37*, 2678–2685.

30 G. ALDRIAN-HERRADA, A. RABIÉ, R. WINTERSTEIGER, J. BRUGIDOU, *J. Pept. Sci.* **1998**, *4*, 266–281.

31 J. K. RUETER, S. O. NORTEY, E. W. BAXTER, G. C. LEO, A. B. REITZ, *Tetrahedron Lett.* **1998**, *39*, 975–978.

32 C. N. SELWAY, N. K. TERRET, *Bioorg. Med. Chem.* **1996**, *4*, 645–654.

33 Y. F. XIE, J. P. WHITTEN, T. Y. CHEN, Z. LIU, J. R. McCARTHY, *Tetrahedron* **1998**, *54*, 4077–4084.

34 M. G. ORGAN, C. E. DIXON, *Biotech. Bioeng.* **2000**, *71*, 71–77.

35 J. HABERMANN, S. V. LEY, J. S. SCOTT, *J. Chem. Soc., Perkin Trans. 1* **1999**, 1253–1256.

36 S. N. FILIHEDDU, S. MASALA, M. TADDEI, *Tetrahedron Lett.* **1999**, *40*, 6503–6506.

37 a) B. L. CHING, A. GANESAN, *Bioorg. Med. Chem. Lett.* **1997**, *7*, 1511–1514; b) R. MALTAIS, D. POIRIER, *Tetrahedron Lett.* **1998**, *39*, 4151–4144.

38 R. MALTAIS, M. R. TREMBLAY, D. POIRIER, *J. Comb. Chem.* **2000**, *2*, 604–614.

39 H.-P. BUCHSTALLER, *Tetrahedron* **1998**, *54*, 3465–3470.

40 W. J. HAAP, D. KAISER, T. B. WALK, G. JUNG, *Tetrahedron* **1998**, *54*, 3705–3724.

41 S. HANESSIAN, R.-Y. YANG, *Tetrahedron Lett.* **1996**, *37*, 5835–5838.

42 A. LOHSE, K. B. JENSEN, K. LUNDGREN, M. BOLS, *Bioorg. Med. Chem.* **1999**, *7*, 1965–1971.

43 H. S. OH, H.-G. HAHN, S. H. CHEON, D.-C. HA, *Tetrahedron Lett.* **2000**, *41*, 5069–5072.

44 B. RAJU, T. P. KOGAN, *Tetrahedron Lett.* **1997**, *38*, 4965–4968.

45 C. SUBRAMANYAM, *Tetrahedron Lett.* **2000**, *41*, 6537–6540.

46 A. A. VIRGILIO, A. A. BRAY, W. ZHANG, L. TRINH, M. SNYDER, M. M. MORRISSEY, J. A. ELLMAN, *Tetrahedron* **1997**, *53*, 6635–6644.

47 C. Y. CHO, R. S. YOUNGQUIST, S. R. PAIKOFF, M. H. BERESINI, A. H. HEBERT, L. T. BERLEAU, C. W. LIU, D. E. WEMMER, T. KEOUGH, P. G. SCHULTZ, *J. Am. Chem. Soc.* **1998**, *120*, 7706–7718.

48 M. UEKI, T. IKEO, M. IWADATE, T. ASAKURA, M. P. WILLIAMSON, J. SLANINOVÁ, *Bioorg. Med. Chem. Lett.* **1999**, *9*, 1767–1772.

49 B. LANGLOIS, L. GILBERT, G. FORAT, *Ind. Chem. Libr.* **1996**, *8*, 244–292.

50 a) J. T. THURSTON, J. R. DUDLEY, D. W. KAISER, I. HECHENBLEIKNER, F. C. SCHAEFER, D. J. HOLM-HANSEN, *J. Am. Chem. Soc.* **1951**, *73*, 2981–2983; b) C. R. JOHNSON, B. ZHANG, P. FANTAUZZI, M. HOCKER, K. M. YAGER, *Tetrahedron* **1998**, *54*, 4097–4106; c) P.

J. Hajduk, J. Dinges, J. M.
Schkeryantz, D. Janowick, M.
Kaminski, M. Tufano, D. J. Augeri,
A. Petros, V. Nienaber, P. Zhong,
R. Hammond, M. Coen, B. Beutel, L.
Katz, S. W. Fesik, *J. Med. Chem.* **1999**,
42, 3852–3859.

51 G. R. Gustafson, C. M. Baldino,
M.-M. E. O'Donnell, A. Sheldon,
R. J. Tarsa, C. J. Verni, D. L. Coffen,
Tetrahedron **1998**, *54*, 4051–4065.

52 M. Stanková, M. Lebl, *Mol. Diversity*
1996, *2*, 75–80.

53 J. L. Silen, A. T. Lu, D. W. Solas,
M. A. Gore, D. Maclean, N. H. Shah,
J. M. Coffin, N. S. Bhinderwala, Y.
Wang, K. T. Tsutsui, G. C. Look, D.
A. Campbell, R. L. Hale, M. Navre,
C. R. DeLuca-Flaherty, *Antimicrob.
Agents Chemother.* **1998**, *42*, 1447–1453.

54 D. Scharn, H. Wenschuh, U.
Reineke, J. Schneider-Mergener, L.
Germeroth, *J. Comb. Chem.* **2000**, *2*,
361–369.

55 M. T. Fiorini, C. Abell, *Tetrahedron
Lett.* **1998**, *39*, 1827–1830.

56 A. S. Fraser, A. M. Kawasaki, P. D.
Cook, *Nucleosides Nucleotides* **1999**, *18*,
1087–1089.

57 T. C. Norman, N. S. Gray, J. T. Koh,
P. G. Schultz, *J. Am. Chem. Soc.*
1996, *111*, 7430–7431.

58 Y.-T. Chang, N. S. Gray, G. R.
Rosania, D. P. Sutherlin, T. C.
Kwon, R. Sarohia, M. Leost, L.
Meijer, P. G. Schultz, *Chem. Biol.*
1999, *6*, 361–375.

59 N. S. Gray, L. Wodicka, A.-M. W. H.
Thunnissen, T. C. Norman, S. Kwon,
F. H. Espinoza, D. O. Morgan, G.
Barnes, S. LeClerc, L. Meijer, S.-H.
Kim, D. J. Lockhart, P. G. Schultz,
Science **1998**, *281*, 533–538.

60 a) C. Chen, R. Dagnino Jr, E. B. De
Souza, D. E. Grigoriadis, C. Q.
Huang, K.-I. Kim, Z. Liu, T. Moran,
T. R. Webb, J. P. Whitten, Y. F. Xie,
J. R. McCarthy, *J. Med. Chem.* **1996**,
39, 4358–4360; b) A. D. Baxter, E. A.
Boyd, P. B. Cox, V. Loh Jr, C.
Monteils, A. Proud, *Tetrahedron Lett.*
2000, *41*, 8177–8181; c) R. Di
Lucrezia, I. H. Gilbert, C. D. Floyd,
J. Comb. Chem. **2000**, *3*, 249–253; d)

M. Legraverend, O. Ludwig, E.
Bisagni, S. LeClerc, L. Meijer, N.
Giocanti, R. Sadri, V. Favaudon,
Bioorg. Med. Chem. **1999**, *7*, 1281–
1293; e) Z. Wu, J. Kim, R. M. Soll, D.
S. Dhanoa, *Biotechnol. Bioeng.* **2000**,
71, 87–90; f) L. F. Hennequin, S.
Piva-Le Blanc, *Tetrahedron Lett.* **1999**,
40, 3881–3884.

61 a) R. W. Sullivan, C. G. Bigam, P. E.
Erdman, M. S. S. Palaski, D. W.
Anderson, M. E. Goldman, L. J.
Ransone, M. J. Suto, *J. Med. Chem.*
1998, *41*, 413–419; b) M. J. Suto, L.
M. Gayo-Fung, M. S. S. Palaski, R.
Sullivan, *Tetrahedron* **1998**, *54*, 4141–
4150.

62 a) D. Obrecht, C. Albrecht, A.
Grieder, M. J. Villalgordo, *Helv.
Chim. Acta* **1997**, *80*, 65–72; b) T.
Masquelin, D. Sprenger, R. Baer, F.
Greber, Y. Mercadal, *Helv. Chim.
Acta* **1998**, *81*, 646–660; c) L. M. Gayo,
M. J. Suto, *Tetrahedron Lett.* **1997**, *38*,
211–214.

63 a) T. Masquelin, N. Meunier, F.
Gerber, G. Rossé, *Heterocycles* **1998**,
48, 2489–2504; b) T. Masquelin, Y.
Delgado, V. Baumlé, *Tetrahedron Lett.*
1998, *39*, 5725–5726.

64 L. Neuville, J. Zhu, *Tetrahedron Lett.*
1997, *38*, 4091–4094.

65 a) B. A. Bunin, J. A. Ellman, *J. Am.
Chem. Soc.* **1992**, *114*, 10997–10998; b)
C. G. Boojamra, K. M. Burow, L. A.
Thompson, J. A. Ellman, *J. Org.
Chem.* **1997**, *62*, 1240–1256.

66 a) J. Lee, D. Gauthier, R. A. Rivero,
J. Org. Chem. **1999**, *64*, 3060–3065; b)
M. K. Schwarz, D. Tumelty, M. A.
Gallop, *Tetrahedron Lett.* **1998**, *39*,
8397–8400.

67 a) J. Lee, W. V. Murray, R. A.
Rivero, *J. Org. Chem.* **1997**, *62*, 3874–
3879; b) G. A. Morales, J. W.
Corbett, W. F. DeGrado, *J. Org.
Chem.* **1998**, *63*, 1172–1177.

68 R. E. TenBrink, W. B. Im, V. H.
Sethy, A. H. Tang, D. B. Carter, *J.
Med. Chem.* **1994**, *37*, 758–768.

69 F. Zaragoza, H. Stephensen, *J. Org.
Chem.* **1999**, *64*, 2555–2557.

70 G. B. Phillips, G. P. Wie, *Tetrahedron
Lett.* **1996**, *37*, 4887–4890.

71 a) D. TUMELTY, M. K. SCHWARZ, M. C. NEEDELS, *Tetrahedron Lett.* **1998**, *39*, 7467–7470; b) J. P. MAYER, G. S. LEWIS, C. MCGEE, D. BANKAITIS-DAVIS, *Tetrahedron Lett.* **1998**, *39*, 6655–6658; c) J. LEE, D. GAUTHIER, R. A. RIVERO, *Tetrahedron Lett.* **1998**, *39*, 201–204; d) Z. WU, P. REA, G. WICKHAM, *Tetrahedron Lett.* **2000**, *41*, 9871–9874.

72 J. P. KILBURN, J. LAU, R. C. F. JONES, *Tetrahedron Lett.* **2000**, *41*, 5419–5421.

73 G. P. WIE, G. B. PHILLIPS, *Tetrahedron Lett.* **1998**, *39*, 179–182.

74 a) P.-C. PAN, C.-M. SUN, *Tetrahedron Lett.* **1999**, *40*, 6443–6446; b) P.-C. PAN, C.-M. SUN, *Biorg. Med. Chem. Lett.* **1999**, *9*, 1537–1540.

75 E. FARRANT, S. S. RAHMAN, *Tetrahedron Lett.* **2000**, *41*, 5383–5386.

76 G. LIU, Y. FAN, J. R. CARLSON, Z.-G. ZHAO, K. S. LAM, *J. Comb. Chem.* **2000**, *2*, 467–474.

77 U. PETERSEN, T. SCHENKE in: Quinolone Antibacterials. KUHLMANN, J., DALHOFF, A., ZEILER, H.-J. (eds), Springer Verlag, Berlin 1998, pp. 63–118.

78 a) Y.-S. OH, S.-H. CHO, *J Heterocyclic Chem.* **1998**, *35*, 17–23; b) K. E. FRANK, P. V. DEVASTHALE, E. J. GENTRY, V. T. RAVIKUMAR, A. KESCHAVARZ-SHOKRI, L. A. MITSCHER, A. NILIUS, L. L. SHEN, R. SHAWAR, W. R. BAKER, *Comb. Chem. High Throughput Screening* **1998**, *1*, 89–99.

79 a) A. A. MACDONALD, S. H. DEWITT, E. M. HOGAN, R. RAMAGE, *Tetrahedron Lett.* **1996**, *37*, 4815–4818; b) A. M. HAY, S. HOBBS-DEWITT, A. A. MACDONALD, R. RAMAGE, *Tetrahedron Lett.* **1998**, *39*, 8721–8724; c) A. M. HAY, S. HOBBS-DEWITT, A. A. MACDONALD, R. RAMAGE, *Synthesis* **1999**, *11*, 1979–1985.

80 R. MOHAN, W. YUN, B. O. BUCKMAN, A. LIANG, L. TRINH, M. M. MORRISSEY, *Biorg. Med. Chem. Lett.* **1998**, *8*, 1877–1882.

81 J. C. H. M. WIJKMANS, A. J. CULSHAW, A. D. BAXTER, *Mol. Diversity* **1998**, *3*, 117–120.

82 W. XU, R. MOHAN, M. M. MORRISSEY, *Tetrahedron Lett.* **1997**, *38*, 7337–7340.

83 a) X. OUYANG, N. TAMAYO, A. S. KISELYOV, *Tetrahedron* **1999**, *55*, 5827–5830; b) X. OUYANG, A. S. KISELYOV, *Tetrahedron* **1999**, *55*, 8295–8302; c) X. OUYANG, Z. CHEN, L. LONGBIN, C. DOMINGUEZ, A. S. KISELYOV, *Tetrahedron* **2000**, *56*, 2369–2377.

84 a) A. FALCHI, M. TADDEI, *Org. Lett.* **2000**, *2*, 3429–3431; b) H. DELEUZE, D. C. SHERRINGTON, *J. Chem. Soc., Perkin Trans. 2* **1995**, 2217–2221; c) C. R. JOHNSON, B. ZHANG, P. FANTAUZZI, M. HOCKER, K. M. YAGER, *Tetrahedron* **1998**, *54*, 4097–4106.

85 a) B. YAN, G. KUMARAVEL, *Tetrahedron* **1996**, *52*, 843–848; b) M. K. SCHWARZ, D. TUMELTY, M. A. GALLOP, *J. Org. Chem.* **1999**, *64*, 2219–2231.

86 a) G. C. MORTON, J. S. M. SALVINO, R. F. LABAUDINIÈRE, T. F. HERPIN, *Tetrahedron Lett.* **2000**, *41*, 3029–3030; b) A. NEFZI, N. A. ONG, M. A. GIULIANOTTI, J. M. OSTRESH, R. A. HOUGHTEN, *Tetrahedron Lett.* **1999**, *40*, 4939–4942.

87 T. S. YOKUM, J. ALSINA, G. BARANY, *J. Comb. Chem.* **2000**, *2*, 282–292.

88 a) S. HOBBS-DEWITT, J. S. KIELY, C. J. STANKOWIC, M. C. SCHROEDER, D. M. REYNOLDS-CODY, M. R. PAVIA, *Proc. Natl. Acad. Sci. USA* **1993**, *90*, 6909–6913; b) S. HOBBS-DEWITT, A. W. CZARNIK, *Acc. Chem. Res.* **1996**, *29*, 114–122.

89 A. K. SZARDENINGS, T. S. BURKOTH, H. H. LU, D. W. TIEN, D. A. CAMPBELL, *Tetrahedron* **1997**, *53*, 6573–6593.

90 E. A. JEFFERSON, E. E. SWAYZE, *Tetrahedron Lett.* **1999**, *40*, 7757–7760.

91 A. S. KISELYOV, S. EISENBERG, Y. LUO, *Tetrahedron* **1998**, *54*, 10635–10640.

92 A. S. KISELYOV, S. EISENBERG, Y. LUO, *Tetrahedron Lett.* **1999**, *40*, 2465–2468.

93 a) Y. FENG, K. BURGESS, *Chem. Eur. J.* **1999**, *5*, 3261–3272; b) Z. WANG, S. JIN, Y. FENG, K. BURGESS, *Chem. Eur. J.* **1999**, *5*, 3273–3278; c) A. J. ZHANG, S. KHARE, K. GOKULAN, D. S. LINTHICUM, K. BURGESS, *Bioorg. Med. Chem. Lett.* **2001**, *11*, 207–210.

94 C. FOTSCH, G. KUMARAVEL, S. K. SHARMA, A. D. WU, J. S. GOUNARIDES, N. R. NIRMALA, R. C. PETTER, *Biorg. Med. Chem. Lett.* **1999**, *9*, 2125–2130.

9
Electrophilic Substitution in Combinatorial and Solid-phase Synthesis

Jan-Gerd Hansel and Stephan Jordan

9.1
Introduction

Reactions in which an electron-deficient reagent attacks a substrate and an electron-deficient leaving group is displaced are called electrophilic substitutions (S_E). The most common leaving group is the proton. While the reaction is typical for aromatic systems, aliphatic substrates only react when hydrogens of sufficient acidity are available. This chapter focuses on the application of S_E in the generation of chemical diversity.

Compared with nucleophilic substitution reactions (see Chapter 8) the use of S_E in the context of combinatorial chemistry is rare. Much still remains to be done in this area of research. Although a number of functional group transformations involving S_E mechanisms have been applied to solution- and solid-phase synthesis, the potential of the reaction to link building blocks in a combinatorial sense is restricted to very special cases. The wide application of acid-labile linkers in solid-phase synthesis often prohibits the use of electrophilic reagents since the linkers are electron-rich aromatics and thus highly susceptible to aromatic S_E.

Reactions involving S_E mechanisms are frequently used in building-block synthesis. Despite its importance in the overall workflow of combinatorial chemistry, this aspect is beyond the scope of this chapter and will not be covered. Reference will only be made to the use of S_E in the preparation of linkers and solid supports (see also Chapter 4). S_E reactions involving organometallic reagents or carbon nucleophiles will be dealt with in the appropriate chapters. Traditional synthetic methods involving S_E have been supplemented by modern transition metal-catalyzed substitution reactions. Please refer to other Chapters for their applications in combinatorial chemistry.

9.2
Electrophilic Substitution at Aliphatic Carbons

9.2.1
Halogen Electrophiles

Halogenated carbonyl compounds are important synthetic intermediates, especially in the formation of heterocycles. Suitable reaction conditions have been devised for the transformation of enolizable carbonyl compounds into the corresponding α-bromo derivatives. The reagent of choice is pyridinium perbromide, which can be employed in both solid-phase [1] and solution-phase synthesis. In the latter case the commercially available polymer-bound version of the reagent has been frequently used [2].

As an example of this transformation the solution-phase bromination of acetophenones to α-bromo-acetophenones will be discussed. The transformation has been incorporated in an oxidation, bromination, and nucleophilic substitution reaction sequence (see Scheme 9.1). Notably, this sequence is performed by adding all the necessary polymer-supported reagents to the starting phenylethanol at the same time in a single reaction vessel. The final yield in this multistep/one-chamber solution-phase synthesis is higher than the combined yields of the three steps performed sequentially [3].

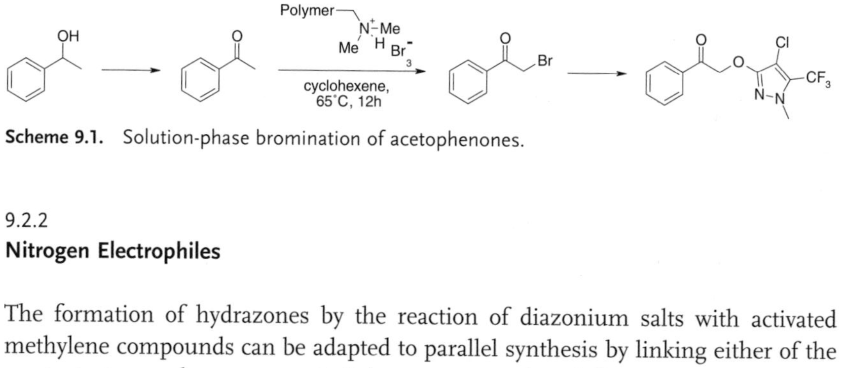

Scheme 9.1. Solution-phase bromination of acetophenones.

9.2.2
Nitrogen Electrophiles

The formation of hydrazones by the reaction of diazonium salts with activated methylene compounds can be adapted to parallel synthesis by linking either of the reactants to a polymer support. Polymer-supported aryl diazonium cations have been treated with the potassium salt of Meldrum's acid at 25 °C to give the corresponding 5-phenylhydrazone derivatives of Meldrum's acid in good yield [4]. Alternatively, polymer-supported Meldrum's acid anion reacts with various aryl diazonium fluoroborates at 25 °C in acetonitrile, yielding the same products [5].

Aliphatic diazo compounds are obtained on solid support from the reaction of tosyl or mesyl azide with immobilized activated methylene compounds. The substrates used are β-ketoesters [6], β-ketoamides [7] (see Scheme 9.2), or malonic acid derivatives [8]. Typically, the reaction is carried out at room temperature using an excess of the azide and an even larger excess of triethylamine or diisopropylethylamine.

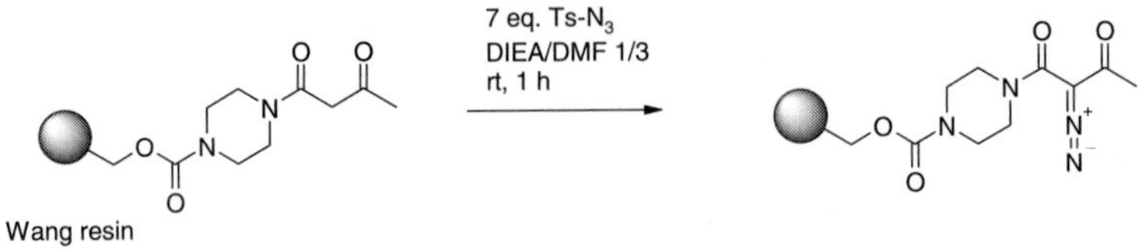

Wang resin

Scheme 9.2. Solid-phase diazotation.

9.2.3
Carbon Electrophiles

Enamines react with a number of electrophiles such as electron-poor olefins. The reaction is used in heterocycle chemistry, as shown in a solid-phase synthesis of dihydropyridines (see Scheme 9.3) [9].

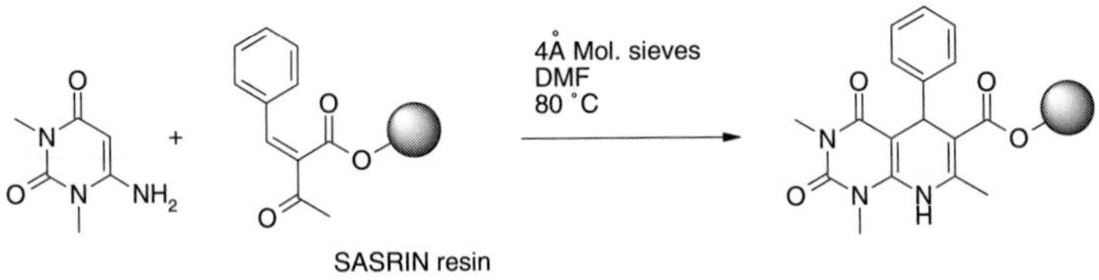

SASRIN resin

Scheme 9.3. Solid-phase enamine substitution and cyclization.

9.3
Electrophilic Substitution at Aromatic Carbons

9.3.1
General Remarks

Almost all kinds of S_E reactions involving arene substrates are used in solution-phase chemistry as a powerful synthetic tool. In solid-phase chemistry the use of this reaction is limited owing to its incompatibility with electron-rich linkers. With the exception of a few examples mentioned below, aromatic S_E is restricted to the functionalization of polystyrene-based supports [10].

Polystyrene can be brominated, nitrated, and acylated or alkylated applying Friedel–Crafts conditions in solvents such as carbon tetrachloride or nitrobenzene. For example, in 1988 Ajayaghosh and Pillai [11] demonstrated the preparation of a photosensitive resin using S_E reactions (see Scheme 9.4). Commercially available

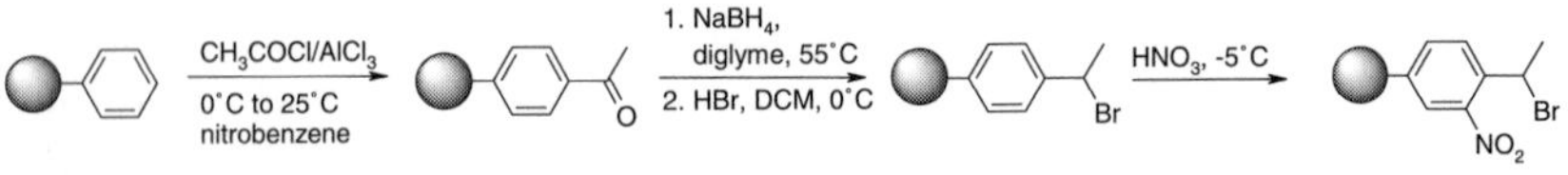

Scheme 9.4. Solid-phase Friedel–Crafts acylation.

polystyrene (crosslinked with 1% divinylbenzene) was first acylated with acetyl chloride under typical Friedel–Crafts conditions. The resulting ketone was reduced to the corresponding alcohol and then halogenated. Using a second S_E reaction, a nitro group was introduced with fuming nitric acid at low temperature.

Higher reaction temperatures applied during acylation of solid-supported material may lead to side-reactions such as partial dealkylation of phenyl groups and hence to soluble polymers. Although Friedel–Crafts alkylations on polystyrene are possible under harsh conditions (strong acids), there are more suitable methods for this kind of C–C bond formation. Reaction of an immobilized organometallic compound (most commonly lithiated by a halogen–lithium exchange reaction) [12] with alkyl halides can be considered as the method of choice.

9.3.2
Halogen Electrophiles

In accordance with their inherent preference for electrophilic substitutions, many heterocycles are easily halogenated. For example, a simple bromination can serve as a starting point for further diversification when followed by palladium-catalyzed C–C bond formation (see Scheme 9.5). In addition to Suzuki couplings, the 6-bromonalidixic acid derivative obtained in the bromination step can also undergo Heck reaction [13].

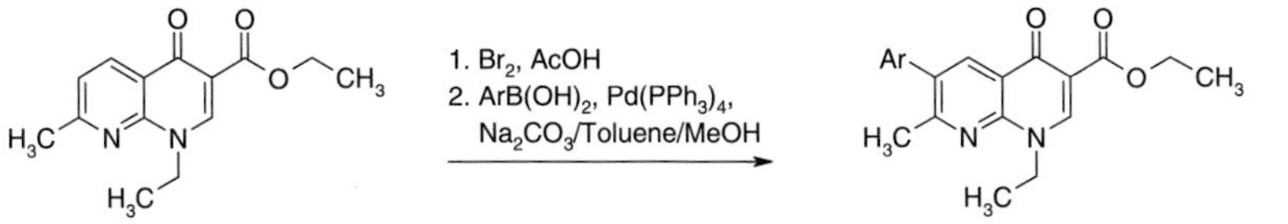

Scheme 9.5. Solution-phase arene bromination and subsequent Suzuki coupling.

There are only a few examples of arene brominations on solid support in the literature. Using N-bromosuccinimide in dimethyl formamide at room temperature, electron-rich arenes such as thiophenes can be brominated (see Scheme 9.6). Combination with a Stille coupling and reiteration of the reaction sequence leads to oligothiophenes – new materials with interesting optical and electronic properties [14].

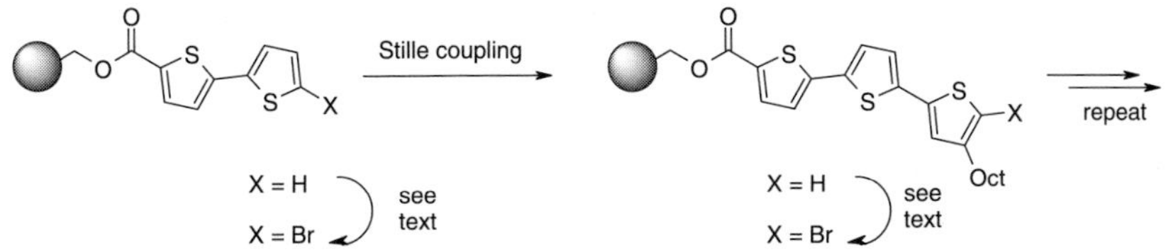

Scheme 9.6. Solid-phase arene bromination and subsequent Stille coupling.

A rare example of direct halogenation on a solid support has been reported for phenols. The phenolic moiety of tyrosine undergoes iodination when treated with bis-(pyridine) iodonium(I) tetrafluoroborate (Ipy_2BF_4) for no more than 10 min (see Scheme 9.7) [15].

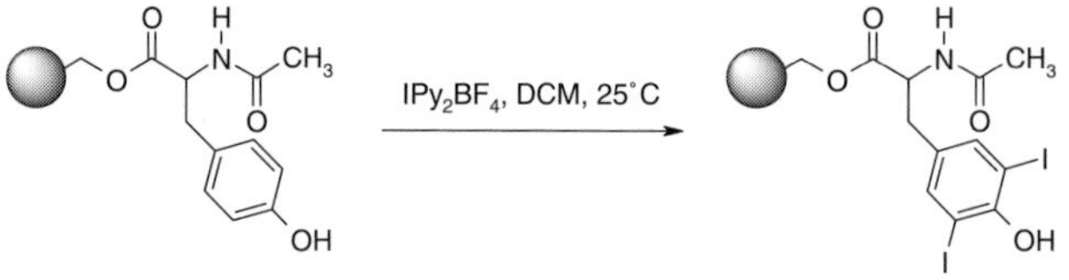

Scheme 9.7. Solid-phase iodination.

The only example in this chapter in which the leaving group in an S_E reaction is not a proton involves germanium-based linkers. These linkers have been developed for solid-phase synthesis as a means of traceless linkage (see Chapter 4). However, reaction of germanium-linked substrates with bromine rapidly releases the corresponding aryl bromides by ipso substitution of the germanium by bromine (see Scheme 9.8). Aryl iodides can be prepared by the same method using iodomonochloride [16].

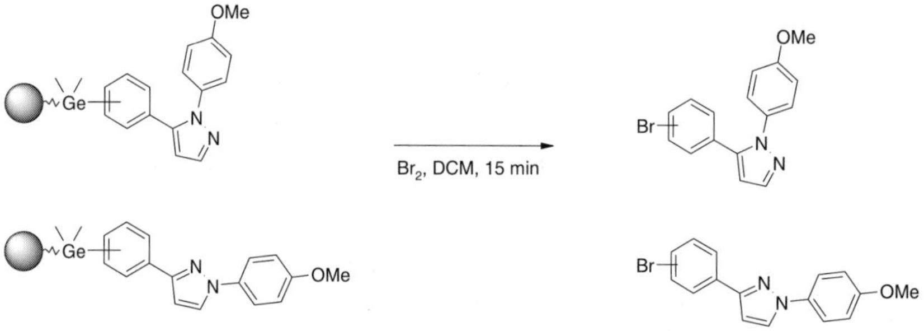

Scheme 9.8. Solid-phase ipso substitution of germanium by bromine.

9.3.3
Nitrogen Electrophiles

Very recently the diazotization of aromatic compounds has found applications in combinatorial chemistry. The formation of diazonium salts and the coupling to electron-rich aromatics to give azo dyes can be performed using polymer-supported reagents (see Scheme 9.9) [17].

Polymer-bound aryl diazonium salts also play a pivotal role in the chemistry of triazene linkers.

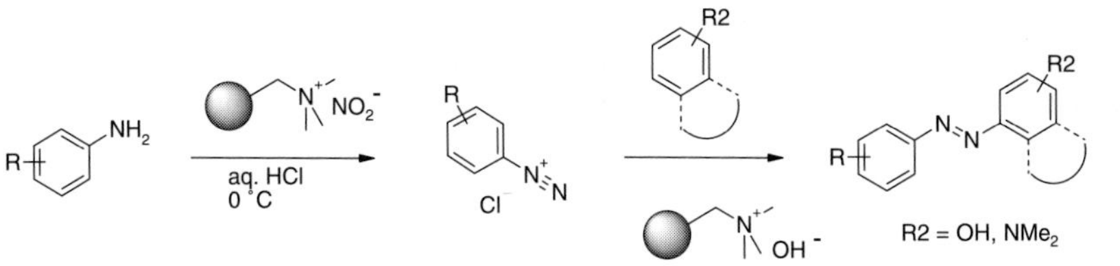

Scheme 9.9. Solution-phase diazotation and azo coupling.

9.3.4
Carbon Electrophiles

Friedel–Crafts chemistry is rarely used to generate diversity. One of the few examples is the superacid-induced solution-phase synthesis of a small library of 3,3-diaryloxindoles (see Scheme 9.10) [18]. The reaction proceeds smoothly in pure triflic acid at room temperature.

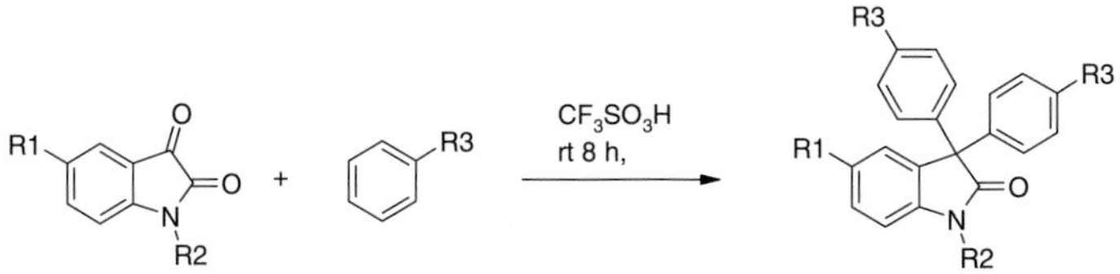

Scheme 9.10. Solution-phase Friedel–Crafts alkylation.

The Pictet–Spengler reaction is a very well-established method for the synthesis of tetrahydroisoquinolines and tetrahydro-β-carbolines both in solution and on solid support [19]. The molecules are easily prepared by intramolecular reaction of an iminium ion with an arene usually under acidic conditions. The imines are typically formed by condensing amines with (aromatic) aldehydes (see Scheme 9.11), cyclohexanones, or aryl methyl ketones [20].

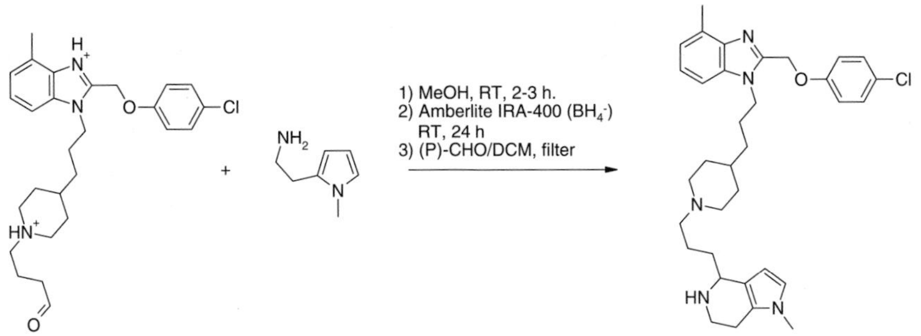

Scheme 9.11. Solution-phase Pictet–Spengler reaction.

Using tryptophan as the starting material, the synthesis of tricyclic carbolines has been reported in numerous examples in the literature. Whereas reactions with aldehydes are normally complete within hours, less reactive ketones require up to several days to reach complete conversion (see Scheme 9.12) [21].

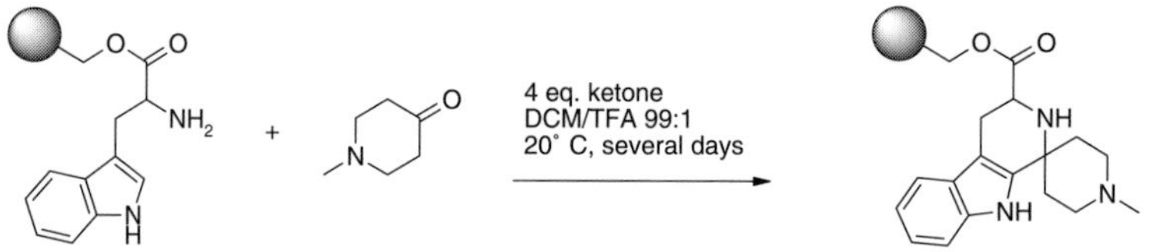

Scheme 9.12. Solid-phase Pictet–Spengler reaction.

A Pictet–Spengler reaction has been used as the key step during the synthesis of indolyl diketopiperazine-based libraries of Fumitremorgin C analogs (see Scheme 9.13). This natural product, isolated by fermentation of the fungi *Aspergillus fumigatus*, appears to be of interest in the area of central nervous system (CNS) and cancer research and has therefore resulted in the preparation of some solid-phase combinatorial libraries [22].

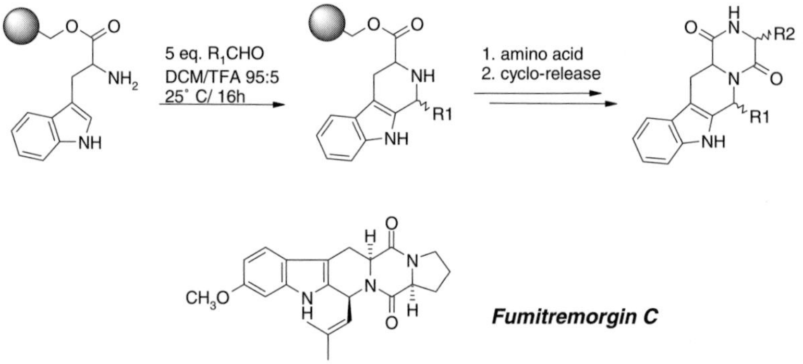

Fumitremorgin C

Scheme 9.13. Solid-phase synthesis of Fumitremorgin C analogs.

The Pictet–Spengler reaction is not limited to the synthesis of six-member ring systems. Also the class of seven-member rings present in the pharmaceutically important diazepine class of compounds is easily accessed by condensing aldehydes with suitably substituted aminomethylfurans (see Scheme 9.14) [23].

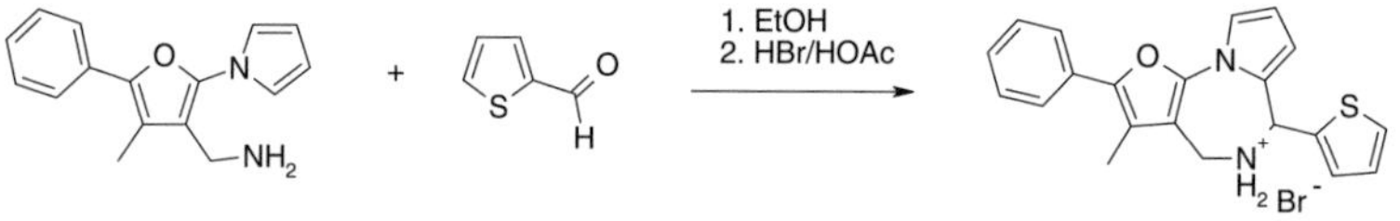

Scheme 9.14. Solution-phase Pictet–Spengler reaction yielding seven-member rings.

In the closely related Bischler–Napieralski reaction dihydroisoquinolines are formed. The intramolecular condensation of an acylated amine with an arene is mediated by the action of a strong dehydrating agent, usually phosphoryl chloride. It leads to dihydroisoquinolines, which can be transformed either into isoquinolines by oxidation or into tetrahydroisoquinolines by reduction (see Scheme 9.15). The harsh reaction conditions are not generally amenable to solid-phase reactions since most linkers are not stable to a large excess of phosphoryl chloride in toluene at 80 °C [24].

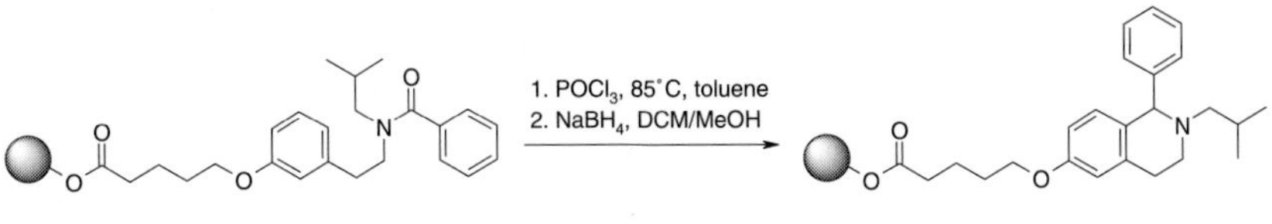

Merrifield resin

Scheme 9.15. Solid-phase Bischler–Napieralski reaction.

The reaction has also been applied to solution-phase synthesis, but has found little application in library production to date [25].

In a Reissert-type reaction involving pyrrole derivatives, C–C bond formation is achieved by aromatic S_E on solid support using N-oxides (see Scheme 9.16). The reaction also works with indoles and with enamines [26].

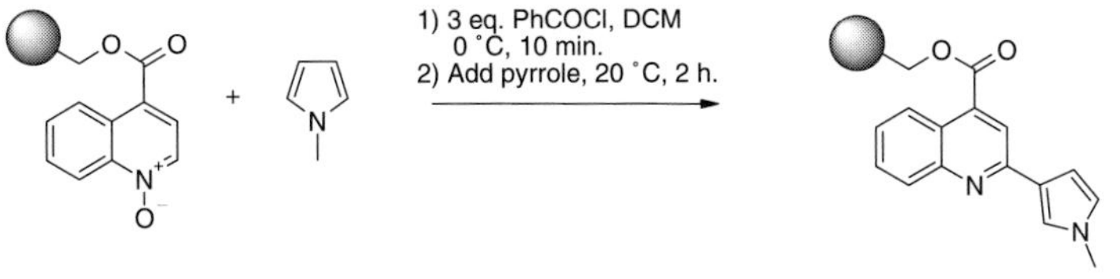

Scheme 9.16. Solid-phase Reissert reaction.

References

1 A. BARCO, S. BENETTI, C. DE RISI, C. MARCHETTI, C. P. POLLINI, V. ZANIRATO, *Tetrahedron Lett.* **1998**, *39*, 7591–7594.

2 a) J. HABERMANN, S. V. LEY, J. S. SCOTT, *J. Chem. Soc. Perkin Trans. 1* **1998**, 3127–3130; b) J. HABERMANN, S. V. LEY, R. SMITS, *J. Chem. Soc. Perkin Trans. 1* **1999**, 2421–2423; c) J. HABERMANN, S. V. LEY, J. J. SCICINSKI, J. S. SCOTT, R. SMITS, A. W. THOMAS, *J. Chem. Soc. Perkin Trans. 1* **1999**, 2425–2427.

3 J. J. PARLOW, *Tetrahedron Lett.* **1995**, *36*, 1395–1396.

4 B. P. BANDGAR, J. V. TOTARE, J. N. NIGAL, *Ind. J. Heterocycle Chem.* **1998**, *8*, 77–78.

5 B. P. BANDGAR, A. M. TAVHARE, S. S. PANDIT, *Ind. J. Chem., Sect. B* **2000**, *38*, 721–723.

6 M. CANO, F. CAMPS, J. JOGLAR,

Tetrahedron Lett. **1998**, *39*, 9819–9822.

7 F. Zaragoza, S. V. Petersen, *Tetrahedron* **1996**, *52*, 5999–6002.

8 a) D. L. Whitehouse, K. H. Nelson Jr, S. N. Savinov, D. J. Austin, *Tetrahedron Lett.* **1997**, *38*, 7139–7142; b) M. R. Gowravaram, M. A. Gallop, *Tetrahedron Lett.* **1997**, *38*, 6973–6976; c) D. L. Whitehouse, K. H. Nelson Jr, S. N. Savinov, R. S. Löwe, D. J. Austin, *Bioorg. Med. Chem.* **1998**, *6*, 1273–1282.

9 M. F. Gordeev, D. V. Patel, J. Wu, E. M. Gordon, *Tetrahedron Lett.* **1996**, *37*, 4643–4646.

10 a) J. H. Adams, R. M. Cook, D. Hudson, V. Jammalamadaka, M. H. Lyttle, M. F. Songster, *J. Org. Chem.* **1998**, *63*, 3706–3716; b) G. Orosz, L. P. Kiss, *Tetrahedron Lett.* **1998**, *39*, 3241–3242.

11 A. Ajayaghosh, V. N. R. Pillai, *Tetrahedron Lett.* **1988**, *21*, 6661–6666.

12 a) S. Havez, M. Begtrup, P. Vedso, *J. Org. Chem.* **1998**, *63*, 7418–7420; b) Z. Li, A. Ganesan, *Synlett* **1998**, 405–406.

13 C. Plisson, J. Chenault, *Heterocycles* **1999**, *51*, 2627–2637.

14 P. R. L. Malenfant, J. M. Fréchet, *Chem. Commun.* **1998**, 2657–2658.

15 G. Arsequell, G. Espuña, G. Valencia, J. Barluenga, R. P. Carlón, J. M. González, *Tetrahedron Lett.* **1998**, *39*, 7393–7396.

16 a) A. C. Spivey, M. C. Diaper, H. Adams, *J. Org. Chem.* **2000**, *65*, 5253–5263; b) M. J. Plunkett, J. A. Ellman, *J. Org. Chem.* **1997**, *62*, 2885–2893.

17 M. Caldarelli, I. R. Baxendale, S. V. Ley, *Green Chem.* **2000**, 43–45.

18 D. A. Klumpp, K. Y. Yeung, G. K. S. Prakash, G. A. Olah, *J. Org. Chem.* **1998**, *63*, 4481–4484.

19 P. P. Fantauzzi, K. M. Yager, *Tetrahedron Lett.* **1998**, *39*, 1291–1294.

20 M. G. Siegel, M. O. Chaney, R. F. Bruns, M. P. Clay, D. A. Schober, A. M. Van Abbema, D. W. Johnson, B. E. Cantrell, P. J. Hahn, D. C. Hunden, D. R. Gehlert, H. Zarrinmayeh, P. L. Ornstein, D. M. Zimmerman, G. A. Koppel, *Tetrahedron* **1999**, *55*, 11619–11639.

21 a) J. P. Mayer, D. Bankaitis-Davis, J. Zhang, G. Beaton, K. Bjergarde, C. M. Andersen, B. A. Goodman, C. J. Herrera, *Tetrahedron Lett.* **1996**, *37*, 5633–5637; b) L. Yang, L. Guo, *Tetrahedron Lett.* **1996**, *37*, 5041–5044; c) L. Yang, *Tetrahedron Lett.* **2000**, *41*, 6981–6984; d) X. Li, L. Zhang, W. Zhang, S. E. Hall, J. P. Tam, *Org. Lett.* **2000**, *2*, 3075–3078.

22 a) A. van Loevezijn, J. H. van Maarseveen, K. Stegman, G. M. Visser, G.-J. Koomen, *Tetrahedron Lett.* **1998**, *39*, 4737–4740; b) H. Wang, A. Ganesan, *Org. Lett.* **1999**, *1*, 1647–1649.

23 a) X. Feng, J. C. Lancelot, A. C. Gillard, H. Landelle, S. Rault, *J. Heterocycle Chem.* **1998**, *35*, 1313–1316; b) S. Vega, M. S. Gil, V. Darias, C. C. Sanchez Mateo, M. A. Exposito, *Pharmazie* **1995**, *50*, 27–33.

24 a) K. Rölfing, M. Thiel, H. Künzer, *Synlett* **1996**, 1036–1038; b) W. D. F. Meutermans, P. F. Alewood, *Tetrahedron Lett.* **1995**, *36*, 7709–7712.

25 a) V. Jullian, J. C. Quirion, H. P. Husson, *Eur. J. Org. Chem.* **2000**, *7*, 1319–1325; b) S. Deprets, G. Kirsch, *Eur. J. Org. Chem.* **2000**, *7*, 1353–1357.

26 M. Z. Hoemann, G. Melikian-Badalian, G. Kumaravel, J. R. Hauske, *Tetrahedron Lett.* **1998**, *39*, 4749–4752.

10

Elimination Chemistry in the Solution- and Solid-phase Synthesis of Combinatorial Libraries

Demosthenes Fokas and Carmen Baldino

10.1
Introduction

One of the challenges for organic chemists involved in the burgeoning field of combinatorial chemistry is to rediscover new uses and applications of old reactions and subsequently adapt them to either solid- or solution-phase chemistry. Elimination reactions, which have been studied thoroughly and used extensively by the chemistry community in the synthesis of several complex molecules, fall into this category [1]. Among them, β-eliminations prevail in organic synthesis with numerous applications in the preparation of olefins. Although olefins are versatile and useful intermediates for combinatorial chemistry, the synthesis of libraries of olefinic substrates by adaptation of the classical β-elimination reaction has not received much attention so far, presumably because of the lack of convergence in the synthesis and diversity of the final products. Instead, β-eliminations along with other elimination reactions are gaining favor in solid-phase synthesis as a release strategy of the desired products from solid support.

In this chapter we will address elimination reactions from a mechanistic rational including: (1) β-eliminations, (2) conjugate eliminations, and (3) addition–elimination reactions. We will also discuss the utility of these transformations in combinatorial chemistry. Olefination reactions such as the Wittig, Horner–Emmons, and ring-closing olefin metathesis, which could fall into the addition–elimination category, will not be discussed herein since they will be addressed in different chapters.

10.2
β-Eliminations in Combinatorial Chemistry

β-Elimination has been used for peptide synthesis since 1967. However, broader applications have been limited, presumably because the cleavage conditions as described needed to be tightly controlled [2]. The expanded utility of β-elimination reactions in combinatorial chemistry and solid-phase synthesis has been realized very recently with the advent of new linkers [3]. The design of traceless linkers that leave no residue on the cleaved product enabled chemists to envision β-elimination reactions as an effective release strategy of the desired products from solid support. The term traceless usually defines linkers that leave no obvious residue on the cleaved molecule. A traceless linker is defined as one where a new C–H or C–C bond is formed at the linkage site of the cleaved molecule. However, this definition has been expanded to linkers that include other cleavage reactions. Although there are several transformations in combinatorial chemistry that involve an intermediate β-elimination step, we will focus on release-based β-eliminations and their application to combinatorial chemistry.

10.2.1
The Hofmann Elimination Solid-phase Synthesis of Tertiary Amines

10.2.1.1 Via a Regenerated Michael Acceptor (REM) Resin
The Hofmann elimination can be traced back to 1851 when Hofmann first reported the elimination of quaternary ammonium compounds [4]. The reaction is usually considered to be a useful method for the synthesis of alkenes and indeed has been previously used by Blettner and Bradley to synthesize dehydroalanine derivatives on resin [5]. The utility of this reaction in combinatorial chemistry was not realized until 1996 when Rees and coworkers introduced this reaction as part of a traceless linker strategy for the solid-phase synthesis of libraries of tertiary amines (Scheme 10.1) [6].

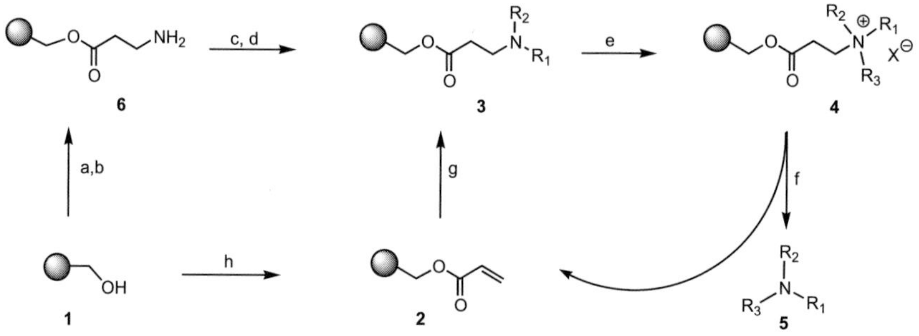

a) Fmoc-β-alanine, DIC, DMAP, DMF; b) 20% piperidine in DMF; c) R_1CHO, NaBH(OAc)$_3$, DMF; d) R_2CHO, NaBH(OAc)$_3$, DMF; e) R_3X, DMF, 20 °C; f) DIEA, DMF or DCM, 20 °C; g) DMF, R_1R_2NH, 20 °C; h) acryloyl chloride, DIEA, DCM, 20 °C.

Scheme 10.1. Synthesis of tertiary amines on a REM resin.

This strategy utilizes a hydroxymethylpolystyrene resin (**1**) derivatized as the acrylate ester (**2**) which upon a Michael addition of a primary or secondary amine gives the secondary or tertiary resin-bound amine (**3**) respectively. In the case where a primary amine is added, the resulting resin-bound secondary amine can be converted into tertiary amine (**3**) by a reductive alkylation, thus introducing a new element of diversity. Quaternization of the tertiary amine with an alkyl halide to give ammonium salt (**4**) introduces another site of diversity and activates the linker for cleavage by a facile Hofmann elimination reaction. Then, NiPr$_2$Et at room temperature liberates the tertiary amine (**5**) into solution and regenerates the acrylate resin (**2**). Similarly, tertiary amines with the general structure **5** and with three sites of diversity can result from a resin-bound equivalent of ammonia (**6**), which can be derived from the coupling of 9-fluorenylmethoxycarbonyl (Fmoc)-β-alanine to hydroxymethylpolystyrene resin followed by Fmoc deprotection. Since the resin linker **2** is regenerated after elimination of the product and is functionalized via a Michael reaction, this resin was referred to as a REM (regenerated Michael acceptor) resin.

The REM resin system can be used in the monoalkylation of diamines without the use of protecting groups (Scheme 10.2). For example, piperazine can be added to resin **2** to give the monoalkylated derivative **7**, which can then be acylated or alkylated cleanly at the second nitrogen [7]. Treatment of resin-bound piperazine (**7**) with isocyanates or alkyl halides can generate compounds of the general structure **8** or **9** respectively. Quaternization and elimination provides disubstituted piperazine adducts **10** and **11** in high purity. A library of 125 piperazines with the general structure **11** was prepared and a few of the compounds were found to be active against δ-opioid receptors [8]. The high purity of substrate **11** demonstrates that although quaternization at the undesired nitrogen in piperazine (**9**) is possible, it does not reduce the purity of the final product since only the desired product can be cleaved from the resin.

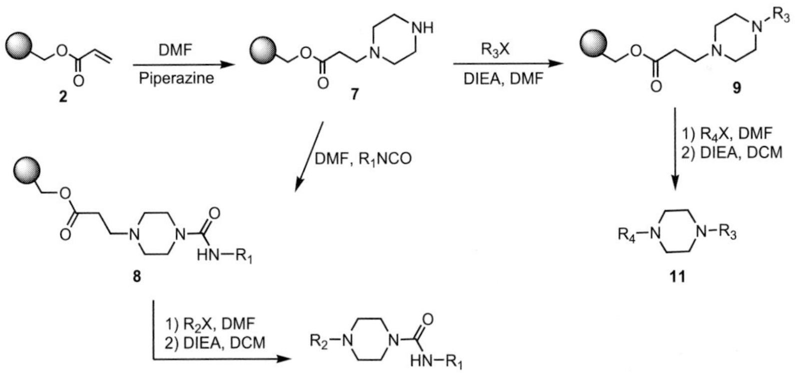

Scheme 10.2. Immobilization and derivatization of diamines on a REM resin.

The overall yield for the three-step sequence ranges from 40% to 80% and is substrate dependent. The purity of the isolated products is good because only the desired tertiary resin-bound amine will be susceptible to the elimination conditions. The method works well when the quaternization step can be conducted at ambient temperature using reactive alkyl halides (i.e. allyl, benzyl, and methyl). The quaternization step only requires heat when less reactive alkylating agents are used. However, quaternization at elevated temperatures can cause the Hofmann elimination to occur prematurely, releasing the tertiary amine into solution where it is susceptible to quaternization by a second equivalent of alkylating agent [7]. Therefore, a more thermally stable linker needs to be developed in order to prevent premature cleavage when less reactive alkyl halides are used. Quaternization can also be problematic with sterically hindered alkylating agents or when sterically hindered resin-bound amines are involved.

Although the Michael addition of primary and secondary alkyl amines to resin **2** works well, there are no examples describing the addition of anilines to resin **2** and, consequently, the synthesis of tertiary anilines. Addition of anilines would presumably require more vigorous reaction conditions.

The REM linker is stable to both mildly acidic and basic conditions, making possible Fmoc protection of amines and *tert*-butyl ester protection of carboxylic acids. Boc protection of amines is also likely to be compatible with the resin, since the cleavage conditions are likely to be similar to those used for deprotecting *tert*-butyl esters. The amount of racemization occurring during the synthesis of *N*-alkylated α-amino ester derivatives on REM resin is likely to be minimal. Furthermore, the REM resin **2** can be recycled and used successfully with no substantial loss of its reactivity. No decrease in yield or purity of products was observed even after five synthesis cycles [7]. However, the ester linkage is not compatible with Grignard reagents, metal hydride reducing agents, or transesterification conditions, which limits the scope of reactions that can be performed on solid phase. However, this problem can be addressed through the use of the more stable sulfone REM resins **12–14**, which have been successfully used in the synthesis of amine libraries. These sulfone REM resins provide enhanced chemical stability and compatibility with a wider range of chemical reagents and reaction conditions [9]. Recently, the amide REM resin derivatives have emerged and two of these resins (**15**, **16**) are currently under evaluation for the automated preparation of amine libraries (Scheme 10.3) [10].

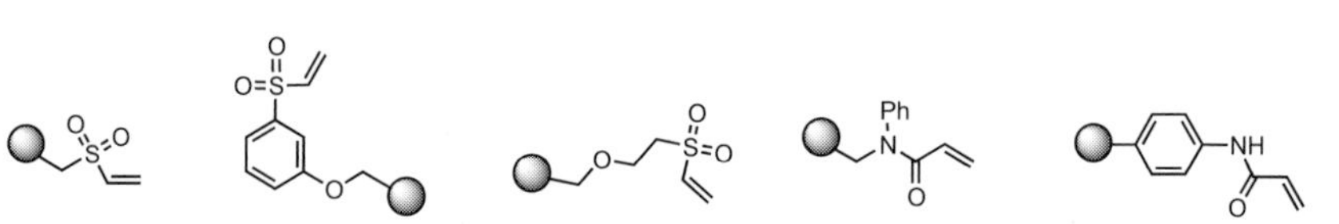

Scheme 10.3. Other REM resins used in the synthesis of tertiary amines.

10.2.1.2 Via a Safety-catch Resin
A small library of 13 G-coupled protein receptor agonists and antagonists was produced by Wade et al. using the sulfide safety-catch resin **17** (Scheme 10.4) [11]. Quaternization of resin **17** followed by oxidation gave resin-bound sulfone **18**, which resembles those derived from the corresponding vinyl sulfone REM resins via a Michael addition–quaternization sequence [9]. Sulfone **18** underwent a Hof-

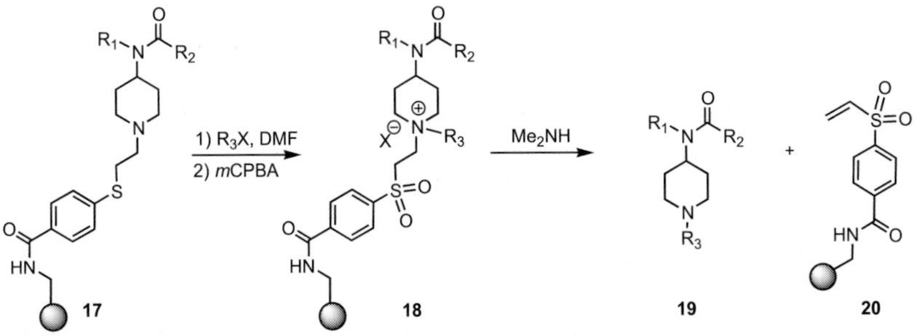

Scheme 10.4. Hofmann elimination on a safety-catch resin.

mann elimination, upon treatment with Me_2NH, to furnish vinyl sulfone resin **20** and release in solution the desired tertiary amine with general structure **19** in a 25% average yield for the six-step sequence. However, all of the final compounds had to be purified by reversed phase high-performance liquid chromatography (HPLC) owing to a minor 3-chlorobenzoic acid impurity. This problem can be circumvented by extended washing of the activated resin with 2 N HCl in tetrahydrofuran (THF), presumably exchanging the anions of the ammonium salt for chloride. The advantage of the sulfide safety-catch resin is the decreased sensitivity of the alkylated product to β-elimination, which could be utilized to increase the yield or extend the scope of alkylating agents to less activated systems. However, oxidation of the sulfide resin prior to amine synthesis would be necessary for more oxidation-sensitive systems.

10.2.1.3 Via a Hydroxylamine Resin
A Hofmann-type elimination has also been utilized independently by both Grigg (Scheme 10.5, route a) and Andersson (Scheme 10.5, route b) in the synthesis of tertiary methylamines using an extremely robust and versatile traceless linker [12]. Polystyrene resin (**21**) with a hydroxylamine linker attached can be converted to resin-bound tertiary hydroxylamine **22**. Quaternization of **22** with MeOTf resulted in the alkoxyammonium intermediate **23**, which upon treatment with NEt_3 in CH_2Cl_2 furnished the resin-bound aldehyde **25** and released into solution the tertiary methylamine with the general structure **24**. An alternative route involving the exposure of **23** to much milder reagents such as lithium iodide (in dioxane or ace-

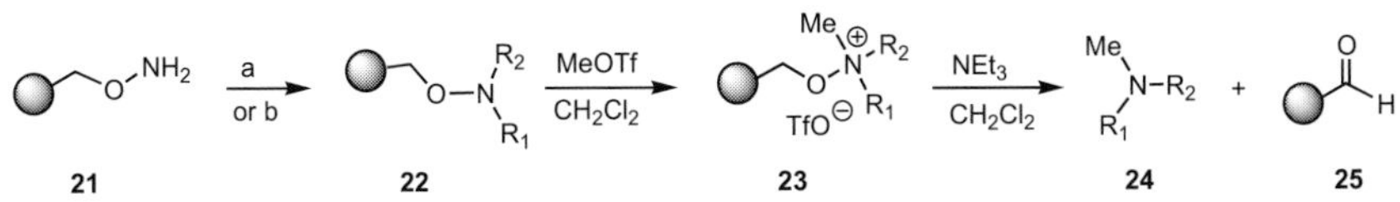

a: 1) BOC anhydride, DIEA, THF; 2) NaH/DMF then R_1Br; 3) 20% TFA/CH_2Cl_2; 4) $R_2CHO/NaBH(OAc)_3$, THF.
b: 1) R_1CHO, AcOH, MeOH; 2) $BH_3.Py$, HCl/Dioxane; 3) R_2CHO, PPTS, $BH_3.Py$, MeOH-THF.

Scheme 10.5. Hofmann elimination on a hydroxylamine resin.

tonitrile) or samarium iodide (in THF) also resulted in highly efficient cleavage delivering tertiary methylamines of high purity in good overall yields.

The quaternization works only with methyl triflate as the alkylating reagent, thus limiting the diversity of the produced library. Although other triflates prepared *in situ* via AgOTf/alkyl halide exchange gave the corresponding quaternary salts in solution, this approach is unsuitable for the solid-phase sequence because of interference of reaction kinetics from the silver halide precipitate. Both routes are amenable to solution-phase parallel synthesis by careful selection of reaction conditions and work-up methods starting from *O*-benzylhydroxylamine. Although this hydroxylamine resin (**22**) is more stable than the ester REM resin (**2**) toward a variety of nucleophilic reagents, it limits a library design to tertiary methylamines. Furthermore, this resin cannot be recycled once it is used – an advantage of the REM resins.

10.2.1.4 **Alternative Cleavage Techniques**

A major challenge associated with the cleavage of amines from resins is the removal of any excess cleavage reagents and byproducts from the reaction mixture. The process generally requires extraction or chromatography, which could render the synthesis of large libraries cumbersome. The pursuit of other cleavage strategies that would allow for the direct isolation of pure compounds has become necessary. Murphy and coworkers introduced a novel two-resin system in which the resin-bound quaternary ammonium compounds were treated with an excess of a second resin-bound amine such as the weakly basic ion-exchange resin Amberlite IRA-95 in dimethylformamide (DMF) and a catalytic amount of Et_3N (Scheme 10.6) [13]. Under these conditions, highly pure products were recovered with good yields after filtration and evaporation of the solvent.

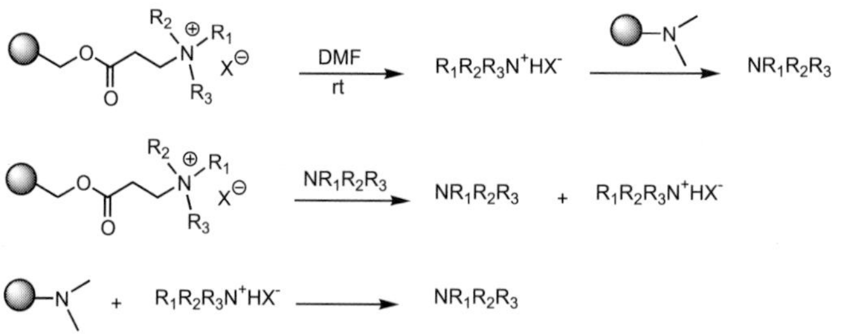

Scheme 10.6. Proposed mechanism for the two-resin system-promoted Hofmann elimination.

Similar results were obtained by treating the resin-bound quaternary ammonium compounds with Amberlite weakly basic ion-exchange resin in DMF in the absence of any additional base. The basic resin is sufficient to achieve cleavage and avoid the need for an aqueous work-up. These surprising two-resin results may be explained by a thermal elimination of the amine from the resin as the HBr salt.

The basic resin then desalts the amine to catalyze the β-elimination. Alternatively, it may also be due to trace amounts of base, present either from previous steps or from trace dimethylamine in the DMF. It was found that yield was a function of reaction time with an optimal length of 18 h for a 14-member library production. Identical results generating highly pure products were obtained using a deprotected Rink amide resin, albeit at lower yield [13].

A novel Wang resin-bound piperazine base (26) that resembles Murphy's two-resin system was introduced by Yamamoto et al. (Scheme 10.7) [14]. It was used successfully (2 equiv., loading 1.62 mmol g^{-1}), in the absence of any other external base, to cleave N-aryl-N'-benzylpiperazines from the resin by treating the quaternary ammonium compounds with resin 26 in CH$_2$Cl$_2$ for 16 h at room temperature. However, the caveat of this reagent and the ion-exchange resin is that they complicate the reusability of the REM resin since at the end of the sequence both resins are mixed together.

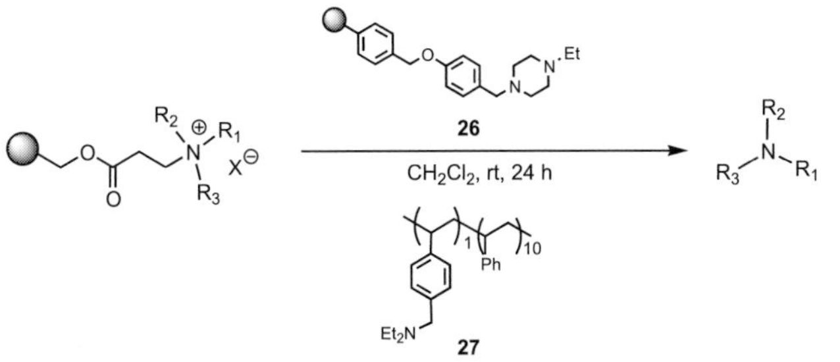

Scheme 10.7. Polymer-supported bases used in Hofmann elimination.

A soluble noncrosslinked polystyrene-bound basic reagent (NCPS-NEt$_2$) (27) has been developed recently by Janda and coworkers (Scheme 10.7) [15]. Use of base 27 (3 equiv., loading 0.85 mmol g^{-1}) in CH$_2$Cl$_2$ eliminates the need for purification and allows the direct isolation of a library of pure tertiary amines through simple filtration and concentration operations. The advantage of this method over the ion-exchange resin and the polymer-supported base methods is that it allows for the recycling of the REM resin by taking advantage of the insolubility of cleavage reagent 27 in methanol. Once the cleavage is complete, filtration of the reaction mixture separates the REM resin from the tertiary amine and the soluble reagent 27. Concentration of the filtrate followed by trituration with cold MeOH results in precipitation of 27. Filtration of the resulting slurry effectively separates the noncrosslinked polystyrene reagent and evaporation of the filtrate leaves behind the tertiary amine in good yields and high purity. However, a large amount of MeOH might be required to triturate the soluble reagent 27, which could render the synthesis of large libraries cumbersome.

Alternatively, Brown has reported a vapor-phase elimination approach as a rapid method for the cleavage of tertiary amines from REM resins 2 and 12 in Irori

MacroKans™ [16]. The MacroKans™ were placed in a glass peptide vessel, which was then sealed under a slight positive pressure of ammonia gas. Products were isolated cleanly in good yields after evaporation, resin sorting, and washing with CH_3CN or dimethylsulfoxide (DMSO). This is a particularly suitable method of parallel processing for the synthesis of large libraries, thus minimizing or eliminating the impurities due to the cleavage reagent.

10.2.2
β-Elimination on Selenyl Resins

The oxidation of selenides to selenoxides and their thermal elimination to alkenes has been studied extensively and has found numerous applications in synthesis [17]. The chemistry was first adapted to solid phase in 1976 when Heitz and co-workers prepared a polymer-supported selenide and oxidatively eliminated it to release an α,β-unsaturated ketone [18]. However, its application in solid-phase synthesis as a cleavage method was realized with the advent of the polymer-bound selenium reagents used in cyclization reactions to construct resin-bound carbocyclic scaffolds [19]. Recently, Nicolaou et al. utilized it in a strategy to generate a 10,000-compound benzopyran library by a solid-phase split-and-pool technique using Irori's NanoKan™ technology [20].

This novel strategy involves immobilization of an o-prenylated phenol (28) through cycloloading with a polystyrene-based selenyl bromide resin [21] to give resin-bound benzopyran scaffold 29 via a precedented 6-endo-trig cyclization (Scheme 10.8) [22]. Further elaboration of 29 to 30 and subsequent cleavage from solid support via oxidation and spontaneous syn-elimination of the selenoxide tether provides benzopyran 31. This is an example of a traceless release from solid support where functionality is generated at the released molecule instead of any linker residue being incorporated at the cleavage site.

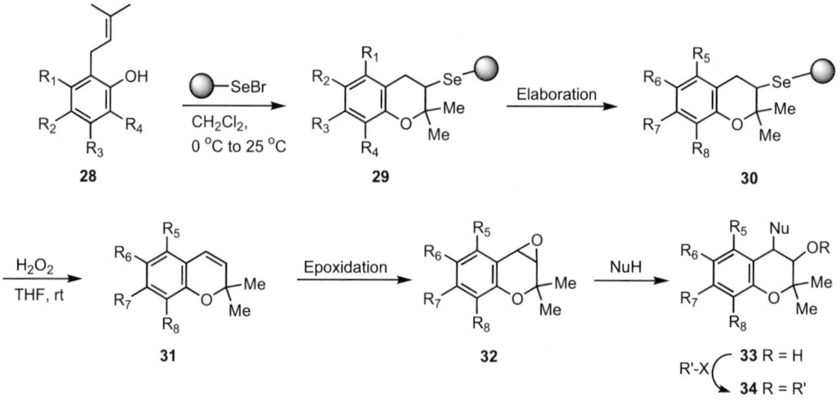

Scheme 10.8. Release of benzopyrans via a resin-bound selenoxide elimination.

Indeed, the newly formed double bond can serve as a starting point for the generation of secondary libraries or more focused libraries, thus introducing additional elements of diversity. For example, epoxidation of the released benzopyran **31** followed by ring opening of the intermediate epoxide **32** with a variety of nucleophiles provides access to a new series of benzopyran derivatives with the general structure **33**. Additionally, benzopyrans with the general structure **34** can result from further elaboration of the secondary hydroxyl group of **33** with a series of electrophiles [20c]. Taking into consideration the current advances in asymmetric epoxidation of olefinic substrates, this sequence could provide entry to chiral benzopyran libraries starting from chiral benzopyran epoxides [23].

The loading to the selenium solid support is compatible with a great variety of prenylated phenols except for substrates with electron-withdrawing groups adjacent to the prenyl group or adjacent to the phenol hydroxyl group participating in the cyclization step. Therefore, these scaffolds have to be loaded with the electron-withdrawing groups masked. The elimination of the selenoxide resin tether proceeds smoothly at room temperature and seems to be independent of the substitution pattern of the benzopyran scaffold.

A similar resin-bound selenoxide elimination on 2-seleno carbohydrates (**35**) was utilized by Nicolaou et al. in the synthesis of a small library of carbohydrate orthoesters representing novel regions of the potent antibiotic everninomicin (Scheme 10.9) [24]. Oxidation of glycoside **35** with *meta*-chloroperbenzoic acid (*m*-CPBA) gave the corresponding resin-bound selenoxide, which underwent a thermal *syn*-elimination. This thermal selenoxide elimination to the intermediate ketene acetal **36** introduces the desired functionality for the formation of orthoesters **37** and **38** and can release the desired products in a traceless manner from solid support. Orthoester **37** is formed from glycoside **35a** and 2,3-allyl orthoester **38** is formed from deprotected glycoside **35b**. However, unlike the solution-phase selenoxides, it was observed that the resin-bound selenoxide was more prone to eliminate at room temperature and therefore necessitated the use of lower temperatures in the oxidation step. Thus, treatment of selenide **35** with *m*-CPBA in CH_2Cl_2 at $-78\ °C$, followed by rapid filtration and transfer to a sealed tube, was found to give the best results. Although the chemistry has been developed in both solution- and solid-phase chemistry, it is well suited to the solid-phase synthesis of novel semisynthetic everninomicins and other carbohydrate libraries.

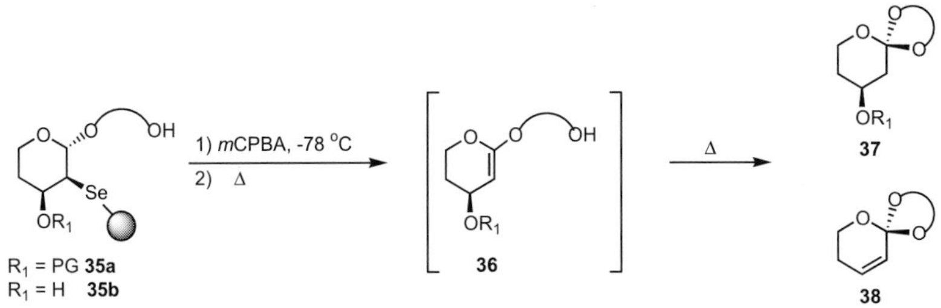

Scheme 10.9. Carbohydrate orthoesters via a resin-bound selenoxide elimination.

10.2.3
β-Elimination on Sulfone Resins

Sulfone elimination has the potential to be a preferred strategy for solid-phase compound cleavage. While no linker and cleavage strategy can be stable to the full range of conditions available to the synthetic chemist, the oxidative activation–elimination strategy promises to increase substantially the variety of options. Schwyzer et al. first described a 2-(4-carboxyphenylsulfonyl)ethanol linker for the synthesis of peptides and oligonucleotides [25]. However, the application and utility of a sulfone-type linker in peptide synthesis was demonstrated in 1992 by Katti et al. with the introduction of a new and readily available linker in the solid-phase synthesis of C-terminal peptides (Scheme 10.10) [26]. For example, Leu-enkephalin (**40**) was released from resin **39** in 54% or 60% overall yields, using either Boc or Fmoc chemistries respectively, after cleavage from the solid support with dioxane/MeOH/4 N NaOH followed by re-acidification.

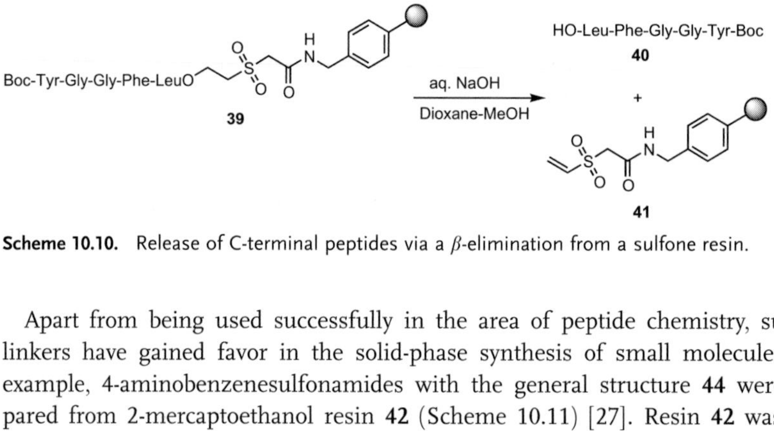

Scheme 10.10. Release of C-terminal peptides via a β-elimination from a sulfone resin.

Apart from being used successfully in the area of peptide chemistry, sulfone linkers have gained favor in the solid-phase synthesis of small molecules. For example, 4-aminobenzenesulfonamides with the general structure **44** were prepared from 2-mercaptoethanol resin **42** (Scheme 10.11) [27]. Resin **42** was converted to sulfone **43**, which underwent a facile β-elimination to release the desired

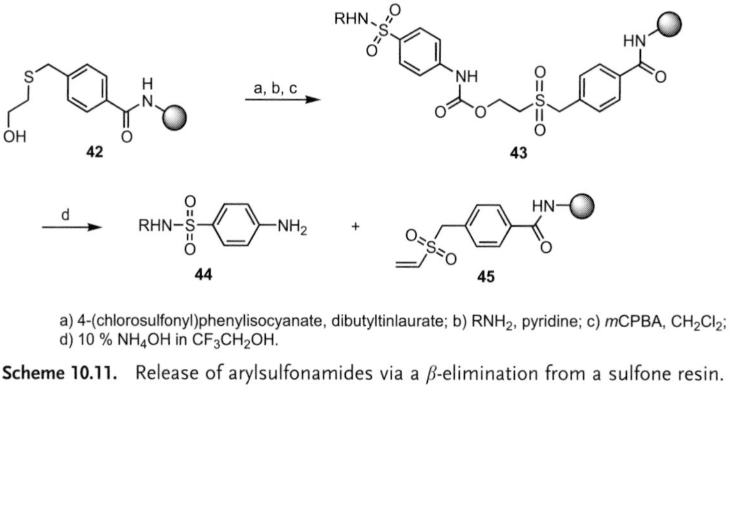

a) 4-(chlorosulfonyl)phenylisocyanate, dibutyltinlaurate; b) RNH$_2$, pyridine; c) *m*CPBA, CH$_2$Cl$_2$;
d) 10 % NH$_4$OH in CF$_3$CH$_2$OH.

Scheme 10.11. Release of arylsulfonamides via a β-elimination from a sulfone resin.

4-aminobenzenesulfonamides, in good yields and purities, upon treatment with aqueous NH$_4$OH.

A small library of seven dehydroalanine derivatives has been prepared by a β-elimination of a sulfinate resin (Scheme 10.12). Anchoring of cysteine onto Merrifield resin through the side-chain thiol group gave a resin-bound sulfide (46). Modification of both C- and N-termini and oxidation of the sulfide with *m*-CPBA followed by a β-elimination of the sulfinate resin 48 furnished the dehydroalanine derivatives 47 in 31–86% yields with high purities after aqueous work-up [28].

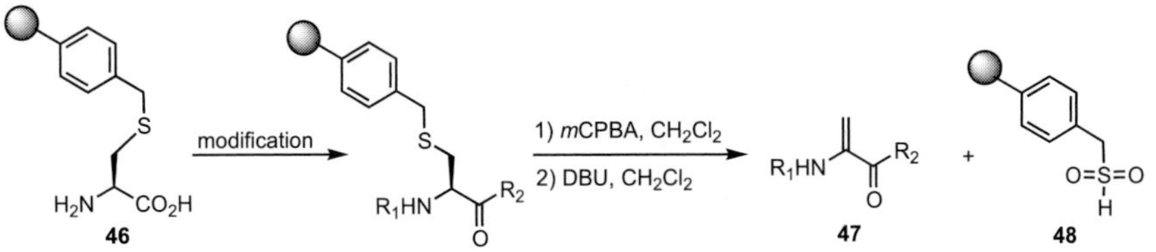

R$_1$: Boc, Cbz-Phe
R$_2$: OMe, OBn, NHBn

Scheme 10.12. Dehydroalanine derivatives via a sulfone elimination.

Elimination of a sulfinate resin has also been utilized in heterocyclization chemistry for the synthesis of a few 2-substituted-4-piperidone derivatives (50) from resin-bound sulfone 49, which serves as a divinyl ketone synthon (Scheme 10.13). The amine reagent can act both as a nucleophile and as a base, thus promoting a Michael addition to resin 49 and inducing elimination of the sulfinate resin followed by a second Michael addition to the newly formed enone [29].

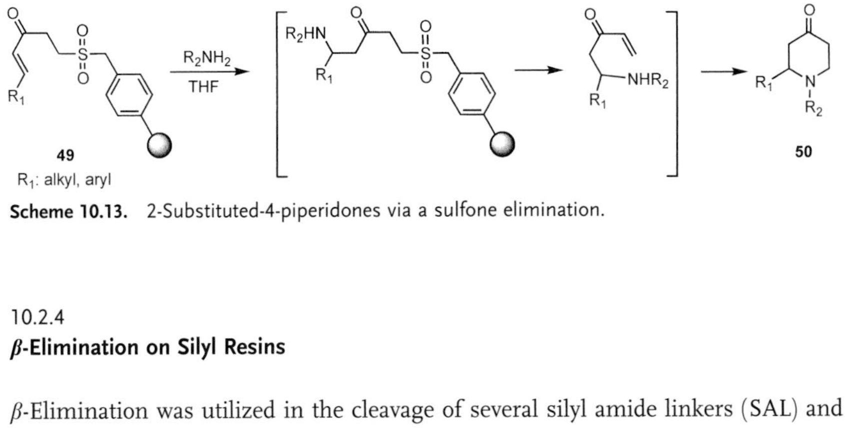

R$_1$: alkyl, aryl

Scheme 10.13. 2-Substituted-4-piperidones via a sulfone elimination.

10.2.4
β-Elimination on Silyl Resins

β-Elimination was utilized in the cleavage of several silyl amide linkers (SAL) and trimethylsilylethyl ester linkers for the facile release of peptide fragments from the solid support (Scheme 10.14). Stabilization of a carbocation by a β-trialkylsilyl group, as shown in intermediate 52, seems to facilitate the release of C-terminal

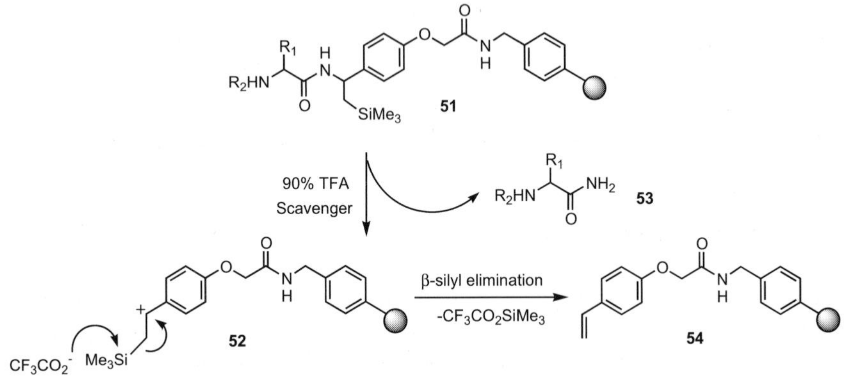

Scheme 10.14. Release of peptides via a β-silyl elimination on silyl amide linkers.

amides (**53**) from silyl resin **51** [30]. β-Elimination of the trialkylsilyl group neutralizes the transient carbocation to give a stable styrene derivative (**54**). These silyl amide linkers gave improved yields of C-terminal tryptophan amides over conventional linkers since an irreversible alkylation of the tryptophan indole nucleus by such carbocations is suppressed. However, acid scavengers (1,2-ethanedithiol/phenol/thioanisole, 5:3:2) were needed as the styrene moiety is sensitive to protonation. Therefore, purification of the final product is required in order to remove the scavenger byproducts.

Similarly, linker **55** was designed to be cleaved by a β-elimination mechanism based on the 2-(trimethylsilyl)ethylester protecting group (Scheme 10.15) [31]. Fluoridolysis or dilute acid cleavage enabled the preparation of protected peptide fragments such as **56**. C-Terminal tryptophans or prolines could be successfully anchored with this linker and no undesired alkylation or diketopiperazine formation was observed upon cleavage.

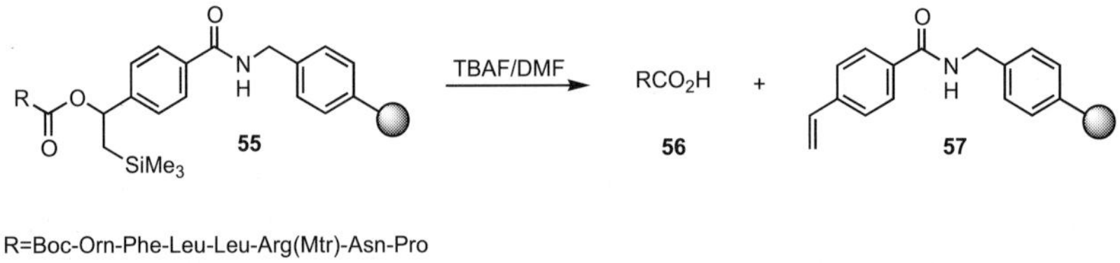

R=Boc-Orn-Phe-Leu-Leu-Arg(Mtr)-Asn-Pro

Scheme 10.15. Release of peptides via β-silyl elimination on 2-(trimethylsilyl)ethyl ester linkers.

A β-elimination mechanism is also involved in the release of olefins **60** from resin-bound allyl silanes such as **59**, the product of a solid-phase cross-olefin metathesis between allylsilane **58** and an olefin (Scheme 10.16) [32].

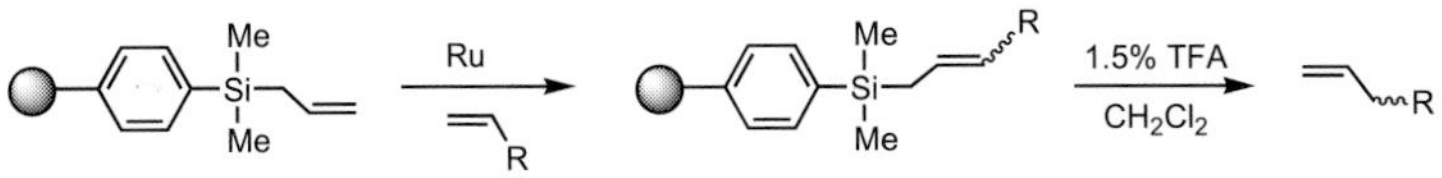

Scheme 10.16. Release of olefins from an allyl silane resin via a β-silyl elimination.

10.2.5
β-Elimination on Fluorenyl Resins

β-Elimination was also implemented in the cleavage of several fluorene-based linkers for the facile release of peptide fragments from solid support (Scheme 10.17) [33]. Quantitative cleavage of the Merrifield peptide **62** and peptide **63**, which corresponds to the sequence 31–38 of uteroglobin, was achieved from the fluorene resin **61** in good yields and high purities with 20% morpholine in DMF or 10% piperidine in DMF. Resin **61** proved to be superior to other fluorene-derived resins where incomplete removal of the protected peptide from the resin has been described [33b]. Also, slight lability to *N,N*-diisopropylethylamine, which was used at the neutralization step after Boc deprotection in peptide synthesis, and basic amino groups of the growing peptide has been detected occasionally in other fluorene-derived resins [33c]. The fluorene nucleus in resin **61** has been conveniently substituted with an electron-donating *N*-amide group to fine-tune its base lability in order to prevent any premature cleavage of the growing peptide chain.

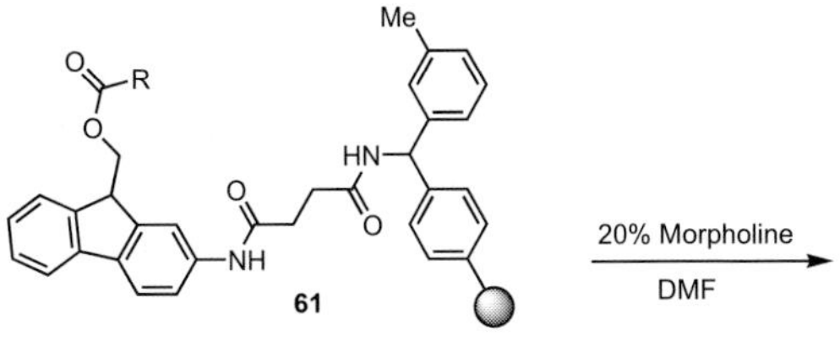

20% Morpholine
DMF

RCO₂H

62 R = Boc-Leu-Ala-Gly-Val
63 R = Boc-Asp(OcHx)-Asp(OcHx)-Thr(Bzl)-Met-Lys(ClZ)-Asp(OcHx)-Ala-Gly

Scheme 10.17. Release of C-terminal peptides via a β-elimination from a fluorenyl resin.

10.2.6
β-Elimination on 2-(2-Nitrophenyl)ethyl Resins

β-Elimination was also utilized with 2-(2-nitrophenyl)ethyl (NPE) linkers. Release of 3′-hydroxy-and 3′-phosphateoligonucleotides **66** and **67** from CPG (controlled pore glass) support was achieved through carbonate and phosphate linkers **64** and **65** respectively (Scheme 10.18) [34]. The conditions used were either 0.5 M 1,8-diazabicyclo[5.4.0]undecene-7 (DBU) in dioxane, pyridine for 1 h, and ammonia for 5 h at 55 °C or 20% piperidine in DMF for 3 h. The linkage was found to be resistant to 40% Et₃N in pyridine for 16 h, conditions commonly used to remove the 2-cyanoethylphosphate protecting group. An important application of 2-(2-nitrophenyl)ethyl linkages is that these supports can be used together with *p*-

nitrophenylethyl-protected nucleoside 2-cyanoethylphosphoramidites for the preparation of oligonucleotides without using ammonia during the final deblocking, because all protecting groups will be cleaved by DBU. This strategy will be of interest for the preparation of oligonucleotides containing ammonia-sensitive compounds such as base analogs, fluorescent compounds, and so on.

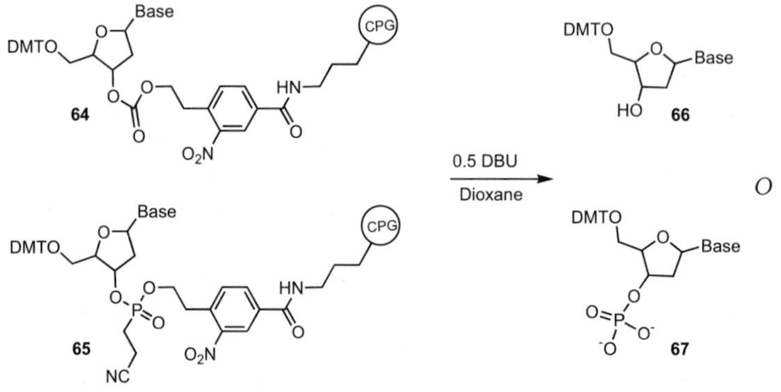

Scheme 10.18. Release of oligonucleotides via a β-elimination.

10.2.7
Radical-based β-Eliminations

10.2.7.1 β-C,O Bond Scission
Peukert and Giese devised the original photolabile linker (**68**) based on the radical-induced β-C,O bond scission of a 2-pivaloylglycerol group for the release of immobilized acids (Scheme 10.19) [35]. Upon irradiation, an α-hydroxyalkyl radical intermediate (**69**) is generated via a Norrish type I reaction with release of carbon monoxide and a *t*-butylradical that leads to isobutene. Elimination then takes place where the glycerol radical is converted into an enolate radical (**70**) and a carboxylic acid (**71**) is released. The reaction is not solvent dependent but selection of the irradiation wavelength is crucial. The pivaloyl linker **68** was found to cleave aromatic

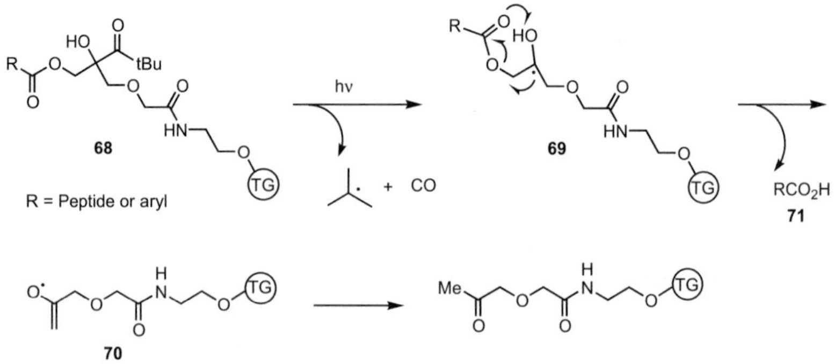

Scheme 10.19. Release of carboxylic acids–peptides by a radical β-C,O scission.

carboxylic acids and peptides with high yields (65–93%) and purities in various solvents such as THF, CH_2Cl_2, dioxane, and DMSO by irradiation with light above 320 nm. The photo byproducts are either volatile (CO and isobutene/isobutane) or inert resin-bound acetone. The linker proved to be stable upon treatment with acids and bases and to be compatible with many reagents and reaction conditions, such as palladium-catalyzed cross-coupling and epoxidation, with broad applicability in combinatorial chemistry.

10.2.7.2 β-C,Se Bond Scission-Release of Olefins

Release of olefins from solid support can also be achieved under mild reductive conditions. Olefins can be loaded onto polymer support by treatment with the polymer-bound selenium bromide resin and released reductively under the influence of Bu_3SnH-AIBN (cat). A radical β-elimination is responsible for producing the release. For example, once olefin **72** is loaded onto solid support as the resin-bound bromoselenide (**73**), reductive debromination of **73** followed by a homolytic cleavage of the C–Se bond and subsequent elimination of resin-bound selenol (**74**) will regenerate the starting olefin (**72**) (Scheme 10.20) [21]. However, purification would be required to isolate the released olefin from tin byproducts. Alternatively, Curran and coworkers' fluorous hydrides in fluorinated solvents could be utilized as well in order to isolate the released olefin in pure form from tin byproducts through a liquid–liquid extraction [36]. In addition, Barton et al.'s buffered N-ethylpiperidine hypophosphite reagent in conjuction with Et_3B as the radical initiator instead of AIBN could provide a viable alternative for this reductive release [37]. Although this radical-based release strategy has not been implemented for library synthesis, it is well suited to the solid-phase synthesis of small molecules.

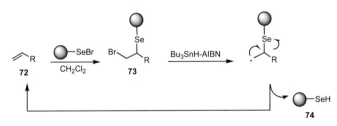

Scheme 10.20. Release of olefins by a radical β-C,Se scission.

10.3
Conjugate Eliminations

10.3.1
1,6-Conjugate Eliminations

Conjugate eliminations were also envisioned as a release strategy of the desired products from solid support, thus leading to the development of new linkers.

Waldmann and coworkers developed an enzyme-labile 4-acyloxybenzyloxy linker system (**75**), which was hydrolyzed with lipase RB 001–05 on TentaGel resin (Scheme 10.21) [38]. The linker is attached through a carboxyl group as an amide to the solid support. The resulting resin contains an acyl group, such as acetate, which can be cleaved by lipases or esterases. A phenolate is thus generated which fragments to give a quinone methide (**76**) and releases compound **77**, the desired product. The quinone methide remains bound to the solid support and is quenched by water or an additional nucleophile.

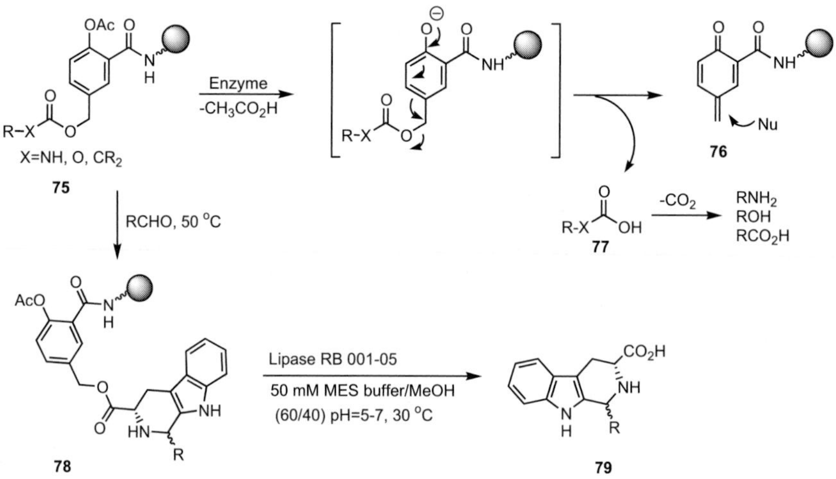

Scheme 10.21. Enzymatic release based on a conjugate 1,6-elimination.

In this way, amines (bound as urethanes), alcohols (bound as carbonates), and carboxylic acids (bound as esters) can be detached from the polymeric carrier under very mild conditions (pH 5–7, room temperature (rt)) and with complete selectivity. The substrate specificity of the enzyme guarantees that only the intended ester is cleaved, and the mild conditions of the biocatalyzed transformations ensure that the compounds constructed on the solid support remain intact during cleavage. The applicability of the enzyme-labile group to multistep synthesis on solid support was proven by the synthesis of tetrahydro-β-carboline carboxylic acids (**79**) by means of the Pictet–Spengler reaction and their subsequent enzyme-mediated release from resin **78**.

A similar 1,6-conjugate elimination mechanism is also involved in the release of carboxylic acids, peptides, amines, and alcohols in good yields by the fluoride-induced cleavage of several labile silyl linkers, such as **80–82** (Scheme 10.22) [39]. Apart from the desired carboxylic acid, linker **82** also releases a quinone methide, which is scavenged by thiophenol. Consequently, the final products have to be purified to remove the thiophenol byproduct.

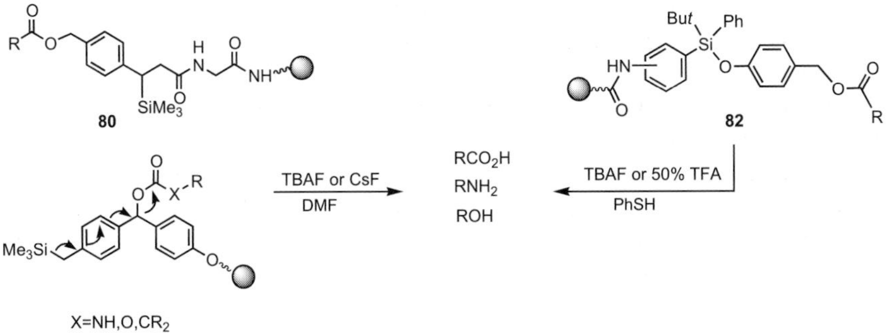

Scheme 10.22. Fluoride-induced release based on a conjugate elimination.

10.3.2
1,4-Conjugate Eliminations

A conjugate 1,4-elimination of a phenylsulfinate anion yielding functionalized 3-arylbenzofurans in a traceless manner was observed in a novel cyclofragmentation pathway involving epoxides (**85**) (Scheme 10.23) [40]. Sulfone anion **85a** undergoes a 5-*exo*-trig cyclization to alkoxide **85b**, which next collapses to benzofuran **86** by a concomitant expulsion of both formaldehyde and a phenylsulfinate anion. This cyclofragmentation-release pathway has been developed in both solution and solid phase. It is well suited to solid-phase synthesis and can lead to the generation of a diverse family of 3-arylbenzofurans starting from commercially available or proprietary 2-hydroxybenzophenones.

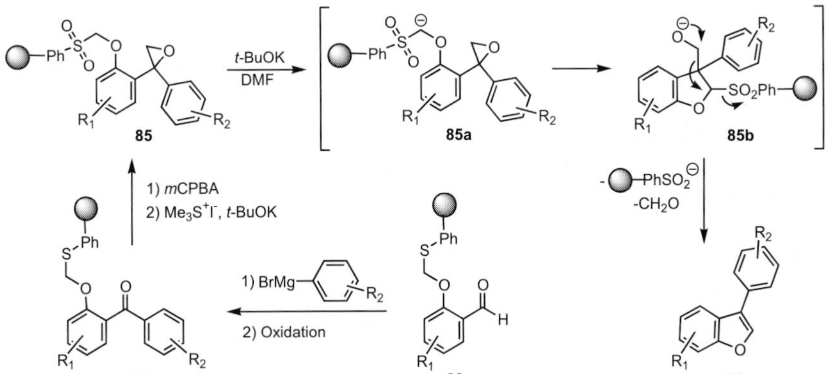

Scheme 10.23. 3-Arylbenzofurans by a 1,4-conjugate cyclofragmentation-release assay.

Greater structural diversity can be introduced by starting from a series of 2-hydroxybenzophenones (**84**) or made directly from the addition of arylmagnesium bromides to resin-bound salicaldehydes (**83**), followed by an oxidation of the intermediate carbinol. All cleavage products are remarkably clean since the mechanism only allows for the release of products generated through intermediate **85**. The limitation of this novel fragmentation pathway lies in the necessity for epoxides derived from 2-hydroxybenzophenones, since both aryl groups are required for the regioselective epoxide opening.

10.4
Addition–Elimination Reactions

Addition–elimination reactions in which a C–C or C–X bond (X = heteroatom) is formed by the combination of two or more reactive centers, in an intra- or inter-molecular fashion, followed by an elimination or extrusion of a small molecule can be useful in combinatorial chemistry for the convergent synthesis of diverse libraries. The structural diversity, which originates directly from the starting materials or building blocks, is retained through the library synthesis. If a polymer support is directly linked to the eliminated molecule, then a traceless release of the desired compounds from solid support can be achieved. Although there are several mechanistically different addition–elimination transformations, here we will identify those utilized in the synthesis of libraries where the addition–elimination step is the final process that completes the synthesis of the desired scaffold.

10.4.1
Addition–Elimination on Vinylogous Systems

10.4.1.1 Entry to Aminomethyleneoxazolones
Aminomethyleneoxazolones are recognized as a class of serine protease inhibitors, and therefore have been the focus of library synthesis [41]. The synthesis involves the addition of a primary or secondary amine to an ethoxymethyleneoxazolone (EMO) (**87**) followed by elimination of a molecule of ethanol (Scheme 10.24). The convergent nature of this scheme greatly facilitates its execution starting from either commercially available or proprietary amines and ethoxymethyleneox-azolones. Baldino et al. reported the automated parallel synthesis of 1600 amino-

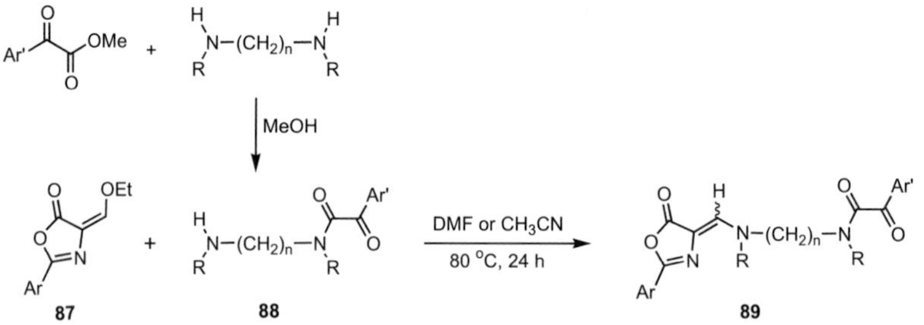

Scheme 10.24. Solution-phase synthesis of aminomethyleneoxazolones.

methyleneoxazolones (**89**) starting from 20 EMOs and 80 secondary amino α-ketoamide intermediates (**88**), prepared *in situ* from ten α-ketoesters and eight diamines under equimolar reagent combinations [42].

The chemistry works well in a wide range of solvents such as CH_3CN, DMF, THF, and dioxane. Ultimately, the choice of the appropriate solvent depends mainly on the solubility of the starting oxazolones. The corresponding products were isolated as E,Z isomers, in good yields and high purities, after evaporation of the solvent. Although there are no examples illustrating the solid-phase implementation of this chemistry, it could be amenable to solid phase as well (i.e. via a resin-bound EMO). However, mild cleavage conditions would be required, since strong acidic or basic conditions might disrupt the oxazolone ring system.

10.4.1.2 Entry to Benzopyrones

Substituted benzopyrones encompass an important class of molecules that possess a wide range of interesting biological activities [43, 44]. Although there are numerous literature methods for the synthesis of the benzopyrone ring system, they are not ideally suited for combinatorial approaches owing to harsh reaction conditions, poor yields, and limited substituent tolerance. A method that is amenable to solution-phase parallel synthesis was developed by Brueggemeier and coworkers for the preparation of a seven-member benzopyrone library (Scheme 10.25) [44]. This sequence utilizes chlorination of a bis-silylated salicylic acid (**90**) with oxalyl chloride to generate the corresponding acid chloride (**91**) followed by a Sonogashira coupling with a terminal alkyne to give cleanly alkynone **92**. Treatment of alkynone **92** with a secondary amine such as diethylamine in refluxing ethanol provides an intermediate enaminone (**94a**) which undergoes a disilylation and an intramolecular Michael addition followed by elimination of volatile diethylamine to give the benzopyrone **95**. Removal of silyl impurities along with the palladium and copper catalysts and any triphenylphosphine requires purification.

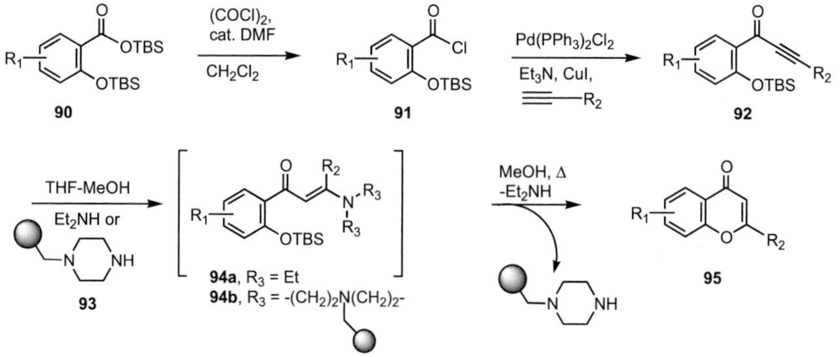

Scheme 10.25. Entry to benzopyrones via an addition–elimination sequence.

The problems identified above were circumvented by a resin capture strategy. Indeed, capturing alkynone **92** with a fivefold excess of piperazinyl Merrifield resin

(**93**) (loading 0.7 mmol g^{-1}) eliminates the need for any further purification [45]. Thus, the resin-bound enaminone **94b** could be easily separated from the excess reagents and byproducts of the reaction mixture by a simple filtration. On-resin cyclization of the enaminone **94b** released the benzopyrone **95** in good yields and good purities and regenerated the piperazinyl resin. This resin capture would facilitate a one-pot conversion of silylated salicylic acids to benzopyrones without requiring any intermediate purification steps – an attractive feature for the synthesis of large libraries. Furthermore, this method can be applied to the synthesis of benzopyrones with no residual functionalities required for linkage to the solid support. The solid-supported capture reagent can be regenerated and recycled for additional rounds of resin capture.

10.4.1.3 2,3-Dihydro-4-pyridone Libraries

The reaction of Danishefsky's diene with imines leading to 2,3-dihydro-4-pyridones has been studied extensively by the chemistry community and has numerous applications in the synthesis of piperidine alkaloids [46]. The reaction is postulated to proceed via a Mannich reaction followed by an intramolecular Michael reaction with a subsequent elimination of a molecule of methanol [46a–c]. The large number of readily available aromatic aldehydes and amines prompted chemists to adapt this reaction to both solid- and solution-phase automated synthesis of large libraries of 2,3-dihydro-4-pyridones. Application of this chemistry to solution-phase parallel synthesis was demonstrated recently by Yu et al. in the construction of a 4320-member library (Scheme 10.26) [47a]. Condensation of benzaldehydes with anilines in the presence of a Yb resin [48] gave the intermediate imine, which upon immediate treatment with Danishefsky's diene gave the corresponding dihydropyridones (**96**) in good yields and purities after filtration of the resin-bound catalyst.

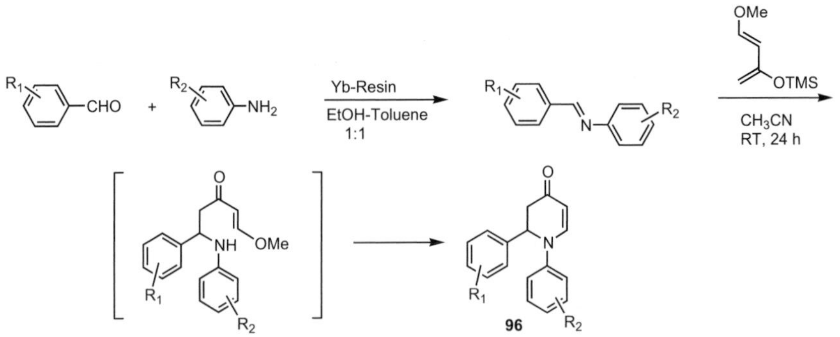

Scheme 10.26. Solution-phase synthesis of a 2,3-dihydro-4-pyridone library.

Adaptation of this chemistry to solid phase was demonstrated by Wang and Wilson starting from Wang resin-bound benzaldehyde and anilines or aliphatic amines [47b]. Several different Lewis acids were found to be compatible with the reaction conditions, including $ZnCl_2$, $AlCl_3$, Et_2AlCl, $TiCl_4$, $BF_3 \cdot Et_2O$, and $Yb(OTf)_3$. However, the water-tolerant $Yb(OTf)_3$ gave the highest yield. The final products were obtained after cleavage with trifluoroacetic acid $(TFA)/CH_2Cl_2$ (1:1) in relatively high yields and purities.

10.4.2
Cycloreversions

10.4.2.1 Pyrrole Libraries

A cycloreversion reaction with loss of CO_2 is involved in the pyrrole synthesis via a 1,3-dipolar cycloaddition of alkynes to münchnones. Mjalli and coworkers and Armstrong and coworkers reported independently the solid-phase synthesis of pyrroles where the münchnone precursors – *N*-acyl-*N*-alkyl-*α*-amino amides – were generated in a single step via an Ugi four-component condensation (U-4CC) instead of relying upon individual acylated amino acids (Scheme 10.27) [49]. Treatment of the *N*-acyl-*N*-alkyl-*α*-amino acid **97a** with Ac_2O or treatment of Armstrong's cyclohexenamide precursor **97b** with HCl in the presence of an acetylenic dipolarophile resulted in the bicyclo intermediates **98a** and **98b**, which rapidly aromatized to pyrroles **99a** and **99b**, respectively, with loss of CO_2.

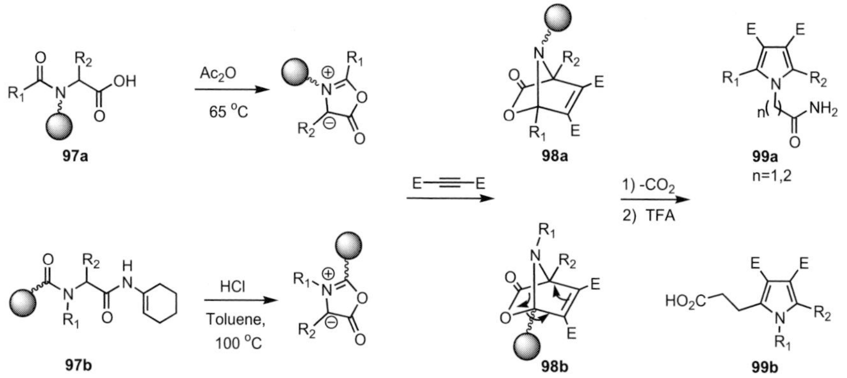

Scheme 10.27. Entry to pyrroles via a dipolar cycloaddition–cycloreversion pathway.

Pyrroles **99a** were isolated in 40% overall yield with high purities, according to Mjalli and coworkers' protocol, after release from solid support. Armstrong and coworkers' four-step protocol, although not optimized, gave pyrroles **99b** with 4–17% yields. Although this chemistry is amenable to solution-phase synthesis, low yields of the tetra-substituted pyrroles are usually observed [50]. This may be attributed to the substitution pattern of the acetylenic dipolarophile and to the tendency of münchnones to self-condense. Furthermore, it would require the use of the corresponding *N*-acyl-*N*-alkyl-*α*-amino acids as building blocks in order to increase the purity of the final products.

10.4.2.2 Imidazole Libraries

Application of the same cycloaddition–cycloreversion strategy described above resulted in the solid-phase synthesis of a 12-member library of 2,4,5-triarylimidazoles by employing an aryltosylimine as the dipolarophile (Scheme 10.28) [51]. Treatment of the resin-bound acid **100** with N'-(3-dimethylamino-propyl)-N-ethylene carbodiimide (EDC) in CH_2Cl_2 at ambient temperature for 24–48 h followed by a cycloaddition of the intermediate münchnone **101** with a tosyli-mine gave bicyclo compound **102**. Cycloreversion of **102** with elimination of tolue-nesulfinic acid and CO_2 provided the polymer-linked imidazole **103**, which was next washed with TFA, without any observed cleavage, to remove any unreacted starting materials. Release from resin upon treatment with AcOH at 100 °C gave the free imidazole **104**. All reagent combinations provided the desired products in good yields and high purities with a 73% average yield over the six-step sequence. However, low yields of imidazoles were observed when this chemistry was carried out in solution phase [52b]. This is at least partly the result of the potential for münchnones to self-condense, which can be suppressed in a solid-phase approach [52].

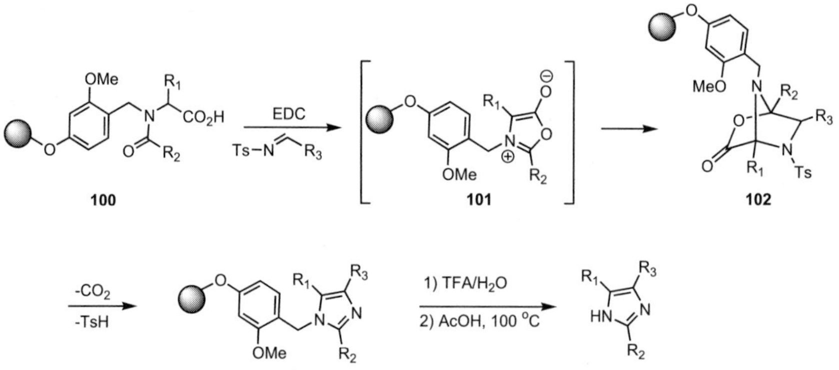

Scheme 10.28. Entry to imidazoles via a dipolar cycloaddition–cycloreversion pathway.

10.4.2.3 Traceless Solid-phase Synthesis of Furans

Isomünchnones readily undergo a [3 + 2] cycloaddition with acetylenes to give bi-cyclo intermediates, which can lead to furans after a cycloreversion and elimina-tion of isocyanate [53]. This strategy was applied to the solid-phase synthesis of furan libraries independently by Gowravaram and Gallop and by Austin and co-workers (Scheme 10.29) [54]. α-Diazocarbonyl **107** can react with Rh(II) catalysts to form a highly reactive rhodium carbenoid that collapses to a mesoionic dipole intermediate – the isomünchnone **108** – which in the presence of an acetylenic dipolarophile can lead to a bicyclo intermediate (**109**). Cycloreversion of **109** and extrusion of the resin-bound isocyanate liberates the tetrasubstituted furan **110** into solution, in good yields and high purities, and leaves no obvious remnant of polymer tethering in the desired product.

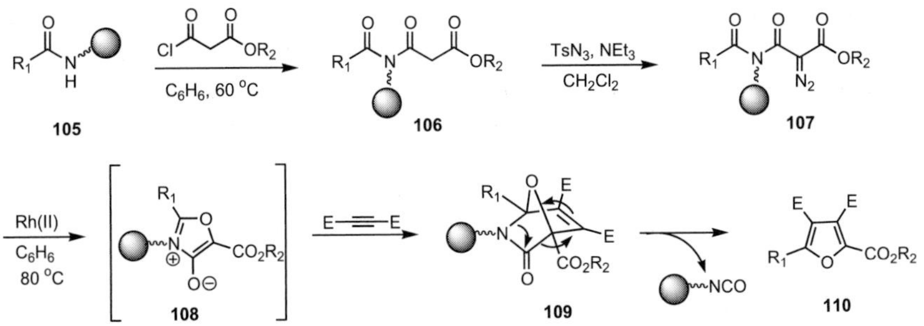

Scheme 10.29. Tetrasubstituted furans via a cycloaddition–cycloreversion pathway.

The cleavage rate from solid support, at similar temperatures, is highest in polar protic solvents, which could allow for the cleavage to be carried out directly in aqueous media. The chemistry could be amenable to solution-phase parallel synthesis since furans were recovered in good yields but would require a special work-up or purification to remove the rhodium catalyst. The released isocyanate could be effectively scavenged with a resin-bound amine.

10.4.2.4 1,2-Diazines

Access to a small library of functionalized 1,2-diazines can be provided by the inverse electron demand Diels–Alder reactions of 3,6-substituted-1,2,4,5-tetrazines on solid phase (Scheme 10.30) [55a]. Treatment of an immobilized azadiene (**111**) with a variety of electron-rich olefins in dioxane at room or elevated temperatures gave bicyclo intermediate **112**, which underwent a cycloreversion with loss of N_2 and concomitant loss of HX to give resin-bound diazine **113**. Removal of the Boc group first, followed by cleavage from solid support under basic conditions, gave the corresponding 3-amino-6-thiomethyl-1,2-pyridazines (**114**) in good to moderate yields. The chemistry can be extended with the azadienes bearing a sulfone group in the 6-position. The sulfone substrate is more reactive than the corresponding thiomethyl substrate in a Diels–Alder reaction as more efficient conversion is generally achieved with the less reactive alkynes. A wide range of electron-rich dienophiles can be used which permits the introduction of two diversity elements on

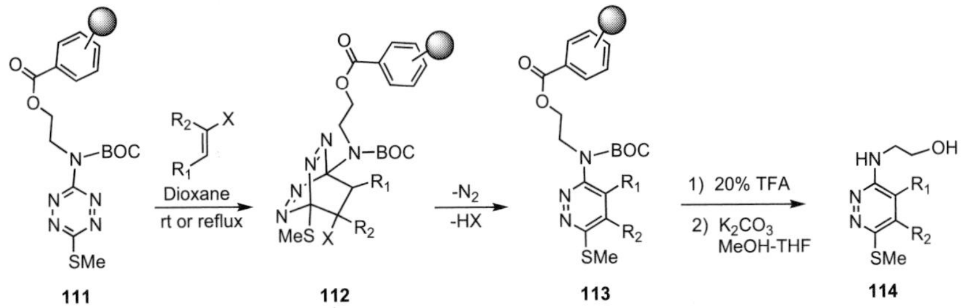

Scheme 10.30. Entry to diazines via a [4 + 2] cycloaddition–cycloreversion pathway.

heteroaromatic scaffolds. Subsequent nucleophilic aromatic substitution of the C-6 methylsulfide/sulfone and acylation/alkylation of the C-3 amine will introduce the third and fourth diversity elements. This chemistry could be amenable to solution phase although library purification might be required for the less reactive enamines or enol ethers derived from acetophenones [55b].

10.5
Summary

β-Elimination has been successfully used in solid-phase synthesis as a release strategy of the desired products from solid support. However, it has not been broadly applied in the synthesis of diverse libraries as the final step that determines the synthesis of the desired scaffold. This is presumably because of the lack of convergence in the synthesis and diversity of the final products. On the contrary, addition–elimination reactions, apart from also being used as a release method of the final products from solid support, can be utilized in the convergent synthesis of diverse libraries. However, β-elimination could be very well utilized in the synthesis of functionalized intermediates where the generated functionality (i.e. double bond) could serve as a handle for the introduction of additional diversity elements. Although elimination reactions can be very useful to the high-throughput synthesis of small molecules by either solid- or solution-phase parallel synthesis, their potential has not been fully realized yet.

References

1 J. MARCH in: Advanced Organic Chemistry, 4th edn, *Wiley-Interscience*, New York 1992, pp. 982–1050.

2 a) G. I. TESSER, J. T. W. A. R. M. BUIS, E. T. M. WOLTERS, E. G. A. M. BOTHÉ-HELMES, *Tetrahedron* 1976, *32*, 1069–1072; b) G. I. TESSER, B. W. J. ELLENBROEK in: Eighth European Peptide Symposium. BEYERMAN, H. C., LINDE, A. v. d., BRINK, W. M. v. d. (eds), Amsterdam 1967.

3 For a review on linkers, see: a) F. GUILLIER, D. ORAIN, M. BRADLEY, *Chem. Rev.* 2000, *100*, 2091–2157; b) I. W. JAMES, *Tetrahedron* 1999, *55*, 4855–4946.

4 A. W. HOFMANN, *Annalen* 1851, *78*, 253–286.

5 C. BLETTNER, M. BRADLEY, *Tetrahedron Lett.* 1994, *35*, 467–470.

6 J. R. MORPHY, Z. RANKOVIC, D. C. REES, *Tetrahedron Lett.* 1996, *37*, 3209–3212.

7 A. R. BROWN, D. C. REES, Z. RANKOVIC, J. R. MORPHY, *J. Am. Chem. Soc.* 1997, *119*, 3288–3295.

8 J. COTTNEY, Z. RANKOVIC, J. R. MORPHY, *Bioorg. Med. Chem. Lett.* 1999, *9*, 1323–1328.

9 For synthesis of tertiary amines via vinyl sulfone REM resins, see: a) F. E. K. KROLL, R. MORPHY, D. REES, D. GANI, *Tetrahedron Lett.* 1997, *38*, 8573–8576; b) P. HEINONEN, H. LÖNNBERG, *Tetrahedron Lett.* 1997, *38*, 8569–8572.

10 M. J. PLATER, A. M. MURDOCH, J. R. MORPHY, Z. RANKOVIC, D. C. REES, *J. Comb. Chem.* 2000, *2*, 508–512.

11 W. S. WADE, F. YANG, T. J. SOWIN, *J. Comb. Chem.* 2000, *2*, 266–275.

12 a) P. Blaney, R. Grigg, Z. Rankovic, M. Thoroughgood, *Tetrahedron Lett.* **2000**, *41*, 6635–6638; b) M. Gustafsson, R. Olsson, C.-M. Andersson, *Tetrahedron Lett.* **2001**, *42*, 133–136.

13 X. Ouyang, R. W. Armstrong, M. M. Murphy, *J. Org. Chem.* **1998**, *63*, 1027–1032.

14 Y. Yamamoto, K. Tanabe, T. Okonogi, *Chem. Lett.* **1999**, 103–104.

15 P. H. Toy, T. S. Reger, K. D. Janda, *Org. Lett.* **2000**, *2*, 2205–2207.

16 A. R. Brown, *J. Comb. Chem.* **1999**, *1*, 283–285.

17 Y. Nishibayashi, S. Uemura in: Topics in Current Chemistry. Wirth, T. (ed.), Springer Verlag, Berlin 2000, vol. 208, pp. 201–214.

18 R. Michels, M. Kato, W. Heitz, *Makromol. Chem.* **1976**, *177*, 2311–2320.

19 For cyclizations mediated by resin-bound selenium reagents, see: a) K. Fujita, K. Watanabe, A. Oishi, Y. Ikeda, Y. Taguchi, *Synlett* **1999**, 1760–1762; b) K. C. Nicolaou, J. A. Pfefferkorn, G.-Q. Cao, S. Kim, J. Kessabi, *Org. Lett.* **1999**, *1*, 807–810.

20 a) K. C. Nicolaou, J. A. Pfefferkorn, A. J. Roecker, G.-Q. Cao, S. Barluenga, H. J. Mitchell, *J. Am. Chem. Soc.* **2000**, *122*, 9939–9953; b) K. C. Nicolaou, J. A. Pfefferkorn, H. J. Mitchell, A. J. Roecker, S. Barluenga, G.-Q. Cao, R. L. Affleck, J. E. Lillig, *J. Am. Chem. Soc.* **2000**, *122*, 9954–9967; c) K. C. Nicolaou, J. A. Pfefferkorn, S. Barluenga, H. J. Mitchell, A. J. Roecker, G.-Q. Cao, *J. Am. Chem. Soc.* **2000**, *122*, 9968–9976.

21 K. C. Nicolaou, J. Pastor, S. Barluenga, N. Winssinger, *Chem. Commun.* **1998**, 1947–1948.

22 For a solution-phase precedent of selenium-mediated 6-*exo*-trig cyclizations of *ortho*-prenylated phenols and related systems, see: a) D. L. J. Clive, G. Chittattu, N. J. Curtis, W. A. Kiel, C. K. Wong, *J. Chem. Soc., Chem. Commun.* **1977**, 725–727; b) P. B. Anzeveno, *J. Org. Chem.* **1979**, *44*, 2578–2580; c) K. C.

Nicolaou, Z. Lysenko, *J. Am. Chem. Soc.* **1977**, *99*, 3185–3187.

23 For asymmetric epoxidation of benzopyran systems with a polymer-bound chiral salen manganese (III) complex, see: C. E. Song, E. J. Roh, B. M. Yu, D. Y. Hi, S. C. Kim, K.-J. Lee, *Chem. Commun.* **2000**, 615–616.

24 K. C. Nicolaou, H. J. Mitchell, K. C. Fylaktakidou, H. Suzuki, R. M. Rodríguez, *Angew. Chem. Int. Ed.* **2000**, *39*, 1089–1093.

25 R. Schwyzer, E. Felder, P. Faili, *Helv. Chim. Acta* **1984**, *67*, 1316–1326.

26 S. B. Katti, P. K. Misra, W. Haq, K. B. Mathur, *J. Chem. Soc., Chem. Commun.* **1992**, 843–844.

27 C. G. Echeverría, *Tetrahedron Lett.* **1997**, *38*, 8933–8934.

28 M. Yamada, T. Miyajima, H. Horikawa, *Tetrahedron Lett.* **1998**, *39*, 289–292.

29 A. Barco, S. Benetti, C. De Risi, P. Marchetti, G. P. Pollini, V. Zanirato, *Tetrahedron Lett.* **1998**, *39*, 7591–7594.

30 H.-G. Chao, M. S. Bernatowicz, G. R. Matsueda, *J. Org. Chem.* **1993**, *58*, 2640–2644.

31 H.-G. Chao, M. S. Bernatowicz, P. D. Reiss, C. E. Klimas, G. R. Matsueda, *J. Am. Chem. Soc.* **1994**, *116*, 1746–1752.

32 M. Schuster, N. Lucas, S. Blechert, *Chem. Commun.* **1997**, 823–824.

33 a) F. Rabanal, E. Giralt, F. Albericio, *Tetrahedron Lett.* **1992**, *33*, 1775–1778; b) M. Mutter, D. Bellof, *Helv. Chim. Acta* **1984**, *67*, 2009–2016; c) Y. Z. Liu, S. H. Ding, J. Y. Chu, A. M. Felix, *Int. J. Pept. Prot. Res.* **1990**, *35*, 95–98.

34 a) R. Eritja, J. Robles, D. Fernandez-Forner, F. Albericio, E. Giralt, E. Pedroso, *Tetrahedron Lett.* **1991**, *32*, 1511–1514; b) F. Albericio, E. Giralt, R. Eritja, *Tetrahedron Lett.* **1991**, *32*, 1515–1518.

35 S. Peukert, B. Giese, *J. Org. Chem.* **1998**, *63*, 9045–9051.

36 D. P. Curran, S. Hadida, S.-Y. Kim, Z. Luo, *J. Am. Chem. Soc.* **1999**, *121*, 6607–6615.

37 D. H. R. Barton, D. O. Jang, J. Cs. Jaszberenyi, *J. Org. Chem.* **1993**, *58*, 6838–6842.

38 B. Sauerbrei, V. Jungmann, H. Waldmann, *Angew. Chem. Int. Ed.* **1998**, *37*, 1143–1146.

39 a) R. Ramage, C. A. Barron, S. Bielecki, D. W. Thomas, *Tetrahedron Lett.* **1987**, *28*, 4105–4108; b) A. Routledge, H. T. Stock, S. L. Flitsch, N. J. Turner, *Tetrahedron Lett.* **1997**, *38*, 8287–8290; c) D. G. Mullen, G. Barany, *J. Org. Chem.* **1988**, *53*, 5240–5248.

40 K. C. Nicolaou, S. A. Snyder, A. Bigot, J. A. Pfefferkorn, *Angew. Chem. Int. Ed.* **2000**, *39*, 1093–1096.

41 N. L. Benoiton, F. Hudesz, F. M. F. Chen, *Int. J. Pept. Prot. Res.* **1995**, *45*, 266–271.

42 C. M. Baldino, D. S. Casebier, J. Caserta, G. Slobodkin, C. Tu, D. L. Coffen, *Synlett* **1997**, 488–490.

43 E. S. C. Wu, J. T. Loch, B. H. Toder, A. R. Borrelli, D. Gawlak, L. A. Radov, N. P. Gensmantel, *J. Med. Chem.* **1992**, *35*, 3519–3525.

44 A. S. Bhat, J. L. Whetstone, R. W. Brueggemeier, *Tetrahedron Lett.* **1999**, *40*, 2469–2472.

45 A. S. Bhat, J. L. Whetstone, R. W. Brueggemeier, *J. Comb. Chem.* **2000**, *2*, 597–599.

46 a) H. Kunz, W. Pfrengle, *Angew. Chem. Int. Ed.* **1989**, *28*, 1067–1068; b) H. Waldmann, M. Braun, M. Drager, *Angew. Chem. Int. Ed.* **1990**, *29*, 1468–1471; c) H. Waldmann, M. Braun, *J. Org. Chem.* **1992**, *57*, 4444–4451; d) H. Yamamoto, K. Hattori, *Tetrahedron* **1993**, *49*, 1749–1760;

e) S. Kobayashi, H. Ishitami, S. Nagayama, *Synthesis* **1995**, 1195–1202.

47 a) L. Yu, C. M. Baldino, M. S. Harris, E. Marler, B. Carr, J. Troth, M. Kearny, J. Mills, J. Brochu, J. Gordon, D. L. Coffen, paper presented at the 217th ACS National Meeting, Anaheim, March 1999, Organic Division, abstract no. 335; b) Y. Wang, S. R. Wilson, *Tetrahedron Lett.* **1997**, *38*, 4021–4024.

48 L. Yu, D. Chen, J. Li, P. G. Wang, *J. Org. Chem.* **1997**, *62*, 3575–3581.

49 a) A. M. M. Mjalli, S. Sarshar, T. J. Baiga, *Tetrahedron Lett.* **1996**, *37*, 2943–2946; b) A. M. Strocker, T. A. Keating, P. A. Tempest, R. W. Armstrong, *Tetrahedron Lett.* **1996**, *37*, 1149–1152.

50 T. A. Keating, R. W. Armstrong, *J. Am. Chem. Soc.* **1996**, *118*, 2574–2583.

51 M. T. Bilodeau, A. M. Cunningham, *J. Org. Chem.* **1998**, *63*, 2800–2801.

52 a) K. T. Potts in: 1,3-Dipolar Cyclo-addition Chemistry. Padwa, A. (ed.), Wiley-Interscience, New York 1984, vol. 2, pp. 1–82; b) R. Consonni, P. D. Croce, R. Ferraccioli, C. La Rosa, *J. Chem. Res., Synop.* **1991**, 188–189.

53 M. H. Osterhout, W. R. Nadler, A. Padwa, *Synthesis* **1994**, 123–141.

54 a) M. R. Gowravaram, M. A. Gallop, *Tetrahedron Lett.* **1997**, *38*, 6973–6976; b) D. L. Whitehouse, K. H. Nelson, Jr, S. N. Savinov, D. J. Austin, *Tetrahedron Lett.* **1997**, *38*, 7139–7142.

55 a) J. S. Panek, B. Zhu, *Tetrahedron Lett.* **1996**, *37*, 8151–8154; b) S. M. Sakya, K. K. Groskopf, D. L. Boger, *Tetrahedron Lett.* **1997**, *38*, 3805–3808.

11
Addition to CC Multiple Bonds
(Except for CC Bond Formation)

Adrian L. Smith

11.1
Introduction

The last decade has witnessed the birth and maturation of combinatorial chemistry as a technique for synthesizing large numbers of compounds. Much of the early emphasis was on synthesizing large (mixtures) libraries for screening and lead generation, and this was almost exclusively based upon solid-phase chemistry. However, the technologies developed during this work were also applicable to the more traditional medicinal chemistry lead optimization process, and this has led to a rethink in the ways in which the medicinal chemist can most efficiently optimize a lead series, both *in vitro* and *in vivo*. Whilst this initially resulted in many people developing solid-phase synthetic routes for the parallel synthesis of arrays of single compounds (whether it be to make 10 or 10,000 compounds), the co-development of postsynthesis sample-handling techniques (particularly in the area of purification) has significantly broadened the scope of chemistries which can be used. In particular, the initial advantages of solid-phase chemistry (primarily related to purification) are now less compelling in many cases where solution-phase chemistry will require less chemistry development time and where crude synthetic products can be purified in an automated fashion.

Today, the term "combinatorial chemistry" is loosely used to describe a very broad range of techniques including solid- and solution-phase chemistry and the synthesis of discrete (single) compound arrays and mixtures. Whilst the different techniques each have their pros and cons, their unifying factor when properly applied is the ability to address a particular problem (e.g. a medicinal chemistry problem) by making and testing a larger and more diverse set of compounds than would have been the case by more traditional methods. Therefore, although a large number of publications have appeared in the area of solid-phase chemistry, the purpose of this chapter is not to act as a comprehensive review of solid-phase chemistry *per se*, but rather to highlight key areas in which chemists may *practically* utilize these techniques in chemical transformations involving overall addition to carbon–carbon multiple bonds. It should be noted that certain transformations such as carbon–carbon bond formation, heterocycle formation, cycloaddition reactions, and transition metal-catalyzed formation of single bonds are covered in other chapters and will not be duplicated here.

11.2
Addition to C=C Double Bonds

Additions to C=C double bonds fall into two broad main categories dependent upon the electronic nature of the double bond. The main focus of this section will deal with electrophilic addition to isolated double bonds, whilst nucleophilic 1,4-addition to conjugated α, β-unsaturated systems will be briefly touched upon at the end.

The reactivity of the isolated C=C double bond arises from the nucleophilic nature of the π-bond, with the majority of reactions involving some form of electrophilic addition. This may result in formal oxidation (e.g. epoxidation) or reduction (e.g. hydrogenation). It may be expected that more electron-rich double bonds will generally exhibit greater reactivity than electron-deficient double bonds in the absence of overriding steric factors. Whilst the C=C double bond is also frequently encountered in radical reactions, this is usually implicated in C–C bond formation and is outside the scope of this chapter. This section aims to summarize some of the key transformations which can be effected by addition to the C=C double bond, emphasizing applicability to combinatorial chemistry.

11.2.1
Epoxidation and Subsequent Epoxide Opening

The epoxidation of an olefin represents one of the more versatile transformations available to the aspiring combinatorial chemist, generating a reactive epoxide intermediate which can be opened by a range of nucleophiles (Scheme 11.1). In particular, opening with amines gives rise to the (hydroxyethyl)amine isostere (**3**), which mimics the tetrahedral intermediate for amide hydrolysis [1], making this a very powerful two-step transformation of an olefin.

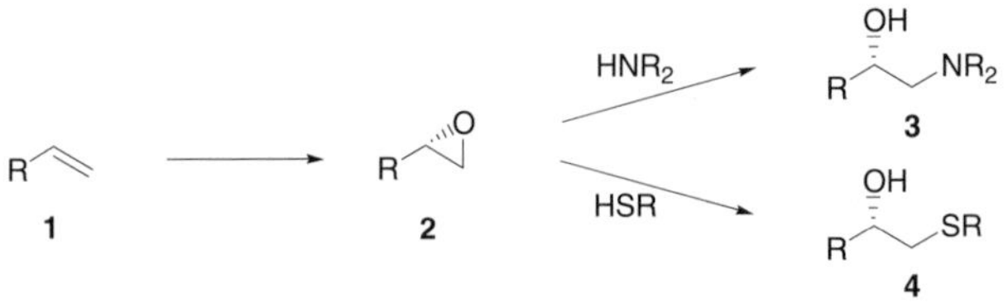

Scheme 11.1. Epoxidation and subsequent epoxide opening of an olefin.

Epoxidation is most usually carried out with *meta*-chloroperbenzoic acid (*m*-CPBA) in a solvent such as dichloromethane and, being a very efficient and mild reaction in the absence of competing functionalities, generally gives clean and full conversion with a minimal excess of reagent. As such, it is suitable for use in both solid- and solution-phase chemistry. It should be noted that there is often little diastereoselectivity with this reagent. Scheme 11.2 shows an example of an epoxidation carried out on solid phase [2]. Here, the author experienced instability of the urethane linker to the generated 3-chlorobenzoic acid byproduct and buffered the reaction with NaHCO₃. Another side-reaction sometimes observed is opening of the epoxide by the generated 3-chlorobenzoic acid, and in these cases more success may be possible with reagents such as dimethyldioxirane [3]. It is also possible to use a range of solid-supported reagents for epoxidation (see Table 11.1) [4].

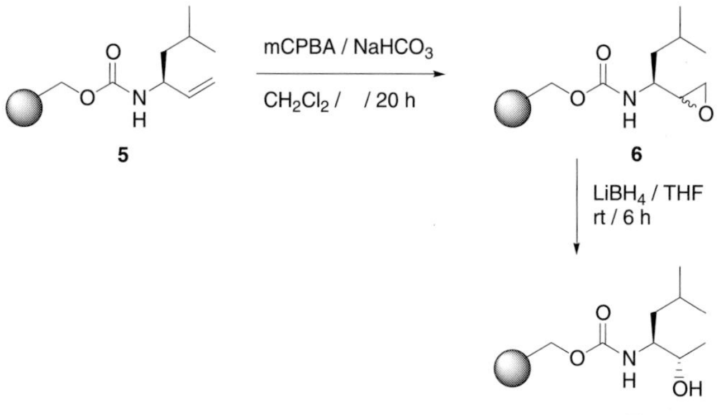

Scheme 11.2. Epoxidation on solid phase.

The two-step processes shown in Scheme 11.1 are very well suited to solution-phase parallel synthesis. Generally, the intermediate (2) is prepared in bulk, purified, and then split into individual reactors for parallel ring opening by a range of nucleophiles. For reactions with amines to give 3, we find a small excess of the amine (typically 1.5 equiv.) in isopropyl alcohol heated at 65 °C for 16 h usually gives clean conversion, even with relatively non-nucleophilic amines such as aniline. It is important that no epoxide remains in final products, since this can give misleading data in biological assays by acting as a suicide inhibitor. It is therefore generally preferred to use the epoxide as the limiting reagent. For reactions with secondary amines to give tertiary amine products, it may be possible to scavenge excess amine with an appropriate resin [9]. Alternatively, we find automated preparative high-performance liquid chromatography (HPLC) to be a convenient method for purification.

Examples of larger solution-phase combinatorial mixtures libraries are known utilizing a similar approach. A library of over 6000 β-amino alcohols was prepared using LiClO₄ in acetonitrile and 1.2 equiv. of amine [10]. In such cases, care is needed to validate the reactivity of reagents properly in order to avoid misleading bioassay data since purification is more difficult. A solid-phase mixtures library of

Tab. 11.1. Polymer-supported epoxidation reagents.

Reagents	Reaction Time	Yield	Comments	Ref.
Oxone®/NaHCO₃/THF/H₂O	48 h	13–97%	TentaGel resin (2 equiv.). 5 equiv. Oxone®. Generates dioxirane resin *in situ*	5
CHCl₃	60 h	70–90%	2 equiv. of dioxirane resin used. Resin is prepared by treatment of polystyrene 2-oxoalkyl resin with Oxone® and NaHCO₃ in THF/H₂O and isolated prior to use. Shelf stability of dioxirane resin not reported	6
—SO₄H THF/Δ	4 h	80%	AG 50W-X8 ion exchange resin (Bio-Rad), converted to persulfonic acid with potassium persulfate	7
—CO₃H THF/40 °C	4 h	50–95%	Resin prepared by treatment of 1% crosslinked carboxypolystyrene with 85% H₂O₂ in MeSO₃H for 16 h. Peracid resin stable to storage at −20 °C. THF is vastly superior to CH₂Cl₂ as solvent	8

5800 phenoxypropanolamines, prepared via a split-and-mix strategy, has also been reported [11]. Here, a resin-bound epoxide was opened with excess amine in acetonitrile at 80 °C for 18 h.

Opening of epoxides on solid phase with alternative nucleophiles to amines is illustrated in Scheme 11.3 [12]. In this example, which includes experimental procedures, azides (**10**) were prepared using buffered sodium azide in dimethylformamide (DMF) at 100 °C for 2 h, whilst thiols (**11**) were prepared by reaction with sodium thiophenoxide in DMF at 0 °C. In these systems, lactonization occurred under the cleavage conditions.

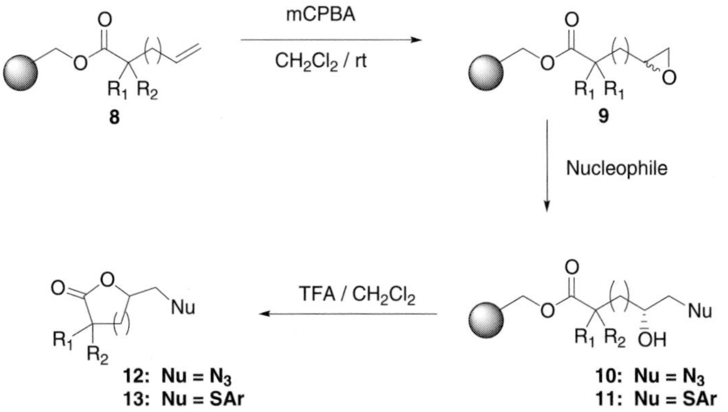

Scheme 11.3. Epoxidation, epoxide opening, and lactonization during cleavage from resin.

An elegant and potentially very versatile extension to the utility of epoxides is demonstrated in Scheme 11.4 [13]. Here, a range of olefins (**14**) was converted in a one-pot procedure to the immobilized α-sulfonated ketones (**17**) by epoxidation with dimethyldioxirane, epoxide opening with a sulfonic acid resin, and subsequent oxidation with Dess–Martin periodinane. The activated sulfonyloxy moiety is an excellent leaving group, and 20 efficient functionalizing cleavage options were demonstrated for resin **17**.

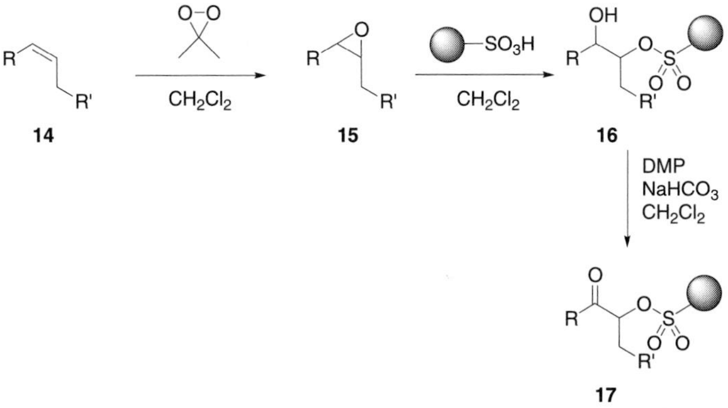

Scheme 11.4. One-pot epoxidation and conversion to α-sulfonated ketones.

11.2.2
Dihydroxylation

The dihydroxylation of an olefin is most efficiently carried out using osmium tetroxide, giving rise to vicinal diols (Scheme 11.5) [14]. The catalytic variant of the reaction is usually employed owing to the toxic and volatile nature of the reagent, and the facile regeneration of the catalyst with oxidants such as *N*-methylmorpholine *N*-oxide (NMO), potassium ferricyanide, hydrogen peroxide, or *tert*-butyl hydroperoxide. Osmium(III) chloride is sometimes used as the osmium source in the catalytic reaction, being less volatile than the tetroxide and therefore easier to weigh out safely. It is converted *in situ* to the tetroxide by the oxidant. High levels of enantioselectivity can be achieved in the reaction using catalytic amounts of osmium tetroxide in the presence of cinchona alkaloid derivatives [15].

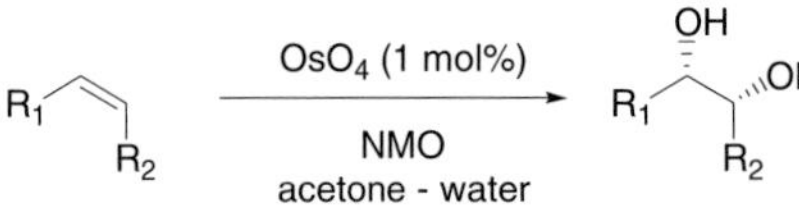

Scheme 11.5. Dihydroxylation of an olefin.

Osmium tetroxide is a very mild reagent and compatible with many functional groups, making it suitable for combinatorial synthesis, at least in principle. There have, however, been few reported uses in library synthesis. For solution-phase library synthesis, perhaps the most practical methods will involve the use of solid-supported osmium tetroxide [4] since this will minimize handling problems for the toxic osmium reagent, allowing recovery by simple filtration. One such example is osmium tetroxide microencapsulated in polystyrene [16]. This work highlighted one potential practical problem associated with such reagents. The dihydroxylation is usually carried out in acetone–water or *tert*-butanol–water mixed solvent systems. Not only does this raise the possibility of solubility issues for reactants, but also solvent/reactant access within the solid-supported reagent may be restricted since polymers such as polystyrene are poorly solvated and not swelled well by these solvents. In this work, acetonitrile was added as a cosolvent in order to achieve good conversion. Good results have been obtained in the area of supported asymmetric dihydroxylations. The OsO$_4$ copolymer of **20** [an analog of (DHQ)$_2$-PHAL] behaves well under standard conditions [*t*-BuOH/H$_2$O (1:1) at 10 °C using K$_3$Fe(CN)$_6$-K$_2$CO$_3$ (3 equiv.) as secondary oxidant], giving high yields and enantioselectivities (Scheme 11.6) [17]. The copolymer is easily recovered by filtration.

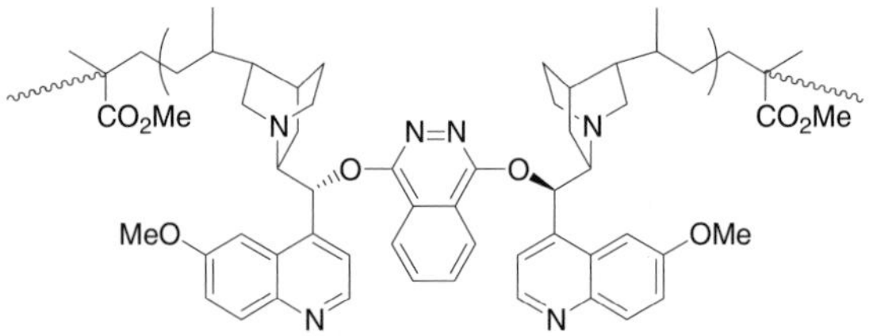

Scheme 11.6. Poly((QN)$_2$-PHAL-*co*-MMA) used in asymmetric dihydroxylation of olefins.

An example of solid-phase parallel synthesis employing asymmetric dihydroxylation is shown in Scheme 11.7 [18]. For this work, ArgoGel Wang resin was employed for the synthesis because of its compatibility with the solvent system, allowing standard Sharpless asymmetric dihydroxylation reaction conditions to be used which would have been less compatible with polystyrene Wang resin as discussed above. A second example demonstrating the use of osmylation in solid-phase chemistry is shown in Scheme 11.8 [19]. Here, a polystyrene-based resin was used and the solvent conditions were modified to tetrahydrofuran (THF)/H_2O (5:1) in order to promote resin swelling.

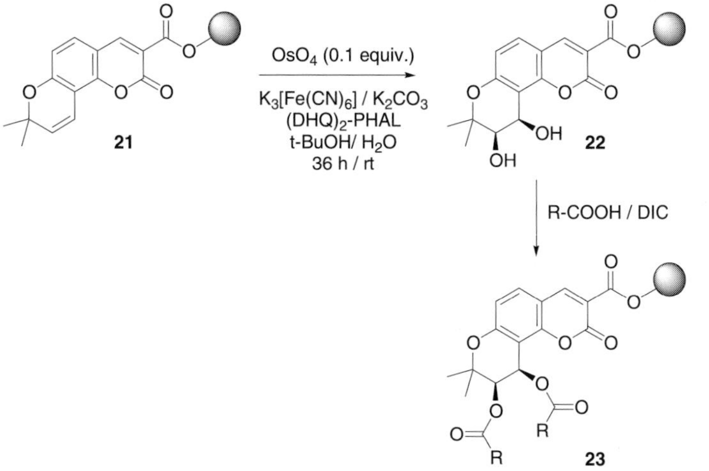

Scheme 11.7. Synthesis of khellactone derivatives.

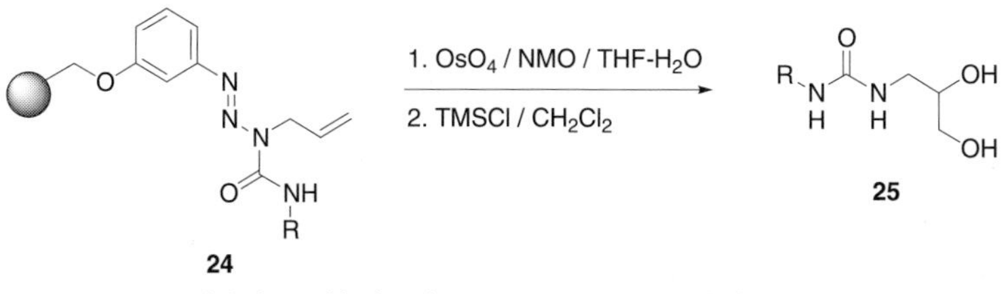

Scheme 11.8. Solid-phase dihydroxylation using T2 triazine linker.

11.2.3
Oxidative Cleavage

In situ oxidative cleavage of olefins to carbonyl compounds can be efficiently achieved by inclusion of sodium periodate as the oxidant in the osmium tetroxide dihydroxylation of olefins (the Lemieux–Johnson reagent) [20]. There have, however, been few if any reports on the use of this reagent combination in combinatorial synthesis, owing to either environmental/handling concerns or possible compatibility problems with resin systems.

More use has been made of ozonolysis (Scheme 11.9) [21], primarily for the solid-phase synthesis of peptide aldehydes (Scheme 11.10) [22–24]. A simple alkenyl ester linkage to Merrifield hydroxy resin may conveniently be used for tethering small molecules [21]; an α, β-unsaturated γ-amino acid linker to polystyrene [22, 24] or direct vinyl linkage to polystyrene (prepared by Wittig reaction) [23] has been used in the preparation of peptide aldehydes.

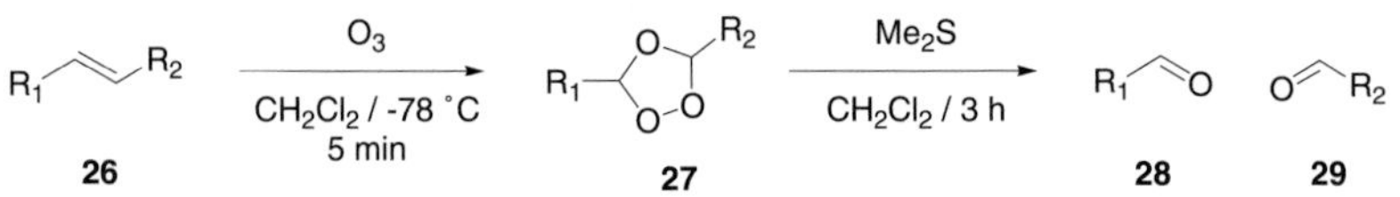

Scheme 11.9. Oxidative cleavage of olefins by ozonolysis.

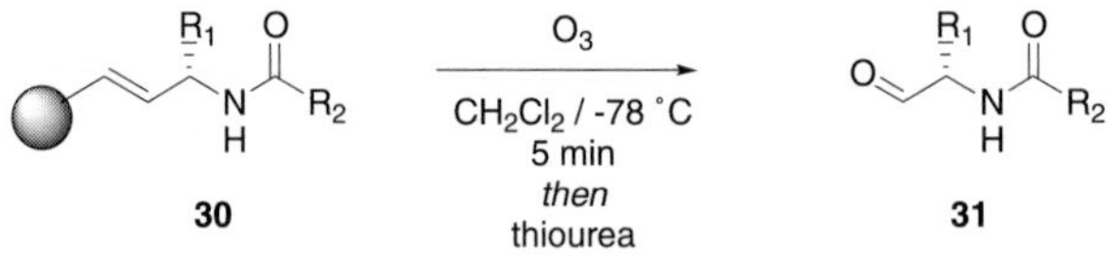

Scheme 11.10. Solid-phase synthesis of peptide aldehydes.

The procedure usually involves bubbling a stream of ozone through a dichloromethane solution of the olefin at −78 °C until a blue coloration remains, followed by quenching of the ozonide with a reducing agent such as dimethylsulfide, triphenylphosphine, or thiourea to give the carbonyl compound. Quenching with a reducing agent such as sodium borohydride gives rise to the corresponding alcohol [21], and a variation on this approach has been used for the preparation of a lactone (Scheme 11.11) [25]. There are certain logistical questions to be addressed in carrying out a large number of such reactions in parallel which may limit the overall utility of the reaction for combinatorial chemistry in the absence of specialist equipment. Additionally, there is a general tendency for swollen polystyrene-based resins to collapse when cooled to the low temperatures usually utilized in this reaction. This may limit reagent diffusion into the resin and will need to be borne in mind.

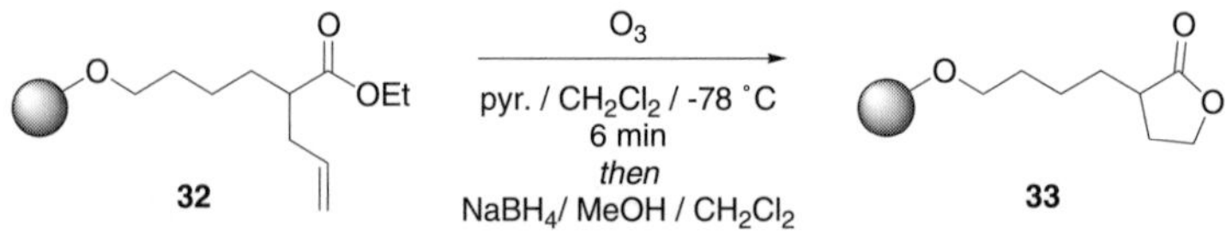

Scheme 11.11. Lactone formation via ozonolysis.

11.2.4
Electrophilic Addition of A–X

The examples above represent electrophilic addition of oxygen to a carbon–carbon double bond. Additionally, other powerful electrophiles – most notably sources of I^+, Br^+, $ArSe^+$, and H^+ – are also able to add (Scheme 11.12). The formation of the intermediate "onium" species such as **35** can then be followed by nucleophilic attack of the counterion X^-, by addition of some external nucleophile (e.g. water), or by an intramolecular process such as lactonization. Examples of such reactions are reported for both solution-phase and solid-phase chemistry and can be very efficient, although their reported use in library synthesis is somewhat limited.

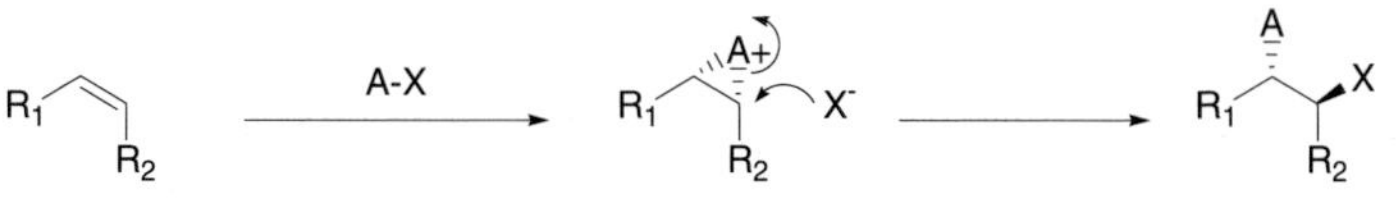

Scheme 11.12. Electrophilic addition of A–X to olefins.

An example of solid-phase iodolactonization and concomitant resin release is shown in Scheme 11.13 [26]. In this case, a modest degree of diastereofacial control (2:1 mixture of enantiomers) was achieved in the iodolactone **38**. Similarly, Scheme 11.14 demonstrates an example of iodoetherification on solid phase with concomitant oxidative cleavage to give the 2,5-disubstituted tetrahydrofuran derivative **40** [27]. Both of these examples provide functional groups which could, in principle, be used for postcleavage derivatization in solution phase.

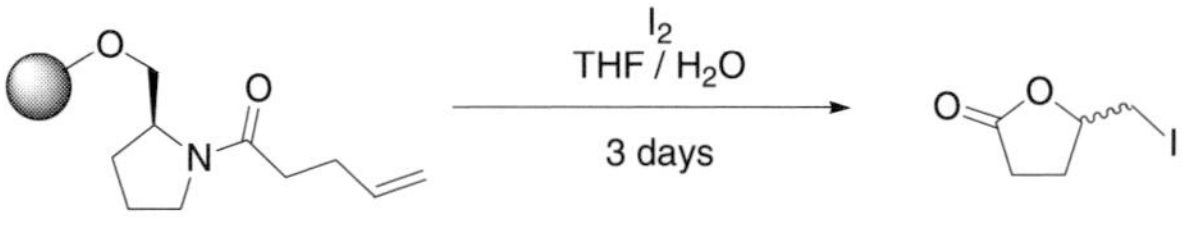

Scheme 11.13. Solid-phase intramolecular iodolactonization.

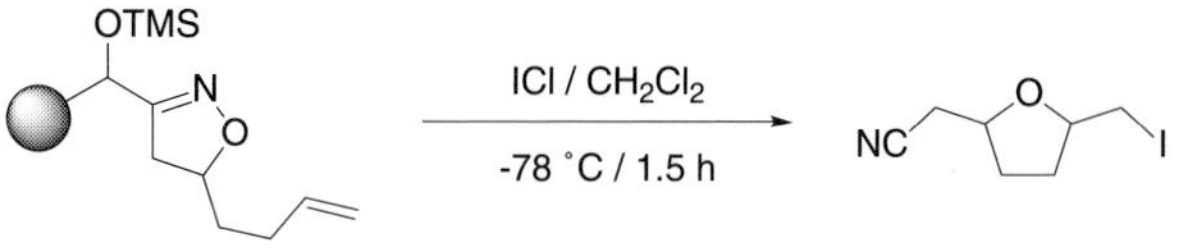

Scheme 11.14. Solid-phase intramolecular iodoetherification.

Perhaps most use has been made of organoselenium reagents in this area of research, providing mild and extremely versatile reactivity [28, 29]. A resin-bound version of phenylselenyl bromide (**43**) has been developed which provides a convenient and odorless way of handling this toxic and smelly reagent (Scheme 11.15) [30]. This reagent can also be converted to polymer-supported phenylselenyl phthalimide (**44**), which provides a convenient method for hydration of olefins (Scheme 11.16) [30]. A polymer-supported version of phenylselenyl cyanide is readily prepared from Merrifield resin and potassium selenocyanate [31].

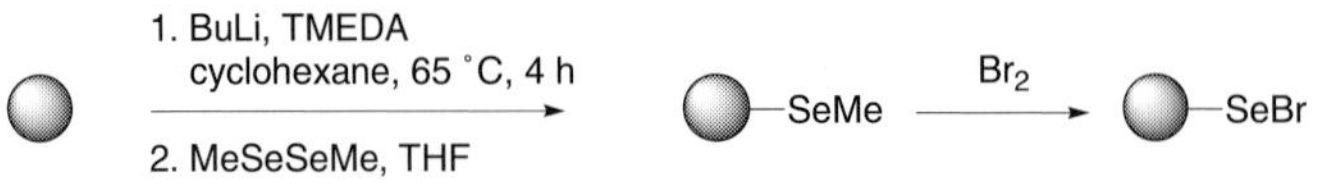

Scheme 11.15. Synthesis of polymer-supported phenylselenyl bromide.

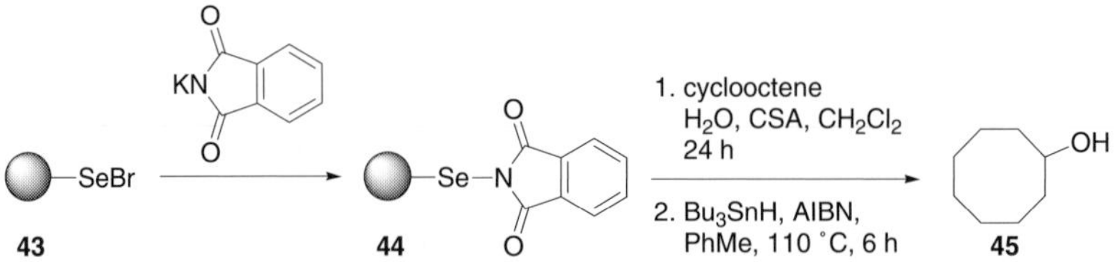

Scheme 11.16. Synthesis and use of polymer-supported phenylselenyl phthalimide.

The polymer-supported phenylselenyl bromide **43** efficiently adds to olefins in the manner depicted in Scheme 11.12 [30]. However, most use has been made of the way in which these selenium reagents facilitate intramolecular cyclizations such as cyclic ether formation (Scheme 11.17) [30], lactonization (Scheme 11.18) [32], and cyclic amine formation (Scheme 11.19) [33]. Reductive hydrodeselenation can be achieved with tributyltinhydride or, alternatively, oxidation results in the formation of an olefin via selenoxide elimination. The use of such reactions in the formation of benzopyran libraries (Scheme 11.20) [34–36] and polycyclic indoline libraries **53** [33] is described in more detail in Chapter X.

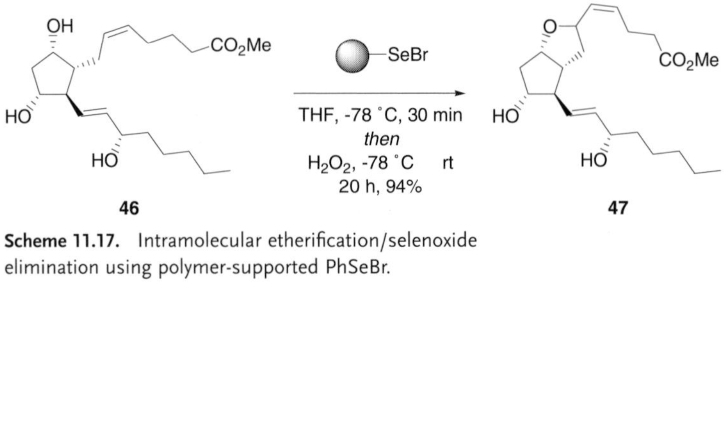

Scheme 11.17. Intramolecular etherification/selenoxide elimination using polymer-supported PhSeBr.

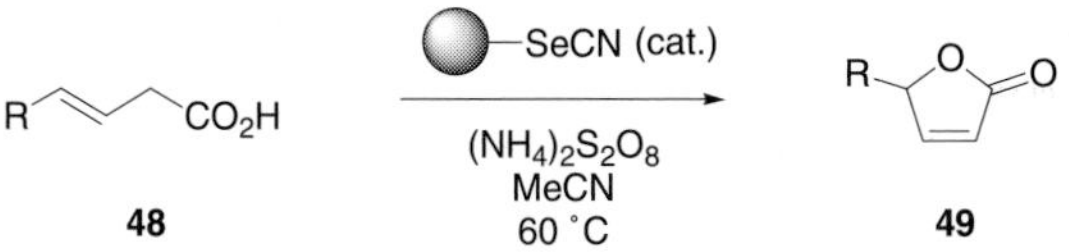

Scheme 11.18. Intramolecular lactonization using polymer-supported PhSeCN.

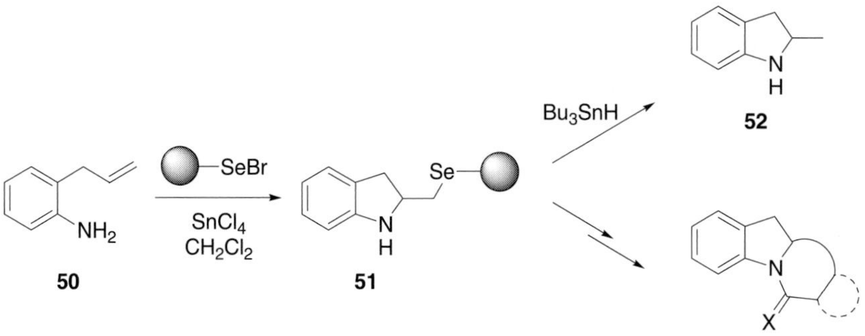

Scheme 11.19. Intramolecular selenoamination of *o*-allylamines in the presence of SnCl₄.

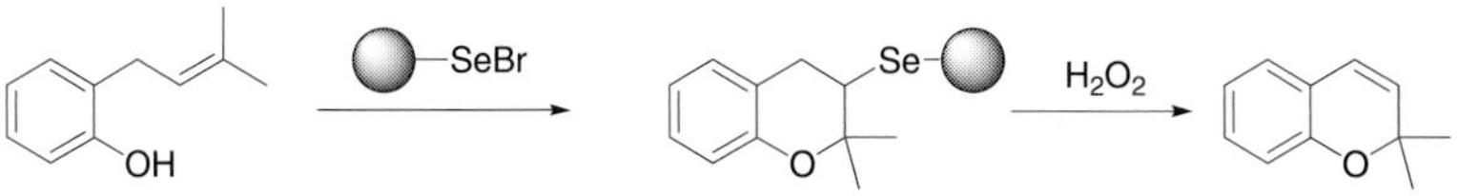

Scheme 11.20. Polymer-supported synthesis of benzopyrans.

11.2.5
Hydrogenation

The catalytic hydrogenation of olefins is, in the absence of sensitive functionalities and with the use of an appropriate catalyst, a very efficient transformation. Catalysts such as 10% Pd on activated carbon were in widespread use long before polymer-supported reagents became fashionable, and are easily removed from reaction mixtures by simple filtration. The parallel processing of multiple samples introduces certain complexities such as addition of hydrogen gas which may be overcome with specialist equipment [37] or, alternatively, catalytic transfer hydrogenation from reagents such as ammonium formate [38] may be considered. Carrying out hydrogenation during a solid-phase synthesis adds yet another layer of complexity since, by necessity, the catalyst must be soluble in order to gain access to the polymer-bound reactant [39]. Additionally, the reaction conditions must be compatible with the resin and linker, which often contain functionalities such as benzyl ethers that are themselves sensitive to hydrogenation. Consequently, this is an underutilized reaction in combinatorial chemistry. Diimide offers an alternative for reduction of alkenes (and alkynes) on solid phase, and excellent conversions have been observed using benzene sulfonylhydrazide in DMF at 100 °C as diimide source [40].

11.2.6
Hydrometallation

Of the hydrometallation processes available, the hydroboration of an olefin represents one of the more useful, not only allowing access to alkylboranes but also the formal hydration of the double bond to an alcohol through a subsequent oxidation step.

The utility of *in situ* hydroborated alkenes as coupling partners in the Suzuki reaction for library generation was demonstrated through the parallel synthesis of an array of 26 prostaglandin analogs on solid phase (Scheme 11.21) [41]. In this work, terminal olefins (**57**) were hydroborated with 9-BBN dimer in vials and then directly transferred to resin **59** contained within the reaction vessels of a parallel synthesizer, efficiently converting it to resin **60** under standard Suzuki conditions. Subsequent elaboration provided the prostaglandin E_1 analogs (**61**) in 18–56% overall yield.

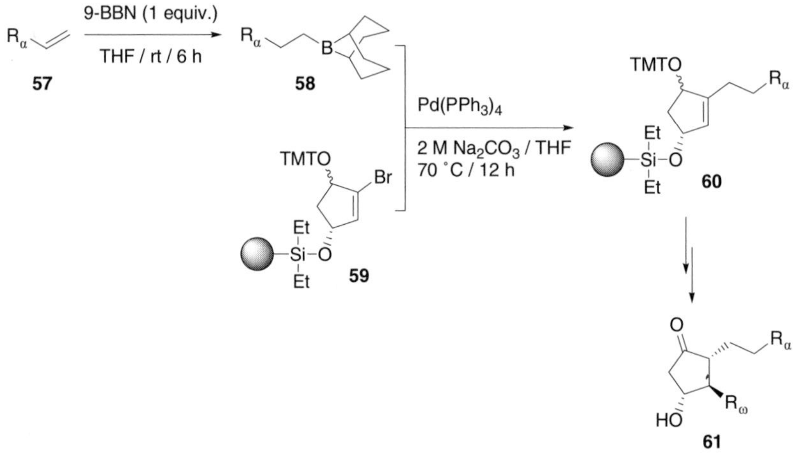

Scheme 11.21. Parallel synthesis of prostaglandin E_1 analogs, utilizing hydroborated alkenes as coupling partners in a Suzuki reaction.

The hydroboration/oxidation of alkenes to alcohols is a potentially useful transformation for the combinatorial chemist, particularly on solid phase. This reaction effectively allows the use of a terminal alkene as a protecting group for the hydroxyethyl moiety, which can then be further functionalized. This was demonstrated during the course of some synthetic studies toward mniopetals (Scheme 11.22) [42]. The triene **62** was selectively hydroborated with 9-BBN to give the primary alcohol **63**. IBX oxidation and a subsequent Bayliss–Hillman reaction using PhSeLi as nucleophile provided the key intermediate **64** which, upon oxidation to the corresponding ketone, underwent an intramolecular Diels–Alder reaction. Cleavage from the resin then gave the polycyclic compound **65** in 35% overall yield.

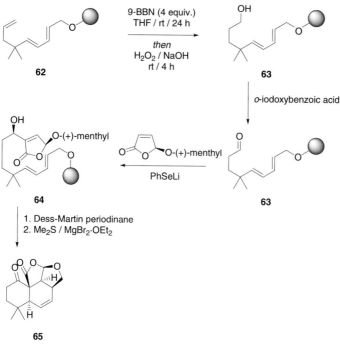

Scheme 11.22. Solid-phase hydroboration/oxidation of a terminal alkene and subsequent elaboration.

11.2.7
1,4-Addition to *α, β*-Unsaturated Carbonyl Systems

The reactions described above all make use of the electron-rich nature of isolated carbon–carbon double bonds and their consequent reactivity toward electrophiles. By contrast, when the carbon–carbon double bond is conjugated to a carbonyl or sulfonyl group, the electronic character of the double bond changes, making it susceptible to 1,4-nucleophilic addition (also known as Michael addition in the case of carbon nucleophiles). Pioneering work in this area of combinatorial chemistry is shown in Scheme 11.23, in which secondary amines **68** add to resin-bound acrylate (**67** → **69**), are quaternized with alkyl halides, and then undergo a Hofmann-type *β*-elimination in the presence of triethylamine to give tertiary amines **72** [43]. When primary amines **68** are used in this sequence (R2 = H), a reductive amination step has been used prior to quaternization to prepare tertiary amine libraries (**72**) with three points of diversity [43, 44]. Subsequent work has shown that replacement of triethylamine with Amberlite weakly basic ion-exchange resin in DMF simplifies work-up procedures since it overcomes the necessity to remove triethylamine salts from cleaved products, and this procedure has been used to generate large libraries of trisubstituted amines (> 10,000 compounds) [44]. A similar approach has been employed utilizing a vinylsulfone resin in place of the acrylate resin **67** [45].

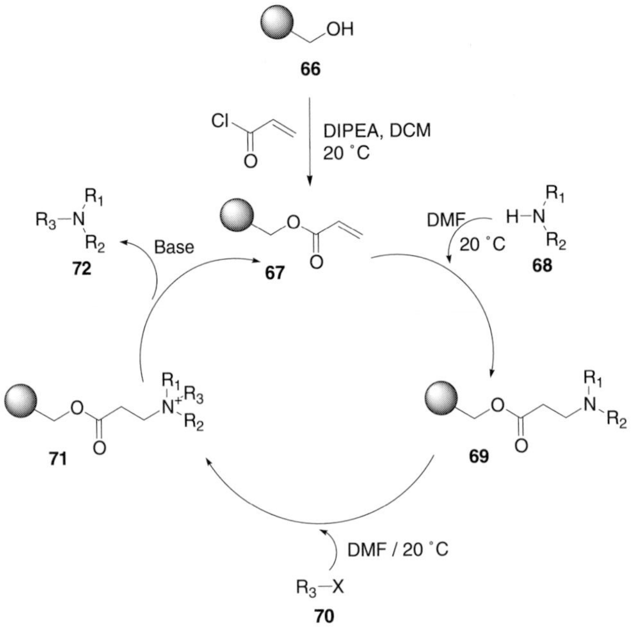

Scheme 11.23. REM linker – 1,4-addition of amines to acrylate, quaternization, and β-elimination to give tertiary amines. DIPEA, *N,N*-diisopropylethylamine.

11.3
Addition to C≡C Triple Bonds

The acetylenic group displays a versatile range of reactivities for use in combinatorial chemistry. Its use in cycloaddition reactions, palladium-mediated reactions, and radical reactions is well documented and covered here in the relevant chapters. Prominent examples on solid phase include their use in the palladium-mediated syntheses of heterocyclic templates such as indoles [46–49] and benzofurans [50].

The discussion above (Sect. 11.2.5) on hydrogenation of olefins applies equally to the hydrogenation of acetylenes – the diimide reduction of an acetylene on solid phase is one of the few reported examples [40]. Hydrometallation is another area where addition to acetylenes could be a very powerful transformation for further functionalization, although there are few reported examples. One such example is the hydrostannylation of a terminal acetylene using a polymer-supported tinhydride to give a vinyl stannane (74 → 75, Scheme 11.24) [51]. The vinyl stannane **75** was obtained as a mixture of *E:Z* isomers and was taken through to compound **76** in order to carry out a cyclorelease palladium-mediated Stille coupling to give **77**. There is, however, relatively little to report of relevance to this chapter at present, although there is scope for future work in this area.

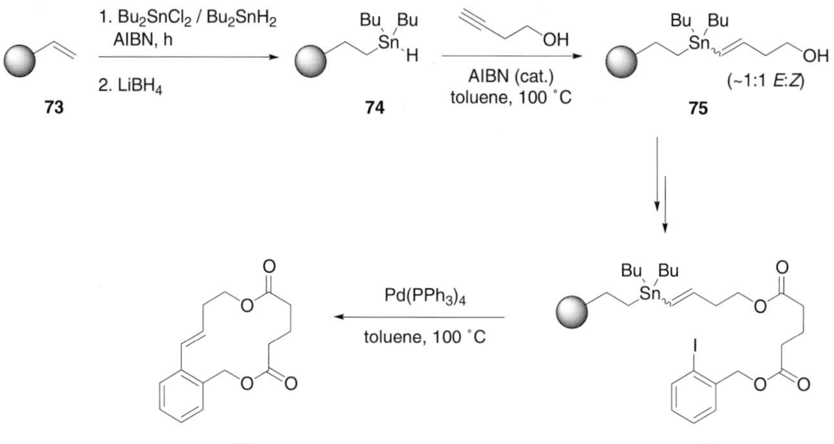

Scheme 11.24. Solid-phase hydrostannylation of an acetylene and subsequent synthesis of a macrocyclic system.

References

1 E. K. Kick, D. C. Roe, A. G. Skillman, G. Liu, T. J. A. Ewing, Y. Sun, I. D. Kuntz, J. A. Ellman, *Chem. Biol.* **1997**, *4*, 297–307.

2 D. P. Rotella, *J. Am. Chem. Soc.* **1996**, *118*, 12246–12247.

3 H. M. C. Ferraz, R. M. Muzzi, T. de O. Vieira, H. Viertler, *Tetrahedron Lett.* **2000**, *41*, 5021–5023.

4 S. V. Ley, I. R. Baxendale, R. N. Bream, P. S. Jackson, A. G. Leach, D. A. Longbottom, M. Nesi, J. S. Scott, R. I. Storer, S. J. Taylor, *J. Chem. Soc., Perkin Trans. I* **2000**, 3815–4196.

5 T. R. Boehlow, P. C. Buxton, E. L. Grocock, B. A. Marples, V. L.

Waddington, *Tetrahedron Lett.* **1998**, *39*, 1839–1842.

6 A. Shiney, P. K. Rajan, K. Sreekumar, *Polym. Int.* **1996**, *41*, 377–381.

7 C. S. Pande, N. Jain, *Synth. Commun.* **1989**, *19*, 1271–1279.

8 C. R. Harrison, P. Hodge, *J. Chem. Soc., Perkin I* **1976**, 605–609.

9 G. M. Coppola, *Tetrahedron Lett.* **1998**, *39*, 8233–8236.

10 B. L. Chng, A. Ganesan, *Bioorg. Med. Chem. Lett.* **1997**, *7*, 1511–1514.

11 W. M. Bryan, W. F. Huffman, P. K. Bhatnagar, *Tetrahedron Lett.* **2000**, *41*, 6997–7000.

12 C. L. Hetet, M. David, F. Carreaux,

B. Carboni, A. Sauleau, *Tetrahedron Lett.* **1997**, *38*, 5153–5156.

13 K. C. Nicolaou, P. S. Baran, Y.-L. Zhong, *J. Am. Chem. Soc.* **2000**, *122*, 10246–10248.

14 M. Schröder, *Chem. Rev.* **1980**, *80*, 187–213.

15 R. A. Johnson, K. B. Sharpless in: Catalytic Asymmetric Synthesis. Ojima, I. (ed.), VCH, New York 1993, pp. 227–272.

16 S. Nagayama, M. Edo, S. Kobayashi, *J. Org. Chem.* **1998**, *63*, 6094–6095.

17 C. E. Song, J. W. Yang, H. J. Ha, S. Lee, *Tetrahedron: Asymmetry* **1996**, *7*, 645–648.

18 Y. Xia, Z.-Y. Yang, A. Brossi, K.-H. Lee, *Org. Lett.* **1999**, *1*, 2113–2115.

19 S. Bräse, S. Dahmen, M. Pfefferkorn, *J. Comb. Chem.* **2000**, *2*, 710–715.

20 R. Pappo, D. S. Allen, R. U. Lemieux, W. S. Johnson, *J. Org. Chem.* **1956**, *21*, 478–479.

21 C. Sylvain, A. Wagner, C. Mioskowski, *Tetrahedron Lett.* **1997**, *38*, 1043–1044.

22 C. Pothion, M. Paris, A. Heitz, L. Rocheblave, F. Rouch, J.-A. Fehrentz, J. Martinez, *Tetrahedron Lett.* **1997**, *38*, 7749–7752.

23 B. J. Hall, J. D. Sutherland, *Tetrahedron Lett.* **1998**, *39*, 6593–6596.

24 M. Paris, A. Heitz, V. Guerlavais, M. Cristau, J.-A. Fehrentz, J. Martinez, *Tetrahedron Lett.* **1998**, *39*, 7287–7290.

25 S. Hanessian, F. Xie, *Tetrahedron Lett.* **1998**, *39*, 737–740.

26 H. S. Moon, N. E. Schore, M. J. Kurth *J. Org. Chem.* **1992**, *57*, 6088–6089.

27 X. Beebe, N. E. Schore, M. J. Kurth, *J. Org. Chem.* **1992**, *57*, 10061–10062.

28 K. C. Nicolaou, N. A. Petasis, *Selenium in Natural Product Synthesis.* CIS Inc., Philadelphia, PA, 1984.

29 D. Liotta, *Organoselenium Chemistry*, Wiley, New York 1986.

30 K. C. Nicolaou, J. Pastor, S. Barluenga, N. Winssinger, *J. Chem.*

Soc., Chem. Commun. **1998**, 1947–1948.

31 K. Fujita, K. Watanabe, A. Oishi, Y. Ikeda, Y. Taguchi, *Synlett* **1999**, 1760–1762.

32 K.-i. Fujita, H. Taka, A. Oishi, Y. Ikeda, Y. Taguchi, K. Fujie, T. Saeki, M. Sakuma, *Synlett* **2000**, 1509–1511.

33 K. C. Nicolaou, A. J. Roecker, J. A. Pfefferkorn, G.-Q. Cao, *J. Am. Chem. Soc.* **2000**, *122*, 2966–2967.

34 K. C. Nicolaou, J. A. Pfefferkorn, G.-Q. Cao, *Angew. Chem., Int. Ed. Engl.* **2000**, *39*, 734–739.

35 K. C. Nicolaou, G.-Q. Cao, J. A. Pfefferkorn, *Angew. Chem., Int. Ed. Engl.* **2000**, *39*, 739–743.

36 K. C. Nicolaou, J. A. Pfefferkorn, A. J. Roecker, G.-Q. Cao, S. Barluenga, H. J. Mitchell, *J. Am. Chem. Soc* **2000**, *122*, 9939–9953.

37 T. Bruckdorfer, H. Linnertz, *GIT Labor-Fachz* **2000**, *44*, 58.

38 B. C. Ranu, A. Sarkar, S. K. Guchhait, K. Ghosh, *J. Indian Chem. Soc.* **1998**, *75*, 690–694.

39 I. Ojima, C.-Y. Tsai, Z. Zhang, *Tetrahedron Lett.* **1994**, *35*, 5785–5788.

40 P. Lacombe, B. Castagner, Y. Gareau, R. Ruel, *Tetrahedron Lett.* **1998**, *39*, 6785–6786.

41 D. R. Dragoli, L. A. Thompson, J. O'Brien, J. A. Ellman, *J. Comb. Chem.* **1999**, *1*, 534–539.

42 U. Reiser, J. Jauch, *Synlett* **2001**, 90–92.

43 A. R. Brown, D. C. Rees, Z. Rankovic, J. R. Morphy, *J. Am. Chem. Soc.* **1997**, *119*, 3288–3295.

44 X. Ouyang, R. W. Armstrong, M. M. Murphy, *J. Org. Chem.* **1998**, *63*, 1027–1032.

45 F. E. K. Kroll, R. Morphy, D. Rees, D. Gani, *Tetrahedron Lett.* **1997**, *38*, 8573–8576.

46 A. L. Smith, G. I. Stevenson, C. J. Swain, J. L. Castro, *Tetrahedron Lett.* **1998**, *39*, 8317–8320.

47 H.-C. Zhang, B. E. Maryanoff, *J. Org. Chem.* **1997**, *62*, 1804–1809.

48 H.-C. ZHANG, K. K. BRUMFIELD, B. E. MARYANOFF, *Tetrahedron Lett.* **1997**, *38*, 2439–2442.

49 M. C. FAGNOLA, I. CANDIANI, G. VISENTIN, W. CABRI, F. ZARINI, N. MONGELLI, A. BEDESCHI, *Tetrahedron Lett.* **1997**, *38*, 2307–2310.

50 D. FANCELLI, M. C. FAGNOLA, D. SEVERINO, A. BEDESCHI, *Tetrahedron Lett.* **1997**, *38*, 2311–2314.

51 K. C. NICOLAOU, N. WINSSINGER, J. PASTOR, F. MURPHY, *Angew. Chem., Int. Ed. Engl.* **1998**, *37*, 2534–2537.

12
Addition to Carbon–Hetero Multiple Bonds

Philipp Grosche, Jörg Rademann, and Günther Jung

12.1
Introduction

Carbon–hetero multiple bonds play a central role in synthetic organic chemistry. They possess a polar character and react with various nucleophiles as well as with electrophiles in numerous addition reactions. They form valuable substrates for the synthesis of structurally diverse and pharmacologically important compounds, such as tetrahydro-β-carbolines, ureas, and guanidines. Many of the addition reactions to CX multiple bonds have found wide application in combinatorial chemistry such as the synthesis of ureas via isocyanates, the Pictet–Spengler reaction, and several Mannich-type reactions. In the literature, most articles in the field of combinatorial chemistry deal with solid-phase reactions, however, solution-phase protocols are in many instances also suited to parallel synthesis.

In this chapter addition reactions to CN double bonds, CS double bonds, and CN triple bonds are discussed. The chapter will not include cycloaddition reactions of CX multiple bond systems and reactions of isocyanides typically employed in multiple component reactions.

This chapter reviews those transformations of the CX multiple bonds that are suitable for combinatorial chemistry, either in solution or on solid phase. To assure clarity and readability, the reactions are grouped according to the number of bonds between the C atom and the heteroatom, the hybridization of the carbon center, and the nature of the attacking agent.

12.2
Additions to CN Double Bonds in sp^2 Systems

CN double bonds are the structural motif of various functional groups. This section refers to CN double bonds in sp^2 systems such as imines, iminium ions, acyliminium ions, and *N*-oxides of nitrogen heterocycles (pyridine, quinoline, isoquinoline). The reacting CN double bond can be isolated; however, the reactions are often performed as a three-component reaction consisting of an amine, a carbonyl compound and the attacking nucleophile, with *in situ* formation of the reactive CN double bond.

12.2.1
Attack by Hydride (Reductive Alkylation)

Reductive alkylation, or reductive amination, is one of the most commonly used reactions in combinatorial chemistry. It is a convenient method for the preparation of secondary and tertiary amines and is one of the easiest ways to generate diversity, as usually the substrate tolerance is very broad. The monoalkylation of primary or secondary amines is affected by condensation with aldehydes or ketones and the subsequent reduction of the imine species. In the case of primary amines, the reaction can be performed either in a two-step procedure by first synthesizing and isolating the imine followed by reduction, or in a one-pot procedure by adding the carbonyl partner together with the reducing agent. Secondary amines are typically converted in a one-pot synthesis because of the low stability of the intermediary iminium ion. The use of primary amines in a one-pot procedure carries the risk of dialkylation as the formed secondary amine can react with another carbonyl, thus forming the tertiary amine. As a reducing agent, NaBH$_3$CN, NaBH(OAc)$_3$, Me$_4$NBH$_4$, LiBH$_4$, NaBH$_4$, BH$_3$·py, and BH$_3$·THF can be employed. In solution phase, both the one-pot synthesis [1] and the reduction of preformed imine species [2] were used. Anilines were, for example, converted upon treatment with aldehydes and NaBH(OAc)$_3$ in dichloromethane at room temperature (rt) to give the respective secondary amines [3]. The degree of dialkylation in the one-pot procedure increases with less-hindered amines and aldehydes and with more electron-rich amines. To avoid double alkylation an excess of amine is often used [4].

Polymer-bound reducing agents have found wide application in solution-phase combinatorial synthesis mostly in the form of ion-exchange resins loaded with borohydride or cyanoborohydride (Scheme 12.1). Polymer-bound ammonium cyanoborohydride was used for the reductive alkylation of primary and secondary amines with aromatic aldehydes in MeOH at rt (Scheme 12.1) [5, 6], or for the reduction of preformed imines in toluene/MeOH under reflux conditions [7]. Polymer-supported borohydride was utilized for the reduction of preformed imines in MeOH at rt. An excess of amine relative to the aldehyde was applied in order to drive the imine formation to completion and to suppress dialkylation. Excess amine was readily scavenged from the desired secondary amine by selective imine formation using a polymer-supported aldehyde [8]. Tertiary amines were formed by treating aldehydes with an excess of secondary amine and polymer-supported

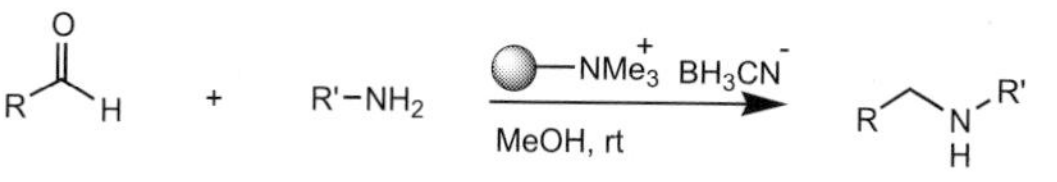

Scheme 12.1

cyanoborohydride in acetic acid/dichloromethane. In this case excess secondary amine was removed using polymer-supported benzoyl chloride [8].

In solid-phase chemistry, both of the components forming the imine species can be linked to the solid phase. Both approaches, the reduction of isolated imines (Scheme 12.2) [9, 10] and the one-pot synthesis [11], can be used. Polymer-bound primary amines can be partly overalkylated using the one-pot strategy, as reported for solution-phase chemistry. Reactions using secondary amines are always carried out as a one-pot procedure. Imines can easily be hydrolyzed, especially under acidic conditions. Thus, with isolated imines the condensation and reduction step can be repeated to achieve complete conversion [12].

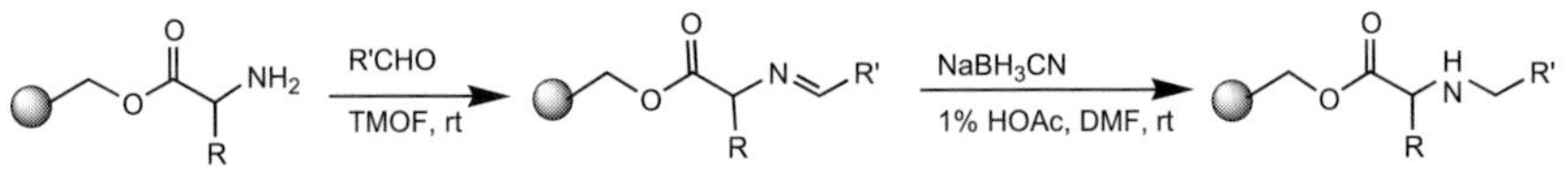

Scheme 12.2

Usually $NaBH_3CN$ in AcOH/dimethylformamide (DMF) is used as the reducing agent. However, several other reagents have also been used in solid-phase chemistry: $NaBH(OAc)_3$ [13], Me_4NBH_4 [14], $LiBH_4$ [15], $NaBH_4$ [10], $BH_3 \cdot py$ [16], and $BH_3 \cdot THF$ [17]. The borane–pyridine complex was found to be an excellent reagent for the *in situ* reduction of iminium ions on solid phase [16]. The reaction was performed in DMF/EtOH (3:1) using an excess of aldehyde or ketone respectively. However, this reagent is not suitable for primary amines as it leads to dialkylation.

In contrast to imines, oximes required harsher conditions for efficient reduction. In a recent example, resin-bound oximes were reduced using $BH_3 \cdot THF$ in 4 M HCl in dioxane, tetrahydrofuran (THF)/MeOH at rt [17].

12.2.2
Addition of Carbon Nucleophiles

A broad range of C-nucleophiles could be added to CN double bonds, namely enolates, ketene acetals, C,H-acidic compounds, electron-rich heterocycles, organometallic compounds, boronic acids, and silanes. The reactions are sorted according to the nature of the nucleophile. In several cases there are only a few examples of solution-phase combinatorial chemistry, which does not necessarily reflect the suitability of solution-phase protocols to library synthesis.

12.2.2.1 Imino Aldol Reaction
Imines react readily with silyl enolates, silyl ketene acetals, and silyl thioketene acetals to afford *N*-substituted *β*-amino ketones, *β*-amino esters, and *β*-amino thioesters. The reaction is catalyzed typically by Lewis acids, preferably with the lanthanide triflates $Ln(OTf)_3$, $Sc(OTf)_3$, and $Yb(OTf)_3$. It can also be performed as a three-component reaction – between amine, aldehyde or ketone, and the silyl

component – where the imine is formed *in situ*. In this case, a dehydration agent such as trimethyl orthoformate is normally added to support imine formation and to prevent degradation of the ketene silyl acetals by H_2O. This Mannich-type reaction has attracted a lot of interest in solution-phase as well as in solid-phase chemistry.

A typical procedure in solution applies equimolar (or nearly equimolar) amounts of imine (or amine and aldehyde) and silyl component and 0.05–0.3 mol% $Ln(OTf)_3$. The reaction is performed in dichloromethane (DCM), acetonitrile (ACN), or mixtures of both solvents at rt. The suitability of this reaction for library synthesis was exemplified by the generation of a small library of 40 2,3-dihydro-4-pyridones using Danishefsky's diene as the silyl component and $Yb(OTf)_3$ as the Lewis acid catalyst [18]. Byproducts and unreacted aldehyde were removed with polyamine resin and the products were obtained in good to excellent purities and yields. Although this reaction was often described as an aza-Diels–Alder reaction in the literature, other authors favored a tandem Mannich–Michael mechanism [19].

Another solution-phase approach utilizes a polymer-bound scandium catalyst (polyallyl-scandium-triflylamide ditriflate) (Scheme 12.3). The reaction is conducted as a three-component reaction, with silyl enolates, ketene silyl acetals, and cyanotrimethyl silane as nucleophiles [20]. Treatment with cyanotrimethyl silane as the nucleophile affords α-aminonitriles in a Lewis acid-catalyzed variation of the classical Strecker synthesis. This reaction is performed in dichloromethane/acetonitrile (2:1) at rt for 19 h. When ketene silyl acetal is used, $MgSO_4$ is added to prevent decomposition of the ketene silyl acetal by traces of water. A cation-exchange resin has also been used for the promotion of the imino-aldol reaction [21].

In the imino-aldol reaction at least one asymmetric C atom is generated. Thus, a procedure for enantioselective addition is of special interest. Catalysis of the reaction with polymer-supported palladium BINAP μ-hydroxo complex (BINAP = 2,2′-bis(diphenylphorphino)-1,1′-binaphthyl) yields an ee of 81% [22].

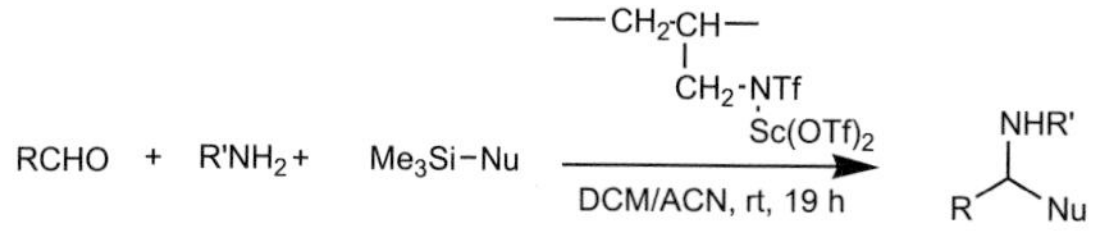

Scheme 12.3

This Mannich-type reaction is very suitable for combinatorial chemistry and has also been adapted to solid-phase chemistry, thus pursuing both possible strategies: the immobilization of the imine and the immobilization of the silyl component. Polymer-supported silyl thioketene acetals obtained by silylation of immobilized thioesters were converted to β-aminoesters by treatment with imines in the presence of $Sc(OTf)_3$ using dichloromethane as solvent [23, 24]. Alternatively, a three-component protocol can be employed (Scheme 12.4) [25]. In both cases reductive cleavage of β-amino acid thioesters with $LiBH_4$ yields β-amino alcohols.

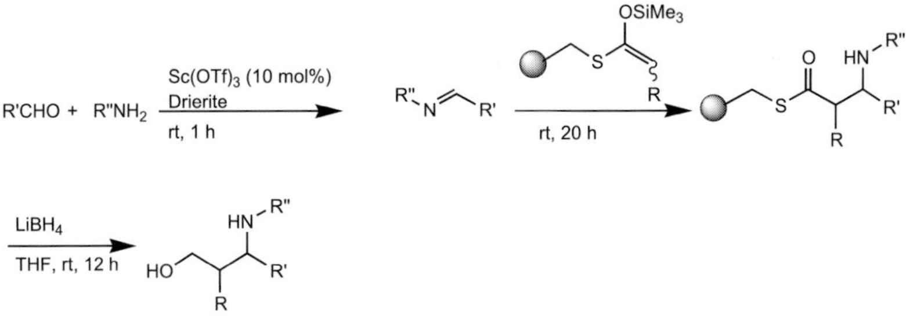

Scheme 12.4

In the second approach, a polymer-supported imine was treated with silyl enol ethers or silyl ketene acetals in the presence of Yb(OTf)$_3$ in dichloromethane to yield β-amino ketones and esters respectively [26, 27]. Instead of imines, polymer-supported acyl hydrazones can also be used as substrates [28].

2,3-Dihydro-4-pyridones were synthesized by tandem Mannich–Michael reaction in good yields. Resin-supported imines were reacted with Danishefsky's diene in THF using Yb(OTf)$_3$ (Scheme 12.5) [29] or ZnCl$_2$ [30] as catalyst.

Scheme 12.5

A more classical example of Mannich chemistry is the solid-phase variant of the Robinson tropanone synthesis. Starting from immobilized primary amines, tropanone derivatives are formed upon treatment with 1,3-acetonedicarboxylic acid and succinic aldehyde under acidic conditions at rt [31].

12.2.2.2 Reaction with Boronic Acids

The Petasis reaction is a three-component condensation reaction between an aldehyde, an amine (usually secondary amine), and a boronic acid. It is also referred to as a boronic acid Mannich reaction. The reaction mechanism is not yet completely elaborated, however, a mechanism involving iminium species is likely and therefore this reaction is discussed here. A wide range of building blocks is accepted in this reaction, but the limited number of suitable aldehydes is remarkable. Besides glyoxylic acid, only a few aldehydes react satisfactorily, e.g. α-hydroxyaldehydes and aldehydes bearing an α-heteroatom. In solution-phase chemistry the condensation is performed using equimolar amounts of the components in THF or acetonitrile at rt. By using 1,2-diamines, glyoxylic acid, and boronic acids, piperazinones and benzopiperazinones are accessible (Scheme 12.6) [32]. The Petasis reaction has also been adapted to solid-phase chemistry. Each of the required components can

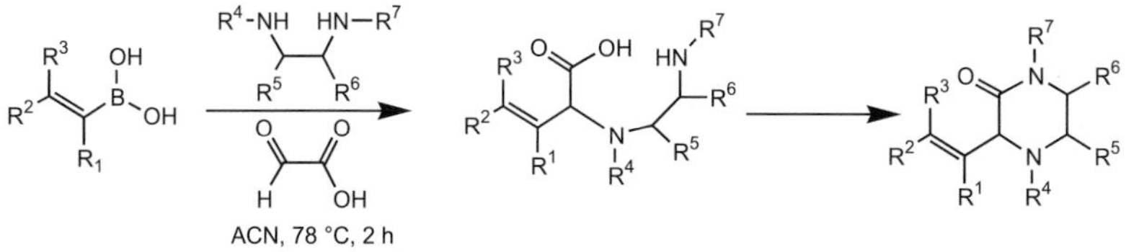

Scheme 12.6

be linked to the solid support. Typically, the components in solution are employed in large excess. Polymer-bound *N*-substituted amino acids have been treated with glyoxylic acid and boronic acids in dichloromethane at rt for 18 h, and the reaction was repeated for additional 60 h. The carboxylic acids obtained were further modified by coupling with amines. After cleavage from the resin, products were obtained in good purity and yield [33, 34]. This reaction was also carried out at 50 °C for 1–2 days in DMF/1,2-dichloroethane (DCE), and these reaction conditions have also been successfully applied to immobilized phenyl boronic acid, glyoxylic acid (Scheme 12.7), and proline or piperazine [35].

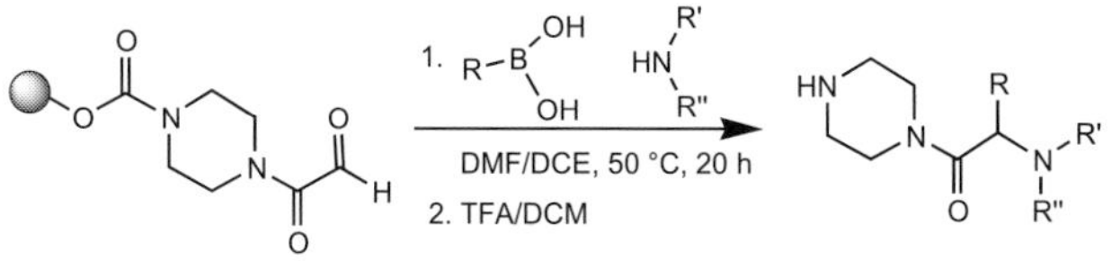

Scheme 12.7

12.2.2.3 Addition of Allylsilanes (Imino-Sakurai Reaction)

Allylsilanes add to CN double bonds in an imino-Sakurai reaction. Both the imine species and the allylsilane can be the immobilized component. Polymer-bound acyl imines have been generated *in situ* by reaction of immobilized carbamate with aromatic and heteroaromatic aldehydes under Lewis acid conditions. These intermediates reacted directly with allylsilanes to yield homoallylic amines. The reaction was performed as a one-pot procedure in acetonitrile using $BF_3 \cdot OEt_2$ as Lewis acid (Scheme 12.8) [36]. A variation employing cyclic *N*-acyliminium ion intermediates has also been performed [37].

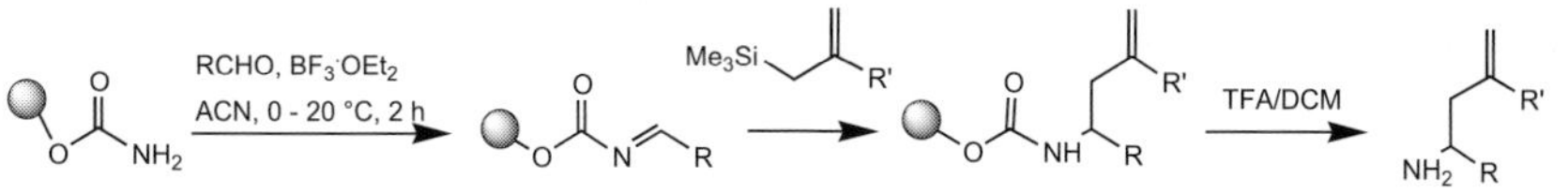

Scheme 12.8

In another approach immobilized allylsilanes were treated with Boc-protected aldimines in dichloromethane using $BF_3 \cdot OEt_2$ as Lewis acid. This reaction yielded Boc-protected homoallylic amines [38].

12.2.2.4 **Reaction with Grignard Reagents, Lithium Organyles, and Zinc Organyles**
Imines and iminium species are converted to amines upon treatment with Grignard reagents, lithium reagents, and zinc organyles. This reaction is often used in solution-phase chemistry and has been adapted to solid-phase chemistry.

Resin-immobilized aldimines, derived from condensation of Rink amide polystyrene resin with aldehydes, reacted with Grignard reagents at 60 °C or lithium reagents at −78 °C to 20 °C to yield after cleavage primary amines in good to excellent purity (Scheme 12.9) [15]. Since the aldimines of ammonia are unstable, similar approaches to primary amines in solution phase need protecting groups at the N atom to improve the stability. In this example, the function is fulfilled by the resin. If the imines are not immobilized via the N atom, secondary amines are obtained after cleavage [39, 40]. Enantiopure immobilized aldimines have been converted diastereoselectively to homoallylamines upon treatment with allyl zinc reagents in THF. The latter were obtained from allylbromide and zinc using CeCl$_3$·7H$_2$O as an additive [41].

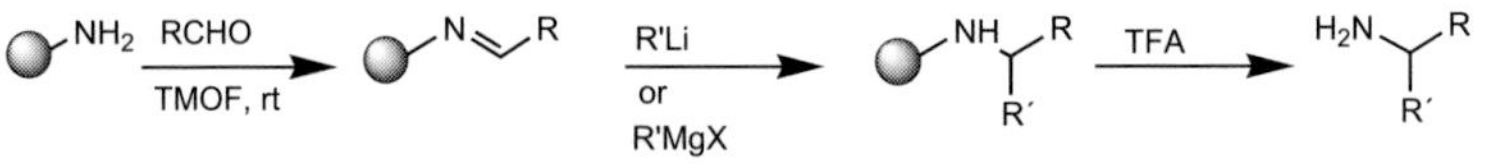

Scheme 12.9

Another kind of CN double bond-containing substrate are the N-acyl-pyridinium or N-acyl-quinoline intermediates. These are generated by the reaction of the heterocycle with acylchlorides or chloroformates. Activation of 4-methoxypyridine can be achieved by reaction with a resin-bound chloroformate. Addition of the Grignard reagent was performed at 20 °C in THF and yielded 2,3-dihydro-4-pyridones in high purity following basic cleavage [42, 43].

Another possibility to generate acyl-pyridinium intermediates is the immobilization of hydroxypyridine and subsequent acylation. Conversion with a Grignard reagent and acidic cleavage affords N-acyl-2,3-dihydro-4-pyridones (Scheme 12.10) [44, 45]. In a similar approach, quinolines were converted to 2-substituted N-acyl-dihydroquinolin-4-ones [46]. A related reaction is the solid-phase variant of the Reissert reaction, which is an efficient activation method for introducing substituents at the C-1 position of isoquinolines. Polymer-bound N-acyl-isoquinolinium intermediates generated *in situ* from immobilized benzoyl chloride react with trimethylsilyl cyanide to form N-acyl-1-cyanodihydroisoquinolines. These Reissert intermediates possess increased acidity at the C-1 position and are smoothly alkylated [47, 48].

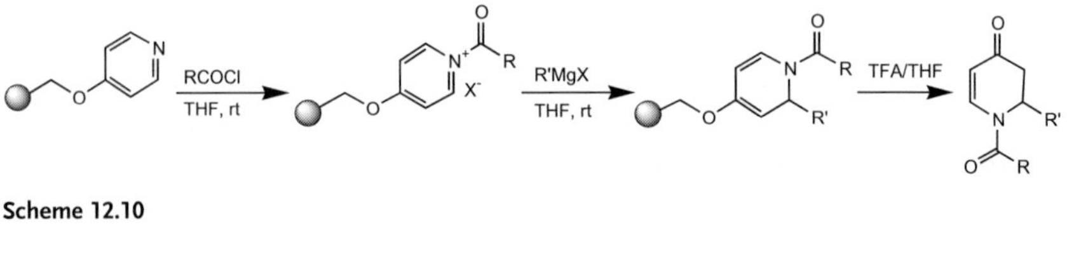

Scheme 12.10

12.2.2.5 **Addition of Copper Alkynes**

Aldehydes, amines, and alkynes react in a Mannich-type reaction yielding prop-argylic amines. As in other examples the reactive imine species is formed *in situ*. The reaction is catalyzed by CuCl and proceeds presumably via copper alkynes. Each of the three components can be linked to the solid support.

Propargylamine was immobilized on 2-chlorotrityl chloride resin and was re-acted with secondary amines and paraformaldehyde in the presence of CuCl to give the aminomethylated products in high purity. The reaction was performed in dioxane at 70–75 °C for 3 h [49]. Solid-supported alkynes can also be prepared via the Sonogashira reaction [50] and have been employed in the addition of alkyne cuprates on solid phase.

Besides the amino alkynes, resin-bound piperazine and benzaldehydes have served as substrates for this reaction (Scheme 12.11) [51].

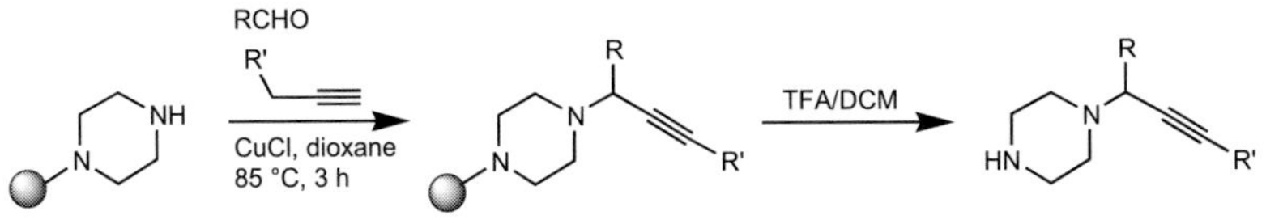

Scheme 12.11

12.2.2.6 **Addition of Electron-rich Aromatic and Heteroaromatic Cycles**

Electron-rich heterocycles can add to CN double bonds. The most famous example in this field is the Pictet–Spengler reaction. In this reaction 3-(2'-aminoethyl)-indoles are condensed with aldehydes or ketones and subsequently cyclized to yield 1,2,3,4-tetrahydro-β-carbolines.

This reaction has often been used in solid-phase chemistry. Tryptophan deriva-tives serve as substrates and are immobilized via the carboxy group. Condensation with the carbonyl compound and cyclization is performed in one pot. Aliphatic and aromatic aldehydes as well as ketones have been used. Neutral or acidic con-ditions are suitable for the Pictet–Spengler reaction. However, the acid-catalyzed route using 1–25% trifluoroacetic acid (TFA) in dichloromethane is most often used, and under neutral conditions the reaction can be conducted at 50–110 °C.

A typical procedure employs tryptophan immobilized on Merrifield resin, with a tenfold excess of aldehyde in 10% TFA/dichloromethane at rt for 16 h (Scheme 12.12). The tetrahydro-β-carbolines were obtained with high purity after cleavage via aminolysis with ethylamine [52]. *N*-Substituted tryptophan derivatives can also be used for the Pictet–Spengler reaction [53]. In an alternative approach the imine obtained from the condensation of various aldehydes with tryptophan immobilized on Wang-polystyrene resin was isolated. Treatment with fluorenylmethoxycarbonyl (Fmoc)-protected amino acid chloride results in the formation of an *N*-acylimi-nium intermediate which undergoes Pictet–Spengler condensation. Fmoc depro-tection leads to resin cleavage through cyclization to yield diketopiperazines in high purity (Scheme 12.13) [54].

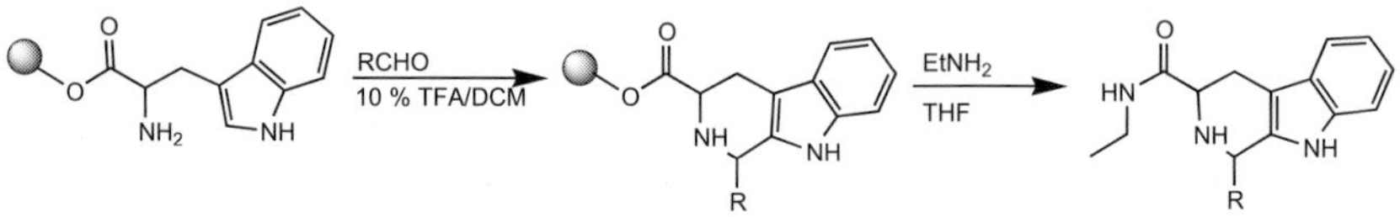

Scheme 12.12

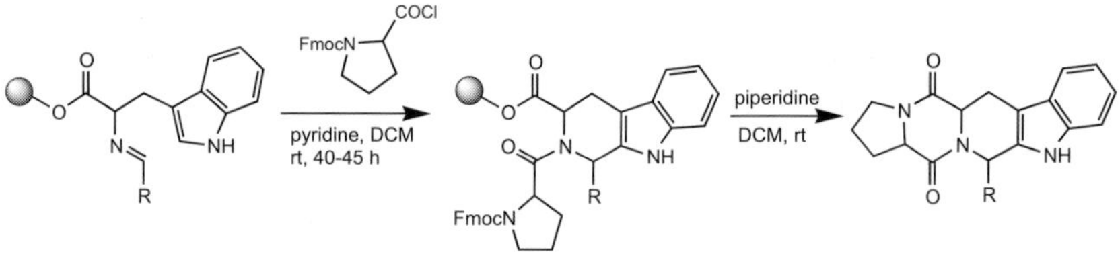

Scheme 12.13

Instead of tryptamine derivatives, electron-rich *m*-tyramine and histamine derivatives can be linked to the support. Condensation and cyclization proceeds well with aliphatic, aromatic, and heteroaromatic aldehydes in pyridine at 100 °C for 14 h. Tetrahydroisoquinolines and tetrahydroimidazopyridines are obtained with high purity after cleavage (Scheme 12.14).

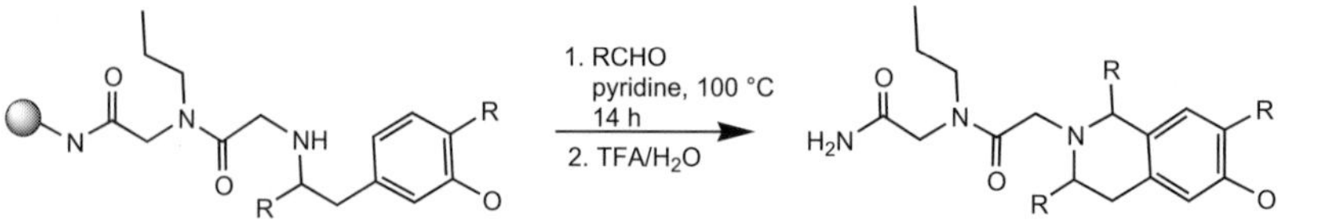

Scheme 12.14

Another reaction of this type is the aminomethylation of resin-supported indoles with formaldehyde and secondary amines, affording 3-aminomethylindole in high yield and purity. The reaction is conducted as a three-component reaction in AcOH/1,4-dioxane (1:4) at rt for 1.5 h. The iminium intermediate is formed *in situ* and is trapped by the indole [55].

Immobilized heterocyclic *N*-oxides, such as quinoline-*N*-oxide or isoquinoline-*N*-oxide react with various nucleophiles, for example indoles, pyrroles, and enamines, in the presence of benzoylchloride to afford 2-substituted quinolines or 1-substituted isoquinolines, respectively [118]. The reaction proceeds via an addition to the CN double bond followed by cleavage of benzoic acid.

12.2.2.7 Radical Reactions

CN double bonds can participate in radical reactions. Radical reactions are covered in detail in Chapter 7, however, some typical examples are also discussed here.

In solid-phase chemistry the intermolecular and intramolecular carbon radical additions to CN double bonds have been established.

In the intermolecular reaction, radicals generated from alkyl iodides in the presence of Et$_3$B and Bu$_3$SnH have been added to resin-bound glyoxylic oxime ethers in dichloromethane at rt to yield α-hydroxylamino acid derivatives in moderate to good yields (Scheme 12.15) [56]. Another approach uses phenylsulfonyl oxime ethers that react with alkyl iodides and hexamethylditin in benzene under irradiation at 300 nm at rt to afford alkylated α-oxime ethers in moderate to good yields [57].

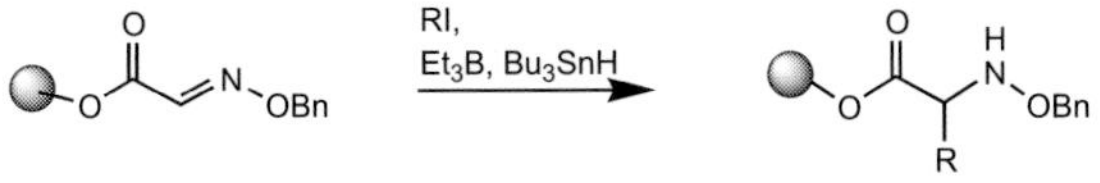

Scheme 12.15

Resin-bound allylamino acetaldoxime and propargylamino acetaldoxime were employed as substrates for an intramolecular radical reaction using Et$_3$B and Bu$_3$SnH in toluene at 80 °C. Functionalized pyrrolidines were formed via a carbon–carbon bond-forming process [58].

12.2.3
Addition of Nitrogen Nucleophiles

The addition of N-nucleophiles to CN double bonds is typically conducted as a three-component reaction between amines, carbonyl compounds, and N-nucleophiles. The active imine species is formed *in situ* and trapped by the N-nucleophile. Benzotriazoles add readily to imines. This approach was adapted to solid-phase chemistry as a linker strategy using polymer-bound benzotriazole and excess aldehyde and amine in THF/trimethyl orthoformate (TMOF) at 20–60 °C (Scheme 12.16) [59, 60]. The adducts can be released in solution with nucleophiles such as hydride ions, Grignard reagents, and organo zinc compounds.

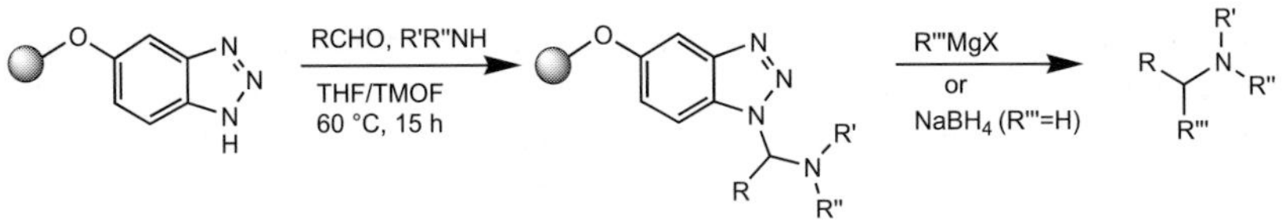

Scheme 12.16

In a second group of addition reactions used in heterocycle synthesis, the reactions proceed presumably via CN double bond intermediates. Phenylenediamines obtained from immobilized nitroanilines via reduction condense with aldehydes to form benzimidazoles. The reaction is performed at 50 °C [61] or with

2,3-dichloro-5,6-dicyanobenzoquinone (DDQ) at rt [62]. The reaction can also be conducted in a one-pot procedure treating nitroaniline with aldehyde and $SnCl_2 \cdot 2H_2O$ [63].

Polymer-bound 2-aminobenzamides (immobilization via the amide nitrogen) react with aldehydes in *N,N*-dimethylacetamide (DMA) in the presence of 5% acetic acid at 100 °C to form dihydroquinazolinones via an imine intermediate. The relatively harsh reaction conditions are required because of the low activity of the N-nucleophile [64].

12.2.4
Addition of Phosphorus Nucleophiles

Various P-nucleophiles react with imines or iminium ions in Mannich-type reactions. Protocols have been developed for solution-phase combinatorial as well as for solid-phase chemistry.

Condensation of a secondary phosphine with an aromatic or heteroaromatic aldehyde and an amine proceeds in THF at rt to yield aminomethylphosphines in good yields (Scheme 12.17). Owing to the air sensitivity of the phosphines the reaction has to be conducted under inert conditions. The protocol was utilized for the synthesis of a 96-member library [65].

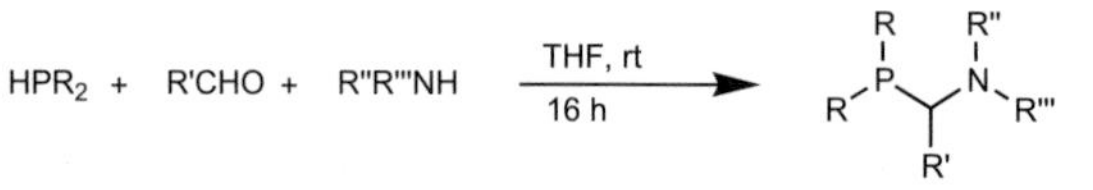

Scheme 12.17

The addition of dialkyl phosphite to imines can be effectively promoted with polymer-supported TBD (1,5,7-triazabicyclo[4.4.0]dec-1-ene) as base to afford α-aminophosphonates [66].

A solid-phase approach to α-amino phosphonates and phosphonic acids starts from polymer-bound H-phosphonates, which were obtained in a three-step procedure. Wang-PS resin was treated with 2-chloro-4*H*-1,3,2-benzodioxaphosphorin-4-one followed by hydrolysis and esterification. Addition of imines using either sonification or $Yb(OTf)_3$ catalysis followed by cleavage afforded the corresponding α-amino phosphonates in high yield and purity [67]. Cleavage of the *p*-nitrophenylethyl group using 1,8-diazabicyclo[5.4.0]undecene-7 (DBU) followed by TFA cleavage from the resin yielded α-amino phosphonic acids.

Aminophosphonites were synthesized by nucleophilic addition of bis-(trimethylsilyl)phosphonite (BTSP) to polymer-supported imine in dichloromethane/DMF. The silyl groups are cleaved upon treatment with methanol to produce amino-phosphinic acid (Scheme 12.18). *N*-Fluorenylmethoxy-carbonyl-9-amino-xanthen-3-yloxymethyl polystyrene was used as support and cleavage was performed using TFA/CH_2Cl_2/triisopropylsilane. Aromatic, heteroaromatic, aliphatic, and sterically hindered aldehydes can be used for imine formation [68].

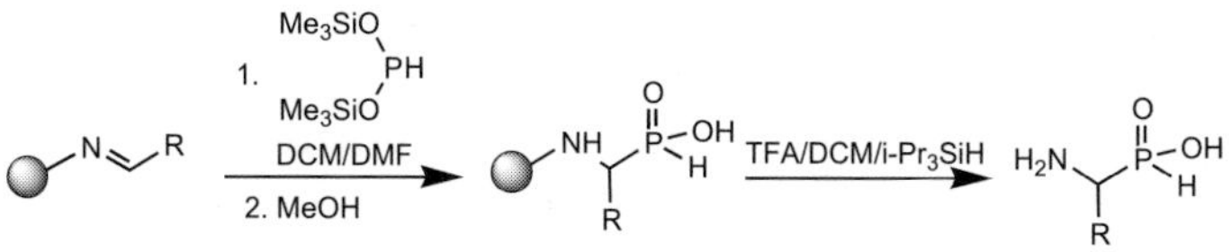

Scheme 12.18

12.2.5
Reactions with Oxygen Nucleophiles

There are only few addition reactions of O-nucleophiles to CN double bonds re-ported in solid-phase chemistry. 1,2-Aminoalcohols were cyclized with aldehydes to form oxazolidines via an imine intermediate. This reaction has been exploited for the immobilization of aldehydes. Polymer-bound serine or threonine was treated with an aldehyde in 1% *N,N*-diisopropylethylamine (DIPEA)/MeOH for 2 h at 60 °C. This oxazolidine linker is stable to the conditions of Fmoc peptide syn-thesis and is cleaved by heating with 5% AcOH/water for 30 min at 60 °C [69].

A similar approach was chosen for the synthesis of disubstituted 1,3-oxazolidines [70]. 1,2-Aminoalcohols linked to the polymeric support were con-densed with aldehydes in TMOF at rt to yield the corresponding imines that were converted to *N*-acyl-1,3-oxazolidines by treatment with acyl chlorides, isocyanates, or isothiocyanates (Scheme 12.19). Cleavage of the Wang-PS resin was performed using DDQ and afforded 1,3-substituted 4-(4'-formylphenoxymethyl)oxazolidines in good yields. In this example, the cleavage using 1% TFA led to the decomposi-tion of the heterocycles. Thus, an oxidative method was required for smooth isola-tion of the products.

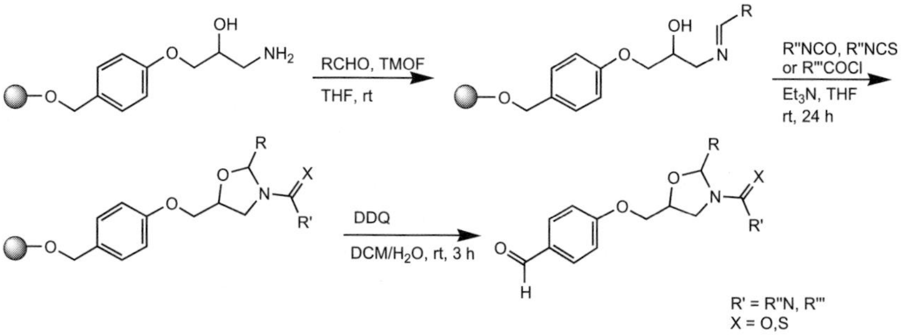

Scheme 12.19

12.2.6
Addition of Sulfur Nucleophiles

Thiols can readily add to imines, which is exploited in combinatorial chemistry for the formation of several heterocycles, namely thiazolidinones, metathiazanones, thiazolines, and benzothiazoles. Imines react with mercapto acids under addition of the thiol followed by acylation of the resulting *N,S*-acetal. α-Mercaptocarboxylic acids are used for the synthesis of thiazolidinones; β-mercaptocarboxylic acids for the synthesis of metathiazanones. The latter reaction is most conveniently performed as a three-component condensation of a primary amine, an aldehyde, and a mercapto carboxylic acid, but a stepwise approach is also possible.

Solid-phase procedures of this reaction [71, 72] start from immobilized amines that are treated with excess aldehyde and mercapto carboxylic acids in THF at 70 °C for 2 h (Scheme 12.20). Molecular sieves or TMOF are used to remove H_2O that formed during the condensation as well as during the cyclization step [72]. Whereas thiazolidinones are obtained in high purities, the formation of metathiazanones is more problematic. Besides immobilized amines, immobilized glyoxylic acid has also been used as a substrate [73].

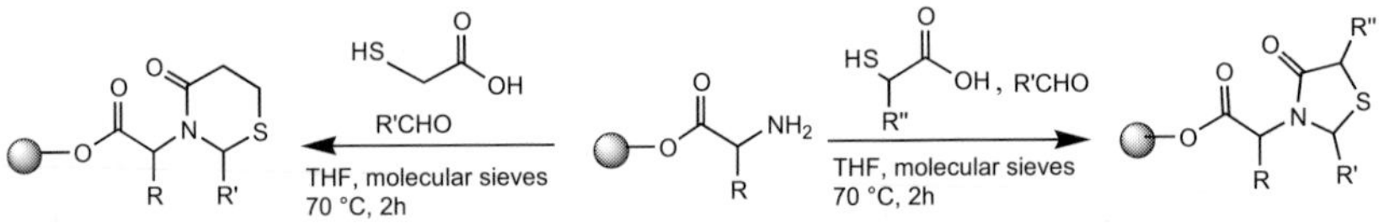

Scheme 12.20

The solid-phase synthesis of thiazolines and benzothiazoles begins from related compounds. 2-Substituted thiazole-4-carboxylic acids were obtained by reaction of aldehydes with unprotected cysteine attached to the polymeric support via an ester bond. The condensation and cyclization step is performed under acidic conditions (toluene/acetonitrile/AcOH; 45:45:10). The thiazolidines formed can be acylated and obtained in good yield after cleavage from the resin [74].

If immobilized 3-amino-4-mercaptobenzoic acid is used instead of cysteine, the reaction with aldehydes in refluxing ethanol for 4 h affords benzothiazoles in good to high purities and similar yields. The reaction proceeds presumably via an imine intermediate, which is subsequently attacked by the mercapto group followed by oxidation. This pathway is analogous to that of the benzimidazole formation from phenylenediamines [75].

12.3
Additions to CN Double Bonds in sp-Systems

In this section, transformations of carbodiimides, isocyanates, and isothiocyanates are discussed. These functional groups have a common feature in that the C atom is highly reactive against nucleophiles because of the two neighboring electron-withdrawing heteroatoms.

12.3.1
Additions to Carbodiimides

Besides their importance as coupling reagents in amide and ester formation, (Chapter 13.3.1) carbodiimides are important synthetic intermediates. They are easily attacked by primary and secondary amines to form guanidines. The carbodiimides are generated commonly by eliminating water or H_2S from ureas or thioureas using suitable dehydrating agents such as Mukaiyama's reagent or *p*-toluenesulfonyl chloride together with Et_3N. Alternatively, the aza-Wittig reaction of iminophosphoranes with isocyanates or isothiocyanates is employed. Often, the carbodiimides are not isolated.

In the solid-phase synthesis of guanidines, the carbodiimide is most often the immobilized component. There are several examples of carbodiimide formation from iminophosphoranes with either isocyanates [76] or with isothiocyanates [77, 78] via an aza-Wittig reaction. Another approach to carbodiimides is the elimination of H_2S from thioureas using Mukaiyama's reagent [79]. The addition of primary and secondary amines to carbodiimides is typically performed at rt. Higher temperatures are used if the formed guanidines are to react further as nucleophiles. This strategy has been exploited for the syntheses of several heterocycles employing a cyclization-release approach. Polymer-supported carbodiimides that are obtained by the reaction of immobilized α-amino acids with isothiocyanates followed by treatment with Mukaiyama's reagent are converted to guanidines upon treatment with amines. These guanidines cyclize to release 2-aminoimidazolinones [79]. A similar cyclization-cleavage approach has been used for the synthesis of 3*H*-quinazolin-4-ones using immobilized 2-azido-benzoic acid as substrate (Scheme 12.21) [80]. If polymer-supported 2′-amino-cinnamic acid is employed, guanidine formation is followed by a Michael addition to yield 3,4-dihydroquinazolines [76].

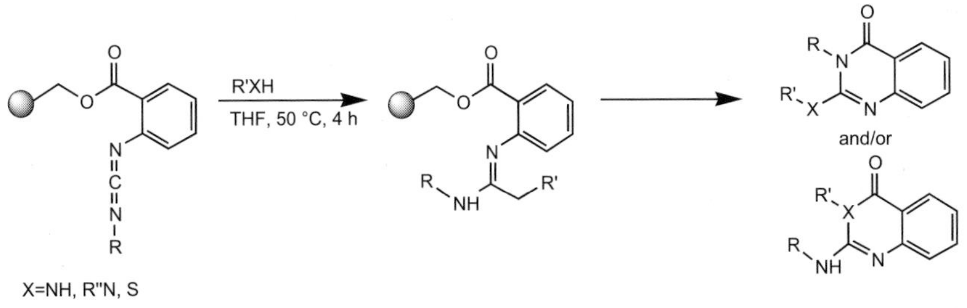

X=NH, R''N, S

Scheme 12.21

12.3.2
Reaction of Isocyanates and Isothiocyanates

The chemistry of these two groups is very similar. There is a vast number of articles describing the addition of nucleophiles to isocyanates and isothiocyanates.

12.3.2.1 Addition of Carbon Nucleophiles

CH acidic compounds react with isothiocyanates to form thioamides. This reaction has been adapted to solid-phase chemistry. Both components required for this synthesis can be linked to the solid support. Resin-bound cyanoacetamide was reacted with aliphatic and aromatic isothiocyanates in DMF at rt using DBU as base to yield thioamides [81]. Also resin-bound cyanoacetic acid has been used as a substrate, employing DIPEA as a base (Scheme 12.22) [82].

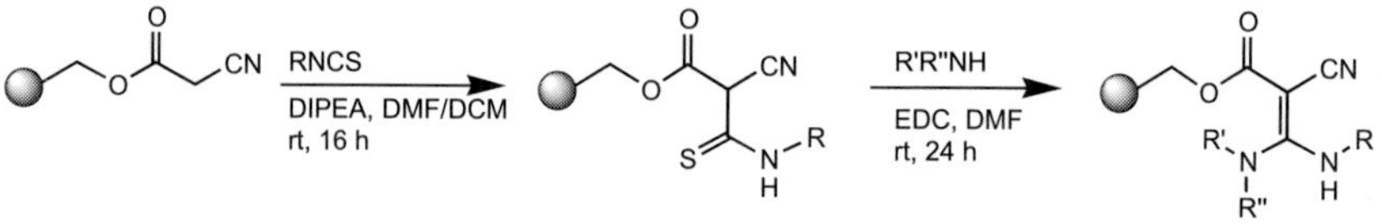

Scheme 12.22

Resin-bound isothiocyanates are alkylated with acceptor-substituted acetonitriles, such as malononitrile and methanesulfonylacetonitrile, under basic conditions to yield the respective thioamides [83].

12.3.2.2 Addition of Nitrogen Nucleophiles

Amines readily react with isocyanates or isothiocyanates to form ureas or thioureas. Both ureas and thioureas are of interest as final products and as precious synthetic intermediates which, among other reactions can be further converted to carbodiimides, guanidines, hydantoins, and quinazoline-2,4-diones. Typically, heterocycle formation proceeds via an intramolecular acylation of the urea. In accordance with their importance, these reactions were soon adapted to combinatorial solution- and solid-phase chemistry.

In solution-phase chemistry, each of the two components may be used in excess. Treatment of the reaction mixture with an appropriate scavenger resin yields pure products (Scheme 12.23). When there is an excess of the amine, a scavenger resin functionalized with isocyanate, aldehyde, or carboxy groups is used. An excess of isocyanate or isothiocyanate has best been removed with amino resins [4, 6]. In solid-phase chemistry either of the components has been linked to the resin; most often, resin-bound primary and secondary aliphatic amines or anilines are used. Typically an excess of the solution component is used in dichloromethane, THF, or DMF at rt. Polymer-bound isocyanates have to be generated on solid support. They are accessible by reaction of immobilized Fmoc-protected amino acids with methyltrichlorosilane [84] or by Curtius rearrangement of acyl azides [85]. The

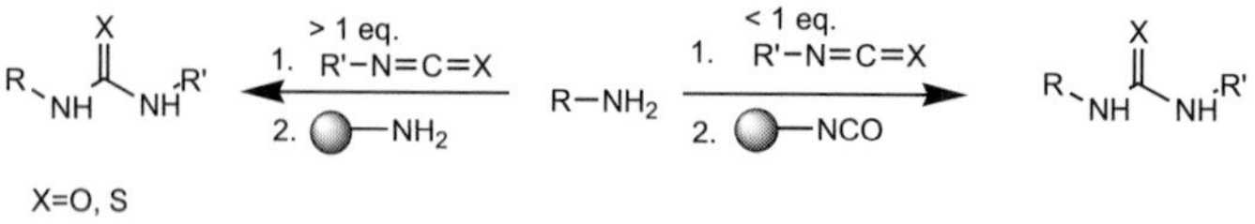

X=O, S

Scheme 12.23

reaction of resin-bound amines with *p*-nitrophenyl chloroformate under basic conditions yields isocyanates via the active carbamate.

Ureas are often formed as intermediates in reaction sequences leading to heterocycles. Several syntheses of this kind have been established on solid phase, namely the syntheses of hydantoins, thiohydantoins, and quinazoline-2,4-diones. Often, a cyclizing cleavage strategy is pursued which guarantees high purities. For example, quinazoline-2,4-diones are obtained in >90% purities if the route depicted in Scheme 12.24 is followed [85].

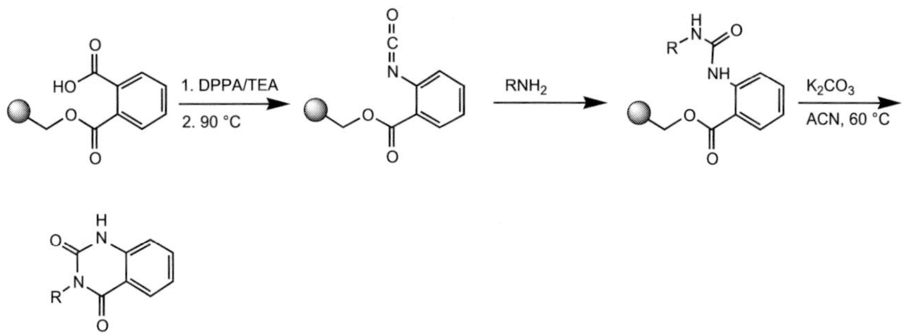

Scheme 12.24   DPPA = Diphenylphosphorylazide; TEA = Triethyl amine.

12.3.2.3 Addition of Oxygen Nucleophiles

Isocyanates form carbamates when treated with alcohols. Alcohols linked to a solid support or to soluble polymers have been reacted with isocyanates at rt using Et$_3$N as a base (Scheme 12.25) [86, 87]. Highly activated isocyanates have been reacted with hydroxy resins without base at 0 °C [88, 89]. The carbamates obtained using this approach are cleaved either under acidic conditions to yield the amines via the carbamic acid or by nucleophiles. Carbamates have also been accessed from immobilized secondary alcohols using isocyanates in DMF in the presence of catalytic amounts of CuCl at rt. This procedure was embedded in a multistep library synthesis, in which the products were obtained in good to excellent purity [90].

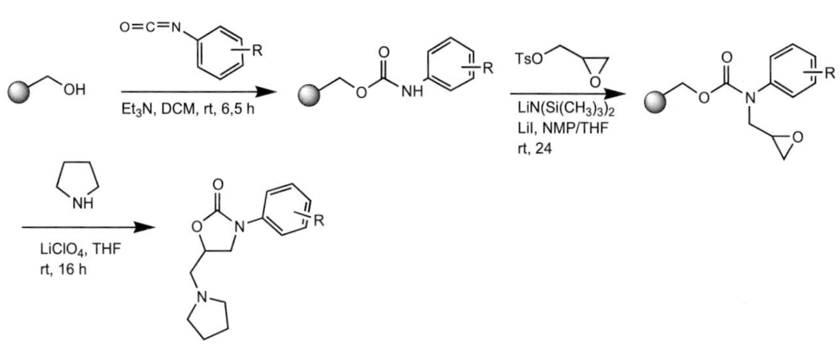

Scheme 12.25

12.3.3
Addition to CS Double Bonds in sp² Systems

Thioamides and thioureas are the most frequently used synthons bearing CS double bonds. Owing to its high nucleophilicity, the sulfur is readily alkylated. The alkylated species is often used as an intermediate to facilitate subsequent substitution with amines to afford amidines and guanidines. Furthermore, the alkylated species can be exploited for the synthesis of heterocycles such as thiazoles from thioamides and α-halo ketones. These reactions are well established in solution and have also been employed successfully on solid phase.

The alkylation of benzimidazol-2-thiones with various alkyl halides was performed using MeO-PEG-OH as soluble polymeric support. The reaction proceeds in dichloromethane using Et₃N as a base and yields the 2-alkylthiobenzimidazoles in good to high purity [91].

S-Alkylation has been conducted on solid phase using, for example, immobilized benzimidazolthiones and quinazoline-2-thioxo-4-ones as substrates. The reaction is typically performed in DMF or N-methyl-2-pyrrolidone using a tertiary base and various alkyl or aryl halides as alkylating agents [92, 93].

As already mentioned, S-alkylations can be a key step in heterocycle syntheses. Thiazoles are formed by the Hantzsch synthesis by reaction of resin-bound thioamides with α-halo ketones [94]. 2,4-Diaminothiophenes are obtained by alkylation of immobilized α-cyano thioamides with α-halo ketones followed by cylization (see Sect. 12.4.1).

Formation of guanidines from thioureas via S-alkylation and substitution with amines has been performed using methyl iodide, Mukaiyama's reagent, or carbodiimides as alkylating agents. Merrifield resin can also act as an alkylation reagent and yields immobilized S-alkyl isothioureas. These are typically cleaved as guanidines when treated with amines [95, 96]. For carbodiimides, the alkylated species is not isolated but is directly substituted with amines [97, 98], whereas the alkylation products of methyl iodide [99, 100] or Mukaiyama's reagent are isolated (Scheme 12.26) [101].

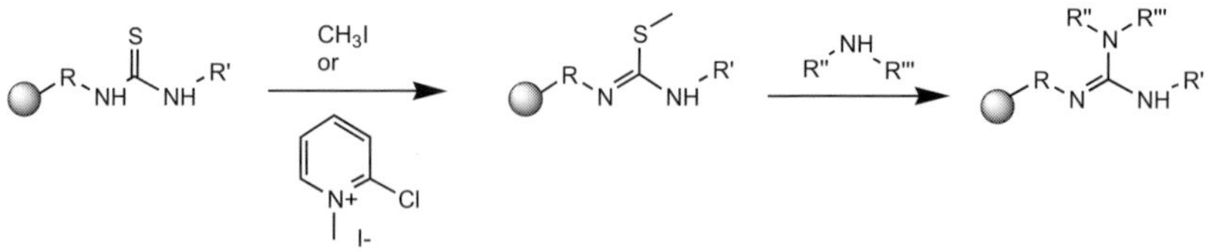

Scheme 12.26

Treatment of thioureas with Mukaiyama's reagent, Et₃N in dichloromethane, or MeCN afforded carbodiimides [79, 102]. Thioamides react in a similar way to thioureas with N'-(3-dimethylaminopropyl)-N-ethylene carbodiimide (EDC) and amines to yield amidine derivatives (Scheme 12.22) [82].

12.3.4
Reaction of CS Double Bonds in sp Systems

CS double bonds occur as part of sp systems in isothiocyanates and in carbon disulfide. Additions of isothiocyanates have been discussed previously together with additions of isocyanates.

Carbon disulfide reacts with amines and Merrifield resin in a three-component reaction to give resin-bound dithiocarbamates. In the first step, the amine adds to carbon disulfide to form dithiocarbamates which are subsequently trapped by the Merrifield resin (Scheme 12.27). The reaction proceeds at rt using DIPEA as the base. The dithiocarbamates obtained with primary amines and anilines can be cleaved with primary and secondary amines in toluene at 60 °C to yield thioureas in good to excellent yield [103].

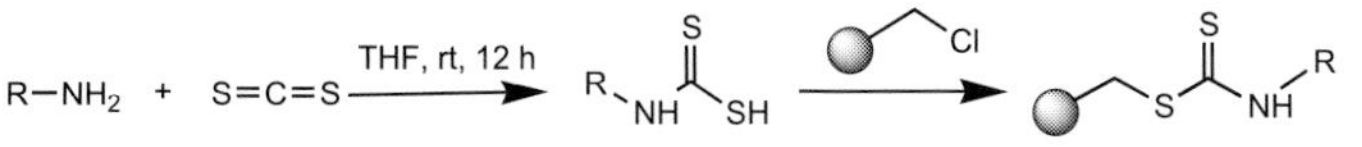

Scheme 12.27

A similar approach to dithiocarbamates uses resin-bound amines. Carbon disulfide was added to the amine in 1,4-dioxane using aqueous KOH as a base. This dithiocarbamate was converted upon treatment with formaldehyde and amino acids to yield tetrahydro-2H-1,3,5-thiadiazin-2-thiones in good to excellent purity and yield [104].

12.4
Additions to CN Triple Bonds (Cyanides, not Isocyanides)

Cyanides and isocyanides participate in numerous inter- and intramolecular addition reactions. Most reactions in which cyanides are transformed lead to amino-substituted heterocycles. Isonitriles are typically converted in multicomponent reactions that are discussed in Chapter 23 and will not be discussed here.

12.4.1
Addition of Carbon Nucleophiles

Nitriles can be attacked by CH acidic compounds. This feature was exploited in the synthesis of aminothienopyridine libraries in a solution-phase approach [105, 106]. Treatment of 3-cyanopyridine-2-thiones with acceptor-substituted bromomethylene compounds and aqueous KOH in DMF at rt resulted in S-alkylation. Cyclization was carried out again with aqueous base to yield aminothienopyridines (Scheme 12.28) [106].

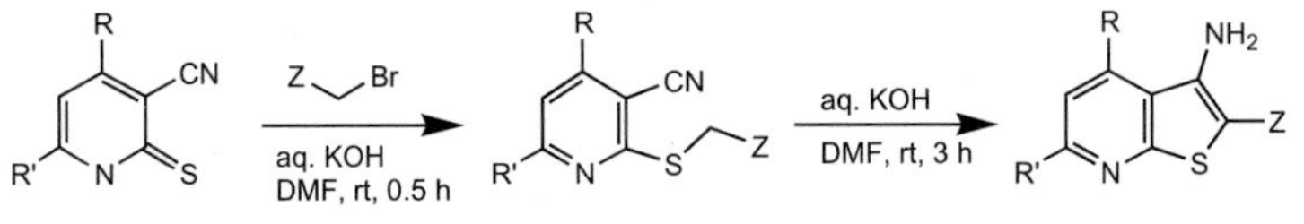

Scheme 12.28

A similar approach in solid-phase chemistry uses α-cyanothioamides as starting material for the formation of aminothiophenes. The α-cyanothioamides are readily obtainable from isothiocyanates and acceptor-substituted acetonitriles. Both of the synthons can be linked to the support. Treatment of the thioamides with α-bromoketones results in S-alkylation. Under basic conditions (DBU in DMF), this intermediate cyclizes with the nitrile and forms 3-aminothiophenes in good purity (Scheme 12.29) [81, 83].

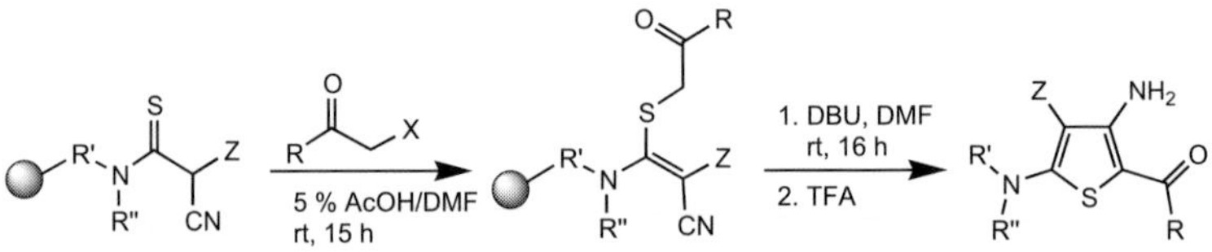

Scheme 12.29

Cyanopyridines are formed upon treatment of resin-bound chalcones with acetonitrile and *t*-BuOK at rt using ultrasonic irradiation. The first step is a Thorpe reaction to yield 3-aminocrotononitriles which react in the subsequent step with the chalcone [107].

12.4.2
Addition of Nitrogen Nucleophiles

Several additions of N-nucleophiles to nitriles were established on solid phase. One of the earliest examples of this concept was the reaction of polymer-bound phthalonitrile with ammonia and NaOMe to yield 1,3-diiminoisoindolines that were further converted to phthalocyanines [108]. The reaction of β-oxo-nitriles with hydrazines afforded 5-aminopyrazoles regioselectively (Scheme 12.30). Both polymer-bound β-keto nitriles [109] and α-formyl nitriles [110] were used as substrates. With the former, a preformed combination between a linker and the β-keto nitrile

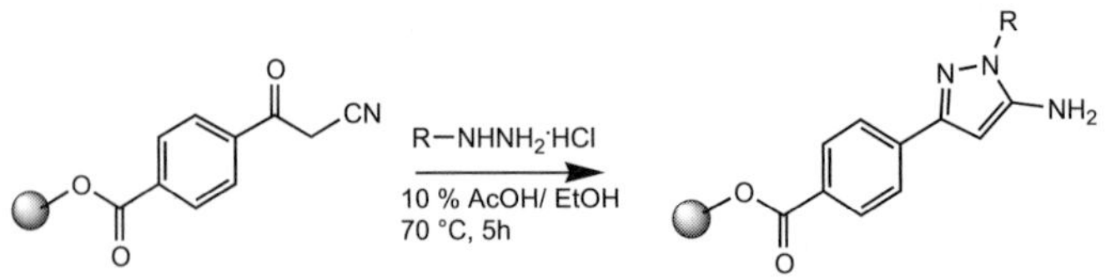

Scheme 12.30

was attached to the support. α-Formyl nitriles are accessible by reaction of immobilized benzylnitrile with Bredereck's reagent [bis-(dimethylamino)-*t*-butyloxymethane]. Cyclization occurs upon treatment with hydrazines and heating at 70 °C in 10% AcOH/EtOH for 5 h. The 5-aminopyrazoles were obtained in high purity.

In a similar approach, α, β-unsaturated nitriles were cyclized with hydrazines using NaOEt as a base to yield 3-amino-2-pyrazolines in high purity. Heating at 70 °C in ethanol for 24 h was necessary for complete conversion [111].

Kaiser's oxime resin was used for the introduction of a hydroxylamine group to cyanofluorobenzenes. Cleavage of the hydroxylamine and cyclization was performed in a one-step procedure using TFA/5 N HCl to yield 3-aminobenzisoxazoles in high purity (Scheme 12.31) [112, 113]. An example with the *in situ* generation of the hydroxylamine nucleophile is the reduction of a polymer-bound 2-nitro-benzylnitrile with tin(II) chloride dihydrate. Hydroxylamine, an intermediate of the reduction, is trapped by cyclization with the nitrile to afford 2-amino-1-hydroxyindoles [114].

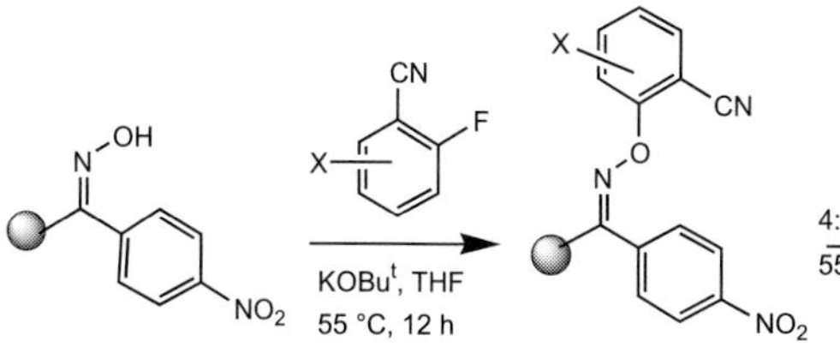

Scheme 12.31

Guanidines can also act as the attacking N-nucleophile. This feature has been exploited for the solid-phase synthesis of 2,4-diaminoquinazolines. The guanidines were synthesized using acylisothiocyanate resin and 2-aminobenzonitriles as building blocks. Cleavage and cyclization occurs during heating with TFA/H$_2$O (95:5) at 80 °C for 16 h. The cleavage step is repeated once. Following this procedure, diaminoquinazolines were obtained in high purity and good yield [115].

There are only a few examples of nucleophilic conversions of nitriles that occur without cyclization. One example is the aminolysis of polymer-bound nitriles with hydroxylamine hydrochloride and DIPEA in 2-methoxyethanol at 85 °C for 16 h to provide amide oximes with a quantitative yield (Scheme 12.32). The products obtained were used in the synthesis of oxadiazoles [116]. Another example is the reaction of α-(benzotriazol-1-yl)acetonitril with amines in 2-methoxyethanol at 75–80 °C. The formed amidines were not isolated but were directly treated with resin-bound chalcones to yield aminopyridines that had high purity after cleavage.

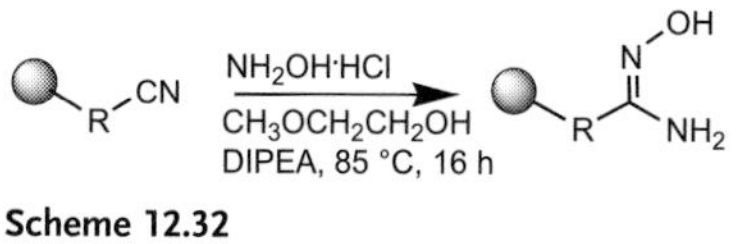

Scheme 12.32

12.4.3
Addition of Sulfur Nucleophiles

Polymer-bound nitriles can be converted to thioamides using dithiophosphoric acid *O,O*-diethylester in THF/H$_2$O at 70 °C (Scheme 12.33) [117]. The thioamides have been used for the synthesis of thiazoles (see Sect. 12.3.3) [94].

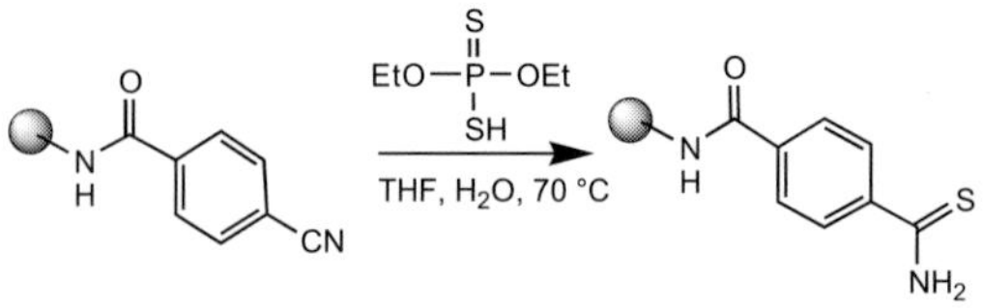

Scheme 12.33

References

1 M. M. Sim, A. Ganesan, *J. Org. Chem.* **1997**, *62*, 3230–3235.

2 A. K. Szardenings, T. S. Burkoth, G. C. Look, D. A. Campbell, *J. Org. Chem.* **1996**, *61*, 6720–6722.

3 B. A. Kulkarni, A. Ganesan, *J. Chem. Soc., Chem. Commun.* **1998**, 785–786.

4 M. M. Sim, C. L. Lee, A. Ganesan, *J. Org. Chem.* **1997**, *62*, 9358–9360.

5 S. V. Ley, M. H. Bolli, B. Hinzen, A.-G. Gervois, B. J. Hall, *J. Chem. Soc., Perkin Trans. 1* **1998**, 2239–2241.

6 J. Habermann, S. V. Ley, J. S. Scott, *J. Chem. Soc., Perkin Trans. 1* **1998**, 3127–3130.

7 K. Hemming, M. J. Bevan, C. Loukou, S. D. Patel, D. Renaudeau, *Synlett* **2000**, 1565–1568.

8 S. W. Kaldor, M. G. Siegel, J. E. Fritz, B. A. Dressman, P. J. Hahn, *Tetrahedron Lett.* **1996**, *37*, 7193–7196.

9 S. W. Kim, S. Y. Ahn, J. S. Koh, J. H. Lee, S. Ro, H. Y. Cho, *Tetrahedron Lett.* **1997**, *38*, 4603–4606.

10 Y. Aoki, S. Kobayashi, *J. Comb. Chem.* **1999**, *1*, 371–372.

11 J. Green, *J. Org. Chem.* **1995**, *60*, 4287–4290.

12 V. Krchnak, A. S. Weichsel, D. Cabel, Z. Flegelova, M. Lebl, *Mol. Diversity* **1995**, *9*, 149.

13 C. G. Boojamra, K. M. Burow, J. A. Ellman, *J. Org. Chem.* **1995**, *60*, 5742–5743.

14 H. Maehr, R. Yang, *Tetrahedron Lett.* **1996**, *37*, 5445–5448.

15 A. R. Katritzky, L. H. Xie, G. F. Zhang, M. Griffith, K. Watson, J. S. Kiely, *Tetrahedron Lett.* **1997**, *38*, 7011–7014.

16 N. M. Khan, V. Arumugam, S. Balasubramanian, *Tetrahedron Lett.* **1996**, *37*, 4819–4822.

17 M. Gustafsson, R. Olsson, C. M. Andersson, *Tetrahedron Lett.* **2001**, *42*, 133–136.

18 M. W. Creswell, G. L. Bolton, J. C. Hodges, M. Meppen, *Tetrahedron* **1998**, *54*, 3983–3998.

19 H. Waldmann, M. Braun, *J. Org. Chem.* **1992**, *57*, 4444.

20 S. Kobayashi, S. Nagayama, T. Busujima, *Tetrahedron Lett.* **1996**, *37*, 9221–9224.

21 M. Shimizu, S. Itohara, *Synlett* **2000**, 1828–1830.

22 A. Fujii, M. Sodeoka, *Tetrahedron Lett.* **1999**, *40*, 8011–8014.

23 S. Kobayashi, I. Hachiya, S. Suzuki,

M. Moriwaki, *Tetrahedron Lett.* **1996**, *37*, 2809–2812.

24 S. Kobayashi, M. Moriwaki, *Tetrahedron Lett.* **1997**, *38*, 4251–4254.

25 S. Kobayashi, M. Moriwaki, R. Akiyama, S. Suzuki, I. Hachiya, *Tetrahedron Lett.* **1996**, *37*, 7783–7786.

26 S. Kobayashi, Y. Aoki, *Tetrahedron Lett.* **1998**, *39*, 7345–7348.

27 S. Kobayashi, R. Akiyama, H. Kitagawa, *J. Comb. Chem.* **2000**, *2*, 438–440.

28 S. Kobayashi, T. Furuta, K. Sugita, O. Okitsu, H. Oyamada, *Tetrahedron Lett.* **1999**, *40*, 1341–1344.

29 Y. H. Wang, S. R. Wilson, *Tetrahedron Lett.* **1997**, *38*, 4021–4024.

30 C. J. Creighton, C. W. Zapf, J. H. Bu, M. Goodman, *Org. Lett.* **1999**, *1*, 1407–1409.

31 D. Jonsson, H. Molin, A. Unden, *Tetrahedron Lett.* **1998**, *39*, 1059–1062.

32 N. A. Petasis, Z. D. Patel, *Tetrahedron Lett.* **2000**, *41*, 9607–9611.

33 S. R. Klopfenstein, J. J. Chen, A. Golebiowski, M. Li, S. X. Peng, X. Shao, *Tetrahedron Lett.* **2000**, *41*, 4835–4839.

34 A. Golebiowski, S. R. Klopfenstein, J. J. Chen, X. Shao, *Tetrahedron Lett.* **2000**, *41*, 4841–4844.

35 N. Schlienger, M. R. Bryce, T. K. Hansen, *Tetrahedron* **2000**, *56*, 10023–10030.

36 W. J. N. Meester, F. P. J. T. Rutjes, P. H. H. Hermkens, H. Hiemstra, *Tetrahedron Lett.* **1999**, *40*, 1601–1604.

37 J. J. N. Veerman, F. P. J. T. Rutjes, J. H. van Maarseveen, H. Hiemstra, *Tetrahedron Lett.* **1999**, *40*, 6079–6082.

38 R. C. D. Brown, M. Fisher, *J. Chem. Soc. Chem. Commun.* **1999**, 1547–1548.

39 B. Chenera, J. A. Finkelstein, D. F. Veber, *J. Am. Chem. Soc.* **1995**, *117*, 11999–12000.

40 M. Schuster, J. Pernerstorfer, S. Blechert, *Angew. Chem. Int. Ed.* **1996**, *35*, 1979–1980.

41 S. Itsuno, A. A. El-Shehawy, M. Y. Abdelaal, K. Ito, *New J. Chem.* **1998**, 775–777.

42 C. X. Chen, B. Munoz, *Tetrahedron Lett.* **1998**, *39*, 3401–3404.

43 C. X. Chen, I. A. McDonald, B. Munoz, *Tetrahedron Lett.* **1998**, *39*, 217–220.

44 C. X. Chen, B. Munoz, *Tetrahedron Lett.* **1998**, *39*, 6781–6784.

45 C. X. Chen, B. Munoz, *Tetrahedron Lett.* **1999**, *40*, 3491–3494.

46 S. Wendeborn, *Synlett* **2000**, 45–48.

47 B. A. Lorsbach, R. B. Miller, M. J. Kurth, *J. Org. Chem.* **1996**, *61*, 8716–8717.

48 B. A. Lorsbach, J. T. Bagdanoff, R. B. Miller, M. J. Kurth, *J. Org. Chem.* **1998**, *63*, 2244–2250.

49 M. A. Youngman, S. L. Dax, *Tetrahedron Lett.* **1997**, *38*, 6347–6350.

50 A. B. Dyatkin, R. A. Rivero, *Tetrahedron Lett.* **1998**, *39*, 3647–3650.

51 J. J. McNally, M. A. Youngman, S. L. Dax, *Tetrahedron Lett.* **1998**, *39*, 967–970.

52 L. H. Yang, L. Q. Guo, *Tetrahedron Lett.* **1996**, *37*, 5041–5044.

53 H. A. Dondas, R. Grigg, W. S. MacLachlan, D. T. MacPherson, J. Markandu, V. Sridharan, S. Suganthan, *Tetrahedron Lett.* **2000**, *41*, 967–970.

54 H. S. Wang, A. Ganesan, *Org. Lett.* **1999**, *1*, 1647–1649.

55 H. C. Zhang, K. K. Brumfield, L. Jaroskova, B. E. Maryanoff, *Tetrahedron Lett.* **1998**, *39*, 4449–4452.

56 H. Miyabe, Y. Fujishima, T. Naito, *J. Org. Chem.* **1999**, *64*, 2174–2175.

57 G. H. Jeon, J. Y. Yoon, S. Kim, S. S. Kim, *Synlett* **2000**, 128–130.

58 H. Miyabe, H. Tanaka, T. Naito, *Tetrahedron Lett.* **1999**, *40*, 8387–8390.

59 A. Paio, A. Zaramella, R. Ferritto, N. Conti, C. Marchioro, P. Seneci, *J. Comb. Chem.* **1999**, *1*, 317–325.

60 A. R. Katritzky, S. A. Belyakov, D. O. Tymoshenko, *J. Comb. Chem.* **1999**, *1*, 173–176.

61 D. Tumelty, M. K. Schwarz, K. Cao, M. C. Needels, *Tetrahedron Lett.* **1999**, *40*, 6185–6188.

62 J. P. Mayer, G. S. Lewis, C. McGee, D. Bankaitis-Davis, *Tetrahedron Lett.* **1998**, *39*, 6655–6658.

63 Z. M. Wu, P. Rea, G. Wickham, *Tetrahedron Lett.* **2000**, *41*, 9871–9874.

64 J. P. Mayer, G. S. Lewis, M. J. Curtis, J. W. Zhang, *Tetrahedron Lett.* **1997**, *38*, 8445–8448.

65 A. M. LaPointe, *J. Comb. Chem.* **1999**, *1*, 101–104.

66 D. Simoni, R. Rondanin, M. Morini, R. Baruchello, F. P. Invidiata, *Tetrahedron Lett.* **2000**, *41*, 1607–1610.

67 C. Z. Zhang, A. M. M. Mjalli, *Tetrahedron Lett.* **1996**, *37*, 5457–5460.

68 E. A. Boyd, W. C. Chan, V. M. Loh, *Tetrahedron Lett.* **1996**, *37*, 1647–1650.

69 N. J. Ede, A. M. Bray, *Tetrahedron Lett.* **1997**, *38*, 7119–7122.

70 H. S. Oh, H. G. Hahn, S. H. Cheon, D. C. Ha, *Tetrahedron Lett.* **2000**, *41*, 5069–5072.

71 M. C. Munson, A. W. Cook, J. A. Josey, C. Rao, *Tetrahedron Lett.* **1998**, *39*, 7223–7226.

72 C. P. Holmes, J. P. Chinn, G. C. Look, E. M. Gordon, M. A. Gallop, *J. Org. Chem.* **1995**, *60*, 7328–7333.

73 N. Schlienger, M. R. Bryce, T. K. Hansen, *Tetrahedron Lett.* **2000**, *41*, 5147–5150.

74 M. Patek, B. Drake, M. Lebl, *Tetrahedron Lett.* **1996**, *36*, 2227–2230.

75 C. L. Lee, Y. L. Lam, S. Y. Lee, *Tetrahedron Lett.* **2001**, *42*, 109–111.

76 F. J. Wang, J. R. Hauske, *Tetrahedron Lett.* **1997**, *38*, 8651–8654.

77 D. H. Drewry, S. W. Gerritz, J. A. Linn, *Tetrahedron Lett.* **1997**, *38*, 3377–3380.

78 C. P. Lopez, P. Molina, E. Aller, A. Lorenzo, *Synlett* **2000**, 1411–1414.

79 D. H. Drewry, C. Ghiron, *Tetrahedron Lett.* **2000**, *41*, 6989–6992.

80 J. M. Villalgordo, D. Obrecht, A. Chucholowsky, *Synlett* **1998**, 1405–1407.

81 F. Zaragoza, *Tetrahedron Lett.* **1996**, *37*, 6213–6216.

82 F. Zaragoza, *Tetrahedron Lett.* **1997**, *38*, 7291–7294.

83 H. Stephensen, F. Zaragoza, *J. Org. Chem.* **1997**, *62*, 6096–6097.

84 P. Y. Chong, P. A. Petillo, *Tetrahedron Lett.* **1999**, *40*, 4501–4504.

85 H. Shao, M. Colucci, S. J. Tong, H. S. Zhang, A. L. Castelhano, *Tetrahedron Lett.* **1998**, *39*, 7235–7238.

86 H. P. Buchstaller, *Tetrahedron* **1998**, *54*, 3465–3470.

87 J. Y. Yoon, C. W. Cho, H. Han, K. D. Janda, *J. Chem. Soc. Chem. Commun.* **1998**, 2703–2704.

88 C. Subramanyam, *Tetrahedron Lett.* **2000**, *41*, 6537–6540.

89 L. J. Fitzpatrick, R. A. Rivero, *Tetrahedron Lett.* **1997**, *38*, 7479–7482.

90 M. J. Sofia, R. Hunter, T. Y. Chan, A. Vaughan, R. Dulina, H. M. Wang, D. Gange, *J. Org. Chem.* **1998**, *63*, 2802–2803.

91 C. M. Yeh, C. M. Sun, *Tetrahedron Lett.* **1999**, *40*, 7247–7250.

92 J. Lee, D. Gauthier, R. A. Rivero, *Tetrahedron Lett.* **1998**, *39*, 201–204.

93 S. Makino, N. Suzuki, E. Nakanishi, T. Tsuji, *Tetrahedron Lett.* **2000**, *41*, 8333–8337.

94 D. Goff, J. Fernandez, *Tetrahedron Lett.* **1999**, *40*, 423–426.

95 D. S. Dodd, O. B. Wallace, *Tetrahedron Lett.* **1998**, *39*, 5701–5704.

96 R. Y. Yang, A. Kaplan, *Tetrahedron Lett.* **2000**, *41*, 7005–7008.

97 L. J. Wilson, S. R. Klopfenstein, M. Li, *Tetrahedron Lett.* **1999**, *40*, 3999–4002.

98 J. A. Josey, C. A. Tarlton, C. E. Payne, *Tetrahedron Lett.* **1998**, *39*, 5899–5902.

99 P. C. Kearney, M. Fernandez, J. A. Flygare, *Tetrahedron Lett.* **1998**, *39*, 2663–2666.

100 A. Gopalsamy, H. Yang, *J. Comb. Chem.* **2000**, *2*, 378–381.

101 S. E. Schneider, P. A. Bishop, M. A. Salazar, O. A. Bishop, E. V. Anslyn, *Tetrahedron* **1998**, *54*, 15063–15086.

102 J. Chen, M. Pattarawarapan, A. J. Zhang, K. Burgess, *J. Comb. Chem.* **2000**, *2*, 276–281.

103 L. Gomez, F. Gellibert, A. Wagner, C. Mioskowski, *J. Comb. Chem.* **2000**, *2*, 75–79.

104 R. Perez, O. Reyes, M. Suarez, H. E. Garay, L. J. Cruz, H. Rodriguez, M. D. Molero-Vilchez, C. Ochoa, *Tetrahedron Lett.* **2000**, *41*, 613–616.

105 S. J. Shuttleworth, M. Quimpere, N. Lee, J. DeLuca, *Mol. Diversity* **1998**, *4*, 183–185.

106 A. M. Shestopalov, V. P. Kislyi, E. Y. Kruglova, K. G. Nikishin, V. V. Semenov, A. C. Buchanan, A. A. Gakh, *J. Comb. Chem.* **2000**, *2*, 24–28.

107 A. L. Marzinzik, E. R. Felder, *J. Org. Chem.* **1998**, *63*, 723–727.

108 C. C. Leznoff, P. I. Svirskaya, B. Khouw, R. L. Cerny, P. Seymour, A. B. P. Lever, *J. Org. Chem.* **1991**, *56*, 82–90.

109 S. P. Watson, R. D. Wilson, D. B. Judd, S. A. Richards, *Tetrahedron Lett.* **1997**, *38*, 9065–9068.

110 R. D. Wilson, S. P. Watson, S. A. Richards, *Tetrahedron Lett.* **1998**, *39*, 2827–2830.

111 L. O. Lyngso, J. Nielsen, *Tetrahedron Lett.* **1998**, *39*, 5845–5848.

112 S. D. Lepore, M. R. Wiley, *J. Org. Chem.* **1999**, *64*, 4547–4550.

113 S. D. Lepore, M. R. Wiley, *J. Org. Chem.* **2000**, *65*, 2924–2932.

114 H. Stephensen, F. Zaragoza, *Tetrahedron Lett.* **1999**, *40*, 5799–5802.

115 L. J. Wilson, *Org. Lett.* **2001**, *3*, 585–588.

116 N. Hebert, A. L. Hannah, S. C. Sutton, *Tetrahedron Lett.* **1999**, *40*, 8547–8550.

117 R. J. Booth, J. C. Hodges, *J. Am. Chem. Soc.* **1998**, *119*, 4882–4883.

118 M. Z. Hoemann, A. Mllikian-Badalian, G. Kumaravel, J. R. Hauske, *Tetrahedron Lett.* **1998**, 4749–4752.

13
Chemistry of the Carbonyl Group

Tobias Wunberg

13.1
Introduction

The carbonyl group is one of the most important functional groups in organic chemistry. It is found in various structural classes, each one having its own characteristic and fascinating chemistry. The wealth of chemical transformations involving carbonyl groups covers virtually all forms of organic reactions, thus making the carbonyl group an unreplacable tool in synthetic organic chemistry. Additionally, this structural unit is extremely important in medicinal chemistry. Seventy per cent of known drugs (CMC database vers. 94.1) contain carbonyl groups in various manifestations [1]. This chapter deals with the application of carbonyl group chemistry to the generation of libraries. It will not include C–C single bond-forming reactions, enolate chemistry, or reductions and oxidations involving carbonyl groups. These reactions are dealt with in other chapters.

13.2
Chemistry of the Carbonyl Group and Combinatorial Chemistry

Since this chapter deals with mechanistically different chemical transformations, a general statement about the application of carbonyl group chemistry for library synthesis cannot be made. Some of the described reactions such as the formation of amides or reductive aminations play a key role in combinatorial chemistry. Others are frequently used, for example Wittig-type olefinations, and some have only rarely been used, e.g. Curtius degradation. Basically, they are all suited for both solution-phase and solid-phase chemistry even though the choice between these two formats has consequences for the reaction conditions. These aspects will be discussed in more detail for each individual reaction.

For some reactions there is a standard procedure which is generally suited to a first test. But this does not guarantee a satisfactory result and does not suggest that other methods may not give better yields and purities. Each reaction may have to be optimized by parameters such as base, solvent, time, temperature, and reagents in order to determine reaction conditions that are applicable to as broad a range of reactions as possible.

13.3
Chemistry of Carboxylic Acids

13.3.1
C(O)–X Bond-forming Reactions: General Remarks

From a mechanistic point of view, the transformation of carboxylic acids into amides or esters might not be the most spectacular reaction, however nicely they demonstrate the fundamental principles and mechanisms of carbonyl group chemistry. For his pioneering work in solid-phase chemistry, Bruce Merrifield chose previously established procedures for the formation of amides (more precisely, peptides) [2]. Since then, the formation of amides has become one of the best-elaborated reactions in solid-phase chemistry [3]. Since peptides usually have pharmacokinetically unfavorable properties, the focus of combinatorial chemistry in medicinal chemistry has recently shifted toward the synthesis of nonpeptidic, more drug-like small molecules (see Chapter 22). Nevertheless, C(O)–N and C(O)–O bond-forming reactions still play a key role in combinatorial chemistry both in solution phase and on solid phase. The reliability and large number of commercially available building blocks explain why this reaction type has maintained its importance.

13.3.1.1   Amides and Ureas

Formation of amides
There is a mechanistic requirement for the transformation of carboxylic acids into amides: the introduction of a suitable leaving group instead of the OH group prior to the reaction with a nucleophile (amine). Two principal ways of activation are feasible: replacement of the OH group with better leaving groups (e.g. acid halides) or transformation of the OH moiety into a suitable leaving group (e.g. active esters, anhydrides). While planning the synthesis of a library one should keep in mind that the choice between solution and solid phase may cause limitations which have their origin in the combination of the specific demands of resin-based chemistry, the efficacy of coupling reagents, and the formation of acid chlorides.

A plethora of methods and conditions has been described for the formation of acid chlorides in solution phase and most of them are applicable to parallel synthesis and library production. Consequently, the reaction of acid chlorides with aliphatic amines is a standard reaction and has been used in numerous preparations of libraries in solution. In our hands, CH_2Cl_2 or tetrahydrofuran (THF) as solvent and NEt_3 or DIEA as base are the standard methods. Even aromatic amines can be smoothly converted into amides under these conditions. Furthermore, this protocol is also suitable for the formation of amides on solid phase with immobilized aliphatic and aromatic amines. For resin-bound carboxylic acids, however, there is an intrinsic problem with the formation of the corresponding acid chloride. Acidic reagents such as $SOCl_2$ or $POCl_3$ are incompatible with acid-sensitive linkers as well as with the resin itself. A few approaches promise a solution to this problem, e.g. Ghosez's reagent [4], Appel's PPh_3/CCl_4 combination [5], oxalyl chloride [6], or alternatively formation of acid fluorides (Scheme 13.1) using cyanuric fluoride [7], DAST [8], or TFFH [9]. However, none of these methods has so far been described for the synthesis of larger libraries.

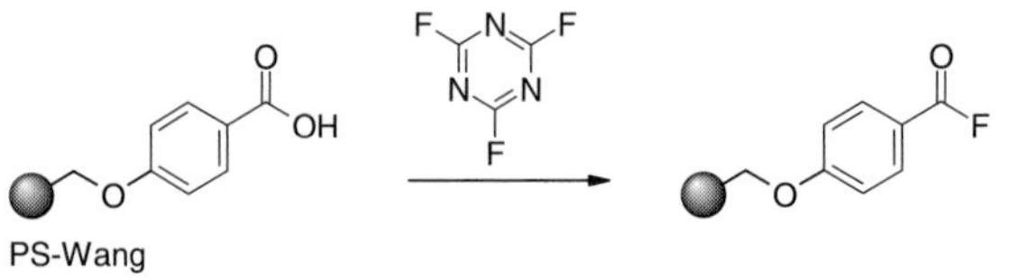

PS-Wang

Scheme 13.1. Transformation of aromatic carboxylic acids into acid fluorides.

In addition to the use of acid chlorides, chemists have been successfully transforming carboxylic acids into amides using coupling reagents. A large variety of reagents has been developed which provide chemists with a tool-box for the synthesis of large and complex structures [10]. Numerous examples demonstrate the efficacy of these methods for library synthesis in solution phase as well as in solid phase. In the latter case, both starting materials can be bound to the resin: activation of a polymer-bound acid followed by addition of the amine, or the addition of an excess acid plus coupling reagent to a polymer-bound amine.

The reaction of active esters with aliphatic amines generally leads to the formation of the desired amide in high yield and purity. However, the formation of anilides using active esters as acylating agents poses difficulties and only the most reactive coupling reagents may give satisfactory results (see Table 13.1). Therefore, the use of acid chlorides is necessary when anilines are used as nucleophiles.

The best conditions for a particular acylation are highly dependent on the steric and electronic nature of both the amine and carboxylic acid. There is no general rule as to which reagent is the best for a particular reaction and each case has to be optimized with respect to coupling reagent, base, solvent, etc. Nevertheless, there are some general guidelines for the choice of an appropriate coupling reagent:

- Carbodiimide-mediated couplings without additional reagents (HOBt) are a source for racemization. Therefore they are not recommended for fluorenylmethoxycarbonyl (Fmoc) amino acids [18].
- DIC/HONSu-mediated couplings occur under slightly acidic conditions, thus avoiding formation of N-acyl ureas.
- HOBt-based active esters suppress racemization of amino acids. This type of activation usually requires activating bases (usually DIEA or NMM).
- Pyrrolidino derivatives of both phosphonium and uronium salts derived from HOBt are slightly more reactive and less toxic than the dimethylamino derivatives (e.g. PyBOP vs. BOP).

Tab. 13.1. Coupling reagents for amide synthesis.

Building blocks		Coupling reagent	Comment	Reference
Acids	**Amines**			
Aliphatic and aromatic (sterically undemanding)	Aliphatic (primary)	DCC or DIC	Potential racemization during coupling of α-amino acids. DIC is preferred for solid-phase chemistry (soluble urea)	11
		DIC/HONSu	Slightly acidic conditions suitable for strongly basic amines (e.g. hydrazine)	12
Aliphatic and aromatic amino acids	Aliphatic (primary)	HOBt/DIC/DIEA	Intermediate formation of HOBt esters avoids racemization	13
Aliphatic and aromatic amino acids	Aliphatic (primary and secondary)	HBTU or TBTU/ DIEA	Preformed HOBt/DIC combination (HBTU: PF_6^- salt; TBTU: BF_4^- salt)	14
		PyBOP/DIEA	Fast coupling makes it suitable for base-sensitive substrates	15
Aliphatic and aromatic amino acids	Aliphatic and aromatic	HOAt/DIC/DIEA	Developed for cyclization of peptides, useful for coupling of hindered amino acids.	16
		HATU/DIEA	Preformed HOAt/DIC combination (PF_6^- salt)	17

- Uronium salt-mediated couplings may cause side-reactions such as the transformation of the N-terminus into a guanidinium residue [19]. Phosphonium reagents do not take part in this reaction [20].
- HBTU and TBTU are very popular for peptide synthesis.
- HATU or HOAt/DIC most likely yield the best results for difficult couplings, e.g. secondary amines [21], cyclizations, or anilines. Unfortunately, compared with other reagents they are rather expensive.

In the early days of combinatorial chemistry, these methods were used for the construction of peptide libraries on solid phase leading to potent, bioactive lead structures or elucidation of binding motifs [22]. Today, the application of coupling reagents for amide synthesis remains an essential part of combinatorial chemistry. Amino acids have gained popularity as readily available, multifunctional templates with a high degree of diversity and biological importance, and they are often used as valuable building blocks for the library synthesis of small molecules [23].

For parallel amide synthesis in solution phase, polymer-bound variations of the well-known coupling reagents have been shown to be powerful tools for avoiding tedious work-up procedures. The first report of modified DIC on a crosslinked polystyrene (PS) resin appeared in the early 1970s [24]. More recently, an N'-(3-dimethylaminopropyl)-N-ethylene carbodiimide (EDC)-based resin (P-EDC) was developed that couples amines and carboxylic acids more efficiently (Scheme 13.2) [25].

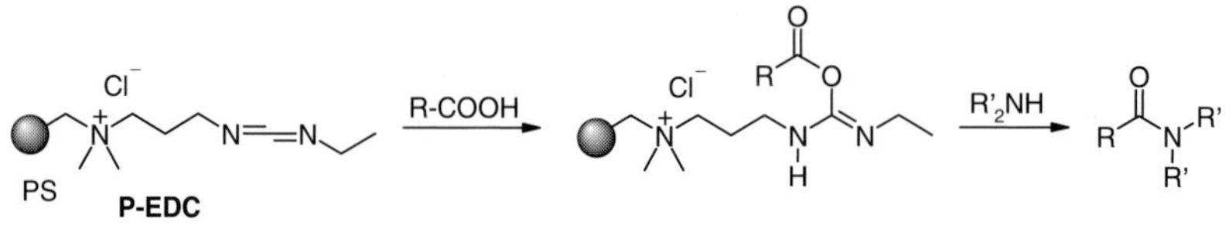

Scheme 13.2. Use of P-EDC in amide bond formation.

Additionally, carboxylic acids can be activated by polymer-bound HOBt [26] or 4-hydroxy-3-nitrobenzophenone. The latter reagent has been used for the synthesis of a library of 8000 amides and esters from which a compound with considerable herbicidal activity was identified (Scheme 13.3) [27].

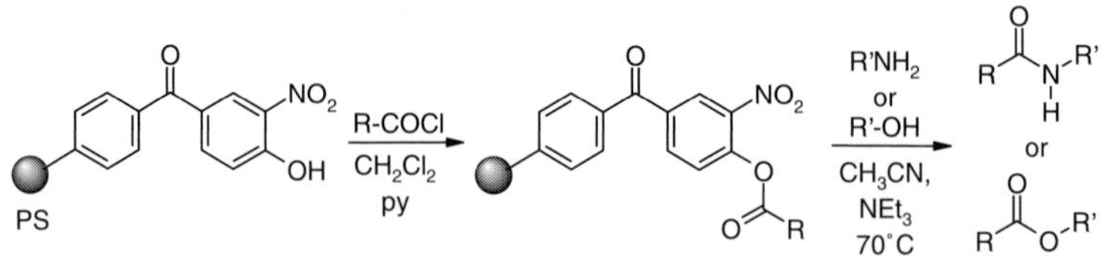

Scheme 13.3. Synthesis of amides using polymer-bound 4-hydroxy-3-nitrobenzophenone.

Shortcomings of these methods include the limited scope of reactivity of N-nucleophiles toward these reagents and problems in determining the absolute loading of the activated resin. The use of polymer-supported tetrafluorophenol promises a solution to both of these problems [28]. In analogy to well-known solution-phase transformations [29], carboxyl and sulfonyl activated esters can be prepared that yield amides and sulfonamides, respectively, with a wide range of N-nucleophiles (Scheme 13.4). Loading of the resin may be quantitatively determined by ^{19}F-nuclear magnetic resonance (NMR) spectroscopy.

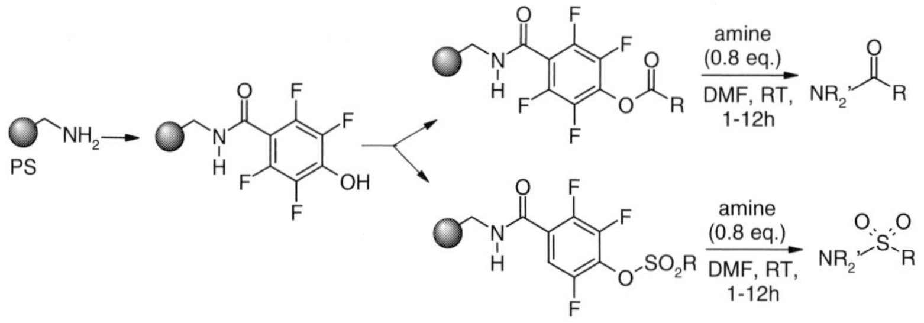

Scheme 13.4. Synthesis of amides using polymer-bound 4-hydroxy-3-nitrobenzophenone.

This methodology is routinely used for the synthesis of large libraries (10,000 members) for lead discovery or for targeted libraries, e.g. optimization of factor Xa inhibitors [30].

Formation of ureas – general aspects

There are two standard procedures that are used in the formation of ureas: The classical reaction of isocyanates with amines, and the treatment of amines with carbonyl insertion compounds such as *p*-NPCF [31] or CDI [32] followed by addition of another amine. Both procedures are suitable for library synthesis in solution phase or on solid support [33].

Formation of ureas using isocyanates. The solvent of choice for the reaction of isocyanates with amines is CH_2Cl_2. The reaction usually is completed within 1–2 h (Scheme 13.5). Prolonged reaction times have been reported to lead to lower product yields [34].

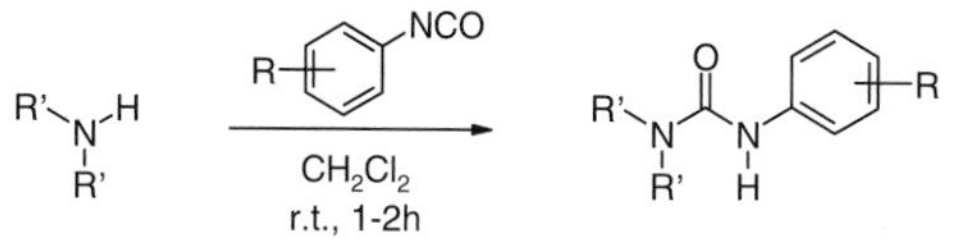

Scheme 13.5. Formation of ureas from amines and isocyanates.

In solid-phase chemistry, either amines or isocyanates can be immobilized. The generation of resin-bound isocyanates may be achieved by treating amines with phosgene or triphosgene [35]. Furthermore, the direct conversion of Fmoc-protected amines into isocyanates has been described previously (Scheme 13.6) [36].

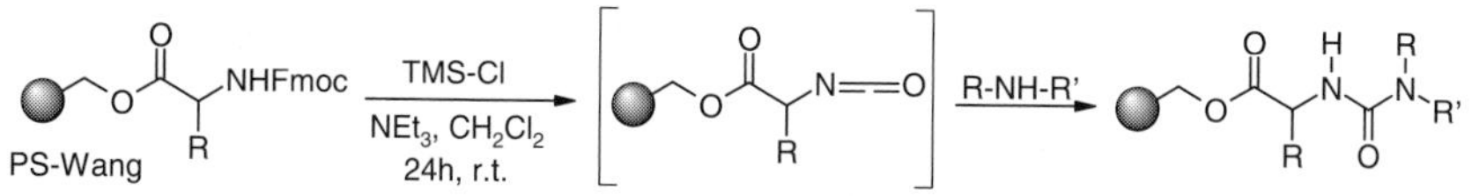

Scheme 13.6. Formation of isocyanates from Fmoc-protected amines on solid phase.

Trapping immobilized isocyanates with excess of amine proceeds rapidly (30 min). Using polymer-bound isocyanates as scavenging reagents, the time required for the complete removal of amines (generally nucleophiles) from the reaction mixture depends on the character of the amine (Scheme 13.7, Table 13.2) [37].

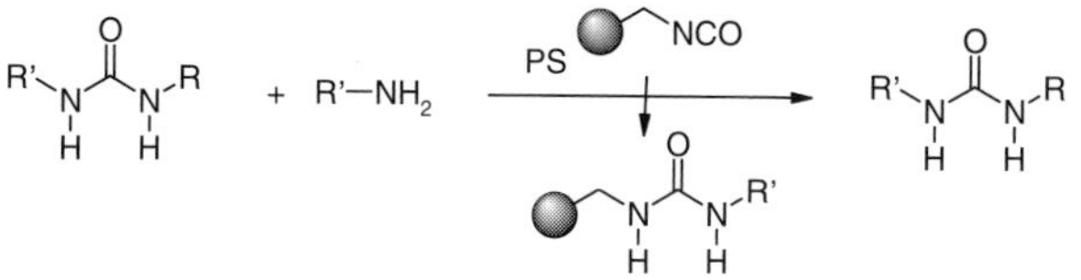

Scheme 13.7. Scavenging of amines using polymer-supported isocyanate.

Tab. 13.2. Comparative scavenging of nucleophiles in DCM (20 °C) or DCE (60 °C).[a]

Nucleophile	PS-isocyanate (equiv.) (1% crosslinked)	T (°C)	% Scavenged	
			1 h	16 h
Piperidine	3.0	20	100	–
Benzyl amine	3.0	20	100	–
Aniline	2.0	20	19	89
Aniline	3.0	60	–	99
2-Aminobenzophenone	3.0	60	–	81
4-Methoxyphenyl-1-butanol	2.0	20	0	68
4-Methoxyphenyl-1-butanol	3.0	60	0	29

[a] Brochure by Argonaut Technologies: *Polymer Reagents and Scavengers.*

Whereas aliphatic amines are completely removed within 1 h, aromatic amines generally require longer reaction times and elevated temperatures.

Formation of ureas using carbonyl insertion compounds
Since a larger number of amines than isocyanates is commercially available, the use of the carbonyl insertion method provides access to a larger and more diverse number of compounds than the reaction of isocyanates (Scheme 13.8).

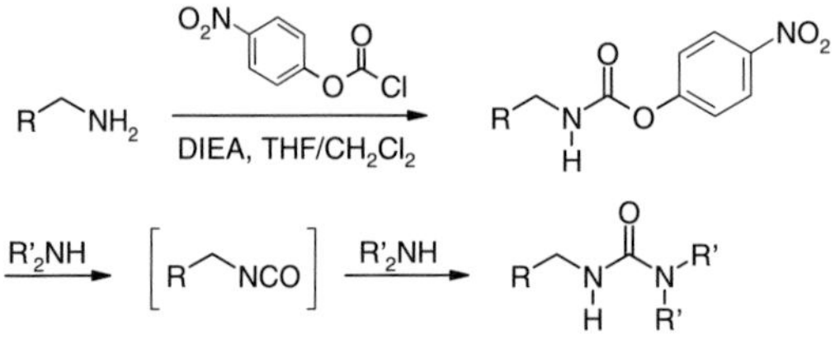

Scheme 13.8. *p*-NPCF-mediated formation of ureas.

This method is limited to the use of electron-rich and sterically undemanding amines as nucleophiles. Even though anilines have also been used, ureas derived from anilines are preferably prepared via transforming them into corresponding isocyanates prior to coupling with the second amine.

The above-mentioned methods for the activation of amines proceed via *in situ* formation of the corresponding isocyanate (Schemes 13.6 and 13.8). Therefore, they are limited to primary amines. However, secondary amines may also be activated via formation of the corresponding carbamoyl chloride using phosgene or triphosgene (Scheme 13.9) [38] or by the CDI analog 1,1'-carbonylbisbenzotriazole [39].

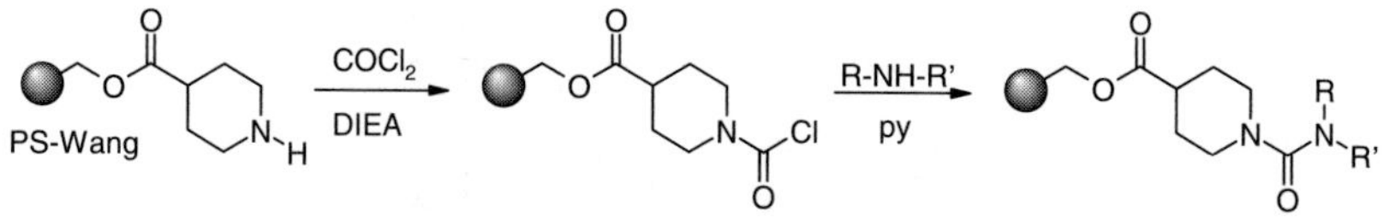

Scheme 13.9. Activation of secondary amines with COCl$_2$ for solid-phase synthesis of ureas.

In summary, the formation of ureas using either isocyanates or carbonyl insertion compounds is a straightforward synthetic transformation suitable for library synthesis both in solution and on solid phase.

Formation of ureas – recent applications

Ureas have been used as precursors of the benzoxazine moiety. Anthranilic acids were treated with isocyanates in solution and the obtained ureas were cyclized with polymeric EDC (Scheme 13.10) [40]. Simultaneously, unreacted excess of anthranilic acid was bound to the carbodiimide and could be filtered off from the reaction mixture yielding products in 80–97% purity.

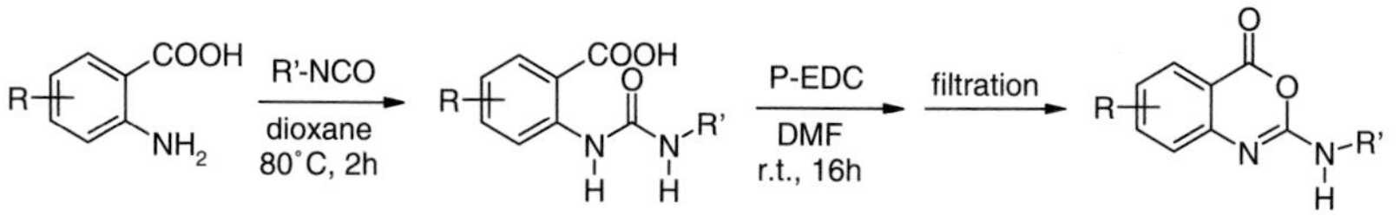

Scheme 13.10. Solution-phase library using ureas as precursors of benzoxazines.

13.3.1.2 Esters and Urethanes

General remarks

For both solution- and solid-phase chemistry, esters and urethanes commonly serve as protecting groups for carboxylic acids and amines, respectively [41]. Additionally, they are one of the most common ways to attach a molecule to a polymeric support. In the context of library synthesis, esters and urethanes may of course occur as part of the targeted molecule [42].

Formation of esters

A wealth of methods exist for the esterification of carboxylic acids. Almost all of them are suitable for parallel synthesis. Among these methods, DIC/DMAP (Steglich esterification) [43], 2,4,6-trichlorobenzoyl chloride (Yamaguchi esterification) [44], or 2,6-dichlorobenzoyl chloride [45] and the reaction of alcohols with acid chlorides or anhydrides are the most prominent ones in solution phase. In particular, the Steglich procedure seems to be suitable for automation and parallelization since polymeric carbodiimides are readily available (Scheme 13.11) [46].

Conversion of polymer-bound alcohols using these methods is as efficient as in solution phase. Owing to the lack of methods for acid chloride formation on solid phase (see Section 13.3.1), esterification of immobilized carboxylic acids is

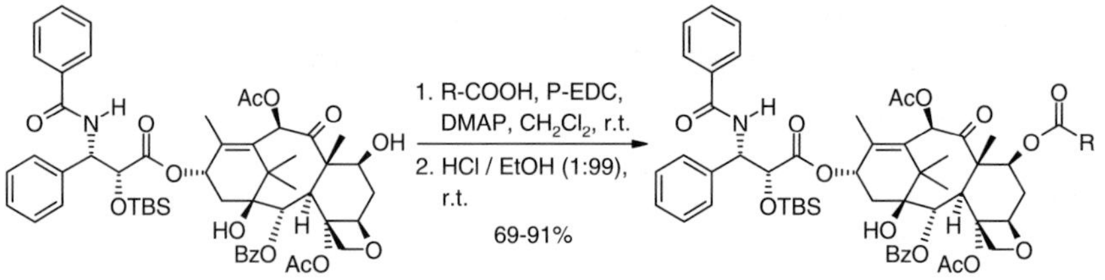

Scheme 13.11. Parallel esterification using polymeric carbodiimides: taxol analogs.

mostly realized using either the Steglich or the Yamaguchi procedure. Nevertheless, the condensation with primary or secondary alcohols using DIC/DMAP or 2,4,6-trichlorobenzoyl chloride has been well described. Additionally, the combination MSNT/NMI has been reported to be superior for the coupling of immobilized aromatic carboxylic acids to primary and secondary alcohols compared with the Steglich or Yamaguchi esterification (Scheme 13.12) [47].

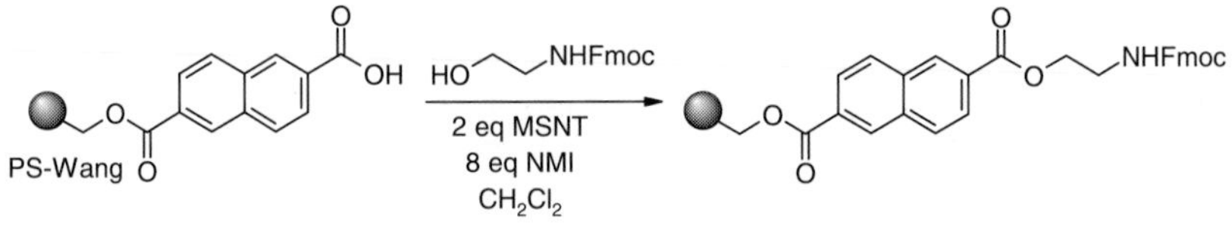

Scheme 13.12. Esterification of aromatic carboxylic acids using MSNT/NMI.

Hydrolysis of esters

In solution, esters are normally hydrolyzed using hydroxides such as LiOH, NaOH, or KOH in aqueous (50%) MeOH or THF [41]. These procedures are suitable for parallelization. Adaptation of these methods to solid phase sometimes requires slight modification of experimental details. Because of the extremely poor swelling properties of PS-based resins in water and alcohols, hydrolysis requires addition of solvents with better swelling properties – usually THF. NaOH or LiOH in THF/H_2O (4:1 v/v) or THF/MeOH/H_2O (3:1:1 v/v) at room temperature (rt) or at 50 °C generally give good conversions (Scheme 13.13) [48].

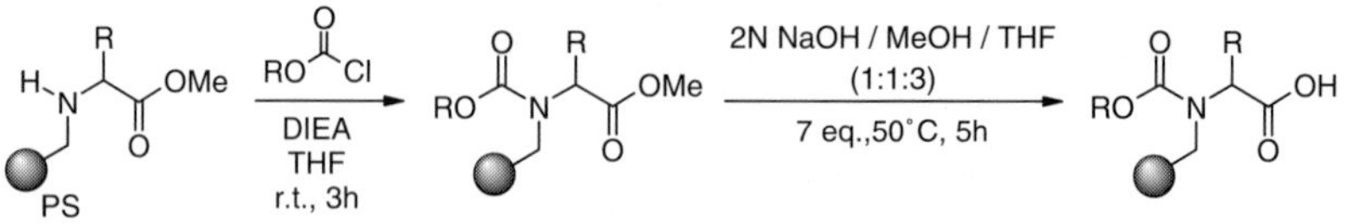

Scheme 13.13. Saponification of esters during library synthesis on solid phase.

Formation of carbamates

There are two standard ways for the formation of carbamates: first, from amines and chloroformates (Scheme 13.13); second, from alcohols and isocyanates or carbonyl insertion compounds. In contrast to the reaction of amines with isocyanates, the reaction of alcohols and isocyanates requires an activating base in order to achieve fast conversion. Usually, NEt$_3$ or DIEA is used (Scheme 13.14).

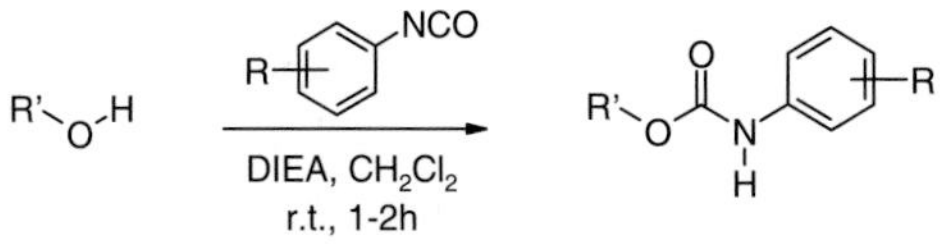

Scheme 13.14. Formation of carbamates from alcohols and isocyanates.

Besides this small modification, experimental details, scope, and limitations as well as applications of this reaction in solution or on solid phase are analogous to the formation of amides and ureas respectively (Section 13.3.1) [49].

13.3.2
Transformation of Carboxylic Acids into Other Functional Groups

The transformation of carboxylic acids into other functional groups is a powerful tool for the generation of new functionality. However, the reactions covered in this chapter have been described mostly as useful transformations but have not been used for library synthesis. The synthesis of larger libraries using these concepts has yet to be realized.

13.3.2.1 Formation of Ketones
Carboxylic acids are often used as precursors in the synthesis of aldehydes and ketones. The standard way to transform a carboxylic acid into a ketone is the two-step procedure of forming the corresponding Weinreb amide and subsequent addition of metalorganic reagents (see Chapter 16). The transformation of a carboxylic acid into the corresponding α-halomethyl ketone can be achieved by adding a diazomethane solution to the previously activated (via a mixed anhydride) carboxylic acid. The resulting diazomethylketone can be transformed into the α-halo ketone by treating it with tetrabutyl ammonium salts, $(n\text{-Bu}_4\text{N}^+\text{X}^-)$ [50].

Using appropriately protected aspartic acid, this reaction sequence was used for the solid-phase synthesis of a highly functionalized template bearing four different functional groups (Scheme 13.15). Substitution of the chlorine followed by further

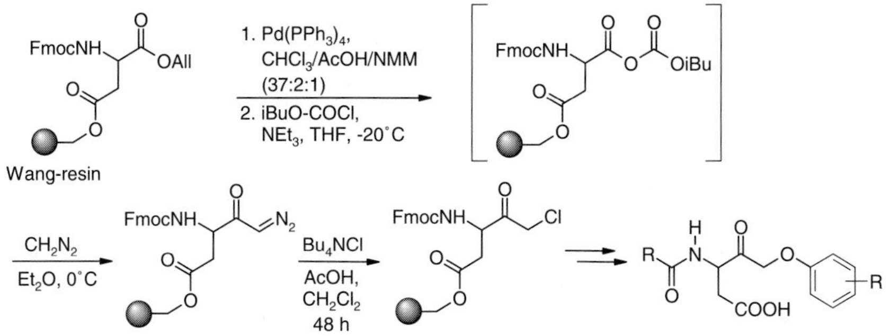

Scheme 13.15. Synthesis of amino acid-derived α-halomethyl ketones on solid phase.

derivatization completed the synthesis of (acyloxy)methylketones as irreversible inhibitors of the cysteine protease ICE.

13.3.2.2 Formation of Amines: Curtius Degradation

The Curtius degradation transforms carboxylic acids via thermal degradation of their corresponding azides into isocyanates. In the presence of alcohols, the reaction yields carbamates, whereas in the presence of water the intermediate isocyanate is directly degraded into an amino group. The standard procedure for solution-phase conversion of the carboxylic acid into the rearranged isocyanate is the two-step procedure forming the acid azide with DPPA in the presence of base (usually NEt₃) followed by heating the bound azide to 90 °C in toluene or xylene [51].

This well-established procedure has been adapted to solid-phase chemistry using polymer-bound aromatic carboxylic acids (Scheme 13.16) [52]. In contrast to solution phase, the presence of water gave impure products, whereas in the presence of alcohols the reaction cleanly yielded the carbamates. However, the use of 9-fluorenylmethanol (Fm-OH) yielded the Fmoc-protected amines from which the unprotected amine can easily be obtained.

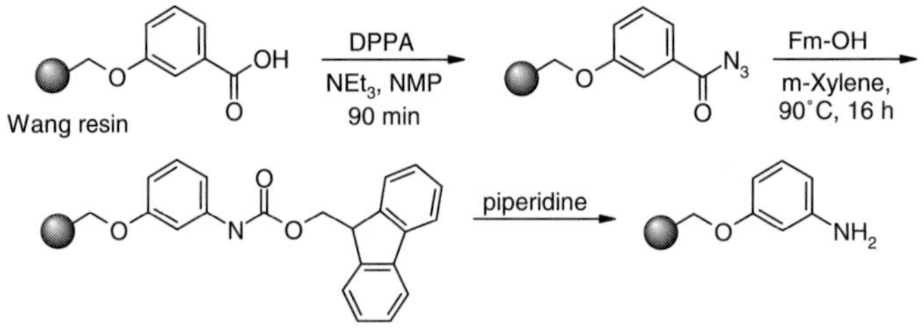

Scheme 13.16. Curtius degradation of carboxylic acids leading to Fmoc-protected amines.

13.3.2.3 Tebbe Olefination

The Tebbe olefination converts esters into the corresponding enol ethers that are the starting points for a variety of chemical transformations leading to such different structures as ketones, amines, thiazoles, and cyclohexanones (Scheme 13.17) [53].

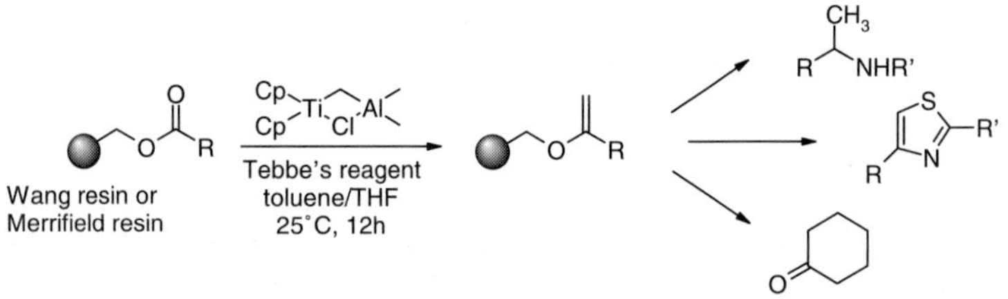

Scheme 13.17. Tebbe olefination of polymer-bound esters.

Despite its synthetic potential, the Tebbe olefination has only rarely been used for parallel synthesis. The starting point for a library of thiazoles on solid phase was the conversion of polymer-bound esters into the corresponding enol ethers using Tebbe reagents in toluene/THF at room temperature. For the following on-resin functionalization, the authors elegantly combined the principles of solid- and solution-phase chemistry (Scheme 13.18): in a first step, the enol ether is brominated on the resin, yielding an α-halo ketone equivalent. Formation of thiazoles via Hantzsch synthesis then simultaneously led to cleavage of the heterocycles from the resin. Removal of excess thiourea was accomplished by applying a polymeric α-halo ketone as a scavenger resin.

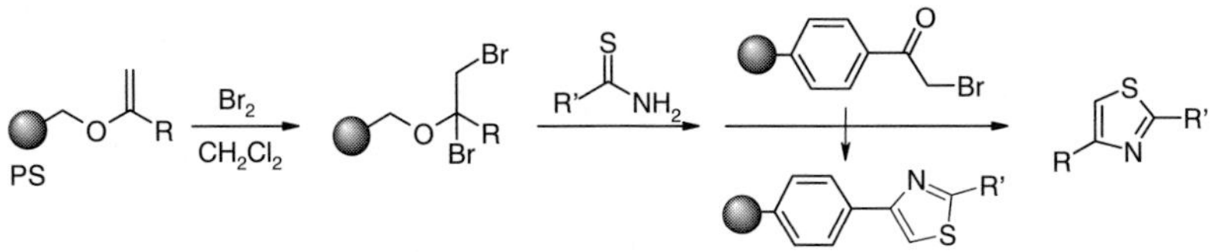

Scheme 13.18. Modification of enol ethers yielding thiazoles.

13.3.2.4 Formation of Thioamides

Thioamides are valuable precursors for heterocycle synthesis, a prevalent structural element in marketed drugs. Formation of thioamides on solid support can be achieved from readily synthesized primary amides of aliphatic and aromatic carboxylic acids by treatment with Lawesson reagent (Scheme 13.19) [54].

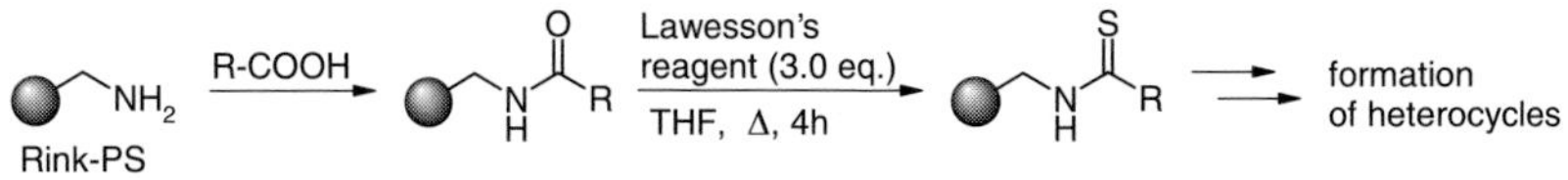

Scheme 13.19. Formation of thioamides on solid support.

13.4
Reactions of Aldehydes and Ketones

13.4.1
Reactions of Carbonyl Groups with C–H Acidic Compounds

13.4.1.1 Wittig and Horner–Emmons Olefinations

General aspects

The Wittig and the Horner–Emmons olefinations are well-established synthetic transformations for the generation of double bonds. Parallelization of these extensively used reactions should be straightforward. However, in the literature there are hardly any descriptions of libraries synthesized in solution. Probably, reactions have been carried out in a parallel fashion but have not specifically been reported as libraries. Some reports deal with ways to simplify the work-up, e.g. by using polymeric triphenylphosphine [55]. The situation is different for solid-phase synthesis. The reaction of aldehydes or ketones with phosphonium ylides yielding C–C double bonds was among the first reactions to be used for the synthesis of nonpeptidic molecules on solid phase [56]. Numerous examples prove the value of Wittig and the Horner–Emmons olefinations for solid-phase synthesis.

Double bonds are versatile starting points for diversification and are, thus, interesting building blocks for combinatorial chemistry. The double bonds obtained by the Wittig reaction have been used for various purposes, e.g. in olefin metathesis reactions [57] or cycloadditions [58] (Scheme 13.20).

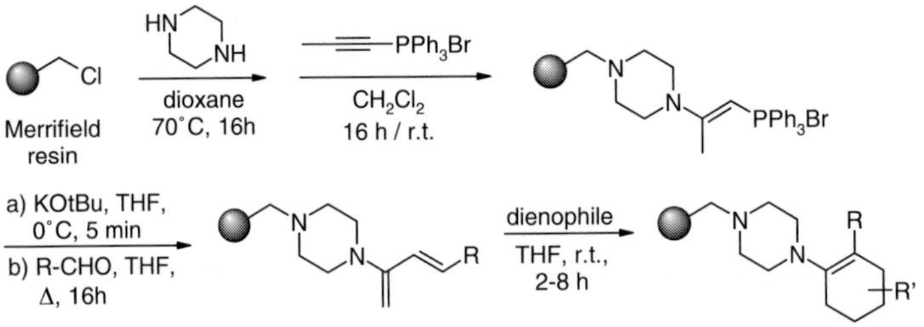

Scheme 13.20. Wittig reaction: diene formation as a precursor for a cycloaddition.

For solid-phase synthesis, two strategies have been pursued: generation of the ylide in solution and adding it to a polymer-bound aldehyde, as well as the opposite direction – generation of the ylide on solid phase and subsequent addition of the aldehyde to the resin. It is important to note that the choice between these two possibilities has consequences for the reaction conditions. If the ylide is generated on solid phase, KOtBu/THF has been reported to be better than NaHMDS/THF for phosphonium salt-derived ylides [58a]. Additionally, the reaction time for the generation of immobilized ylides is shorter (5–60 min) than in solution phase (up to 3 h). Hydrolysis of these sensitive intermediates can be a significant problem unless the resin is rigorously dried prior to use, e.g. azeotroping it with benzene.

Owing to the highly basic conditions of the reaction, the choice of the linker is also crucial. Even though the Wang linker has been employed previously [see 65], most examples use more base-stable linkers, e.g. amide linkage [59], trityl-ethers [56], and carbamates [60]. To avoid ester hydrolysis of substrates bound to Wang resin, the following procedure has been recommended (Scheme 13.21): generation of the phosphonate anion, removal of excess base under inert conditions, and addition of the aldehyde dissolved in 60% cyclohexane in THF [61].

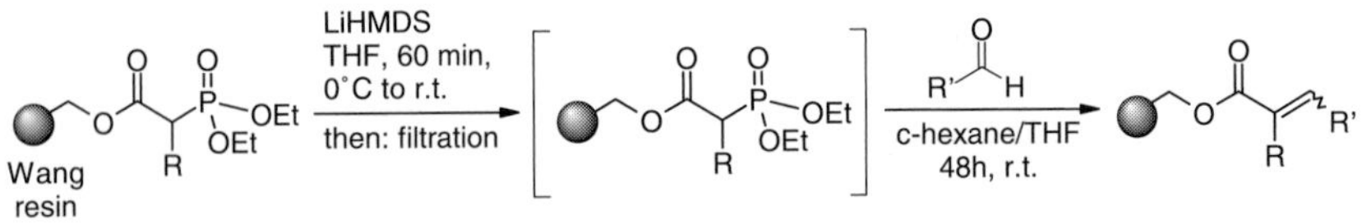

Scheme 13.21. Horner–Emmons reaction on Wang resin.

The reasoning behind the cyclohexane/THF mixture was that the less-polar solvent mixture should suppress ester hydrolysis, while allowing the Horner–Emmons condensation to proceed. The authors successfully synthesized acrylic acid derivatives from aldehydes containing aliphatic, aromatic, and basic functionalities. The reaction generally was completed within 2 days.

The undesired cleavage from the resin during ylide generation can also be avoided by attaching the ylide to the resin via the phosphonate moiety. In this case, the Horner–Emmons reaction has been used for cleaving the molecule from the resin (Scheme 13.22) [62]. This approach was used for the library synthesis of cyclic ketones such as (DL)-muscone: a polymer-bound methylphosphonate was deprotonated with *n*-BuLi and quenched with an alkenyl ester. The double bond was used for subsequent introduction of an aldehyde moiety. The following intramolecular olefination ("cyclorelease") led to the cyclic enones which were further derivatized in solution phase.

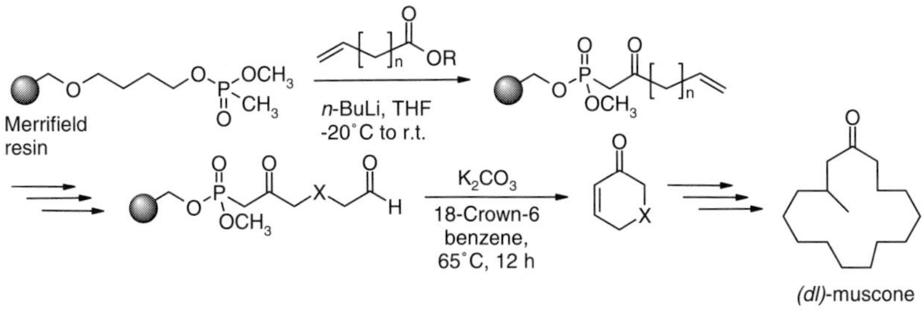

Scheme 13.22. Intramolecular Horner–Emmons reaction.

Recent examples

The use of commercially available polymer-bound PPh$_3$ for Wittig reactions combines ideas from solid-phase synthesis and the application of polymer-supported reagents (Scheme 13.23) [63]. On the one hand (as a polymer-supported reagent) it offers the advantage that the byproduct, triphenylphosphine oxide, remains at-

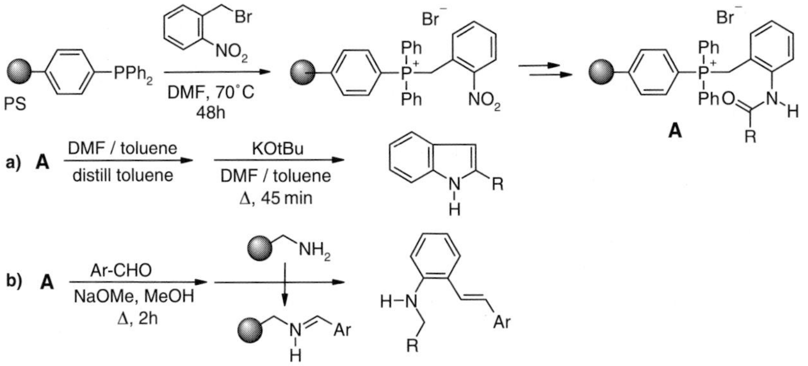

Scheme 13.23. Wittig reaction using polymer-supported triphenylphosphine.

tached to the resin. On the other hand (similar to the function of a linker in solid-phase synthesis) it immobilizes benzyl halides.

Starting from 2-nitrobenzylbromide, after immobilization and chemical modifications, the products may be cleaved off from the support by an intra- or intermolecular Wittig reaction. The intramolecular Wittig reaction yields indoles, whereas the intermolecular pathway gives access to stilbenes [64].

Double bonds have also been used for the synthesis of peptidomimetics. In this context, they serve as nonhydrolyzable rigid mimetics of the three-dimensional structure of the amide bond. The Horner–Emmons reaction has been employed as part of a two-step procedure for solid-phase synthesis of β, γ-unsaturated δ-amino acids that mimic dipeptides (Scheme 13.24) [65]. This concept combines both the use of the double bond as an amide bond mimetic as well as the use of the double bond for diversification.

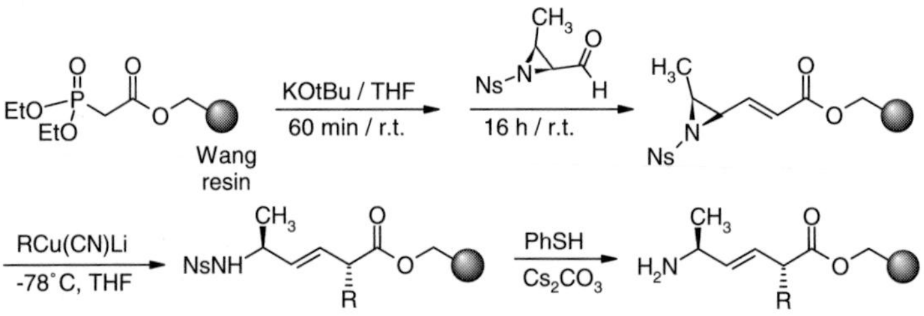

Scheme 13.24. Horner–Emmons reaction: solid-phase derivatization of the double bond.

13.4.2
Reductive Amination

13.4.2.1 General Aspects

Reductive amination is one of the most frequently used reactions for library generation in solution phase as well as on solid phase. Starting materials are diverse, inexpensive, and readily available from commercial sources. Additionally, the secondary amines obtained may be widely applied to the generation of novel structures, e.g. heterocycles, sulfonamides, and amides [66]. Last but not least, the reaction is well suited to automated synthesis. Not surprisingly, a large number of methods has been elaborated for the multitude of electronic and steric demands of the building blocks.

13.4.2.2 Formation of Imines

Imines are formed by the condensation of an amine with a ketone or an aldehyde. Classical methods in solution phase shift the equilibrium toward the imine by removing the water generated during the reaction, e.g. via azeotropic distillation or applying drying reagents such as molecular sieves or inorganic sulfates. These methods have also been used for solid-phase chemistry [67], but one should bear in mind that the distribution of these reagents into a parallel array is not very convenient. Therefore, trimethylorthoformate (TMOF) has gained popularity as a universal and resin-compatible desiccant (Scheme 13.25) [68]. Additional methods for imine formation include the simple use of a large excess of amine [69] or using ultrasound in the presence of Na_2SO_4 [70].

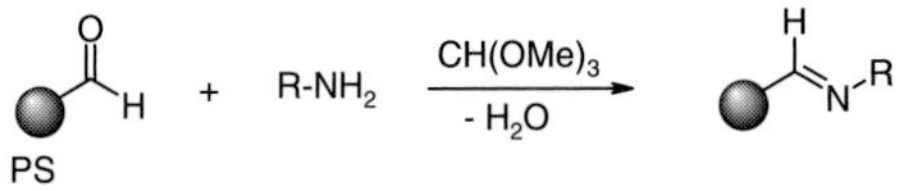

Scheme 13.25. Imine formation on solid support using TMOF as dehydrating reagent.

Imines formed in the presence of TMOF may be isolated and hence can be used for various purposes besides reduction to amines, e.g. for cycloadditions [71]. In cases where the solubility of the amine component in neat TMOF is low, cosolvents such as THF or dimethylformamide (DMF) can be used. As a rule of thumb, the reaction is slower with polymer-bound aldehydes and the amine in solution than with the opposite situation.

A given combination of building blocks has consequences for the reaction conditions depending on the reactivity of the amine and carbonyl components (Table 13.3). The reaction of aliphatic amines and aromatic aldehydes reliably yields the desired imine within short reaction times at room temperature. Other scenarios may result in side-reactions or incomplete conversions. Aromatic amines also form imines, but, depending on their electronic nature, this reaction may require elevated temperatures and elongated reaction times. Imines derived from aliphatic aldehydes tend to form the tautomeric enamines. The reactivity of ketones is strongly influenced by their structure. Cyclic ketones react nicely whereas acyclic ketones often show hardly any conversion [72].

13.4.2.3 Reduction of Imines/Enamines
A number of reducing agents has been used for the reduction of imines to secondary amines. The most frequently used reagents on solid phase are: $NaCNBH_3$ in 1% HOAc/DMF [73], freshly prepared $NaBH(OAc)_3$ or $(NBu_4)BH(OAc)_3$ [74], and $BH_3 \cdot py$ [69a, 75]. They all reduce imines and enamines (e.g. derived from secondary amines and aldehydes) equally well. Generally, the reducing agent is added *in situ* without isolation of the imines or the enamines.

Tab. 13.3. Imine formation depending on amine and carbonyl component.

Carbonyl	Amine	Comment
Aromatic aldehyde	Aliphatic	Clean reaction (3 h, rt)
Aliphatic aldehyde	Aliphatic	Equilibrium with the tautomeric enamine
Aromatic aldehyde	Aromatic	Requires elevated temperatures and longer reaction times (e.g. 50 °C, 16 h)
Cyclic ketone	Aliphatic	Mostly a clean reaction
Acyclic ketone	Aliphatic	Low conversion

Tab. 13.4. Optimized reaction conditions for reductive aminations using $NaCNBH_3$.

Aldehyde/Amine[a]	Format	Experimental detail
Aliphatic aldehyde/aliphatic amine	Solution phase	Simply requires $NaCNBH_3$ in TMOF, large excess of aldehyde, no additional proton source
Aliphatic aldehyde/aliphatic amine[pb]	Solid phase (resin)	Simply requires $NaCNBH_3$ in TMOF, large excess of aldehyde, no additional proton source
Sterically hindered aliphatic aldehydes/aliphatic amine	Solution phase	Requires premixing (30 min) of amine and aldehyde prior to addition of $NaCNBH_3$, no additional proton source
Sterically hindered aliphatic aldehydes/aliphatic amine[pb]	Solid phase (resin)	Requires premixing (30 min) of amine and aldehyde prior to addition of $NaCNBH_3$, requires additional proton source (MeOH !!!, HOAc leads to overalkylation)
Aromatic aldehyde/ aliphatic amine	Solution phase	Premixing with the amine in TMOF followed by the addition of $NaCNBH_3$ in the presence of 1% HOAc
Aromatic aldehyde/ aliphatic amine[pb]	Solid phase (resin)	Premixing with the amine in TMOF followed by the addition of $NaCNBH_3$ in the presence of 1% HOAc
Aliphatic amine/ketone[pb]	Solid phase (pin)	pH = 5 in MeOH (HOAc)
Aromatic amine/ketone[pb]	Solid phase (pin)	pH = 7 in MeOH (HOAc, NMM)

[a] p.b., (= polymer bound) – indicates which part is on solid phase.

Reaction conditions have been investigated for all reagents in order to achieve maximum conversion and to suppress overalkylation. The classical reagent for reduction, $NaCNBH_3$ [76], has been optimized with respect to aldehydes and amines for both solution- and solid-phase chemistry (Table 13.4) [77].

13.4.2.4 Applications

Reductive aminations using polymeric reagents

Reductive aminations in solution phase sometimes require the tedious removal of boron-containing by-products generated from the reduction step. Solid-supported borohydrides circumvent this problem. For the synthesis of secondary amines from primary aliphatic amines an excess of primary amines relative to the aldehyde component has been used. After reduction of the corresponding imine with resin-bound borohydride, excess primary amine was removed by selective imine formation using a polymer-supported aldehyde (Scheme 13.26) [78].

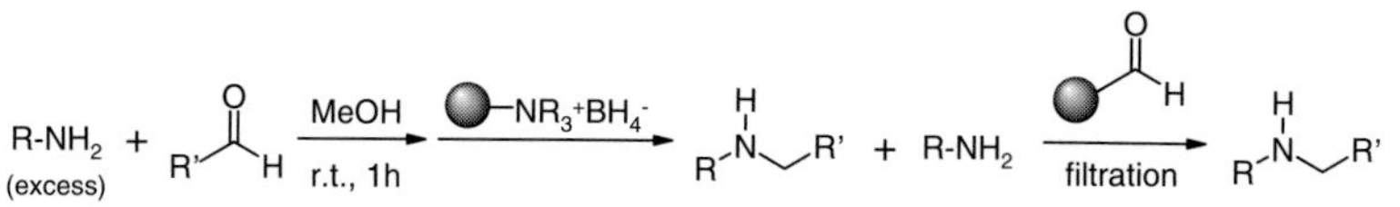

Scheme 13.26. Reductive amination in solution phase using polymeric borohydride.

Favorable relative kinetics of the competing reactions allowed simultaneous addition of both resins to the reaction mixture, thus simplifying the experimental procedure.

Reductive aminations as a starting point for library synthesis

For library synthesis on solid phase, a major application of reductive aminations is their use as a starting point for library synthesis. During the first step, large diversity has already been introduced from a multitude of commercially available amines as building blocks. Subsequent acylation or sulfonylation of the generated secondary amines is the next step during the synthesis of the desired compounds.

Most commonly, secondary amines are generated from a polymer-bound, electron-rich aromatic aldehyde (Scheme 13.27). The choice of the appropriate reducing agent does not seem to be crucial, e.g. 2 equiv. $RNH_2/NaBH(OAc)_3$ and 3 equiv. $RNH_2/CH(OMe)_3/BH_3 \cdot py$ have been successfully applied [79].

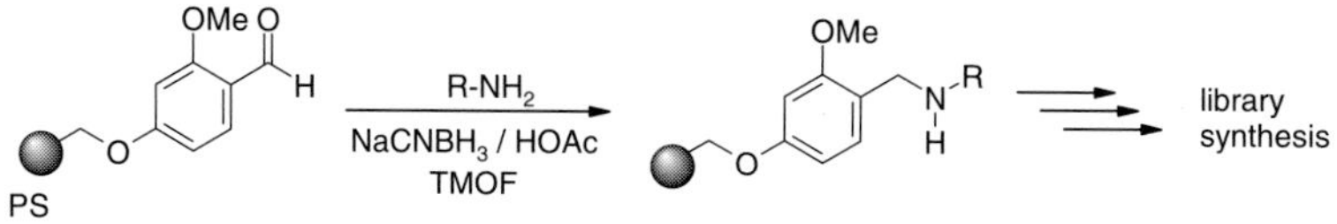

Scheme 13.27. Reductive amination as starting point for library synthesis.

Rink's amide linker has been used as a polymer-bound source of amines (Scheme 13.28) [80]. It became obvious that clean conversion of the sterically demanding benzhydrylamine was not straightforward and that the choice of the

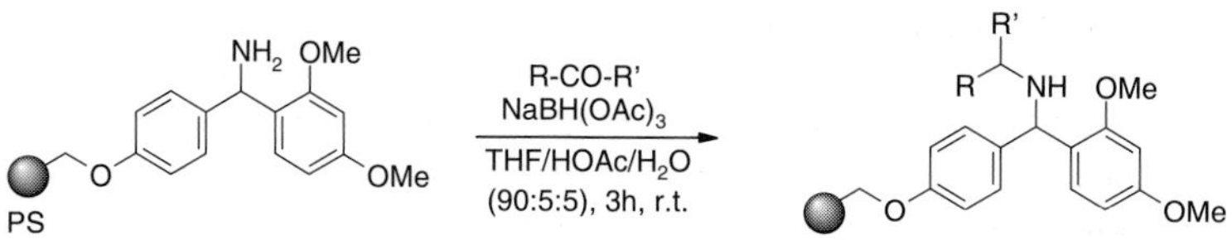

Scheme 13.28. Reductive amination of Rink's amide linker.

reaction conditions for reducing the imine was crucial. Standard methods failed (e.g. 1–10 equiv. R-CHO/NaCNBH$_3$/1% HOAc in DMF). However, adjusting the solvent system from anhydrous DMF to aqueous THF gave clean monoalkylation with aliphatic and aromatic aldehydes as well as with ketones.

Reductive amination of ketones: benzopyran library

Even though reductive amination of ketones is often difficult, thorough optimization of the reaction conditions can still lead to suitable synthetic protocols, as demonstrated by the conversion of benzopyranones (Scheme 13.29). For the reduction of imines derived from these relatively unreactive ketones, combinations of complex boron hydrides with standard proton sources such as HOAc or MeOH (Table 13.4) failed. Furthermore, heating or sonication led to significant cleavage from the resin. However, the use of Ti(OiPr)$_4$ in combination with NaBH(OAc)$_3$ resulted in complete conversion to the desired amine.

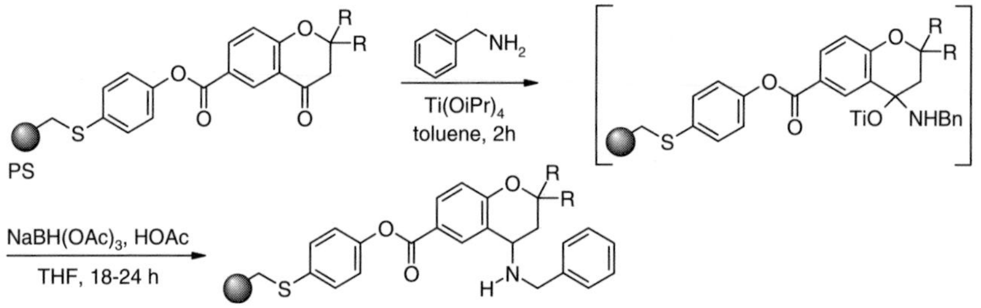

Scheme 13.29. Reductive amination of ketones using Ti(OiPr)$_4$/NaBH(OAc)$_3$.

The presence of Ti(OiPr)$_4$ requires that the reaction is carefully run under inert conditions in order to avoid precipitation of TiO$_2$. The feasibility of this protocol was demonstrated by the synthesis of a 8448-member library of benzopyrans [81].

Serendipitous reactions as an opportunity for affinity breakthroughs

Neuropeptide Y is believed to be involved in the regulation of feeding, energy metabolism, vascular tone, learning and memory, and the release of pituitary hormones. Researchers have discovered that benzimidazoles act as potent antagonists of the NPY-1 receptor. Further optimization of the side-chain has been addressed by a solution-phase library. In order to optimize interaction with the receptor, the distal piperidine has been modified via reductive amination using polymeric reagents and scavenger resins. Only one compound from this library appeared to be more active than the lead for the library. However, upon resynthesis it became clear that the anticipated product was not consistent with the spectroscopic data and that the intermediate imine underwent a spontaneous Pictet–Spengler cyclization, even in the absence of any reducing agent (Scheme 13.30) [82].

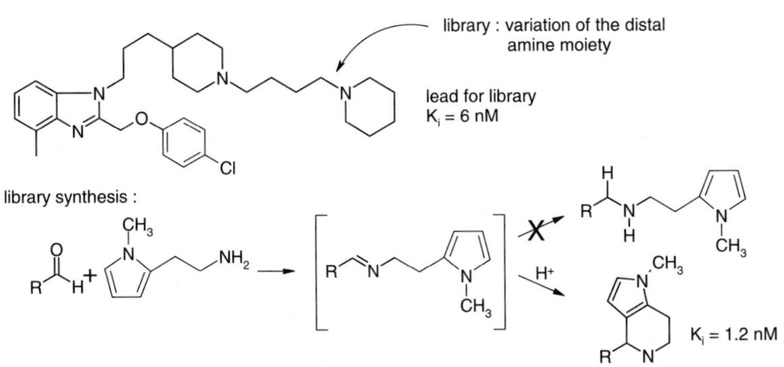

Scheme 13.30. Unexpected side-reaction yielding biologically active compound.

References

1 G. W. BEMIS, M. A. MURCKO, *J. Med. Chem.* **1999**, *42*, 5095–5099.

2 B. MERRIFIELD, *J. Am. Chem. Soc.* **1963**, *85*, 2149–2154.

3 G. JUNG, A. G. BECK-SICKINGER, *Angew. Chem., Int. Ed. Engl.* **1992**, *31*, 367–383.

4 L. GHOSEZ, I. GEORGE-KOCH, L. PATINY, M. HOUTEKIE, P. BOVIE, P. NSHIMYUMUKIZA, T. PHAN, *Tetrahedron* **1998**, *54*, 9207–9222.

5 a) R. APPEL, *Angew. Chem.* **1975**, *87*, 863–874; b) J. B. LEE, *J. Am. Chem. Soc.* **1966**, *33*, 3440.

6 B. C. HAMPER, S. A. KOLODZIELJ, A. M. SCATES, *Tetrahedron Lett.* **1998**, *39*, 2047–2050.

7 C. K. SAMS, J. LAU, *Tetrahedron Lett.* **1999**, *40*, 9359–9362.

8 C. KADUK, H. WENSCHUH, M. BEYERMANN, K. FORNER, L. A. CARPINO, M. BIENERT, *Lett. Peptide Sci.* **1996**, *2*, 285.

9 L. A. CARPINO, A. EL-FAHAM, *J. Am. Chem. Soc.* **1995**, *117*, 5401–5402.

10 G. B. FIELDS, R. L. NOBLE, *Int. J. Pept. Protein Res.* **1990**, *35*, 161–214.

11 a) J. C. SHEEHAN, G. P. HESS, *J. Am. Chem. Soc.* **1955**, *77*, 1067–1068; b) J. C. SHEEHAN, M. GOODMAN, G. P. HESS, *J. Am. Chem. Soc.* **1956**, *78*, 1367–1369.

12 a) D. SARANTAKIS, J. TEICHMAN, E. L. LIEN, R. L. FENICHEL, *Biochem. Biophys. Res. Commun.* **1976**, *73*, 336–342; b) G. B. FIELDS, C. G. FIELDS, J. PETEFISH, H. E. VAN WART, T. A. CROSS, *Proc. Natl. Acad. Sci. USA* **1988**, *85*, 1384–1388.

13 W. KÖNIG, R. GEIGER, *Chem. Ber.* **1970**, *103*, 788–798.

14 R. KNORR, A. TRECIAK, W. BANNWARTH, D. GILLESSEN, *Tetrahedron Lett.* **1989**, *30*, 1927–1930.

15 a) J. MARTINEZ, J. BALI, M. RODRIGUEZ, B. CASTRO, M. MAGOUS, J. LAUR, M. F. LIGNON, *J. Med. Chem.* **1985**, *28*, 1874–1875; b) J. COSTE, D. LE-NGUYEN, B. CASTRO, *Tetrahedron Lett.* **1990**, *31*, 205–208; c) B. CASTRO, J. R. DORMOY, G. EVIN, C. SELVE, *Tetrahedron Lett.* **1975**, *16*, 1219–1222.

16 a) L. A. CARPINO, *J. Am. Chem. Soc.* **1993**, *115*, 4397–4398; b) L. A. CARPINO, A. EL-FAHAM, *Tetrahedron* **1999**, *55*, 6813–6830.

17 L. A. CARPINO, A. EL-FAHAM, C. A. MINOR, F. ALBERICIO, *J. Chem. Soc., Chem. Commun.* **1994**, 201–203.

18 a) E. ATHERTON, R. C. SHEPPARD, *J. Chem. Soc. Chem. Commun.* **1985**, 165–166; b) H. GROSS, L. BILK, *Tetrahedron Lett.* **1968**, *24*, 6935–6939.

19 a) H. Gausepohl, U. Pieles, R. W. Frank, in: Proceedings of the 12th American Peptide Symposium. Smith, J. A., Rivier, J. E. (eds), ESCOM, Leiden 1992, pp. 523–524; b) S. C. Story, J. V. Aldrich, *Int. J. Pept. Protein Res.* **1994**, *43*, 292–296.

20 F. Albericio, M. Cases, J. Alsina, S. A. Triolo, L. A. Carpino, S. A. Kates, *Tetrahedron Lett.* **1997**, *38*, 4853–4856.

21 a) A. A. Virgilio, J. A. Ellman, *J. Am. Chem. Soc.* **1994**, *116*, 11580–11581; b) A. Nefzi, J. M. Ostresh, R. A. Houghten, *Tetrahedron Lett.* **1997**, *38*, 4943–4946.

22 a) C. P. Holmes, C. L. Adams, L. M. Kochersperger, R. B. Mortensen, L. A. Aldwin, *Biopolymers* **1995**, *37*, 199–211; b) Z. M. Ruggeri, R. A. Houghten, S. R. Russel, T. S. Zimmerman, *Proc. Natl. Acad. Sci. USA* **1986**, *83*, 5708–5712; c) A. G. Beck-Sickinger, W. Gaida, G. Schnorrenberg, R. Lang, G. Jung, *Int. J. Pept. Protein Res.* **1990**, *36*, 88–94.

23 a) B. A. Bunin, J. A. Ellman, *J. Am. Chem. Soc.* **1992**, *114*, 10997–10998; b) M. A. Marx, A.-L. Grillot, C. T. Louer, K. A. Beaver, P. A. Bartlett, *J. Am. Chem. Soc.* **1997**, *119*, 6153–6167; c) E. E. Swayze, *Tetrahedron Lett.* **1997**, *38*, 8643–8636.

24 N. M. Weinshenker, C.-M. Shen, *Tetrahedron Lett.* **1972**, *13*, 3281–3284.

25 M. C. Desai, L. M. S. Stramiello, *Tetrahedron Lett.* **1993**, *34*, 7685–7688.

26 a) I. E. Pop, B. P. Deprez, A. L. Tartar, *J. Org. Chem.* **1997**, *62*, 2594–2603; b) K. Dendrinos, J. Jeong, W. Huang, A. G. Kalivrentenos, *Chem. Commun.* **1998**, 449–500.

27 J. J. Parlow, J. E. Normansell, *Mol. Diversity* **1995**, *1*, 266–269.

28 J. M. Salvino, N. V. Kumar, E. Orton, J. Airey, T. Kiesow, K. Crawford, R. Mathew, P. Krolikowski, M. Drew, D. Engers, D. Krolikowski, T. Heroin, M. Gardyan, G. McGeehaan, R. Labaudiniere, *J. Comb. Chem.* **2000**, *2*, 691–697.

29 L. KisFaludy, I. Schon, O. Nyeki, M. Low, *Tetrahedron Lett.* **1974**, *39*, 1785.

30 Y. Gong, M. Becker, Y. M. Choi-Sledeski, R. S. Davis, J. M. Salvino, V. Chu, K. D. Brown, H. W. Pauls, *Bioorg. Med. Chem. Lett.* **2000**, *10*, 1033–1036.

31 a) S. M. Hutchins, K. T. Chapman, *Tetrahedron Lett.* **1994**, *35*, 4055–4058; b) S. M. Hutchins, K. T. Chapman, *Tetrahedron Lett.* **1995**, *36*, 2583–2586.

32 H. A. Staab, *Liebigs Ann. Chem.* **1957**, *609*, 75–83.

33 a) S. W. Kaldor, J. E. Fritz, J. Tang, E. R. McKinney, *Bioorg. Med. Chem. Lett.* **1998**, *8*, 3041–3044; b) J. W. Nieuwenhuijzen, P. G. M. Conti, H. C. J. Ottenheijm, J. T. M. Linders, *Tetrahedron Lett.* **1998**, *39*, 7811–7814.

34 K. Burgess, J. Ibarzo, S. Linthicum, D. H. Russel, H. Shin, A. Shitangkoon, R. R. Totani, A. J. Zhang, *J. Am. Chem. Soc.* **1997**, *119*, 1556–1564.

35 D. Limal, V. Semetey, P. Dalbon, M. Jolivet, J.-P. Briand, *Tetrahedron Lett.* **1999**, *40*, 2749–2752.

36 P. Y. Chong, P. A. Petillo, *Tetrahedron Lett.* **1999**, *40*, 4501–4504.

37 a) J. R. Booth, J. C. Hodges, *J. Am. Chem. Soc.* **1997**, *119*, 4882; b) M. W. Creswell, G. L. Bolton, J. C. Hodges, M. Meppa, *Tetrahedron* **1998**, *54*, 3983.

38 G. T. Wang, Y. Chen, S. Wang, R. Sciotti, T. Sowin, *Tetrahedron Lett.* **1997**, *38*, 1895–1898.

39 a) H. A. Staab, G. Seel, *Liebigs Ann. Chem.* **1958**, *612*, 187–193; b) A. R. Katritzky, D. P. M. Pleynet, B. Yang, *J. Org. Chem.* **1997**, *62*, 4155–4158.

40 B. O. Buckman, M. M. Morissey, R. Mohan, *Tetrahedron Lett.* **1998**, *39*, 1487–1488.

41 a) P. J. Kocienski, *Protecting Groups*. Thieme, Stuttgart 1994; b) T. W. Green, P. G. M. Wuts, *Protective Groups in Organic Synthesis*, 2nd edn, Wiley, New York 1991.

42 M. F. Gordeev, D. V. Patel, B. P. England, S. Jonnalagadda, J. D.

COMBS, E. M. GORDON, *Bioorg. Med. Chem. Lett.* **1998**, *6*, 883–889.

43 B. NEISES, W. STEGLICH, *Angew. Chem. Int. Ed. Engl.* **1978**, *17*, 522–523.

44 J. INANAGA, K. HIRATA, H. SAEKI, T. KATSUKI, M. YAMAGUCHI, *Bull. Chem. Soc. Jpn.* **1979**, *52*, 1989–1993.

45 P. SIEBER, *Tetrahedron Lett.* **1987**, *28*, 6147–6150.

46 L. BHAT, Y. LIU, S. F. VICTORY, R. H. HIMES, G. I. GEORG, *Bioorg. Med. Chem. Lett.* **1998**, *8*, 3181–3186.

47 J. NIELSEN, L. O. LYNGSO, *Tetrahedron Lett.* **1996**, *37*, 8439–8442.

48 N. GRIEBENOW, T. WUNBERG, unpublished results.

49 a) B. A. DRESSMAN, L. A. SPANGLE, S. W. KALDOR, *Tetrahedron Lett.* **1996**, *37*, 937–940; b) H. P. BUCHSTALLER, *Tetrahedron* **1998**, *54*, 3465–3470; c) D. P. ROTELLA, *J. Am. Chem. Soc.* **1996**, *118*, 12246–12247; d) T. WUNBERG, C. KALLUS, T. OPATZ, S. HENKE, W. SCHMIDT, H. KUNZ, *Angew. Chem. Int. Ed.* **1998**, *37*, 2503–2505.

50 M. T. MUJICA, G. JUNG, *Synlett* **1999**, 1933–1935.

51 a) R. BONJOUKLIAN, R. A. RUDEN, *J. Org. Chem.* **1977**, *42*, 4095–4103; b) B. CHANTEGREL, S. GELIN, *Synthesis* **1981**, 315–316; c) H. SHAO, M. COLUCCI, S. TONG, H. ZHANG, A. L. CASTELHANO, *Tetrahedron Lett.* **1998**, *39*, 7235–7238.

52 L. S. RICHTER, S. ANDERSEN, *Tetrahedron Lett.* **1998**, *39*, 8747–8750.

53 C. P. BALL, A. G. M. BARRETT, A. COMMERCON, D. CAMPERE, Y. KUHN, R. S. ROBERTS, M. L. SMITH, O. VENIER, *Chem. Commun.* **1998**, 2019–2020.

54 J.-F. PONS, Q. MISHIR, A. NOUVET, F. BROOKFIELD, *Tetrahedron Lett.* **2000**, *41*, 4965–4968.

55 D. H. DREWRY, D. M. COE, S. POON, *Med. Chem. Rev.* **1999**, *19*, 97–148.

56 C. CHEN, L. A. AHLBERG RANDALL, R. B. MILLER, A. J. JONES, M. J. KURTH, *J. Am. Chem. Soc.* **1994**, *116*, 2661–2662.

57 K. C. NICOLAOU, N. WINSSINGER, J. PASTOR, S. NINKOVIC, F. SARABIA, Y. HE, D. VOURLOUMIS, Z. YANG, T. LI,

P. GIANNAKAKOU, E. HAMEL, *Nature* **1997**, *387*, 268–272.

58 a) M. CRAWSHAW, N. W. HIRD, K. IRIE, K. NAGAI, *Tetrahedron Lett.* **1997**, *40*, 7115–7118; b) N. W. HIRD, K. IRIE, K. NAGAI, *Tetrahedron Lett.* **1997**, *38*, 7111–7114; c) P. GROSCHE, A. HÖLTZEL, T. B. WALK, A. W. TRAUTWEIN, G. JUNG, *Synthesis* **1999**, *11*, 1961–1970.

59 D. A. CAMPBELL, J. C. BERMAK, *J. Org. Chem.* **1994**, *59*, 658–660.

60 D. P. ROTELLA, *J. Am. Chem. Soc.* **1996**, *118*, 12246–12247.

61 J. M. SALVINO, T. J. KIESOW, S. DARNBROUGH, R. LABAUDINIERE, *J. Comb. Chem.* **1999**, *1*, 134–139.

62 K. C. NICOLAOU, J. PASTOR, N. WINSSINGER, F. MURPHY, *J. Am. Chem. Soc.* **1998**, *120*, 5132–5133.

63 W. T. FORD, in: ACS Symposium Series 308: Polymeric Reagents and Catalysts. FORD, W. T. (ed.), ACS, Washington 1986, pp. 155–185.

64 I. HUGHES, *Tetrahedron Lett.* **1996**, *37*, 7595–7598.

65 P. WIPF, T. C. HENNINGER, *J. Org. Chem.* **1997**, *62*, 1586–1597.

66 a) Heterocycle formation: P.-P. KUNG, E. SWAYZE, *Tetrahedron Lett.* **1999**, *40*, 5651–5654; b) 684-membered library of MMP inhibitors: A. K. SZARDENINGS, D. HARRIS, S. LAM, L. SHI, D. TIEN, Y. WANG, D. V. PATEL, M. NAVRE, D. A. CAMPBELL, *J. Med. Chem.* **1998**, *41*, 2194–2200; c) benzo-diazepinediones: C. G. BOOJAMRA, K. M. BUROW, J. A. ELLMAN, *J. Org. Chem.* **1995**, *60*, 5742–5743.

67 C. P. HOLMES, J. P. CHINN, G. C. LOOK, E. M. GORDON, M. A. GALLOP, *J. Org. Chem.* **1995**, *60*, 7328–7333.

68 G. C. LOOK, M. M. MURPHY, D. A. CAMPBELL, M. A. GALLOP, *Tetrahedron Lett.* **1995**, *36*, 2937–2940.

69 a) N. M. KHAN, V. ARUMUGAM, S. BALASUBRAMANIAN, *Tetrahedron Lett.* **1996**, *37*, 4819–4822; b) C. T. BUI, F. A. RASOUL, F. ERCOLE, Y. PHAM, N. J. MAEJI, *Tetrahedron Lett.* **1998**, *39*, 9279–9282.

70 S. V. LEY, D. M. MYNETT, W.-J. KOOT, *Synlett* **1995**, 1017–1019.

71 a) G. C. LOOK, J. R. SCHULLEK, C. P.

HOLMES, J. P. CHINN, E. M. GORDON, M. A. GALLOP, *Bioorg. Med. Chem. Lett.* **1996**, *6*, 707–712; b) R. SINGH, J. M. NUSS, *Tetrahedron Lett.* **1999**, *40*, 1249–1252.

72 C. KALLUS, T. WUNBERG, unpublished results.

73 a) G. C. LOOK, M. M. MURPHY, D. A. CAMPBELL, M. A. GALLOP, *Tetrahedron Lett.* **1995**, *36*, 2937–2940; b) M. MOTOKOFF, K. REN, L. K. WONG, *J. Med. Chem.* **1992**, *35*, 4696–4703.

74 a) A. F. ABDEL-MAGID, C. A. MARYANOFF, K. G. CARSON, *Tetrahedron Lett.* **1990**, *31*, 5595–5598; b) D. W. GORDON, J. STEELE, *Bioorg. Med. Chem. Lett.* **1995**, *5*, 47–50.

75 E. E. SWAYZE, *Tetrahedron Lett.* **1997**, *38*, 8643–8646.

76 R. F. BORCH, M. D. BERNSTEIN, H. D. DURST, *J. Am. Chem. Soc.* **1971**, *93*, 2897–2904.

77 a) A. K. SZARDENINGS, T. S. BURKOTH, G. C. LOOK, D. A. CAMPBELL, *J. Org. Chem.* **1996**, *61*, 6720–6722; b) A. M.

BRAY, D. S. CHIEFRAI, R. M. VALERIO, N. J. MAEJI, *Tetrahedron Lett.* **1995**, *36*, 5081–5084.

78 S. W. KALDOR, M. G. SIEGEL, J. E. FRITZ, B. A. DRESSMAN, P. J. HAHN, *Tetrahedron Lett.* **1996**, *37*, 7193–7196.

79 a) E. E. SWAYZE, *Tetrahedron Lett.* **1997**, *38*, 8465–8468; b) A. M. FIVUSH, T. M. WILLSON, *Tetrahedron Lett.* **1997**, *38*, 7151–7154; c) commerical sources with high loadings, e.g. Rapp Polymere or NovaBiochem.

80 E. G. BROWN, J. M. NUSS, *Tetrahedron Lett.* **1997**, *38*, 8457–8460.

81 J. G. BREITANBACHER, H. C. HUI, *Tetrahedron Lett.* **1998**, *39*, 8207–8210.

82 M. G. SIEGEL, M. O. CHANEY, R. F. BRUNS, M. P. CLAY, D. A. SCHOBER, A. M. ABBEMA, D. W. JOHNSON, B. E. CANTRELL, P. J. HAHN, D. C. HUNDEN, D. R. GEHLERT, H. ZARRINMAYEH, P. L. ORNSTEIN, D. M. ZIMMERMANN, G. A. KOPPEL, *Tetrahedron* **1999**, *55*, 11619–11639.

14
Oxidation Except CC Double Bonds

Henning Steinhagen

14.1
Introduction

Oxidations belong to the very fundamental set of transformations in organic chemistry. Comparing the rich repertoire of modern oxidation reactions in solution [1], relatively few reaction types and principles have been transferred to solid phase and consequently to combinatorial chemistry. The main reason for this fact is that oxidations usually do not add diversity in the construction of compound libraries. Also, manipulations of the oxidation stage in combinatorial (multistep) synthesis are often unwanted and can be avoided by choosing the right building blocks of correct oxidation states. Nevertheless, oxidations on solid support are sometimes very useful and reliable transformations. By far the most applied transformation is the oxidation of primary and secondary alcohol functions to aldehydes and ketones. Another important class of oxidation reactions are the formation of sulfoxides and sulfones from sulfides. But also more complex oxidation types such as heterocyclic synthesis and oxidative coupling reactions are becoming more popular.

The aim of this chapter is to give an overview of useful oxidation reactions on solid phase. The reactions can be principally divided into two classes: (1) polymer-bound substrate and (2) polymer-bound oxidant. Being applied to combinatorial chemistry these two classes often correspond to solid-phase (class 1) and to solution-phase (class 2) combinatorial chemistry. Since the field of oxidation reactions on solid support [2] and especially the use of polymer-bound reagents [3] has been reviewed and published extensively, the focus of this chapter is more to present some interesting examples rather than giving a complete literature survey of all published procedures. Nevertheless, it has been an aim to supply the reader with the most relevant literature in the references.

14.2
Oxidation of Alcohols to Aldehydes and Ketones

Conversion of alcohols are among the most used and reliable oxidation reactions on solid phase. In general, many protocols from the classic repertoire of solution-phase oxidation reactions have been transferred successfully to solid phase. By choosing the appropriate reagent, synthetic limitations of the substrate (e.g. temperature instability, chemoselectivity, solvent) and also of the linker/polymer support system should be carefully considered. The most common and reliable reagents for oxidation of polymer-bound alcohols include SO_3/pyridine (Scheme 14.1) [4], Swern conditions (Scheme 14.2) [5], Dess–Martin periodinane (DMP; Scheme 14.3) [6], perruthenate (TPAP) [7] and Cr reagents (PDC [8], CrO_2Cl_2 [9], $CrO_2(Ot\text{-}Bu)_2$ [9b]; Scheme 14.4). Other published methods involve NCS [10] and IBX [11] as oxidants. There have also been some reports for the oxidation of polymer-bound phenols by various conditions mostly for coupling reactions [12, see 57, 59, 60a] (Scheme 14.14). In comparison, there are many polymer-

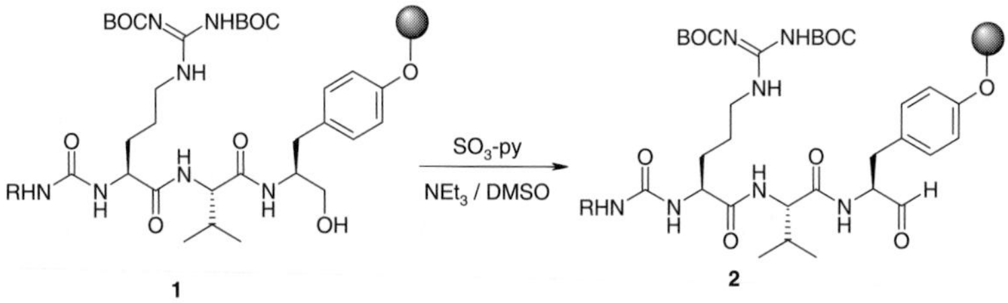

Scheme 14.1. Oxidation step in the synthesis of peptide aldehydes of the (S)-MAPI type [4f].

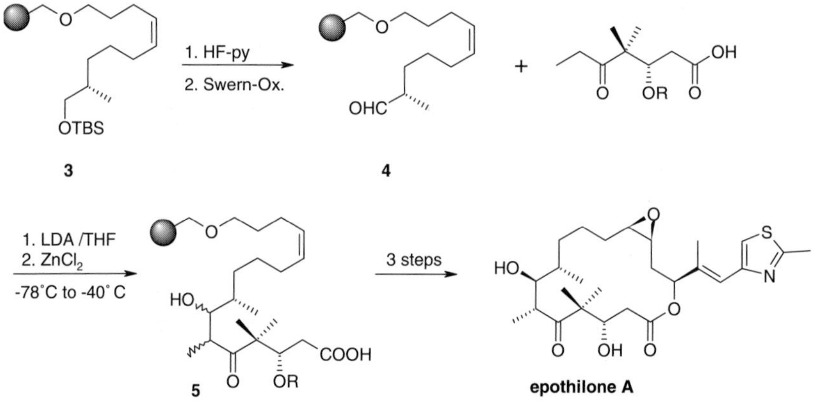

Scheme 14.2. Oxidation using the aldol strategy in the solid-phase synthesis of epothilone A [5].

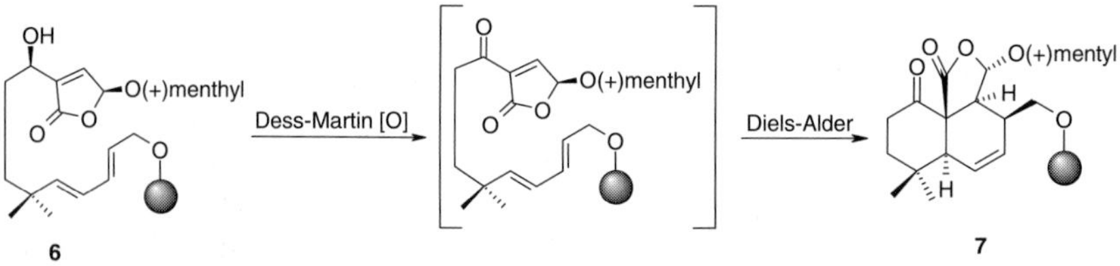

Scheme 14.3. Domino oxidation/Diels–Alder pathway to mniopetals [6e].

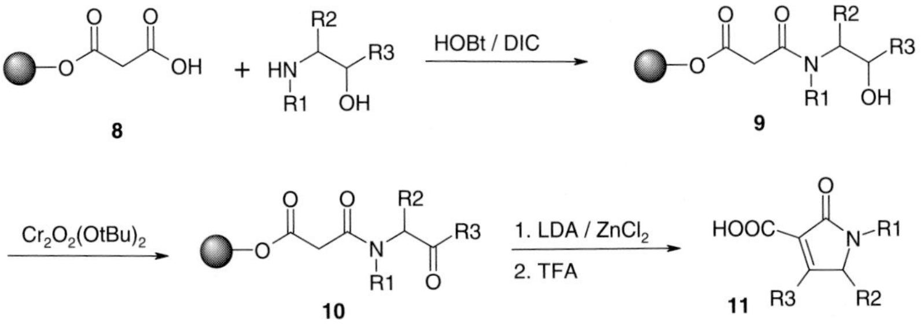

Scheme 14.4. Oxidation–cyclization strategy for the solid-phase synthesis of carboxypyrrolinones [9a].

bound oxidants available that are suitable for oxidation of alcohols in solution to the corresponding aldehydes and ketones. Poly(vinyl pyridinium dichromate) (PVPDC) is an inexpensive, easy-to-use, recyclable reagent for the oxidation of different alcohols, including primary, secondary and allylic alcohols [13]. The latter can be also oxidized using polymer-supported ammonium perchlorate, although the reagent fails with saturated alcohols [14]. Silica-based chromium reagents have also been reported as easy to prepare and use for the oxidation of several alcohols [15]. Many other convenient-to-use reagents such as KMnO$_4$ on kieselguhr [16] and ammonium chlorochromate on silica [17] have been reported to oxidize a variety of alcohols. Ammonium chlorochromate proved especially useful for the oxidation of benzoins to benzils. Nevertheless, the main problems of chromium-based reagents remain selectivity and toxicity issues and the mostly stoichiometric use. This draw back could be overcome by a very useful reagent for the oxidation of several alcohols, including primary, benzylic and secondary alcohols: polymer-supported perruthenate (PSP) [18]. The reagent can be used stoichiometrically or in catalytic amounts with molecular oxygen or an N-oxide as cooxidant. Using molecular oxygen as a cooxidant also avoids the need for conventional work-up (Scheme 14.5) [18b]. Another recently introduced useful oxidant is a TEMPO-derived oxammonium resin [19]. This powerful oxidant cleanly converts primary alcohols to aldehydes and also secondary alcohols to ketones (Scheme 14.6).

14.2.1
Examples of the Oxidation of Polymer-bound Primary Alcohols to Aldehydes

Bradley and coworkers [4f] have described an interesting example using SO$_3$/pyridine for the oxidation of polymer-bound primary alcohols in the synthesis of peptide aldehydes based on the human immunodeficiency virus (HIV) protease inhibitor (S)-MAPI. Peptide aldehydes are an important class of transition-state analogs of different proteases and have been of considerable interest since their initial discovery in natural products [20]. The efficiency and reliability of the oxidation step was dramatically improved by the incorporation of a small polyethylene glycol (PEG) spacer between the linker and the solid support (polystyrene resin).

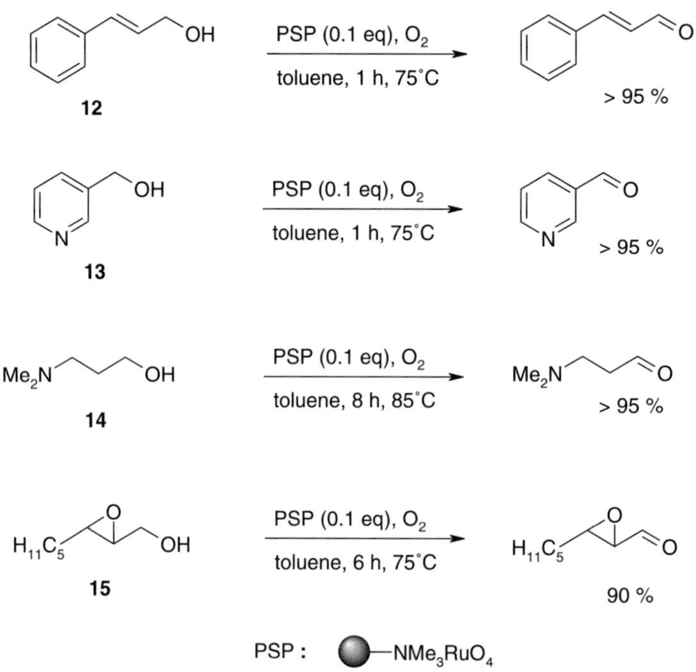

Scheme 14.5. Examples of oxidation using PSP [18].

In the course of the synthesis, the polymer-bound primary alcohol **1** was oxidized by SO_3/pyridine to the peptide aldehyde **2**, which could be readily cleaved using trifluoroacetic acid (TFA).

An interesting example of the Swern oxidation of a primary alcohol in a multi-step sequence on solid support (Merrifield resin) was demonstrated by Nicolaou et al. [5] in the synthesis of the anticancer agents epothilones A and B. In the course of the synthesis, the TBS-protected alcohol **3** was deprotected with HF/pyridine and subsequently oxidized to the aldehyde **4** (> 95% yield). The aldehyde was essential for further conversion to **5** in a stereoselective aldol reaction. After three additional steps [1, esterification; 2, metathesis upon cleavage; 3, epoxidation with methyl-(trifluoro-methyl)dioxirane], epothilone A was obtained.

14.2.2
Examples of the Oxidation of Polymer-bound Secondary Alcohols to Ketones

An interesting example using the oxidation of a secondary alcohol was demonstrated in the synthesis of a building block for a library of analogs of mniopetal [6e]. The Dess–Martin oxidation was used as an electronic activation of the di-enophile **6** for the domino oxidation/inverse intramolecular Diels–Alder reaction attached to Wang resin. The final product **7** was cleaved under oxidative conditions using 2,3-dichloro-5,6-dicyanobenzoquinone (DDQ) [21].

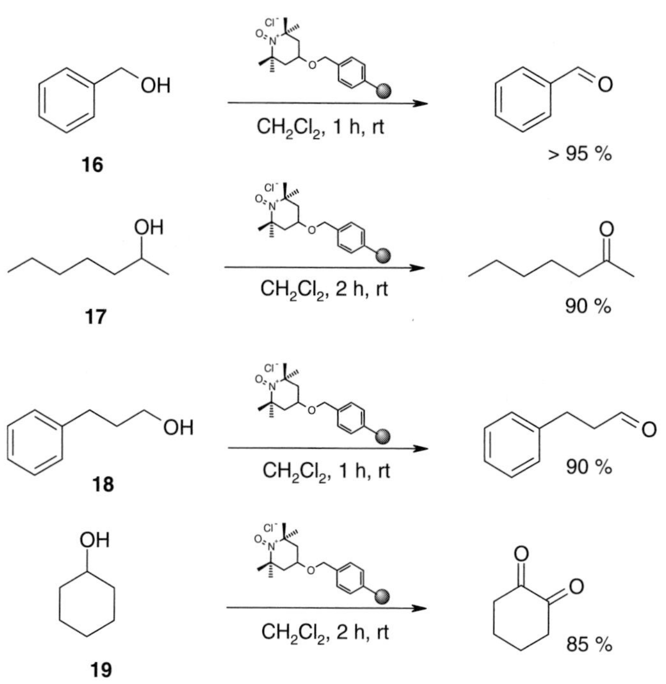

Scheme 14.6. Examples of oxidation with oxammonium resin [19].

Miller at al. presented a simple and efficient way for the preparation of carboxy-pyrrolinones on solid phase [9a]. The pyrrole system is an interesting potentially bioactive motif for both agrochemicals and pharmaceutical drugs. In this approach, Wang resin-bound malonic acid [22] 8 was coupled with an amino alcohol in the presence of 1-hydroxybenzotriazole hydrate (HOBt)/DIC, providing the hydroxyamide 9 in high yield. For the oxidation step to 10, several oxidizing reagents were tested. Under Swern conditions the conversion was not complete because the Swern reagent is unstable at temperatures above 10 °C. SO$_3$/pyridine is stable at room temperature, but the reagent was not effective for all of the examples shown, especially N-aryl-substituted compounds. The most broadly applicable reagent was CrO$_2$(Ot-Bu)$_2$ generated *in situ* according to Leznoff et al. [9b]. The oxidation of 9 to 10 proceeded successfully for most substrates, and after ring closure with lithium diisopropylamide (LDA)/ZnCl$_2$ and subsequent cleavage with TFA, the carboxypyrrolinones 11 could be obtained.

14.2.3
Examples of the Oxidation of Alcohols by Polymer-bound Reagents

Ley and Hinzen introduced polymer-supported perruthenate (PSP), which can be easily prepared by washing an anion exchange resin with an aqueous solution of potassium perruthenate [18]. The obtained reagent can be used stoichiometrically or in catalytic amounts using molecular oxygen or *N*-oxides [*N*-methylmorpholine *N*-oxide (NMO), TMAO] as cooxidant. Especially in the catalytic version (0.1 equiv. PSP, O_2 as cooxidant), this versatile reagent is very useful.

PSP cleanly oxidizes α,β-unsaturated (e.g. **12**) and benzylic alcohols (e.g. **13**) in 1 h to the corresponding ketones. Also nonactivated primary alcohols (e.g. **14**) and epoxy alcohols (e.g. **15**) are converted, but reaction times are somewhat longer. Secondary alcohols are unreactive, providing chemoselectivity with this reagent. Another useful polymer-bound oxidant is the oxammonium TEMPO resin that was introduced by Rademann and coworkers [19]. In contrast to PSP, this strong oxidant reacts with both primary alcohols to aldehydes (e.g. **16**, **18**) and secondary alcohols (e.g. **17**) to the corresponding ketones.

In substrates such as **19**, where the initial product is easy enolizable, further oxidation directly leads to the 1,2-diketo compounds. The mechanism of oxidation (Scheme 14.7) can be rationalized by the intermediate formation of an alkoxy-TEMPO salt **22** that cleaves to **23** and the corresponding oxidized compound **24**. Prior to the oxidation, the TEMPO polystyrene **20** needs to be activated by NCS, Br_2 or Cl_2, yielding the oxammonium salt **21**.

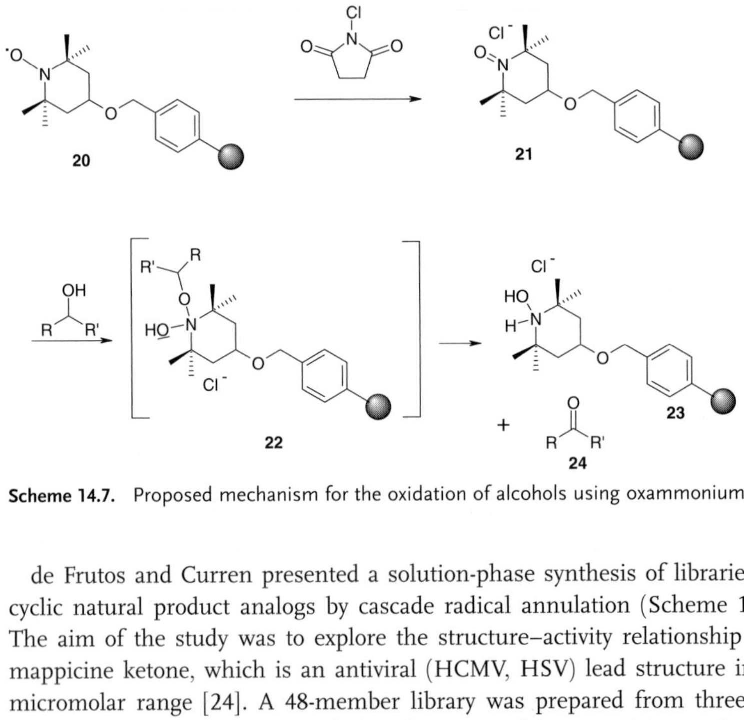

Scheme 14.7. Proposed mechanism for the oxidation of alcohols using oxammonium resin [19].

de Frutos and Curren presented a solution-phase synthesis of libraries of polycyclic natural product analogs by cascade radical annulation (Scheme 14.8) [23]. The aim of the study was to explore the structure–activity relationship (SAR) of mappicine ketone, which is an antiviral (HCMV, HSV) lead structure in the low micromolar range [24]. A 48-member library was prepared from three building blocks and the library was purified at the stage of the mappicine analogs. Therefore, it was anticipated that a clean and easily removable oxidant would be obtained with the goal of having sufficiently clean products for biological testing without

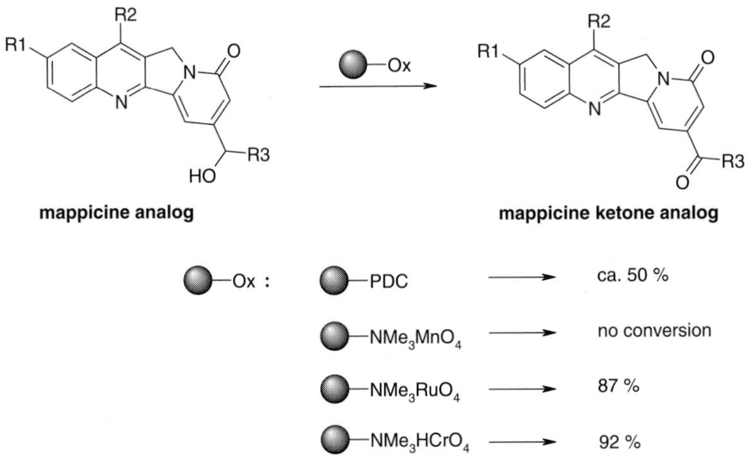

mappicine analog　　　　　　　　**mappicine ketone analog**

Scheme 14.8. Screening of polymer-bound oxidants in the synthesis of mappicine ketone analogs [23].

further chromatography. From the screened oxidants, only the polymer-supported perruthenate (PSP) and chromic acid proved to be useful, whereas other reagents such as Dess–Martin periodinane (DMP), polymer-bound PDC, and permanganate oxidation did not give good results.

14.3
Oxidation of Polymer-bound Aldehydes to Carboxylic Acids

Compared with the oxidation of alcohols to aldehydes and ketones, there are relatively few reports of the direct oxidation of aldehydes to carboxylic acids under solid-phase conditions. This is because the corresponding alcohols serve as more stable and easily protectable functional groups in both solid and liquid phase. Nevertheless, there have been some reports of the oxidation of polymer-bound aldehydes to carboxylic acids including the use of $NaClO_2$ [25] and MCPBA [26] as oxidants. An impressive example was demonstrated by Nicolaou et al. in the solid-phase synthesis of a combinatorial library of sarcodictyins (Scheme 14.9) [25a]. Sarcodictyins were discovered in 1987 in a Mediterranean coral [27a] and their potent antitumor activity (Taxol-like mode of action) was recognized by Ciomei et al. in 1997 [27b]. In the example given, the tricyclic core system **25** was constructed via multistep synthesis on Merrifield resin. After deprotection of **25** using TBAF, the primary alcohol was oxidized stepwise first to the unsaturated aldehyde **26** with DMP. After subsequent oxidation to the corresponding carboxylic acid with $NaClO_2$, esterification using MeOH/DCC led to the methyl ester **27**. The stepwise procedure gave optimal chemical yields (~95%).

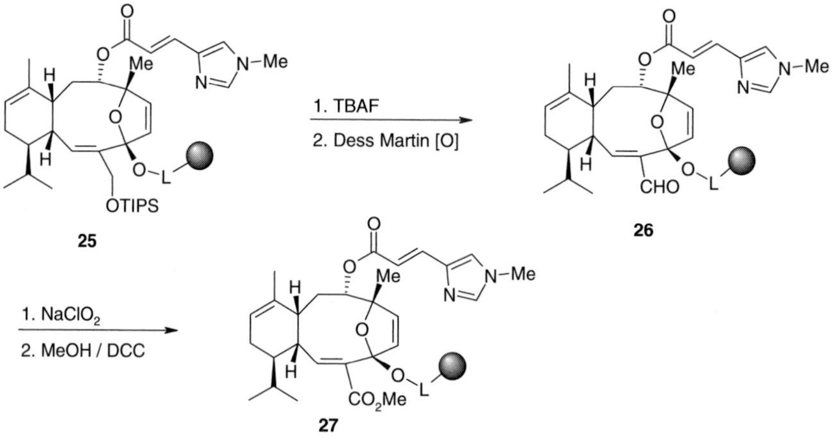

Scheme 14.9. Oxidation sequence in the solid-phase synthesis of sarcodictyins [25a].

14.4
Oxidation of Sulfur-containing Compounds

Oxidation of sulfur-containing compounds on solid phase is a quite common and usually high-yielding reaction. Most typically sulfides are converted to their corresponding sulfoxides or sulfones. Typical oxidants for polymer-bound sulfur-containing compounds are MCPBA [28, 29] and H_2O_2 [30, 31], but also less common reagents such as $NaBrO_2$ [32], OsO_4/NMO [33], and ozone [29e] have been reported. In the case of MCPBA, the oxidation can usually be controlled by the number of equivalents used (see Scheme 14.11) [28a]. Besides the oxidation of sulfides and sulfoxides, polymer-bound sulfinamides have also been oxidized to sulfonamides using different conditions [34].

Other oxidation procedures of polymer-bound sulfur-containing compounds include thioacetals, which have been oxidatively cleaved from solid support using $PhI(Tfa)_2$ [35] or H_5IO_6 [36].

14.4.1
Examples of the Oxidation of Polymer-bound Sulfides to Sulfoxides and Sulfones

There are in general two reasons for the oxidation of a sulfur-containing compound on solid phase. Either the oxidized form displays some interesting biological or chemical activity, or the oxidation procedure is part of a cleavage strategy in which, for example, the sulfoxide function is cleaved by elimination or the sulfone is directly displaced with a nucleophile. A typical example of a cleavage procedure using a sulfone was presented by Gayo and Suto in the synthesis of substituted aminopyrimidines (Scheme 14.10) [29b]. The solid-supported (TG Thiol resin)

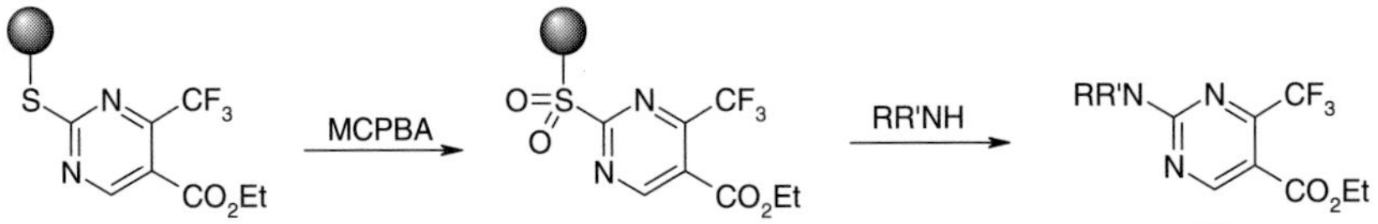

Scheme 14.10. Oxidation–cleavage strategy for the synthesis of aminopyrimidines [29b].

pyrimidine **28** was oxidized to the corresponding sulfone **29** with MCPBA and subsequently cleaved through nucleophilic displacement by an amine, yielding the substituted 2-amino pyrimidine **30**.

An interesting example for the oxidation of sulfides to either sulfoxides or sulfones with MCPBA was demonstrated by Mata in the synthetic approach to β-lactams on solid support (Merrifield and Wang resin) (Scheme 14.11) [28a]. The carboxylic acid **31** was loaded onto Merrifield resin yielding **32**, and then either oxidized to the sulfoxide **33** with 1.4 equiv. MCPBA at 0 °C or to the sulfone **34** using a fivefold excess of MCPBA at room temperature. Eventually, the β-lactams could be cleaved using AlCl₃.

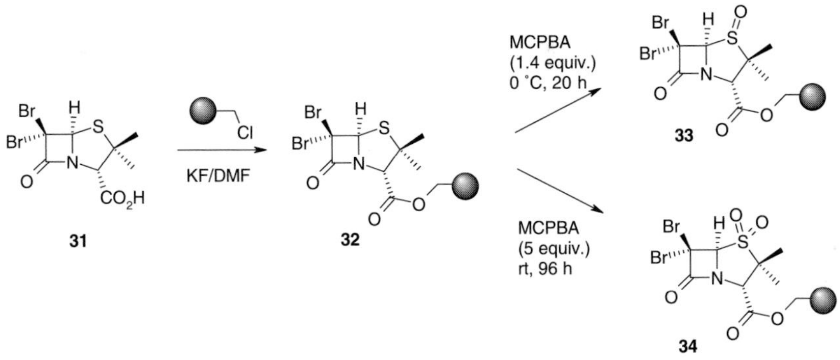

Scheme 14.11. Oxidation sequence of β-lactams to sulfoxides and sulfones [28a].

Grimstrup and Zaragoza presented an interesting synthetic route to highly functionalized benzamides on solid phase (Wang resin) by a facile nucleophilic substitution pathway (Scheme 14.12) [29c]. The thiol **35**, which was derived from the corresponding fluoro compound and RSH, was first oxidized to the sulfone **36**. This highly electron-deficient aryl fluoride was now easily substituted by a secondary amine R'R''NH, yielding **37**, which could be readily cleaved to **38** from solid support upon treatment with TFA. The order of the oxidation step and the displacement step were in this case very important, since MCPBA also oxidizes the anilide amine if the nucleophilic displacement with the amine was performed directly on **35** followed by MCPBA oxidation.

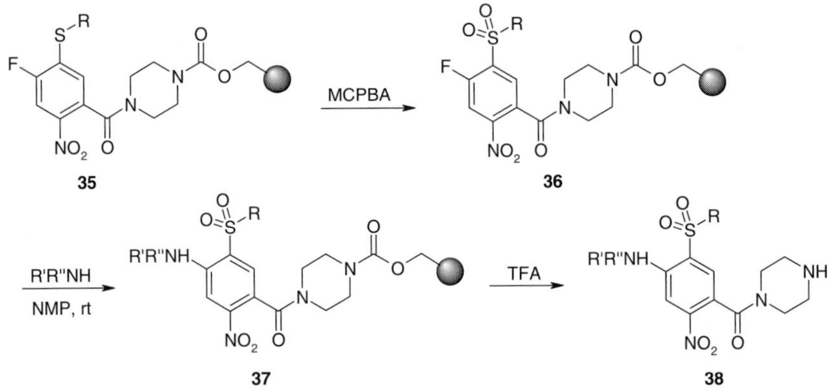

Scheme 14.12. Nucleophilic substitution strategy to highly functionalized benzamides on solid support [29c].

14.5
Oxidation of Selenium- and Phosphorus-containing Compounds

Compared with oxidations of sulfur-containing compounds on solid support, there have been fewer examples of oxidation of selenium- or phosphorus-containing compounds. Nevertheless, selenium compounds have attracted some attention mainly because of their use as linker systems which can be easily cleaved under oxidative conditions (see Chapter 4.3.7.3). Compared with nonpolymer-bound selenium compounds [37], which are usually problematic because of their odor and toxicity, the solid-phase-bound analogs are both totally odorless and convenient to use. The oxidation of selenides to the corresponding selenoxides is usually performed using H_2O_2 [38], but also with other reagents such as $NaIO_4$ [39], have been reported. Oxidation of phosphorus-containing compounds is relatively rare and often linked to the synthesis of oligonucleotides [40]. Phosphites have been oxidized to phosphates using NMO [41] and iodine [42]. Phosphonates have also been oxidized to phosphates using iodine [40].

14.5.1
Examples of the Oxidation (Cleavage) of Selenides to Selenoxides on Solid Support

A useful and versatile application of selenium-based solid-phase synthesis was presented by Nicolaou et al. in the combinatorial synthesis of benzopyran-based natural products (Scheme 14.13) [38c,d]. Various substituted *ortho*-prenylated phenols **39** were directly loaded onto the selenium-based solid support (polystyrene) leading to formation of the 2,2-dimethylbenzopyran system **40**. This very mild and selective procedure exhibited a broad tolerance toward a wide range of poly-functionalized aromatic compounds. In all cases, the resin could be quantitatively

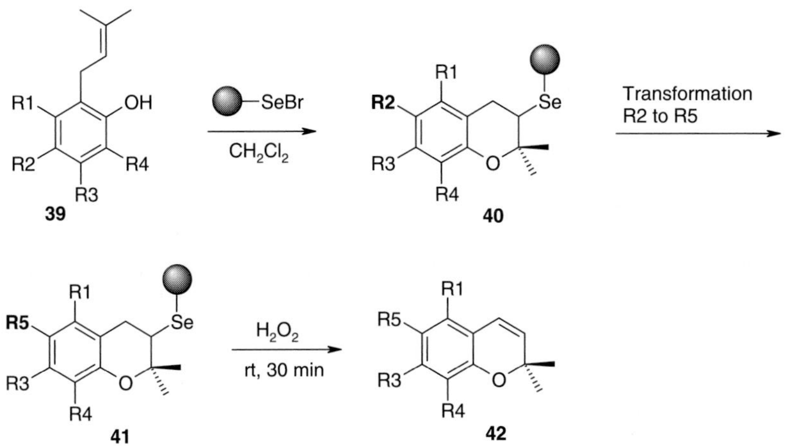

Scheme 14.13. Combinatorial synthesis of benzopyran-based natural products using a selenium linker system [38c,d].

loaded using a threefold excess of **39**. The further strategy in the course of constructing a combinatorial library contained various chemical manipulations (in general transformation of **R2** to **R5**). These transformations included annulations, glycosidations, aldol condensations, and various coupling reactions, yielding the products represented by formula **41**. The selenium oxidation step readily cleaved the products, yielding **42** in high purity and efficiency with the introduction of a double bond.

14.6
Oxidative Formation of Heterocycles on Solid Support

Solid-phase synthesis of heterocycles has become a central part of research in the field of combinatorial chemistry mainly because of the great structural variety and potential biological activity. Familiar examples include quinolones, β-lactams and dihydropyridines. There are many methods for the synthesis of heterocycles using oxidations as important synthesis steps (compare Chapter 22). In general, the main synthetic approaches can be divided into three categories represented by: A, oxidative aromatization; B, oxidative cyclization or condensation; and C, formation of heterocycles upon oxidative cleavage. Examples of A include the synthesis of pyridines [43], quinazolinones [44] and isoquinolines [45]. Examples of B include the synthesis of indolizines [46], pyrimidines [47], pyridazines [48], isoxazolines [49], benzodiazepines [50], indazoles [51] and benzimidazoles [52]. Polymer-bound quinolines have been activated for coupling by oxidation to the corresponding N-oxides using MCPBA [53]. Examples of C include the synthesis of isoxazolines [54], γ-butyrolactones [55] and phenanthridines [56].

14.7
Oxidative Coupling and Cleavage Reactions on Solid Support

There are relatively few reports of oxidative coupling reactions on solid support. Nevertheless, some quite complex reaction sequences have been successfully performed. Phenols have been coupled to cyclic biarylethers using phenolic oxidation employing thallium(III) salts [57]. Biaryls have been formed by oxidative coupling using 1,3-dinitrobenzene (see Scheme 14.15) [58] and vanadium-based oxidants [59]. Oxidative homocouplings of *ortho*-hydroxystyrenes yielding carpanones have been performed using PhI(OAc)$_2$, as described in Scheme 14.14 [60a].

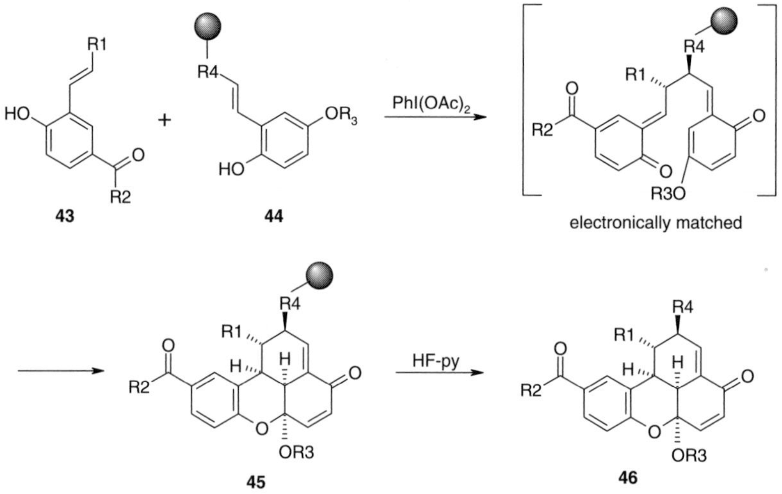

Scheme 14.14. Oxidative coupling strategy for the synthesis of carpanones [60].

In contrast to oxidative couplings, there have been many reports of cleavage reactions under oxidative conditions (compare Section 14.7.2). Most of these reports directly refer to linker cleavage strategies (see Chapter 4.5). An interesting application is the preparation of peptide aldehydes using the cleavage of amino alcohols [61] and diols [62] with NaIO$_4$ (see Scheme 14.16).

Other applications of oxidative cleavage procedures include arylethers which have been cleaved using cerium ammonium nitrate (CAN) as an oxidant [63]. Benzylethers have been cleaved from solid support (Wang resin) using DDQ oxidation [64] and benzylamines which have been cleaved from Merrifield *p*-benzyloxybenzylamine (BOBA) resin by DDQ [65]. Aryl hydrazides have been used as linker systems and can be oxidatively cleaved using different conditions including Cu(OAc)$_2$ and NBS [66].

14.7.1
Examples of Oxidative Coupling Reactions on Solid Support

Shair and coworkers presented a biomimetic approach to carpanone-like molecules through oxidative heterocoupling on solid support (Scheme 14.14, PS-DES resin, silyl and trityl linkers) [60a] based on the previous synthesis in solution phase by Chapman et al. [67]. In a first step, phenols **43** and **44** were oxidatively coupled with PhI(OAc)$_2$ and directly transformed to the Diels–Alder adduct **45**. The choice of the oxidant was crucial to this step, since the reacting phenols had to be electronically differentiated and coupled in the anticipated way. The inverse electron demand Diels–Alder reaction proceeded via an electronically matched transition state. Cleavage from solid support with HF/pyridine yielded the carpanone-like products **46**. Experiments are under way to generate a split-and-pool synthesis of a 100,000-member library and high-throughput biological screens [60b]. Schreiber and coworkers presented a stereoselective synthesis of a biaryl-containing medium ring system using the oxidation of biaryl cuprates on solid support (polystyrene resin, silicon linker) (Scheme 14.15) [58a].

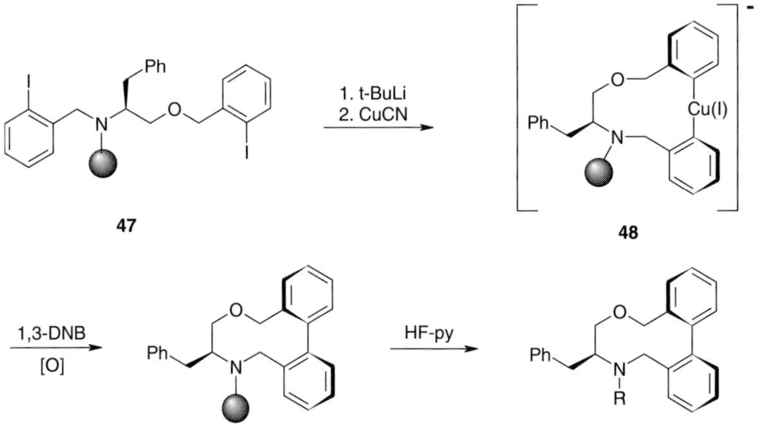

Scheme 14.15. Intramolecular biaryl synthesis using oxidation of cuprates [58a].

In the first step, the attached biarylic compound **47** was converted to the intermediary cyclic cuprate **48** using lithiation and treatment with copper cyanide. Subsequently, the biaryl ten-membered ring **49** was formed upon exposure to the oxidant 1,3-dinitrobenzene (1,3-DNB). After cleavage from solid support using HF/pyridine the product **50** could be isolated in a diastereomeric ratio of 6:1 (P/M). A disadvantage of the solid-phase approach was the decreased diastereomeric ratio, which was better (20:1) in the analog reaction in solution phase.

14.7.2
Examples of Oxidative Cleavage Reactions on Solid Support

A useful route to the synthesis of C-terminal α-oxo aldehydes on solid support was presented by Malnyk and coworkers [62]. As already described in Section 14.2.1, peptide aldehydes are an important class of potentially bioactive compounds [20]. Since peptide aldehydes are prone to epimerization, considerable attention has been focused on the construction of a new appropriate linker system to allow mild reaction conditions. The strategy that was chosen (Scheme 14.16) contained a new linker system as a key element that could be removed in a two-step procedure using TFA and NaIO$_4$ as oxidants. The peptide to be converted into the peptide aldehyde was first loaded onto the linker moiety **51** using standard solid-phase peptide synthesis (SPPS), yielding **52**, and then subsequently transferred to diol **53**. Mild cleavage of the diol using NaIO$_4$ yielded the desired peptide aldehyde **54**.

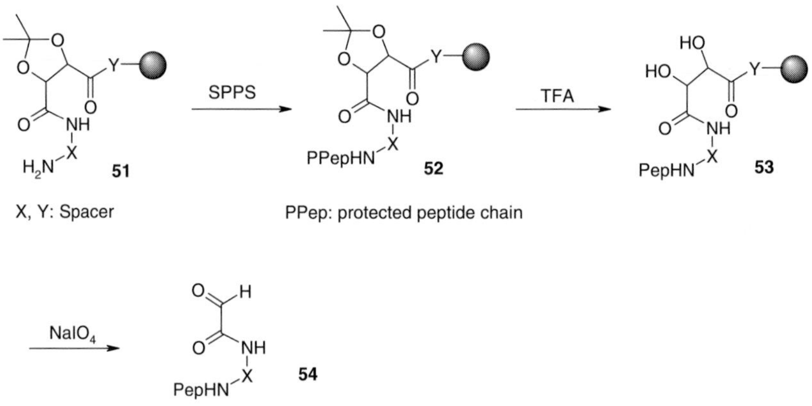

Scheme 14.16. Synthesis of α-oxo aldehydes using formyl transfer from the linker system [62].

References

1 a) R. C. LAROCK, *Comprehensive Organic Transformations*, 2nd edn. Wiley-VCH, New York 1999; b) M. HUDLICKY, Oxidations in Organic Chemistry, ACS Monograph 186, Washington, DC 1990.

2 a) F. Z. DÖRWALD, *Organic Synthesis on Solid Phase*. Wiley-VCH, Weinheim 2000; b) G. JUNG in: Combinatorial Chemistry. BANNWARTH, W. and FELDER, E. (eds), Wiley-VCH, Weinheim 1999; c) W. BANNWARTH, E. FELDER in: Combinatorial Chemistry. BANNWARTH, W. and FELDER, E. (eds), Wiley-VCH, Weinheim 2000; d) B. A. BUNIN, *The Combinatorial Index*. Academic Press, San Diego 1998; e) D. OBRECHT, J. M. VILLALGORDO, *Solid-Supported Combinatorial and Parallel Synthesis of Small-Molecular-Weight Compound Libraries*. Pergamon, Oxford 1998; dynamic internet databases of references: f) http://www.combinatorial.com/; g) http://www.5z.com/divinfo/.

3 a) Key reference including literature coverage until 2000: S. V. LEY, I. R. BAXENDALE, R. N. BREAM, P. S. JACKSON, A. G. LEACH, D. A. LONGBOTTOM, M. NESI, J. S. SCOTT, R. I. STORER, S. J. TAYLOR, *J. Chem. Soc., Perkin Trans 1* **2000**, 3815–4195; b) D. H. DREWRY, D. M. COE, S. POON, *Med. Res. Rev.* **1999**, *19*, 97–148; c) R. T. TAYLOR, *Polymer-Bound Oxidizing Agents, ACS Symposium Series* **1986**, *308*, 42; d) Novabiochem-Catalogue, Polymer Supported Reagents Handbook 2001; e) B. HINZEN in: Combinatorial Chemistry. BANNWARTH, W. and FELDER, E. (eds), Wiley-VCH, Weinheim 2000.

4 a) C. CHEN, L. A. AHLBERG RANDALL, R. B. MILLER, A. D. JONES, M. J. KURTH, *J. Am. Chem. Soc.* **1994**, *116*, 2661; b) B. BORHAN, J. A. WILSON, M. J. GASCH, Y. KO, D. M. KURTH, M. J. KURTH, *J. Org. Chem.* **1994**, *60*, 7375; c) C. CHEN, L. A. AHLBERG RANDALL, R. B. MILLER, A. D. JONES, M. J. KURTH, *Tetrahedron* **1997**, *53*, 6595; d) M. REGGELIN, V. BRENIG, R. WELCKER, *Tetrahedron Lett.* **1998**, *39*, 4801; e) D. P. ROTELLA, *J. Am. Chem. Soc.* **1996**, *118*, 12246; f) P. PAGE, M. BRADLEY, I. WALTERS, S. TEAGUE, *J. Org. Chem.* **1999**, *64*, 794; g) secondary alcohol: P. S. FURTH, M. S. REITMANN, A. F. COOK, *Tetrahedron Lett.* **1997**, *38*, 5403; h) benzylic alcohol: M. FIVUSH, T. M. WILSON, *Tetrahedron Lett.* **1997**, *38*, 7151.

5 a) K. C. NICOLAOU, N. WINSSINGER, J. PASTOR, S. NINKOVIC, F. SARABIA, Y. HE, D. VOURLOUMIS, Z. YANG, T. LI, P. GIANNAKAKOU, E. HAMEL, *Nature* **1997**, *387*, 268; b) K. C. NICOLAOU, D. VOURLOUMIS, J. PASTOR, N. WINSSINGER, Y. HE, S. NINKOVIC, F. SARABIA, H. VALLBERG, F. ROHSCHANGAR, M. P. KING, M. R. V. FINLEY, P. GIANNAKAKOU, P. VERDIER-PINARD, E. HAMEL, *Angew. Chem. Int. Ed. Engl.* **1997**, *36*, 2097; c) M. MARX, A. L. GRILLOT, C. T. LOUER, K. A. BEAVER, P. A. BARTLETT, *J. Am. Chem. Soc.* **1997**, *119*, 6153.

6 a) K. C. NICOLAOU, N. WINSSINGER, D. VOURLOUMIS, T. OHSHIMA, S.

KIM, J. PFEFFERKORN, T. LI, *J. Am. Chem. Soc.* **1998**, *120*, 10814; b) K. C. NICOLAOU, J. PASTOR, N. WINSSINGER, F. MURPHY, *J. Am. Chem. Soc.* **1998**, *120*, 5132; c) M. REGGELIN, V. BRENIG, R. WELCKER, *Tetrahedron Lett.* **1998**, *39*, 4801; d) secondary alcohol: L. A. THOMPSON, F. L. MOORE, Y.-C. MOON, J. A. ELLMANN, *J. Org. Chem.* **1998**, *63*, 2066; e) secondary alcohol: U. REISER, J. JAUCH, *Synlett* **2001**, *1*, 90.

7 a) Benzylic alcohol: W. LI, B. YAN, *J. Org. Chem.* **1998**, *63*, 4092; b) secondary benzylic alcohol: B. YAN, Q. SUN, J. R. WAREING, C. F. JEWELL, *J. Org. Chem.* **1996**, *61*, 8765.

8 a) A. M. BRAY, D. S. CHIEFARI, R. M. VALERIO, N. JOE MAEJI, *Tetrahedron Lett.* **1995**, *28*, 5081; b) J. A. ELLMANN, *Proceedings of the Solid-Phase Synthesis of Prostaglandins Conference.* 209th ACS National Meeting, Anaheim, CA, 1995.

9 a) Secondary alcohol: P. C. MILLER, T. J. OWEN, J. M. MOLYNEAUX, J. M. CURTIS, C. R. JONES, *J. Comb. Chem.* **1999**, *1*, 223; b) C. C. LEZNOFF, T. M. FYLES, J. WEATHERSTON, *Can. J. Chem.* **1977**, *55*, 1143.

10 a) K. C. NICOLAOU, N. WINSSINGER, J. PASTOR, F. MURPHY, *Angew. Chem., Int. Ed. Engl.* **1998**, *37*, 2534.

11 a) W. H. PEARSON, R. B. CLARK, *Tetrahedron Lett.* **1997**, *38*, 7669; b) U. REISER, J. JAUCH, *Synlett* **2001**, *1*, 90; c) secondary alcohol: T. L. BOEHM, H. D. H. SHOWALTER, *J. Org. Chem.* **1996**, *61*, 6498.

12 P. A. TEMPEST, R. W. ARMSTRONG, *J. Am. Chem. Soc.* **1997**, *119*, 7607.

13 J. M. J. FRECHET, P. DARLING, M. J. FARRALL, *J. Org. Chem.* **1981**, *46*, 1728.

14 H. YANG, B. LI, *Synth. Comm.* **1991**, *21*, 1521.

15 a) B. KHADILKAR, A. CHITNAVIS, A. KHARE, *Synth. Comm.* **1996**, *26*, 205; b) J. G. LEE, J. A. LEE, S. Y. SOHN, *Synth. Comm.* **1996**, *2*, 543.

16 L. D. LOU, W.-X. LOU, *Synth. Comm.* **1997**, *27*, 3697.

17 G. S. ZHANG, Q.-Z. SHI, M.-F. CHEM, K. CAI, *Synth. Comm.* **1997**, *27*, 3691.

18 a) B. Hinzen, S. V. Ley, *J. Chem. Soc., Perkin 1* **1997**, 1907; b) B. Hinzen, R. Lenz, S. V. Ley, *Synthesis* **1998**, 977.

19 S. Weik, G. Nicholson, G. Jung, J. Rademann, *Angew. Chem., Int. Ed. Engl.* **2001**, *40*, 1436.

20 a) T. Aoyagi, T. Takeuchi, A. Matsuzaki, M. Kawamura, M. Kondo, M. Hamada, K. Maeda, H. Umezawa, *J. Antibiot.* **1969**, *22*, 283; b) J. O. Westerik, R. Wolfenden, *J. Biol. Chem.* **1972**, *247*, 8195.

21 a) S. Kobayashi, R. Akiyama, *Tetrahedron Lett.* **1998**, *39*, 9211; b) T. L. Deegan, O. W. Gooding, S. Baudart, J. A. Proco, *Tetrahedron Lett.* **1997**, *38*, 4973.

22 B. C. Hamper, S. A. Kolodziej, A. M. Scates, *Tetrahedron Lett.* **1998**, *39*, 2047.

23 O. de Frutos, D. P. Curran, *J. Comb. Chem.* **2000**, *2*, 639.

24 a) T. R. Govindachari, K. R. Ravidranath, N. Viswanathan, *J. Chem. Soc., Perkin Trans 1* **1974**, 1215; b) I. Pendrak, R. Wittrock, W. D. Kingsbury, *J. Org. Chem.* **1995**, *60*, 2912.

25 a) K. C. Nicolaou, N. Winssinger, D. Vourloumis, T. Ohshima, S. Kim, J. Pfefferkorn, T. Li, *J. Am. Chem. Soc.* **1998**, *120*, 10814; b) J. C. H. M. Wijkmans, A. J. Culshaw, A. D. Baxter, *Mol. Diversity* **1998**, *3*, 117.

26 a) X. Beebe, N. E. Schore, M. J. Kurth, *J. Am. Chem. Soc.* **1992**, *114*, 10061; b) X. Beebe, C. L. Chiappari, M. J. Kurth, N. E. Schore, *J. Org. Chem.* **1993**, *58*, 7320; c) J. T. Ayres, C. K. Mann, *J. Polym. Sci., Polym. Lett. Ed.* **1965**, *3*, 505.

27 a) M. D'Ambrosio, A. Guerriero, F. Pietra, *Helv. Chim. Acta* **1987**, *70*, 2019. b) M. Ciomei, C. Albanese, W. Pastori, M. Grandi, F. Pietra, M. D'Ambrosio, A. Guerriero, C. Battistini, *Proc. Am. Assoc. Cancer Res.* **1997**, *38*, 5, *Abstract 30*.

28 Oxidation to sulfoxides: a) E. G. Mata, *Tetrahedron Lett.* **1998**, *39*, 5287; b) T. Masquelin, D. Sprenger, R. Baer, F. Gerber, Y. Mercadal, *Helv. Chim. Acta* **1998**, *81*, 646; c) M. Patek, B. Drake, M. Lebl, *Tetrahedron Lett.*

1995, *36*, 2227; d) C. P. Holmes, J. P. Chinn, G. C. Look, E. M. Gordon, M. A. Gallop, *J. Org. Chem.* **1995**, *60*, 7328.

29 Oxidation to sulfones: a) J. S. Patek, B. Zhu, *Tetrahedron Lett.* **1996**, *37*, 8151; b) L. M. Gayo, M. J. Suto, *Tetrahedron Lett.* **1997**, *38*, 211; c) M. Grimstrup, F. Zaragoza, *Eur. J. Org. Chem.* **2001**, 3233; d) U. Grabowska, A. Rizzo, K. Farnell, M. Quibell, *J. Comb. Chem.* **2000**, *2*, 475; e) F. E. K. Kroll, R. Morphy, D. Rees, D. Gani, *Tetrahedron Lett.* **1997**, *38*, 8573; f) M. Yamada, T. Miyajima, H. Horikawa, *Tetrahedron Lett.* **1998**, *39*, 289; g) B. A. Kulkarni, A. Ganesan, *Tetrahedron Lett.* **1999**, *40*, 5633; h) D. Obrecht, C. Abrecht, A. Grieder, J. M. Villalgordo, *Helv. Chim. Acta* **1997**, *80*, 65; i) C. Garcia-Echeverria, *Tetrahedron Lett.* **1997**, *38*, 8933; k) A. Barco, S. Benneti, C. De Risi, P. Machetti, G. P. Pollini, V. Zanirato, *Tetrahedron Lett.* **1998**, *39*, 7591; l) C. J. Burns, D. Robert, L. M. Salvino, G. McGeehan, S. M. Condon, R. Morris, M. Morrissette, R. Mathew, S. Darnbrough, K. Neuenschwander, A. Scotese, S. W. Djuric, J. Ullrich, R. Labaudiniere, *Angew. Chem., Int. Ed. Engl.* **1998**, *37*, 2848; m) F. Gosselin, M. Di Renzo, T. H. Ellis, W. B. Lubell, *J. Org. Chem.* **1996**, *61*, 7980; n) E. Flanigan, G. R. Marshal, *Tetrahedron Lett.* **1970**, *11*, 2403; PEG-bound thioethers: o) X. Y. Zhao, K. W. Jung, K. D. Janda, *Tetrahedron Lett.* **1997**, *38*, 977; p) X. Y. Zhao, K. D. Janda, *Tetrahedron Lett.* **1997**, *38*, 5437. q) K. C. Nicolaou, S. A. Snyder, A. Bigot, J. A. Pfefferkorn, *Angew. Chem.* **2000**, *112*, 1135. *Angew Chem. Int. Ed. Engl.* **2000**, *39*, 1036.

30 Oxidation to sulfoxides: H. R. Russel, R. W. A. Luke, M. Bradley, *Tetrahedron Lett.* **2000**, *41*, 5287.

31 Oxidation to sulfones: D. L. Marshall, I. E. Liener, *J. Org. Chem.* **1970**, *35*, 867.

32 Oxidation to sulfoxides: D. M. R. Cody, S. H. DeVitt, J. C. Hodges,

J. S. Kiely, W. H. Moss, M. R. Pavia,
B. D. Roth, M. C. Schroeder, C. J.
Stankovic, US patent no. 5324483,
1994.

33 Oxidation to sulfones: H. S. Han,
K. D. Janda, *Tetrahedron Lett.* **1997**,
38, 1527.

34 D. B. D. De Bont, W. J. Moree, R. M.
J. Liskamp, *Bioorg. Med. Chem.* **1996**,
4, 667.

35 C. M. Huwe, H. Künzer, *Tetrahedron
Lett.* **1999**, *40*, 683.

36 a) H. B. Lee, S. Balasubramanian,
J. Org. Chem. **1999**, *64*, 3454; b) V.
Bertini, F. Lucchesini, M. Pocci,
A. De Munno, *Tetrahedron Lett.* **1998**,
39, 9263.

37 a) K. C. Nicolaou, N. A. Petasis,
Selenium in Natural Product Synthesis.
CIS, Philadelphia, PA 1984; b) D.
Liotta, *Organoselenium Chemistry.*
Wiley, New York 1986.

38 a) K. C. Nicolaou, J. Pastor, S.
Barluenga, N. Winssinger, *Chem.
Comm.* **1998**, 1974; b) T. Ruhland,
K. Anderson, H. Pedersen, *J. Org.
Chem.* **1998**, *63*, 9204; c) K. C.
Nicolaou, J. A. Pfefferkorn, G.-Q.
Cao, *Angew. Chem.* **2000**, *112*, 750;
d) K. C. Nicolaou, G.-Q. Cao, J. A.
Pfefferkorn, *Angew. Chem.* **2000**,
112, 755; e) R. Michels, M. Kato,
W. Heitz, *Makromol. Chem.* **1976**,
177, 2311; f) H. Russell, R. W. A.
Luke, M. Bradley, *Tetrahedron Lett.*
2000, *41*, 5287.

39 M. J. Kurth, L. A. Ahlberg Randall,
K. Takenouchi, *J. Org. Chem.* **1996**,
61, 8755.

40 F. Eckstein, *Oligonucleotides and
Analogues; A Practical Approach.*
Oxford University Press, Oxford 1991.

41 C. A. Metcalf III, C. B. Vu, R.
Sundaramoorthi, V. A. Jacobsen,
E. A. Laborde, J. Green, Y. Green,
K. J. Macek, T. J. Merry, S. G.
Pradeepan, M. Uesugi, V. M.
Varkhedkar, D. A. Holt, *Tetrahedron
Lett.* **1998**, *39*, 3435.

42 a) K. K. Ogelvie, M. J. Nemer,
Tetrahedron Lett. **1980**, *21*, 4159; b) M.
D. Matteucci, M. H. Caruthers,
Tetrahedron Lett. **1980**, *21*, 3185.

43 a) M. F. Gordeev, D. V. Patel, J. Wu,

E. M. Gordon, *Tetrahedron Lett.* **1996**,
37, 4643; b) S. Tadasse, A. Bhandari,
M. A. Gallop, *J. Comb. Chem.* **1999**,
1, 184.

44 J. P. Mayer, G. S. Lewis, M. J.
Curtis, J. W. Zhang, *Tetrahedron
Lett.* **1997**, *38*, 8445.

45 S. Berteina, A. De Maesmaeker,
Tetrahedron Lett. **1998**, *39*, 5759.

46 D. A. Goff, *Tetrahedron Lett.* **1999**, *40*,
8741.

47 a) A. L. Marzinzik, E. R. Felder,
J. Org. Chem. **1998**, *63*, 723; b) B. C.
Hamper, K. Z. Gan, T. J. Owen,
Tetrahedron Lett. **1999**, *40*, 4973.

48 a) A. M. Boldi, C. R. Johnson, H. O.
Eissa, *Tetrahedron Lett.* **1999**, *40*, 619;
b) C. O. Ogbu, M. N. Qabar, P. D.
Boatman, J. Urban, J. P. Meara, M.
D. Ferguson, J. Tulinsky, C. Lum,
S. Babu, M. A. Blaskovich, H.
Nakanishi, F. Q. Ruan, B. L. Cao,
R. Minarik, T. Little, S. Nelson,
M. Nguyen, A. Gall, M. Kahn,
Bioorg. Med. Chem. Lett. **1998**, *8*, 2321.

49 a) R. C. Johnson, B. Zhang, PCT
WO 0116116 A1; b) J.-F. Cheng, A.
M. M. Mjalli, *Tetrahedron Lett.* **1998**,
39, 939; c) B. B. Shankar, D. Y.
Yang, S. Girton, A. K. Ganguly,
Tetrahedron Lett. **1998**, *39*, 2447.

50 J. M. Berry, P. W. Howard, D. E.
Thurston, *Tetrahedron Lett.* **2000**, *41*,
6171.

51 B. Yan, H. Gstach, *Tetrahedron Lett.*
1996, *37*, 8325.

52 D. Tumelty, M. K. Schwarz, K. Cao,
M. C. Needels, *Tetrahedron Lett.* **1999**,
40, 6185.

53 M. Z. Hoemann, A. Melikian-
Badalian, G. Kumaravel, J. R. Hauske,
Tetrahedron Lett. **1998**, *39*, 4749.

54 S. Kobayashi, R. Akiyama,
Tetrahedron Lett. **1998**, *39*, 9211.

55 Y. Watanabe, S. Ishikawa, G. Takao,
T. Toru, *Tetrahedron Lett.* **1999**, *40*,
3411.

56 W.-R. Li, N.-M. Hsu, H.-H. Chou, S.
T. Lin, Y.-S. Lin, *Chem. Comm.* **2000**,
401.

57 a) K. Nakamura, H. Nishiya, S.
Nishiyama, *Tetrahedron Lett.* **2001**, *42*,
6311; b) S. Yamamura, S. Nishiyama,
J. Synth. Org. Chem. Jpn. **1997**, *55*, 1029.

58 a) D. R. Spring, S. Krishnan, S. L. Schreiber, *J. Am. Chem. Soc.* **2000**, *122*, 5656; b) F. Ullmann, J. Bielecki, *Chem. Ber.* **1901**, *34*, 2174.

59 J. W. ApSimon, D. M. Dixit, *Can. J. Chem.* **1982**, *60*, 368.

60 a) C. W. Lindsey, L. K. Chan, B. C. Goess, R. Joseph, M. D. Shair, *J. Am. Chem. Soc.* **2000**, *122*, 422; b) biological activity of carpanone-like molecules: S. Maeda, H. Masuda, T. Tokoroyama, *Chem. Pharm. Bull.* **1995**, *43*, 935.

61 J. Rademann, M. Meldal, K. Bock, *Chem. Eur. J.* **1999**, *5*, 1218.

62 a) O. Melnyk, J.-S. Fruchart, C. Grandjean, H. Gras-Masse, *J. Org. Chem.* **2001**, *66*, 4153; b) J.-S. Fruchart, H. Gras-Masse, O. Melnyk, *Tetrahedron Lett.* **1999**, *40*, 6225.

63 a) A. Peyman, C. Weiser, E. Uhlmann, *Bioorg. Med. Chem. Lett.* **1995**, *5*, 2469; b) K. Fukase, K. Egusa, Y. Nakai, S. Kusumoto, *Mol. Diversity* **1997**, *2*, 182; c) H. P. Nestler, P. A. Bartlett, W. C. Still, *J. Org. Chem.* **1994**, *59*, 4723.

64 a) K. Fukase, Y. Nakai, K. Egusa, J. A. Porco, S. Kusumoto, *Synlett* **1999**, 1074; b) Y. Ito, T. Ogawa, *J. Am. Chem. Soc.* **1997**, *119*, 5562.

65 S. Kobayashi, Y. Aoki, *Tetrahedron Lett.* **1998**, *39*, 7345.

66 a) C. Millington, R. Quarrell, G. Lowe, *Tetrahedron Lett.* **1998**, *39*, 7201; b) F. Stieber, U. Grether, H. Waldmann, *Angew. Chem., Int. Ed. Engl.* **1999**, *38*, 1073.

67 O. L. Chapman, M. R. Engel, J. P. Springer, J. C. Clardy, *J. Am. Chem. Soc.* **1971**, *93*, 6696.

15
Reductions in Combinatorial Synthesis

Christopher P. Corrette and Conrad W. Hummel

15.1
Introduction

Reductions are one of the most fundamental and widely used transformations in organic chemistry. The reduction of a wide variety of functional groups has been exploited in both solid- and solution-phase combinatorial library synthesis. There are a number of reviews that cover combinatorial chemistry, including reductions [1]. This chapter is divided into two main sections. The first section covers reductions where the substrate is support bound. The second section covers reductions where the substrate remains in solution (resin-bound reagents and catalysts, catch-and-release purification, and fluorous chemistry). In general, the combinatorial chemistry of peptides has been omitted in our coverage.

15.2
Solid-phase Reductions

15.2.1
Aldehyde Reductions

The reduction of solid-supported aldehydes to alcohols is a straightforward reaction. As is the case with solution-phase chemistry, the popular borohydride reagent $NaBH_4$ chemoselectively reduces aldehydes in the presence of other reducible functionalities, including esters [2], acetals [3], and nitro groups [4]. Typical conditions use excess $NaBH_4$ in tetrahydrofuran (THF) at ambient temperatures. The use of an alcoholic co-solvent such as ethanol or methanol is often required. In most cases, upon completion of the reaction, a dilute acid wash (AcOH or HCl) is needed to completely remove the unreacted reagent. The progress of these reactions, as well as other carbonyl reductions, can be followed by monitoring the disappearance of the carbonyl band (about 1700 cm^{-1}) in the FTIR spectrum of the resin [5].

Aldehyde reductions have been performed on a number of resins, including the Merrifield resin [2a], the Wang resin [2b], and an arylsulfonate ester resin [6]. They have also been used as a key step in the preparation of a variety of other resins. The preparation of the acid-labile resins 2 [7] and 4 [8] each required NaBH$_4$-mediated aldehyde reductions (Scheme 15.1). The improved synthesis of SASRIN resin, described by Katritzky et al. [9], and the synthesis of a photolabile 6-nitrovanillin-based resin, described by Zehavi and Patchornik [4], also required aldehyde reductions. The use of an acid wash to remove excess reagent is inappropriate in the case of the acid-labile resin preparations. In some of these cases, an alcoholic solvent wash suffices.

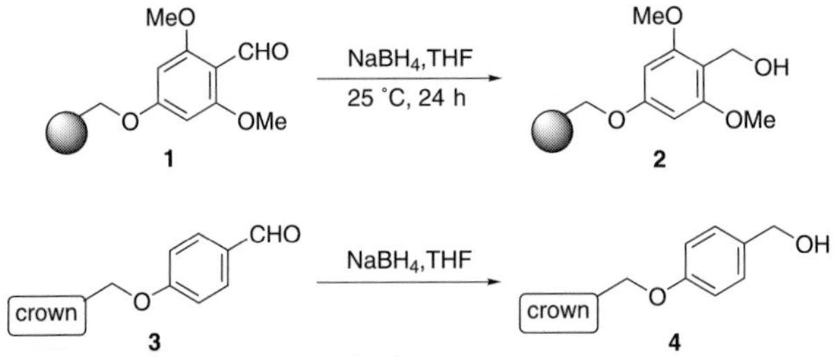

Scheme 15.1. Examples of aldehyde reductions in linker preparations.

15.2.2
Ketone Reductions

As with aldehyde reductions, the reagent of choice for simple ketone reductions on a solid support is NaBH$_4$ [10]. However, various other reducing agents have been described in the literature to carry out this transformation. These include the borohydride reducing agents diisoamylborane [11] and LiBH$_4$ [12]. The latter was used in the preparation of an acid-labile xanthone-based resin described by Sieber. NaBH$_4$ in combination with diethyl methoxyborane has also been used to reduce the diketone 5 [13] to diol 6 (Scheme 15.2). Following boron oxidation and removal, this diol was cyclized with dilute acid to δ-lactone 7.

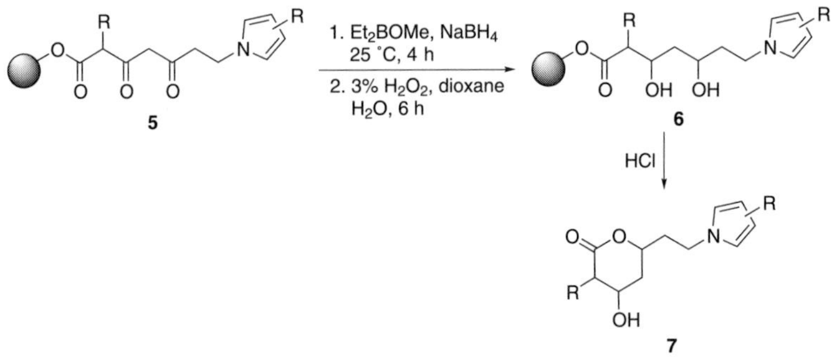

Scheme 15.2. Diketone reduction on solid phase.

Asymmetric reductions of ketones bound to resin are also possible. The silane re-agents diphenylsilane and α-naphthylphenylsilane in the presence of the catalyst prepared from μ-dichlorotetraethylenedirhodium(I) and (+) or (−)-DIOP asymmet-rically reduced a polymer-bound ketone [14]. The resulting secondary alcohol, of unknown optical purity, was part of a resin-bound asymmetric rhodium-containing hydrogenation catalyst that was used to reduce the double bond of α-N-acylami-noacrylic acids. However, the optical purities (∼ 70% ee) of the resulting amino acids were comparable to those obtained from the analogous homogeneous catalyst.

Diastereoselective ketone reductions can be achieved by a variety of methods. An example of a diastereoselective ketone reduction is found in Paterson's library syn-thesis of polyketides [15]. To increase the degree of diversity in this library, the re-duction of β-hydroxy ketone intermediates employed two different sets of condi-tions, allowing access to both syn- and anti-1,3-diols. The former was obtained with 95% diastereoselectivity (ds). Treatment of the β-hydroxy ketone with (c-Hex)$_2$BCl and triethylamine delivered a boron aldolate, which could be reduced with LiBH$_4$ affording the syn derivative with 95% ds. The 1,3-anti-diol was generated with 97% ds via the anti-selective Evans–Tishchenko reduction protocol, using stoichiometric quantities of SmI$_2$ and propanal, followed by a LiBH$_4$-mediated ester cleavage. Pat-erson also points out that attempts to effect the anti reduction with Me$_4$NBH(OAc)$_3$ resulted in lower diastereoselectivity (75%).

Other examples of both chelation- and nonchelation-controlled diastereoselec-tive ketone reductions were described by Ellman in his synthesis of an aspartyl protease inhibitor library [16]. To achieve chelation-controlled reduction, the α,β-dialkoxy ketone 8 was treated with Zn(BH$_4$)$_2$ in diethyl ether and THF at −20 °C (Scheme 15.3). Following deprotection and resin cleavage, the triol product was peracetylated. Analysis by gas chromatography (GC) showed that the syn product had been delivered in 60–80% ds. Following separation of this mixture by prepa-rative high-performance liquid chromatography (HPLC), the major diastereomer was found to have a 90% ee, which demonstrated that only a small degree of race-mization had occurred during an eight-step sequence that included reduction. The anti compound was obtained with 74% ds, utilizing the nonchelation-controlled conditions of L-selectride in THF at −75 °C. In another example of substrate-controlled reduction, Schlessinger was able to reduce ketone 10 stereoselectively with NaBH$_4$ (Scheme 15.4) [17].

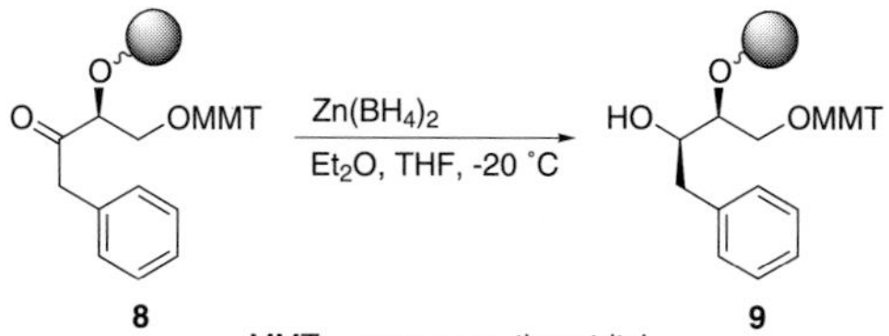

MMT = mono-p-methoxytrityl

Scheme 15.3. Diastereoselective reduction of an α,β-dialkoxy ketone.

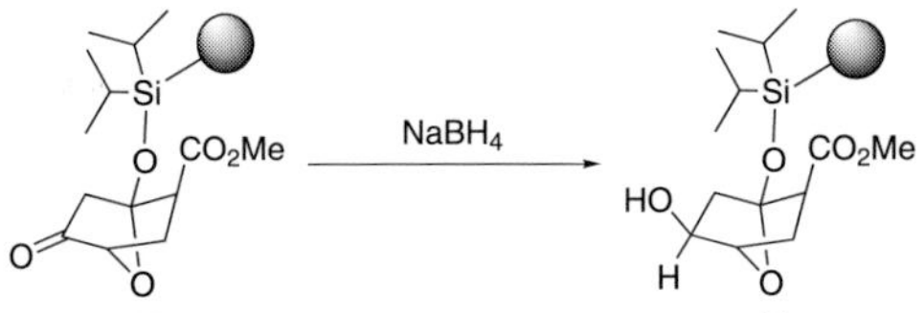

Scheme 15.4. Diastereoselective reduction of an oxabicyclo ketone.

15.2.3
Ester Reductions

A number of hydride-based reagents can reduce solid-supported esters. For all practical purposes, alcohols have been the usual targets of these reductions. Despite the ability to stop the reduction at the aldehyde oxidation state with the use of stoichiometric reagents such as DIBAL in solution, this transformation has not been reported for solid-phase-supported esters. The difficulty of this transformation is not surprising, as chemists usually use excess reagents to drive reactions to completion on solid phase and the use of stoichiometric reagents is very difficult to achieve. The transformation from a carboxylic acid oxidation state to an aldehyde has been successfully addressed on solid phase by utilizing Weinreb amides instead of esters. These amides are known reliably to produce aldehydes upon reduction (see Section 15.2.6).

Of the reducing reagents used to carry out ester reductions to the corresponding alcohol, DIBAL [2b, 18] and LAH [4, 13, 19] have been used most frequently. An example of a solid-phase DIBAL reduction is shown in Scheme 15.5, where the resin-bound α,β-unsaturated *tert*-butyl ester is reduced to the allylic alcohol **13**. DIBAL reductions have been run in various solvents, including THF, Et$_2$O, CH$_2$Cl$_2$, and toluene, from $-78\,°C$ to ambient temperature, while LAH reductions typically utilize THF or ether as the solvent at ambient to reflux reaction temperatures. The quench of the excess reagent is usually carried out using ethyl acetate/water or dilute aqueous HCl. Merrifield, Wang, and trityl resins have all proven to be stable to this strong reducing agent. Examples of LAH reductions are shown in Scheme 15.6. Example 15.6A shows that an amide is concomitantly reduced with an ester of a resin-bound amino acid to give the amino alcohol **15** [13]. LiBH$_4$ has also been used to reduce resin-bound esters, as shown in Scheme 15.7 [20]. A

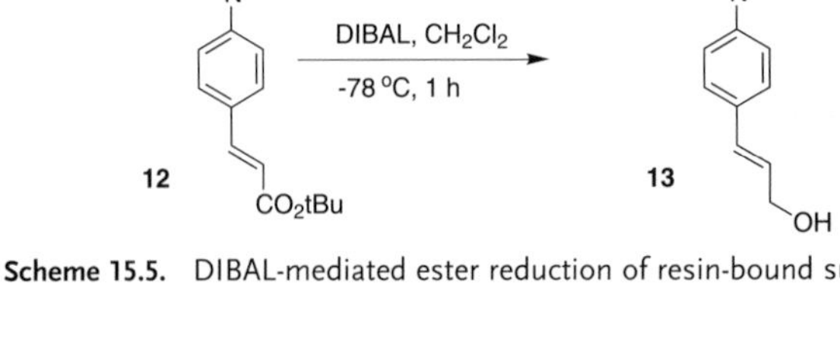

Scheme 15.5. DIBAL-mediated ester reduction of resin-bound substrate.

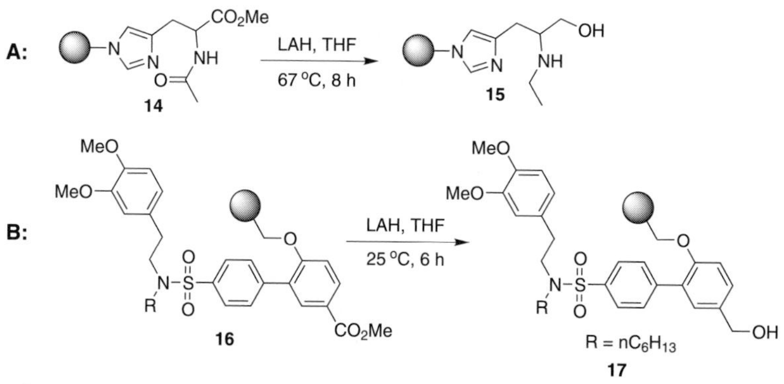

A:

B:

R = nC6H13

Scheme 15.6. LAH reductions on solid support.

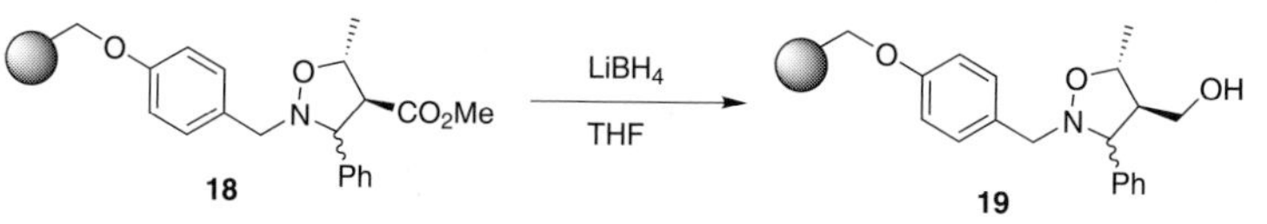

Scheme 15.7. Ester reduction on solid phase with LiBH$_4$.

NaBH$_4$-mediated ester reduction in ethanol and water has also been demonstrated; however, this protocol may not be general as esters are typically inert to this reagent.

Esters are common points of attachment for substrates to solid supports. Cleavage can be effected by saponification or by reduction, affording acids or alcohols, respectively, as the final products. For reductions, both DIBAL [21] and LAH [22] have been described in the literature. Tietze and coworkers have demonstrated the use of DIBAL in the two examples shown in Scheme 15.8, where concomitant reduction of a diester [21c] and a β-ketoester [21d] have led to diols **21** and **23**, respectively. Kurth and coworkers have also used this reagent to reduce the resin-bound β-hydroxy ester **24**, thus providing diol **25** [21a], which is a representative of a 27-compound library (Scheme 15.9). Kuster and Scheeren have reductively cleaved

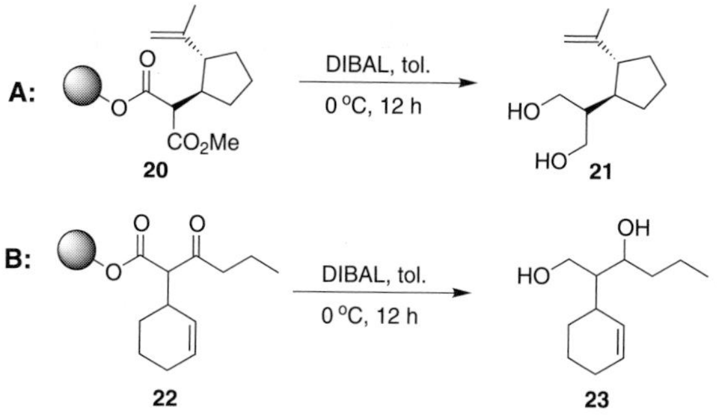

A:

B:

Scheme 15.8. Reductive cleavage of ester-linked substrates.

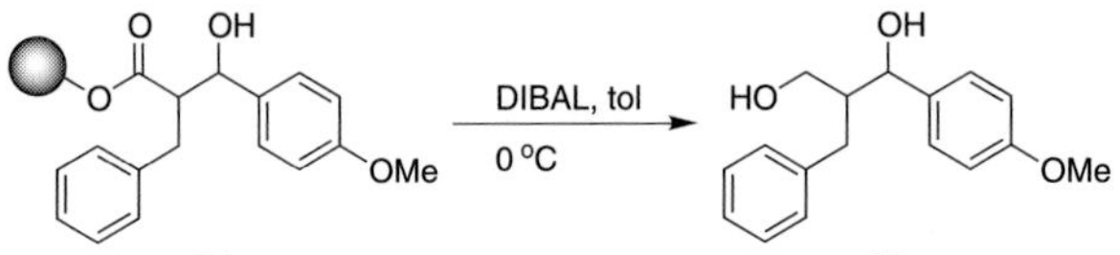

Scheme 15.9. Kurth's DIBAL-mediated reduction of a β-hydroxy ester.

the α-amino ester **26** from Wang resin to obtain the primary amino alcohol **27** using LAH (Scheme 15.10) [22].

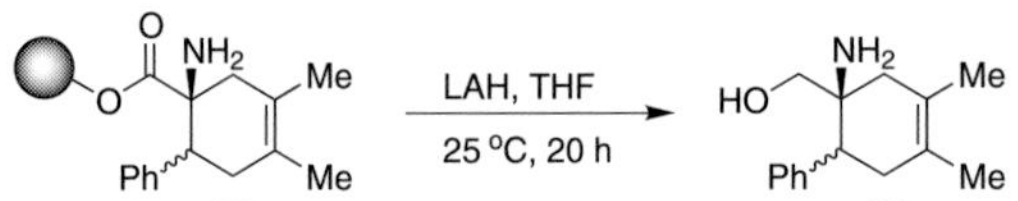

Scheme 15.10. Reductive cleavage of an amino ester with LAH.

Resin-bound acid chlorides have also been reduced to alcohols. Goldwasser and Leznoff have shown that diacid chlorides can be loaded onto 2% crosslinked polystyrene resin to give ester-linked acid chlorides. The resins can then be treated with amines or hydride reducing agents to give amides or alcohols, respectively [23]. In the latter case, the most successful reducing reagents were NaBH$_4$ and NaCNBH$_3$ (4 equiv., in THF at ambient temperature), which delivered the desired products in 67% and 65% yields, respectively. However, following basic cleavage and esterification, it was found that these products were contaminated with the dimethylester byproduct arising from incomplete reduction.

15.2.4
Mixed Anhydride Reductions

A mild method employed to obtain alcohols from carboxylic acids consists of using a mixed anhydride intermediate. In their search for an optimized phosphomannose isomerase inhibitor, researchers at Affymax coupled ten different symmetrical anhydrides and diacids to an immobilized 2-aminoindane-2-carboxamide to deliver acids of the general structure **29** (Scheme 15.11). This set of acids was treated with isobutylchloroformate and triethylamine in THF, followed by NaBH$_4$ in water, to deliver the primary alcohols **30** [24]. Further diversification led to a library of 600 analogs (60 pools of ten compounds). Chemists at Signal have also utilized this reduction protocol to deliver pyrimidine **33** following Mitsunobu alkylation and subsequent cleavage (Scheme 15.12) [25].

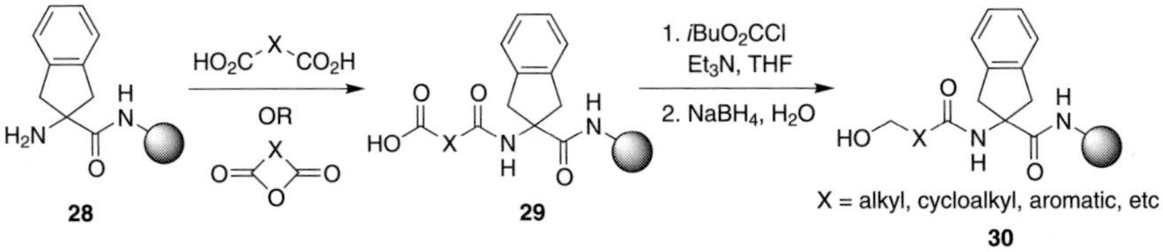

Scheme 15.11. Solid-phase reduction of acid via mixed anhydride.

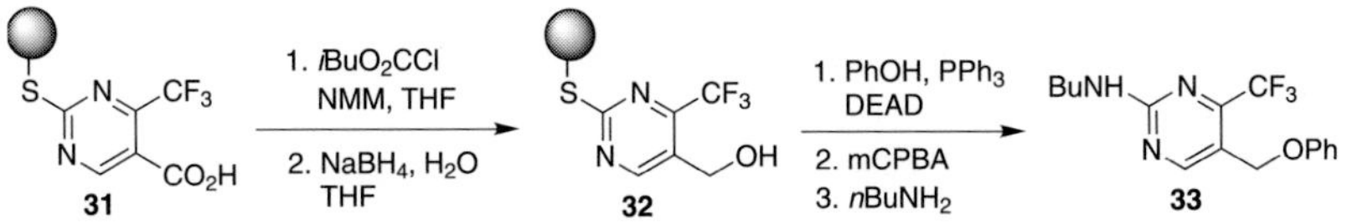

Scheme 15.12. Pyrimidine synthesis using mixed anhydride reduction.

15.2.5
Thioester Reductions

In 1996, Kobayashi reported the first synthesis of polymer-supported silyl enol ethers (thioketene silyl acetals) [26]. These reactive intermediates were converted to small libraries of β-amino alcohols [26a,c] and 1,3-diols [26b] according to Scheme 15.13. Treatment of compound **34** with either an imine or an aldehyde in the presence of catalytic amounts of Sc(OTf)$_3$ gave the resin-bound β-amino thioester **35** or β-hydroxy thioester **37**, respectively. Following reductive cleavage of the thioester with LiBH$_4$, the desired β-amino alcohols and 1,3-diols were obtained in good yields. However, a chromatography step was required to obtain pure compounds in both cases. The authors also demonstrated that the resin-bound β-

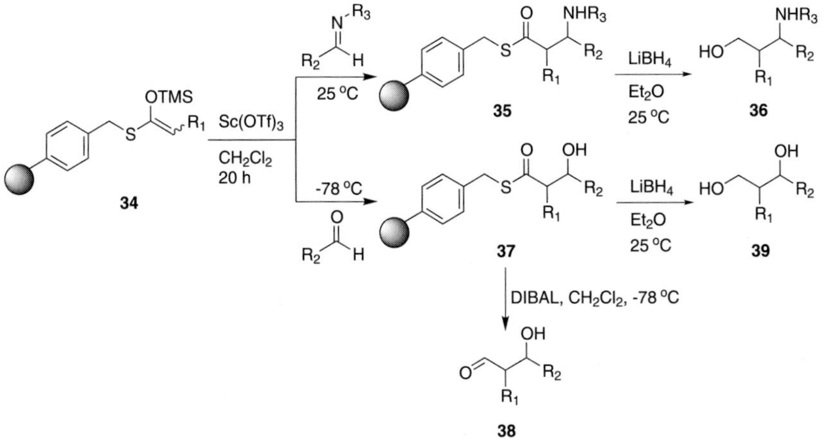

Scheme 15.13. Reductive cleavage of thioesters.

hydroxy thioester could be reduced to the β-hydroxy aldehyde **38** with DIBAL in CH$_2$Cl$_2$, at −78 °C [26b].

Thioester reduction on solid phase has also been successfully carried out using LAH in THF. Scheme 15.14 shows the reduction of dithioester **40** to the propane-1,3-dithiol derivative **41**, which was used to produce ketones via dithiane chemistry [27].

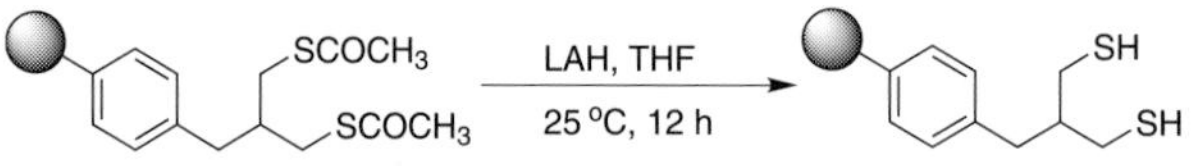

Scheme 15.14. LAH reduction of 1,3-dithioester.

In their Evans aldol approach to polyketide libraries, Reggelin et al. have also utilized thioesters and their reduction as part of the synthetic strategy [28]. They first converted resin-bound oxazolidinones **42** to thioesters **43** via a three-step protocol involving hydrolysis, thioester formation, and alcohol protection. They next reduced the thioester with LiBH$_4$ and reoxidized the resulting alcohol to aldehyde **45** with Dess–Martin reagent (Scheme 15.15). This aldehyde could then be used in a subsequent Evans aldol reaction. Unlike the examples given above, this route was designed to leave the substrate on the resin following thioester reduction.

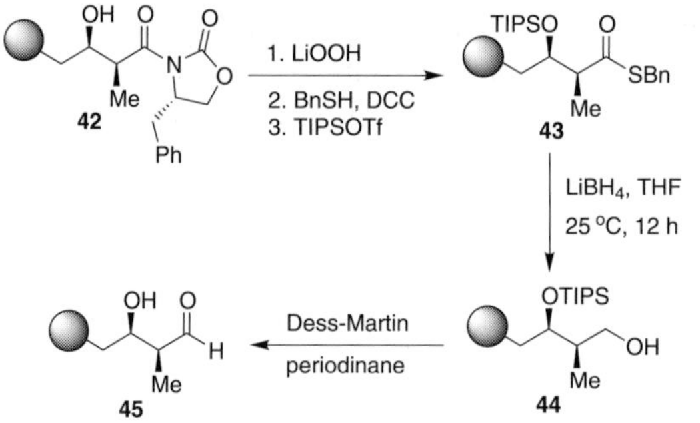

Scheme 15.15. Resin-bound thioester reduction with LiBH$_4$.

15.2.6
Weinreb Amide Reductions

As mentioned in Section 15.2.3, the partial reduction of an ester to an aldehyde is a difficult transformation on solid phase, since excess reagents are typically used to drive reactions to completion. If reactions fail to go to completion and purifications are required, then one of the major advantages of solid-phase synthesis is forfeited. In the case of ester reductions, alcohols are normally obtained owing to the use of excess reagents. Reggelin et al. solved this problem in the solid phase by making use of Weinreb amides, which deliver aldehydes upon reduction in standard solution-phase chemistry. In their approach to polyketide libraries, they demonstrated that the imides **46** can be converted to Weinreb amides under standard conditions (Scheme 15.16). Following alcohol protection, these amides can be selectively reduced with DIBAL to deliver the support-bound aldehydes **48** [29]. In 1998, Reggelin et al. replaced this sequence with another (see Section 15.2.5) that goes through a thioester instead of the Weinreb amide. However, this change was due to difficulties arising from steps other than reduction [28].

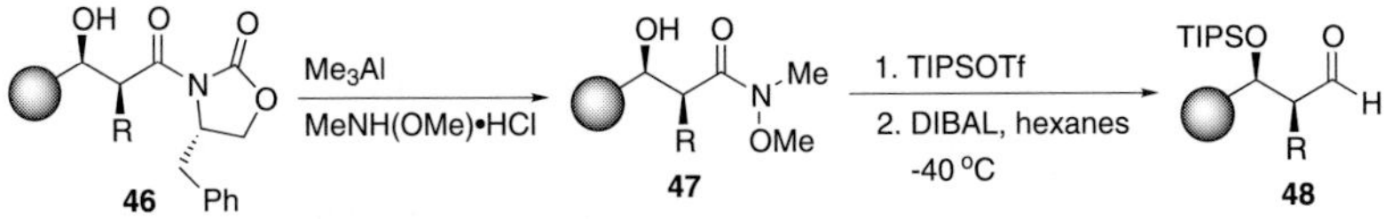

Scheme 15.16. Weinreb amide approach to support-bound aldehyde.

Two other groups have developed Weinreb amide-based linkers that give aldehydes in solution upon reductive cleavage [30]. Salvino et al. modified a resin that had been used for the synthesis of hydroxamic acids to give an *O*-linked Weinreb amide linker [30d], whereas Martinez and coworkers synthesized an *N*-linked version [30a]. The synthesis of C-terminal amino acid aldehydes has been successfully demonstrated by both groups, providing clean products in low to moderate yields using LAH as the reductant. Schemes 15.17 and 15.18 show the synthesis of Boc-Phe-H using both linkers. In the case of the Martinez linker, this support has also been applied to the synthesis of C-terminal peptide aldehydes [30b] and side-chain aldehydes [30c]. The peptide analogs were produced without noticeable epimeri-

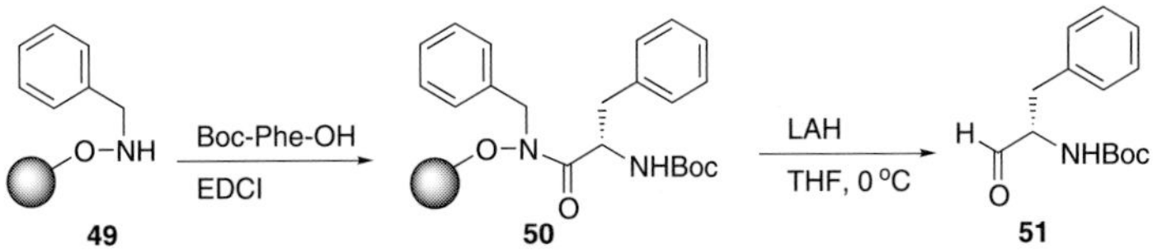

Scheme 15.17. Synthesis of C-terminal amino acid aldehyde using an *O*-linked Weinreb amide resin.

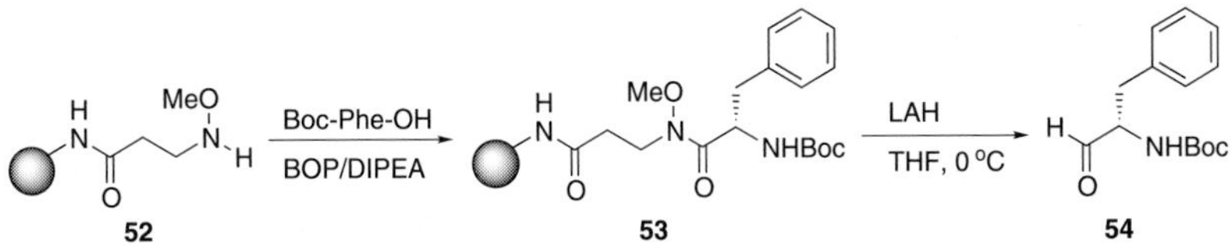

Scheme 15.18. Synthesis of C-terminal amino acid aldehyde using an *N*-linked Weinreb amide resin.

zation and the side-chains were reduced in high yield with LiAl(O*t*Bu)$_3$H. The
O-linked resin was also employed to deliver ethyl ketones by treating intermediate Weinreb amides with ethyl Grignard [30d]. Dinh and Armstrong also introduced an *N*-linked Weinreb amide resin, which delivered ketones in low to
moderate yields upon treatment with Grignard reagents [31]. However, attempts to
reductively cleave these amides with LAH produced aldehydes in very low yields
(< 20%).

15.2.7
Sulfur Reductions

There have been a number of different solid-phase applications involving disulfide and sulfone reductions. These include examples of simple functional group
manipulation, linker preparation, and "traceless" linker cleavage. Hummel and
Hindsgaul have shown that thio-oligosaccharides can be synthesized from polymer-
supported sugar derivatives functionalized with an anomeric disulfide (Scheme
15.19) [32]. Reduction of the disulfide with dithiothreitol (DTT) and subsequent
sodium thiolate formation leads to a thio-sugar capable of reacting with triflated
glycosides. Following deprotection and resin cleavage, thio-oligosaccharides 57
were formed. In another example, Ellman and coworkers synthesized a series of
cyclic 9- and 10-member thioethers, utilizing a PBu$_3$-mediated reduction of a *tert*-
butyl-protected disulfide as a key step [33]. They later modified this approach by
linking the substrate through a disulfide bond, as shown in Scheme 15.20 [34].
In this case, the disulfide bond was cleaved with tris-(2-carboxyethyl)phosphine
(TCEP), which gave the free thiol in solution along with excess TCEP and the phosphine oxide byproduct. The contaminants were of little consequence, as support-
bound *N*,*N*,*N'*,*N'*-tetramethylguanidine (TMG) was not only able to induce cyclization to the thioether, but was also able to scavenge these byproducts. In a recent

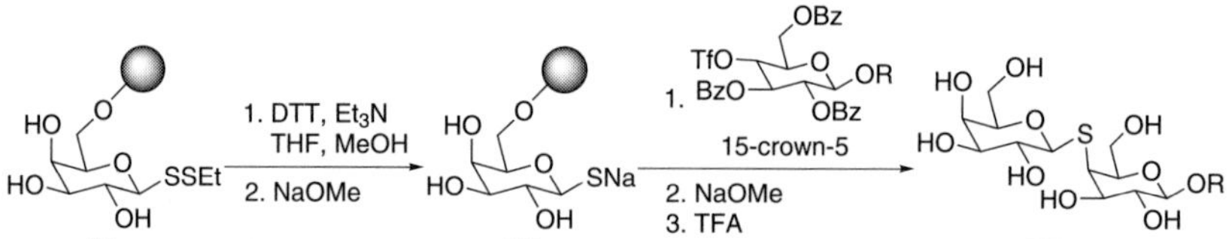

Scheme 15.19. Anomeric disulfide reduction in the preparation of thio-oligosaccharides.

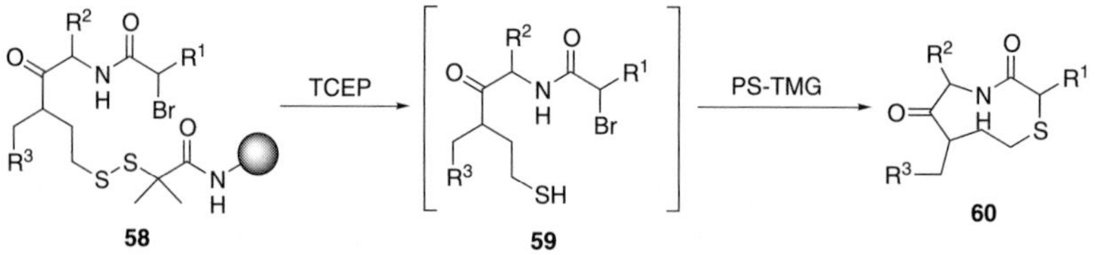

Scheme 15.20. TCEP-mediated disulfide reduction leading to a *β*-turn mimetic library.

application, Lam and coworkers used $NaBH_4$ to reduce a disulfide bond in their solid-phase approach to 1,4-benzothiazin-3(4H)-one derivatives [35].

In an example of linker preparation, Sucholeiki and coworkers have utilized a β-mercaptoethanol-mediated reduction to synthesize the photolabile resin **62** (Scheme 15.21) [36]. Zhao and Janda also utilized a disulfide reduction in the synthesis of the thiol linker shown in Scheme 15.22 [37]. Treatment of a mixture of disulfides generated in a previous alkylation step with DTT in water led to the desired resin **64** in excellent yield. Following construction of the targeted substrate on solid phase, oxidation of the thiol with $KHSO_5$ in water gave a sulfone that was cleaved with 5% Na/Hg in methanol and DMF to yield **66** [38]. The solvent mixture of methanol and DMF (1:8) was crucial for this successful reductive cleavage [39]. The use of additional methanol or THF as a replacement for DMF gave lower yields and required the use of larger quantities of Na/Hg. Zhao and Janda have also pointed out that, for reactions using soluble polyethylene glycol (PEG)-based polymers, isopropanol is an excellent choice for polymer precipitation, and leading to products of higher purity [38].

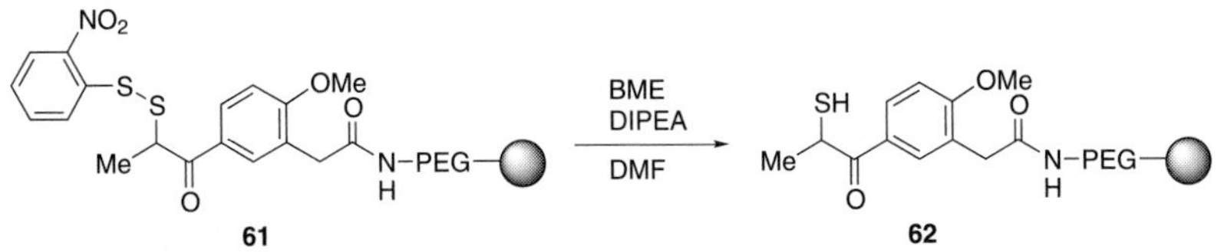

Scheme 15.21. Linker preparation utilizing β-mercaptoethanol-mediated disulfide reduction.

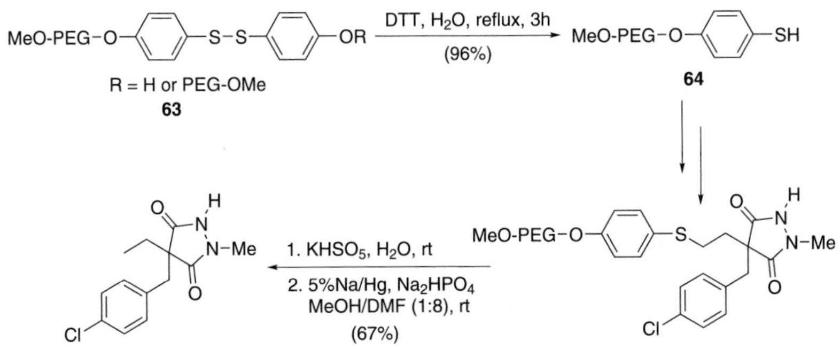

Scheme 15.22. Preparation and use of Janda's soluble polymeric thiol linker.

15.2.8
Selenium Reductions

The ease with which selenium can be reduced and oxidized is well understood, however the toxicity and odor of selenium compounds has limited their use. Nicolaou et al. recognized that by attaching selenium to a solid support these drawbacks would be mitigated, and the positive attributes of selenium reagents and substrates could then be exploited [40]. They have shown that supported selenium reagents make excellent "traceless" linkers because loading of substrates and subsequent cleavage to simple alkyl groups is carried out with equal ease. As shown in Scheme 15.23, reduction of the selenium bromide **67** with $LiBH_4$ delivers lithium selenide **68**, which then reacts with substrates containing an alkyl halide. Following substrate modification (not shown), the carbon–selenium bond can be reduced with Bu_3SnH to give an alkyl group with no vestige of the linker remaining. Alternatively, Nicolaou et al. have shown that these selenides can be treated with oxidizing reagents to give an alkene.

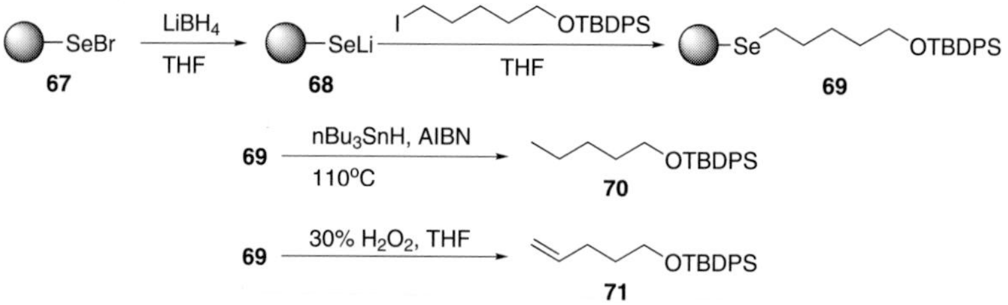

Scheme 15.23. Preparation and use of Nicolaou's selenium linker.

In another example of a selenium-based traceless linking strategy, Ruhland et al. prepared and reacted $PS\text{-}SeB(OEt)_3^- Na^+$ with hydroxy-containing alkyl halides. The resulting intermediates were then alkylated with phenols under Mitsunobu conditions and cleaved with Bu_3SnH to give alkylated phenols [41].

In an example that does not involve selenium linkage to a solid support, Pearson and Clark used a vinyl selenium reagent as an anionophile in a [3 + 2] cycloaddition with a solid-supported 2-azaallyl anion (Scheme 15.24) [42]. Following condensation, the phenylselenide group was then reduced with Bu_3SnH giving a 1,2-disubstituted pyrrolidine.

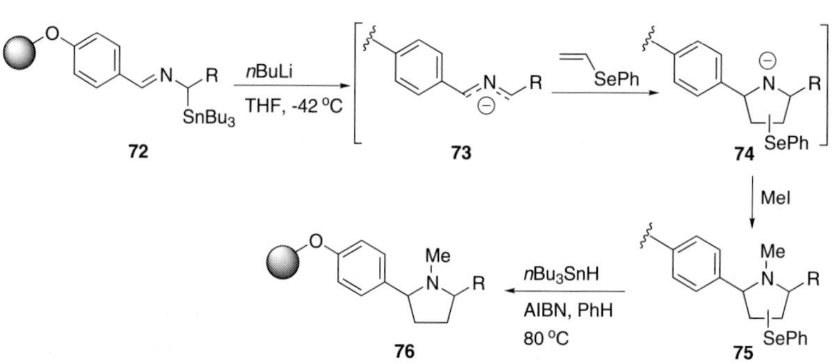

Scheme 15.24. Use of phenylvinylselenium as an "anionophile" and selenium removal with $n\text{-}Bu_3SnH$.

15.2.9
Quinone Reductions

The redox chemistry of quinones has been used to prepare solid-phase linkers for peptide synthesis [43]. The quinone can be easily reduced with NaBH$_4$ or Na$_2$S$_2$O$_3$ to provide the corresponding dihydroquinone. Treatment of the reduced product with TBAF results in phenoxide displacement at the C-terminus of the attached peptide, thus releasing the molecule from the resin. In initial efforts, the phenolic oxygen forms a lactone and releases the peptide functionalized at the C-terminus (Scheme 15.25A) [43a]. An improved approach is shown in Scheme 15.25B, in which the peptide is provided without functionalization at the C-terminus via S$_N$2 displacement by the phenol [43b].

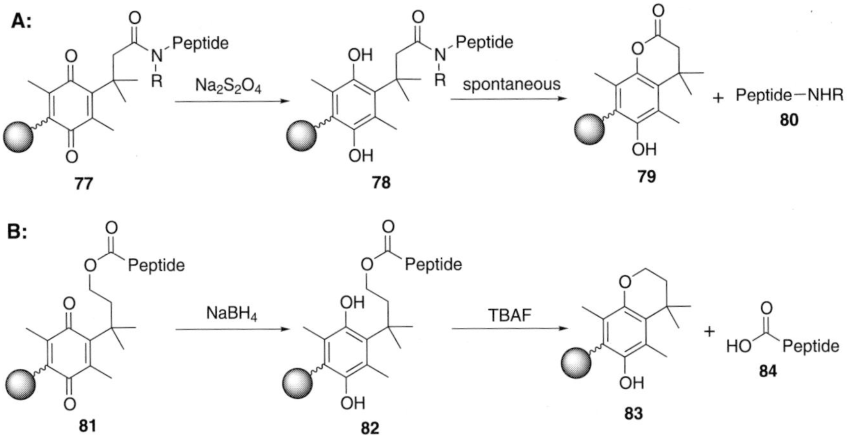

Scheme 15.25. Peptide release via quinone reduction and phenol or phenoxide-mediated cleavage.

15.2.10
Amide Reductions

The reduction of amides to amines is a useful technique for the generation of diversity in combinatorial synthesis. Either separately, or in combination with the complementary reductive amination techniques of aldehydes described in Section 15.2.12.4, an enormous number of amines and polyamines can be accessed from the diverse set of amides and polyamides that can be generated via a combinatorial approach. Further utility arises from the ability to increase diversity by modification of the product amines through any number of common approaches.

The reducing agents most routinely employed are BH$_3$·THF [44], BH$_3$·pyr [45], or BH$_3$·SMe$_2$ [46], although Red-Al [47] and LiBH$_4$ [48] have also been used. Borane reductions are easily applied to resin-bound amides since any excess reducing agent and resulting inorganic byproducts are easily washed away from the polymer support. An important factor for such reductions is the cleavage of the borane–

amine adducts. Acidic cleavage is incompatible with a variety of polymer supports, therefore heating with piperidine has been used to remove boron from the desired products [44b]. An iodine-mediated cleavage of the borane–amine adduct has also been reported (Scheme 15.26) [45b, 49]. This method is compatible with acid-sensitive solid supports such as Wang or Rink resins.

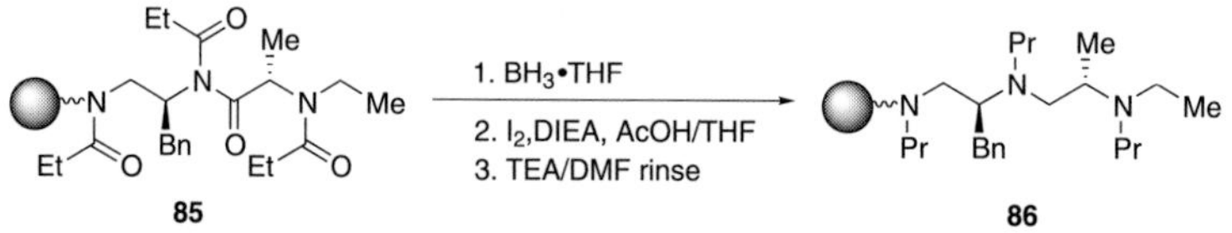

Scheme 15.26. Iodine work-up for borane reduction of amides on solid phase.

Several articles have been published on the use of amide reductions for the generation of combinatorial libraries [44, 45a,c]. Dipeptides are typically constructed on resin, acylated at the N-terminus, and exhaustively reduced with $BH_3 \cdot THF$ or $BH_3 \cdot pyr$ to give polyamines. This methodology has been used to generate hydantoins and cyclic ureas, thus generating "libraries from libraries" [44a]. This sequence has also been used to construct a 100,000-compound library of polyamines and bicyclic guanidines (Scheme 15.27) [44b].

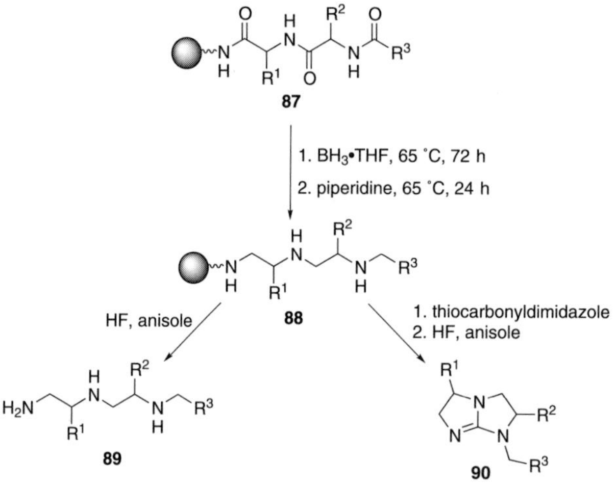

Scheme 15.27. Synthesis of polyamines and bicyclic guanadines via borane reduction of dipeptides.

15.2.11
Carbamate Reductions

There are several examples of carbamate reductions wherein the carbonyl is exhaustively reduced to an N-methylamine. All of these examples use aluminum hydride reagents. For example, Liu and Ellman used Red-Al-mediated carbamate reduction to prepare chiral *N*-methyl-2-pyrrolidinemethanol ligands [47]. Carbamate linkers have been used as latent methylamines in solid-phase synthesis. Formation of the carbamate on resin followed by LAH reduction generated the desired primary and secondary amines [50]. A carbamate linker has also been used to generate substituted *N*-methylpyrrolidines via reductive cleavage with LAH (Scheme 15.28) [51]. Xiao et al. used $Na_2S_2O_4$ to reduce an *O*-piperidinyl carbamate linker, thus revealing a secondary amine that underwent a cyclizative cleavage with a pendant ester to release a desired alcohol [52].

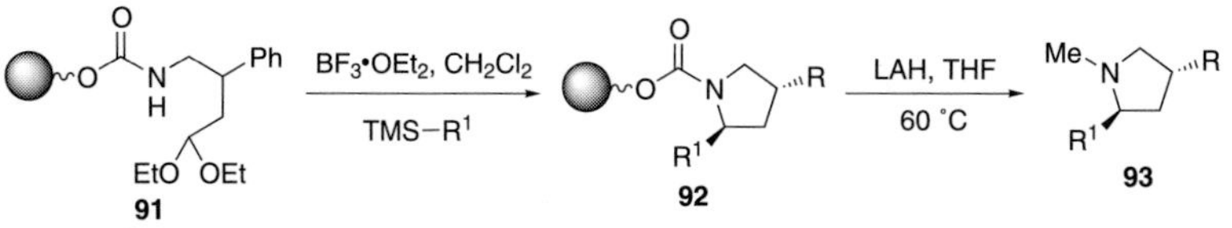

Scheme 15.28. Synthesis of subsituted *N*-methylpyrrolidines via reductive cleavage.

15.2.12
Reductive Amination

15.2.12.1 General Considerations
Reductive amination is one of the most widely applied techniques in combinatorial chemistry and is an excellent method for generating diversity (Scheme 15.29) [53]. Amines, aldehydes, and ketones as inputs for reductive amination are plentiful, commercially available, and typically inexpensive. The technique has been well established in routine organic chemistry [54], and has therefore provided a strong foundation for explorations in both solid- and solution-phase library synthesis. There is a wide variety of protocols for both imine formation and reduction, thus providing a large set of choices for reaction conditions and reagents that can be screened for a given set of inputs (e.g. primary vs. secondary amines).

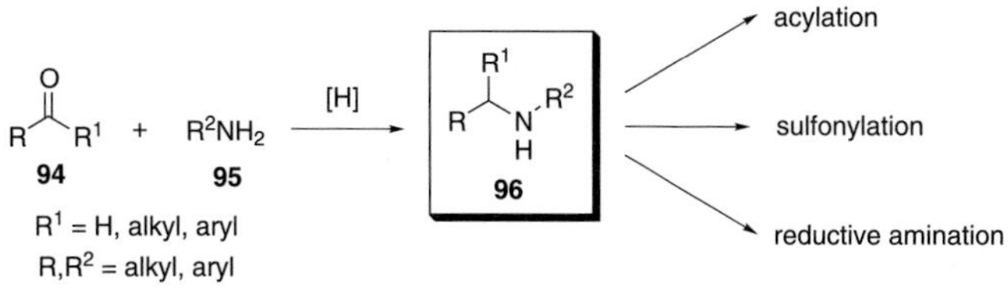

Scheme 15.29. Diversity from secondary amines prepared by reductive amination.

Generally carbonyl reactivity follows the order aldehyde > cyclic ketone ≫ acyclic ketone. Steric considerations are important in reductive amination reactions, thus hindered inputs may require forcing conditions for adequate results. For example, hindered amino acid esters such as valine [53a] have been reported to be poor inputs for reductive amination reactions as have methyl amines [55]. Secondary amines can require repetitive cycles [56] of reductive amination, or very long reaction times [56, 57] to obtain good conversion.

Electronic factors can also be important in reductive amination reactions, as evidenced by the higher reactivity of alkyl amines versus anilines. The more reactive nature of alkyl amines and aldehydes can render them prone to bis-alkylation [44b, 58]. Reductive aminations can also be sensitive to solvent choice. Brown and co-workers found that the use of aqueous THF was crucial for the successful reductive amination of a Rink amide resin [59].

Racemization of amino acid esters as amine inputs during reductive amination is another concern for library synthesis [60]. Racemization is expected to occur by equilibration during imine formation [61]. In order to suppress racemization, imines can be formed in the presence of a reducing agent, thus avoiding equilibration [58a, 61].

There are several methods for monitoring the progress of solid-phase reductive amination. Qualitative nitrogen analyses such as a (Kaiser) ninhydrin test can quickly and easily assess reaction progress [62]. Infrared spectroscopy can also be used to detect consumption of the carbonyl and/or imine components in a reductive amination process by their characteristic C=O (1740–1720 cm^{-1}) and C=N ($\sim$1650 cm^{-1}) stretches [63].

15.2.12.2 Imine Formation

Imines are formed by condensation of amines and aldehydes or ketones with loss of water. Typically, aliphatic aldehyde-derived imines tautomerize to their corresponding enamines. The classic methods for water removal such as azeotrope formation or trapping with molecular sieves are neither practical nor economical for application in a solid-phase library format [64]. A number of different approaches have therefore been developed for imine formation in combinatorial chemistry to supplant these techniques.

The most widely used reagent for imine formation in solid-phase organic synthesis has been trimethylorthoformate (TMOF) [53b, 65]. TMOF is used as the reaction solvent, or in conjunction with other solvents as solubility dictates, and reacts with water as it is formed, thus driving the equilibrium toward imine formation. TMOF is compatible with most resins and common reducing agents such as NaBH$_3$CN, NaBH(OAc)$_3$, or BH$_3$·pyridine (BAP).

The imine equilibrium is often shifted by using an excess of the input not loaded on resin (anywhere from 2 to 20 equiv.). Imines can be formed *in situ* in the presence of a reducing agent with or without added acids. The typical additive for imine formation is 1–10% acetic acid (by volume) but *p*-TsOH [66] and PPTS [67] have also been used to catalyze the condensation. Anhydrous inorganic salts such as Na$_2$SO$_4$ (with ultrasound) have been used to drive formation of imines in solid-phase applications [68]. Titanium reagents such as Ti(O*i*Pr)$_4$ have been used to facilitate reductive aminations via carbonyl activation and amine addition [69]. Use of Ti(O*i*Pr)$_4$ made possible the reductive amination of a relatively unreactive ketone **97**, as shown in Scheme 15.30 [69a].

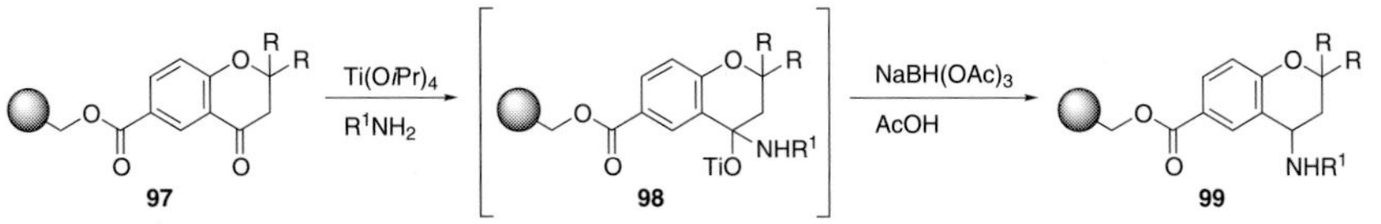

Scheme 15.30. Reductive amination with Ti(O*i*PR)$_4$.

15.2.12.3 Reducing Agents for Reductive Amination

There is a myriad of reducing agents that can be used for reductive aminations in solid-phase applications. The most common are NaBH$_3$CN (Borch reduction) [53, 56, 70], NaBH(OAc)$_3$ [63a, 21b, 71], and BAP [57, 72], but Me$_4$NBH(OAc)$_3$ [73], NaBH$_4$ [74], and LiBH$_4$ [75] have also been used. Me$_4$NBH(OAc)$_3$ and NaBH$_3$CN have been used sequentially in a reductive amination protocol in which the individual reductants provided incomplete or impure reaction products respectively [76].

The borohydride reagents listed above typically reduce both imines and tautomeric enamines (formed from aliphatic aldehydes) and can be used in large excess as they are easily washed from the resin [77]. NaBH(OAc)$_3$ has become a favored reducing agent as it is not toxic and displays excellent chemoselectivity, but NaBH$_3$CN is still frequently used. Borohydride reagents such as NaBH$_4$ and LiBH$_4$ can be problematic as they may reduce starting aldehydes and ketones to alcohols. The usual solution in such cases is to preform the imine before application of the reductant in a two-step process.

15.2.12.4 Reductive Aminations as the Entry Point for Library Preparation

Frequently, reductive amination is used to load a variety of inputs onto a resin, thus introducing diversity in the first step of a library synthesis. A typical approach is to react primary amines (e.g. amino acid esters) with electron-rich, resin-bound aldehydes (Scheme 15.31) [63a, 78]. The acetophenone-containing resin **103** has been loaded with amines via reductive amination for entry into parallel amide and sulfonamide preparations (Scheme 15.32) [79].

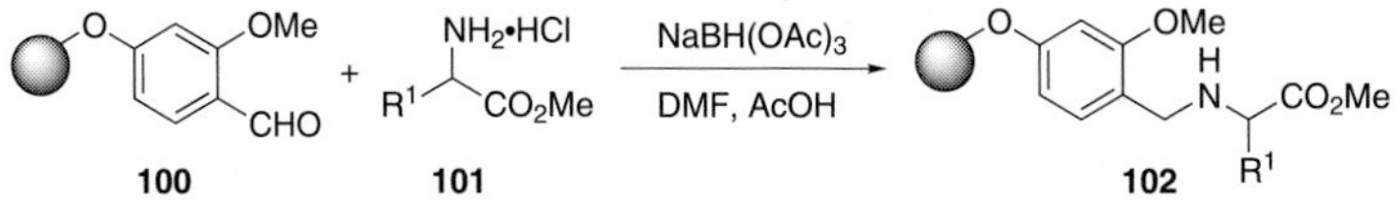

Scheme 15.31. Loading of amino acid esters onto AMEBA-type resin.

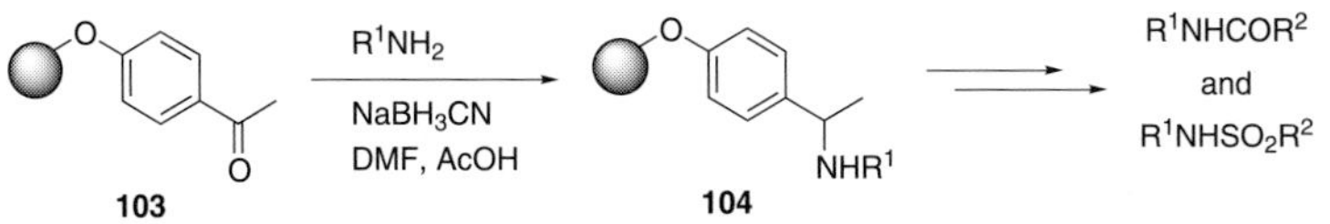

Scheme 15.32. Reductive amination on an acetophenone-containing resin.

The combination of primary amines and electron-rich, resin-bound aldehydes usually works well under a number of conditions, but hindered aldehydes have occasionally been problematic. For example, a 3,5-dimethoxy-4-benzaldehyde-containing resin was shown to be less reactive toward reductive amination than the monomethoxy analog [80]. Schwarz et al. also observed that di-ortho (e.g. dimethyl)-substituted aldehydes can be poor substrates for reductive amination [81]. However, Ellman and coworkers reported reductive amination conditions that work well with a polymer-supported 3,5-dimethoxy-4-benzaldehyde (Scheme 15.33) [78a].

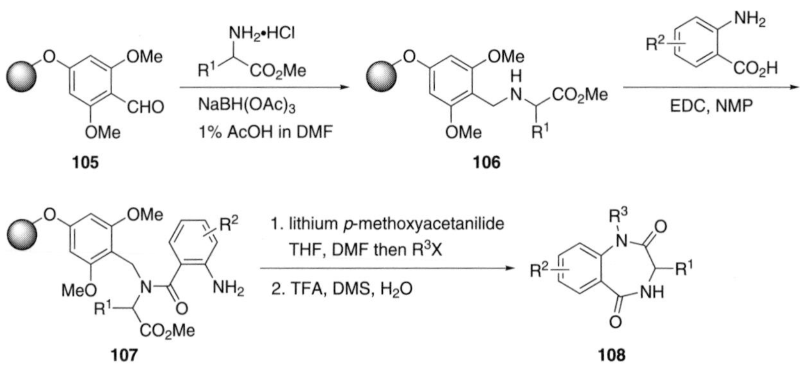

Scheme 15.33. Ellman's 1,4-benzodiazepine-2,5-dione synthesis.

Polymer-supported amines have also been used for reductive aminations with aldehydes. There are several examples where Rink amide resin has been reductively alkylated [59b, 62b, 75a]. Katritzky et al. formed a variety of imines (both electron-rich and -poor aromatic aldehydes) with TMOF and reduced them with LiBH$_4$ [75a]. Rivero and coworkers also used a two-step reductive amination procedure to prepare a library of 500 macrocycles (Scheme 15.34) [62b]. Benzyloxy-

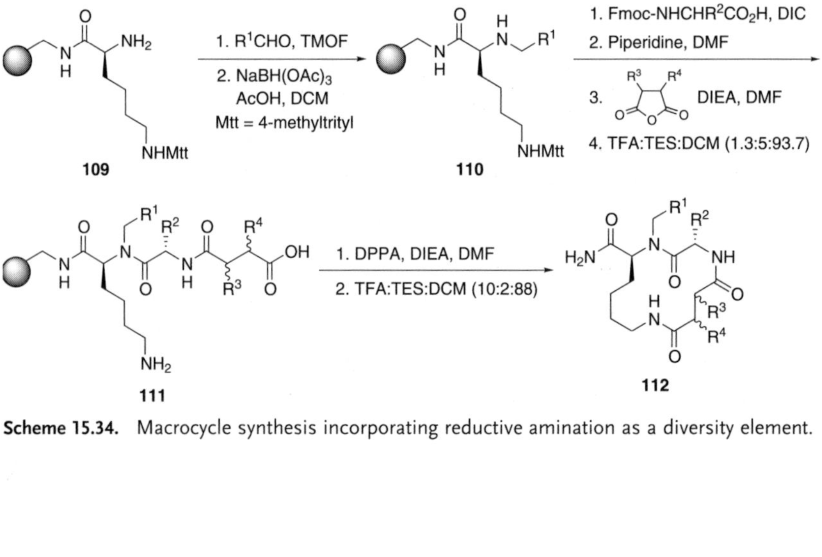

Scheme 15.34. Macrocycle synthesis incorporating reductive amination as a diversity element.

Tab. 15.1. Recent libraries using reductive amination (since January 2000).

Library	Reactants[a]	Reducing agent	Method[b]	Reference
Secondary amides	Aniline/aldehydes	$NaBH_4$	B	74e
Macrolides	Aldehyde/primary amines Secondary amines/aldehydes	$NaBH_3CN$	B	58b
Tertiary amines	Hydroxylamine/aldehydes	BAP	B	67
Tetrahydro-quinoxalines	Aldehydes/amino alcohol	$NaBH(OAc)_3$	A	78f
Peptide aldehydes	Aldehydes/primary amines	$NaBH(OAc)_3$	A	83
Macrocycles	Primary amine/aldehydes	$NaBH(OAc)_3$	B	62b, 85
3,4-Dihydro-1,4-benzothiazines	Anilines/ketones (intramolecular)	$NaBH_3CN$	A	84
Quinoxalinones	Aldehydes/amino acid	$NaBH(OAc)_3$	A	86
Pyrrolidines	Primary amine/aldehydes	$NaBH_3CN$	A	87
2-Carboxindoles	5-Aminoindole/aldehydes	$NaBH_3CN$	A	88
β-Ketoamides/imidazoles	Aldehydes/primary amines	$LiBH_4$	B	75b
Amines/amides	Aldehydes/primary amines	BAP	B	72c
Tricyclics	Primary amine/cinnamaldehydes	$NaBH(OAc)_3$	A	89
Piperidines	Aldehydes/primary amines	BAP	B	72b
Tertiary methylamines	Hydroxylamine/aldehydes	$NaBH(OAc)_3$	A	90
Phenolic amino acids	Primary amines/aldehydes	$NaBH(OAc)_3$	B	91
Neoglycopeptides	Amino acids/β-glycoside aldehydes	$NaBH_3CN$	A	92
Lysine/glutamic acid derivatives	Amino acids/aldehydes	$NaBH(OAc)_3$	B	93

[a] First entry is resin-bound component.
[b] A, one-step procedure, *in situ* imine formation/reduction; B, two-step procedure, imine formation then reduction.

aniline and *p*-benzyloxybenzylamine (BOBA) resins have also been reductively alkylated with aldehydes and $NaBH_4$ using two-step procedures [74c,e].

15.2.12.5 Recent Examples of Reductive Amination on Resin

There is a large number of libraries that have incorporated a reductive amination step. This fact highlights the power and reliability of reductive amination for library synthesis. Recently, a number of libraries have been prepared using reductive amination chemistry. These library syntheses are listed in Table 15.1.

Specific examples include a diketopiperizine library effort at Affymax that relies on reductive amination for a key diversity step (Scheme 15.35) [65a, 82]. Groth and Meldal reported a combinatorial approach to *N*-terminal peptide aldehydes and diketopiperazines using reductive amination [83]. They found that $NaBH(OAc)_3$ in dimethyl sulfoxide (DMSO)/CH_2Cl_2/AcOH (50:50:1) was optimal after an exten-

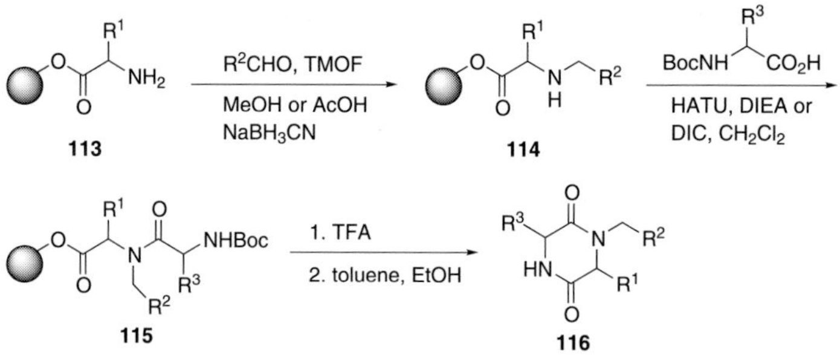

Scheme 15.35. Affymax's solid-phase approach to diketopiperazines.

sive study of conditions, thus emphasizing the importance of varying reaction conditions for successful reductive aminations. Barany and coworkers prepared a set of 3,4-dihydro-1,4-benzothiazines using a one-pot alkylation–intramolecular reductive amination to prepare the thiazine core [84]. A group at Abbott has prepared an antibiotic screening library of 70,000 macrolides [58b]. The synthesis involved three reductive amination steps (aldehydes with a primary amine and two secondary amines) to introduce diversity and began with a preconstructed macrolide core.

15.2.13
Azide Reductions

15.2.13.1 General Considerations
Reductions of aromatic azides provide anilines that are handles for diversification and may be incorporated into benzo-fused heterocycles. Tin ($SnCl_2$), phosphine, or sulfur reagents are commonly used to carry out this transformation. Under some tin reduction conditions, side-reactions, such as azide displacement or *N*-acetylation, can be problematic. In these cases, phosphine- or sulfur-mediated azide reductions can be used as replacements.

15.2.13.2 Azide Reductions in Glycopeptide Preparations
Solid-phase azide reductions are heavily used in glycopeptide preparations. The azide is frequently used as a point of attachment for the peptide, but may also be a handle for diversification. The most commonly used reductants are sulfur based, such as DTT or 1,3-propanedithiol, although phosphines have been used in several examples.

Peters et al. [94] in an early example, and later Rademann and Schmidt [95], obtained *N*-acetates from azide reductions using thioacetic acid and pyridine in solid-phase glycopeptide preparations. Danishefsky and coworkers used both thiophenol and 1,3-propanedithiol with Hünigs base in THF to effect azide reduction in solid-supported trisaccharide- and disaccharide-containing glycopeptides, thus avoiding *N*-acetate formation [96]. Glycopeptides have also been prepared on solid support via azide reduction using DTT and 1,8-diazabicyclo[5.4.0]undecene-7 (DBU) in

DMF [97]. A modified Staudinger reaction has been applied to the preparation of amides using a solid-supported glycoazide in a one-pot procedure (Scheme 15.36) [98].

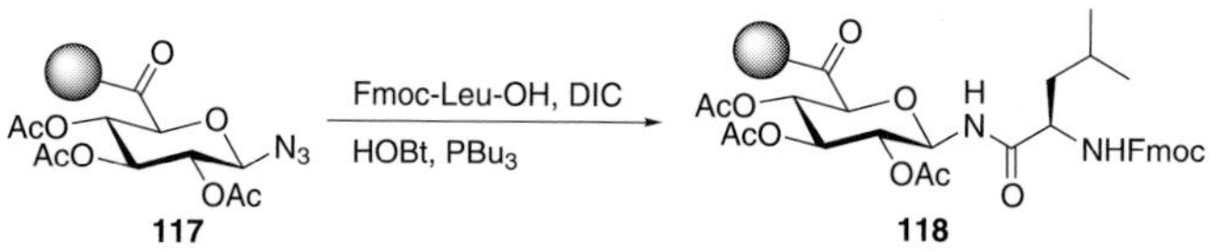

Scheme 15.36. Modified Staudinger reaction for the preparation of glycopeptides.

15.2.13.3 Small Molecule Libraries Incorporating Azide Reduction

A variety of small molecule libraries has been synthesized that incorporate azide reductions using primarily $SnCl_2$ and triarylphosphines. Ellman's group has prepared several small molecule arrays by generating diversity at the amine prepared by a tin-mediated azide reduction (Scheme 15.37) [16, 99]. Kim and coworkers have also prepared a group of oligoureas using a tin-mediated azide reduction [100].

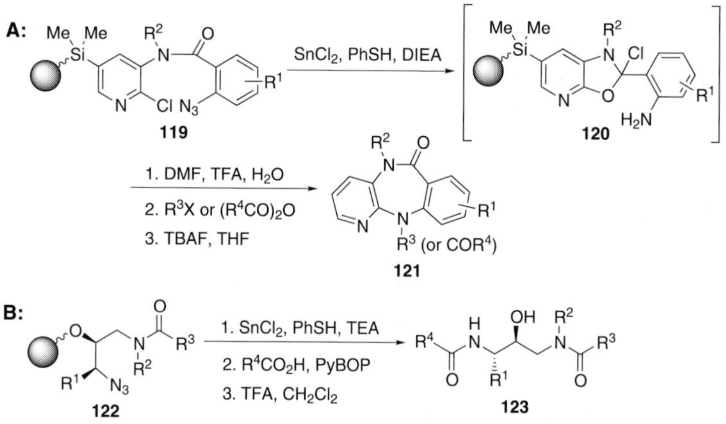

Scheme 15.37. Examples of small-molecule synthesis using azide reductions.

Another example from Chiron is the preparation of a small group of 1,4-benzo-diazepine-2,5-diones by a PBu_3-mediated azide reduction [101]. Reaction of the resulting aniline with a pendant ester formed the diazepine heterocycle. Trifluoroacetic acid cleavage provided 21 benzodiazepines in good yields (Scheme 15.38). Kahne and coworkers also used aqueous PMe_3 to reduce azides on a TentaGel-supported carbohydrate [102].

It has been reported by Zhou and coworkers that the reduction of a primary azide with $SnCl_2$ and thiophenol provided significant amounts of an azide displacement product (resulting from attack by thiophenol) [103]. To circumvent this

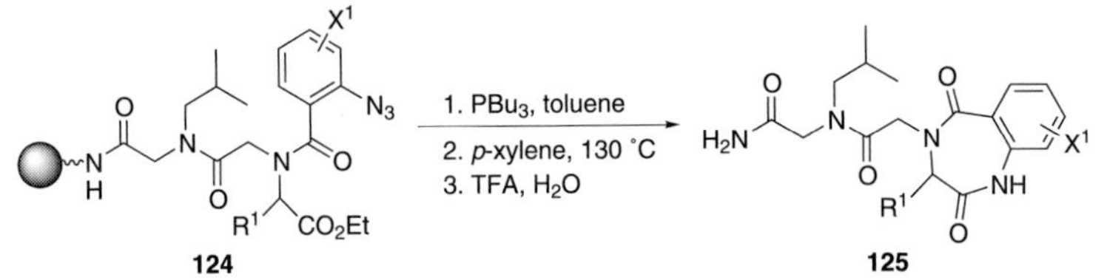

Scheme 15.38. Synthesis of 1,4-benzodiazepine-2,5-diones using an azide reduction.

problem, aqueous PPh₃ provided the amine in good yield without any azide displacement and allowed the synthesis of an array of hydroxybisamides. Nicolaou et al. also used an aqueous PPh₃-mediated azide reduction methodology to provide an amine for diversification in the preparation of a library of 50 sarcodictyins [104].

15.2.13.4 Recent Examples of Azide Reduction on Resin

A variety of recent libraries has incorporated azide reductions in their synthetic approach and are summarized in Table 15.2. For example, a library of phenolic steroids has been prepared by Poirier and coworkers using both tin- and phosphine-mediated azide reductions on various solid supports [105]. The phosphine reduction was employed when an *o*-nitrobenzyl ether linkage was used to avoid reduction of the nitro group on the linker [105b]. A library of substituted oxazoles has been prepared via derivatization of an amine generated from an azide reduction with DTT and Hünigs base (Scheme 15.39) [106]. A small library of 1,3-oxazolidines has been prepared using a tin-mediated azide reduction on solid support [107]. An azide on solid support has also been reduced with TMSI. This acidic protocol was used to avoid base-induced formation of a lactam side-product when using DTT/DBU [108].

A library of 1300 disaccharides has been prepared on solid phase via azide reduction with aqueous PMe₃ followed by amine derivatization with isocyanates and

Tab. 15.2. Recent libraries utilizing azide reduction (since January 1999).

Library	Reducing agent	Reference
Glycopeptides	PhSH or 1,3-propanedithiol, DIEA	96b
	DTT, DBU	97b
	PBu₃	98
	DTT, DBU or TMSI	108
Hydroxybisamides	Aqueous PPh₃	103
Estradiols	Aqueous PPh₃	105a
	SnCl₂, PhSH, TEA or aqueous PPh₃	105b
Oxazoles	DTT, DIEA	106
1,3-Oxazolidines	SnCl₂, PhSH, TEA	107
Disaccharides	Aqueous PMe₃	109
1,3-Bis(acylamino)-2-butanones	SnCl₂, PhSH, TEA	78e
Phenylglycinnamides	Aqueous PMe₃	110

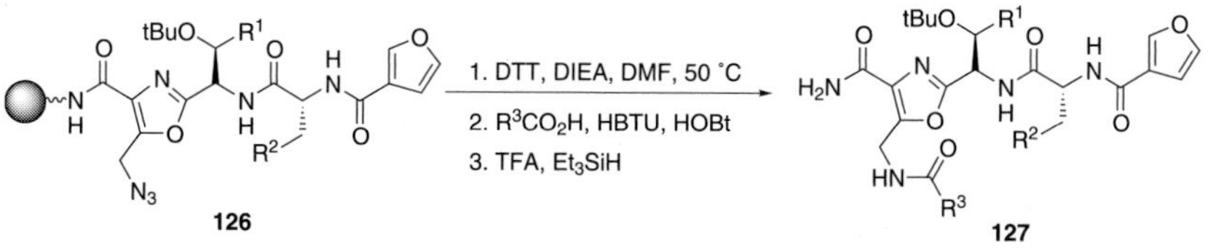

Scheme 15.39. Synthesis of oxazoles incorporating an azide reduction.

acids (Scheme 15.40) [109]. An array of 18 1,3-bis-(acylamino)-2-butanones has been prepared in which diversity was generated at an amine prepared by a $SnCl_2$ azide reduction [78e]. Notably, a dimethyl ketal survived the conditions of this tin-mediated reduction. A library of phenylglycinnamides has been prepared by generating a galactosylamine by means of an azide reduction with 1,3-propanedithiol [110]. The galactosylamine was used in a series of Ugi reactions and cleaved from the resin to generate eight different phenylglycinnamides.

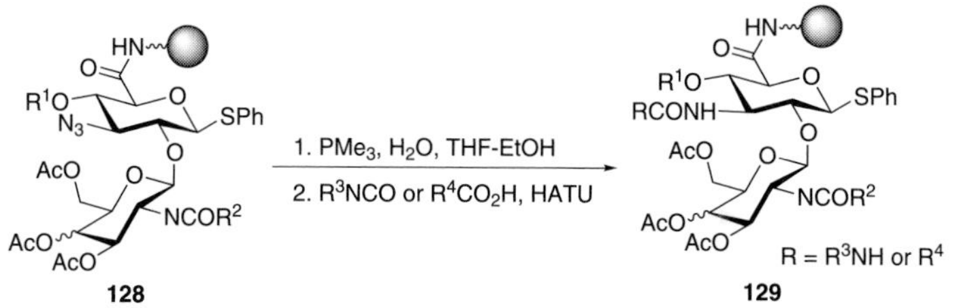

Scheme 15.40. Disaccharide derivatization at an amine generated by an azide reduction.

15.2.14
Nitro Group Reductions

15.2.14.1 General Considerations
The reduction of aromatic nitro groups to anilines is an often-used transformation in combinatorial chemistry. The aromatic nitro group serves two important functions: it facilitates S_NAr reactions and provides an amine for further manipulation following reduction. A frequent use of the resulting aniline in library synthesis has been in the preparation of various benzo-fused heterocyclic compounds (Scheme 15.41) [111].

There are a variety of protocols that have been developed for the reduction of nitro groups, and each offers different advantages and disadvantages. Some experimentation may be required to find suitable conditions for the system under study since the most frequently used reducing reagent ($SnCl_2$), while quite reliable, does not always provide consistent results [112].

15.2.14.2 Tin-mediated Nitro Reductions
The reduction of aromatic nitro groups is often carried out using a tin reagent (usually an aqueous solution of $SnCl_2$ in DMF) [113, 114]. Acidic conditions typi-

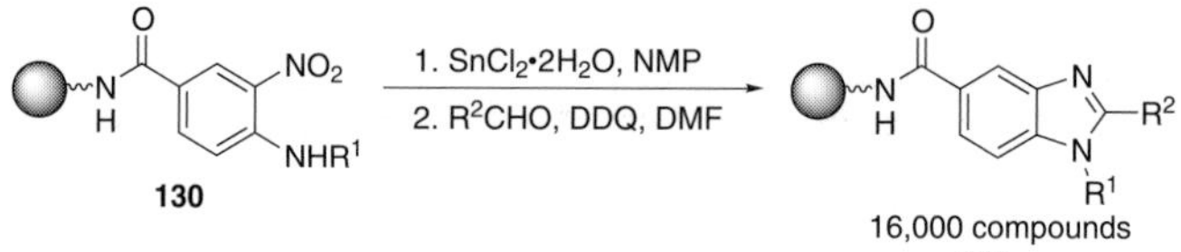

Scheme 15.41. Synthesis of a benzimidazole library via tin-mediated nitro reduction.

cally accompany nitro reduction with $SnCl_2$, which presents a potential problem with acid-sensitive polymer supports. Addition of a small amount of buffer such as sodium acetate often remedies this situation [35, 115]. DMF is the solvent of choice for tin reductions, but N-formylation of o-diaminobenzenes generated from 2-amino-substituted nitrobenzenes and subsequent cyclization to benzimidazoles has been observed as a side-reaction [116]. This result can be avoided by employing other solvents such as NMP or N-methylmorpholine (NMM). Tin reductions often require heating and can benefit from exclusion of oxygen [117].

Tin reductions have been used in a number of library syntheses. In an early example, the tin-mediated reduction of solid-supported substituted nitrobenzenes led to anilines that were derivatized to provide a small library of phenols following cleavage (Scheme 15.42) [118]. The synthesis of a library of 3,4,5-substituted 1,5-benzodiazepin-2-ones began with a tin-mediated nitro reduction on polymer support (Scheme 15.43) [119]. Sequential hydrolysis and intramolecular amide coupling provided the benzodiazepine core.

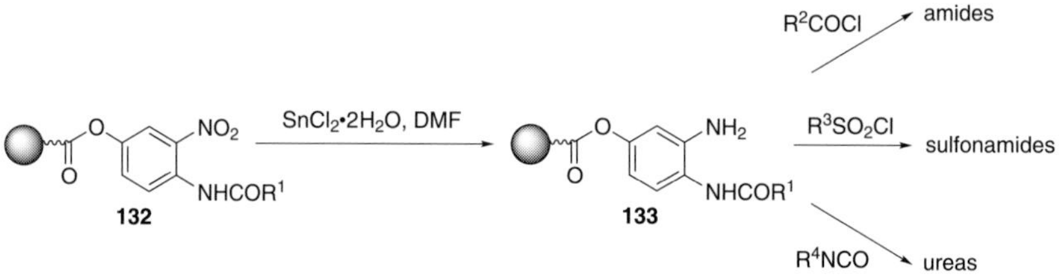

Scheme 15.42. Meyer's approach to a phenolic library via a tin-mediated nitro reduction.

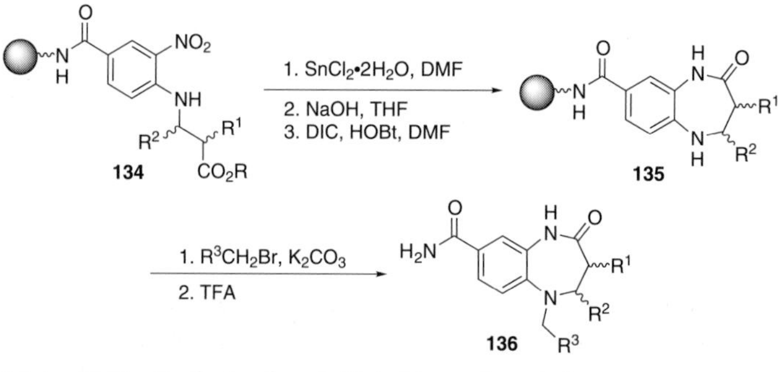

Scheme 15.43. Synthesis of a substituted benzodiazepin-2-one library via a tin-mediated nitro reduction.

15.2.14.3 **Nitro Reductions with Alternative Reagents**

Tin-mediated reductions of nitro groups can occasionally give inconsistent results or suffer from incomplete reactions [112, 120], a serious problem in library synthesis where reliability and purity are essential. Furthermore, tin impurities are known to be problematic in many drug-screening assays, especially cellular assays [121]. Therefore, a variety of reagents and conditions has been developed as alternatives to tin-mediated nitro reductions.

Sodium borohydride with $Cu(acac)_2$ was used in the preparation of a benzimidazole library in which $SnCl_2$ gave inconsistent results (Scheme 15.44) [112]. A comparative study of nitro reductions with $Na_2S_2O_4$ versus $SnCl_2$ on a set of 74 compounds has been performed [121]. The results show that $Na_2S_2O_4$ is as effective as $SnCl_2$ in nitro reductions, although resins compatible with aqueous solutions must be used with $Na_2S_2O_4$. A set of sixteen different conditions for the solid-phase reduction of a nitropyrimidine has also been explored [120]. In this study, $SnCl_2$ gave only 50–65% conversion to the aminopyrimidine, while the best results were obtained with $LAH/AlCl_3$, although the final products were contaminated with aluminum salts. Aromatic nitro groups have been successfully reduced with $CrCl_2$ at room temperature [122]. Other metal-mediated nitro reductions have also been applied to library synthesis in both solid phase (Zn, NH_4Cl [123]) and solution phase (Fe, HCl [124]).

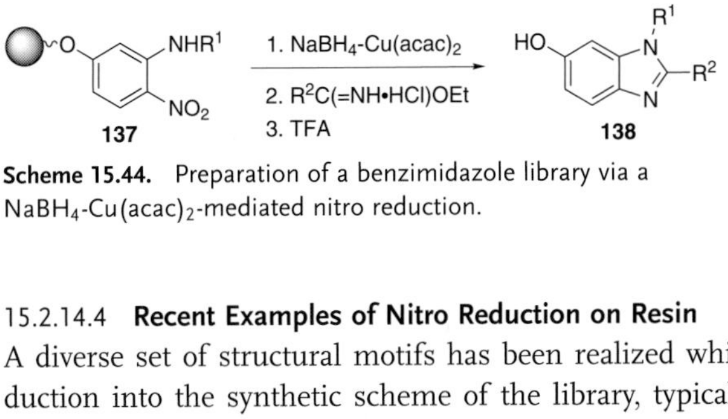

Scheme 15.44. Preparation of a benzimidazole library via a $NaBH_4$-$Cu(acac)_2$-mediated nitro reduction.

15.2.14.4 **Recent Examples of Nitro Reduction on Resin**

A diverse set of structural motifs has been realized which incorporate a nitro reduction into the synthetic scheme of the library, typically mediated by tin. Table 15.3 lists recent libraries synthesized with incorporation of a nitro reduction. Included in this set are libraries of 1,4-benzoxa- and benzothiazin-3(4*H*)-ones as well as benzimidazoles. A library of 56 1,4-benzoxa- and benzothiazin-3(4*H*)-ones was prepared via a reduction, cyclization, and derivatization approach (Scheme 15.45) [35]. A traceless solid-phase approach to a diverse group of substituted benzimidazoles incorporated a tin-mediated nitro reduction (Scheme 15.46) [125].

15.2.15
Imine Reductions (not Reductive Amination)

There are relatively few examples of imine reductions in combinatorial chemistry that do not involve imines formed from carbonyls and amines. For the reduction of imines generated from carbonyl compounds, see Sections 15.2.12 and 15.3.1.2 (reductive amination). For an example of imine formation via an aza-Wittig reac-

Tab. 15.3. Recent libraries utilizing nitro reduction (since January 1999).

Library	Reducing agent	Reference
2-Alkylthioimidazoles	Zn, NH$_4$Cl, methanol	123b
Benzimidazoles	SnCl$_2$, NMP	125
	SnCl$_2$, NMP	126
	SnCl$_2$, NMM	116b
	Zn, NH$_4$Cl, methanol	123c
Benzimidazolones	Zn, NH$_4$Cl, methanol	123a
Benzo[c]isoxazoles	SnCl$_2$, DMF	127
1,5-Benzodiazepin-2-ones	SnCl$_2$, DMF	119
1,4-Benzothiazepin-5-ones	SnCl$_2$	128
1,5-Benzothiazepin-4-ones	SnCl$_2$, DMF	129
Benzothiazines	SnCl$_2$, DMF	84
Benzothiazoles	SnCl$_2$, DMF, NaOAc	115
1,4-Benzoxa/thiazin-3(4H)-ones	SnCl$_2$, DMF, NaOAc	35
2-Carboxyindoles	SnCl$_2$, DMF	88
Diaminobenzamides	Fe, HCl	124
Dibenzo[b,f]oxazocines	SnCl$_2$, DMF	130
2,3-Dihydro-[1,5]-benzothiazepines	SnCl$_2$, DMF, NaOAc	115
2,3-Dihydro-[1,5]-benzothiazepine-4(5H)-ones	SnCl$_2$, DMF	81
3,4-Dihydro-2(1H)-quinazolinones	SnCl$_2$, DMF	131
3,4-Dihydro-1H-quinazolin-2-thiones	SnCl$_2$, DMF	131
1,2,3,4-Tetrahydroquinoxalin-2-ones	SnCl$_2$, NMP	132

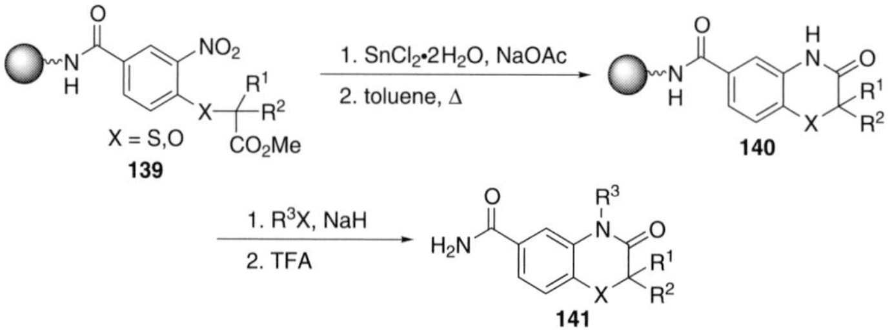

Scheme 15.45. Synthesis of 1,4-benzoxa- and thiazin-3(4H)-ones via a tin-mediated reduction.

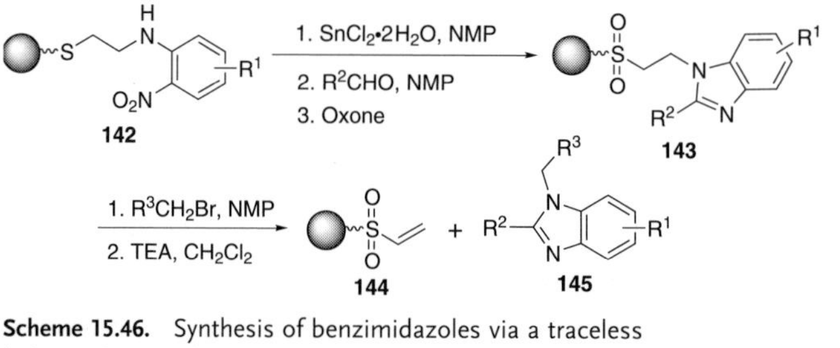

Scheme 15.46. Synthesis of benzimidazoles via a traceless linker and tin-mediated nitro reduction.

tion and subsequent reduction, see Section 15.3.1.2. Imine reductions have been used in the synthesis of compounds on solid support as well as in linker activation prior to cleavage (see below).

Bischler–Napieralski cyclization products have been prepared on solid phase, and the resulting cyclic imines were reduced with NaBH$_3$CN to provide tetrahydroisoquinolines (Scheme 15.47) [133]. When NaBH$_4$ was used in this application, the dihydroisoquinoline was cleaved at the ester–resin linkage. An indolenine intermediate, generated via a Fischer indole reaction, was reduced with NaBH$_4$ in the synthesis of a small library of spiroindolines [134]. Resin-bound imines of amino acids, prepared from transimination with N-H ketimines, have been reduced with NaBH$_3$CN in an approach to a library of hydantoins [135]. In a new linker application, a phenanthridine was reduced with NaBH$_4$/BH$_3$·THF [136]. The desired acid was subsequently released via oxidative cleavage.

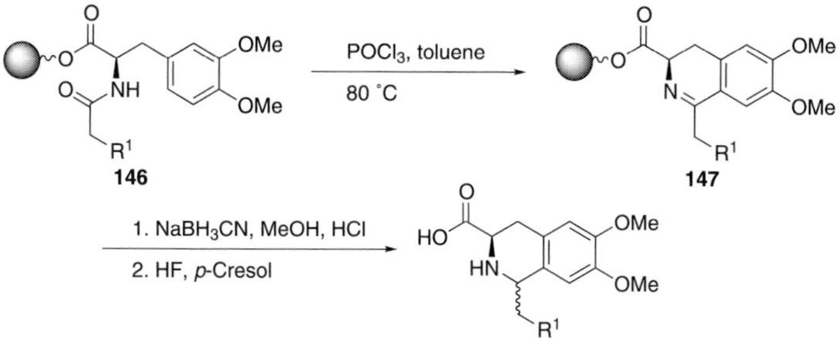

Scheme 15.47. Synthesis of tetrahydroisoquinolines involving an imine reduction.

15.2.16
Nitrile Reduction

Conti and coworkers have reported the reduction of a nitrile on solid support [137]. An aromatic nitrile was reduced with BH$_3$·SMe$_2$ in diglyme at 80 °C to provide a benzylamine. The resulting molecule was then released from the resin using an α-chloroethyl chloroformate methanol activation-cleavage strategy.

15.2.17
N–N and N–O Bond Reductions

Samarium diiodide has been used to cleave N–O bonds in a hydroxylamine traceless linker application [138]. Recently, a report was published that described both nitrosamine and hydrazine reductions on solid phase for the preparation of an array of α-substituted primary amines (Scheme 15.48) [139]. DIBAL reduction of the nitrosamine to the corresponding hydrazine followed by addition of an aldehyde gave the resin-bound hydrazone. Nucleophilic addition and subsequent borane reduction of the resulting derivatized hydrazine provided the target amines in mod-

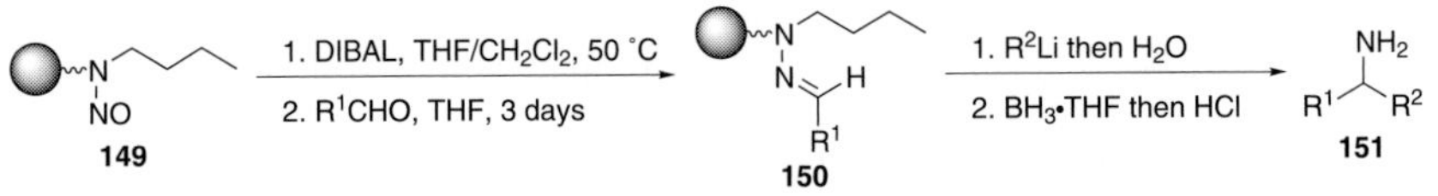

Scheme 15.48. Preparation of α-substituted primary amines via hydrazine reductive cleavage.

est yields. This approach has also been used to prepare chiral hydrazones and the corresponding chiral amine products with modest enantioselectivity (50–86% ee) [140].

15.2.18
Miscellaneous Reductions

There are a number of reductions performed on solid-supported functional groups for which there are relatively few examples. These reductions can be categorized as those in which the substrate remains attached to the resin, and those where it is released. Wustrow and coworkers have used a reductive cleavage performed with Pd(OAc)$_2$ and formic acid to provide benzoate esters and benzamides from aryl sulfonates [141]. Reductions where substrates remain attached are listed in Table 15.4.

15.3
Solution-phase Reductions

15.3.1
Supported Reagents

15.3.1.1 Asymmetric Reagents
A number of different research groups have shown that polymer-bound amino alcohols can act as chiral ligands in asymmetric hydride reductions of various func-

Tab. 15.4. Miscellaneous reductions on solid-supported substrates.

Reduction	Reducing agent	Reference
Ozonide to alcohol	NaBH$_4$, sonication	142
Ozonide to aldehyde	PPh$_3$, sonication	142
Epoxide to alcohol	LiBH$_4$	143
Peroxide to alcohol	(EtCO$_2$)$_3$P	144
Acetal to hydroxyether	DIBAL	145
Lactone to diol	LAH	43b
Alkyl chloride to alkane	NaI then Bu$_3$SnH	146
Isoxazole to aldehyde	LAH	147
Hydroxybenzotriazole to benzotriazole	PCl$_3$ or SmI$_2$	148
Tin chloride to tin hydride	LiBH$_4$	149
Phosphine sulfide to phosphine	TfOMe then HMPT	150

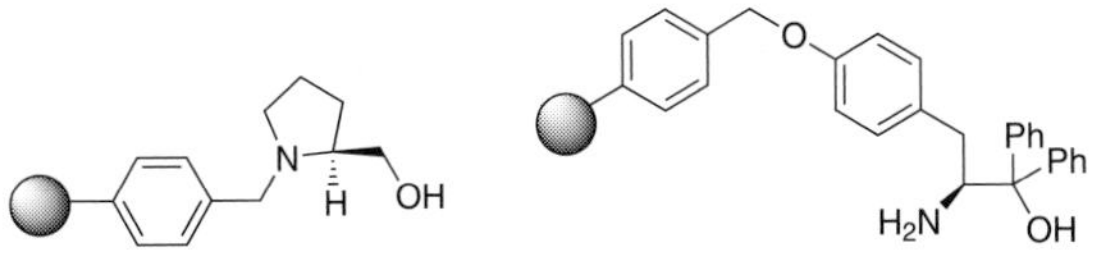

Fig. 15.1. Itsuno's asymmetric reduction ligands.

tional groups. Itsuno and coworkers attached optically active prolinol to polystyrene to give **152** and treated this product with $BH_3 \cdot THF$ to derive an enantioselective reducing agent (Figure 15.1) [151]. This reagent reduced prochiral ketones to secondary alcohols in good optical purity. The highest optical yield (80%) was obtained with a 1% crosslinked reagent (with 50% functionalization), which was 20% higher than that obtained by the solution-phase control. Following hydrolysis of the reaction mixture with 2 M HCl, the polymer was collected via filtration. Borane regeneration allowed this reagent to be used two more times. Itsuno and coworkers also attached amino alcohols to a polymer through a pendant aromatic group (Figure 15.1) [152]. An acetophenone oxime was reduced with the reagent derived from this polymer-bound amino alcohol and $NaBH_4/ZrCl_4$ or $BH_3 \cdot THF$ [153]. The optical purity of the product was only 35% ee; however, the reagent could be recycled.

Adjidjonou and Caze have also synthesized polymer-bound amino alcohols that were combined with $NaBH_4$ to reduce acetophenone [154]. These reagents delivered the product with modest enantioselectivity (up to 75% ee), which was much more enantioselective than the product obtained from a solution-phase control experiment (12% ee). Frechet et al. derived ligands from ephedrine and polystyrene resin and utilized them in the LAH-mediated reduction of acetophenone [155]. The enantiomeric excess of the product was 79% when a lightly loaded insoluble polymer species was utilized in the presence of an achiral phenol. The minimally substituted resin allows the chiral amino alcohols to act independently from one another and allows the hydride to access all of these units fully, thus providing higher enantioselectivity.

15.3.1.2 Non-asymmetric Reagents

Borane-based reagents

In 1977, Gibson and Bailey introduced the first solid-supported borohydride exchange resin (BER) [156]. It should can be noted that, following use, this reagent can be collected by filtration and regenerated. Early studies with this reagent focused on carbonyl reductions and related chemoselectivities, which were found to be better than those produced by $NaBH_4$ in solution [157]. It was understood that this difference in selectivity was due in part to the slower reaction kinetics of the support-bound reagent.

Fig. 15.2. Macroporous polystyrene-supported borohydride and cyanoborohydride.

Despite the recognized benefits of BER, improvements have been continually sought. As has been demonstrated with solution-phase $NaBH_4$ reductions, the addition of catalytic quantities of transition metal salts ($CuSO_4$ [158], Ni_2B [159], and $Ni(OAc)_2$ [160]) enhances reactivity and provides the ability to reduce a broader spectrum of functional groups. This area of research has also seen the introduction of zinc [161] and zirconium [162] borohydride polymers.

Reductive amination. Commercially available solid-supported reducing agents such as BER and $NaBH_3CN$ on exchange resin ($PS-BH_3CN$) are useful for solution-phase reductive aminations [163]. Recently macroporous polystyrene versions of $NaBH_4$ ($MP-BH_4$) and $NaBH_3CN$ ($MP-BH_3CN$) have also become commercially available (Figure 15.2) [164]. All of these reagents have the same advantage: they are easily removed from the reaction mixture via filtration. $PS-BH_3CN$ and $MP-BH_3CN$ have the added advantages of avoiding contamination of final products with cyanide and providing enhanced chemoselectivities (relative to BER and $MP-BH_4$).

Typically, reductive aminations with BER and $MP-BH_4$ are two-step procedures, usually performed in methanol (Scheme 15.49). The imine is preformed with 3-Å molecular sieves followed by addition of the reducing agent [165]. Any unreacted amine can be scavenged with an appropriate polymer-supported scavenger (e.g. Wang resin or PS-carboxaldehyde) [166].

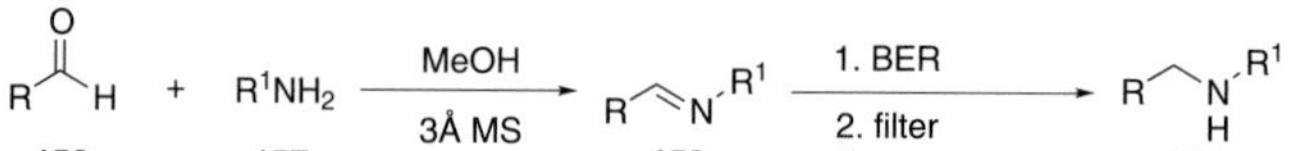

Scheme 15.49. Two-step solution-phase reductive amination with BER.

Ley et al. have contributed a number of papers on the subject of polymer-supported reagents, including reductive amination with BER [165c, 167] and $PS-BH_3CN$ in conjunction with scavenger resins [168]. Ley et al. recently described the reductive amination of substituted bicyclo[2.2.2]octanes with BER and amine scavenging with Wang resin (Scheme 15.50) [165c]. Kaldor et al. also reported the use of BER and scavenger resins in the parallel preparation of small molecules [166a]. They used PS-NCO, PS-CHO, and PS-COCl to scavenge excess primary and secondary amines from crude reaction mixtures and isolated products with purities exceeding 90% (HPLC).

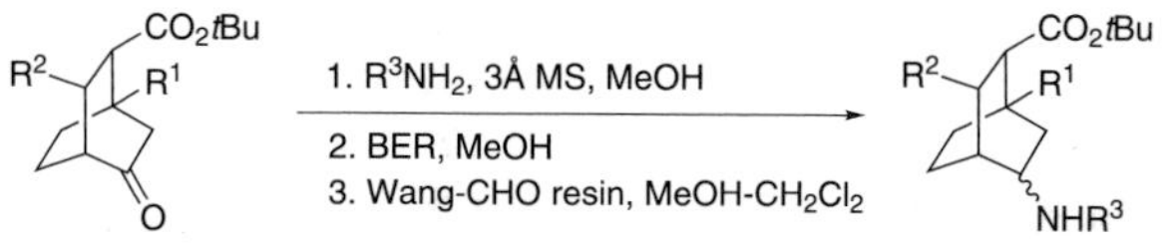

Scheme 15.50. Ley's reductive amination/amine scavenging approach to subsituted bicyclo[2.2.2]octanes.

Aldehyde and ketone reductions. In 1983, Yoon et al. studied the chemoselectivity of carbonyl reductions in a series of competitive reduction experiments with BER (no additives) [157a]. Their results showed that aldehydes were reduced in preference to ketones. More interesting were their observations that there was selectivity between aldehydes and between ketones. Aromatic aldehydes were preferentially reduced in the presence of aliphatic aldehydes. Benzaldehydes with para-substituted electron-withdrawing groups were reduced preferentially to those with para electron-donating groups. It was also shown that unhindered ketones were reduced in preference to hindered ketones. In a separate study by Yoon et al., it was also shown that the addition of $CuSO_4$ to BER increased the diastereoselectivity of the reduction of norcamphor to norborneol (endo/exo = 94:6 vs. 82:18) [158]. The reduction of ketones and aldehydes has also been carried out using zinc [161] and zirconium [162] borohydride reagents immobilized on polyvinylpyridine. The zinc-based reagent is completely inert toward ketones; however, addition of $FeCl_3$ gives low to moderate yields of ketone reduction products. The solid-supported zirconium borohydride reduces both aldehydes and ketones in the absence of an additive. Further, it has been shown that the BER-Ni(OAc)$_2$ system fully reduces aromatic aldehydes to toluene derivatives in high yield regardless of aromatic substitution [160b]. A hindered equivalent of BER, which diastereoselectively reduces ketones to secondary alcohols, has recently been introduced by Smith et al. [169].

Studies on the reduction of α,β-unsaturated aldehydes and ketones have also been carried out using these reducing agents. BER selectively adds hydride in a 1,2-fashion to these substrates, delivering allylic alcohols in high yield [157b]. The same properties are exhibited by the zirconium reagent [162]; however, the zinc reagent [161] shows chemoselectivity in that it reduces aldehydes without affecting ketones. Sim and Yoon showed that addition of 0.1 equiv. of $CuSO_4$ to BER under standard conditions (5 equiv. BER, methanol, room temperature) fully reduced α,β-unsaturated systems to saturated alcohols [158]. However, if the amount of BER was reduced to 2 equiv., the saturated ketone was isolated [158]. Despite these results, Ley et al. recently published a report describing the isolation of the allylic alcohol from a BER-$CuSO_4$-mediated α,β-unsaturated ketone reduction [167a]. In their synthesis of ($\pm$)-epimaritidine, Ley et al. successfully utilized BER-$CuSO_4$ and BER-NiCl$_2$ to carry out the 1,2-reduction of an α,β-unsaturated ketone [167a]. It should be pointed out that the structural complexity of the substrate in the Ley synthesis is much greater than that of Sim and Yoon. In the report of the synthesis of ($\pm$)-epibatidine, Ley and coworkers also used the parent BER to carry out ketone reductions in high yield [170].

Ester and acid chloride reductions.   The reduction of fully oxidized carbons has also been studied, but to a much lesser extent. Esters, for example, seem to be inert to these exchange resins even when transition metal salts are employed. Acid chlorides, on the other hand, have been reduced to both aldehydes and alcohols depending on the resin used. Simple long-chain acid chlorides have been selectively reduced to aldehydes in high yield by passage through a column of BER [171]. Depending on the reaction conditions, Tamami and Goudarzian have shown that polymeric $Zn(BH_4)_2$ can deliver either the alcohol or the aldehyde, however the products are not obtained cleanly [161b]. For example, if phenylacetyl chloride is treated with $Zn(BH_4)_2$ in hot THF, a 70:20 mixture of the alcohol and aldehyde is recovered. If the reaction is run at room temperature in CH_2Cl_2, a 25:65 mixture is obtained. Tamani and Lakouraj have also demonstrated that high yields of clean alcohol can be obtained by using another polymeric zinc borohydride, poly-η-(pyrazine)zinc borohydride, in THF at ambient temperature [172]. In Ley and coworkers' synthesis of ($\pm$)-epibatidine, the first step involved an aromatic acid chloride reduction to an alcohol mediated by BER [170].

Epoxide reductions.   The reduction of epoxides has also been studied. BER with $CuSO_4$ does not react with aliphatic epoxides, yet cleanly reduces styrene oxide to ethylbenzene [158]. Despite requiring additional quantities of reagents (10 equiv. BER and 0.5 equiv. $CuSO_4$), α-methylstyrene oxide and β-methylstyrene oxide also gave the fully saturated alkylphenyl derivatives upon reduction. Supported $Zn(BH_4)_2$ was capable of reducing both aliphatic and styrenyl derivatives, however this reagent did not give fully reduced products. Instead, a mixture of the more and less substituted alcohols was obtained, with the former predominating [161b]. The poly-pyrazine zinc reagent was inert toward both types of epoxides [172].

Halide reductions.   Sim and Yoon looked at the reduction of alkyl and aryl halides in detail. BER-$CuSO_4$ was found to be inert toward simple alkyl and aryl chlorides, while readily reducing primary and secondary alkyl bromides as well as aryl bromides and iodides [158]. It should be noted that activated chlorides (benzylic or α to an ester) can be reduced by this system. These chemoselectivities were demonstrated by performing competition experiments. For instance, 1-bromo-4-chlorobutane was readily reduced to 1-chlorobutane (95%) and *p*-bromochlorobenzene was cleanly reduced to chlorobenzene (99%). Since aryl bromides required heat to be effectively reduced, while aryl iodides did not, it was possible to selectively reduce *p*-bromoiodobenzene to bromobenzene at ambient temperature with a 97% yield. Yoon et al. have also shown that BER-Ni(OAc)$_2$ has nearly the same selectivity profile as BER-$CuSO_4$, and that this nickel-based system can be used to reduce 1-octyl tosylate to octane in 95% yield provided that NaI is present [160a].

Disulfide reductions. Attempts to reduce disulfides with polymer-supported reagents has given variable results. BER-CuSO$_4$ quantitatively reduces diphenyl disulfide, yet fails to convert *n*-butyl disulfide to *n*-butylthiol [158]. On the other hand, polymeric Zn(BH$_4$)$_2$ has been successful in reducing both substrates (100% and 40% respectively), as well as others [161b]. The parent BER quantitatively reduces diphenyl disulfide [173].

Azide reductions. BERs and combinations with nickel or copper salts are effective at reducing alkyl and aryl azides [158, 174]. In an early application of BERs, both aryl and arylsulfonyl azides were reduced in methanol to amines and sulfonamides [175]. BER-Ni(OAc)$_2$ has been used to reduce a variety of azides [174]. Tamami and Lakouraj's piperazine-based zinc reagent can reduce both aryl and alkyl azides to amines [172]. Tamami and Goudarzian's pyridine-based version reduces aryl and arylsulfonyl azides but does not react with alkyl azides [161b].

Nitro reductions. A number of support-bound borohydride reagents has been used to reduce nitro groups [176]. BER-Ni(OAc)$_2$ reduces aromatic and aliphatic nitro groups and can be easily removed via filtration in a solution-phase approach [177]. The BER-CuSO$_4$ reagent couple also reduces aromatic and aliphatic nitro groups [158]. BER-NiCl$_2$ was used by Ley and coworkers to reduce a nitro group in their synthesis of epibatidine [170].

Reductive cyclizations. The reductive addition of alkyl iodides to electron-deficient alkenes has been demonstrated utilizing the BER-Ni$_2$B system [159a]. Examples of radical additions to α,β-unsaturated esters, nitriles, and ketones have been shown to occur in high yields. It has been demonstrated that the same reagent affects aliphatic alkene and vinyl ether reactions with α-bromo esters [159b].

Miscellaneous reductions. BER-Ni(OAc)$_2$ also has been reported to reduce aldehyde oximes to amines [178].

Tin-based reagents

Polymeric tin hydrides are capable of reducing a number of functional groups, including carbonyls, alkyl halides, and alcohols [179]. The last are reduced through the intermediacy of a phenylthionocarbonate, according to the methodology set forth by Barton [179b,c]. The main advantage of these reagents over tributyltinhydride (TBTH) is in the work-up. Separations to remove toxic tin byproducts are avoided as the tin species can be easily removed by filtration.

In 1975, Crosby and coworkers introduced the first of the supported tin reagents, a polystyrene-based *n*-butyldihydridotin species [179a]. This reagent directly links a tin atom to the phenyl ring of the polystyrene backbone. In 1993, Neumann and Petersheim published an optimized preparation for a polystyrene-based monohydridotin reagent that utilizes a two-carbon linker between the tin and aromatic backbone of the polymer [180]. Since aromatic tin bonds can be labile, this aliphatic carbon-linked tin reagent was believed to be more stable than Crosby's reagent. Dumartin et al. introduced tin reagents with 3- and 4-carbon linkers that more closely resemble the structure and reactivity of TBTH [179d].

These tin reagents have been used to carry out carbonyl reductions in high yield, including the reduction of both aliphatic and aromatic aldehydes and ketones. It has also been shown that chemoselectivity can be achieved with these reagents, as alkyl halides have been reduced in the presence of ketones [179a]. Neumann and coworkers demonstrated the feasibility of alcohol deoxygenation by utilizing the Barton protocol. This methodology required the conversion of an alcohol to a phenylthionocarbonate, which was then reduced with a solid-supported tin reagent to give the saturated alkyl compound [179b,c]. Neumann and coworkers have also applied this reagent to the reductive cyclization of ω-alkenyl halides [181].

Trialkylsilane-mediated reduction of carbonyls
A polymer-supported trialkylsilane has been used to hydrosilylate carbonyl aldehydes and ketones [182]. Treatment of the carbonyl compounds with the trialkylsilane and Wilkinson's catalyst generated resin-bound alkoxysilanes (Scheme 15.51). Cleavage of the resulting alkoxysilane with HF provided the desired alcohols in fair to good yields.

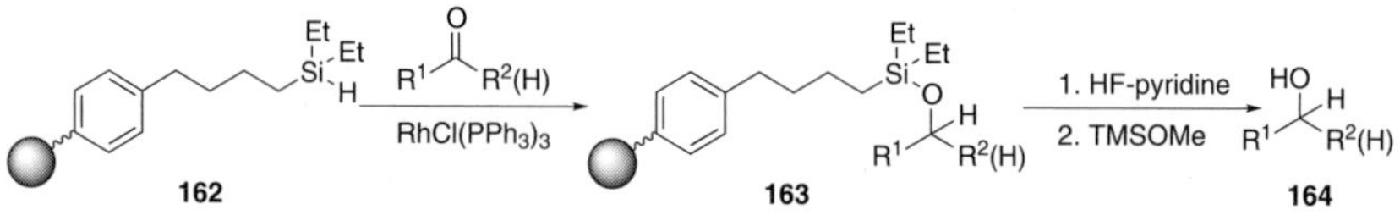

Scheme 15.51. Reduction of aldehydes and ketones via hydrosilylation.

Polymer-supported dihydrolipoic acid-mediated reduction of disulfides
Disulfides of cystamine, cysteine, 2-hydroxyethyl disulfide, and oxidized glutathione have been reduced with polymer-bound dihydrolipoic acid [183]. The polymer is prepared via $NaBH_4$ reduction of lipoic acid on polymer (Scheme 15.52). The best results for disulfide reduction were obtained with a polyacrylamide solid support in a pH range of 7.5–8.5.

Scheme 15.52. Preparation of polymer-supported dihydrolipoic acid.

Polymer-supported dihydropyridine-mediated reductions
Polymer-supported 1,4-dihydropyridines (PS-DHPs) have been used as NADH-type reducing agents [184]. A divalent cation, typically magnesium, is required for reducing activity and the reactions can be run in either organic or aqueous systems.

Bourguignon and coworkers used 1,4-dihydronicotinamide on Merrifield resin to reduce C=O, C=N, C=S, and C=C double bonds [184a]. Obika and coworkers have developed a chiral sulfinyl-containing DHP on Merrifield resin that was used to reduce methyl benzoylformate to the corresponding hydroxy ester (Scheme 15.53) [184c]. Quantitative chemical yields and high optical yields (96% ee) were obtained when the reaction was run in acetonitrile–benzene (1:1) with 2.5 equiv. of supported DHP and Mg(ClO$_4$)$_2$, respectively. The oxidized supported DHP could be regenerated by treatment with propyl-1,4-dihydronicotinamide (PNAH).

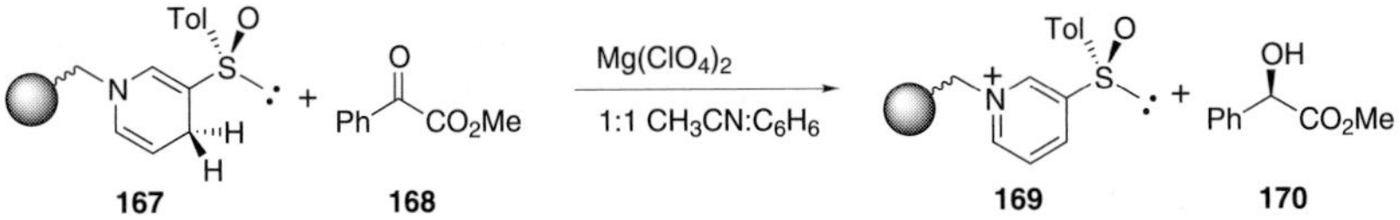

Scheme 15.53. Polymer-supported chiral NADH model ketoester reduction.

Polymer-supported sulfide reductions of ozonides

Ozonide reductions have been performed with solid-supported triphenylphosphine [185] and sulfides [186]. Appell and coworkers have prepared 3,3′-thiodipropionic acid and its sodium salts as supported analogues of dimethylsulfide for reductive quenching of ozonides [186]. The best results were obtained in ozonolysis reactions with the monosodium salt; as such, a polymer-supported version **172** was prepared. The corresponding dialdehyde of ethyl 3-cyclopentenecarboxylate was generated in a 92% yield after quenching the ozonide with this polymer-supported reagent (Scheme 15.54).

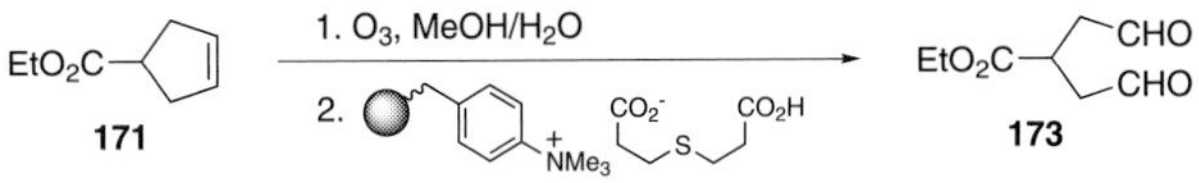

Scheme 15.54. Polymer-supported sulfide for reductive ozonolysis work-up.

Polymer-supported triphenylphosphines for the reduction of azides

Polymer-supported triphenylphosphine (PS-PPh$_2$) is similar to unsupported triphenylphosphine in solution-phase azide reductions. An added advantage of PS-PPh$_2$ is that the phosphine oxide generated is left on the polymer and is easily removed from the product by filtration.

Polystyrene-supported triphenylphosphine has been used to reduce azides in a series of azido nucleosides [187]. Yields were nearly quantitative and were similar to those obtained with unsupported triphenylphosphine. Polyethylene glycol-supported triphenylphosphine (PEG-PPh$_2$) has been successfully applied to azide reductions, providing amines in shorter reaction times than with PS-PPh$_2$ [188].

Reaction of PS-PPh$_2$ and azides provides iminophosphoranes that in turn can react with aldehydes to provide imines (aza-Wittig reaction). This approach has been used to generate a set of 20 imines which were reduced with PS-BH$_3$CN or NaBH$_3$CN to give amines in good to excellent yields (Scheme 15.55) [189]. Imines have also been prepared in a similar fashion using 1 equiv. of a noncrosslinked polystyrene-supported triphenylphosphine [190]. This resin-bound phosphine has a higher loading (1 mmol g^{-1}) than PEG-PPh$_2$ (0.5 mmol g^{-1}) and can be used in stoichiometric quantities (PS-PPh$_2$ is typically used in excess).

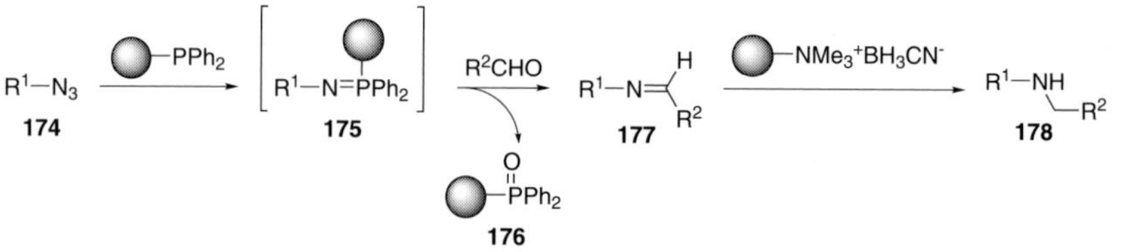

Scheme 15.55. Synthesis of amines via aza-Wittig reaction and imine reduction.

15.3.2
Supported Catalysts

15.3.2.1 Asymmetric Catalysis
Homogeneous asymmetric catalysis has been widely studied in both academic and industrial settings. A subset of this research involves the reduction of prochiral ketones to chirally enriched secondary alcohols. Two of the more efficient methods of carrying out this transformation have been described by the research groups of Noyori [191] and Corey [192]. Despite the advantages of the catalyst systems introduced by these groups, the cost of catalyst preparation is high, thereby making reuse desirable. The recovery and purification can be a difficult process; therefore, a number of research groups have pursued the preparation and use of heterogeneous analogs of these catalysts. By attaching these compounds to a solid support, it is believed that the ease with which a catalyst could be recovered and reused would be increased. However, catalyst-recycling improvements cannot come about at the expense of catalyst activity and stereoselectivity. Polymeric catalyst design has therefore taken into account the issues of active site symmetry, accessibility, and flexibility. The three major areas of research in this field include hydrogenations, transfer hydrogenations, and borane-mediated reductions.

Hydrogenations
Of the homogeneous asymmetric catalysts designed to carry out the reduction of prochiral ketones with molecular hydrogen, perhaps none has garnered more attention than the BINAP-Ru catalyst designed by Noyori [191b]. It should not be surprising therefore that this catalyst system has been chosen for exploitation by a number of research groups interested in heterogeneous catalysis. At least two

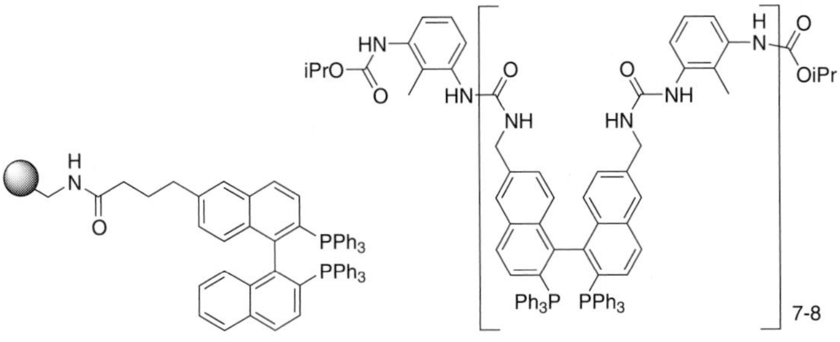

Fig. 15.3. Polymeric BINAP ligands.

different approaches have been used to incorporate the BINAP structure into a polymer.

An approach chosen by a group from Oxford Asymmetry involved attaching this C-2-symmetric ligand to an existing polymer with the attachment point distal from the active site phosphine atoms [193]. This goal was accomplished by mono-functionalizing the ligand at the 6-position with an alkyl carboxylic acid and then coupling this group to aminomethyl polystyrene resin. The resulting non-C-2-symmetric resin-bound ligand **179** (Figure 15.3) was then treated with a ruthenium(II) complex and methanolic HBr in acetone to give the active hydrogenation catalyst. The catalyst (1.7 mol%) was added to a methanol/THF solution of the substrate, which was then treated with 10 atm of hydrogen and heated to 70 °C. Reduction of the β-ketoester, methyl propionylacetate, was complete in 18 h with an enantioselectivity of 97%. This heterogeneous catalyst was similar in activity and selectivity to the parent homogeneous BINAP-Ru catalyst. Further, these data show that the loss of C-2 symmetry is not detrimental to the parent catalyst's selectivity. Perhaps more important is that this catalyst was easily recovered and reused two more times with only minimal losses in yield and enantioselectivity.

Another approach that has been used to incorporate BINAP into a polymer was carried out by the Lemaire group [194]. This approach involved copolymerization of a 6,6'-dimethylamine BINAP ligand with 2,6-tolylene diisocyanate to give a C-2-symmetric BINAP polymer **180** (Figure 15.3). This noncrosslinked polymer was soluble in DMF and DMSO, yet insoluble in the typical hydrogenation solvent – methanol. Utilizing conditions similar to those described above, Lemaire and coworkers were able to completely reduce methyl propionylacetate in 14 h (0.1 mol% catalyst, 40 atm., 50 °C) to the desired β-hydroxyester in 98% ee. This catalyst was also recovered and reused up to four times without any loss in activity or selectivity. Lemaire and coworkers utilized the same polymer in the presence of chiral diamines [191c] to reduce "simple" ketones (lacking proximal heteroatoms), such as substituted acetophenones to alcohols [195]. However, the enantiomeric excesses of the products varied between 58% and 96%. It was also shown that the absolute configuration of the added diamine is crucial to retain good enantioselectivity.

Chan and coworkers described another example of catalytic asymmetric hydrogenation in 1999 [196]. Although the prepared catalyst was used in an olefin reduction, which is beyond the scope of this chapter, it is worthy of note. The polymer formation was conceptually similar to that described by Lemaire, in that the C-2 symmetry was retained by copolymerizing either enantiomer of a 5,5'-difunctionalized BINAP with (2S,4S)-pentanediol and terephthaloyl chloride. These polymers contained a polyester backbone, which imparted solubility in the reaction solvent mixture of methanol and toluene (2:3, v/v). It was also possible to precipitate these catalysts with excess methanol following reaction completion. Utilization of either polymer to reduce 2-(6'-methoxy-2'-naphthyl)acrylic acid was complete within 18 h, giving nearly equal and opposite enantiomeric excesses (about 93% ee) of naproxen. These polymers were recycled up to ten times without any loss of activity or selectivity.

Transfer hydrogenations

The replacement of molecular hydrogen by hydrogen donors is an issue of practical importance in the field of catalytic asymmetric reduction (because of safety concerns). As was the case for standard homogeneous hydrogenations, the Noyori laboratory has made some of the most significant contributions in this area. Noyori and coworkers introduced the (1S,2S)- and (1R,2R)-N-(p-toluenesulfonyl)-1,2-diphenylethylenediamine (TsDPEN) ligands, which carry out hydride transfer-mediated ketone reductions in high yields and enantioselectivities when complexed with ruthenium [191d].

Both the Oxford Asymmetry [197] and Lemaire [198] groups have incorporated this ligand into polymers, using handles on the aromatic sulfonyl portion of the ligand as the linkage point to the resin (Figure 15.4). Each group adopted a strategy similar to the one they took in forming the BINAP polymers, described above. The Oxford group attached the ligand via an amide bond to preformed polymers (PS and PEG-PS) whereas the Lemaire group took a copolymerization approach. The Lemaire group did not concern itself with producing a linear C-2-symmetric polymer as they had previously, because the parent TsDPEN ligand is not C-2 symmetric. They copolymerized styrene and a TsDPEN ligand, equipped with a vinyl group, in both the presence and absence of divinylbenzene, thus producing both a crosslinked and a linear polymer.

Both groups studied the reduction of acetophenone; however, each group took their own approach to optimize the reaction conditions. The Oxford group focused on the variation of the hydride source, polymer, and solvent, while keeping the

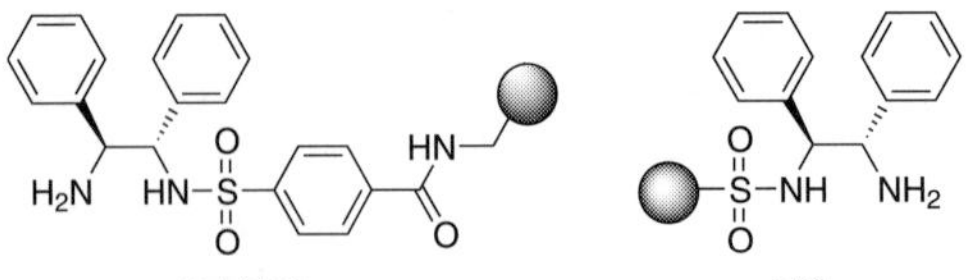

181 (PS)
182 (PS-PEG)

183

Fig. 15.4. Polymeric TsDPEN ligands.

transition metal constant [197]. The Lemaire group varied the polymer and transition metal, while keeping the hydride source and solvent constant [198]. Regardless of which polymeric ligand (PS or PEG-PS) was used in the catalyst preparation with [RuCl$_2$(p-cymene)]$_2$, the Oxford group encountered difficulties with isopropanol as the hydrogen donor. In the case of ligand **181** (Figure 15.4), the activity of the catalyst and the optical purity of the products were acceptable; however, catalyst recycling failed. In the case of ligand **182**, both the conversion and the enantioselectivity observed were low with the initial use of the catalyst. To circumvent these problems a switch was made from isopropanol to a mixture of formic acid and triethylamine (5:2). This combination led to successful reductions using either ligand. The catalyst formed from ligand **182**, in neat HCO$_2$H:Et$_3$N, gave product in 97% ee with 95% conversion in 28 h and could be reused once without any loss in ee. The catalyst formed from ligand **181** required a cosolvent to deliver favorable results. Addition of either DMF or CH$_2$Cl$_2$ resulted in enantiomer excesses of 94% or better with a reasonable degree of conversion (> 60% at 18 h). This catalyst was also successfully subjected to recycling.

Although the Lemaire group varied both the transition metal and the polymer in their efforts to find a heterogeneous transfer hydrogenation catalyst, there was little difference in activity and selectivity between their crosslinked and non-crosslinked polymers. From these results, they chose to focus on the significance of the transition metal [198]. Both Ir(I) and Ru(II) complexes were used in the preparation of the catalysts. The iridium catalyst was prepared by combination of the polymeric TsDPEN ligand **183** and [Ir(I)(COD)Cl]$_2$ in an isopropanolic solution of KOH, whereas the preparation of two ruthenium catalysts (from either [Ru(benzene)Cl$_2$]$_2$ or [Ru(p-cymene)Cl$_2$]$_2$) required heat (70 °C) and the replacement of KOH with triethylamine. Of these, the best results were found utilizing the iridium-based catalyst, which gave 96% conversion to the S-alcohol with 94% ee after 72 h. Unfortunately, the reuse of this catalyst led to poor results in terms of activity and selectivity. The ruthenium-based catalysts, on the other hand, were much less selective (31–64% ee), but were able to be reused up to four times.

For comparative purposes it is interesting to note that when both groups employed their ligand with [RuCl$_2$(p-cymene)]$_2$, a crosslinked polymer, and isopropanol (as solvent and hydride source), the optical purity of the alcohol produced was similar (84% ee for Lemaire and 90.5% ee for Oxford); however, the activities were quite different. After 48 h, the former group saw just 23% conversion while the latter group saw 88% conversion after 18 h. It must be noted that the catalyst load (2.5% vs. 1% respectively) and usage of KOH (presence vs. absence respectively) were different.

In another effort to identify a heterogeneous catalyst system capable of carrying out asymmetric reductions, the Lemaire group has copolymerized dialdimine ligands **184**, **185** (Figure 15.5) with varying amounts of polystyrene and/or DVB [199]. The iridium-based catalysts formed from the resulting ligands were used in the isopropanol-mediated transfer hydrogenation of acetophenone. Although the level of activity for these catalysts was high, the enantiomeric excess of the products obtained were never greater than 70%. Catalyst recycling suffered losses in

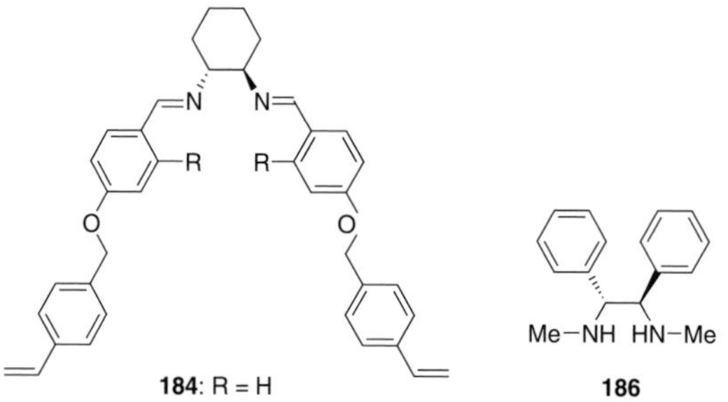

Fig. 15.5. Dialdimines and diamine used in the preparation of transfer hydrogenation catalysts.

both activity and selectivity. It is interesting to note that ruthenium and cobalt failed to catalyze the reduction, and that a 71% crosslinked polymer gave higher optical purity than both 15% and 100% crosslinked polymers. In another example from the Lemaire group, diamine **186** (Figure 15.5) was copolymerized with both a diacid chloride and a diisocyanate to give a poly(amide) and a poly(urea), respectively [200]. Utilizing the rhodium-based catalysts prepared from these ligands, the reduction results were less than optimal. The poly(amide) gave product in only 28% ee and the poly(urea) resulted in a product of only 60% ee. The latter catalyst could be recycled at least twice.

Also worthy of mention are the Lemaire group's efforts directed toward catalyst formation using "molecular imprinting" [201]. In an application of this methodology, this group copolymerized a preformed diamine–rhodium complex with diisocyanate in the presence of the compound to be imprinted (the alcohol product) – optically pure 1-(S)-phenylethanol. Once the polymer was formed and the alcohol was washed away, an "imprint" of the product was left in the catalytic site, which allowed for binding of acetophenone (or a similar substrate) and "biased" reduction to the desired products. In practice, the "imprinting" effect was found to be real, yet small. The enantiomeric excess of the product from acetophenone reduction increased modestly, from 33% to 43%, from the polymer catalyst formed in the absence of the template to the one formed in the presence of the template. Both of these optical purities were lower than those obtained using the diamine in a homogeneous control reaction (55%). A drawback to this method is that it does not allow for the reduction of a diverse set of ketones as the substrates must have a similar structure to the imprinted molecule.

Borane-based reductions

A third major area of research directed at the heterogeneous asymmetric catalysis of prochiral ketone reductions is focused on borane-based catalysts. Successful solution-phase asymmetric reductions using chiral oxazaborolidines, described by Itsuno et al. [151–153] and Corey et al. [192], have prompted much of this research.

Some of the early work carried out by Itsuno et al. involved covalent attachment

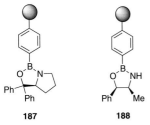

187   **188**

Fig. 15.6. Oxazaborolidine catalysts.

of an amino alcohol to a polystyrene backbone followed by carbonyl reduction with the amino alcohol based-borane reagent to give products with moderate enantioselectivity [151–153]. Although this work allowed for the recovery of the ligand, it did not allow for the recovery of the intact boron catalyst for reuse. Some of the more recent work has addressed this issue by covalently linking the boron atom of the catalyst directly to the aromatic ring of polystyrene.

In an effort to capture some of the success of Corey and coworkers' CBS catalyst and apply it to the solid phase, a group from Sandoz derived a catalyst from (*S*)-α,α-diphenyl-2-pyrrolidinemethanol and a crosslinked polystyrene boronic acid [202]. Once in hand, this catalyst **187** was used to reduce acetophenone and cyclohexylmethyl ketone (Figure 15.6). In both cases, 10 mol% of the catalyst was sufficient when used in combination with a stoichiometric reductant in THF at 40 °C. In the case of the reduction of the aromatic ketone acetophenone, an excellent enantioselectivity of 98% was obtained for the product when $BH_3 \cdot SMe_2$ was used as the stoichiometric reductant and care was taken to add the ketone slowly. This result was in line with that obtained when the monomeric catalyst was employed. The reduction of cyclohexylmethyl ketone also gave product with the same selectivity (about 80% ee) as was obtained by the monomer catalyst. This reduction required the use of either $BH_3 \cdot SMe_2$ or $BH_3 \cdot 1,4$-oxathiane as the stoichiometric reductant. Following a methanol quench, it was shown that at least one round of recycling was possible with this catalyst.

In a conceptually similar approach, Caze et al. have prepared catalysts from (1*R*,2*S*)-(−)-norephedrine and two polystyrene boronic acids with differing degrees of crosslinking [203]. The optimized reduction conditions involved premixing 30 mol% of $BH_3 \cdot SMe_2$ (the stoichiometric reductant) with 30 mol% of the lesser crosslinked catalyst **188** in THF at 20 °C, and after 30 min gradually adding all of the ketone and the remainder of the catalyst. This procedure delivered the product of propiophenone reduction in 89% ee and in high yield. Recycling of this catalyst was carried out up to three times. The more highly crosslinked catalyst, as well as a thiophene-linked catalyst [204], gave inferior results to those obtained by the less crosslinked polymer mentioned above.

More recently, Wandrey and coworkers attached a modified CBS ligand to a siloxane-based copolymer via Pt-catalyzed hydrosilylation [205]. The resulting soluble polymer is similar to the original Itsuno polymers in that the chiral amino

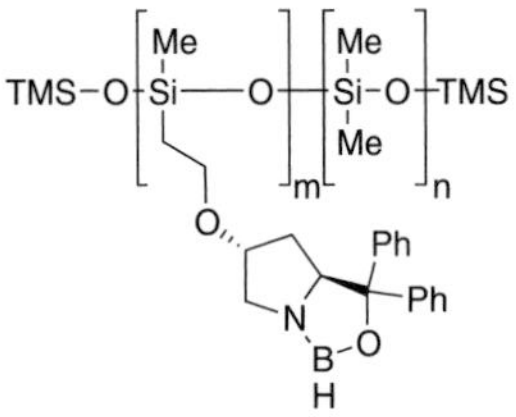

Fig. 15.7. Wandrey's oxazaborolidine catalyst.

alcohol, not the boron atom, acts as the point of attachment to the polymer. The catalyst **189** is formed by combination of the polymer amino alcohol with BH$_3$·SMe$_2$ in THF (Figure 15.7). Aryl ketone reduction is then carried out by treatment with the catalyst and stoichiometric quantities of BH$_3$·SMe$_2$ in THF. The resulting secondary alcohols are obtained in enantiomeric excesses ranging from 89% to 98%, which compares favorably with nonpolymeric results. Unfortunately, the products still have to be purified by distillation or chromatography.

In early 2001, Zhao and coworkers reported the preparation of catalyst **190** (Figure 15.8) [206]. Unlike the amino alcohol described above, the nitrogen of this ligand is attached to the resin via a sulfonamide bond. Product ee values were good to excellent for the reduction of aromatic ketones and moderate for alkyl ketones when this catalyst was employed. The combination of NaBH$_4$ and Me$_3$SiCl (or BF$_3$·OEt$_2$) was used as the stoichiometric reductant. Although the catalyst could be reused up to three times, a regeneration step was required.

The asymmetric reduction of ketones has also been carried out using zinc complexes of polynaphthyl ligands. These catalysts have been shown by the Pu group to mediate the catecholborane reduction of prochiral ketones [207]. Although the reduction of arylmethyl ketones gave products in good yield with ee values as high as 80%, the reductions of alkyl and branched methyl ketones were much less successful. After quenching the reaction, the homogeneous polymer was precipitated by addition of methanol. Reuse of this catalyst also required a regeneration step.

15.3.2.2 Non-Asymmetric Catalysis

One of the main drawbacks of tributyltinhydride-mediated reductive dehalogenations is the tin waste stream that is created. Utilization of polymeric tin reagents reduces the difficulties associated with their removal. A further improvement has

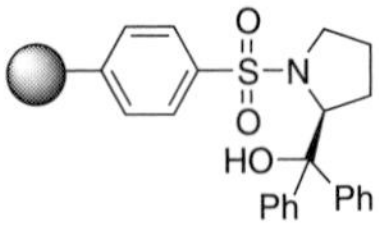

Fig. 15.8. Zhao's polymer-supported sulfonamide.

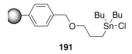

191

Fig. 15.9. Enholm's tin catalyst.

been introduced which uses these tin reagents in catalytic quantities in the presence of stoichiometric amounts of sodium borohydride.

Enholm and Schulte developed a noncrosslinked polymer (**191**) that is soluble in a number of organic solvents (Figure 15.9) [208]. This reagent can easily be removed from a reaction mixture by precipitation with methanol. Alkyl and aryl halide reductions have been carried out in N,N-dimethylacetamide (DMA) with 1.5 equiv. of NaBH$_4$, 0.1 equiv. of **191**, and AIBN, as initiator. Because the reactions are homogeneous, the reaction rates are faster than those found with insoluble polymeric catalysts, with completion typically observed in just a few hours at 80 °C.

Dumartin and coworkers demonstrated the utility of a polymeric tin iodide (**192**) and compared it with the reducing capabilities of Neumann's tin chloride reagent (**193**) (Figure 15.10, see Section 15.3.1.2) [181, 209]. In the comparative analysis of 1-bromoadamantane reduction, 0.05–0.9 equiv. of polymer **192** or **193** was combined with NaBH$_4$ (2.5 equiv.), AIBN (0.1 equiv.), and substrate in ethanol and heated to 65 °C for 12 h. When 0.2 equiv. of the catalyst **192** was used, adamantane was obtained in 93% GC yield, while 0.5 equiv. of Neumann's reagent gave only 40% of the same product. Dumartin's group also showed that catalyst **192** produces very low levels of tin pollution and can be reused.

Bergbreiter and Walker introduced a catalytic tin halide polymer that reduced alkyl and aryl bromides and iodides when combined with NaBH$_4$ and catalytic quantities of a crown ether [210]. Blanton and Salley extended this methodology by attachment of both the crown ether and tin chloride to the same lightly crosslinked polymer [211]. Although this polymeric co-catalyst showed lower activity than the soluble catalyst controls, it showed a marked increase in activity (48%) over controls with one catalyst supported and the other in solution. It appears that this lightly crosslinked polymer was sufficiently mobile to allow the two catalysts to interact. More recently, Deleuze and coworkers introduced an insoluble maleimide-based polymer for catalytic tin reductions [212]. The reduction of 1-bromoadamantane was successfully demonstrated, but the high reaction temperature required (95 °C) caused significant leaching of tin.

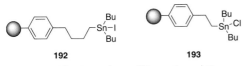

192　　　　**193**

Fig. 15.10. The tin catalysts of Dumartin and Neumann.

15.3.3
Unsupported Reagents Using Catch-and-release Purification

15.3.3.1 Reductive Amination

A catch-and-release approach has been used for purification of reductive aminations on acid-containing products [213]. The reductive amination was performed
with $NaBH_4$ and the crude reaction was mixed with DOWEX 1×8–400 formate
resin. The solution was drained and the resin treated with TFA to provide the
clean product. A catch-and-release strategy has also been used to prepare small
groups of ureas and amides via solution-phase reductive amination with $Ti(OiPr)_4/$
$NaBH_4$ (Scheme 15.56) [69c].

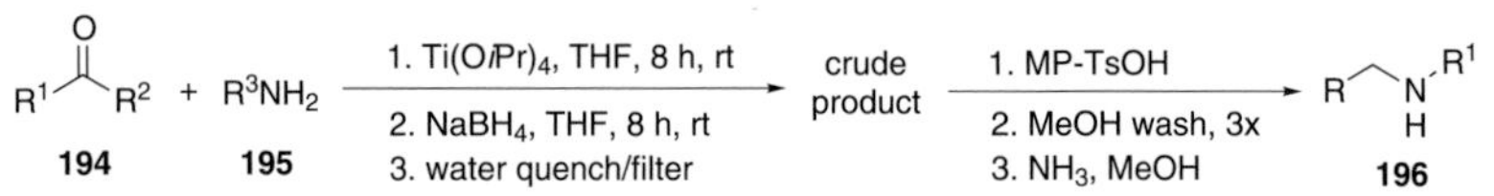

Scheme 15.56. Capture and release for rapid purification of a
solution-phase reductive amination reaction.

15.3.3.2 Amide Reductions

Bussolari and coworkers have also reported a resin quench-capture method for the
work-up of solution-phase amide reductions with $BH_3 \cdot THF$ [214]. Borane–amine
adducts were quenched by acidic AG 50W-X2 resin and boron-containing salts
were washed away while the desired amine was captured by the resin. Subsequent
treatment of the resin with ammonia released the desired amine products with
excellent purities ($> 95\%$ by LCMS). This approach was used to prepare a 300-
member library of 2-alkoxy- and 2-acyloxyphenylpropyl amines.

15.3.4
Fluorous Chemistry

All of the reductions described above required that either the substrate or the reagent be attached to a polymer support; however, a new method is emerging that
allows both reactants to remain in solution which takes advantage of the fact that
highly fluorinated reagents are immiscible in both standard organic and aqueous
phases at ambient temperature, yet are miscible in organic solvents at elevated
temperatures. This solubility profile simplifies product isolation and purification
by making it possible to separate products from byproducts by straightforward ex-
tractive work-ups. Curran and coworkers has shown that "fluorous" chemistry is
ideally suited to carry out tin-based reductions [215]. They demonstrated that per-
fluorinated tin reagents can reduce a number of functional groups including selenides, alkyl halides, nitro groups, xanthates, and aldehydes. Alkyl halides have
been reduced with both stoichiometric and catalytic quantities of the tin hydride
reagent using $NaCNBH_3$ as the stoichiometric reductant in the latter case. Curran

and coworkers also described reductive additions and cyclizations of alkyl and aryl halides to alkenes.

15.4
Conclusions

Reductions have been of enormous synthetic utility in both supported and unsupported combinatorial applications. While solid-phase organic synthesis has provided many examples of reductions over the last few decades, the area of solution-phase combinatorial synthesis has emerged and has grown rapidly more recently. As the introduction of new solid-supported reagents and catalysts continues, the ability of those involved in the drug discovery process to both generate and optimize lead compounds should increase.

References

1 a) P. H. H. Hermkens, H. C. J. Ottenheijm, D. Rees, *Tetrahedron* **1996**, *52*, 4527–4554; b) P. H. H. Hermkens, H. C. J. Ottenheijm, D. Rees, *Tetrahedron* **1997**, *53*, 5643–5678; c) S. Booth, P. H. H. Hermkens, H. C. J. Ottenheijm, D. Rees, *Tetrahedron* **1998**, *54*, 15385–15443; d) R. E. Dolle, K. H. Nelson, Jr, *J. Comb. Chem.* **1999**, *1*, 235–282; e) D. H. Drewry, D. M. Coe, S. Poon, *Med. Res. Rev.* **1999**, *19*, 97–148; f) R. G. Franzen, *J. Comb. Chem.* **2000**, *2*, 195–214; g) R. E. Dolle, *J. Comb. Chem.* **2000**, *2*, 383–433; h) S. V. Ley, I. R. Baxendale, R. N. Bream, P. S. Jackson, A. G. Leach, D. A. Longbottom, M. Nesi, J. S. Scott, R. I. Storer, S. J. Taylor, *J. Chem. Soc. Perkin 1* **2000**, 3815–4195.

2 a) J.-U. Peters, S. Blechert, *Synlett* **1997**, 348–350; b) S. Hanessian, F. Xie, *Tetrahedron Lett.* **1998**, *39*, 737–740.

3 Q.-S. Ren, W.-Q. Huang, P.-L. Ho, *Reactive Polymers* **1989**, *11*, 237–244.

4 U. Zehavi, A. Patchornik, *J. Am. Chem. Soc.* **1973**, *95*, 5673–5677.

5 J. M. Frechet, C. Schuerch, *J. Am. Chem. Soc.* **1971**, *93*, 492–496.

6 E. W. Baxter, J. K. Rueter, S. O. Nortey, A. B. Reitz, *Tetrahedron Lett.* **1998**, *39*, 979–982.

7 K. M. Brummond, J. Lu, *J. Org. Chem.* **1999**, *64*, 1723–1726.

8 C. T. Bui, N. J. Maeji, F. Rasoul, A. M. Bray, *Tetrahedron Lett.* **1999**, *40*, 5383–5386.

9 A. R. Katritzky, D. Toader, K. Watson, J. S. Kiely, *Tetrahedron Lett.* **1997**, *38*, 7849–7850.

10 a) P. F. Alewood, R. I. Brinkworth, R. J. Dancer, B. Garnham, A. Jones, S. B. H. Kent, *Tetrahedron Lett.* **1992**, *33*, 977–980; b) J. J. Burbaum, M. H. J. Ohlmeyer, J. C. Reader, I. Henderson, L. W. Dillard, G. Li, T. L. Randle, N. H. Sigal, D. Chelsky, J. J. Baldwin, *Proc. Natl. Acad. Sci. USA* **1995**, *92*, 6027–6031; c) P. S. Furth, M. S. Reitman, A. F. Cook, *Tetrahedron Lett.* **1997**, *38*, 5403–5406; d) E. B. Akerblom, A. S. Nygren, K. H. Agback, *Mol. Diversity* **1998**, *3*, 137–148; e) A. V. Purandare, M. A. Poss, *Tetrahedron Lett.* **1998**, *39*, 935–938.

11 M. R. Pavia, H. V. Meyers, G. Milot, M. E. Hediger, US Patent No. WO 99/33431, July 8, **1999**.

12 P. Sieber, *Tetrahedron Lett.* **1987**, *28*, 2107–2110.

13 D. R. Cody, S. H. H. DeWitt, J. C. Hodges, J. S. Kiely, W. H. Moos, M. R. Pavia, B. D. Roth, M. C.

SCHROEDER, C. J. STANKOVIC, US Patent No. 5,324,483, June 28, **1994**.

14 T. MASUDA, J. K. STILLE, *J. Am. Chem. Soc.* **1978**, *100*, 268–272.

15 I. PATERSON, M. DONGHI, K. GERLACH, *Angew. Chem. Int. Ed. Engl.* **2000**, *39*, 3315–3319.

16 C. E. LEE, E. K. KICK, J. A. ELLMAN, *J. Am. Chem. Soc.* **1998**, *120*, 9735–9747.

17 R. H. SCHLESSINGER, C. P. BERGSTROM, *Tetrahedron Lett.* **1996**, *37*, 2133–2136.

18 a) S. BRÄSE, D. ENDERS, J. KÖBBERLING, F. AVEMARIA, *Angew. Chem. Int. Ed. Engl.* **1998**, *37*, 3413–3415; b) B. FURMAN, R. THURMER, Z. KALUZA, R. LYSEK, W. VOELTER, M. CHMIELEWSKI, *Angew. Chem. Int. Ed. Engl.* **1999**, *38*, 1121–1123.

19 a) S.-S. WANG, *J. Am. Chem. Soc.* **1973**, *95*, 1328–1333; b) E. GIRALT, J. RIZO, E. PEDROSO, *Tetrahedron* **1984**, *40*, 4141–4152; c) R. S. GARIGIPATI, B. ADAMS, J. L. ADAMS, S. K. SARKAR, *J. Org. Chem.* **1996**, *61*, 2911–2914; d) J. D. WINKLER, W. McCOULL, *Tetrahedron Lett.* **1998**, *39*, 4935–4936.

20 S. KOBAYASHI, R. AKIYAMA, *Tetrahedron Lett.* **1998**, *39*, 9211–9214.

21 a) M. J. KURTH, L. A. A. RANDALL, C. CHEN, C. MELANDER, R. B. MILLER, K. McALISTER, G. REITZ, R. KANG, T. NAKATSU, C. GREEN, *J. Org. Chem.* **1994**, *59*, 5862–5864; b) S. V. LEY, D. M. MYNETT, W.-J. KOOT, *Synlett* **1995**, 1017–1020; c) L. F. TIETZE, A. STEINMETZ, *Angew. Chem. Int. Ed. Engl.* **1996**, *35*, 651–652; d) L. F. TIETZE, T. HIPPE, A. STEINMETZ, *Chem. Commun.* **1998**, 793–794.

22 G. J. KUSTER, H. W. SCHEEREN, *Tetrahedron Lett.* **2000**, *41*, 515–519.

23 J. M. GOLDWASSER, C. C. LEZNOFF, *Can. J. Chem.* **1978**, *56*, 1562–1568.

24 A. BHANDARI, D. G. JONES, J. R. SCHULLEK, K. VO, C. A. SCHUNK, L. L. TAMANAHA, D. CHEN, Z. YUAN, M. C. NEEDELS, M. A. GALLOP, *Bioorg. Med. Chem. Lett.* **1998**, *8*, 2303–2308.

25 L. M. GAYO, M. J. SUTO, *Tetrahedron Lett.* **1997**, *38*, 211–214.

26 a) S. KOBAYASHI, I. HACHIYA, S. SUZUKI, M. MORIWAKI, *Tetrahedron Lett.* **1996**, *37*, 2809–2812; b) S.

KOBAYASHI, I. HACHIYA, M. YASUDA, *Tetrahedron Lett.* **1996**, *37*, 5569–5572; c) S. KOBAYASHI, M. MORIWAKI, R. AKIYAMA, S. SUZUKI, I. HACHIYA, *Tetrahedron Lett.* **1996**, *37*, 7783–7786.

27 V. BERTINI, F. LUCCHESINI, M. POCCI, A. DE MUNNO, *Tetrahedron Lett.* **1998**, *39*, 9263–9266.

28 M. REGGELIN, V. BRENIG, R. WELCKER, *Tetrahedron Lett.* **1998**, *39*, 4801–4804.

29 M. REGGELIN, V. BRENIG, *Tetrahedron Lett.* **1996**, *37*, 6851–6852.

30 a) J.-A. FEHRENTZ, M. PARIS, A. HEITZ, J. VELEK, C.-F. LIU, *Tetrahedron Lett.* **1995**, *36*, 7871–7874; b) J.-A. FEHRENTZ, M. PARIS, A. HEITZ, J. VELEK, F. WINTERNITZ, J. MARTINEZ, *J. Org. Chem.* **1997**, *62*, 6792–6796; c) M. PARIS, C. DOUAT, A. HEITZ, W. GIBBONS, J. MARTINEZ, J.-A. FEHRENTZ, *Tetrahedron Lett.* **1999**, *40*, 5179–5182; d) J. M. SALVINO, M. MERVIC, H. J. MASON, T. KIESOW, D. TEAGER, J. AIREY, R. LABAUDINIERE, *J. Org. Chem.* **1999**, *64*, 1823–1830.

31 T. Q. DINH, R. W. ARMSTRONG, *Tetrahedron Lett.* **1996**, *37*, 1161–1164.

32 G. HUMMEL, O. HINDSGAUL, *Angew. Chem. Int. Ed. Engl.* **1999**, *38*, 1782–1784.

33 a) A. A. VIRGILIO, J. A. ELLMAN, *J. Am. Chem. Soc.* **1994**, *116*, 11580–11581; b) A. A. VIRGILIO, S. C. SCHÜRER, J. A. ELLMAN, *Tetrahedron Lett.* **1996**, *37*, 6961–6964.

34 A. J. SOUERS, A. A. VIRGILIO, A. ROSENQUIST, W. FENUIK, J. A. ELLMAN, *J. Am. Chem. Soc.* **1999**, *121*, 1817–1825.

35 C. L. LEE, K. P. CHAN, Y. LAM, S.-Y. LEE, *Tetrahedron Lett.* **2001**, *42*, 1167–1169.

36 a) I. SUCHOLEIKI, *Tetrahedron Lett.* **1994**, *35*, 7307–7310; b) F. W. FORMAN, I. SUCHOLEICKI, *J. Org. Chem.* **1995**, *60*, 523–528.

37 X.-Y. ZHAO, K. D. JANDA, *Tetrahedron Lett.* **1997**, *38*, 5437–5440.

38 X.-Y. ZHAO, *Tetrahedron Lett.* **1998**, *39*, 8433–8436.

39 X.-Y. ZHAO, *Bioorg. Med. Chem. Lett.* **1998**, *8*, 2439–2442.

40 K. C. Nicolaou, J. Pastor, S. Barluenga, N. Wissinger, *Chem. Commun.* **1998**, 1947–1948.

41 T. Ruhland, K. Andersen, H. Pedersen, *J. Org. Chem.* **1998**, *63*, 9204–9211.

42 W. H. Pearson, R. B. Clark, *Tetrahedron Lett.* **1997**, *38*, 7669–7672.

43 a) A. Zheng, D. Shan, B. Wang, *J. Org. Chem.* **1999**, *64*, 156–161; b) A. Zheng, D. Shan, X. Shi, B. Wang, *J. Org. Chem.* **1999**, *64*, 7459–7466.

44 a) A. Nefzi, J. M. Ostresh, J.-P. Meyer, R. A. Houghten, *Tetrahedron Lett.* **1997**, *38*, 931–934; b) P. J. Brown, K. P. Hurley, L. W. Stuart, T. M. Wilson, *Synthesis* **1997**, 778–782; c) J. M. Ostresh, C. C. Schoner, V. T. Hamashin, A. Nefzi, J.-P. Meyer, R. A. Houghten, *J. Org. Chem.* **1998**, *63*, 8622–8623; d) A. Nefzi, J. M. Ostresh, R. A. Houghten, *Tetrahedron* **1999**, *55*, 335–344; e) A. Nefzi, N. A. Ong, R. A. Houghten, *Tetrahedron Lett.* **2000**, *41*, 5441–5446; f) B. Yan, N. Nguyen, L. Liu, G. Holland, B. Raju, *J. Comb. Chem.* **2000**, *2*, 66–74; g) A. N. Acharya, A. Nefzi, J. M. Ostresh, R. A. Houghten, *J. Comb. Chem.* **2001**, *3*, 189–195.

45 a) A. Nefzi, J. M. Ostresh, M. A. Giulianotti, R. A. Houghten, *J. Comb. Chem.* **1999**, *1*, 195–198; b) S. Manku, C. Laplante, D. Kopac, T. Chan, D. G. Hall, *J. Org. Chem.* **2001**, *66*, 874–885; c) A. Nefzi, M. A. Giulianotti, R. A. Houghten, *J. Comb. Chem.* **2001**, *3*, 68–70.

46 S. Kobayashi, Y. Aoki, *Tetrahedron Lett.* **1998**, *39*, 7345–7348.

47 G. Liu, J. A. Ellman, *J. Org. Chem.* **1995**, *60*, 7712–7713.

48 K. Burgess, D. Lim, *Chem. Commun.* **1997**, 785–86.

49 D. G. Hall, C. Laplante, S. Manku, J. Nagendran, *J. Org. Chem.* **1999**, *64*, 698–699.

50 C. Y. Ho, M. J. Kukla, *Tetrahedron Lett.* **1997**, *38*, 2799–2802.

51 J. J. N. Veerman, F. P. J. T. Rutjes, J. H. vanMaarseveen, H. Hiemstra, *Tetrahedron Lett.* **1999**, *40*, 6079–6082.

52 X.-Y. Xiao, M. P. Nova, A. W. Czarnik, *J. Comb. Chem.* **1999**, *1*, 379–382.

53 a) D. W. Gordon, J. Steele, *Bioorg. Med. Chem. Lett.* **1995**, *5*, 47–50; b) G. C. Look, M. M. Murphy, D. A. Campbell, M. A. Gallop, *Tetrahedron Lett.* **1995**, *36*, 2937–2940.

54 a) S. Kim, C. H. Oh, J. S. Ko, K. H. Ahn, Y. J. Kim, *J. Org. Chem.* **1985**, *50*, 1927–1932; b) M. D. Bomann, I. C. Guch, M. DiMare, *J. Org. Chem.* **1995**, *60*, 5995–5996; c) A. F. Abdel-Magid, K. G. Carson, B. D. Harris, C. A. Maryanoff, R. D. Shah, *J. Org. Chem.* **1996**, *61*, 3849–3862.

55 A. R. Brown, D. C. Rees, Z. Rankovic, J. R. Morphy, *J. Am. Chem. Soc.* **1997**, *119*, 3288–3295.

56 R. Devraj, M. Cushman, *J. Org. Chem.* **1996**, *61*, 9368–9373.

57 N. W. Kahn, V. Arumugam, S. Balasubramanian, *Tetrahedron Lett.* **1996**, *37*, 4819–4822.

58 a) A. Nefzi, J. M. Ostresh, R. A. Houghten, *Tetrahedron Lett.* **1997**, *38*, 4943–4946; b) I. Akritopoulou-Zanze, T. J. Sowin, *J. Comb. Chem.* **2001**, *3*, 301–311.

59 a) E. G. Brown, J. M. Nuss, *Tetrahedron Lett.* **1997**, *38*, 8457–8460; b) E. G. Brown, J. M. Nuss, US Patent No. 5,861,532, January 19, **1999**.

60 a) D. H. Coy, S. J. Hocart, Y. Sasaki, *Tetrahedron* **1988**, *44*, 835–841; b) P. T. Ho, D. Chang, J. W. X. Joyce, G. F. Musso, *Peptide Res.* **1993**, *6*, 10–12.

61 C. G. Boojamra, K. M. Burow, L. A. Thompson, J. A. Ellman, *J. Org. Chem.* **1997**, *62*, 1240–1256.

62 a) J. Matthews, R. A. Rivero, *J. Org. Chem.* **1997**, *62*, 6090–6092; b) C. L. Lanter, J. W. Guiles, R. A. Rivero, *Mol. Diversity* **1999**, *4*, 149–153.

63 a) M. T. Bilodeau, A. M. Cunningham, *J. Org. Chem.* **1998**, *63*, 2800–2801; b) L. Weber, P. Iaiza, G. Biringer, P. Barbier, *Synlett* **1998**, 1156–1157; c) E. Farrant, S. S. Rahman, *Tetrahedron Lett.* **2000**, *41*, 5383–5386.

64 C. P. Holmes, J. P. Chinn, G. C. Look, E. M. Gordon, M. A. Gallop, *J. Org. Chem.* **1995**, *60*, 7328–7333.

65 a) A. K. Szardenings, T. S. Burkoth, G. C. Look, D. A. Campbell, *J. Org. Chem.* **1996**, *61*, 6720–6722; b) E. M. Gordon, M. A. Gallop, D. V. Patel, *Acc. Chem. Res.* **1996**, *29*, 144–154.

66 A. R. Katritzky, S. A. Blyakov, D. O. Tymoshenko, *J. Comb. Chem.* **1999**, *1*, 173–176.

67 M. Gustafsson, R. Olsson, C.-M. Andersson, *Tetrahedron Lett.* **2001**, *42*, 133–136.

68 F. Stieber, W. Grether, H. Waldmann, *Angew. Chem. Int. Ed. Engl.* **1999**, *38*, 1073–1077.

69 a) J. G. Breitenbucher, H. C. Hui, *Tetrahedron Lett.* **1998**, *39*, 8207–8210; b) J. G. Breitenbucher, H. C. Hui, Figliozzi, World Patent Application WO00003681, January 27, **2000**; c) S. Bhattacharyya, L. Fan, L. Vo, J. Labadie, *Comb. Chem. High Throughput Screening* **2000**, *3*, 117–124.

70 A. M. Bray, D. S. Chiefari, R. M. Valerio, N. J. Maeji, *Tetrahedron Lett.* **1995**, *36*, 5081–5084.

71 a) J. S. Koh, J. A. Ellman, *J. Org. Chem.* **1996**, *61*, 4494–4495; b) V. Krchnak, A. S. Weichsel, *Tetrahedron Lett.* **1997**, *38*, 7299–7302.

72 a) S. Carrington, J. Renault, S. Tomasi, J.-C. Corbel, P. Uriac, I. S. Blagbrough, *Chem. Commun.* **1999**, 1341–1342; b) E. A. Jefferson, K. G. Sprankle, E. E. Swayze, *J. Comb. Chem.* **2000**, *2*, 441–444; c) P. W. Davis, E. E. Swayze, *Biotechnol. and Bioeng. (Comb. Chem.)* **2000**, *71*, 19–27.

73 a) S. Caddick, D. Hamza, S. N. Wadman, *Tetrahedron Lett.* **1999**, *40*, 7285–7288; b) R. C. D. Brown, M. Fisher, *Chem. Commun.* **1999**, 1547–1548.

74 a) D. Sarantakis, J. J. Bicksler, *Tetrahedron Lett.* **1997**, *38*, 7325–7328; b) P. C. Kearney, M. Fernandez, J. A. Flygare, *J. Org. Chem.* **1998**, *63*, 196–200; c) Y. Aoki, S. Kobayashi, *J. Comb. Chem.* **1999**, *1*, 371–372; d) K. Paulvannan, T. Chen, J. W. Jacobs, *Synlett* **1999**, 1609–1611; e) K. H. Gordon, S. Balasubramanian, *Org. Lett.* **2001**, *3*, 53–56.

75 a) A. R. Katritzky, L. Xie, G. Zhang, M. Griffith, K. Watson, J. S. Kiely, *Tetrahedron Lett.* **1997**, *38*, 7011–7014; b) H. B. Lee, S. Balasubramanian, *Org. Lett.* **2000**, *2*, 323–326.

76 K. G. Estep, C. E. Neipp, L. M. S. Stramiello, M. D. Adam, M. P. Allen, S. Robinson, E. J. Roskamp, *J. Org. Chem.* **1998**, *63*, 5300–5301.

77 B. Barlaam, P. Koza, J. Berriot, *Tetrahedron* **1999**, *55*, 7221–7232.

78 a) C. G. Boojamra, K. M. Burow, J. A. Ellman, *J. Org. Chem.* **1995**, *60*, 5742–5743; b) A. M. Fivush, T. M. Wilson, *Tetrahedron Lett.* **1997**, *38*, 7151–7154; c) E. E. Swayze, *Tetrahedron Lett.* **1997**, *38*, 8465–8468; d) K.-L. Yu, R. Civiello, D. G. M. Roberts, S. M. Seiler, N. A. Meanwell, *Bioorg. Med. Chem. Lett.* **1999**, *9*, 663–666; e) D. S. Yamashita, X. Dong, H.-J. Oh, C. S. Brook, T. A. Tomaszek, L. Szewczuk, D. G. Tew, D. F. Veber, *J. Comb. Chem.* **1999**, *1*, 207–215; f) V. Krchnak, J. Smith, J. Vagner, *Tetrahedron Lett.* **2001**, *42*, 2443–2446.

79 C. T. Bui, A. M. Bray, F. Ercole, Y. Pham, F. A. Rasoul, N. J. Maeji, *Tetrahedron Lett.* **1999**, *40*, 3471–3474.

80 a) C. T. Bui, A. M. Bray, Y. Pham, R. Campbell, F. Ercole, F. A. Rasoul, N. J. Maeji, *Tetrahedron Lett.* **1998**, *4*, 155–163; b) C. T. Bui, F. A. Rasoul, F. Ercole, Y. Pham, N. J. Maeji, *Tetrahedron Lett.* **1998**, *39*, 9279–9282.

81 M. K. Schwarz, D. Tumelty, M. A. Gallop, *J. Org. Chem.* **1999**, *64*, 2219–2231.

82 a) A. K. Szardenings, T. S. Burkoth, *Tetrahedron* **1997**, *53*, 6573–6593; b) A. K. Szardenings, D. Harris, S. Lam, L. S. Shi, D. Tien, Y. Wang, D. V. Patel, M. Navre, D. A. Campbell, *J. Med. Chem.* **1998**, *41*, 2194–2200; c) A. K. Szardenings, V. Antonenko, D. A. Campbell, N. DeFrancisco, S. Ida, L. Shi, N. Sharkov, D. Tien, Y. Wang, M. Navre, *J. Med. Chem.* **1999**, *42*, 1348–1357.

83 T. Groth, M. Meldal, *J. Comb. Chem.* **2001**, *3*, 45–63.

84 T. S. Yokum, J. Alsina, G. Barany, *J. Comb. Chem.* **2000**, *2*, 282–292.

85 C. L. Lanter, J. W. Guiles, R. A. Rivero, US Patent No. 6,228,986, May 8, **2001**.

86 V. Krchnak, L. Szabo, J. Vagner, *Tetrahedron Lett.* **2000**, *41*, 2835–2838.

87 J. M. Alvarez-Gutierrez, A. Nefzi, R. A. Houghten, *Tetrahedron Lett.* **2000**, *41*, 851–854.

88 J. Tois, R. Franzen, O. Aitio, K. Huikko, J. Taskinen, *Tetrahedron Lett.* **2000**, *41*, 2443–2446.

89 S. Sun, I. J. Turchi, D. Xu, W. V. Murray, *J. Org. Chem.* **2000**, *65*, 2555–2559.

90 P. Blaney, R. Grigg, Z. Rankovic, M. Thoroughgood, *Tetrahedron Lett.* **2000**, *41*, 6635–6638.

91 A. D. Morely, *Tetrahedron Lett.* **2000**, *41*, 7405–7408.

92 P. Arya, K. M. K. Kutterer, A. Barkley, *J. Comb. Chem.* **2000**, *2*, 120–126.

93 P. J. Connolly, K. N. Beers, S. K. Wetter, W. V. Murray, *Tetrahedron Lett.* **2000**, *41*, 5187–5191.

94 S. Peters, T. Bielfeldt, M. Meldal, K. Bock, H. Paulsen, *Tetrahedron Lett.* **1992**, *33*, 6445–6448.

95 J. Rademann, R. R. Schmidt, *Carbohydrate Research* **1995**, *269*, 217–225.

96 a) J. Y. Roberge, X. Beebe, S. J. Danishefsky, *Science* **1995**, *269*, 202–204; b) K. A. Savin, J. C. G. Woo, S. J. Danishefsky, *J. Org. Chem.* **1999**, *64*, 4183–4186.

97 a) E. Meinjohanns, M. Meldal, T. Jensen, O. Werdelin, L. Galli-Stampino, S. Mouritsen, K. Bock, *J. Chem. Soc., Perkin Trans. 1* **1997**, 871–884; b) P. M. St. Hilaire, L. Cipolla, A. Franco, U. Tedebark, D. A. Tilly, M. Meldal, *J. Chem. Soc., Perkin Trans. 1* **1999**, 3559–3564.

98 J. P. Malkinson, R. A. Falconer, I. Toth, *J. Org. Chem.* **2000**, *65*, 5249–5252.

99 a) E. K. Kick, J. A. Ellman, *J. Med. Chem.* **1995**, *38*, 1427–1430; b) F. X. Woolard, J. Paetsch, J. A. Ellman, *J. Org. Chem.* **1997**, *62*, 6102–6103; c) A. Lee, L. Huang, J. A. Ellman, *J. Am. Chem. Soc.* **1998**, *121*, 9907–9914.

100 J. M. Kim, Y. Bi, S. J. Paikoff, P. G. Schultz, *Tetrahedron Lett.* **1996**, *37*, 5305–5308.

101 D. A. Goff, R. N. Zuckermann, *J. Org. Chem.* **1995**, *60*, 5744–5745.

102 R. Liang, L. Yan, J. Loebach, M. Ge, Y. Uozumi, K. Sekanina, N. Horan, J. Gildersleeve, C. Thompson, A. Smith, K. Biswas, W. C. Still, D. Kahne, *Science* **1996**, *274*, 1520–1522.

103 J. Zhou, A. Termin, M. Wayland, C. M. Tarby, *Tetrahedron Lett.* **1999**, *40*, 2729–2732.

104 K. C. Nicolaou, N. Winssinger, D. Vourloumis, T. Ohshima, S. Kim, J. Pfefferhorn, J.-Y. Xu, T. Li, *J. Am. Chem. Soc.* **1998**, *120*, 10814–10826.

105 a) M. R. Tremblay, J. Simard, D. Poirier, *Bioorg. Med. Chem. Lett.* **1999**, *9*, 2827–2832; b) M. R. Tremblay, D. Poirier, *J. Comb. Chem.* **2000**, *2*, 48–65.

106 U. Grabowska, A. Rizzo, K. Farnell, M. Quibell, *J. Comb. Chem.* **2000**, *2*, 475–490.

107 H. S. Oh, H.-G. Hahn, S. H. Cheon, D.-C. Ha, *Tetrahedron Lett.* **2000**, *41*, 5069–5072.

108 K. M. Halkes, P. M. St. Hilaire, A. M. Jansson, C. H. Gotfredsen, M. Meldal, *J. Chem. Soc., Perkin Trans. 1* **2000**, 2127–2133.

109 a) M. J. Sofia, N. Allanson, N. T. Hatzenbuhler, R. Jain, R. Kakarla, N. Kogan, R. Liang, D. Liu, D. J. Silva, H. Wang, D. Gange, J. Anderson, A. Chen, F. Chi, R. Dulian, B. Huang, M. Kamau, C. Wang, E. Baizman, A. Branstrom, N. Bristol, R. Goldman, K. Han, C. Longley, S. Midha, H. R. Axelrod, *J. Med. Chem.* **1999**, *42*, 3193–3198; b) D. J. Silva, H. Wang, N. M. Allanson, R. K. Jain, M. J. Sofia, *J. Org. Chem.* **1999**, *64*, 5926–5929.

110 K. Oertel, G. Zech, H. Kunz, *Angew. Chem., Int. Ed. Engl.* **2000**, *39*, 1431–1433.

111 J. P. Mayer, G. S. Lewis, C. McGee, D. Bankaitis-Davis, *Tetrahedron Lett.* **1998**, *39*, 6655–6658.

112 G. B. Phillips, G. P. Wei, *Tetrahedron Lett.* **1996**, *37*, 4887–4890.

113 For a non-combinatorial example, see:
M. Bartra, P. Romea, F. Urpi, J.
Vilarrasa, *Tetrahedron* **1990**, *46*, 587–
594.

114 a) J. P. Mayer, J. Zhang, K.
Biergarde, D. M. Lenz, J. J.
Gaudino, *Tetrahedron Lett.* **1996**, *37*,
8081–8084; b) Y. Pei, R. A.
Houghten, J. S. Kiely, *Tetrahedron
Lett.* **1997**, *38*, 3349–3352; c) J. Lee,
W. V. Murray, R. A. Rivero, *J. Org.
Chem.* **1997**, *62*, 3874–3879; d) A. S.
Kiselyov, R. W. Armstrong,
Tetrahedron Lett. **1997**, *38*, 6163–6166;
e) J. P. Mayer, G. S. Lewis, M. J.
Curtis, J. Zhang, *Tetrahedron Lett.*
1997, *38*, 8445–8448; f) A. S.
Kiselyov, L. Smith II, R. W.
Armstrong, *Tetrahedron* **1998**, *54*,
5089–5096.

115 C. L. Lee, Y. Lam, S.-Y. Lee,
Tetrahedron Lett. **2001**, *42*, 109–111.

116 a) D. Tumelty, M. K. Schwarz, M. C.
Needles, *Tetrahedron Lett.* **1998**, *39*,
7467–7470; b) A. Mazurov, *Bioorg.
Med. Chem. Lett.* **2000**, *10*, 67–70; c) Z.
Wu, P. Rea, G. Wickham, *Tetrahedron
Lett.* **2000**, *41*, 9871–9874.

117 G. A. Morales, J. W. Corbett, W. F.
DeGrado, *J. Org. Chem.* **1998**, *63*,
1172–1177.

118 H. V. Meyers, G. J. Dilley, T. L.
Durgin, T. S. Powers, N. A.
Winssinger, H. Zhu, M. R. Pavia,
Mol. Diversity **1995**, *1*, 13–20.

119 J. Lee, D. Gauthier, R. A. Rivero,
J. Org. Chem. **1999**, *64*, 3060–3065.

120 R. Di Lucrezia, I. H. Gilbert, C. D.
Floyd, *J. Comb. Chem.* **2000**, *2*, 249–
253.

121 R. A. Scheuerman, D. Tumelty,
Tetrahedron Lett. **2000**, *41*, 6531–6535.

122 a) A. Hari, B. L. Miller, *Tetrahedron
Lett.* **1999**, *40*, 245–248; b) A. Hari, B.
L. Miller, *Angew. Chem., Int. Ed. Engl.*
1999, *38*, 2777.

123 a) P.-C. Pan, C.-M. Sun, *Tetrahedron
Lett.* **1999**, *40*, 6443–6446; b) C.-M.
Yeh, C.-M. Sun, *Tetrahedron Lett.*
1999, *40*, 7247–7250; c) Y.-C. Chi,
C.-M. Sun, *Synlett* **2000**, 591–594.

124 M. S. South, B. L. Case, T. A. Dice,
G. W. Franklin, M. J. Hayes, D. E.
Jones, R. J. Lindmark, Q. Zeng,

J. J. Parlow, *Comb. Chem. High
Throughput Screening* **2000**, *3*, 139–
151.

125 D. Tumelty, K. Cao, C. P. Holmes,
Org. Lett. **2001**, *3*, 83–86.

126 J. M. Smith, J. Gard, W. Cummings,
A. Kanizsai, V. Krchnak, *J. Comb.
Chem.* **1999**, *1*, 368–370.

127 H. Stephensen, F. Zaragoza,
Tetrahedron Lett. **1999**, *40*, 5799–5802.

128 A. Nefzi, N. A. Ong, M. A.
Giulianotti, J. M. Ostresh, R. A.
Houghten, *Tetrahedron Lett.* **1999**, *40*,
4939–4942.

129 G. C. Morton, J. M. Salvino, R. F.
Labaudiniere, T. F. Herpin,
Tetrahedron Lett. **2000**, *41*, 3029–3033.

130 X. Ouyang, A. S. Kiselyov,
Tetrahedron **1999**, *55*, 8295–8302.

131 Q. Sun, X. Zhou, D. J. Kyle,
Tetrahedron Lett. **2001**, *42*, 4119–4121.

132 F. Zaragoza, H. Stephensen, *J. Org.
Chem.* **1999**, *64*, 2555–2557.

133 W. D. F. Meutermans, P. F.
Alewood, *Tetrahedron Lett.* **1995**, *36*,
7709–7712.

134 Y. Cheng, K. T. Chapman,
Tetrahedron Lett. **1997**, *38*, 1497–1500.

135 S.-H. Lee, S.-H. Chung, Y.-S. Lee,
Tetrahedron Lett. **1998**, *39*, 9469–9472.

136 W.-R. Li, N.-M. Hsu, H.-H. Chou, S.
T. Lin, Y.-S. Lin, *Chem. Commun.*
2000, 401–402.

137 P. Conti, D. Demont, J. Cals, H. C.
J. Ottenheijm, D. Leysen, *Tetrahedron
Lett.* **1997**, *38*, 2915–2918.

138 R. M. Myers, S. P. Langston, S. P.
Conway, C. Abell, *Org. Lett.* **2000**, *2*,
1349–1352.

139 J. H. Kirchhoff, S. Brase, D. Enders,
J. Comb. Chem. **2001**, *3*, 71–77.

140 D. Enders, J. H. Kirchhoff, J.
Kobberling, T. H. Peiffer, *Org. Lett.*
2001, *3*, 1241–1244.

141 S. Jin, D. P. Holub, D. J. Wustrow,
Tetrahedron Lett. **1998**, *39*, 3651–3654.

142 C. Sylvain, A. Wagner, C.
Mioskowski, *Tetrahedron Lett.* **1997**,
38, 1043–1044.

143 D. P. Rotella, *J. Am. Chem. Soc.* **1996**,
118, 12246–12247.

144 D. Bonnet, C. Rommens, H. Gras-
Masse, O. Melnyk, *Tetrahedron Lett.*
1999, *40*, 7315–7318.

145 B. Furman, R. Thürmer, Z. Kaluza, W. Voelter, M. Chmielewski, *Tetrahedron Lett.* **1999**, *40*, 5909–5912.

146 a) C. R. McArthur, P. M. Worster, J. Jiang, C. C. Leznoff, *Can. J. Chem.* **1982**, *60*, 1836–1841; b) P. M. Worster, C. R. McArthur, C. C. Leznoff, *Angew. Chem. Int. Ed. Engl.* **1979**, *18*, 221–222.

147 F. Gosselin, J. V. Betsbrugge, M. Hatam, W. D. Lubell, *J. Org. Chem.* **1999**, *64*, 2486–2493.

148 K. Schiemann, H. D. H. Showalter, *J. Org. Chem.* **1999**, *64*, 4972–4975.

149 K. C. Nicolaou, N. Winssinger, J. Pastor, F. Murphy, *Angew. Chem. Int. Ed. Engl.* **1998**, *37*, 2534–2537.

150 S. R. Gilbertson, X. Wang, *Tetrahedron Lett.* **1996**, *37*, 6475–6478.

151 S. Itsuno, K. Ito, A. Hirao, N. Nakahama, *J. Chem. Soc., Perkin Trans. 1* **1984**, 2887–2893.

152 S. Itsuno, Y. Sakurai, K. Shimizu, K. Ito, *J. Chem. Soc., Perkin Trans. 1* **1990**, 1859–1863.

153 S. Itsuno, M. Nakano, K. Ito, A. Hirao, M. Owa, N. Kanda, S. Nakahama, *J. Chem. Soc., Perkin Trans. 1* **1985**, 2615–2619.

154 K. Adjidjonou, C. Caze, *Eur. Polym. J.* **1995**, *31*, 749–754.

155 J. M. J. Frechet, E. Bald, P. Lecavalier, *J. Org. Chem.* **1986**, *51*, 3465–3467.

156 H. W. Gibson, F. C. Bailey, *J. Chem. Soc., Chem. Commun.* **1977**, 815.

157 a) N. M. Yoon, K. B. Park, Y. S. Gyoung, *Tetrahedron Lett.* **1983**, *24*, 5367–5370; b) A. R. Sande, M. H. Jagadale, R. B. Mane, M. M. Salunkhe, *Tetrahedron Lett.* **1984**, *25*, 3501–3504.

158 T. B. Sim, N. M. Yoon, *Bull. Chem. Soc. Jpn.* **1997**, *70*, 1101–1107.

159 a) T. B. Sim, J. Choi, M. J. Joung, N. M. Yoon, *J. Org. Chem.* **1997**, *62*, 2357–2361; b) M. J. Young, J. H. Ahn, D. W. Lee, N. M. Yoon, *J. Org. Chem.* **1998**, *63*, 2755–2757.

160 a) N. M. Yoon, H. J. Lee, J. H. Ahn, J. Choi, *J. Org. Chem.* **1994**, *59*, 4687–4688; b) B. P. Bandgar, S. H. Kshirsagar, P. P. Wadgaonkar, *Syn. Commun.* **1995**, *25*, 941–945.

161 a) H. Firouzbadi, B. Tamami, N. Goudarzian, *Syn. Commun.* **1991**, *21*, 2275–2285; b) B. Tamami, N. Goudarzian, *Iran. J. Chem., Chem. Eng.* **1996**, *15*, 63–71.

162 B. Tamami, N. Goudarzian, *J. Chem. Soc., Chem. Commun.* **1994**, 1079.

163 a) R. O. Hutchins, N. R. Natale, I. M. Taffer, *J. Chem. Soc., Chem. Commun.* **1978**, 1088–1089; b) N. M. Yoon, E. G. Kim, H. S. Son, J. Choi, *Synth. Commun.* **1993**, *23*, 1595–1599.

164 Commercially available from NovaBiochem and Argonaut Technologies. Both provide experimental details on their websites: www.NovaBiochem.com and www.Argotech.com.

165 a) M. W. Creswell, G. L. Bolton, J. C. Hodges, M. Meppen, *Tetrahedron* **1998**, *54*, 3983–3998; b) C. L. Nunns, L. A. Spence, M. J. Slater, D. J. Berrisford, *Tetrahedron Lett.* **1999**, *40*, 9341–9345; c) S. V. Ley, A. Massi, *J. Chem. Soc., Perkin Trans. 1* **2000**, 3645–3654.

166 a) S. W. Kaldor, M. G. Siegel, J. E. Fritz, B. A. Dressman, P. J. Hahn, *Tetrahedron Lett.* **1996**, *37*, 7193–7196; b) M. G. Siegel, M. O. Chaney, R. F. Bruns, M. P. Clay, D. A. Schober, A. M. V. Abbema, D. W. Johnson, B. E. Cantrell, P. J. Hahn, D. C. Hunden, D. R. G. Zarrinmayeh, P. L. Ornstein, D. M. Zimmerman, G. A. Koppel, *Tetrahedron* **1999**, *55*, 11619–11639.

167 a) S. V. Ley, O. Schucht, A. W. Thomas, P. J. Murray, *J. Chem. Soc., Perkin Trans. 1* **1999**, 1251–1252; b) S. V. Ley, A. Massi, *J. Comb. Chem.* **2000**, *2*, 104–107.

168 a) J. Habermann, S. V. Ley, J. S. Scott, *J. Chem. Soc., Perkin Trans. 1* **1998**, 3127–3130; b) S. V. Ley, M. H. Bolli, B. Hinzen, A.-G. Gervois, B. J. Hall, *J. Chem. Soc., Perkin Trans. 1* **1998**, 2239–2241.

169 K. Smith, G. E. El-Hiti, D. Hou, G. A. DeBoos, *J. Chem. Soc., Perkin Trans. 1* **1999**, 2807–2812.

170 J. Habermann, S. V. Ley, J. S. Scott, *J. Chem. Soc., Perkin Trans. 1* **1999**, 1253–1255.

171 K. Y. GORDEEV, G. A. SEREBREN-
NIKOVA, R. P. EVSTIGNEEVA, *J. Org.
Chem.* **1986**, *21*, 2615–2616.

172 B. TAMAMI, M. M. LAKOURAJ, *Synth.
Commun.* **1995**, *25*, 3089–3096.

173 N. M. YOON, J. CHOI, J. H. AHN, *J.
Am. Chem. Soc.* **1994**, *59*, 3490–3493.

174 N. M. YOON, J. CHOI, Y. S. SHON,
Synth. Commun. **1993**, *23*, 3047–3053.

175 G. W. KABALKA, P. P. WADGAONKAR,
N. CHATLA, *Synth. Commun.* **1990**, *20*,
293–299.

176 For a comprehensive review of solid-
supported reagents, see [1h].

177 N. M. YOON, J. CHOI, *Synlett* **1993**,
135–136.

178 B. P. BANDGAR, S. M. NIKAT, P. P.
WADGAONKAR, *Synth. Commun.* **1995**,
863–869.

179 a) N. M. WEINSHENKER, G. A.
CROSBY, J. Y. WONG, *J. Org. Chem.*
1975, *40*, 1966–1971; b) M. GERLACH,
F. JÖRDENS, H. KUHN, W. P.
NEUMANN, M. PETERSEIM, *J. Org.
Chem.* **1991**, *56*, 5971–5972; c) W. P.
NEUMANN, M. PETERSEIM, *Synlett*,
1992, 801–802; d) G. DUMARTIN, G.
RUEL, J. KHARBOUTLI, B. DELMOND,
M.-F. CONNIL, B. JOUSSEAUME, M.
PEREYRE, *Synlett*, **1994**, 952–954.

180 W. P. NEUMANN, M. PETERSEIM, *React.
Polym.* **1993**, *20*, 198–205.

181 U. GERIGK, M. GERLACH, W. P.
NEUMANN, R. VIELER, V. WEINTRITT,
Synthesis **1990**, 448–452.

182 Y. HU, J. A. PORCO, JR, *Tetrahedron
Lett.* **1998**, *39*, 2711–2714.

183 M. GORECKI, A. PATCHORNIK, *Biochim
Biophys Acta* **1973**, *303*, 36–43.

184 a) G. DUPAS, A. DECOMEILLE, J.
BOURGUIGNON, G. QUEGUINER,
Tetrahedron **1989**, *45*, 2579–2590; b) F.
M. MENGER, C. A. WEST, J. DING, *J.
Am. Chem. Soc.* **1997**, 633–634; c) S.
OBIKA, T. NISHIYAMA, S. TATEMATSU,
M. NISHIMOTO, K. MIYASHITA, T.
IMANISHI, *Heterocycles* **1998**, *49*, 261–
267.

185 P. FERRABOSCHI, C. GAMBERO, M. N.
AZADANI, E. SANTANIELLO, *Synth.
Commun.* **1986**, *16*, 667–672.

186 R. B. APPELL, I. A. TOMLINSON, I.
HILL, *Synth. Commun.* **1995**, *25*, 3589–
3595.

187 T. HOLLETZ, D. CECH, *Synthesis* **1994**,
789–791.

188 P. WENTWORTH, JR, A. M.
VANDERSTEEN, K. D. JANDA, *Chem.
Commun.* **1997**, 759–760.

189 K. HEMMING, M. J. BEVAN, C.
LOUKOU, S. D. PATEL, D. RENAUDEAU,
Synlett **2000**, 1565–1568.

190 A. B. CHARETTE, A. A. BOEZIO, M. K.
JANES, *Org. Lett.* **2000**, *2*, 3777–3779.

191 a) R. NOYORI, T. OHKUMA, M.
KITAMURA, *J. Am. Chem. Soc.* **1987**,
109, 5856–5858; b) R. NOYORI in:
Asymmetric Catalysis in Organic
Synthesis, Wiley-Interscience, New
York **1994**; c) T. OHKUMA, H. OOKA, S.
HASHIGUCHI, T. IKARIYA, R. NOYORI,
J. Am. Chem. Soc. **1995**, *117*, 2675–
2676; d) S. HASHIGUCHI, A. FUJII, J.
TAKEHARA, T. IKARIYA, R. NOYORI, *J.
Am. Chem. Soc.* **1995**, *117*, 7562–7563.

192 E. J. COREY, R. K. BAKSHI, S. SHIBATA,
J. Am. Chem. Soc. **1987**, *109*, 5551–
5553.

193 D. J. BAYSTON, J. L. FRASER, M. R.
ASHTON, A. D. BAXTER, M. E. C.
POLYWKA, E. MOSES, *J. Org. Chem.*
1998, *63*, 3137–3140.

194 R. HALLE, B. COLASSON, E. SCHULZ,
M. SPAGNOL, M. LEMAIRE, *Tetrahedron
Lett.* **2000**, *41*, 643–646.

195 R. HALLE, E. SCHULZ, M. SPAGNOL,
M. LEMAIRE, *Synlett* **2000**, *5*, 680–682.

196 Q.-H. FAN, C.-Y. REN, C.-H. YEUNG,
W.-H. HU, A. S. C. CHAN, *J. Am.
Chem. Soc.* **1999**, *121*, 7407–7408.

197 D. J. BAYSTON, C. B. TRACERS, M. E. C.
POLYWKA, *Tetrahedron: Asym.* **1998**, *9*,
2015–2018.

198 R. HALLE, E. SCHULZ, M. LEMAIRE,
Synlett **1997**, 1257–1258.

199 E. BREYSSE, C. PINEL, M. LEMAIRE,
Tetrahedron: Asym. **1998**, *9*, 897–900.

200 P. GAMEZ, B. DUNJIC, F. FACHE, M.
LEMAIRE, *J. Chem. Soc., Chem.
Commun.* **1994**, 1417–1418.

201 a) P. GAMEZ, B. DUNJIC, C. PINEL,
M. LEMAIRE, *Tetrahedron Lett.* **1995**,
36, 8779–8782; b) F. LOCATELLI,
P. GAMEZ, M. LEMAIRE, *J. Mol. Cat.
A.* **1998**, *135*, 89–98.

202 C. FRANOT, G. B. STONE, P. ENGELI,
C. SPÖNDLIN, E. WALDVOGEL,
Tetrahedron: Asym. **1995**, *6*, 2755–2766.

203 C. Caze, N. E. Moualij, P. Hodge, C. J. Lock, J. Ma, *J. Chem. Soc., Perkin Trans. 1* **1995**, 345–349.

204 C. Caze, N. El Moualij, P. Hodge, C. J. Lock, *Polymer* **1995**, *36*, 621–629.

205 M. Felder, G. Giffels, C. Wandrey, *Tetrahedron: Asym.* **1997**, *8*, 1975–1977.

206 J.-B. Hu, G. Zhao, Z.-D. Ding, *Angew. Chem. Int. Ed. Engl.* **2001**, *40*, 1109–1111.

207 a) L. Pu, *Chem. Rev.* **1998**, *98*, 2405–2494; b) W.-S. Huang, Q.-S. Hu, L. Pu, *J. Org. Chem.* **1999**, *64*, 7940–7956.

208 E. J. Enholm, J. P. Schulte II, *Org. Lett.* **1999**, *8*, 1275–1277.

209 G. Dumartin, M. Pourcel, B. Delmond, O. Donard, M. Pereyre, *Tetrahedron Lett.* **1998**, *39*, 4663–4666.

210 D. E. Bergbreiter, S. A. Walker, *J. Org. Chem.* **1989**, *54*, 5138–5141.

211 J. R. Blanton, J. M. Salley, *J. Org. Chem.* **1991**, *56*, 490–491.

212 A. Chemin, H. Deleuze, B. Maillard, *J. Chem. Soc., Perkin Trans. 1* **1999**, 137–142.

213 B. C. Bookser, S. Zhu, *J. Comb. Chem.* **2001**, *3*, 205–215.

214 J. C. Bussolari, D. C. Rehborn, D. W. Combs, *Tetrahedron Lett.* **1999**, *40*, 1241–1244.

215 a) D. P. Curran, S. Hadida, *J. Am. Chem. Soc.* **1996**, *118*, 2531–2532; b) K. Olofsson, S.-Y. Kim, M. Larhed, D. P. Curran, A. Hallberg, *J. Org. Chem.* **1999**, *64*, 4539–4541; c) D. P. Curran, S. Hadida, M. Hoshino, A. Studer, P. Wipf, P. Jeger, S.-Y. Kim, R. Ferritto, US Patent No. 4,859,247, January 12, **1999**; d) D. P. Curran, S. Hadida, S.-Y. Kim, Z. Luo, *J. Am. Chem. Soc.* **1999**, *121*, 6607–6615.

16
Cycloadditions in Combinatorial and Solid-phase Synthesis

Markus Albers and Thorsten Meyer

16.1
Introduction

Cycloadditions are one of the most efficient reactions for the synthesis of isocyclic and heterocyclic compounds in organic chemistry. The two most widespread are the Diels–Alder reaction ($[4+2]$) and the 1,3-dipolar cycloaddition ($[3+2]$). The Diels–Alder reaction is not only suitable for synthesis of carbocyclic but also for *N*- or *O*-heterocyclic six-member rings (hetero-Diels–Alder). The 1,3-dipolar cycloadditions are often used to synthesize five-member aza- or azoxaheterocycles. In addition to these, there is a whole string of other, less common cycloadditions such as the $[2+2]$ cycloaddition for the formation of four-member rings or the $[6+3]$ cycloaddition. The reactions usually proceed smoothly and only in some cases is a moderate application of heat required. Owing to the pericyclic mechanism two bonds are formed simultaneously in a usually stereo- and regiospecific way, which can be promoted by Lewis acid catalysis [1]. Therefore, cycloadditions are extremely valuable for the generation of stereogenic centers, especially during natural product synthesis.

   In this chapter the application of cycloadditions to combinatorial chemistry is discussed. Almost all concepts and strategies are based on solid phase, with only a few based on solution-phase chemistry. One reason behind this is the requirement for full conversions. Transformations on solid support have significant advantages over those in solution. Most important is the simple removal of nonresin-bound byproducts and excess of reagents, which is necessary to allow completion of the cycloaddition within a reasonable time, by simply washing the resin with an appropriate solvent. In this way the products have high purity without a time-consuming and expensive chromatographic purification. Solution-phase combinatorial chemistry is only applicable if the excess component is volatile, which allows its removal *in vacuo*, or it carries an additional functional group, such as an amine, alcohol, ketone, or aldehyde, so that scavengers can be used for its capture.

   Many types of cycloadditions have been applied to solid phase, including the use of a plethora of different dienes, dienophiles, dipoles, dipolarophiles, or olefins. These are either resin-bound or used as reagents, and are inter- or intramolecular variants. Normal or inverse electron demand, and ordinary or hetero-Diels–Alder reactions, complete the set of combinatorial methods available to chemistry [2]. In this chapter, the [4 + 2], [3 + 2], [2 + 2], and [6 + 3] cycloadditions are described in more detail; additionally, sigmatropic rearrangements as pericyclic reactions are described.

16.2
[4 + 2] Cycloadditions

The Diels–Alder reaction is the most synthetically efficient method for the preparation of six-member rings and is a well-established reaction in combinatorial chemistry. The typical [4 + 2] cycloaddition is a Diels–Alder reaction with normal electron demand between an electron-rich diene and an electron-poor dienophile. There have been many examples of this reaction, especially in solid-phase synthesis [2].

16.2.1
Diels–Alder Reaction with Resin-bound Dienes

A variation on the typical reaction is to start with resin-bound dienes. One of the most reactive, because it is electron rich, and well-known dienes is the Danishefsky diene. Scheme 16.1 shows a direct route for the generation of a polymer-supported version. For this purpose a polystyrene diethylsilane resin reacts first with trifluoromethanesulfonic acid, forming a silyl triflate resin. Subsequent treatment with diverse α,β-unsaturated ketones and aldehydes produces silyl enol ether dienes very easily. These reactive dienes are trapped with dienophiles such as N-phenylmaleimide, yielding the bicyclic cycloadducts. Cleavage off the resin with

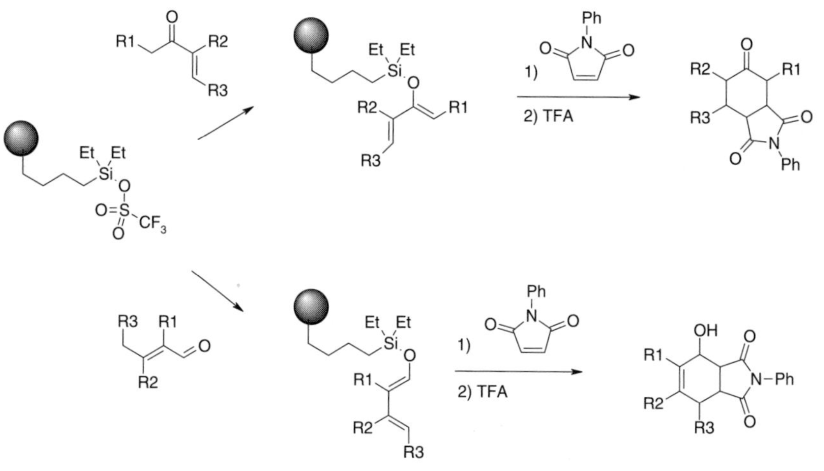

Scheme 16.1. Diels–Alder reaction with resin-bound siloxy dienes.

trifluoroacetic acid forms the corresponding ketones and alcohols in 62–100% yield and 81–98% purity [3].

Another possible route for the synthesis of cyclohexanone derivatives can be realized by the attachment of α,β-unsaturated ketones to a piperazine resin as enamines (Scheme 16.2) [4]. The electron-rich dienes react with electron-poor E-nitrostyrene derivatives to the nitro-substituted cyclohexanones with moderate yields but high purities. However, an almost equal mixture of diastereomers is obtained, which may be explained by a nonconcerted cycloaddition mechanism reflecting the enamine character of the diene used. It is known from solution phase that [4 + 2] cycloaddition with aminobutadienes can proceed via a step-wise process [5]. This conclusion is supported by the observation of small amounts of an open-chain enamine addition product.

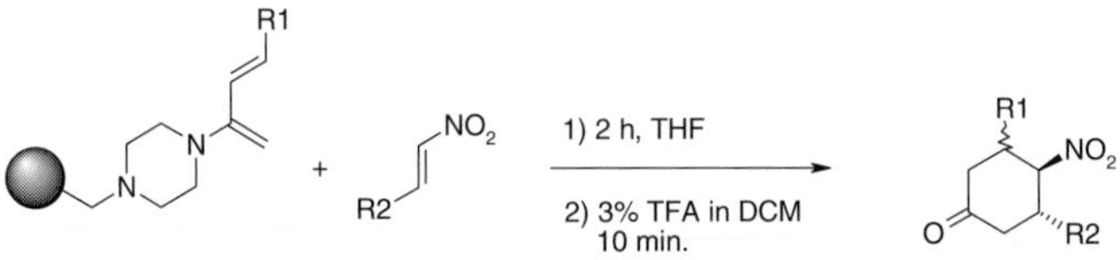

Scheme 16.2. Diels–Alder reaction with resin-bound amino dienes. THF, tetrahydrofuran; TFA, trifluoroacetic acid.

Using different substituted resin-bound cyclic dienamines, as depicted in Scheme 16.3, the cycloaddition with a set of dienophiles (maleimides, nitro styrenes, diazo derivatives) occurs in a stereospecific way [6]. The endo but racemic adduct is obtained by an α-face attack of phenylmaleimide giving 61% yield and purity > 85%. The cyclic dienamine is built up by acylation of resin-bound 4-hydroxypyridine followed by treatment with a Grignard reagent.

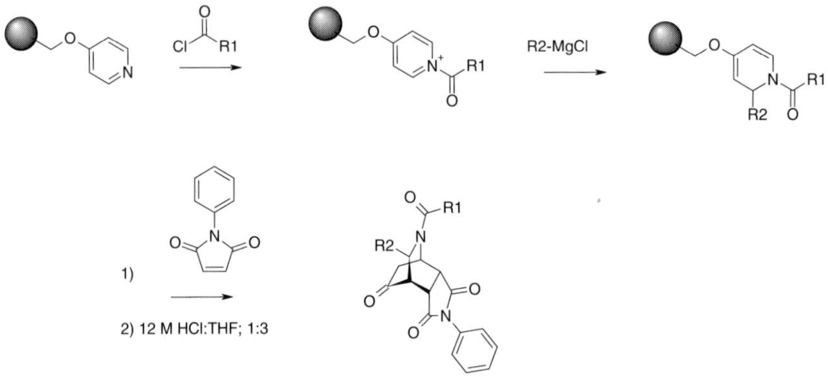

Scheme 16.3. Solid-phase-supported Diels–Alder reaction with cyclic amino dienes.

Tebbe olefination of supported α,β-unsaturated esters is another elegant method used to build up dienyl ethers (Scheme 16.4). The cycloaddition to diverse dienophiles in toluene produces cyclohexenes, which are converted to the corresponding cyclohexanone library after cleavage with trifluoroacetic acid. High endoselectivity is observed at reaction temperatures of 80–100 °C, except for N-methylmaleimide, for which room temperature is found to be sufficient [7].

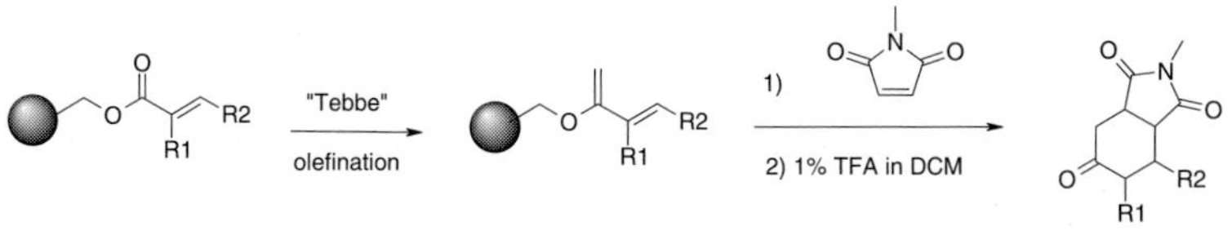

Scheme 16.4. Resin-bound oxydienes by Tebbe olefination. (dichloromethane (DCM))

An interesting intramolecular ruthenium-catalyzed olefin/alkyne metathesis reaction is used to produce different cyclic dienes suitable for Diels–Alder reactions on solid support (Scheme 16.5) [8]. The synthesis starts with resin-bound allylic amides, which are deprotonated with lithium *tert*-butoxide, followed by reaction with acetylene methanesulfonate derivatives in dimethylsulfoxide. The alkynylated product obtained undergoes metathesis reaction with Grubb's ruthenium catalyst and the diene thus formed is treated with maleimide in refluxing toluene to yield the desired cycloadduct as a single diastereomer. The reaction sequence is used in a "split-and-mix" fashion to prepare a $10 \times 4 \times 5 \times 16$-member isoindoline combinatorial library.

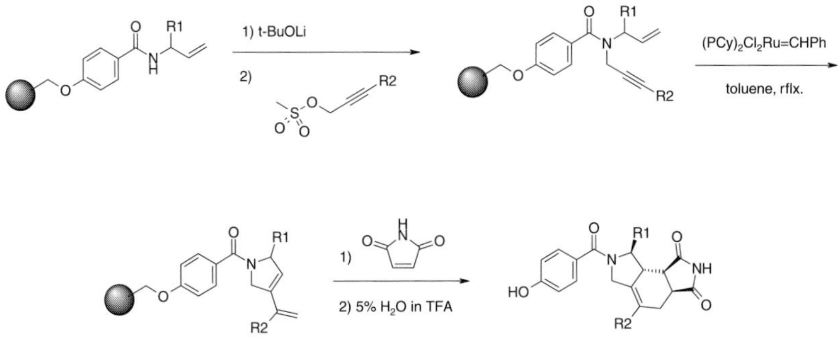

Scheme 16.5. Formation of resin-bound dienes by metathesis.

Scheme 16.6 gives a very attractive example of a completely stereoselective synthesis of a tricyclic core starting from optically pure 3-bromo-3,5-cyclohexadiene-1,2-diol linked to solid support via a ketal [9]. Subsequent epoxidation with dimethyldioxirane takes place with complete facial selectivity, and epoxide opening is achieved with amines followed by acylation of the intermediate alcohol. The use of

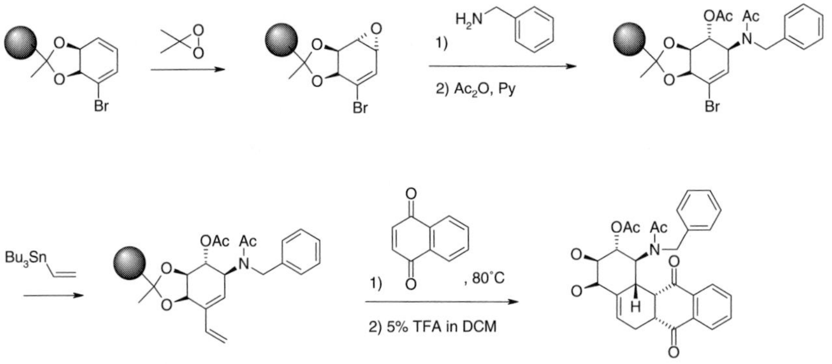

Scheme 16.6. Stereoselective Diels–Alder reaction via a chiral cyclic diene.

a Stille coupling with vinyl stannanes generates the diene for the Diels–Alder reaction with different dienophiles, which proceeds with complete facial selectivity and endoselectivity. The high facial selectivity is rationalized by the sterically demanding ketal group which is efficiently shielding the β-face of the molecule. In sum, 16 cycloadducts are synthesized in high yields and purities.

Another example of an enantioselective cycloaddition is the reaction between polymer-supported chiral amino furans and a variety of dienophiles, which can be applied to the formation of sugar derivatives in natural product synthesis (Scheme 16.7) [10]. In order to generate the resin-bound furans, silylchloride resin [11] is treated with the potassium enolate of the optically pure amino furanone. Reaction with methyl acrylate provides the oxabicycloheptene adduct in a regio- and stereoselective way and, because of the steric demand of the pyrrolidine residue, the dienophile reacts at the less hampered α-face. Cleavage off the resin with tetrabutylammoniumfluoride resumes the synthesis of α-substituted cyclohexenones.

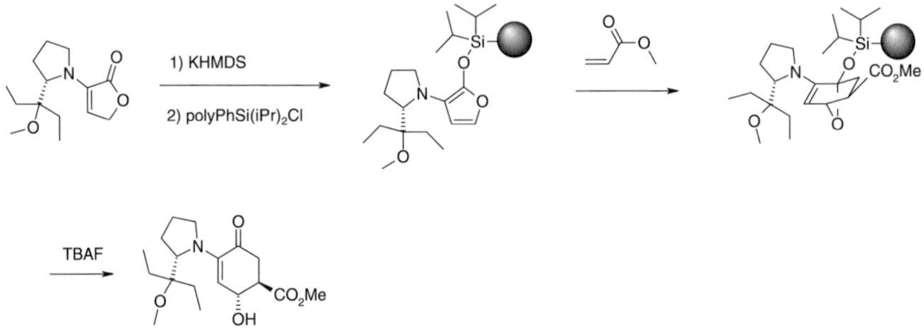

Scheme 16.7. Enantioselective cycloaddition via optically pure furan as diene.

A synthetically interesting methodology is the Diels–Alder/retro-Diels–Alder reaction as a safety-catch procedure. A resin-bound diene temporarily catches a dienophile by Diels–Alder reaction, this is then modified further on solid support

[12]. Finally, a retro-Diels–Alder reaction releases the transformed dienophiles from the resin by regeneration of the resin-bound diene. Obviously, a major advantage of this method lies in the high chemical and stereochemical purity of the released compounds. Scheme 16.8 displays this procedure with a resin-bound furan in a cycloaddition reaction to an electron-deficient alkyne. The resulting resin-bound bicyclic Michael system undergoes stereoselective addition of thiophenol, while the desired olefinic product is released from the support by a retro-Diels–Alder reaction. It is noteworthy that the Michael addition carried out in solution shows an E/Z ratio of 81:19.

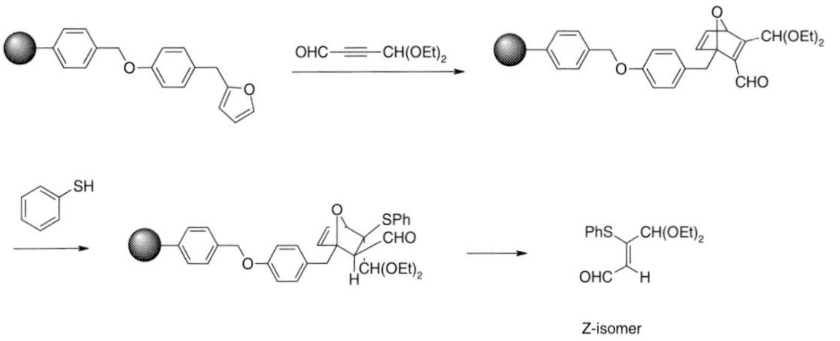

Z-isomer

Scheme 16.8. Solid-phase Diels–Alder/retro-Diels–Alder as a safety-catch procedure.

16.2.2
Diels–Alder Reaction with Resin-bound Dienophiles

In contrast to the polymer-supported dienes, examples with resin-bound dienophiles are not so well represented. This may be due to the fact that more dienophiles than dienes are commercially available. Therefore, in solid-phase Diels–Alder reactions dienophiles are usually used as excess reagents. Nevertheless, few reports with resin-bound dienophiles are published.

One of the first Diels–Alder cycloadditions applied to solid-phase combinatorial chemistry is the reaction between polymer-supported acrylates and butadiene derivatives (Scheme 16.9) [13]. Acrylic acids are first attached to a polystyrene resin via esterification, and the subsequent cycloaddition reactions take place in hot toluene or xylene. After cleavage off the resin with tetrabutylammonium hydroxide and esterification with diazomethane, mixtures of cis/trans isomers are obtained

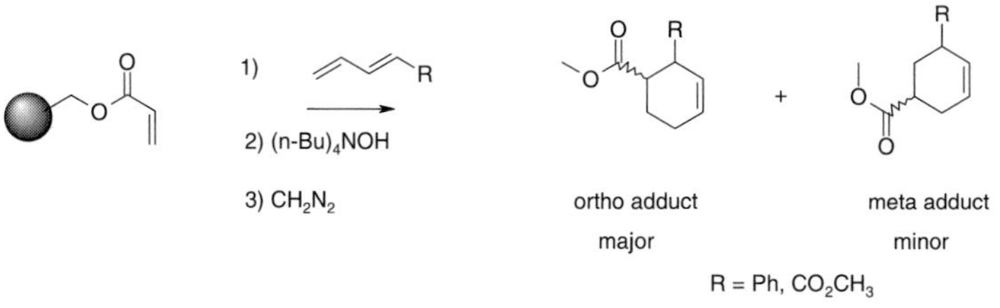

| ortho adduct | meta adduct |
| major | minor |

$R = Ph, CO_2CH_3$

Scheme 16.9. Diels–Alder reaction with resin-bound acrylates.

with high regioselectivities. The observed *ortho/meta* ratio is in accordance with the results obtained in solution phase. It is well documented that 1-substituted-1,3-butadienes react with 1-substituted alkenes containing an electron-withdrawing group, forming predominantly the *ortho* adduct.

Polymer-supported dehydroalanine derivatives are used as dienophiles in Diels–Alder reactions (Scheme 16.10) [14]. The dehydroalanines are generated by coupling of *N*- and *S*-protected cysteines to the resin, oxidation of the sulfides to the sulfones, and elimination to the desired olefins. After cycloaddition with cyclopentadiene at 80 °C and cleavage with 20% trifluoroacetic acid in dichloromethane, mixtures of endo/exo isomers (1:2 to 1:4, 51–81% yield) are detected (determined by ¹H-NMR). The selectivities are similar to those reported in solution phase.

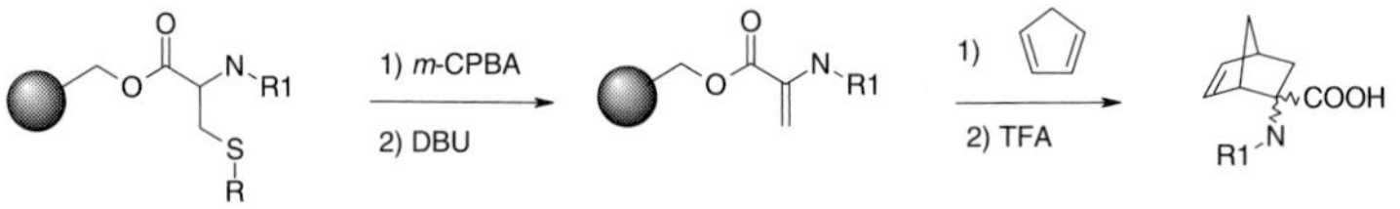

Scheme 16.10. Diels–Alder reaction with resin-bound dehydroalanine derivatives.

The stereoselectivity of Diels–Alder reactions can be increased by use of chiral auxiliaries, as shown with Evans oxazolidinone. In order to introduce the stereochemical information, enantiomerically pure Boc-L-tyrosine methylester is coupled to hydroxymethyl Merrifield resin using Mitsunobu methodology (Scheme 16.11) [15]. Reduction of the ester and treatment with thionylchloride gives the chiral oxazolidinone, which is then acylated with *trans*-crotonic anhydride, triethylamine, and dimethylaminopyridine, thus forming the desired resin-bound dienophile. The subsequent cycloaddition with cyclopentadiene is catalyzed by diethylaluminum chloride as Lewis acid and the cycloadduct is cleaved from the resin by lithium benzyloxide. The endo/exo ratio of 21:1 (86% ee) compares nicely with the results obtained in solution phase.

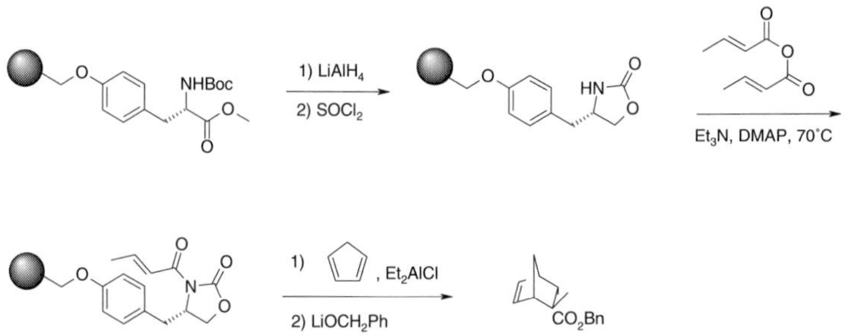

Scheme 16.11. Optically pure oxazolidinones as chiral auxiliaries in cycloaddition.

Besides Lewis acids, the application of high pressure also facilitates the cycloaddition reaction. Especially highly substituted and unreactive starting materials, which show no reaction under normal conditions, can be forced to react. In a microwave-assisted Knoevenagel reaction between resin-bound nitroalkenes and aldehydes, E/Z mixtures of trisubstituted dienophiles are generated (Scheme 16.12) [16]. Treatment with 2,3-dimethylbutadiene under high pressure conditions (15 kbar, 25 °C) yields the cycloadducts, whereas stereoselective reduction with lithium aluminumhydride gives the cyclic amines via a traceless linker strategy. The stereoselective formation of one diastereomer from a diastereomeric mixture of two cycloadducts (from a E/Z mixture of nitroalkenes) is rationalized by an aci-nitro intermediate.

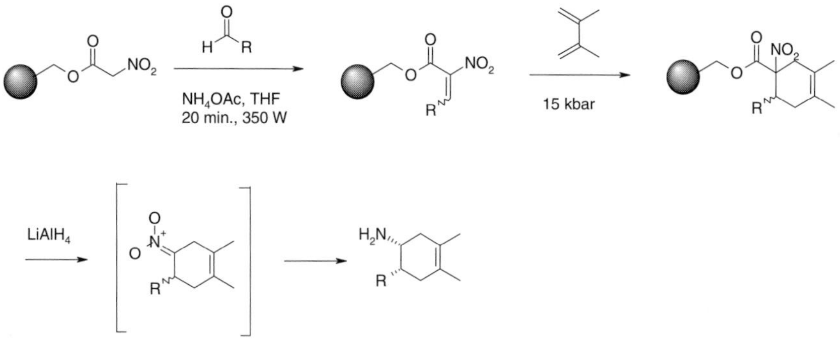

Scheme 16.12. Diels–Alder reaction under high-pressure conditions.

During the synthesis of oligomers via tandem Diels–Alder reactions, solid-phase methodology is superior to solution-phase chemistry (Scheme 16.13) [17], whereas under homogeneous reaction conditions, the reaction of an acrylate and a bisdiene would lead to mixtures of oligomers and polymers. Oligomerization can be pre-

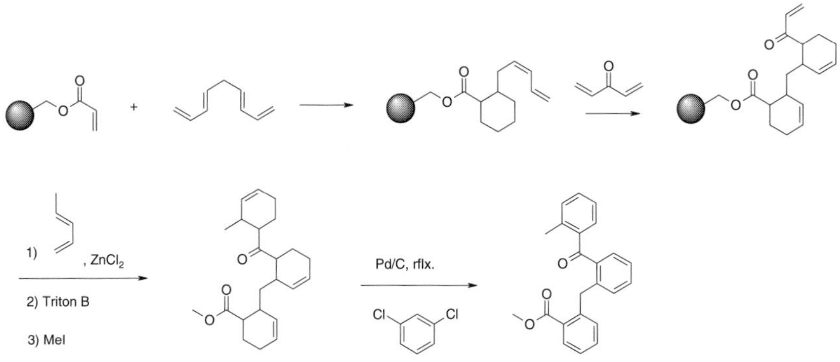

Scheme 16.13. Oligomerization via iterative Diels–Alder reaction on solid support.

cisely controlled on solid support. Starting with a resin-bound acrylate derivative, the Diels–Alder reaction with an excess of bisdiene gives only one cycloadduct. The second cycloaddition is performed with the bisdienophile divinyl ketone in the presence of 15 equiv. of $ZnCl_2$ and the final cycloaddition is achieved by capping the resin-bound dienophile with butadiene. Treatment with Triton B® and iodomethane then reveals the tricyclic ester. All cycloadditions reported proceed in a regioselective way and, after aromatization with palladium/carbon in boiling dichlorobenzene, the desired acetophenones are obtained.

16.2.3
Intramolecular Diels–Alder Reaction on Solid Support

In cycloaddition chemistry the intramolecular strategy is an elegant method of reaction management. By tethering both components it is possible to bring the reactive centers spatially together, so that the reaction proceeds under milder conditions and the turnovers are also improved.

The intramolecular Diels–Alder reaction (IMDA) with furan as the diene partner has been widely used to prepare rigid oxygenated tricyclic compounds. A resin-bound furan can easily be prepared through reductive alkylation of resin-bound glycine and a set of furaldehydes (Scheme 16.14) [18]. Introduction of an activated dienophile is then realized by acylation with different acrylic acids and cleavage with trifluoroacetic acid provides the tricyclic lactams in high yields and purities (> 90%). Owing to the pericyclic and intramolecular reaction pathway just the exo isomers are obtained. In order to overcome the limitation of the commercial availability of activated dienophiles, maleic anhydride is used, which can be hydrolyzed and further reacted with amines.

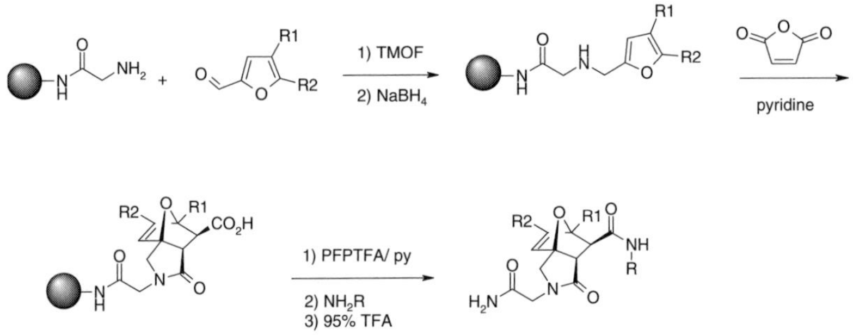

Scheme 16.14. Intramolecular Diels–Alder reaction with furan derivatives. TMOF, trimethylorthoformate. (PFPTFA, trifluoroacetic acid pentafluorophenylester).

In order to increase the stereoselectivity, novel unsaturated amino acids are used as dienophiles (Scheme 16.15) [19]. Phosphonoacetyl Wang resin [20] is hence treated with an optically pure fluorenylmethoxycarbonyl (Fmoc)-protected amino acid aldehyde, forming the electron-deficient dienophile with the intention of in-

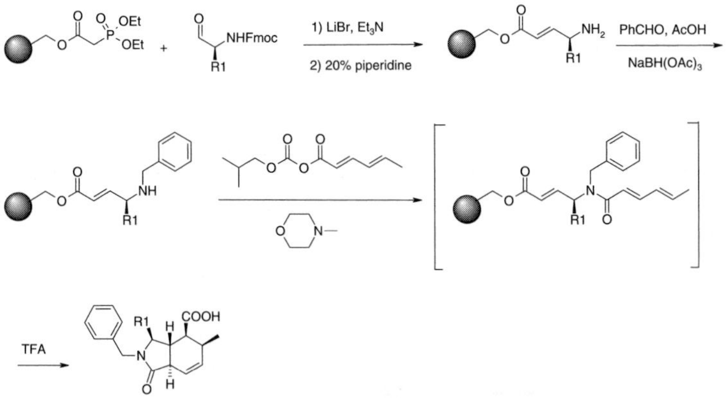

Scheme 16.15. Enantioselective cycloaddition with unsaturated amino acid derivatives.

troducing the required stereochemical information into the cycloaddition process. Deprotection of the amine allows the introduction of the diene component using different methods. The first example shows the acylation with 2,4-hexadionic acid using isobutyl chloroformate as an activator. The following reductive alkylation with benzaldehyde gives the benzylated precursor for the intramolecular Diels–Alder reaction, which is complete within 30 h at room temperature. Cleavage from the resin provides a major diastereomer (> 90%) derived from the endo transition state, which is influenced by the 1,3-allylic interaction of the dienophile and steric effect of the substituent R1.

An alternative transformation of the primary unsaturated amines is to connect them directly to a diene by reductive alkylation with 2,4-hexadienal [19] (Scheme 16.16). Coupling to the resin-bound amine is realized by reductive alkylation of the aldehyde or by acylation of the corresponding furanacrylic acids. Again, during the cycloaddition only one stereoisomer is formed but after prolonged reaction times of up to 2 days. The side-chain double bond of the vinylfuran acts as a part of the diene participating in the Diels–Alder reaction and the aromaticity of the furan ring is restored through final rearrangement.

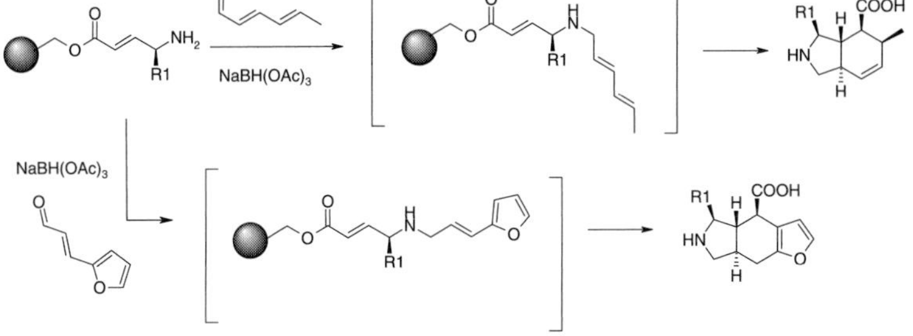

Scheme 16.16. Formation of tricyclic products by intramolecular cycloaddition.

16.2.4
Hetero-Diels–Alder Reaction on Solid Support

The hetero-Diels–Alder reaction is a well-established method for the synthesis of six-member heterocycles. An efficient way to access dihydropyrans lies in the reaction of α,β-unsaturated ketones with enols. Starting with resin-bound acetoacetate, the unsaturated ketones are formed by Knoevenagel reactions with different aliphatic aldehydes (Scheme 16.17) [21]. Treatment with a variety of enol ethers at 60 °C for 3 days sets up a hetero-Diels–Alder reaction with inverse electron demand. Final cleavage with sodium methanolate provides the cycloadducts with excellent purities of 90%. However, the diastereoselectivities are not very satisfying, displaying a ratio of 1:1 to 5:1 for the endo and exo products.

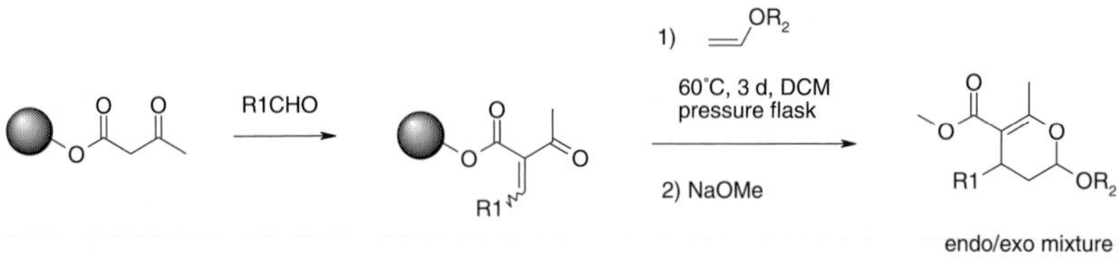

Scheme 16.17. Formation of dihydropyrans by hetero-Diels–Alder reaction.

Diastereoselectivity can be increased by the use of chiral Lewis acids. Scheme 16.18 shows another cycloaddition to dihydropyrans with inverse electron demand. The synthesis commences with resin-bound benzylidenepyruvic acid [22] and the hetero-Diels–Alder reactions with electron-rich enol ethers are catalyzed by Eu(fod)$_3$. Cleavage off the resin is achieved by reduction with lithium aluminumhydride and the corresponding alcohols are formed in high yields with an endo/exo selectivity of up to >97:3. The results are similar to those obtained under homogeneous liquid-phase conditions.

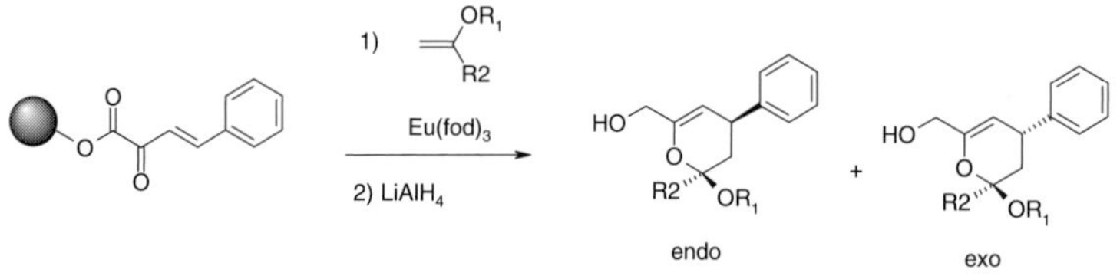

Scheme 16.18. Chiral Lewis acids as catalyst in a hetero-Diels–Alder reaction.

The aza-Diels–Alder reaction allows one of the most convenient and versatile approaches to nitrogen-containing six-member heterocycles. In a typical reaction imines act as dienophiles and react with dienes to the precious cycloadducts. In most cases the imines can be prepared *in situ* from amines and aldehydes, and even simply mixing all three components (amine, aldehyde, and olefin) together gives good results using Lewis acids for catalysis. A very short synthesis is the one-pot reaction between amino-methylated polystyrene resin, aldehydes, and diene catalyzed by ytterbium triflate (Scheme 16.19) [23]. The cycloaddition proceeds smoothly at room temperature over 12–48 h and yields and purities are >90%. The cycloadducts are released from the resin utilizing a tertiary amine N-dealkylation method which involves chloroethyl chloroformate treatment and methanolic decomposition of the resulting carbamates [24]. Only the desired [4 + 2] products are released from the solid support, enabling a clean resin cleavage.

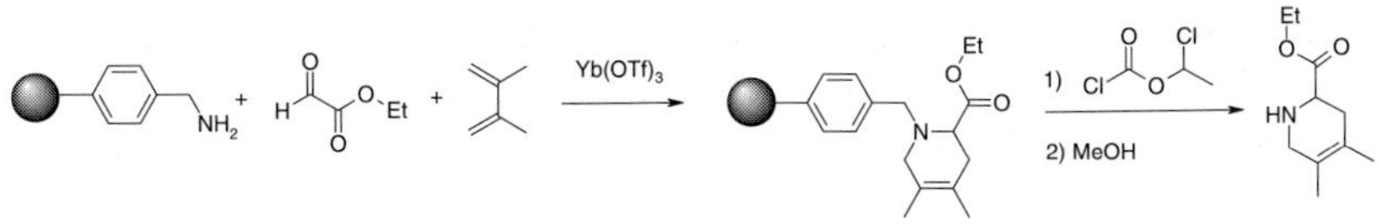

Scheme 16.19. Aza-Diels–Alder reaction using Lewis acids for catalysis.

In order to generate triazolopyridazines, urazines as electron-poor diaza dieno-philes are used in a hetero-Diels–Alder reaction with dienes [25]. Therefore, di-ethylphosphonoacetic acid is coupled to resin-bound amino acids and submitted to Horner–Wadsworth–Emmons reaction conditions with different α,β-unsaturated aldehydes (Scheme 16.20) [26]. The dienes thus obtained then react with different urazines, generated *in situ* from urazoles and iodobenzene diacetate.

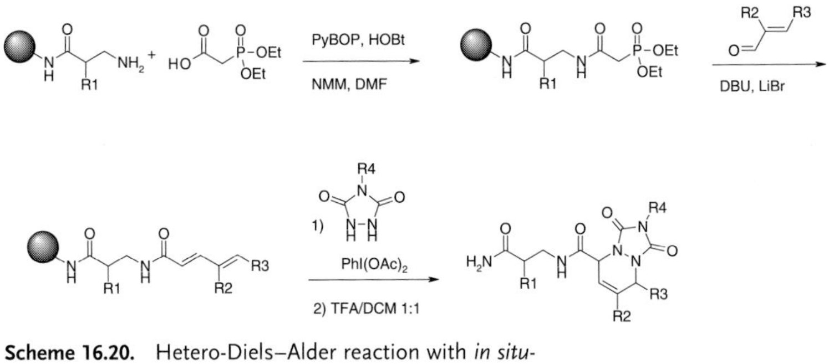

Scheme 16.20. Hetero-Diels–Alder reaction with *in situ-*generated urazines. (PyBOP, benzotriazole-1-yl-oxy-tis-pyrrolidino-phosphonium hexafluorophosphate, NMM, N-methylmorpholine).

16.2.5
Diels–Alder Reaction in Solution Phase

It is true that transformations on solid support have some advantages over those in solution, such as the use of an excess of reagents and the ease of removal of non-resin-bound byproducts; in contrast, solution-phase chemistry often requires min-imal investment of time during method development, has feasible scale-up, has easy reaction monitoring, and no attachment points are required. A variety of imaginative techniques have been developed to rapidly purify the many reactions of a solution-phase library in a parallel fashion. Some of these techniques include acid/ base extraction [27], fluorous-phase extraction [28], polymer-supported reagents [29] and catalysts [30], solid-phase extraction [31], or polymer-supported quench/ scavenging reagents [32].

The reaction with imines and dienes is well suited to solution-phase combinatorial chemistry using the scavenger methodology. Scheme 16.21 outlines the cyclo- addition to 2,3-dihydro-4-pyridones under Lewis acid catalysis [33]. An equimolar mixture of aldehydes and primary amines in trimethylorthoformate reacts to form the imines. After evaporation of the solvent the cycloadducts are formed using an excess of Danishefsky diene under ytterbium triflate catalysis, which finally hydro- lyzes both the cycloadducts to the desired pyridones and the excess of the diene to the corresponding ketone. The diene decomposition product and, if the reaction does not go to completion, any unreacted imine are removed with a polyamine resin. After simple filtration followed by an acidic aqueous work-up dihydropyri- done products are obtained with good yields (up to 90%) and high purities (80– 90%). A variety of different imines, derived from alkyl, alkylaryl, pyridyl amines, and from substituted anilines, undergo the cyclization.

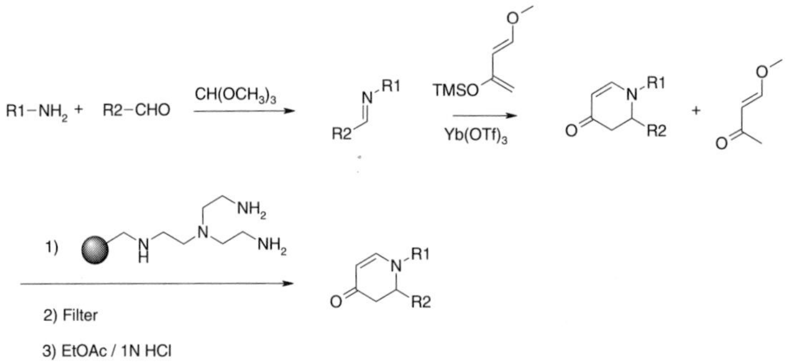

Scheme 16.21. Diels–Alder reaction using polymer-supported scavengers.

Even whole natural products are synthesized using an organized array of polymer-supported reagents. Scheme 16.22 shows the synthesis of epibatidine with a purity > 90%, avoiding the use of chromatographic purification steps [34]. A key step in the synthesis is a cycloaddition reaction between a nitro alkene derivative and an excess of a silyl-protected 2-oxadiene. Beginning with chloronicotinic acid, chloride – the dienophile – is obtained by a reduction/oxidation sequence to the aldehyde and addition of nitromethane with a subsequent elimination step. Treat- ment with an excess of the volatile silyl-protected 2-oxadiene at 120 °C provides the cycloadduct in a quantitative yield. Hydrolysis to the corresponding ketone, reduc- tion to the alcohol, mesylation, and reduction of the nitro group to the amine with final cyclization forms the endo isomer of epibatidine.

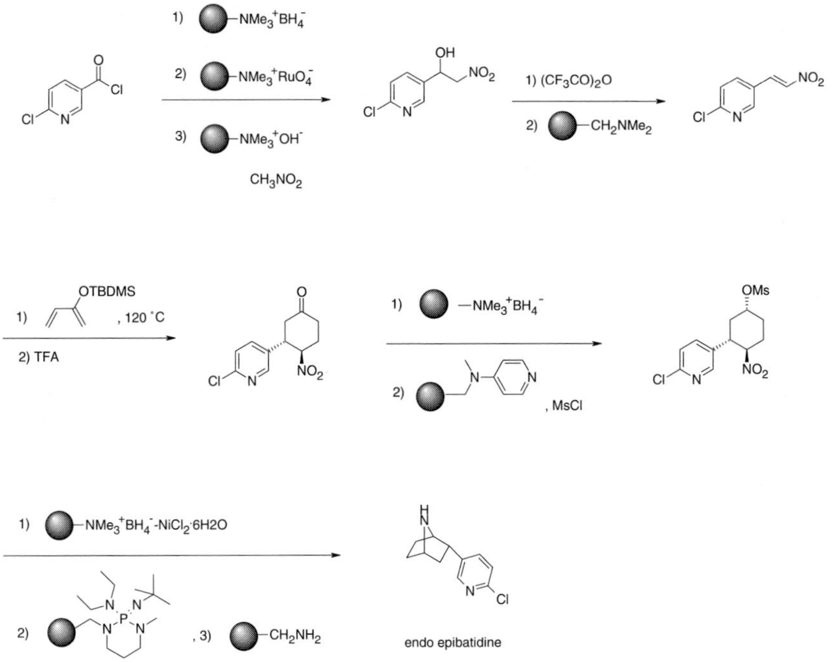

Scheme 16.22. Synthesis of epibatidine by Diels–Alder reaction.

16.3
[3 + 2] Cycloadditions

The most widely studied cycloadditions in solid-phase combinatorial synthesis are [3 + 2] cycloadditions, which have been shown to comprise a wide range of dipoles (nitrones, nitrile oxides, pyridinium salts, azomethine ylides, etc.) and dipolarophiles (alkenes, dienes, and alkynes). Depending on the nature of the 1,3-dipoles employed in the transformations, various heterocycles such as isoxozazoles, isoxazolines, pyrrolidines, indolizines, and pyrrazoles are obtained [35]. These five-member ring systems represent a branch of unique pharmacophores and some are also versatile synthetic intermediates in further functional group interconversions.

As mentioned in the introduction to this chapter, most applications of these transformations are aimed at solid-phase combinatorial chemistry, while only one solution-phase example has been reported to date [36].

16.3.1
Formation of Isoxazoles, Isoxazolines, and Isoxazolidines

Isoxazoles and isoxazolines are obtained by [3 + 2] cycloaddition of nitrile oxides to alkynes or alkenes [37], while isoxazolidines are formed through reactions of nitrones and olefins (Scheme 16.23). As nitrile oxides often suffer from decomposition and dimerization in solution [38], these transformations should be carried out on solid phase on which either the dipolarophile or the 1,3-dipole can be immobilized.

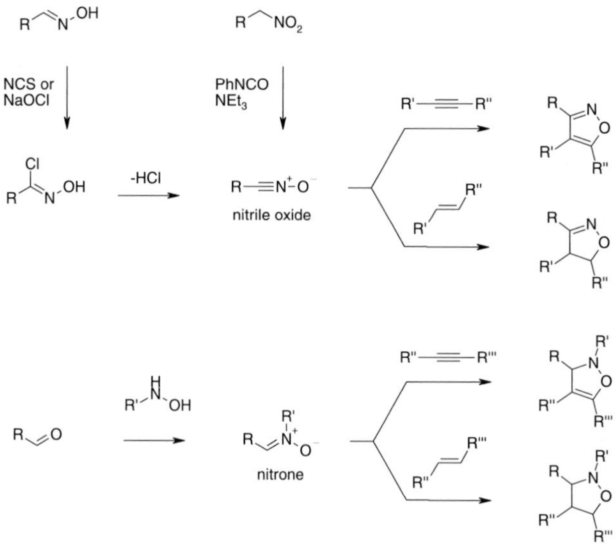

Scheme 16.23. Synthesis strategies for the preparation of isoxazoles, isoxazolines, and isoxazolidines.

In a representative example of the preparation of isoxazoles and oxazolines [39], a polymer-bound olefin or alkyne is treated with nitrile oxides (Scheme 16.24), which are typically generated *in situ* either by using Mukaiyama's method utilizing phenyl isocyanate and triethylamine [40] or by oxidizing oximes with sodium hyporchloride [41]. The conversions observed are generally high, although in some cases the cycloaddition step has to be repeated up to three times when less stable nitrile oxides are used.

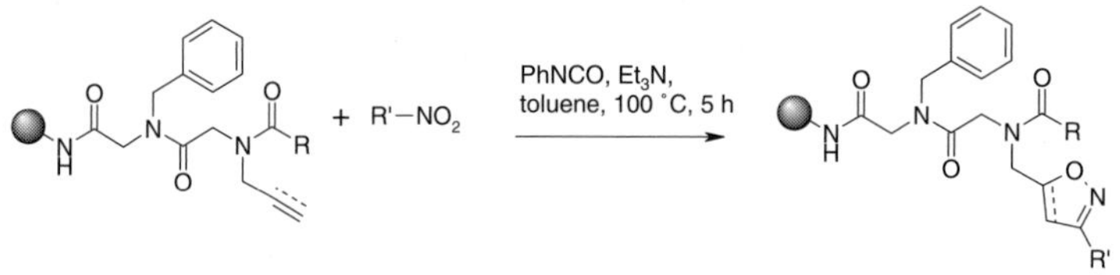

Scheme 16.24. Formation of isoxazoles from resin-bound alkynes.

Intramolecular modifications of the above-mentioned [3 + 2] process have also been well established on solid-phase [42]. In a generic example, polymer-supported nitro olefins undergo 1,3-dipolar cycloadditions, giving three stereogenic centers in the resulting tetrahydrofuroisoxazolines (Scheme 16.25). This highly stereoselective process proceeded after obtaining the nitro olefins from Michael additions of dienol alkoxides to β-nitrostyrene.

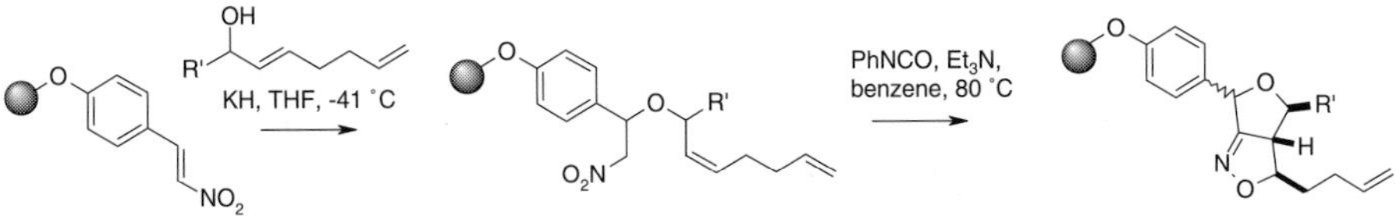

Scheme 16.25. Intramolecular addition of a polymer-bound nitrile oxide to an olefin.

During the synthesis of a natural product-like library (see also Chapter 21), an intramolecular cycloaddition is used as the key step in building up a polycyclic template [43]. This product is formed by a Tamura tandem reaction [44] of a polymer-bound epoxycyclohexanol and a set of nitrone carboxylic acids (Scheme 16.26). After initial coupling of the 1,3-dipoles to a shikimic acid-derived alcohol, the subsequent [3 + 2] cycloaddition proceeds with high stereo- and regioselectivity. A variety of reagents and conditions have been screened for further manipulations of the tetracyclic core thus formed and a split-and-mix library of more than 2 million compounds has been synthesized.

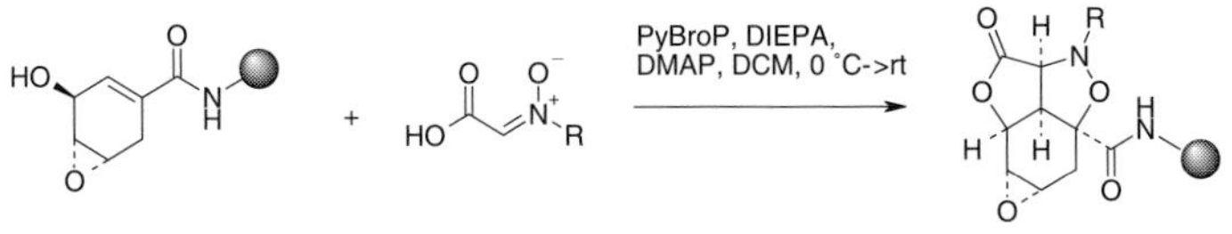

Scheme 16.26. Syntheses of tetracyclic cores by tandem transesterification cycloadditions of epoxycyclohexanols and nitrone carboxylic acids.

On the other hand, when nitrones are prepared *in situ* from 2-bromobenzaldehyde and methyl hydroxylamine and consequently reacted with polymer-bound acrylates through a nontethered transition state (Scheme 16.27), they were found to be less suitable for solid-phase combinatorial synthesis [45]. The yields recorded are in the range of 24–45% and even boosting the excess of reagent up to 40 equiv. and extending the reaction times does not improve the results. In the latter case, cleavage of the acids from the 2-chlorotrityl resin is observed owing to the prolonged exposure to heat.

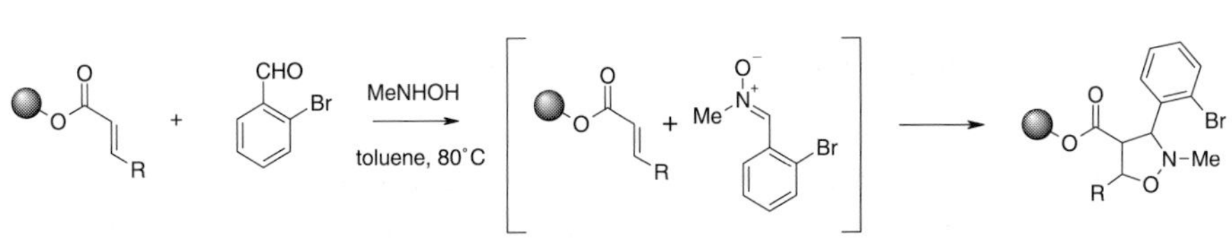

Scheme 16.27. Syntheses of isoxazolidines by reactions of polymer-bound olefins and nitrones.

More success is encountered by the same research group when polymer-bound hydroxylamines are reacted with aldehydes and electron-deficient olefins such as vinylsulfones and *N*-substituted maleimides (Scheme 16.28). Moreover, it is worth mentioning that the electronic nature of the aldehydes employed has little impact on the synthesis of isoxazolidines, whereas nitrones derived from ketones do not react at all. The superiority of this alternative approach over the route involving immobilized olefins is demonstrated by the synthesis of a small split-and-mix library comprising ten compounds.

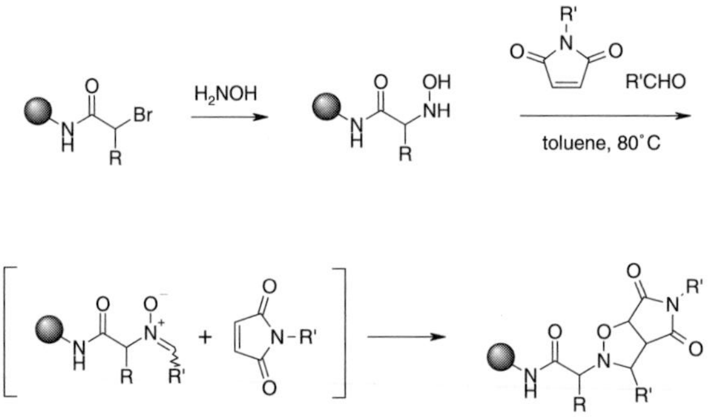

Scheme 16.28. Syntheses of isoxazolidines by reactions of polymer-bound nitrones and olefins.

In order to avoid elevated temperatures, ytterbium triflate was successfully introduced to [3 + 2] cycloaddition reactions, which were then found to proceed at room temperature (Scheme 16.29) [46]. Acrylates of 1,3-oxazolin-2-ones are the best olefinic reaction partners, which can be attributed to their favorable electronics. When other dipolarophiles such as methyl vinyl ketones or substituted acetylenes are used, reduced yields are observed, which can be attributed to the reduced ability of Lewis acid coordination of the unsaturated systems screened.

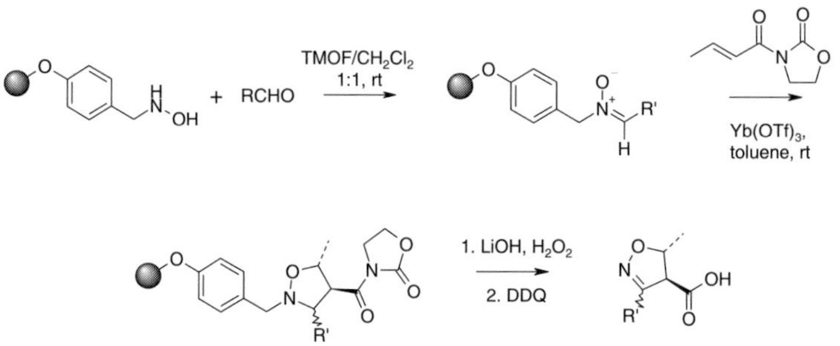

Scheme 16.29. Yb(Otf)$_3$-catalyzed 1,3-dipolar cycloadditions of polymer-bound nitrones.

The isoxazolidines thus obtained are then derivatized further and consequently converted to their corresponding isoxazolines by oxidative cleavage using 2,3-dichloro-5,6-dicyanobenzoquinone (DDQ). During the initial work-up, ascorbic acid is added in order to reduce the amount of remaining oxidant, but chromatography on silica gel is still found to be necessary.

Isoxazolidines can also be prepared from immobilized nitrile oxides, which are easily generated through oxidation of polymer-bound aldoximes. In one case, N-chlorosuccinimide (NCS) was used as the oxidant and the corresponding hydroximoyl chlorides were converted to the reactive species by the dropwise addition of triethylamine (Scheme 16.30) [47]. Immediate trapping with an excess of olefins gave the desired heterocycles with yields of 60–80% and purities of >90%. Isoxazoles can also be prepared using this methodology.

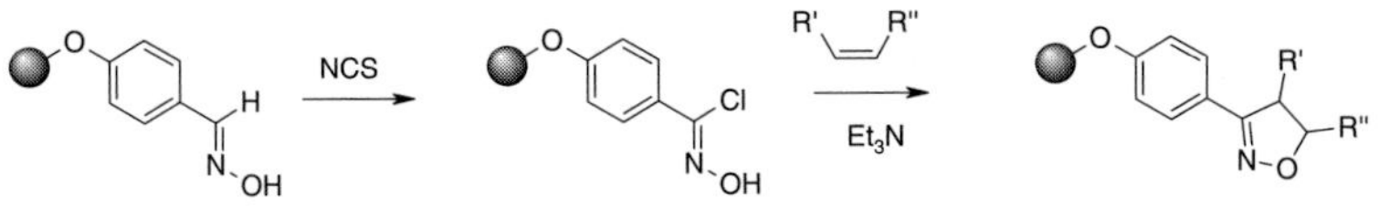

Scheme 16.30. Syntheses of isoxazolines by oxidation of a polymer-bound oxime with NCS and subsequent olefinic trapping.

In another example, the use of an additional amine base can be avoided when a polymer-bound aldoxime is oxidized with commercially available household bleach. After elimination of hydrogen chloride, the corresponding nitrile oxides are obtained [48]. The generality and ease of this protocol is demonstrated when the inverse strategy is pursued and resin-bound acrylates are successfully converted to isoxazolines.

Solid-supported reagents can also be used for the *in situ* preparation of nitrones with regard to the solution-phase synthesis of isoxazolidines (Scheme 16.31) [49]. This process has been carried out using polymer-supported perruthenate (PPS) as the oxidant, but this procedure is limited to symmetrical hydroxylamines only. In order to circumvent this limitation, aldehydes are normally condensed with primary hydroxylamines in the presence of solid-supported acetate. After removal of the polymer-bound reagent and transfer of the crude nitrone to a solution of methyl acrylate, the desired cycloaddition product is isolated with an 81% yield.

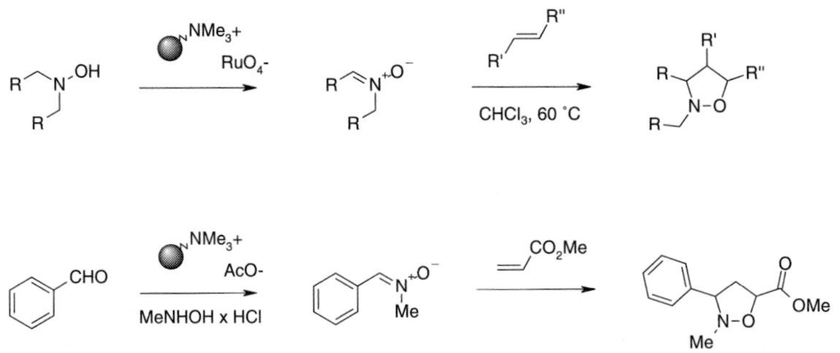

Scheme 16.31. Synthesis of isoxazolidines using polymer-supported reagents.

16.3.2
Formation of Pyrrolidines

Pyrrolidines are typically formed by [3 + 2] cycloaddition reactions of stabilized azomethine ylides and alkenes – a well-documented process using solid support (Scheme 16.32) [50]. Generation of the reactive dipoles can be achieved thermally by Lewis acid activation or under basic conditions. The neighboring effects of a stabilizing electron-withdrawing group are thereby required and, for that reason, amino acids are the preferred building blocks as they are commercially available in large numbers. Strategically, either the azomethine ylide or the dipolarophile can be immobilized on solid support and both strategies have been used successfully. Less commonly, ylides have been formed through transmetalation processes.

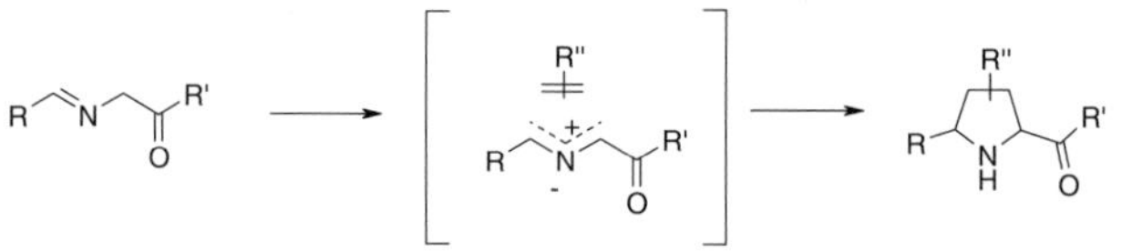

Scheme 16.32. Formation of pyrrolidines by [3 + 2] cycloaddition of azomethine ylides to olefins.

In a representative example of a heat-induced cycloaddition, a polymer-bound amino acid was first condensed with aldehydes and then reacted with *N*-substituted maleimides (Scheme 16.33) [51]. The resulting prolines were obtained with high diastereoselectivities and with satisfactory yields and purities of >72%.

Alternatively, the process can be carried out as a multicomponent procedure [52] when amino acids and maleimides are reacted together with polymer-bound aldehydes.

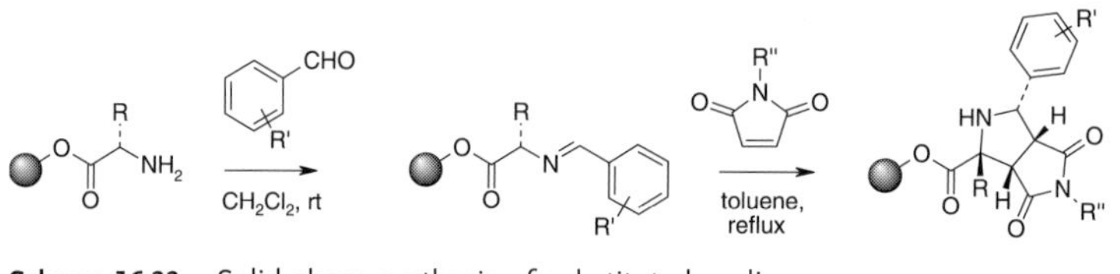

Scheme 16.33. Solid-phase synthesis of substituted prolines.

Pursuing the inverse strategy, resin-bound dipolarophiles can also be reacted with azomethine ylides [53], but the introduction of a base and a Lewis acid is vital for the success of the transformations. After condensation of aromatic aldehydes to 3-hydroxyacetophenone attached to Wang resin, the resulting α,β-unsaturated ketones are treated with *N*-metalated azomethine ylides in the presence of 1,8-diazabicyclo[5.4.0]undecene-7 (DBU) and LiBr (Scheme 16.34). Highly substituted pyrrolidines are obtained with satisfying regio- and diastereoselectivities, but chalcones derived from sterically demanding aldehydes, for example 2,6-dichlorobenz-

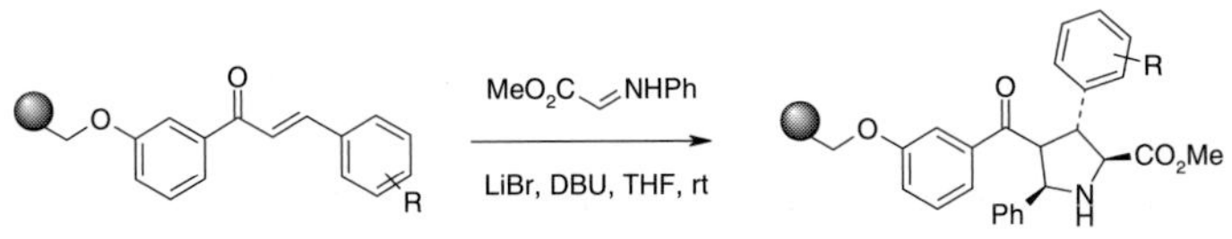

Scheme 16.34. Formation of pyrrolidines from polymer-supported chalcones. rt, room temperature.

aldehyde, do not yield any products. There has been no success using silver(I) acetate as an additive – the catalyst most often used in imine cycloaddition reactions [54].

Other examples that make use of silver salts include the silver acetate-induced cycloaddition of tryptophan-derived imines to polymer-bound acrylates (Scheme 16.35) [55] and synthetic efforts toward a split-and-mix library of mercaptoacyl proline-based inhibitors of angiotensin-converting enzyme (ACE) [56]. Silver nitrate has also been used in the synthesis of fully substituted prolines derived from histidine precursors [57], while intramolecular cycloadditions have yielded polycyclic cores when both the imine and the enone are immobilized on solid support [58]. It is worth mentioning that, during the syntheses of hydantoin-containing

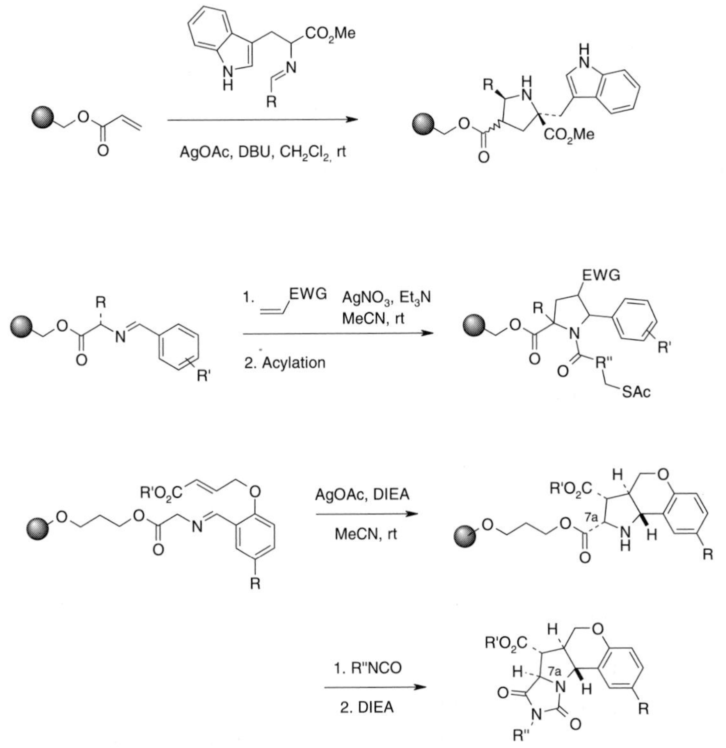

Scheme 16.35. Silver(I) salt-induced formation of substituted prolines.

heterocycles, the insertion of 1,3-propanediol as a spacer moiety between the polymeric backbone and the glycinate generally facilitates the cycloaddition (Scheme 16.35). The desired tetracyclic cores are released from the resin after treatment with isocyanates and base, whereas the latter reagent epimerizes the stereogenic center at C7a. Another example of the tandem azomethine ylide cycloaddition and carbanilide cyclization strategy uses zinc acetate and DBU as the catalytic system [59].

Other conventionally used Lewis acids such as cesium fluoride [60], silver triflate [61], or cobalt dichloride [62] have not yet been adapted to solid-phase combinatorial synthesis.

More reactive dipoles can be generated by transmetalation of 2-aza-allyl-stannanes and butyl lithium (Scheme 16.36) [63], and the resulting unstabilized anions are able to undergo $[2 + 3]$ cycloadditions with electron-rich alkenes. Although mixtures of regio- and stereoisomers are generally obtained, this protocol complements the related azomethine ylide transformations which usually require electron-poor olefins.

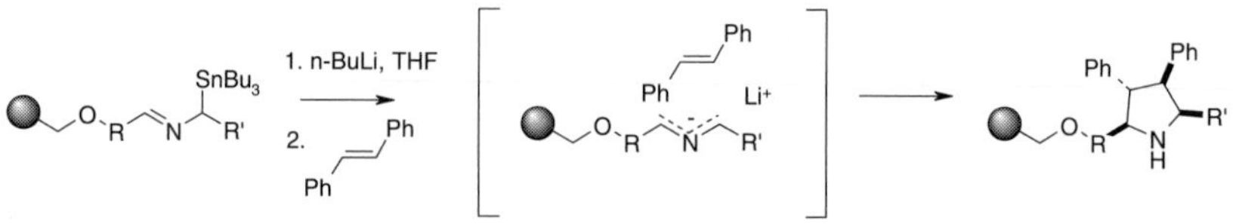

Scheme 16.36. Synthesis of pyrrolidines via aza-allyl anion cycloadditions.

16.3.3
Formation of Furans

Efficient traceless solid-phase syntheses of furans derived from polymer-supported isomünchnones have been reported [64]. The highly reactive 1,3-dipolar intermediates which participate in the cycloaddition reactions with electron-deficient acetylenes are generated *in situ* by the decomposition of diazoesters with Rh(II) catalysts. Upon heating, the intermediate bicyclic cycloadduct rearranges to the desired furans and leaves polymer-supported isocyanate behind (Scheme 16.37).

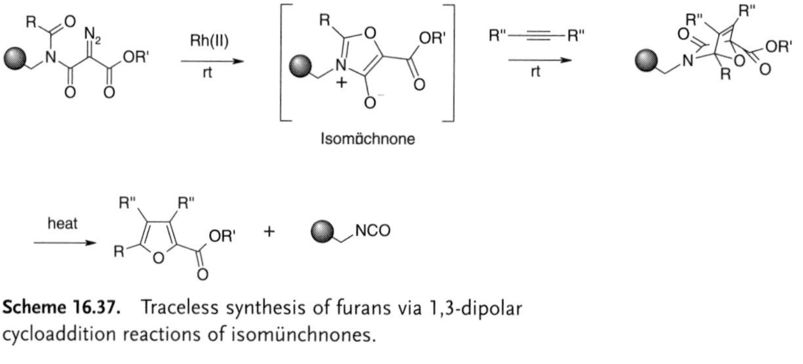

Scheme 16.37. Traceless synthesis of furans via 1,3-dipolar cycloaddition reactions of isomünchnones.

16.3.4
Formation of Imidazoles, Pyrroles, Pyrazoles, and Other Nitrogen-containing Heterocycles

Imidazoles have also been synthesized on solid support utilizing a münchnone [3 + 2] cycloaddition reaction with aryltosylimines as the key bond-forming step (Scheme 16.38) [65]. This methodology has been successfully executed in solution phase before, but the reaction yields are reduced by the tendency of münchnones to undergo self-condensation [66]. This problem can be readily circumvented by attaching the dipoles to AgroGel™-MB-CHO.

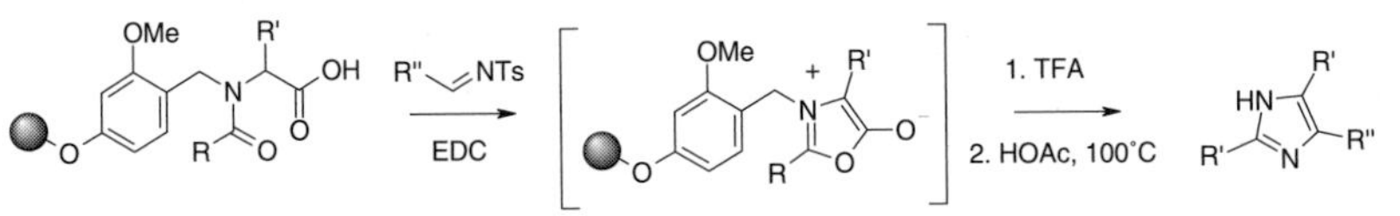

Scheme 16.38. Formation of imidazoles via münchnone intermediates.

Münchnones are prepared by reaction of an acylated polymer-bound amino acid and N'-(3-dimethylaminopropyl)-N-ethylene carbodiimide (EDC) and should immediately be reacted with tosylimines in one pot. It is difficult to cleave the immobilized imidazoles thus obtained from the polymeric support but their release can be achieved by boiling the resins in neat acetic acid, which takes advantage of the robustness of the polymer-bound heterocycles, and unreacted starting materials or nonimidazole byproducts are removed through simple washing with trifluoroacetic acid (TFA) prior to the cleavage step. This new linking strategy has allowed the preparation of an exploratory library containing 12 heterocycles (Scheme 16.38).

When münchnones are combined with electron-deficient acetylenes, pyrroles are obtained (Scheme 16.39) [67]. The precursors for the 1,3-dipolar cycloaddition are available through the Ugi four-component condensation (4UCC) (see Chapter 23.7.5) and undergo an acid-catalyzed cycloelimination step. The resulting

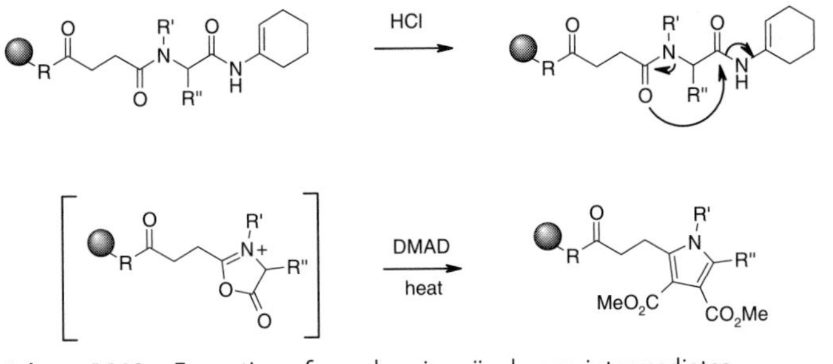

Scheme 16.39. Formation of pyrroles via münchnone intermediates.

1,3-oxazolinium-2-ones are then trapped with dimethyl acetylenedicarboxylate (DMAD) or other electron-deficient acetylenes [68] and yield pyrroles after aromatization and loss of carbon dioxide.

In another example of a [2 + 3] cycloaddition reaction involving electron-deficient acetylenes, DMAD reacts with polymer-bound azomethine imines, forming pyrazoles (Scheme 16.40) [69]. The 1,3-dipoles employed are generated from α-silylnitrosoamides by a 1,4-silatropic shift and give heterocyclic Michael adducts in up to 70% yield. The ratio of the regioisomers obtained is highly dependent on the size of the adjacent substituents, whereas in the case of R = H only one isomer can be detected. Another interesting aspect of this strategy is the cyclization-release methodology, avoiding the need for the cleavage operation. However, purification by silica gel chromatography was still found to be necessary.

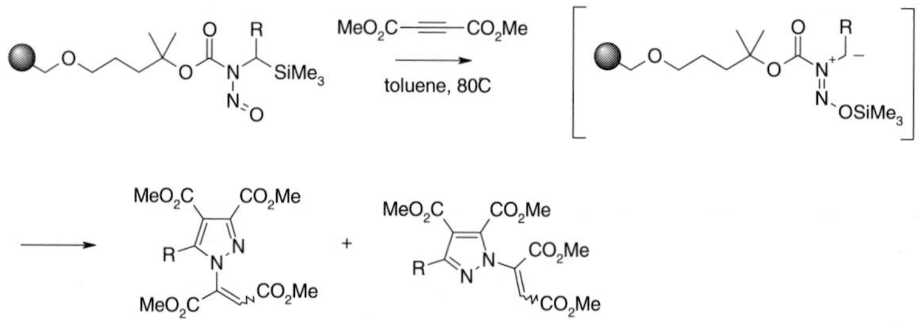

Scheme 16.40. Traceless synthesis of pyrazoles.

Indolizines have also been synthesized on solid support by [3 + 2] cycloaddition reactions of pyridinium ylides with electron-deficient olefins [70]. After alkylation of polymer-bound isonicotinic acid with 2-bromoacetophenones, the resulting pyridinium salts are treated with α,β-unsaturated ketones at elevated temperatures (Scheme 16.41). However, the resulting tetrahydroindolizines rearrange upon

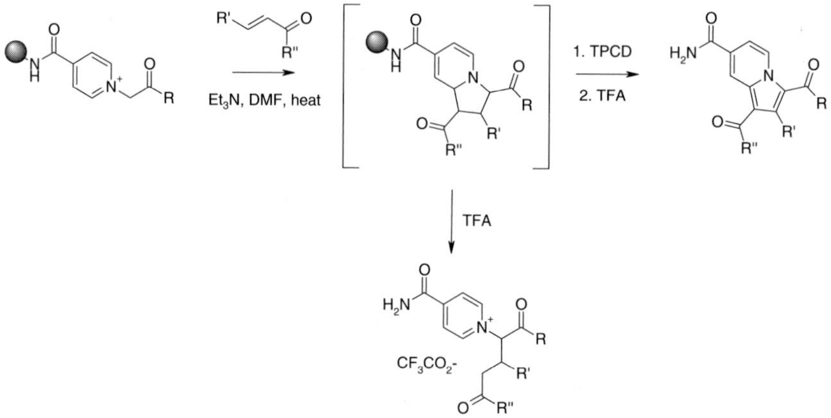

Scheme 16.41. Solid-phase synthesis of indolizines.

acidic cleavage with TFA, a phenomenon also observed during the transfer of the Tsuge reaction to solid-phase chemistry [71].

The formation of the open-chain pyridinium salts is suppressed through oxidation of the bicyclic core with the bimetallic complex TPCD [Co(pyridine)$_4$-(HCrO$_4$)$_2$]. After treatment with TFA, aromatic indolizines are obtained and an exploratory library of nine members has been prepared.

16.4
[2 + 2] Cycloadditions

The [2 + 2] cycloaddition reaction is one of the most synthetically efficient methods used for the preparation of four-member rings. However, only a limited number of protocols have been adapted to solid-phase combinatorial chemistry, while particular focus has been turned toward the synthesis of mono-cyclic β-lactams via the venerable Staudinger reaction [72]. In a representative example (Scheme 16.42), the cycloaddition reaction is initiated through the slow addition of acid chlorides to a suspension of the imine resins in the presence of triethylamine [73]. Owing to the high reactivity and the accompanying tendency to undergo polymerization reactions, the use of a multifold excess of the reagent is required. However, even cycloadditions to imines derived from highly hindered amino acids usually give satisfying results and the scope of the reaction can be extended to amino, O-protected and vinyl ketenes.

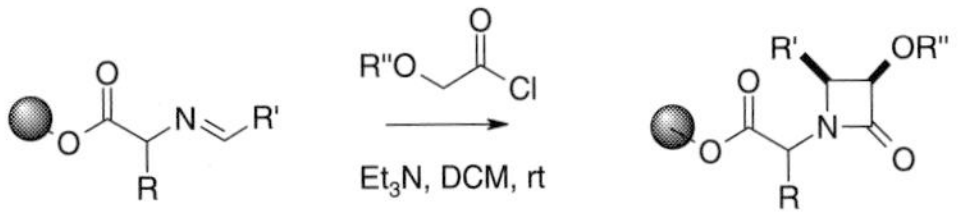

Scheme 16.42. β-Lactams through [2 + 2] cycloaddition reactions of ketenes to resin-bound imines.

The thus formed highly functionalized four-member ring heterocycles are also valuable precursors for further chemical manipulations, particularly, when the β-lactam strain is used to facilitate ring-opening reactions. A striking example of β-lactams as versatile intermediates was given en route to a split-and-mix library of 4140 dihydroquinolinones (Scheme 16.43) [74]. Here, the nitro group of a [2 + 2] cycloadduct is reduced and used as an internal nucleophile for the ring expansion reactions.

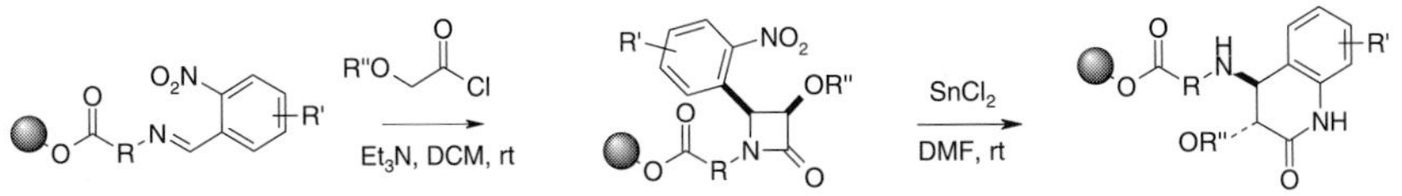

Scheme 16.43. Dihydroquinolinones via polymer-supported β-lactam intermediates.

Another method for the solid-phase preparation of β-lactams from imines involves titanium ester enolates derived from 2-pyridinethiols (Scheme 16.44) [75].

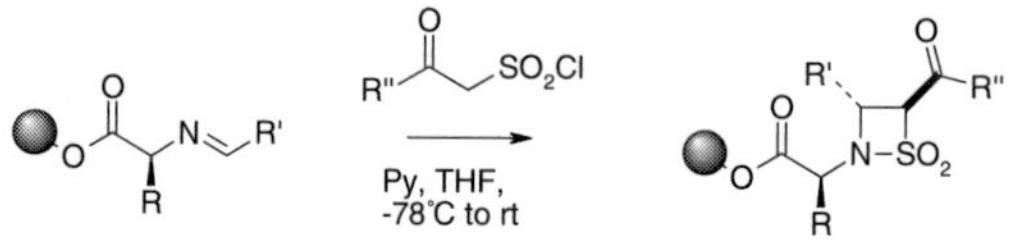

Scheme 16.44. Solid-phase synthesis of β-sultams.

Moreover, when sulfenes are used in the cycloaddition reactions to polymer-supported imines, structurally analogous β-sultams are obtained [76]. Both reactive species are generated *in situ* and smoothly react at −78 °C. While the imines are prepared by the condensation of aldehydes to immobilized amino acids, the sulfenes are formed by the addition of pyridine to methylchlorosulfonyl acetates.

In agreement with the solution-phase synthesis of sultams, two trans diastereomers are obtained [77], but the success of the reaction is reduced when sterically more demanding amino acids such as aspartic acid *tert*-butyl ester are used. On the other hand, the utility of this method is indisputably high – the entire reaction sequence can be carried out with an acid-labile and a photolabile linker. It nicely allows for the tiered release of compounds from polymeric beads onto live cells during high-throughput screening (HTS). Further chemical modifications of the thiazetidine core also extend the scope of this strategy.

A [2 + 2] keteneiminium cycloaddition reaction has been used to prepare cyclobutanones on solid support (Scheme 16.45) [78]. The alkene resins are thereby added to a fivefold excess of the keteneiminium salts generated *in situ* from N,N-dialkylamides according to the method of Ghosez and coworkers [79]. The resulting iminium salts are then hydrolyzed to the corresponding ketones with aqueous sodium bicarbonate solution and further converted to γ-lactams and γ-lactones. This solid-phase protocol is superior to the analogous chemistry carried out in solution, as generally higher conversions are obtained and the purification of the cyclobutanone iminium salts is facilitated by the immobilization on solid phase.

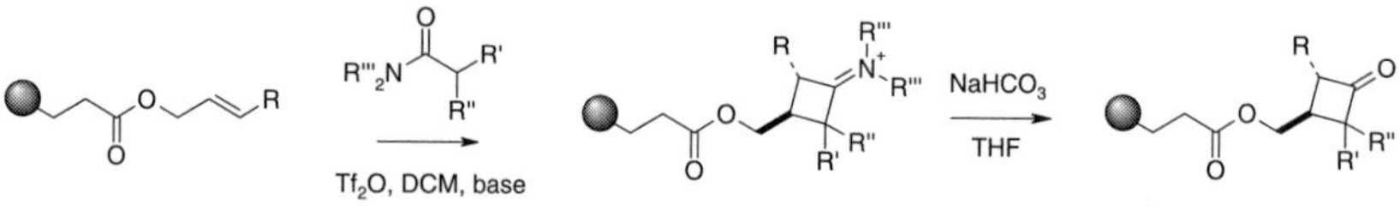

Scheme 16.45. Solid-phase synthesis of cyclobutanes.

16.5
[6 + 3] Cycloadditions on Solid Support

The [6 + 3] cycloaddition is an example of a more exotic reaction in combinatorial solid-phase chemistry. One example of a [6 + 3] cycloaddition is the reaction between fulvenes and benzoquinones forming heterosteroid frameworks (Scheme 16.46) [80]. In order to build up the resin-bound fulvene derivatives, different acids are attached to polystyrene amino resin employing standard conditions (dicyclo-hexyl carbodiimide (DCC), hydroxybenzotriazole (HOBt), dimethylaminopyridine (DMAP)). Treatment with Meerwein salt and different sodium cyclopentadienides provides the desired resin-bound fulvenes. Through cycloaddition with benzoqui-nones the tricyclic adduct is released in a traceless fashion from the resin, which can be recovered and used again. After purification by filtration through a short pad of silica gel, the products are isolated in good yields and purity. In addition to benzoquinones, iminobenzoquinones are also used in this type of cycloaddition.

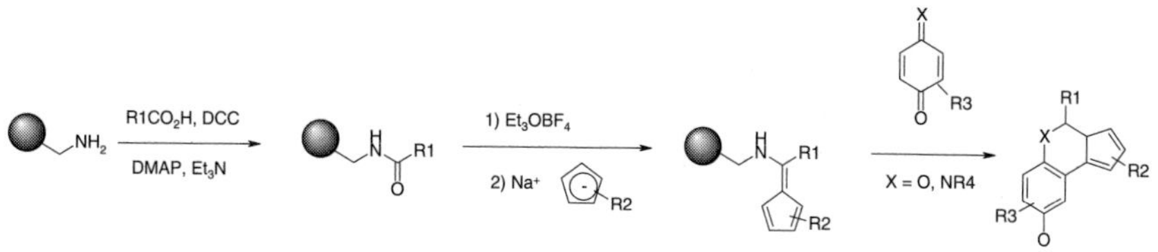

Scheme 16.46. [6 + 3] cycloaddition on solid support.

16.6
Rearrangements

Among cycloadditions, sigmatropic rearrangements also belong to the group of pericyclic reactions. New carbon–carbon bonds are formed and, owing to the peri-cyclic mechanism, there is the possibility of building up stereogenic centers in a stereoselective fashion using chiral induction. Until now only Claisen rearrange-ments have been applied to combinatorial chemistry. A typical example is the polymer-supported Ireland–Claisen rearrangement (Scheme 16.47) [81]. In this solid-phase synthesis, a trialkylsilane linker is used that is first converted to the more reactive silyl triflate. Treatment with different enolizable allylic esters pro-vides the resin-bound silyl enol ethers as the reactive precursors for rearrange-

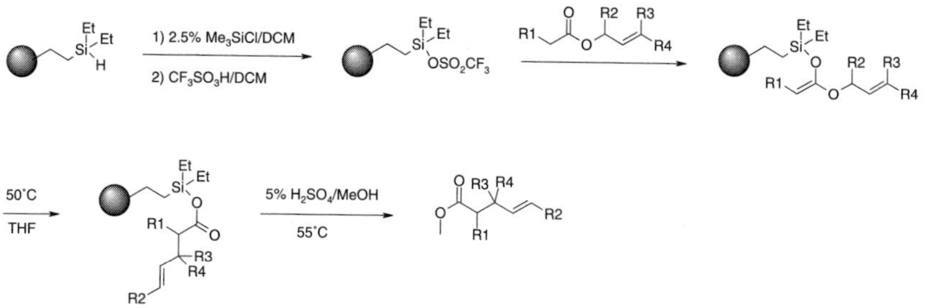

Scheme 16.47. Polymer-supported Ireland–Claisen rearrangement.

ment. After completion of the reaction at 50 °C in tetrahydrofuran, cleavage from the resin is realized by methanolysis of the resin-bound silyl esters. The products are isolated in good yields and high purity.

Not only polystyrene-based resins but also silica gel or mesoporous molecular sieves have been used as solid support. Their thermal resistance at high temperatures and the opportunity of using polar solvents such as methanol or water make these materials superior to traditional resins. Several different silica gels and molecular sieves, which are capped with aminopropyltriethoxysilane and which vary in their average mean pore size, are employed in a Claisen rearrangement (Scheme 16.48) [82]. Attachment of hydroxymethylbenzoic acid with diisopropylcarbodiimide gives the hydroxy-methylated support, which is coupled to further acids bearing allylic side-chains. The Claisen rearrangement is then performed at 225 °C without using any solvent and the products are cleaved as their methyl esters after treatment with methanolic sodium methanolate. When silica gel is used as the solid support, three major products have been isolated by column chromatography that have been identified as two Claisen products (ratio 1.6:1) and a phenol. In contrast, using the mesoporous molecular sieves gave only the desired Claisen product. It is therefore concluded that higher selectivity correlates with the greater distance between the molecules attached to the mesoporous molecular sieves.

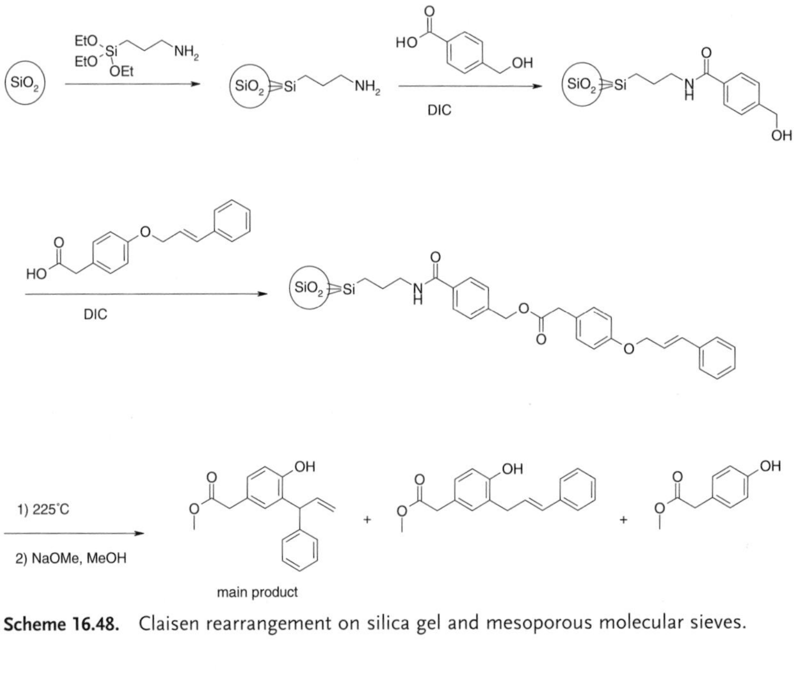

Scheme 16.48. Claisen rearrangement on silica gel and mesoporous molecular sieves.

References

1 a) J. SAUER, R. SUSTMANN, *Angew. Chem.* **1980**, *92*, 773–801; b) W. C. HERNDON, *Chem. Rev.* **1972**, *72*, 157–179; c) K. N. HOUK, A. P. MARCHAND, R. E. LEHR, *Pericyclic Reactions*, Vol. 2, Academic Press, New York 1977; d) P. V. ALSTON, R. M. OTTENBRITE, *J. Org. Chem.* **1975**, *40*, 1111–1116; e) B. M. TROST, J. IPPEN, W. C. VLADUCHIK, *J. Am. Chem. Soc.* **1977**, *99*, 8116–8118.

2 a) R. E. SAMMELSON, M. J. KURTH, *Chem. Rev.* **2001**, *101*, 137–202; b) S. BOOTH, *Tetrahedron*, **1998**, *54*, 15385–15443; c) B. A. BUNIN, *The Combinatorial Index*, Academic Press, New York 1998.

3 E. M. SMITH, *Tetrahedron Lett.* **1999**, *40*, 3285–3288.

4 M. CRAWSHAW, N. W. HIRD, K. IRIE, K. NAGAI, *Tetrahedron Lett.* **1997**, *40*, 7115–7118.

5 J. BARLUENGA, F. AZNAR, M.-P. CABAL, C. VALDES, *J. Chem. Soc. Perkin Trans. 1* **1990**, 633.

6 C. CHEN, B. MUNOZ, *Tetrahedron Lett.* **1999**, *40*, 3491–3494.

7 C. P. BALL, A. G. M. BARRETT, A. COMMERCON, D. COMPÈRE, C. KUHN, R. S. ROBERTS, M. L. SMITH, O. VENIER, *Chem. Commun.* **1998**, 2019–2020.

8 a) D. A. HEERDING, D. T. TAKATA, C. KWON, W. F. HUFFMAN, J. SAMANEN, *Tetrahedron Lett.* **1998**, *39*, 6815–6818; b) S. C. SCHÜRER, S. BLECHERT, *Synlett*, **1999**, *12*, 1879–1882.

9 S. WENDEBORN, A. DE MESMAEKER, W. K.-D. BRILL, *Synlett*, **1998**, 865–868.

10 R. H. SCHLESSINGER, C. P. BERGSTROM, *Tetrahedron Lett.* **1996**, *37*, 2133–2136.

11 a) M. J. FARRALL, J. M. J. FRECHET, *J. Org. Chem.* **1976**, *41*, 3877; b) J. T. RANDOLF, K. F. MCCLURE, S. J. DANISHEFSKY, *J. Am. Chem. Soc.* **1995**, *117*, 5712.

12 L. BLANCO, R. BLOCH, E. BUGNET, S. DELOISY, *Tetrahedron Lett.* **2000**, *41*, 7875–7878.

13 V. YEDIDA, C. C. LEZNOFF, *Can. J. Chem.* **1980**, *58*, 1144–1150.

14 B. A. BURKETT, C. L. L. CHAI, *Tetrahedron Lett.* **1999**, *40*, 7035–7038.

15 J. D. WINKLER, W. MCCOULL, *Tetrahedron Lett.* **1998**, *39*, 4935–4936.

16 G. J. KUSTER, H. W. SCHEEREN, *Tetrahedron Lett.* **2000**, *41*, 515–519.

17 J. D. WINKLER, Y.-S. KWAK, *J. Org. Chem.* **1998**, *63*, 8634–8635.

18 K. PAULVANNAN, T. CHEN, J. W. JACOBS, *Synlett*, **1999**, *10*, 1609–1611.

19 a) S. SUN, W. V. MURRAY, *J. Org. Chem.* **1999**, *64*, 5941–5945; b) S. SUN, I. J. TURCHI, D. XU, W. V. MURRAY, *J. Org. Chem.* **2000**, *65*, 2555–2559.

20 P. WIPF, T. C. HENNINGER, *J. Org. Chem.* **1997**, *62*, 1586–1588.

21 L. F. TIETZE, T. HIPPE, A. STEINMETZ, *Synlett*, **1996**, 1043–1044.

22 S. LECONTE, G. DUJARDIN, E. BROWN, *Eur. J. Org. Chem.* **2000**, *65*, 639–643.

23 W. ZHANG, W. XIE, J. FANG, P. G. WANG, *Tetraderon Lett.* **1999**, *40*, 7929–7933.

24 P. CONTI, D. DEMONT, J. CALS, C. J. OTTENHEIJM, D. LEYSEN, *Tetrahedron Lett.* **1997**, *38*, 2915.

25 C. O. OGBU, M. N. QUABAR, P. D. BOATMAN, J. URBAN, J. P. MEARA, M. D. FERGUSON, J. TULINSKY, C. LUM, S. BABU, M. A. BLASKOVICH, *Bioorg. Med. Chem. Lett.* **1998**, *8*, 2321–2326.

26 A. M. BOLDI, C. R. JOHNSON, H. O. EISSA, *Tetrahedron Lett.* **1999**, *40*, 619–622.

27 a) D. L. BOGER, W. CHAI, R. S. OZER, C. ANDERSSON, *Bioorg. Med. Chem. Lett.* **1997**, *7*, 463–466; b) D. L. BOGER, C. TARBY, P. L. MYERS, L. H. CAPORALE, *J. Am. Chem. Soc.* **1996**, *118*, 2109–2112; c) S. CHENG, D. D. COMER, J. P. WILLIAMS, P. L. MYERS, D. L. BOGER, *J. Am. Chem. Soc.* **1996**, *118*, 2567–2570.

28 a) A. STUDER, D. P. CURRAN, *Tetrahedron* **1997**, *53*, 6681–6688; b) A. STUDER, S. HADID, R. FERRITTO, S. KIM, P. JEGER, P. WIPF, D. P. CURRAN, *Science* **1997**, *275*, 823–830.

29 a) J. J. PARLOW, *Tetrahedron Lett.* **1996**, *37*, 5257–5261; b) J. J. PARLOW, J. E.

NORMANSELL, *Mol. Diversity* **1995**, *1*, 266–270; c) D. C. SHERRINGTON, P. HODGE, *Synthesis and Separations Using Functional Polymers*, Chichester, Wiley 1988; d) A. AKELAH, D. C. SHERRINGTON, *Chem. Rev.* **1981**, *81*, 557–569.

30 a) S. KOBAYASHI, S. NAGAYAMA. *J. Am. Chem. Soc.* **1996**, *118*, 8977–8978; b) S. B. JANG, *Tetrahedron Lett.* **1997**, *38*, 1793–1796.

31 a) M. G. SIEGEL, P. J. HAHN, B. A. DRESSMAN, J. E. FRITZ, J. R. GRUNWELL, S. W. KALDOR, *Tetrahedron Lett.* **1997**, *38*, 3357–3360; b) L. M. GAYO, M. J. SUTO, *Tetrahedron Lett.* **1997**, *38*, 513–516.

32 a) R. J. BOOTH, J. C. HODGES, *J. Am. Chem. Soc.* **1997**, *119*, 4882–4883; b) S. W. KALDOR, M. G. SIEGEL. B. A. DRESSMAN, P. J. HAHN, *Tetrahedron Lett.* **1996**, *37*, 7193–7195.

33 M. W. CRESWELL, G. L. BOLTON, J. C. HODGES, M. MEPPEN, *Tetrahedron*, **1998**, *54*, 3983–3998.

34 J. HABERMANN, S. V. LEY, J. S. SCOTT, *J. Chem. Soc., Perkin Trans. 1*, **1999**, 1253–1255.

35 E. J. KANTOROWSKI, M. J. KURTH, *Mol. Diversity* **1996**, *2*, 207–216.

36 B. HINZEN, S. V. LEY, *J. Chem. Soc., Perkin Trans. 1*, **1998**, 1–2.

37 K. B. G. TORSSEL, *Nitrile Oxides, Nitrones and Nitronates in Organic Synthesis*, VCH Publishers, Weinheim 1988.

38 R. A. WHITNEY, E. S. NICHOLAS, *Tetrahedron Lett.* **1981**, *22*, 3371–3374.

39 Y. PEI, W. H. MOOS, *Tetrahedron Lett.* **1994**, *35*, 5825–5828.

40 T. MUKAIYAMA, T. HOSHINO *J. Am. Chem. Soc.* **1960**, *62*, 5339–5342.

41 a) P. N. CONFALONE, S. S. KO, *Tetrahedron Lett.* **1984**, *25*, 947; b) A. P. KOZIKOWSKI, J. G. SCRIPKO, *J. Am. Chem. Soc.* **1984**, *106*, 253.

42 X. BEEBE, C. L. CHIAPPARI, M. M. OLMSTEAD, M. J. KURTH, N. E. SCHORE, *J. Org. Chem.* **1995**, *60*, 4204–4212.

43 D. S. TAN, M. A. FOLEY, B. R. STOCKWELL, M. D. SHAIR, S. L. SCHREIBER, *J. Am. Chem. Soc.* **1999**, *121*, 9073–9087.

44 a) O. TAMURA, T. OKABE, T. YAMAGUCHI, K. GOTANDA, K. NOE, M. SAKAMOTO, *Tetrahedron* **1995**, *51*, 107–118; b) O. TAMURA, T. OKABE, T. YAMAGUCHI, J. KOTANI, K. GOTANDA, M. SAKAMOTO, *Tetrahedron* **1995**, *51*, 119–128.

45 W. J. HAAP, D. KAISER, T. B. WALK, G. JUNG, *Tetrahedron* **1998**, *54*, 3705–3724.

46 S. KOBAYASHI, R. AKIYAMA, *Tetrahedron Lett.* **1998**, *39*, 9211–9214.

47 B. B. SHANKAR, D. Y. YANG, S. GIRTON, A. K. GANGULY, *Tetrahedron Lett.* **1998**, *39*, 2447–2448.

48 J.-F. CHENG, A. M. M. MJALLI, *Tetrahedron Lett.* **1998**, *39*, 939–942.

49 B. HINZEN, S. V. LEY, *J. Chem. Soc., Perkin Trans. 1*, **1998**, 1–2.

50 E. J. KANTOROWSKI, M. J. KURTH, *Mol. Diversity* **1996**, *2*, 207–216.

51 A. J. BRICKNELL, N. W. HIRD, *Bioorg. Med. Chem. Lett.* **1996**, *6*, 2441–2444.

52 B. C. HAMPER, D. R. DUKESHERER, M. S. SOUTH, *Tetrahedron Lett.* **1996**, *37*, 3671–3674.

53 S. P. HOLLINSHEAD, *Tetrahedron Lett.* **1996**, *37*, 9157–9160.

54 D. A. BARR, M. J. DORRITY, R. GRIGG, S. HARGREAVES, J. F. MALONE, J. MONTGOMERY, J. REDPATH, P. STEVENSON, M. THORNTON-PETT, *Tetrahedron* **1995**, *51*, 273–294.

55 H. A. DONDAS, R. GRIGG, W. S. MACLACHLAN, D. T. MACPHERSON, J. MARKANDU, V. SRIDHARAN, S. SUGANTHAN, *Tetrahedron Lett.* **2000**, *41*, 967–970.

56 M. M. MURPHY, J. R. SCHULLEK, E. M. GORDON, M. A. GALLOP, *J. Am. Chem. Soc.* **1995**, *117*, 7029–7030.

57 B. HENKEL, W. STENZEL, T. SCHOTTEN, *Bioorg. Med. Chem. Lett.* **2000**, *10*, 975–977.

58 Y.-D. GONG, S. NAJDI, M. M. OLMSTEAD, M. J. KURTH, *J. Org. Chem.* **1998**, *63*, 3081–3086.

59 G. PENG, A. SOHN, M. A. GALLOP, *J. Org. Chem.* **1999**, *64*, 8342–8349.

60 R. N. BUTLER, D. M. FARRELL, *J. Chem. Res. Synop.* **1998**, 82–83.

61 D. M. COOPER, R. GRIGG, S. HARGREAVES, P. KENNEWELL, J.

Redpath, *Tetrahedron* **1995**, *51*, 7791–7808.

62 P. Allway, R. Grigg, *Tetrahedron Lett.* **1991**, *32*, 5817–5820.

63 W. H. Pearson, R. B. Clark, *Tetrahedron Lett.* **1997**, *38*, 7669–7672.

64 a) M. R. Gowravaram, M. A. Gallop, *Tetrahedron Lett.* **1997**, *38*, 6973–6976; b) D. L. Whitehouse, K. H. Nelson Jr, S. N. Savinov, D. J. Austin, *Tetrahedron Lett.* **1997**, *38*, 7139–7142.

65 M. T. Bilodeau, A. M. Cunningham, *J. Org. Chem.* **1998**, *63*, 2800–2801.

66 R. Consonni, P. D. Croce, R. Ferraccioli, C. La Rosa, *J. Chem. Res. Synop.* **1991**, 188–189.

67 A. M. M. Mjalli, S. Sarshar, T. J. Baiga, *Tetrahedron Lett.* **1996**, *37*, 2943–2946.

68 A. M. Strocker, T. A. Keating, P. A. Tempest, R. W. Armstrong, *Tetrahedron Lett.* **1996**, *37*, 1149–1152.

69 K.-I. Washizuka, K. Nagai, S. Minakata, I. Ryu, M. Komatsu, *Tetrahedron Lett.* **2000**, *41*, 691–695.

70 D. A. Goff, *Tetrahedron Lett.* **1999**, *40*, 8741–8745.

71 A. J. Bricknell, N. W. Hird, S. A. Readshaw, *Tetrahedron Lett.* **1998**, *39*, 5869–5872.

72 a) B. Ruhland, A. Bombrun, M. A. Gallop, *J. Org. Chem.* **1997**, *62*, 7820–7826; b) Z.-J. Ni, D. Maclean, C. P. Holmes, M. M. Murphy, B. Ruhland, J. W. Jacobs, E. M. Gordon, M. A. Gallop, *J. Med. Chem.* **1996**, *39*, 1601–1608; c) R. Singh, J. M. Nuss, *Tetrahedron Lett.* **1999**, *40*, 1249–1252.

73 B. Ruhland, A. Bhandari, E. M. Gordon, M. A. Gallop, *J. Am. Chem. Soc.* **1996**, *118*, 253–254.

74 Y. Pei, R. A. Houghten, J. S. Kiely, *Tetrahedron Lett.* **1997**, *38*, 3349–3352.

75 V. Molteni, R. Annunziata, M. Cijnquini, F. Cozzi, M. Benaglia, *Tetrahedron Lett.* **1998**, *39*, 1257–1260.

76 M. F. Gordeev, E. M. Gordon, D. V. Patel, *J. Org. Chem.* **1997**, *62*, 8177–8181.

77 M. J. Szymonifka, J. V. Heck, *Tetrahedron Lett.* **1989**, *30*, 2869–2872.

78 R. C. D. Brown, J. Keily, R. Karim, *Tetrahedron Lett.* **2000**, *41*, 3247–3251.

79 J. B. Falmagne, C. Schmit, J. Escudero, H. Vanlierde, L. Ghosez, *Org. Synth.* **1990**, *69*, 199.

80 B.-C. Hong, Z.-Y. Chen, W.-H. Chen, *Org. Lett.* **2000**, *17*, 2647–2648.

81 Y. Hu, J. A. Porco Jr, *Tetrahedron Lett.* **1999**, *40*, 3289–3292.

82 I. Sucholeiki, M. P. Pavia, C. T. Kresge, S. B. McCullen, A. Malek, S. Schramm, *Mol. Diversity* **1998**, *3*, 161–171.

17
Main Group Organometallics

Christopher Kallus

17.1
Introduction

Main group organometallics represent a class of powerful carbon nucleophiles that allow the construction of C–C single bonds. Moreover, they can act as strong bases and metalating agents, but also possess chelating and Lewis acid characteristics. They are indispensable tools for the construction of organic molecular frameworks which receive strong interest in classical syntheses of physiologically active substances. In contrast, the use of main group organometallics in combinatorial chemistry is underrepresented. This situation may be due to two major factors. First, the number of commercially available compounds is limited, as is their "diversity" in terms of structural variety. Second, the organometallic species are rather moisture-sensitive and decompose rapidly in the presence of air. On the other hand, almost no combinatorial equipment provides fully inert reaction conditions, making many combinatorial syntheses involving organometallic reagents difficult to carry out. Reagents carrying more sophisticated residues have to be freshly prepared, but not every desirable chemical functionality is compatible with the high reactivity of this reagent class. As a consequence, only robust anchoring groups such as Ellman's tetrahydropyranyl linker or Wang ethers can be applied in solid-phase chemistry, whereas standard linkers such as Wang esters or Rink amides are mostly excluded. The resins used have to be dried carefully and the reactions need to be carried out in dry glassware where deep cooling can be applied. In practice, batches of resins for library synthesis have been prepared simultaneously prior to further diversification by easy reaction steps carried out in common parallel synthesis equipment. Novel synthetic technologies such as Chemspeed™ or the Irori™ system may lead to new trends.

All these factors in a very small number of combinatorial protocols in the area in question. Moreover, not a single large library has been prepared with main group organometallics, but only small collections of single compounds. In the context of combinatorial chemistry, organometallic reagents have also been used for the synthesis of novel resin–anchor conjugates. Generally, the published protocols describe solid-phase syntheses. However, an increasing number of interesting solutions to this enormously challenging field have been presented and will be re-viewed in this chapter.

C–C bonds are formed by the attack of metalated carbon nucleophiles onto electrophiles such as alkyl and aryl halides, carbonyl groups such as aldehydes, ketones, and their heteroanalogs, as well as carbonic acid derivatives. Among meta-lated species, organomagnesium and -lithium reagents are most commonly used in combinatorial chemistry, whereas cuprates have drawn less attention. Only oc-casionally and in special cases have organoaluminum, -boron, or -indium species been employed.

Since the chemistry of the different metalated species is quite often similar, the structure of this chapter is not guided by the various main group elements. In-stead, it is organized by the type of chemical transformation, thus avoiding double citations of the literature examples, where different organometallics are used for the same transformation. Rather than dividing the subject into a strictly mecha-nistic sense, the order is completely practical, following the synthetic intention focused on a diversifiable substrate (organonucleophile or electrophile) which can be modified in a combinatorial sense. For instance, metalated aromatics are used to make structurally diverse aromatic compounds by reaction with different types of electrophiles while resin-bound ketones give diverse alcohols when reacted with organometallics.

17.2
Reactions of Metalated Aromatics

Aryl lithium species are versatile precursors in the preparation of substituted aromatics. Major applications include the syntheses of different functionalized resins in polymer-assisted synthesis and of "mini-libraries" of substituted hetero-cycles. Metalated aromatics are generally prepared from the corresponding aryl halides with n-BuLi in tetrahydrofuran (THF) at low temperatures. Their reactions with electrophiles are among the most frequently applied in combinatorial synthe-sis owing to the fact that not only a broad range of electrophiles can be applied under formation of different functional motifs, but also a large number of aryl bromides as direct precursors of aryl lithiums are commercially available or can be obtained by bromination during the combinatorial synthesis. Additionally, some heterocyclic systems can be directly lithiated, which makes this type of reaction even more interesting in a combinatorial sense. This feature represents an elegant alternative to the functionalization of aromatics by Suzuki reactions, avoiding the preparation of boronic acid building blocks.

Metalation of aromatic rings is one of the most fundamental reactions in solid-phase synthesis. It has become the standard way of functionalizing simple poly-styrene in order to couple handles, linkers, and reagents for further modifications. A typical reaction sequence consists of bromination of polystyrene in the presence of $Tl(OAc)_3$ or with ferric(III) chloride as a catalyst, and subsequent lithiation using n-BuLi. Direct lithiation can be performed using n-BuLi and tetramethylethylene diamine or triethylene diamine. Lithiated polystyrene can serve as the starting material for the preparation of polymer-bound carboxylic acids, thiols, sulfides, boronic acids, amides, silyl chlorides, phosphines, alkyl bromides, aldehydes, alcohols, or trityl functional groups for applications in polymer-assisted syntheses using the corresponding electrophiles (Scheme 17.1) [1–6]. It also serves as the starting material for sodium tris-ethoxyborane on selenylpolystyrene, one of the first selenium-based linker systems that allows a "traceless" cleavage [7]. The exchange was achieved by treating the lithiated polystyrene with selenium powder in dry THF. This material also reacts with SO₂ in THF to give polymer-bound lithium phenyl sulfinate, a well-established linker for the synthesis of trisubstituted olefins (see below) [8].

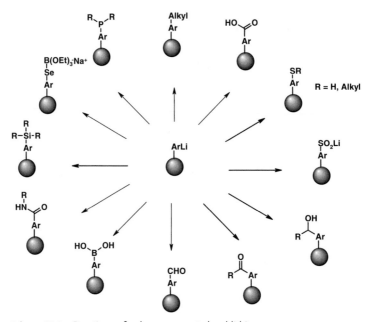

Scheme 17.1. Reactions of polymer-supported aryl lithiums.

A direct lithiation-substitution sequence on solid support has been described with several five-member ring heterocycles. Substituted hydroxyimidazoles are obtained from *O*-imidazolyl-hydroxypolystyrene and *n*-BuLi followed by reactions with various electrophiles such as alkyl halides, amides, aldehydes, carbon tetrachloride, disulfides, or acid chlorides (Scheme 17.2). Interestingly, in contrast to the solution-phase procedure, solid-phase lithiation acylation occurs without any formation of tertiary alcohols. The compounds thus synthesized have been purified by crystallization or chromatography [9]. In a similar manner, 2- and 2,5-functionalizations of 3-polystyrenyl-*O*-trityl-hydroxymethylfurans and -thiophenes can be achieved. The first substitution takes place at the least hindered α-position, presumably because of the steric bulk of the trityl linker. A subsequent lithiation

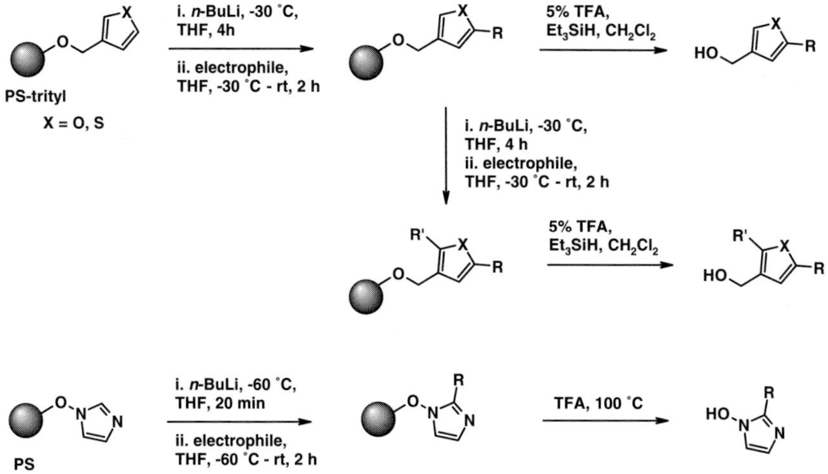

Scheme 17.2. Lithiation-substitution sequences of resin-bound heteroaromatics.

provides access to an attack on the position between the heteroatom and the hydroxymethyl group [10].

Solid-supported phenyl lithium and thiophenyl lithium serve as versatile precursors to the preparation of the corresponding resin-bound aryl isopropylsquarene by reaction with diisopropyl squarate. The reactive intermediates provide a reaction platform to generate several completely different cores, the so-called Multiple Core Structure Libraries (MCSLs). From various possible compound shapes, quinones, hydroquinones, and vinylogous amines derived from arylsquarenes have been realized in small libraries. Further transformations at the squarene carbonyl group are also possible (see below) [11].

In a similar fashion, the replacement of bromine with Grignard reagents leads to magnesiated heterocycles as useful intermediates for further diversification. This procedure has successfully been applied to the synthesis of several functionalized thiophenes (Scheme 17.3). In contrast to reactions with the very polar organolithiums, the ester linkage and other functional groups are stable below −20 °C in the presence of an excess of Grignard reagent. Moreover, selective exchanges on thiophene dibromides can be achieved at low temperatures. It should be mentioned that, in contrast to the alkoxymethyl thiophenyl lithium species, the first exchange and reaction with an electrophile takes place at the sterically more hindered position between the heteroatom and the linking ester group. The directing effect of the anchoring group is based on a complexation between the magnesium and the ester group. The reaction is typically performed in the presence of CuCN·2LiCl in THF [12, 13].

Furthermore, organozinc bromides serve as building blocks for the solid-phase synthesis of substituted aromatics. Since this reaction is catalyzed by Pd, it will be discussed in Chapter 19.

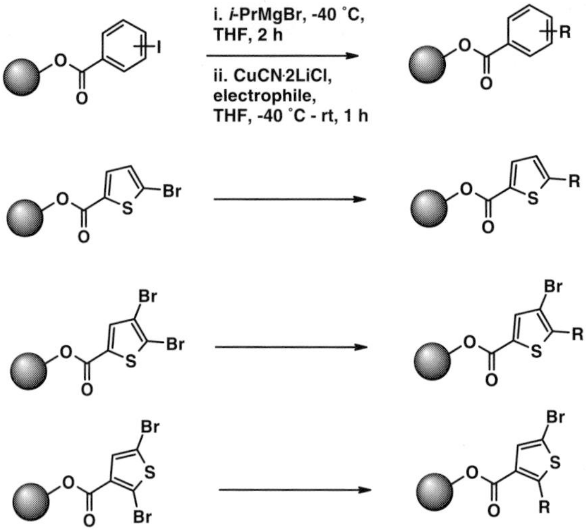

PS-Wang

Scheme 17.3. Halogen exchange with Grignard reagents and subsequent reactions with electrophiles.

17.3
1,2-Additions to C=X Groups

17.3.1
Reactions with Aldehydes

The reaction of resin-bound aldehydes with Grignard reagents is one of the most commonly performed main group organometallic transformations on solid support. The reaction has been applied to the synthesis of a fluoride-labile linker system which is suitable for immobilization of substrates such as esters, carbamates, and carbonates. Polystyrene methoxybenzaldehyde reacts smoothly with *p*-TMSCH$_2$C$_6$H$_4$MgBr, forming the secondary diphenyl alcohol [14] (Scheme 17.4).

A Grignard reaction with an aldehyde has also been applied during the solid-phase synthesis of (*S*)-zearalenone [15]. A stannane linker derived from polymer-supported tin chloride in a reaction carries the aldehyde component which reacted with a chiral TBS-protected organomagnesium alcohol leading to both diastereomers of the secondary alcohol (Scheme 17.5).

In another example, polymer-supported fluorenyl aminoaldehydes were treated with various aryl magnesium reagents (Scheme 17.6). The norephedrines were obtained after trifluoroacetic acid (TFA)-induced cleavage of the fluorenyl linker [16].

One of the few examples dealing with resin-bound alkyl Grignard reagents is the synthesis of 2,5-dihydrofurans and 1,3-dihydroisobenzofurans [17]. Hydroxy aryl or alkenyl iodides were coupled on Wang resin as *p*-alkoxybenzyl ethers. The

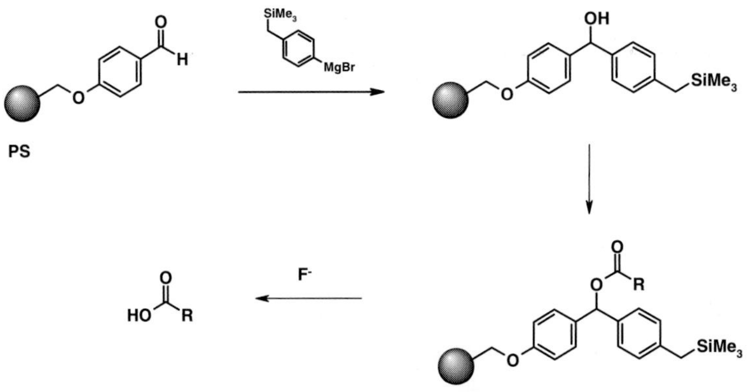

Scheme 17.4. Solid-phase synthesis of a secondary alcohol as fluorine-labile linker via addition of a Grignard reagent to an immobilized benzaldehyde derivative.

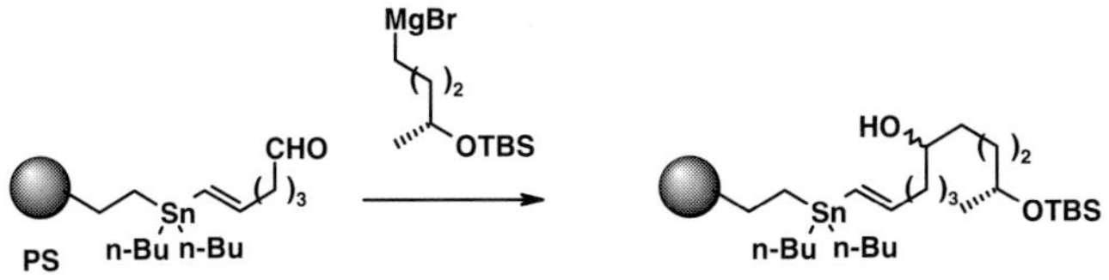

Scheme 17.5. Addition of a TBS-protected Grignard reagent to a resin-bound aldehyde during synthesis of (S)-zearalenone.

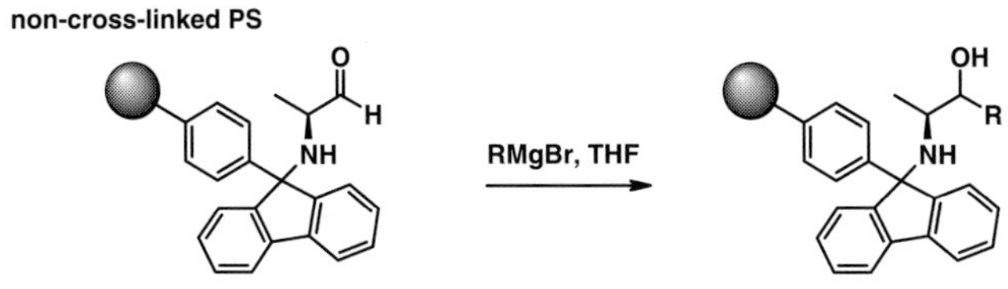

non-cross-linked PS

Scheme 17.6. Formation of secondary alcohols from amino acid derivatives in the synthesis of norephedrines.

subsequent iodine–magnesium exchange with *i*-PrMgBr gave the corresponding Grignard reagent on beads, which reacts with several aldehydes to form secondary alcohols (Scheme 17.7). Finally, cyclization during cleavage from the resin with TFA occurs, yielding the desired furans.

Secondary alcohols can also be prepared by reacting organolithium species with aldehydes. The reaction of polymer-supported *m*-alkoxy benzaldehyde with deprotonated 2-phenyl-1,3-dithiane formed a dithiane-protected 3-alkoxy benzoin as the key step in the synthesis of a photolabile safety-catch linker [18]. Since the dithiane group is easily removed by either bis-((trifluoroacetoxy)iodo)benzene, mercury(II) perchlorate, or periodic acid, this example may serve as a general protocol for the preparation of α-hydroxy ketones. The success of the reaction is controlled

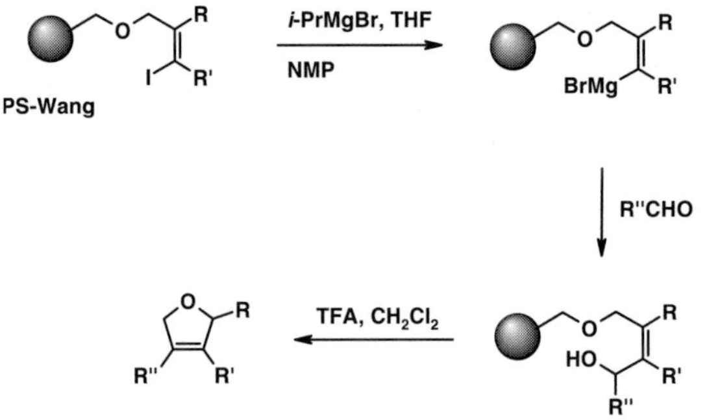

Scheme 17.7. Formation of resin-bound Grignard reagents and addition of aldehydes providing secondary alcohols as precursors to dihydrofurans.

by elemental analysis of sulfur and characteristic gel-phase ^{13}C-NMR signals of the dithiane methylenes (Scheme 17.8).

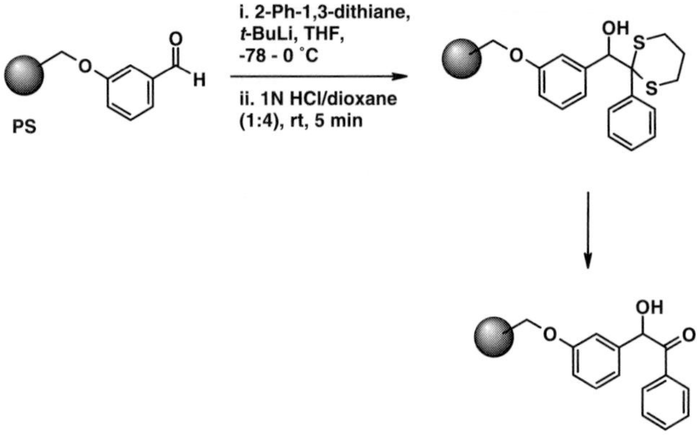

Scheme 17.8. α-Hydroxy ketones by reaction of polymer-bound aldehydes with lithiated dithiane.

Similarly, secondary amines can be lithiated when immobilized as amidines and transformed into secondary α-amino alcohols by reaction with aromatic aldehydes [19]. Subsequent Williamson ether synthesis with benzyl halides and cleavage from the resin by treatment with hydrazine/acetic acid provides a small library of aminoethers (Scheme 17.9).

Allyl indium and allyl boronic acid pinacolates have been proven to be very mild reagents for the nucleophilic attack of aldehydes. They are compatible even with base and photolabile nitrobenzylic linker groups and are, in some cases, superior to Grignard reagents or organolithiums [20]. In recent examples, resin-bound aryl or amino acid aldehydes have been transformed into their homoallylic alcohols (Scheme 17.10). The organoindium species have been prepared *in situ* from allyl

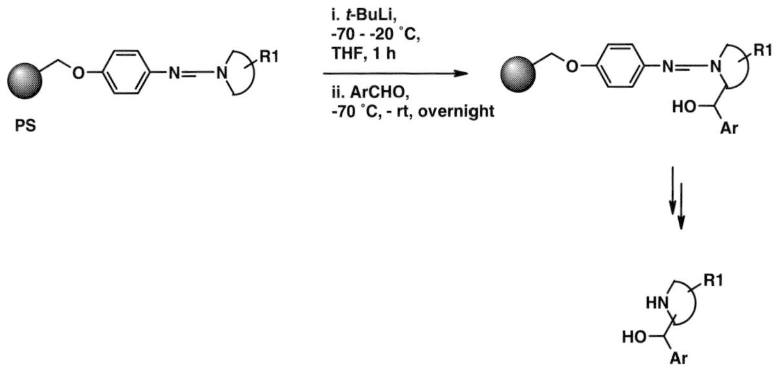

Scheme 17.9. Preparation of secondary amino alcohols from lithiated amidines.

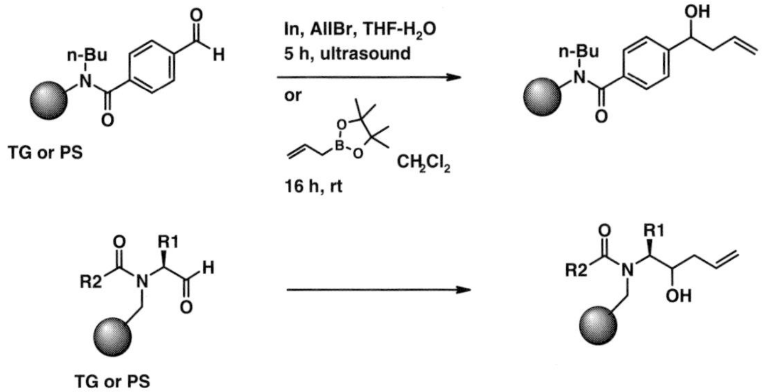

Scheme 17.10. Formation of homoallylic alcohols on solid support from allyl indium or allyl boronic acid pinacolate and aldehydes.

bromide and indium powder in aqueous THF under ultrasonication. Alternatively, the transformation can be carried out with boronic acid pinacolates in dichloromethane under anhydrous conditions. The reactions were successfully conducted on both Tentagel® resin and polystyrene resins.

17.3.2
Reaction with Ketones

The reaction of ketones as electrophiles with organometallic reagents leads to tertiary alcohols. As in the preceding chapters, the solid-phase application of this reaction has been used in the synthesis of linkers and mini-libraries. In a representative example, a novel acid-labile linker for solid-phase synthesis has been prepared by the reaction of MeMgBr with resin-bound methyl phenylethyl ketone [21]. Amines can then be immobilized on solid support by a functional group equivalent to the Boc group, when coupled with *p*-nitrophenyl chloroformate onto this resin (Scheme 17.11).

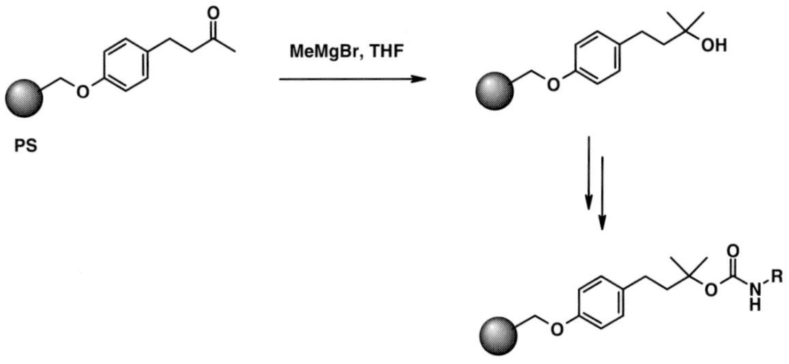

Scheme 17.11. Solid-phase synthesis of tertiary alcohols for the preparation of t-alkyloxycarbonyl-linked amines.

3-Substituted 2-cyclohexenones can be synthesized by 1,2-addition of Grignard reagents to polymer-bound alkoxy cyclohexenones, which were primarily prepared from hydroxymethyl resin and 1,3-cyclohexandione [22]. During cleavage from the resin with TFA, cyclohexenones were formed by elimination of water (Scheme 17.12). Organolithium compounds are also feasible but give somewhat lower yields, while the reaction with more hindered carbon nucleophiles generally suffers from low yields and unsatisfactory purities.

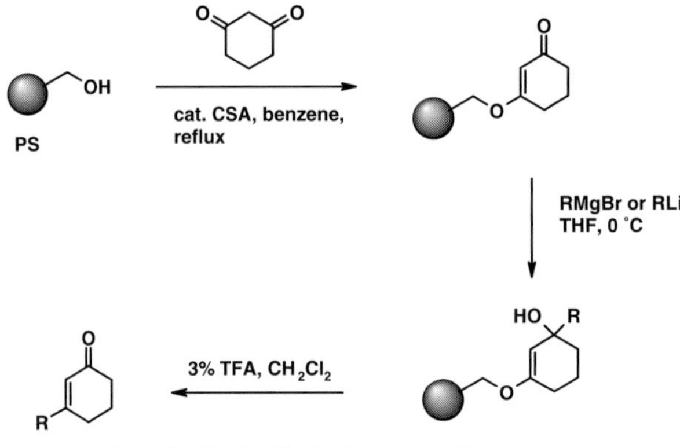

Scheme 17.12. Synthesis of substituted cyclohexenones.

For the preparation of tertiary alcohols from aryl squarenes in MCSLs, two strategies using phenyl lithium reagents have been considered [23]. One involves the use of a polymer-bound aryl lithium species, generated from an aryl bromide with n-BuLi, which was reacted with alkyl 4-isopropoxysquarenes to yield benzo-fused quinones after thermal cyclization and cleavage from the resin (Scheme 17.13, see also Section 17.2). The other pathway starts with a resin-bound squarene which is then converted to a tertiary alcohol with phenyl lithium. If the hydroxyl

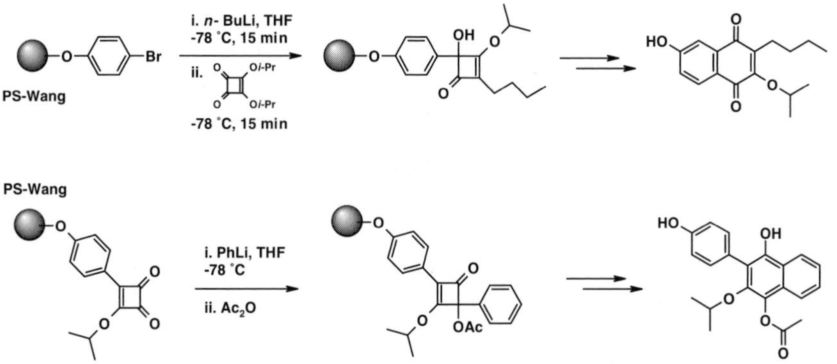

Scheme 17.13. MCSLs by addition of alkyl lithiums on squarenic acid derivatives.

group is protected, the subsequent cyclization affords aryl-substituted benzohy-droquinones.

Treatment of resin-bound 2-substituted dihydropyridones with Grignard species in the presence of CeCl₃ led to 1,2-addition and the formation of tertiary alcohols [24]. TFA-induced cleavage of the compounds from the resin produced 2,4-disubstituted pyridines under oxidative conditions (O₂ or air) or 2,4-disubstituted tetrahydropyridines under reductive conditions (triethylsilane) (Scheme 17.14). The starting material for this reaction sequence was obtained by the resin activation/capture strategy (REACAP, see below).

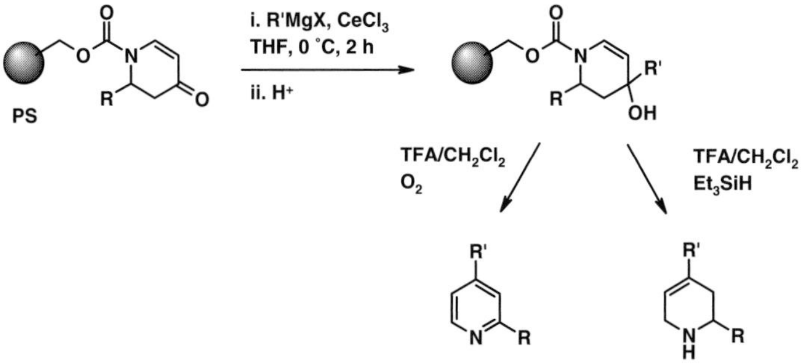

Scheme 17.14. Reaction of Grignard reagents with piperidinones to tertiary alcohols during the preparation of pyridines and tetrahydropyridines.

17.3.3
Reaction with Imines

The conversion of imines with organometallic reagents is slightly harder because their reactivity is lower than the carbonyl groups. While organolithium compounds undergo complete conversions at −78 °C, Grignard reagents do react with imines at elevated temperatures. The scope of the solid-phase version of this reaction has been demonstrated on aldimines prepared from the amino group of Rink resin and aldehydes making use of the tea-bag method [25] (Scheme 17.15). In principle, aldimines derived from aromatic aldehydes work well, independent of their electronic nature. The polymer support functions as a suitable NH-protecting group, while other "masked" imines such as sulfenimines, sulfonimines, and TMS-protected imines were not applicable in solid-phase chemistry. The cleavage of the products thus formed from the resin yielded primary amines. Similarly, Schiff bases from polymer-bound benzaldehydes were reacted with allyl magnesium bromide, providing secondary α-allyl amines [26].

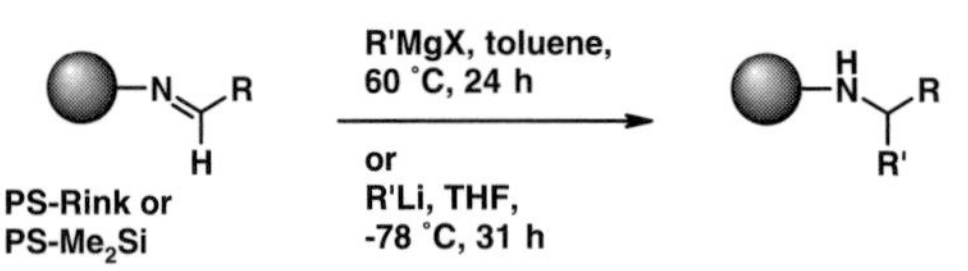

Scheme 17.15. Substituted primary amines from solid-phase synthesis of aldimines with organonucleophiles.

A convenient approach toward 2-substituted dihydropyridones that is an alternative to the tandem Mannich–Michael reaction with Danishefsky's diene uses 4-methoxypyridine or 4-hydroxypyridine attached to polystyrene-hydroxymethyl chloroformate resin [27, 28]. The activated resin reacts with 4-methoxypyridine giving the acylpyridinium species, which is now accessible to a nucleophilic attack by the Grignard reagent, a methodology which has been named "resin activation/capture approach" (REACAP) by the authors. The reaction can principally be considered as an addition of a Grignard reagent to an activated C=N double bond. It leads to the direct formation of the resin-bound dihydropyridinone, which can be cleaved off with sodium hydroxide in methanol. The same reaction type was applied after linking 4-hydroxypyridine to hydroxymethyl polystyrene under Mitsunobu conditions. Activation of the pyridine by acylation with acid chlorides allowed the addition of Grignard reagents. The desired dihydropyridone was formed during the cleavage with TFA from the resin (Scheme 17.16).

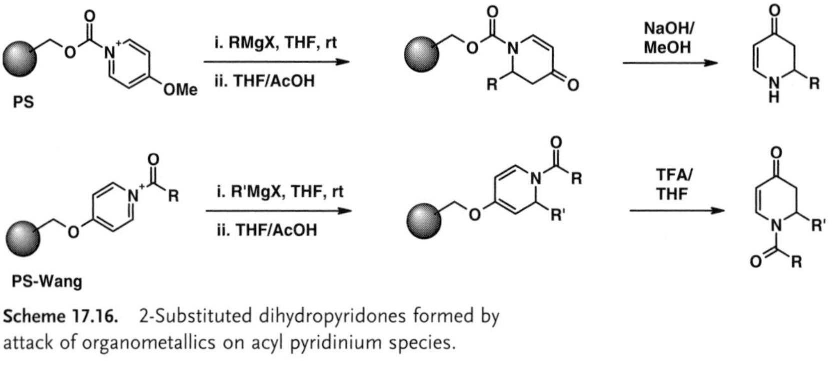

Scheme 17.16. 2-Substituted dihydropyridones formed by attack of organometallics on acyl pyridinium species.

Any unreacted acylpyridinium intermediate remained on the solid support, forming polymer-bound pyridine during the cleavage step. The carboxylic acid thus released as a byproduct was removed by a scavenger resin.

The reactivity of terminal alkynes in the presence of catalytic amounts of copper(I) chloride has been demonstrated by reaction of aldimines in Mannich reactions [29]. The solid-phase three-component synthesis of α-substituted propargylic amines is an application of this reaction type carried out with resin-bound alkynes, with secondary amines, or with aldehydes in hot dioxane (Scheme 17.17). In very recent examples, the milder and more flexible Petasis variant of the solid-supported Mannich reaction has been presented [30–32]. Immobilized secondary amines have been reacted with aldehydes and boronic acids to give the Mannich products in a very simple manner.

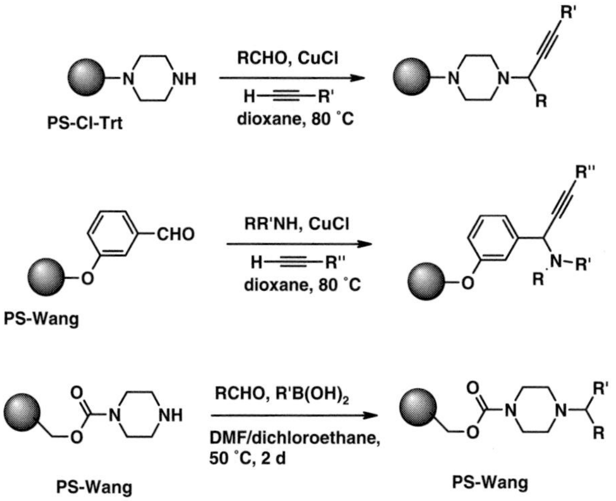

Scheme 17.17. Solid-supported Mannich reactions with metalated species. DMF, dimethylformamide.

17.3.4
Reaction with Enolates

α-Substituted ketones can be easily prepared by reaction of polymer-bound TMS enolethers with triflates after the addition of methyl lithium, as demonstrated during the synthesis of prostaglandin E$_2$ methyl esters on noncrosslinked polystyrene (Scheme 17.18) [33]. Presumably, the organonucleophile cleaves the relatively stable O–TMS bond, forming the highly reactive lithium enolate and tetramethylsilane. This reaction has not yet been applied to the synthesis of a series of compounds or libraries.

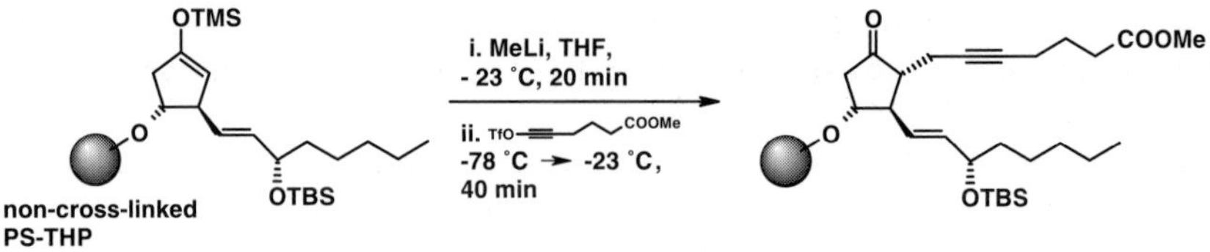

Scheme 17.18. α-Alkylated ketones by reaction of TMS enolates with alkynyl lithium reagent.

17.4
Conjugate Addition to α,β-Unsaturated Carbonyls and Related Systems

Because of their particularly soft character, cuprates are the reagents of choice for conjugate additions, but only a few examples of this reaction type in solid-phase syntheses or combinatorial chemistry have yet been described. An application has been demonstrated by the 1,4-addition to resin-bound dihydropyridones (obtained from a Grignard reaction, see above) [27]. Cuprates have been prepared in a classical manner by the reaction of copper(I) iodide and Grignard reagents in the presence of borontrifluoride etherate (Scheme 17.19).

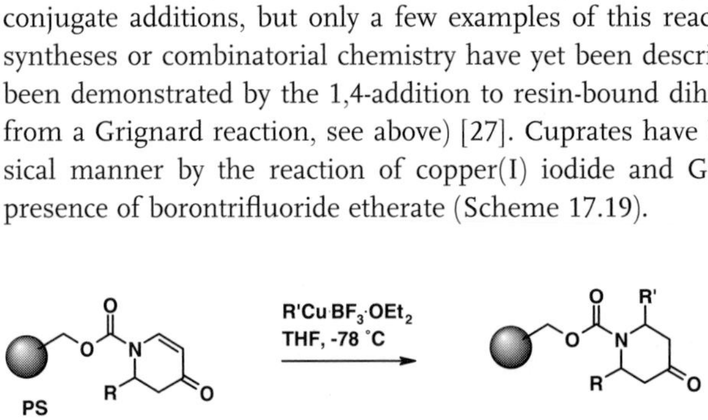

Scheme 17.19. Synthesis of 3-substituted piperinones by reaction of cuprates with dihydropyridones.

Alkenyl-substituted ketones can be prepared by the reaction of α,β-unsaturated ketones, as demonstrated during the synthesis of prostaglandins either by use of solid or soluble supports (Scheme 17.20) [33, 34]. The vinyl cuprates employed in

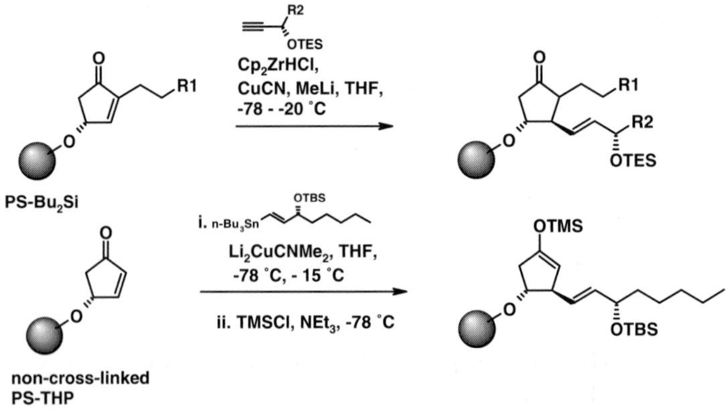

Scheme 17.20. 1,4-Additions of unsaturated metalated species to cyclopentenones.

this reaction can be prepared *in situ* from readily accessible terminal alkynes by hydrozirconation or stannylation followed by transmetalation using Lipshutz's or Babiak's protocol [35, 36].

The synthesis of β-substituted esters can be performed in a similar manner. High enantioselectivities have been obtained in addition reactions of Me_2CuLi or $Ph_2CuMgBr$ to chiral unsaturated esters immobilized on Wang Resin [37]. The resulting trisubstituted hydroxy ester cyclizes upon cleavage from the resin forming δ-lactones in excellent yields and enantiopurities (Scheme 17.21).

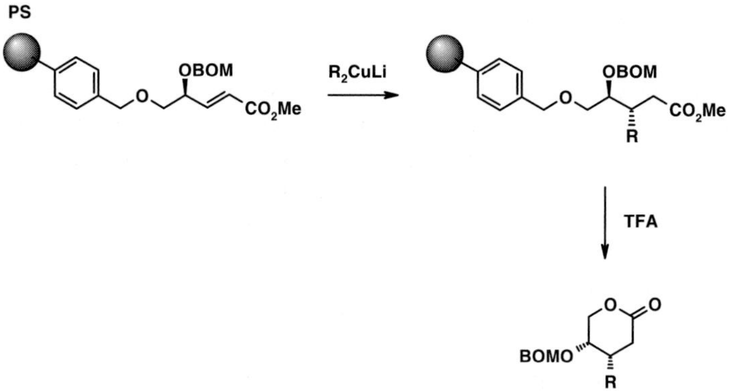

Scheme 17.21. 1,4-Addition of cuprates on immobilized unsaturated esters.

17.5
Nucleophilic Substitutions

N-Nosylated alkenylaziridines are suitable substrates for S_N2' substitutions, where the nosylate serves as a leaving group, expelled initially by the attack of cuprates on the double bond [38]. Typical products are (E)-alkene amino acids, which are useful isosteric nonhydrolyzable peptide bond replacements. In selected examples, copper reagents prepared from organolithium or Grignard reagents were used for the preparation of these α-substituted dipeptide isosteres on Wang resin (Scheme 17.22).

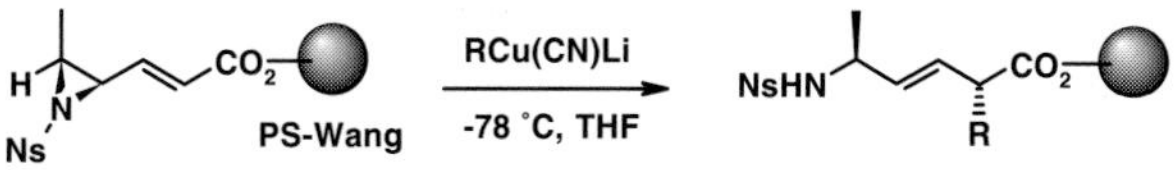

Scheme 17.22. Addition of cuprates to vinyl nosylate aziridines.

Grignard reagents can also serve as agents for functionalizing cleavage products synthesized on the benzotriazole linker [39]. In an extremely short reaction sequence, consisting of resin capture of aldimines in a Mannich reaction and subsequent cleavage by nucleophilic substitution with Grignard reagents in refluxing toluene, primary and secondary amines have been obtained in variable yields and excellent purities. As in all similar cleavage reactions, an aqueous extraction step is necessary (Scheme 17.23).

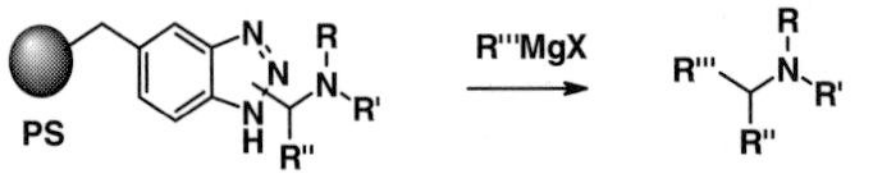

Scheme 17.23. Cleavage of highly substituted amines from benzotriazole linker with Grignard reagents.

Finally, the application of cuprates as cleavage reagents has been demonstrated in related solid-phase reactions [8]. Trisubstituted olefins were synthesized by the reaction of terminal allyl species immobilized on a phenylsulfone linker with cuprates prepared *in situ* from a range of Grignard reagents and copper(I) iodide (Scheme 17.24). Purification by column chromatography of the products thus obtained was found to be necessary.

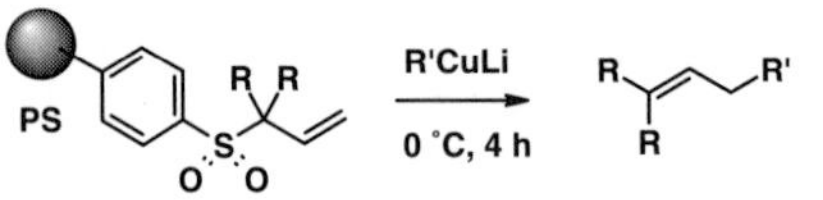

Scheme 17.24. Trisubstituted olefins from additions of cuprates on vinyl sulfones.

17.6
Reactions on Carboxylic Acid Derivatives and Related Systems

From a mechanistic point of view, reaction of Grignard compounds with carboxylic acid derivatives is quite similar to the 1,2-additions mentioned previously. The initial step is again the nucleophilic attack on the C=O double bond, leading to a tetrahedral intermediate. The subsequent restitution of the carbonyl group is accompanied by the simultaneous ejection of a leaving group.

17.6.1
Reaction with Esters

The solid-phase synthesis of α,α-dialkyl-substituted alcohols has been accomplished by the reaction of resin-bound esters with Grignard reagents. After two successive additions, tertiary alcohols are obtained. In contrast to the previously mentioned additions to ketones, only symmetrically disubstituted products are accessible. This type of reaction has been carried out with methyl-2-hydroxy-3-phenyl-propionate immobilized on Wang resin by an ether linkage giving the corresponding dimethyl alcohol [40]. It has also been used for the preparation of a small library of N-alkyl-2-pyrrolidine-dialkyl-methanol ligands as potential catalysts in diethylzinc additions from hydroxy proline, attached by the THP linker onto solid support [41] (Scheme 17.25).

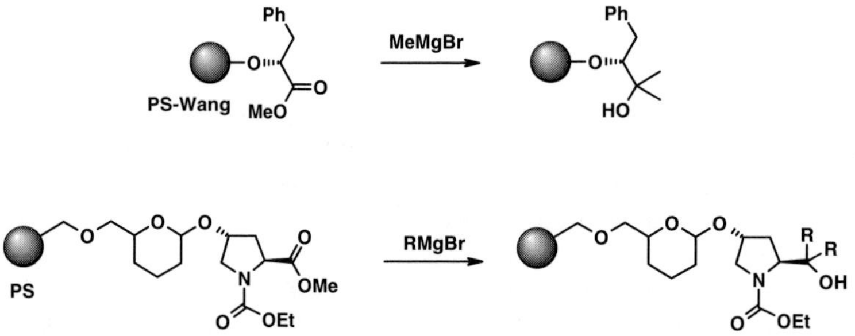

Scheme 17.25. Formation of α,α-distributed alcohols by the addition of Grignard reagents to resin-bound esters.

In many examples, Grignard reagents have been applied for the release of products from the solid support. During the solid-phase synthesis of a PDE4 inhibitor, MeMgBr was used to cleave the ester linkage leading to a dimethylphenyl alcohol after treatment with ammonium chloride in ethyl acetate (Scheme 17.26) [42].

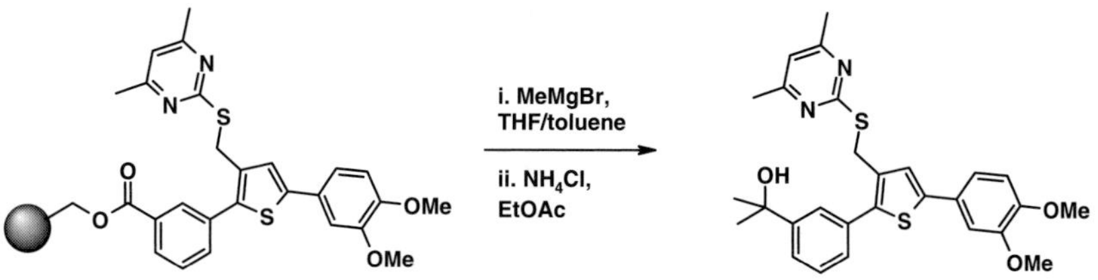

Scheme 17.26. Tertiary alcohols by cleavage of ester linkage with Grignard species.

17.6.2
Reactions with Weinreb Amides and Related Systems

One of the most popular transformations with main group organometallics on solid supports is the reaction of Weinreb amides with kryptobases. These N-methoxy-N-methylamides react with Grignard species to give ketones and with hydrides to give aldehydes, which contain desirable functional groups either for further diversification (reductive amination) or as recognizing moieties in libraries of serine/cysteine protease inhibitors. Little or no overalkylation occurs, presumably because of the formation of a chelate of the metal ion with the carbonyl oxygen and the N-methoxy oxygen. This stable intermediate decomposes readily to give a carbonyl group upon treatment with acid (Scheme 17.27).

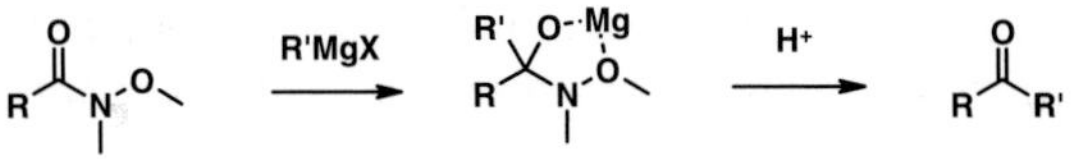

Scheme 17.27. Formation and breakdown of the tetrahedral bidentate during reaction of Grignard reagents with Weinreb amides.

For the synthesis of ketone or aldehyde libraries, the use of Weinreb amides as anchoring groups is preferable. Although additional efforts during the work-up are necessary, this strategy is advantageous because the reactive keto group is released only at the stage of the cleavage step. In this context, the N-alkyoxyamide function of β-alanin Weinreb amide on aminomethyl polystyrene has been used as a linker group [43]. Since preformation of the linker conjugate with the substrate has to be carried out in solution phase, this methodology appears to be somewhat laborious (Scheme 17.28).

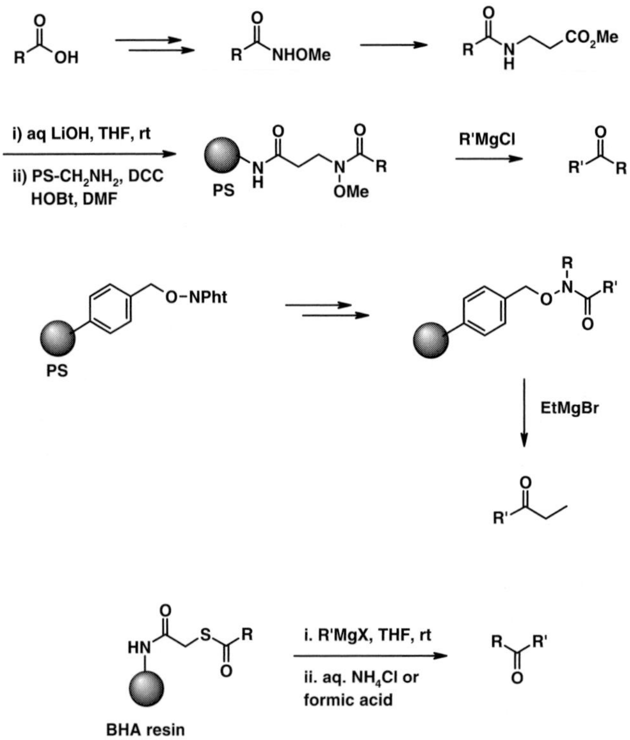

Scheme 17.28. Syntheses of ketones by cleavage of Weinreb-type linkers.

More conveniently, solid-supported Wang O-hydroxyl-N-alkyl amine, obtained by reaction of N-hydroxyphthalimide with Wang resin and subsequent cleavage of the phthalimide with methylamine, has been used in the immobilization of substrates by a Weinreb amide linker [44]. The cleavage was carried out using EtMgBr to provide ethyl ketones. Another successful method for the preparation of ketone libraries on solid phase uses mercaptoacetylamide on polystyrene resin, which acts in the same way as a Weinreb amide [45]. The starting carboxylic acid building blocks were coupled as thioamides onto this linker. After treatment with Grignard reagents, analytically pure products were released from the resin.

The combination of a Weinreb amide and an isonitrile functionality in one single synthetic building block allows the preparation of a dipeptide methyl ketone library by application of methyl Grignard reagent after the Ugi 4CC reaction (Scheme 17.29). A 96-member library around a known anticonvulsant structure has been synthesized according to this strategy [46].

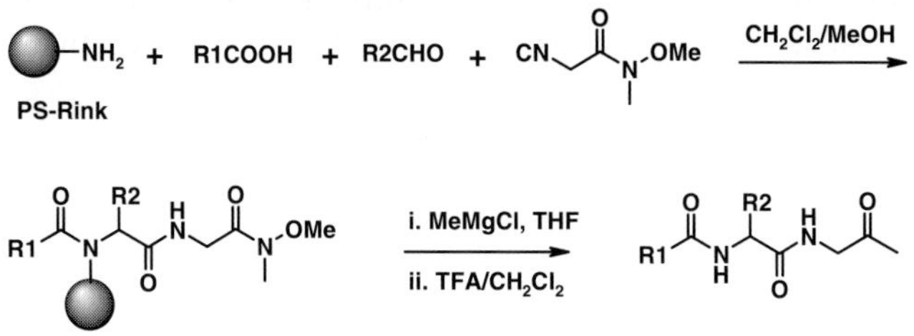

Scheme 17.29. Synthesis of dipeptide alkyl ketones.

Another solid-phase protocol using the Weinreb methodology leads to α-hydroxy-ketones, which have been shown to be efficient inhibitors of the aspartate protease renin [47]. Since the hydroxy group is the recognizing motif within this enzyme family, fluorenylmethoxycarbonyl (Fmoc)-hydroxy-β-amino acids are suitable templates for this purpose, preferably attached via the hydroxy group to the solid support. The immobilized amino acids have been converted into ketones by formation of the Weinreb amide with HCl·NH(OMe)Me and dropwise addition of Grignard reagent in a one-pot procedure (Scheme 17.30). It is noteworthy that an acetic acid/dichloromethane washing step is of crucial importance, presumably because magnesium salts from the bisdentate complex have to be removed prior to cleavage from the polystyrene resin. This method worked very well in the case of primary unhindered Grignard reagents, while attempts with organolithium species resulted in complex product mixtures.

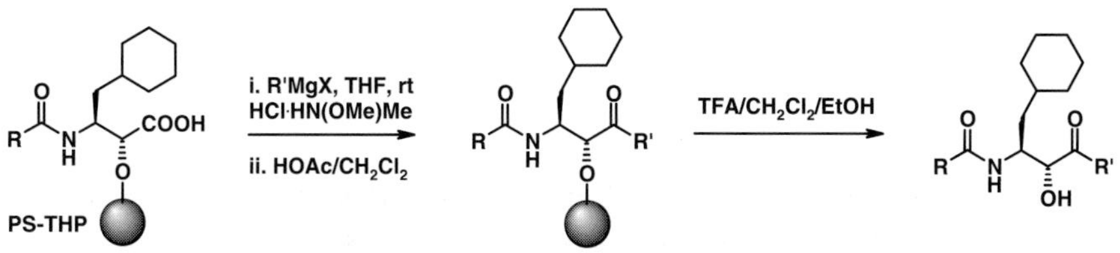

Scheme 17.30. Solid-phase synthesis of an α-hydroxy ketone library.

A highly sophisticated application of the Weinreb amide transformation was part of the synthesis of aspartyl protease inhibitor libraries targeting cathepsin D [48, 49]. 2,3-Dihydroxy propionic acid served as a suitable template, which was attached to Wang resin via its α-hydroxy group, as in the previous example. The carboxylic acid had to be transformed into a ketone for further chemical modifications. The standard Weinreb methodology that was chosen initially to accomplish this task led to unwanted side-products for two reasons. First, overalkylation occurred to a remarkable degree since the excess Grignard reagent was not completely destroyed, while the breakdown of the tetrahedral intermediate had already delivered the free ketone susceptible to further attack. Second, a competitive N–O bond cleavage leads to N-methyl products in substantial amounts, depending on the nature of the Grignard reagent used. While the problem of overalkylation has been countered by addition of acetone to the reaction mixture to quench excess Grignard, the problem of side-products occurring as a result of N–O bond cleavage has remained unsolved. Therefore, the N-methoxy-N-methyl amide was replaced by a simple pyrrolidine amide. The Grignard addition to this substrate provided the desired ketones with no detectable overalkylation product (Scheme 17.31). Herein, the α-alkoxy group serves as a Lewis base to stabilize the tetrahedral intermediate instead of the N-methoxy group of the standard Weinreb amide. For the library synthesis, a diverse set of commercially unavailable Grignard reagents has been prepared from activated magnesium turnings (alkyl Grignards) or Mg(anthracene)(THF)$_3$ complex (activated benzyl-type Grignards). Using the aforementioned chemistry, a library of more than 1000 compounds has been synthesized, delivering several highly potent cathepsin D inhibitors. Later, the same concept was applied to the synthesis of libraries targeting plasmepsin II, which also yielded nanomolar inhibitors [50].

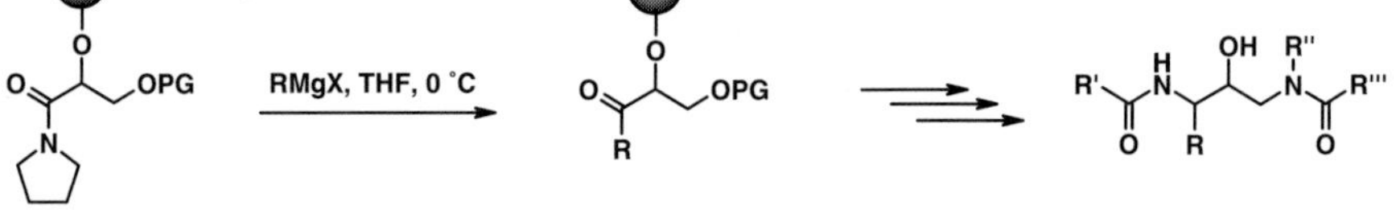

Scheme 17.31. Side-reactions of standard Weinreb amides and synthesis of polymer-bound ketones from α-alkoxypyrrolidine amides.

The solution-phase preparation of ketones from Weinreb amides is also feasible using polymer-bound sulfonylhydrazide as a resin-capture reagent to isolate the desired products [51]. In a short reaction sequence, this strategy was applied to synthesize a small library of 1,2,3-thiadiazoles, released by treating the resin-bound hydrazones with SOCl$_2$. The tetrahedral intermediates from the Grignard reaction were decomposed by addition of macroporous polystyrene-sulfonic acid resin (MP-TsOH) and the ketones thus produced were captured by gentle heating with PS-TsNHNH$_2$ in acetic acid/THF (Scheme 17.32).

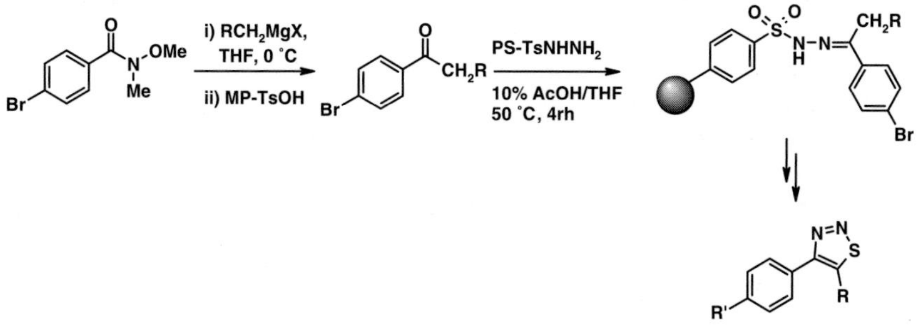

Scheme 17.32. Solution phase synthesis of ketones from Weinreb amides during preparation of 1,2,3-thiadiazoles.

17.7
Aminolysis of Esters

Aminolysis of esters has been applied in combinatorial chemistry in order to generate the overwhelming diversity of amides in a single step. Organoaluminums are the reagents of choice for inducing this reaction owing to their strong Lewis acidity. This protocol has mostly been used for the derivatizing cleavage from hydroxyethyl polystyrene resin. Published examples include the synthesis of tetrahydrochinoline amides by Pictet–Spengler reaction [52] and the formation of 3-hydroxypropionamides utilizing the Baylis–Hillman reaction [53] (Scheme 17.33).

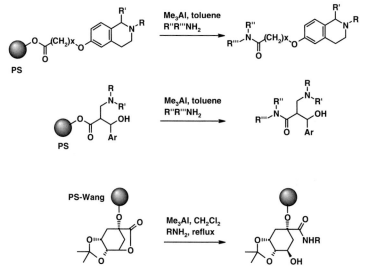

Scheme 17.33. Aminolysis of ester bonds with organoaluminium reagents on solid support.

The filtrates must be carefully quenched and additional work-up was found to be necessary. This reaction type has also been used for the simultaneous lactone opening and amidation on solid support [37].

References

1 T. M. Fyles, C. C. Leznoff, *Can. J. Chem.* **1976**, *54*, 935.

2 F. Camps, J. Castells, M. J. Ferrando, J. Font, *Tetrahedron Lett.* **1971**, 1713–1714.

3 W. Heitz, R. Michels, *Makromol. Chem.* **1971**, *148*, 9.

4 M. J. Farall, J. M. Fréchet, *J. Org. Chem.* **1976**, *41*, 3877–3882.

5 F. G. Bordwell, B. M. Pitt, *J. Am. Chem. Soc.* **1955**, *77*, 572–577.

6 G. A. Crosby, N. M. Weinshenker, H.-S. Uh, *J. Am. Chem. Soc.* **1975**, *97*, 2232–2235.

7 T. Ruhland, K. Andersen, H. Pedersen, *J. Org. Chem.* **1998**, *63*, 9204–9211.

8 C. Halm, J. Evarts, M. J. Kurth, *Tetrahedron Lett.* **1997**, *38*, 7709–7712.

9 S. Havez, M. Begtrup, K. Andersen, T. Ruhland, *J. Org. Chem.* **1998**, *63*, 7418–7420.

10 Z. Li, A. Ganesan, *Synlett* **1998**, 405–406.

11 P. A. Tempest, R. W. Armstrong, *J. Am. Chem. Soc.* **1997**, *119*, 7607–7608.

12 L. Boymond, M. Rottländer, G. Cahiez, P. Knochel, *Angew. Chem. Int. Ed. Engl.* **1998**, *37*, 1701–1703.

13 M. Abarbi, J. Thibonnet, L. Bérillon, F. Dehmel, M. Rottländer, P. Knochel, *J. Org. Chem.* **2000**, *65*, 4618–4634.

14 A. Routledge, H. T. Stock, S. L. Flitsch, N. J. Turner, *Tetrahedron Lett.* **1997**, *38*, 8287–8290.

15 K. C. Nicolaou, N. Winssinger, J. Pastor, F. Murphy, *Angew. Chem. Int. Ed. Engl.* **1998**, *37*, 2534–2537.

16 F. Gosselin, J. v. Betsbrugge, M. Hatam, W. D. Lubell, *J. Org. Chem.* **1999**, *64*, 2486–2493.

17 M. Rottländer, P. Knochel, *J. Comb. Chem.* **1999**, *1*, 181–183.

18 A. Routledge, C. Abell, S. Balasubramanian, *Tetrahedron Lett.* **1997**, *38*, 1227–1230.

19 P. S. Furth, M. S. Reitman, R. Gentles, A. F. Cook, *Tetrahedron Lett.* **1997**, *38*, 6643–6646.

20 C. L. Cavallaro, T. Herpin, B. F. McGuiness, Y. C. Shimshock, R. E. Dolle, *Tetrahedron Lett.* **1999**, *40*, 2711–2714.

21 R. Léger, R. Yen, M. W. She, V. J. Lee, S. C. Hecker, *Tetrahedron Lett.* **1998**, *39*, 4171–4174.

22 M. E. Fraley, R. S. Rubino, *Tetrahedron Lett.* **1997**, *38*, 3365–3368.

23 P. A. Tempest, R. W. Armstrong, *J. Am. Chem. Soc.* **1997**, *119*, 7607–7608.

24 C. Chen, B. Munoz, *Tetrahedron Lett.* **1998**, *39*, 3401–3404.

25 A. R. Katritzky, L. Xie, G. Zhang, *Tetrahedron Lett.* **1997**, *38*, 7011–7014.

26 B. Chenera, J. A. Finkelstein, D. F. Veber, *J. Am. Chem. Soc.* **1995**, *117*, 11999–12000.

27 C. Chen, I. A. McDonald, B. Munoz, *Tetrahedron Lett.* **1998**, *39*, 217–220.

28 C. Chen, B. Munoz, *Tetrahedron Lett.* **1998**, *39*, 6781–6784.

29 J. J. McNally, M. A. Youngman, S. L. Dax, *Tetrahedron Lett.* **1998**, *39*, 967–970.

30 N. Schlienger, M. R. Bryce, T. K. Hansen, *Tetrahedron* **2000**, *56*, 10023–10030.

31 S. R. Klopfenstein, J. J. Chen, A. Golebiowski, M. Li, S. X. Peng, X. Shao, *Tetrahedron Lett.* **2000**, *41*, 4835–4839.

32 A. Golebiowski, S. R. Klopfenstein, J. J. Chen, X. Shao, *Tetrahedron Lett.* **2000**, *41*, 4841–4844.

33 S. Chen, K. D. Janda, *J. Am. Chem. Soc.* **1997**, *119*, 8724–8725.

34 L. A. Thompson, F. L. Moore, Y.-C. Moon, J. A. Ellman, *J. Org. Chem.* **1998**, *63*, 2066–2067.

35 B. H. Lipshutz, E. L. Ellsworth, *J. Am. Chem. Soc.* **1990**, *112*, 7440.

36 K. A. Babiak, J. R. Behling, J. H. Dygos, K. T. McLaughlin, J. S. Ng, V. J. Kalish, S. W. Kramer, R. L. Shone, *J. Am. Chem. Soc.* **1990**, *112*, 7441.

37 S. Hanessian, J. Ma, W. Wang, *Tetrahedron Lett.* **1999**, 4631–4634.

38 P. Wipf, T. C. Henninger, *J. Org. Chem.* **1997**, *62*, 1586–1587.

39 K. Schiemann, H. D. H. Showalter, *J. Org. Chem.* **1999**, *64*, 4972–4975.

40 S. Hanessian, F. Xie, *Tetrahedron Lett.* **1998**, *39*, 737–740.

41 G. Liu, J. A. Ellman, *J. Org. Chem.* **1995**, *60*, 7712–7713.

42 Y. Han, A. Giroux, C. Lépine, F. Laliberté, Z. Huang, H. Perrier, C. I. Bayly, R. N. Young, *Tetrahedron* **1999**, *55*, 11669–11685.

43 T. Q. Dinh, R. W. Armstrong, *Tetrahedron Lett.* **1996**, *37*, 1161–1164.

44 J. M. Salvino, H. Mervic, H. J. Mason, T. Kiesow, D. Teager, J. Airey, R. Labudiniere, *J. Comb. Chem.* **1999**, *1*, 134–139.

45 I. Vlattas, J. Dellureficio, R. Dunn, I. I. Sytwu, J. Stanton, *Tetrahedron Lett.* **1997**, *38*, 7321–7324.

46 S. W. Kim, S. M. Bauer, R. W. Armstrong, *Tetrahedron Lett.* **1998**, *39*, 6993–6996.

47 O. B. Wallace, *Tetrahedron Lett.* **1997**, 4939–4942.

48 E. K. Kick, D. C. Roe, A. G. Skillman, G. Liu, T. J. A. Ewing, Y. Sun, I. D. Kuntz, J. A. Ellman, *Chem. Biol.* **1997**, *4*, 297–307.

49 C. E. Lee, E. K. Kick, J. A. Ellman, *J. Am. Chem. Soc.* **1998**, *120*, 9735–9747.

50 T. S. Haque, A. G. Skillman, C. E. Lee, H. Habashita, I. Y. Gluzman, T. J. A. Ewing, D. E. Goldberg, I. D. Kuntz, J. A. Ellman, *J. Med. Chem.* **1999**, *42*, 1428–1440.

51 Y. Hu, S. Baudart, J. A. Porco, Jr. *J. Org. Chem.* **1999**, *64*, 1049–1051.

52 K. Rölfing, M. Thiel, H. Künzer, *Synlett* **1996**, 1036–1038.

53 O. Prien, K. Rölfing, M. Thiel, H. Künzer, *Synlett* **1997**, 325–326.

18
Enolates and Related Species in Combinatorial and Solid-phase Synthesis

Jochen Krüger

18.1
Introduction

The formation of carbon–carbon bonds utilizing enolates and related species is among the most prominent and useful processes in organic synthesis. Accordingly, the extension of the established solution-phase repertoire to combinatorial synthesis onto liquid phase and solid phase represents both an ambitious as well as a rewarding goal.

Comprehensive compilations of the latest achievements in solid- and liquid-phase enolate chemistry have been published recently [1].

In this chapter, we give a detailed account of the use of enolates and related species in combinatorial chemistry. In addition to the available literature, we provide an extensive comparison of solution- and solid-phase strategies in each section. Moreover, we focus on the practical aspects and assess the suitability of a given method for automation and library synthesis. We also discuss the scope and limitations of each method, as well as its convenience for practical use.

18.2
Aldol Reactions

18.2.1
General Aspects

The aldol reaction between an aldehyde and an enolate or an enolate equivalent delivers β-hydroxy carbonyls or, after dehydration, α,β-unsaturated carbonyls. Both structural motifs represent valuable functionalities in natural product and heterocycle synthesis. Thus, it seems conceivable that well-established methods for classical solution-phase aldol chemistry could be adapted to the requirements of combinatorial chemistry. However, the emphasis in this field was clearly placed on solid-phase strategies.

The direct condensation between an aldehyde and a C–H acidic ketone was successfully employed in solid-phase synthesis and library construction (see Section 18.2.2). In particular, Na and Zn enolates have played a dominant role in this field. Although the basic reaction conditions entail limitations arising from the base sensitivity of some polymeric linkers or starting materials, this method presents a powerful tool to perform aldol reactions on solid support.

For stereoselective aldol protocols, we discuss recent progress in the application of boron enolates in solid-phase chemistry (see Section 18.2.3). In general, the efficiency of a boron aldol process on solid support is highly dependent on the nature of the polymer, the polymeric linker, and the reactivity of the enolate. Thus, for a given synthetic problem in solid-phase chemistry, the boron aldol protocol has to be adjusted to the requirements of the desired transformation. To the best of our knowledge, multicomponent library synthesis utilizing stereoselective boron aldol reactions has not yet been reported.

In Section 18.2.4 we detail the latest achievements in the field of liquid- and solid-phase Mukaiyama aldol chemistry. This mild version of the aldol reaction can serve as an attractive alternative for the direct aldol condensation employing enolizable carbonyl compounds.

18.2.2
Li, Na, K, and Zn Enolates in Aldol Reactions

The synthesis of β-hydroxy carbonyls via a crossed aldol reaction between an aldehyde and an enolizable carbonyl compound has been adopted to solid phase, whereas liquid-phase protocols in this field are rather scarce.

Ruhland and Künzer described the formation of aldol intermediates (2) upon reaction of resin-bound aldehyde 1 with acetophenone in the presence of K_2CO_3 at elevated temperature (Scheme 18.1) [2]. Six aromatic and heteroaromatic methylketones were reported to be suitable substrates for this transformation, however, yields and purities were not given. The intermediates (2) were used in the synthesis of a diverse quinoline library of 12 products. Thereby, the effectiveness of the aldol transformation as a key step in this sequence was demonstrated.

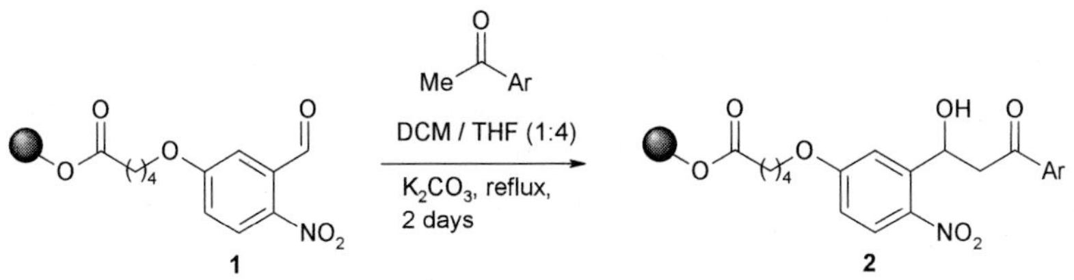

Scheme 18.1. Aldol reactions on solid phase employing K enolates.

A major obstacle in aldol chemistry on solid support is incomplete or sluggish reactions due to variable extents of retro-aldol processes. In this regard, the use of

Zn enolates has proved to be advantageous, since the amounts of retro-aldol products being formed during the process were minimized. As Kurth et al. [3] demonstrated, immobilized Zn enolates react smoothly with aromatic aldehydes and give the desired β-hydroxy ketones in good yields and purities. Accordingly, ester **3** loaded on Merrifield resin was deprotonated at −78 °C using an excess of lithium diisopropylamide (LDA) followed by the addition of anhydrous $ZnCl_2$ to trigger the transmetalation to the corresponding Zn enolate. Finally, 27 aromatic aldehydes were added and the aldol reaction proceeded smoothly at 0 °C within 30 min (Scheme 18.2) [3]. Cleavage from the resin was accomplished by reduction of the ester linkage with diisobutylaluminium-hydride (DIBAL-H) to yield the crude diols (**4**) as mixtures of syn and anti products which were purified by preparative thin layer chromatography (TLC). In the case of *p*-methoxybenzaldehyde as electrophile in this reaction, an overall isolated yield of 26% was reported.

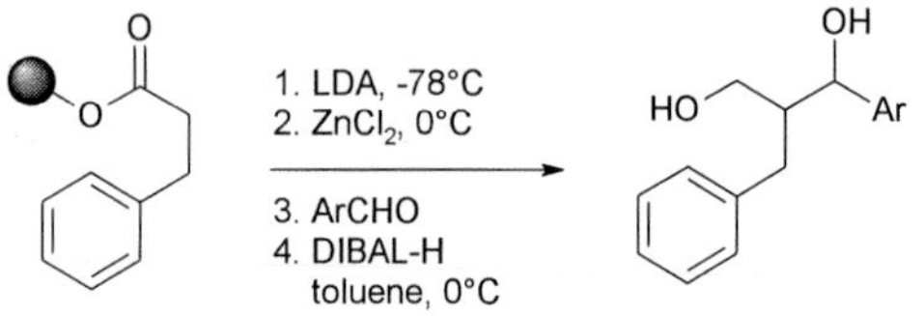

Scheme 18.2. Aldol reactions on solid phase employing Zn enolates.

A related strategy was applied by Nicolaou et al. for the synthesis of an epothilone library [4]. A key step in this route leading to diversity with concomitant construction of the C–C skeleton was an aldol reaction mediated by *in situ*-generated Zn enolates (see Scheme 21.14). Both contributions demonstrate that even more sophisticated reaction conditions including inert gas techniques and low-temperature protocols are routinely applied in solid-phase synthesis today. The only drawback in terms of manipulative convenience arose from the fact that the products were purified by preparative TLC after cleavage. This aspect might be difficult to realize in an automated library synthesis process.

The synthesis of α,β-unsaturated ketones using a crossed aldol strategy on solid support is well established and has successfully been applied to solid-phase organic synthesis. Either the aldehyde moiety or the C–H acidic ketone can be linked to the polymer via an ether or an amide linkage, whereas linkage via an ester proved to be incompatible with the strongly basic reaction conditions.

In general, a solution of NaOMe as a solution in methanol is routinely used as a standard base to promote the desired condensation event. A typical example of this reaction is illustrated in Scheme 18.3. Here, the immobilized acetophenone (**5**) was treated with 12 equiv. of NaOMe (0.5 M solution in MeOH) and 12 equiv. of aromatic aldehydes at room temperature to yield enones (**6**) (five examples) [5]. The reaction was conducted in tetrahydrofuran (THF) to effect sufficient swelling of the polymer. Alternatively, trimethylorthoformate (TMOF), which facilitates the dehydration step, was used as the solvent instead [6]. This type of aldol condensa-

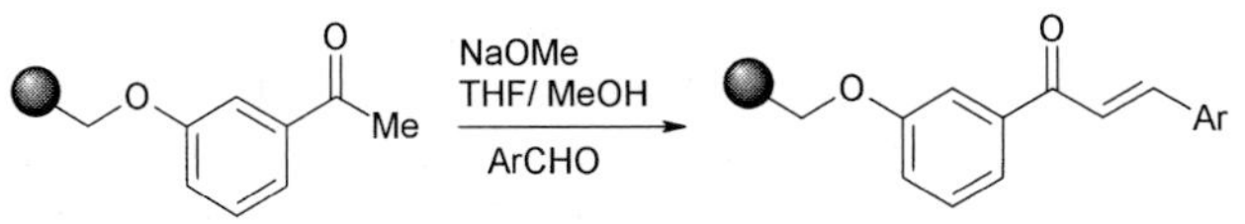

Scheme 18.3. Aldol condensations employing Na enolates.

tion of ketones has also been carried out employing Li enolates which were generated *in situ* using an excess of LiOH·H$_2$O (20 equiv.) in dioxane [7] or dimethoxyethane (DME) [8] as solvent.

In conclusion, reliable aldol protocols for the solid-phase synthesis of α,β-unsaturated ketones are available; however, these processes are primarily restricted to the condensation of methyl-arylketones with aromatic aldehydes, thus limiting the options for diversification in library synthesis.

18.2.3
Boron Enolates in Aldol Reactions

The introduction of boron enolates has provided a significant impetus in aldol chemistry, especially with regard to stereoselective protocols. Therefore, the adaptation of solution-phase boron aldol strategies to solid-phase chemistry has been an intense field of investigation.

Gennari et al. reported an enantioselective aldol addition on solid support mediated by a chiral boron enolate (Scheme 18.4) [9]. Accordingly, 4-(hydroxymethyl)-benzaldehyde was bound to trityl resin and the resulting polymer (**7**) was treated three times with an excess (~ 3 equiv.) of chiral boron enolate (**8**) in dichloromethane at −78 °C. The resulting product was cleaved from the resin and purified by chromatography to yield the corresponding β-hydroxythiolester (**9**) in 60% yield with 88% ee.

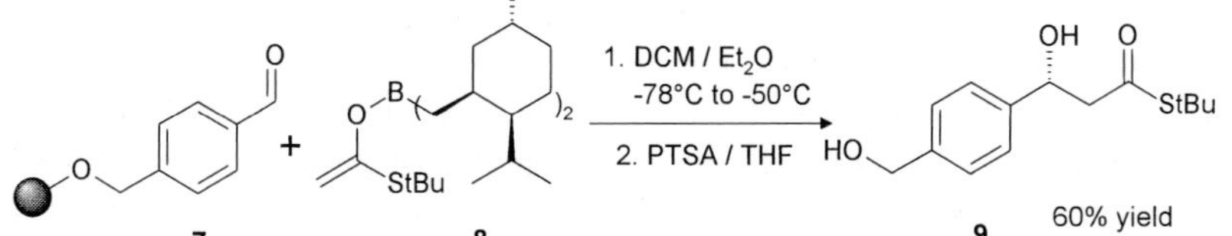

Scheme 18.4. Enantioselective aldol reactions using chiral B enolates.

Paterson and coworkers demonstrated a powerful methodology that permits the construction of polyketide libraries employing stereoselective propionate additions to aldehydes (Scheme 18.5) [9, 10]. Thus, aldehyde **10** was reacted twice with an excess of preformed (*E*)-propionate (**11**) in an argon atmosphere at low temperature to yield the corresponding aldol products on the polymer (Scheme 18.5, path

A). The resulting target molecules were cleaved from the resin using HF-pyridine to yield the anti-configured diols (**12**) after flash chromatography with high levels of diastereoselectivity (five examples: 90–97% de) and good yields. Complementary to this, the syn-configured product (**14**) could be accessed using the (Z)-enolate (**13**) (Scheme 18.5. path B). To achieve complete conversion in this reaction, aldehyde **10** was successively incubated three times, each time with an excess of enolate **13**. The aldol product was cleaved from the resin and the resulting diol was isolated after chromatography with 99% yield with 95% de in favor of the syn configuration.

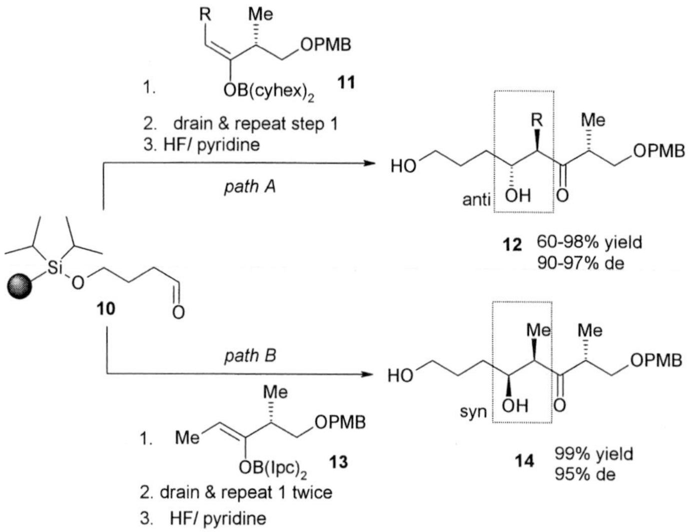

Scheme 18.5. Stereoselective synthesis of anti and syn aldols using chiral B enolates.

Eventually, this concept of stereoregulated aldol chemistry was extended to the construction of more complex polyketide sequences on solid support [10].

Evans oxazolidinone presumably represents the most prominent chiral auxiliary in aldol chemistry. Accordingly, two strategies have been described to transfer this concept to solid-phase chemistry. In the initial report by Reggelin and Brennig, aldehyde **16**, immobilized on Wang resin, was reacted with preformed boron enolate (**15**) (Scheme 18.6, path A) [11]. The aldol addition proceeded as expected and delivered the syn product **17** after cleavage from the resin (BCl$_3$) in 74% de (the authors could demonstrate that slight epimerization occurred under the harsh cleavage conditions; no yield was reported). However, further manipulations of the resin-bound aldol product (e.g. OH protection) proved difficult. These problems were attributed to the Wang linker present in **16**.

Accordingly, a more flexible Si-based linker was introduced for this reaction (Scheme 18.6, path B) [12]. With **10** in place, the aldol reaction as well as further manipulations proceeded without complications, even though lactone **18**, which

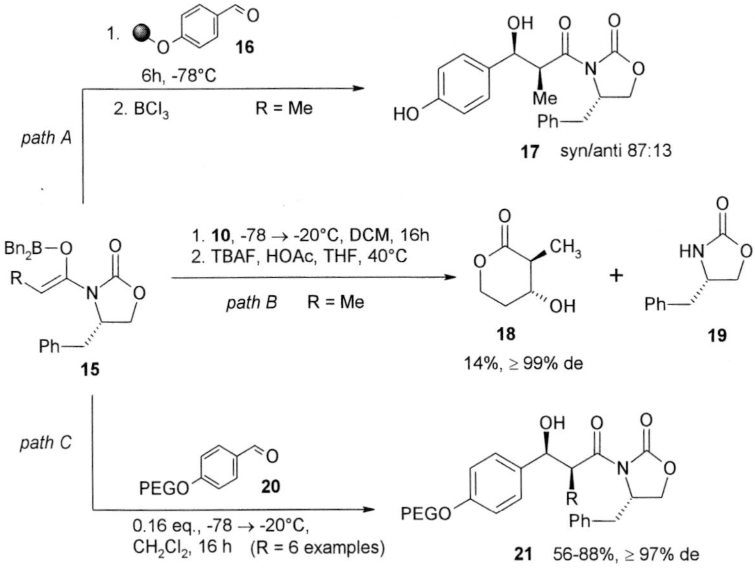

Scheme 18.6. Stereoselective addition of chiral B enolates to resin-bound aldehydes.

formed upon cleavage from the resin was obtained in poor yield. Moreover, isolation of the final product required chromatography since the crude reaction mixture was contaminated by oxazolidinone (**19**). Optimal results for the solid-supported Evans aldol protocol were elaborated for soluble polyethylene glycol (PEG)-bound aldehyde **20**. This scenario comes closest to solution-like conditions having a favorable impact on reaction kinetics (Scheme 18.6, path C) [13]. The polymeric aldol adducts were precipitated from solution to obtain **21** (six examples, 56–88% yield) and ^{1}H-NMR measurements revealed that each product was formed as a pure stereoisomer.

The three examples depicted in Scheme 18.6 clearly demonstrate that the choice of the polymeric linker plays a crucial role in this solid-supported aldol process. Furthermore, the scope of these reactions is limited to aldehydes bearing additional functional groups which allow the electrophile to be anchored to the polymer. After cleavage, these functional groups reside as integral parts in the target molecules.

An alternative strategy is based on resin-bound oxazolidinone **22**, which was synthesized on polymer support using L-tyrosine as the source of chirality [14]. As depicted in Scheme 18.7 the (Z)-configurated boron enolate was generated by reaction of **22** with an access of Bu$_2$BOTf in the presence of a tertiary amine such as triethylamine using methylene chloride as the solvent of choice.

The excess of Bu$_2$BOTf had to be drained away prior to the addition of the aldehyde since deleterious effects with respect to the stereochemical outcome were reported if additional Bu$_2$BOTf was present in the reaction mixture [15]. After completion of the aldol reaction, the resin was subjected to an aqueous LiOH solution

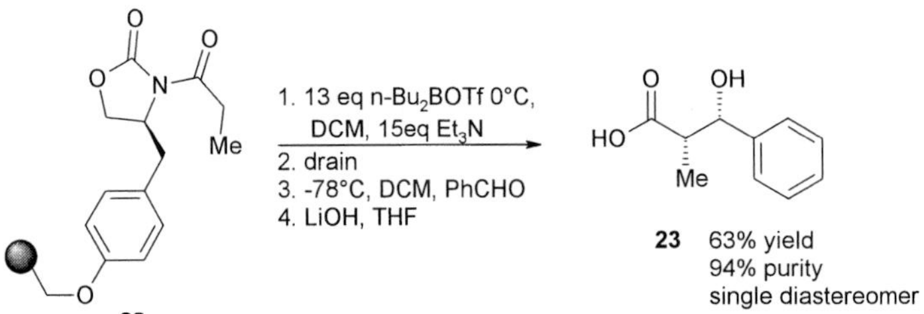

Scheme 18.7. Stereoselective addition of polymer-bound chiral B enolate **22** to benzaldehyde.

in THF to effect cleavage of carboxylic acid **23**. Alternatively, NaOMe was used to form the corresponding methyl ester instead. Finally, product **23** was obtained as the optically pure syn diastereomer with a chemical purity of 94%, as concluded from high-performance liquid chromatography (HPLC).

In summary, a number of solution-phase protocols for the addition of boron enolates to aldehydes have been successfully adapted to solid-phase synthesis. These reactions proceed with similar efficiency to the well-established solution-phase reactions. However, a relatively small number of examples have been reported to date, rendering an assessment of the scope of solid-supported boron aldol reactions exceedingly difficult. Moreover, product isolation is routinely achieved by aqueous work-up in combination with chromatographic methods. This aspect could certainly hamper the application of such a protocol for automated synthesis and the construction of multicomponent libraries still remains a challenging goal.

18.2.4
The Mukaiyama Aldol Reaction

The Mukaiyama protocol describes the Lewis acid-mediated reaction between an aldehyde and a silyl enol ether or ketene silyl acetal to yield aldol adducts. Much effort has been expended in adapting this reaction to the needs of combinatorial chemistry. The research activities in this area can be divided into two major categories: (1) the solution-phase approach employing immobilized catalysts and (2) the solid-phase strategy comprising soluble catalysts and polymer-bound aldehydes and silyl enol ethers.

18.2.4.1 Solution-phase Protocols Using Polymer-bound Reagents
As outlined in Table 18.1, a variety of polymeric catalysts have been reported to effect the Mukaiyama aldol reaction in solution. For instance, immobilized trityl cation **24** can function as a Lewis acid to activate dimethylacetals or benzaldehyde toward addition of a range of silyl enol ethers (Table 18.1, entries 1 and 2). The corresponding aldol products were isolated after aqueous work-up and preparative TLC in good yields.

Tab. 18.1. Polymer-supported reagents promoting the Mukaiyama reaction.

Entry	Silyl enol ether	Electrophile	Catalyst	Product	Conditions and comments	Reference
1	Me$_3$SiO–C(R2)=C(R3)R1	R4–CH(OMe)$_2$; R4 = aromatic and aliphatic	$\mathbf{24}$ (polymer-supported triphenylcarbenium ClO$_4^-$)	R1–C(O)–C(R2)(R3)–CH(R4)–OMe	5 examples; 72–95% yield; all products purified by TLC	16
2	R$_3$SiO–C(R2)=CH–R1	Ph–CHO	24	R1–C(O)–CH(R2)–CH(OSiR$_3$)–Ph	6 examples; 72–96% yield; products isolated by aqueous work-up and preparative TLC	16
3	cyclohexenyl–OSiMe$_3$	CH$_2$O	Yb-Amberlyst 15	2-(hydroxymethyl)cyclohexanone	Reaction carried out in THF/H$_2$O 4:1; 82% yield; product chromatographed	17
4	OSiMe$_3$–C(R2)=CH–R1	R3–CHO; R3 = aromatic and aliphatic	Al-Montmorillonite	R1–C(O)–CH(R2)–CH(OSiMe$_3$)–R3	11 examples; 68–97% yield; products isolated by filtration and Kugelrohr destillation	18
5	OSiMe$_3$–C(R2)=C(R3)–R1	R4–CHO; R4 = aromatic and aliphatic	Montmorillonite K10	R1–C(O)–C(R2)(R3)–CH(OH)–R4	15 examples; 62–87% yield; reactions carried out in water; all products chromatographed	19

Tab. 18.1. (continued)

Entry	Silyl enol ether	Electrophile	Catalyst	Product	Conditions and comments	Reference
6	OSiMe₃ / MeO / Me / Me	R1–CHO; R1 = aromatic and aliphatic	Montmorillonite K10	MeO / O / OH / R1 / Me Me	5 examples; 52–89% yield; water as solvent	19
7	OSiMe₃ / R1 / R2	R3–CHO; R3 = aromatic and aliphatic	Neutral Al₂O₃	R1 / O / OH / R2 / R3	13 examples; 70–90% yield; sonication necessary; reagents used neat; all products chromatographed	20
8	OSiMe₃ / R1–O / R3 / R2	R4–CHO; R4 = aromatic and aliphatic	Al₂O₃–ZnCl₂	R1–O / O / OH / R4 / R2 R3	19 examples; 70–90% yield; sonication essential; products isolated by aqueous work-up and chromatography	21
9	OSiMe₃ / RX / Me / Me ; X = O,S	Ph–CHO	**25** (Sc(OTf)₂ SO₃– on resin)	RX / O / OH / Ph / Me Me	1 example; reaction in water; 98% yield	22
10	OSiMe₃ / R1 / Me / Me	R3–CHO; R3 = aromatic	**26** (resin–CF₂–SO₂–O–Si–Me₃)	R1 / O / R2 / R2	10 examples; CH₂Cl₂ at −78 °C; 8–85% yield; pure products obtained by filtration	23

11	OSiMe₃ / EtO / Me Me	Ph–CHO	COOH NH–Tos / O / + BH₃	OH Ph / EtO O Me Me	24	1 example stoichiometric amount of BH₃ was used; at −78 °C: 28% yield; 90% ee; at −10 °C: 70% yield; 69% ee; best solvent: THF; product chromatographed

Entry 11: $OSiMe_3$, EtO, Me, Me; Ph–CHO; reagent: COOH, NH-Tos, + BH_3; product: OH, Ph, EtO, O, Me, Me; 24.

1 example stoichiometric amount of BH_3 was used; at −78 °C: 28% yield; 90% ee; at −10 °C: 70% yield; 69% ee; best solvent: THF; product chromatographed

Entry 12: $OSiMe_3$, BnO; Ph–CHO; reagent: polymer-bound BINOL-Ti(OR) complex, Br; product: OH, Ph, BnO, O; 25.

1 example 32% conversion after 36 h, 26% ee; 2 mol% polymer used; catalyst removed by precipitation from pentane product obtained by filtration

Entry 14: $OSiMe_3$, Ph; Ph–CHO; reagent: polymer-bound $Pd(OH_2)_2^{2+}$, $2\ BF_4^-$; product: OH, Ph, O, Ph; 26.

1 example; 5 mol% catalyst in DMF at rt; 94% yield; 74% ee

Commercially available ion-exchange resins such as amberlysts or montmorillonite were shown to promote the Mukaiyama reaction in aqueous solvent mixtures (Table 18.1, entries 3–6). The products were obtained after filtration and distillation or chromatography. Neutral Al_2O_3 or $ZnCl_2$-doped Al_2O_3 (Table 18.1, entries 7 and 8) have also been described as catalysts for the aldol addition, however, tedious reaction conditions, including sonication, render these methods less attractive for automation and library synthesis. A polymeric Sc(OTf)$_3$ (**25**) was introduced by Kobayashi et al. [27] which cleanly effected the synthesis of aldol adducts, especially in aqueous solution.

A major advantage of polymer-supported reagents consists in simplification of the work-up procedure. Under optimal conditions, the product isolation is reduced to a filtration and washing process, thus avoiding tedious chromatography. In this regard, Ley and coworkers reported the formation of α,β-unsaturated ketones via a Mukaiyama protocol (Table 18.1, entry 10). Accordingly, the reaction of aromatic aldehydes and silyl enol ethers was catalyzed by immobilized trimethylsilyltriflate (TMSOTf) (**26**) and the desired aldol products were isolated in excellent yields with high purities by mere filtration.

Finally, enantioselective protocols were elaborated for the Mukaiyama reaction (Table 18.1, entries 11–14). The authors describe the synthesis of polymeric Lewis acids and their application in asymmetric aldol reactions. In general, only benzaldehyde was used as a model system for these studies and moderate levels of enantioselectivities with variable yields were observed.

18.2.4.2 Solid-phase Protocols

Kobayashi et al. [27] reported the synthesis of immobilized thioketene silyl acetals and their use in Mukaiyama reactions mediated by rare earth metal salts. Thus, silylated propionate **27** was reacted with benzaldehyde in the presence of 20 mol% Sc(OTf)$_3$ to effect conversion to the corresponding aldol adduct (Scheme 18.8) [27]. The latter was cleaved from the resin upon reduction with LiBH$_4$ to yield diol **28**, which was isolated after aqueous work-up and preparative TLC in 82% yield. Moreover, heteroaromatic and aliphatic aldehydes as well as dimethylacetals work equally well as substrates for this sequence (in total, 12 examples). However, if unsubstituted trimethylsilyl ketene acetals derived from acetic acid were employed, the aldol adducts were isolated in poor yields. Better results were obtained if the TMS group was replaced by *t*-butyldimethylsilyl (TBDMS). Accordingly, TBDMS ketene silyl acetal (**29**) reacted smoothly with aldehyde **30** under BF$_3$ catalysis (Scheme 18.9) [28]. The resulting aldol intermediate was subjected to a tetrabutyl-

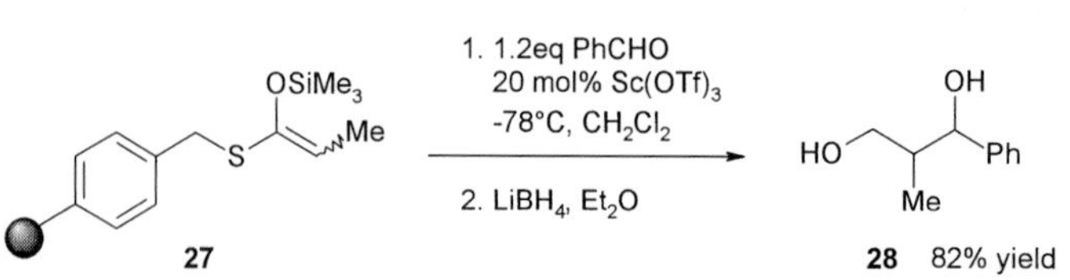

Scheme 18.8. Mukaiyama reactions on solid support using Sc(Otf)$_3$ as catalyst.

ammonium-fluoride (TBAF) solution in THF buffered by acetic acid to effect removal of the TBDMS group. The deprotected alcohol induced cyclative cleavage to
give lactone **31**, which served as a precursor for the synthesis of 2-deoxy-L-glucose.

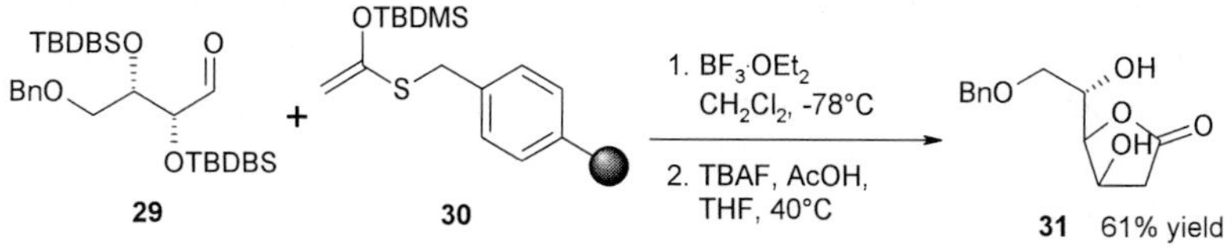

Scheme 18.9. Synthesis of 2-deoxy-L-glucose precursor using a solid-phase Mukaiyama reaction.

The Mukaiyama protocol has also been successfully conducted with immobilized aldehydes. As depicted in Scheme 18.10, cyclic TMS enolether **33** was reacted
with aldehyde **32** mediated by Yb(OTf)$_3$ to yield α,β-unsaturated ketone **34** after
cleavage in 91% yield [29]. The salient features of this process include the use of a
MeCN/H$_2$O mixture as the solvent for this reaction. Since conventional polystyrene resins were not compatible with these hydrophilic conditions, a resin
based on a polar polyoxyethylene/polyoxypropylene matrix was introduced for this
reaction.

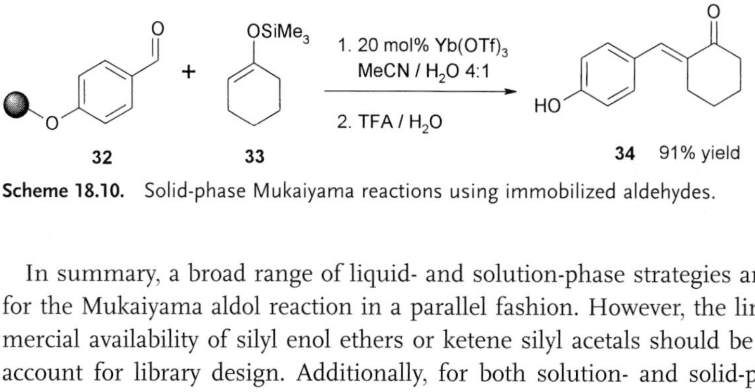

Scheme 18.10. Solid-phase Mukaiyama reactions using immobilized aldehydes.

In summary, a broad range of liquid- and solution-phase strategies are available
for the Mukaiyama aldol reaction in a parallel fashion. However, the limited commercial availability of silyl enol ethers or ketene silyl acetals should be taken into
account for library design. Additionally, for both solution- and solid-phase strategies, it has to be evaluated whether the work-up and purification protocols suit
the individual requirements for automation and library synthesis.

18.3
1,4-Addition of Enolates to Michael Acceptors

The 1,4-addition of enolates to Michael acceptors can be effected by base catalysis.
Consequently, method development for solution-phase parallel synthesis focused
on the use of immobilized bases to promote the addition step and simplify catalyst
separation and product isolation. As early as 1958, Bergmann and Corett reported
the use of basic exchange resins as catalysts for the Michael addition of enolates to

α,β-unsaturated carbonyls and acrylonitriles (Scheme 18.11) [30]. The reaction conditions were elaborated for a variety of Michael donors such as ketones and α-branched aldehydes (15 examples in total). To date, a range of novel polymeric catalysts has been introduced. Of particular interest are immobilized quaternary ammonium salts that proved to be effective for the addition of soft enolate nucleophiles to Michael acceptors [31]. Along these lines, asymmetric Michael additions have been reported based on immobilized optically active alkaloids [32]. However, applications of these methods for the synthesis of versatile libraries in solution phase have not yet been demonstrated.

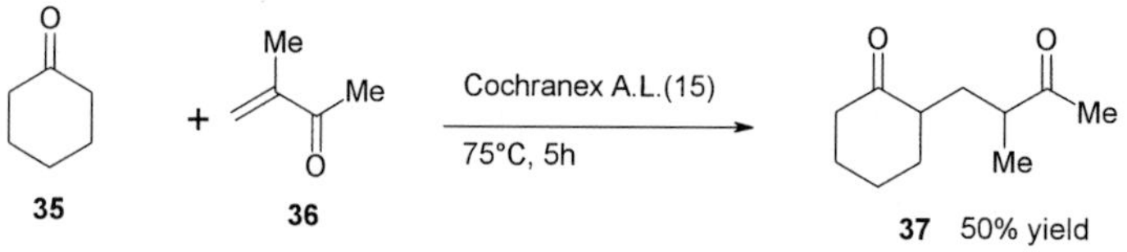

Scheme 18.11. Enolate additions to Michael acceptors mediated by a basic exchange resin.

More recently, Ley and Massi reported a tandem Michael protocol for solution-phase combinatorial synthesis leading to the bicyclo[2.2.2]octanone skeleton **40** that was further elaborated in subsequent amination reactions [33]. The bicyclic scaffold was constructed upon reaction of 2 equiv. of conjugated dianions (**39**) with *tert*-butyl acrylates (**38**). As outlined in Scheme 18.12, the reaction mixture was quenched with Amberlyst 15 (path A) and a simple filtration process provided pure endo products (**40**) (with R2 = OEt or O*i*Bu), whereas the crude product (with R2 = methyl) was contaminated by hydrolyzed 3-methyl-2-cyclohexen-1-one. Therefore, this mixture was treated with polymer-supported thiophenol and diisopropylethylamine (DIPEA) to scavenge the excess of hydrolyzed starting material (**39c**) followed by neutralization of ammonium salts using MP carbonate (path B). After filtration the desired products (**40**) were finally isolated in pure form. These carefully orchestrated reaction and work-up conditions formed the basis for a rapid

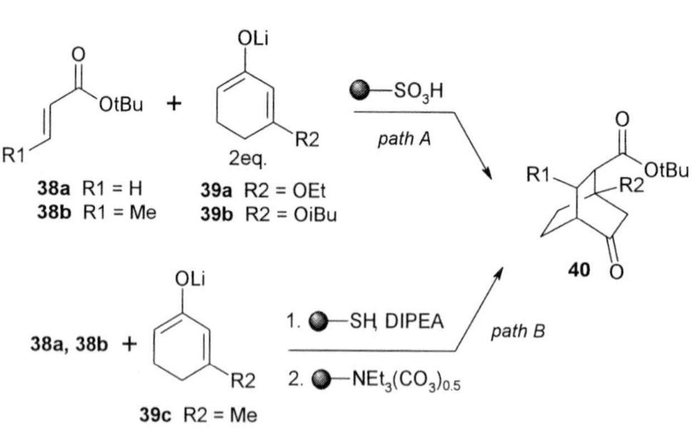

Scheme 18.12. Tandem Michael reactions using polymer-supported reagents.

synthesis of a larger library, since the reaction protocol can easily be automated and avoids tedious chromatography.

The synthesis of the bicyclo[2.2.2]octanone framework via a tandem Michael reaction has also been adapted to solid phase [34].

The 1,4-addition of enolates to Michael acceptors in solid-phase chemistry has been examined by Domínguez et al. [35]. These authors investigated the addition of resin-bound glycinates (41) to various α,β-unsaturated acceptors (Scheme 18.13).

The standard conditions for typical Michael acceptors such as acrylates, include 3 equiv. of Schwesinger-type base BEMP and 5 equiv. of acceptor 42 in NMP at room temperature (rt) for 16 h to yield the Michael adducts 43 in high purity. For more reactive acceptors (e.g. acrylonitriles) only 1.5 equiv. of 42 were used to avoid double addition, while less reactive electrophiles 42 (e.g. substituted acrylates) required additional base and acceptor 42 (each 10 equiv.). In total, 15 examples are given composing a diverse set of enolate additions to α,β-unsaturated Michael acceptors. Additionally, the resin-bound intermediates 42 were further modified on the polymer, including deprotection of the imine, acylation of the amino group with quinaldic acid, and cleavage using trifluoroacetic acid (TFA). The final products were obtained in good yields (61–88%) and purities (74–87%), indicating that the Michael addition as depicted in Scheme 18.13 proceeded on the resin without incident and with high efficiency.

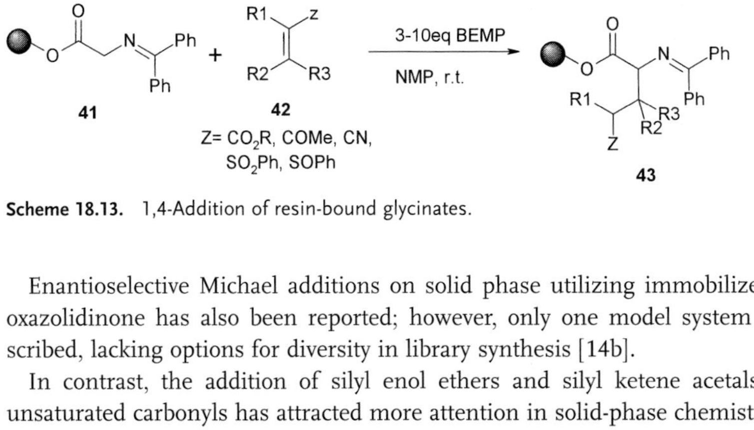

Scheme 18.13. 1,4-Addition of resin-bound glycinates.

Enantioselective Michael additions on solid phase utilizing immobilized Evans oxazolidinone has also been reported; however, only one model system was described, lacking options for diversity in library synthesis [14b].

In contrast, the addition of silyl enol ethers and silyl ketene acetals to α,β-unsaturated carbonyls has attracted more attention in solid-phase chemistry.

As depicted in Scheme 18.14, solid-supported thiosilyl ketene acetal 44 reacted smoothly with α,β-unsaturated ketones in the presence of 20 mol% Sc(OTf)$_3$ to yield the corresponding adducts on solid phase. The products were cleaved by the

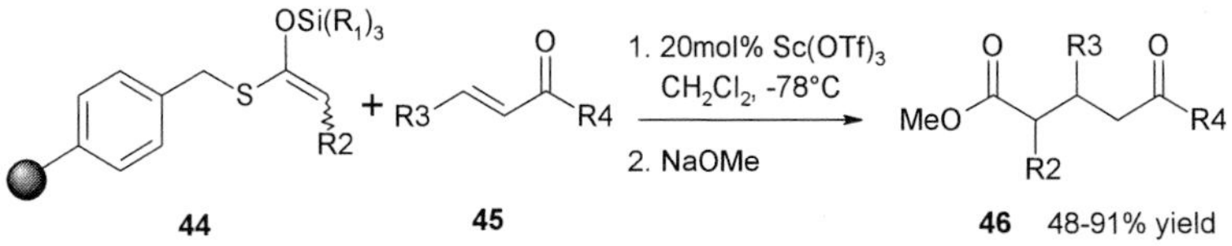

Scheme 18.14. 1,4-Additions of silyl ketene acetals on solid phase.

action of NaOMe and were subjected to chromatography to yield 1,5-dicarbonyls (**46**) in good to excellent yields (48–91%, nine examples) [17a].

An alternative approach was based on resin-bound Michael acceptors which were reacted with soluble silyl enol ethers. This concept was realized by Ellingboe and coworkers, who reported the synthesis of immobilized 1,5-diketones employing polymeric enone **47** and 4 equiv. of TMS enol ether (**48**). The Michael addition was effected by CsF in dimethyl sulfoxide (DMSO) to yield **49**. The best results were obtained if R1 was an aromatic or nonenolizable aliphatic residue, while R2 should preferably be aromatic or heteroaromatic. The target molecules **49** represented important intermediates for the construction of a pyridine library (Scheme 18.15) [36].

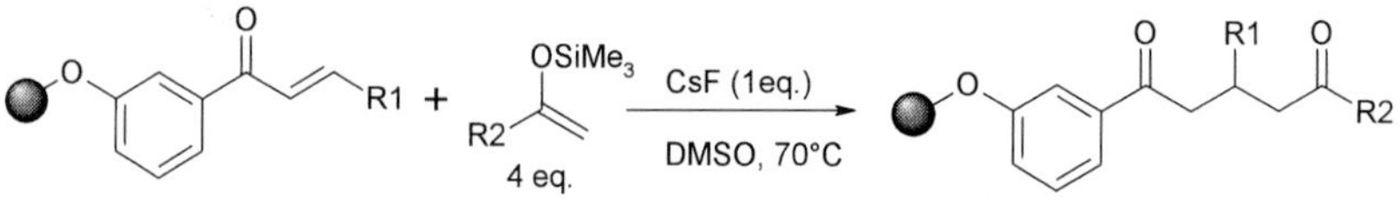

Scheme 18.15. Michael addition of silyl enol ether to polymer-supported α,β-unsaturated ketones.

In summary, a diverse repertoire of combinatorial methods for Michael additions is available. The application of these methods to library synthesis has demonstrated that 1,4-additions of enolates are useful transformations for creating diversity with concomitant formation of C–C bonds.

18.4
Alkylation of Enolates

The alkylation of C–H acidic carbonyls represents a valuable transformation to gain an additional diversity step for a given library. Thus, extensive efforts have been devoted to optimizing alkylation protocols especially for solid-phase synthesis.

18.4.1
a-Alkylation of Carbonyl Compounds

The alkylation of ketone-derived enolates has been realized in solution-phase chemistry employing solid-supported reagents. In this regard, Montanari and coworkers reported the reaction of benzylbromide with benzylmethylketone using an immobilized quaternary phosphonium salt as a phase transfer catalyst [37]. The product could be obtained in excellent yield (95%) after filtration and extraction. However, the scope of this reaction has not yet been extended to additional examples or library synthesis.

Silyl enol ethers have also been described to function as suitable substrates for the alkylation process. Accordingly, **56** was reacted with prenylbromide (**51**) in the presence of ZnCl$_2$-doped acidic alumina to give ketone **52** in 72% yield (Scheme 18.16) [38]. This reaction worked well with allylic, benzylic, and tertiary halides, leading to the alkylated products after a simple extraction step with CH$_2$Cl$_2$ in good yields (60–72%).

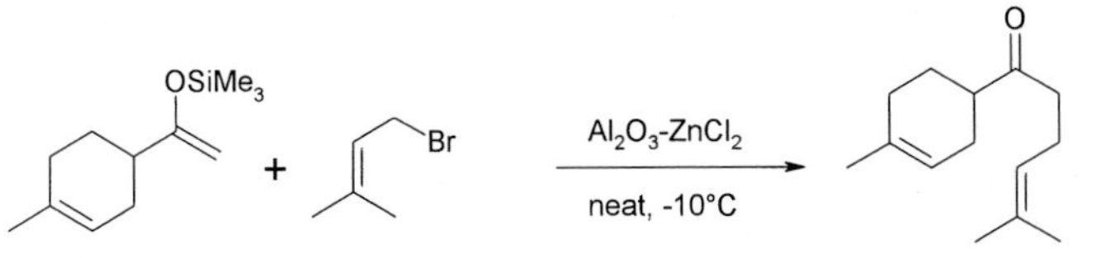

Scheme 18.16. Alkylation of silyl enol ethers using ZnCl$_2$-doped alumina.

The elegant studies of Ellman and coworkers toward the synthesis of arylacetic acid libraries represented classical solid-phase synthesis [39, 40]. This concept called for a reliable enolate-alkylation protocol that proved to be feasible for the construction of a diverse set of target molecules (Scheme 18.17). Accordingly, Kenner et al.'s sulfonamide linker [41] was chosen to bind the aryl acetic acid precursor to solid phase to yield polymer **53**. The resin was subsequently treated with 15 equiv. of LDA to generate the corresponding trianion, which was trapped by quenching with an excess of alkylhalide to selectively yield the monoalkylated product **54**. This alkylation reaction works equally well for a range of halides, including benzylic and aliphatic halides (five examples). The intermediates **54** were further manipulated to give a diverse set of arylacetic acids which were isolated in excellent yields (88–100%) and high purities after cleavage. From these results, it could be concluded that the alkylation event further upstream in this sequence proceeded smoothly and with high efficiency.

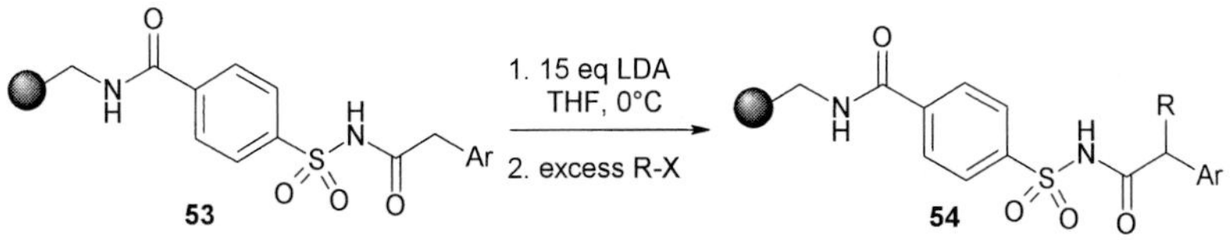

Scheme 18.17. α-Alkylation on solid support using Kenner's sulfonamide linker.

The alkylation of amide and ester enolates derived from deprotonation of the corresponding α-acidic precursor has been reported in solid-phase synthesis using LiHMDS [42] and KHMDS [43], respectively.

18.4.2
α- and γ-Alkylation of 1,3-Dicarbonyl Compounds

The 1,3-dicarbonyl moiety represents a versatile template for the synthesis of heterocycle libraries, especially if alkylation protocols can provide an additional anchor for diversity. Accordingly, the development of methods which selectively target the mono- or dialkylation of α,β-dicarbonyls has attracted recent attention.

In this context, a variety of solution-phase strategies has been elaborated. Most strategies are based on the use of polymer-supported bases such as Amberlite anion-exchange resins [44] or immobilized fluoride sources [45]. A typical example is illustrated in Scheme 18.18 [46]. Celite coated with KF cleanly effected the monoalkylation of acetylacetone to yield **56** in 96% yield. In contrast, the system Al_2O_3–KF predominantly delivered the corresponding disubstituted product under otherwise identical conditions.

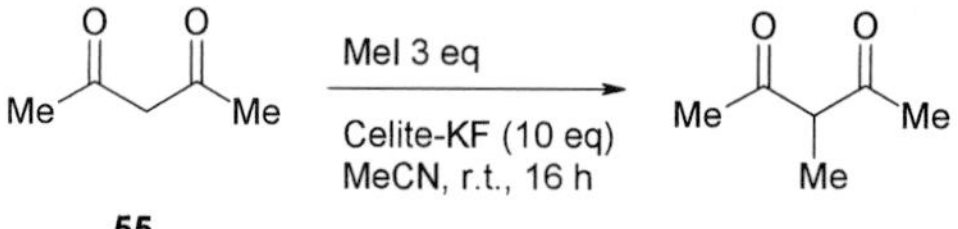

Scheme 18.18. Alkylation of β-dicarbonyls using immobilized KF.

In general, the scope of the solution-phase strategies is limited to a small number of examples, and to the best of our knowledge no application in library synthesis has been published to date.

In contrast, methods that were developed for the alkylation of 1,3-dicarbonyls on solid support have had a significant impact on library synthesis.

The optimal conditions for the α-alkylation of β-diketones and β-ketoesters include the use of excess TBAF and alkyl halide in THF at rt (Table 18.2, entries 1 and 2). Under these conditions, the oxygen atoms in **57** and **58** are shielded by complexation with the sterically demanding tetrabutylammonium counterion, thereby preventing O-alkylations with concomitant nucleophilic activation toward α-alkylation. Importantly, this protocol required strictly anhydrous conditions, otherwise the yields were poor. This methodology has successfully been applied in the synthesis of pyrazole and isoxazole libraries. However, a major drawback arises from the incompatibility of this process with the presence of N-heteroaryl or furyl-functionalities, and, furthermore, the use of benzylbromide as alkylating agent gave only sluggish results. As an alternative to TBAF, other bases such as 1,8-diazabicyclo[5.4.0]undecene-7 (DBU) were used to effect the desired α-alkylation of C–H acidic β-dicarbonyl compounds (Table 18.2, entry 3). The latter example also illustrates that electrophiles bound to solid phase can be efficiently employed in these alkylation reactions.

Similar processes involved the use of Merrifield resin or the more reactive (bromo-methyl)-polystyrene/divinylbenzene (PS/DVB) resin as polymeric electrophile to covalently bind 1,3-dicarbonyls. The requisite enolates were generated in solution by utilizing NaH [51], Na in ethanol [52], or KOH under phase transfer conditions as base [53]. Contributions in this field mainly focused on the development of new linkers and will not be discussed in detail.

A slightly different concept relies on the well-established Pd-catalyzed allylic alkylation of C–H acidic 1,3-dicarbonyls with allylic acetates and carbonates. This

Tab. 18.2. Alkylation of resin-bound β-dicarbonyls.

Entry	β-Dicarbonyl	R'X/electrophile	Product	Conditions	Comment	Reference
1	**57** R = Me, Bn, CH₂CH₂CO₂Me	R' = aliphatic X = Br, I	**58**	20 equiv. TBAF 36 equiv. R'X THF, 3 h, rt	6 examples no γ- or O-alkylation observed	47
2	**59** R = Ph	R' = allyl, Et, CH₂Z R' ≠ benzyl X = Br, I	**60**	10 equiv. TBAF 44 equiv. R'X THF, 2 h, rt	5 examples conditions incompatible with R = N-heteroaryl or furyl	48 49
3	**61**	**62**		10% DBU/DMF 20 °C, 16 h	3 examples	50

Z = electron-withdrawing group.

chemistry has been adapted to solid phase and it has been shown that resin-bound malonates could be cleanly α-dialkylated by this method [54].

An important modification of 1,3-dicarbonyls is the γ-alkylation. This has been accomplished on solid phase commencing from resin-bound β-ketoester (**63**), which was deprotonated with an excess of LDA to generate the corresponding dienolate (Scheme 18.19) [55].

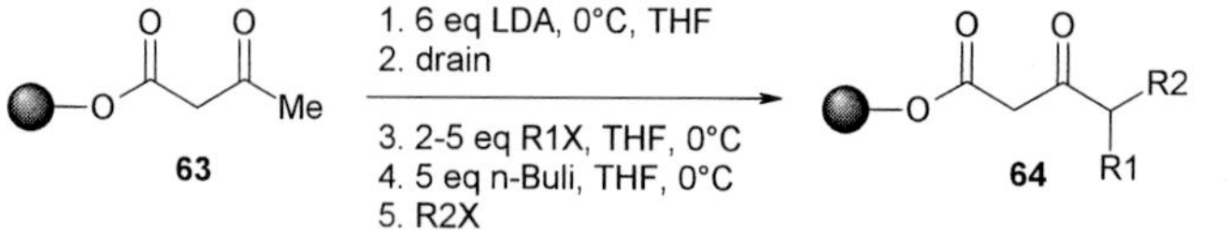

Scheme 18.19. γ-Alkylation of resin-bound β-ketoesters.

Subsequently, the excess LDA was drained away and 2–5 equiv. of alkyl halide (R1X) were added to effect mono-γ-alkylation. The resulting monoenolate could serve as starting material for a second alkylation step. Thus, treatment of the enolate intermediate with *n*-BuLi regenerated the dianion which was quenched with R2X to yield dialkylated products (**64**) in good yields. The γ-alkylation worked well with aliphatic, benzylic, and allylic halides (X = Br or I; nine examples).

18.4.3
Stereoselective Alkylations of Enolates

The diastereoselective alkylation of enolates controlled by chiral auxiliaries is one of the most reliable processes in asymmetric solution-phase chemistry. The transfer of this concept to solid phase would offer manipulative convenience since excess reagents could be removed by simple washing procedures, and, finally, the auxiliary could be recovered by filtration.

Table 18.3 summarizes the progress in this field that has been achieved to date. Leznoff and coworkers [56, 57] reported the application of immobilized amino alcohols as chiral auxiliaries to effect stereoselective alkylation of ketones (Table 18.3, entry 1).

The levels of asymmetric induction were found to be in the same range as in solution. The stereoselective alkylation of resin-bound amide **67** was described by Kurth and coworkers (Table 18.3, entry 2) [58, 59]. The alkylation occurred with good diastereomeric excess and cleavage from the resin was accomplished by an iodo-lactonization protocol. The adoption of the Meyers oxazoline methodology to solid phase (Table 18.3, entry 3) was accompanied by drawbacks, mainly arising from low chemical yields due to incomplete hydrolysis of the final product. Allin [61] and Burgess [62] independently reported their endeavors toward a solid-phase version of an Evans oxazolidinone auxiliary (Table 18.3, entry 4). Both groups could demonstrate that strong electrophilic halides react readily with pregenerated oxazolidinone enolates to yield the corresponding alkylated products with medium to high levels of optical purity. Cleavage from the resin was accomplished by hy-

Tab. 18.3. Stereoselective enolate alkylations on solid phase.

Entry	Resin	R'X	Product	Conditions	Remarks	Reference
1	**65** R= Me, Bn	MeI, iPrI	**66**	1. LDA, THF, 0 °C 2. R'X, rt 3. Cleavage (H⁺)	2 examples; 60–95% ee; 80–87% yield	56 57
2	**67** R= Me, allyl	MeI, allyl-I	**68**	1. 2 equiv. LDA, 0 °C 2. 3 equiv. R'X, 0 °C → rt 24 h	2 examples; >87% de	58 59
3	**69**	BnCl	**70**	1. n-BuLi THF 2. 650 equiv. BnCl 3. Cleavage (H₂SO₄, EtOH)	1 example; 56% ee; 43–45% yield	60
4	**71** R= CH₂; CH₂OCH₂	BnBr BnOCH₂Cl allyl-Br	**72** R2= CO₂H or CH₂OH	1. 2–3 equiv. LDA, THF, 0 °C 2. 2–5 equiv. BnBr 3. LiOH or LiBH₄ (cleavage)	4 examples; 76–96% ee; 20–66% yield	61 62

drolysis (LiOH) to liberate the carboxylic acid or, alternatively, by reduction (LiBH$_4$) leading to the corresponding alcohols.

For all examples depicted in Table 18.3, convenient recycling of the chiral auxiliary has been demonstrated and the polymers could be reused for further applications. However, the majority of protocols required rather drastic reaction conditions for the cleavage of the final product, which implies tedious work-up and purification procedures rendering this chemistry less amenable for automation.

Diastereoselective alkylations have also been successfully applied in the synthesis of more complex molecules on solid support. For instance, Hanessian et al. showed that polymeric ester 73 could be deprotonated by the action of KHMDS followed by the addition of reactive electrophiles such as allylic iodides (three examples) to yield the corresponding α-alkylated product on solid support (Scheme 18.20) [63].

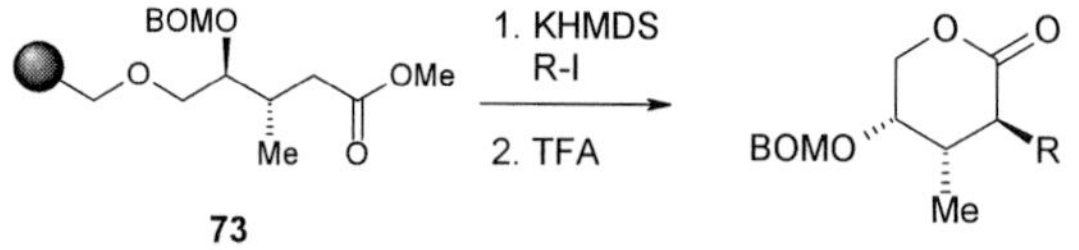

Scheme 18.20. Diastereoselective alkylation of esters on solid support.

This resin was treated with TFA to effect cleavage and lactonization leading to enantiopure lactones (74) which were isolated in 75% yield (using two steps). The stereochemical outcome of this reaction was rationalized via a Felkin-type transition state model.

In Chen and Janda's elegant studies toward a resin-supported synthesis of prostaglandin F$_{2\alpha}$, the α-alkylation of the central cyclopentanone core represented a key step [64].

As depicted in Scheme 18.21, the requisite enolate was generated by lithiation of the TMS enol ether (75) using excess methyl lithium. Upon addition of triflate 76 the desired product 77 was isolated in pure form. For these studies a soluble noncrosslinked polystyrene was selected as polymer support, since this choice guaranteed solution-like kinetics for each step of this sequence. Additionally, the soluble

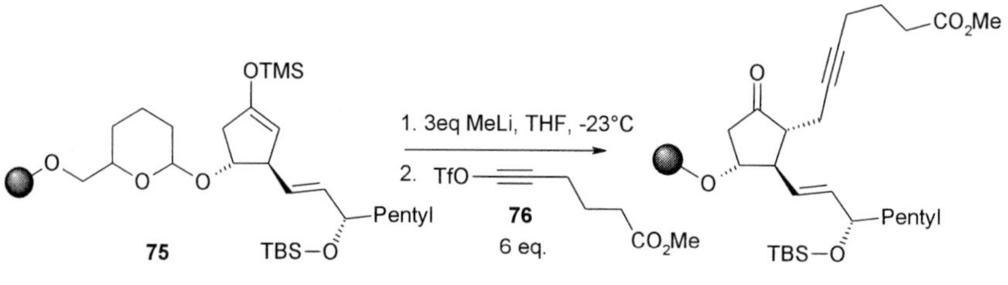

Scheme 18.21. Enolate alkylation as a key step in solid-phase synthesis of prostaglandin F$_{2\alpha}$.

polymer allowed the monitoring of reaction intermediates by nuclear magnetic resonance (NMR).

18.4.4
Alkylation of Protected Glycines

Suitably protected glycine esters are readily alkylated to yield protected α-amino acids. Accordingly, this route offers convenient access to tailor-made unnatural amino acids which are valuable templates in organic synthesis.

Benzophenone imines derived from glycine esters serve as the starting materials of choice for the alkylation of glycine. This is the result of the convenient protocols that are available for imine formation as well as the high selectivity of these imines toward monoalkylation.

Polymer-supported reagents have been developed to produce the above-mentioned transformation. For instance, Palacios et al. reported the use of immobilized *P*-ylides as strong non-nucleophilic bases [65]. Moreover, a stereoselective alkylation procedure has been elaborated previously [66].

As depicted in Scheme 18.22, protected glycine **78** was reacted with an excess of reactive halides, such as benzyl or allylic bromide under phase transfer conditions (seven examples). As chiral catalyst, the immobilized tertiary ammonium salt **80** was used, and for a range of products medium to good levels of induction were observed. The products **79** were isolated after filtration and extraction into the organic layer without the need for chromatographic purification.

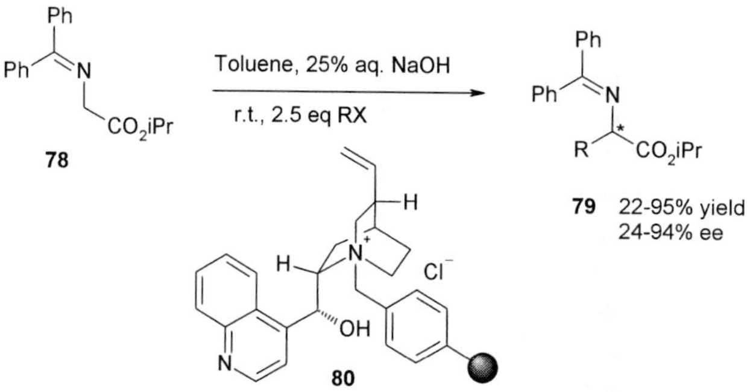

Scheme 18.22. Stereoselective alkylation of glycine imine **78**.

The alkylation of glycine Schiff bases has also been established in solid-phase chemistry (Table 18.4). The most frequently reported conditions include the use of Schwesinger-type bases in combination with an excess of alkyl halides in polar solvents. For instance, imine **81** immobilized on Wang resin was treated with BEMP and alkyl halide to selectively yield the monoalkylated products **82** (Table 18.4, entry 1). These products were further modified on solid support by cleavage of the benzophenone imine followed by acylation of the amino group. Finally, the resulting peptides were cleaved from the resin with TFA to deliver the corre-

Tab. 18.4. Alkylation of glycine derivatives on solid support.

Entry	Polymer	R'X	Product	Conditions	Remarks	Reference
1	**81**	R' = aliphtic, benzylic X = Br, I	**82**	2–10 equiv. R'X 20 equiv. BEMP NMP, rt, 24 h	10 examples; exclusively mono-alkylation	67 68
2	**83**	R' = benzyl, allyl, aliphatic X = Br	**84**	MeCN, excess K$_2$CO$_3$ 1.5 equiv. R'X, reflux	5 examples; products obtained by precipitation from Et$_2$O	69
3	**85**	R' = benzylic, allylic X = Br	**86**	2 equiv. BEMP 3 equiv. R'X	36 examples; for R = iPr no complete reaction	70
4	**87**	R' = benzyl, allyl, aliphatic X = Br, I	**88**	1. 5 equiv. BEMP, 5 equiv. R'X, DCM 1 equiv. **89** 2. NH$_2$OH 3. TFA	17 examples; 61–100% yield; 51–89% ee	71

sponding products in excellent yields (77–100%). These results indicate that the alkylation event at the outset of the sequence must have proceeded with high efficiency.

A related process was described for PEG-bound glycine **83** (Table 18.4, entry 2). The authors showed that the PEG polymer not only simplified product isolation, but also accelerated the reaction significantly. This catalytic effect was attributed to the close relationship of the PEG polymers with crown ethers, which are well known to be beneficial in phase transfer reactions.

Dialkylated amino acids were accessed starting from natural α-amino acids which were transformed to aldimines of type **85**, and subsequent alkylation with halides (Table 18.4, entry 3) led to the expected products **86**. This concept has been validated for a broad range of amino acids (36 examples).

Finally, a stereoselective alkylation of protected glycine on polymer support has been reported by O'Donnell et al. (Table 18.4, entry 4) [71]. Optimal results were obtained if Corey's quaternary ammonium catalyst **89** was used as the source of chirality. The immobilized intermediates were deprotected (NH$_4$OH) and cleaved from the resin (TFA) to yield the free amino acids (**88**) in good chemical and optical purity.

18.5
Claisen-type Condensations

The reactions between enolates and esters deliver 1,3-dicarbonyl compounds which are valuable intermediates in heterocyclic chemistry. An early study by Cohen et al. demonstrated that this kind of transformation could be realized utilizing polymer reagents. Accordingly, benzophenone was deprotonated with immobilized trityl lithium (**91**) and the resulting enolate was trapped with polymeric o-nitrophenyl benzoate (**90**) to yield β-diketone (**92**) with a 96% yield after filtration (Scheme 18.23) [72]. The yield by far exceeded the results obtained for the corresponding reaction in solution phase (≤ 50%). This result is due to the fact that immobilized trityl lithium could be used in great excess to drive the enolate addition to completion. More importantly, the presence of the alkyl lithium base proved to be compatible with the o-nitrophenyl benzoate (**90**) since both reactive sites were shielded by a polymeric matrix, thus avoiding deleterious interactions. This "wolf–lamb" strategy has clearly shown that polymer reagents not only simplify work-up but can also provide an innovative effect in optimizing reaction conditions.

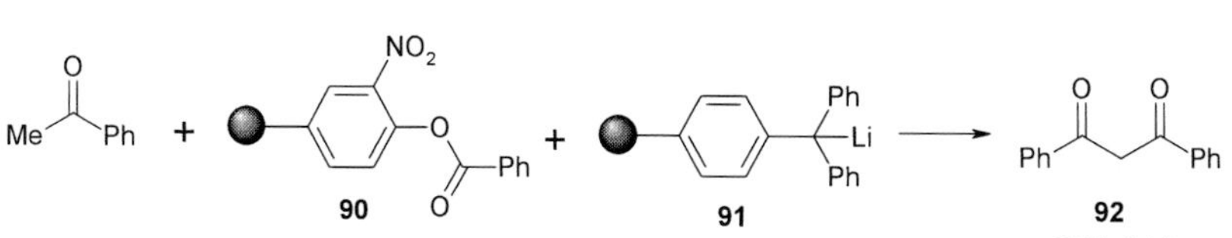

Scheme 18.23. "Wolf–lamb" reaction for the synthesis of β-ketoesters.

The synthesis of β-diketones has also been reported in solid-phase chemistry. In a standard protocol NaH is used as the base of choice to generate the requisite ketone enolate, which reacts with an excess of ester to yield the desired products. A typical example is illustrated in Scheme 18.24. Here, methylketone (**93**) was attached to the polymer by a Rink amide linker which proved stable toward the strongly basic reaction conditions. Upon reaction with aromatic esters in the presence of NaH, the β-diketones (**95**) were obtained in excellent yields [48, 49, 73]. However, the scope of the reaction was limited to a narrow set of aromatic and heteroaromatic esters.

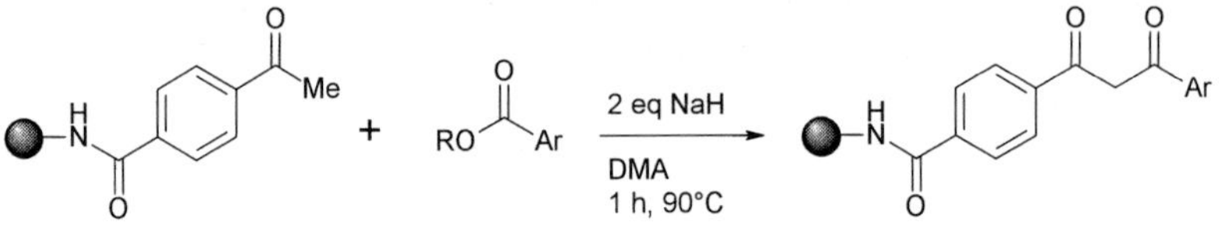

93 **94** **95** 95-100% yield

Scheme 18.24. 1,3-β-Diketone synthesis on solid support.

A more recent contribution in this field reported a novel selenium linker which was used to immobilize methylketone (**96**). Enolate generation was effected by LiHMDS followed by addition of acylcyanides as electrophiles to produce 1,3-diketones (**97**) in good yields (Scheme 18.25) [74]. Alternatively, the use of Weinreb amides [75] or β-ketoester [76] as electrophiles in this Claisen-type reaction have been described.

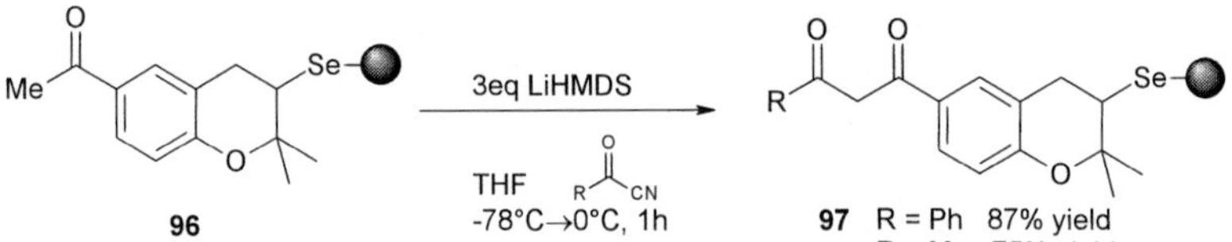

96 **97** R = Ph 87% yield
R = Me 75% yield

Scheme 18.25. Solid-phase synthesis of 1,3-diketones using acylcyanides.

In conclusion, the construction of β-dicarbonyls, especially in solid-phase synthesis, is a well-established process. However, the scope of this reaction is primarily restricted to nonenolizable electrophiles such as aromatic esters, acylcyanides, or Weinreb amides.

18.6 Dieckmann Condensations

The Dieckmann reaction as an intramolecular version of the Claisen condensation offers a valuable means of construction of ring systems with the concomitant formation of a C–C bond. If nonsymmetric dicarbonyl precursors are used in this reaction, fine-tuning of the conditions is often needed to selectively obtain the desired product. For instance, amino acid derivatives (**98**) have been used in solution-phase chemistry as open-chain cyclization precursors. Chemoselective deprotonation of the C–H acidic amide was achieved by preferably allowing electron withdrawing groups for R3 (Scheme 18.26) [77, 78]. Amberlyst A26 (OH⁻ form) was selected as base for this reaction and the resulting tetramic acids (**99**) were first obtained as salts immobilized on the basic exchange resin. Subsequently, the resin was washed to remove byproducts and excess reagents followed by incubation with TFA to liberate the final products (**99**) as free acids. In total, this concept of solid-phase resin capture delivered a range of tetramic acids in good yields and high purities (ten examples), including a convenient work-up protocol which should be suitable for automation.

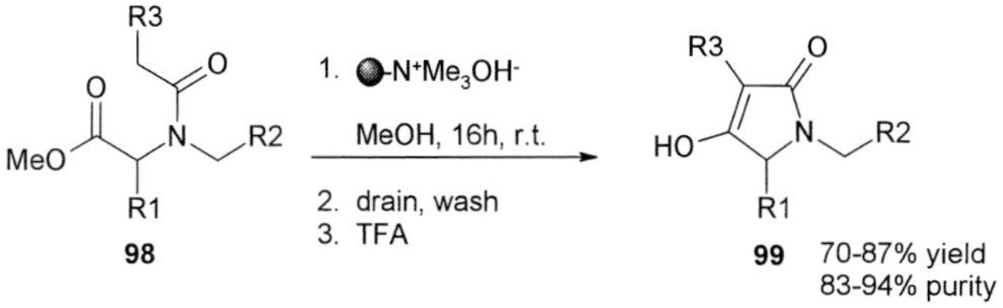

Scheme 18.26. Parallel synthesis of tetramic acids in solution phase.

Following the seminal contributions of Rapoport in the field of solid-phase-supported unidirectional Dieckmann reactions [79], the vast majority of more recent communications have dealt with the solid-phase synthesis of tetramic acids. Accordingly, OH-functionalized polymers such as Wang resin loaded with amino acid derivatives were exposed to basic reaction conditions, as outlined in Scheme 18.27.

A variety of bases were employed for this reaction, including NaOEt [80], KOH [81], DIPEA [82], or Bu₄NOH [83]. The cyclization occurred as expected with concomitant release of the anionic form of the tetramic acid into solution. The free acid (**101**) was finally isolated by treatment with a strong acid or acidic ion-exchange resin. More importantly, the cyclative cleavage strategy employed in these examples guaranteed high purities for the products since the latter were only released into solution if the desired cyclization event took place first.

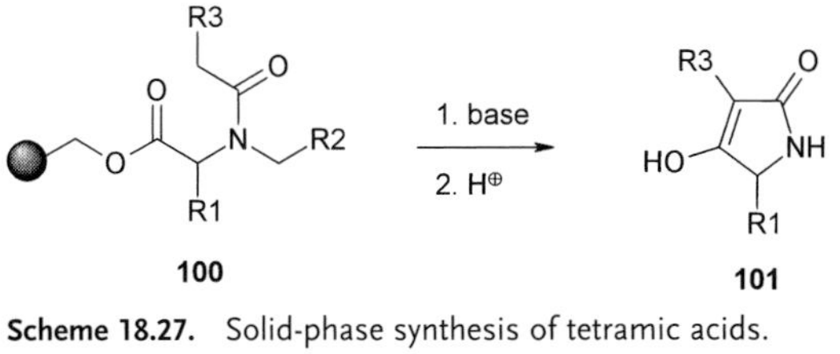

Scheme 18.27. Solid-phase synthesis of tetramic acids.

Owing to the convenient reaction protocols both the liquid- and solid-phase synthesis of tetramic acids appears to be an attractive tool for multicomponent library synthesis.

18.7
Knoevenagel Condensations

The condensation of aldehydes with carbonyl compounds bearing an additional electron withdrawing group (e.g. β-ketoesters or malonates) is usually designated as a Knoevenagel condensation. The resulting alkylidene products serve as ideal precursors for the synthesis of a range of heterocycles such as dihydropyrimidines, pyrimidines, pyrimidinones, dihydropyridines, pyrazoles, dihydropyrans, or coumarines. Libraries prepared in solution phase as well as on solid phase with a broad range of templates underscore the value of this reaction for combinatorial purposes.

Owing to the high acidity of the activated carbonyl moiety, weak bases such as ammonia, ammonium salts, primary or secondary amines, and salts thereof suffice to promote the condensation. Hence, polymer-supported Brönsted bases have been applied in solution-phase Knoevenagel reactions. A typical example is shown in Scheme 18.28. A range of malonates, malonitriles, cyanoacetates, and oxoesters (**102**) were condensed with nonenolizable aromatic aldehydes in the presence of polymeric amine (**103**) under continuous flow conditions. The alkylidene products (**104**) were isolated after removal of the solvent and purification by chromatography or recrystallization in acceptable to excellent yields (29–98%) [84]. Other polymeric reagents reported for the Knoevenagel process include the use of ion-exchange resins [85], immobilized tetraalkylammonium hydroxides [86], Al_2O_3 [87], Al_2O_3–KF [88], or molecular sieves [89]. However, these solution-phase methods have not yet been applied in library synthesis.

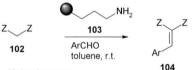

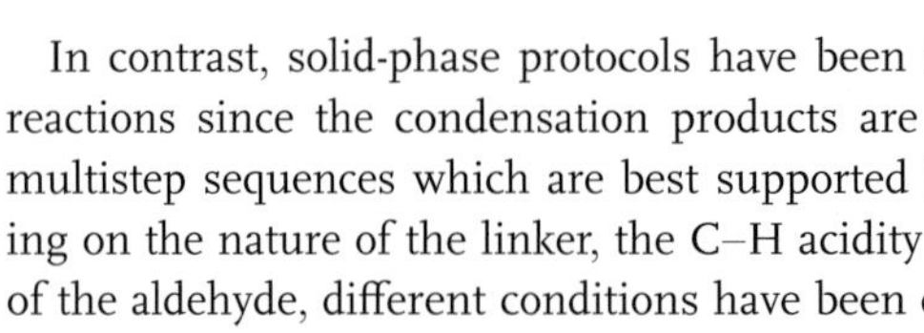

Z = electron withdrawing group

Scheme 18.28. Knoevenagel condensations using polymer reagents.

In contrast, solid-phase protocols have been far more popular for Knoevenagel reactions since the condensation products are often important intermediates in multistep sequences which are best supported by solid-phase chemistry. Depending on the nature of the linker, the C–H acidity of the substrate, and the structure of the aldehyde, different conditions have been employed. Piperidine or piperidine acetate have been successfully used as catalysts for the Knoevenagel reaction and seem to be the methods of choice. For instance, Gallop, Bhandari, and coworkers reported that malonate (**105**) immobilized on TentaGel resin cleanly reacts with aromatic aldehydes (eight examples) under the conditions depicted in Scheme 18.29 [90, 91]. The Knoevenagel products (**106**) were further transformed to pyrrolopyridines utilizing a Hantzsch protocol for construction of the pyridine heterocycle. The final products in this sequence were obtained in decent yields (20–50%) with excellent purities (> 90%), reflecting almost quantitative conversion for each step of the synthesis. Additionally, this methodology was extended to a library synthesis of 4800 compounds, thereby demonstrating the broad scope of the Knoevenagel reaction. Instead of TMOF, molecular sieves were also used to remove water from the reaction mixture [92]. Under similar reaction conditions, salicylaldehydes were effectively condensed with resin-bound malonates to yield the coumarin skeleton on solid support [93].

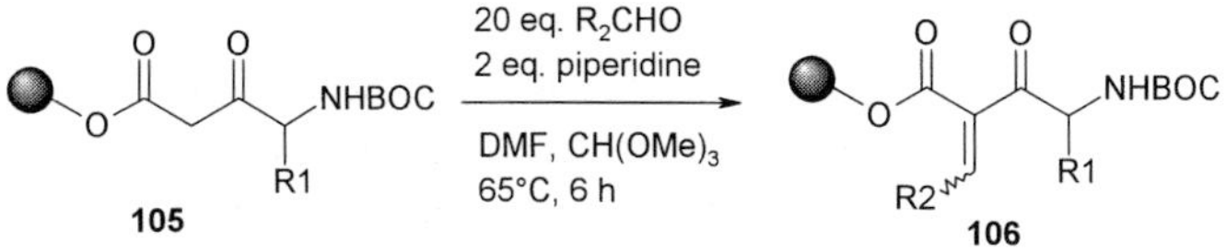

Scheme 18.29. Knoevenagel condensations on solid support.

If the basic piperidine is buffered by acetic acid, even milder conditions for the Knoevenagel condensation are obtained. As illustrated in Scheme 18.30, malonates (**107**) loaded onto Wang resin were reacted with a range of aromatic and aliphatic aldehydes to yield the corresponding Knoevenagel products on solid support. After cleavage, the carboxylic acids (**108**) were obtained as mixtures of E and Z isomers in good yields (45–98%) and high purities (conversion 92–98%). This protocol was further applied in a library synthesis comprising 96 products of type **108** and the majority of products were obtained with purities > 75% as determined by HPLC [94]. Moreover, this procedure was especially suited for the use of enolizable aliphatic aldehydes in the Knoevenagel condensation on solid support [95]. In this context, it was also reported that ethylenediamine diacetate (EDDA) very effectively catalyzed the condensation of both aromatic and aliphatic aldehydes with resin-bound malonates [5c].

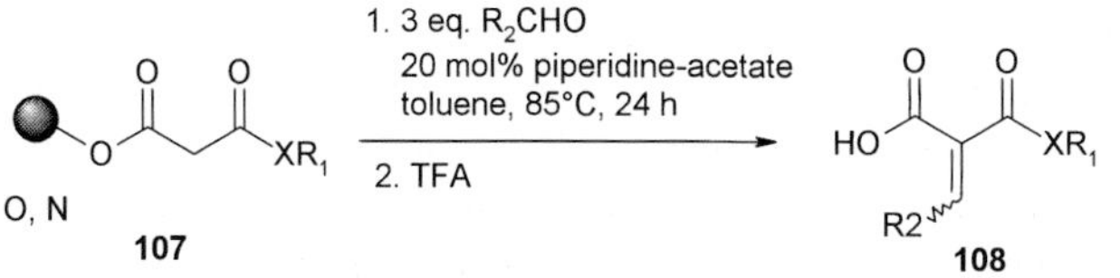

Scheme 18.30. Knoevenagel condensations on solid support.

18.8
Addition of Enolates to Imines

18.8.1
Synthesis of β-Amino Esters and Alcohols via Enolate Addition to Imines

Mannich-type reactions include the addition of enolates to imines and deliver valuable β-amino acid derivatives. Mild reaction conditions were reported employing silyl enol ether or ketene silyl acetals which add to preformed or *in situ*-generated aldimines in the presence of catalytic amounts of Lewis acids. In this regard, a variety of polymeric rare earth metal salts were reported to promote the aforementioned process efficiently [96]. For instance, Kobayashi et al. introduced a microencapsulated scandium triflate which exhibited high catalytic activity in the Mannich reaction [22a]. As outlined in Scheme 18.31, the imino-aldol reaction between aldimine **109** and propiophenone-derived silyl enol ether **110** proceeded smoothly to produce β-aminoketone **111** in excellent yield. Furthermore, the catalyst was recycled and reused up to seven times in the Mannich process without there being any deleterious effects on the yield. These reactions work best with aldimines originating from aromatic or heteroaromatic aldehydes.

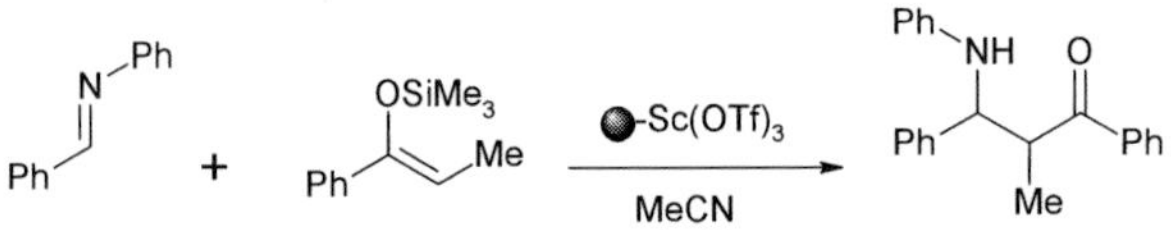

Scheme 18.31. Mannich-type reactions promoted by microencapsulated Sc(Otf)$_3$.

Complementary, polymer-supported catalysts such as π-acidic dicyanoketeneacetals [97] or optically active palladium–BINAP complexes [26] have also been used to effect the addition of silyl enol ethers to imines.

The use of polymeric catalysts for the Mannich reaction has dramatically simplified the isolation and recycling of the catalysts by mere filtration and washing operations, however, the isolation of the reaction products, such as **111**, was routinely achieved by aqueous work-up procedures followed by chromatography. Therefore, the application of this methodology to the synthesis of libraries would require the work-up and purification protocols to be adapted to automation.

An interesting multicomponent Mannich reaction has been recently reported by Prabhakaran and Iqbal (Scheme 18.32) [98]. Ketones or β-ketoesters (**112**) were condensed with aromatic aldehydes (**113**) and acetylchloride in acetonitrile in the presence of a polyaniline-supported cobalt catalyst. The resulting *N*-acetyl-β-amino acid derivatives (**114**) (eight examples given) were isolated by a simple work-up protocol comprising a filtration and a washing step to deliver **114** in acceptable yields (46–68%) and excellent HPLC purities (95–100%). Here, the anticonfigurated products were formed predominantly. The mechanism of this reaction presumably involves the addition of *in situ*-generated cobalt enolates to a cationic iminium species [99].

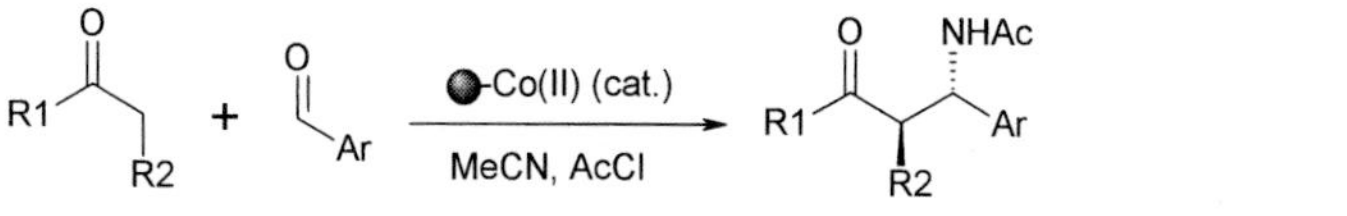

Scheme 18.32. Multicomponent Mannich reactions using a polymer-supported cobalt catalyst.

Mannich reactions have also been established in solid-phase chemistry. Kobayashi et al. introduced resin-bound thiosilyl ketene acetal **115** and elaborated an efficient Mannich protocol on solid support (Scheme 18.33) [100]. Thus, reaction of **115** with imines **116** in the presence of 10 mol% Sc(OTf)$_3$ delivered the corresponding β-amino thioesters which were subsequently cleaved from the resin upon reduction with LiBH$_4$ to yield the β-amino alcohols **117** in good yields (42–78%; 19 examples) after aqueous work-up and preparative TLC. A broad range of substrates were used in this reaction with R1 being hydrogen, methyl, or benzyloxy, and R2 and R3 preferably being aromatic or heteroaromatic groups. This reaction can also be conducted as a one-pot, three-component condensation starting from amine, aldehyde, and silyl enol ether comprising *in situ* generation of the requisite aldimine [101]. Furthermore, the scope of this methodology has been extended by using the Sc(OTf)$_3$-mediated Mannich reaction as a key step in the solid-phase synthesis of amino sugar derivatives [28].

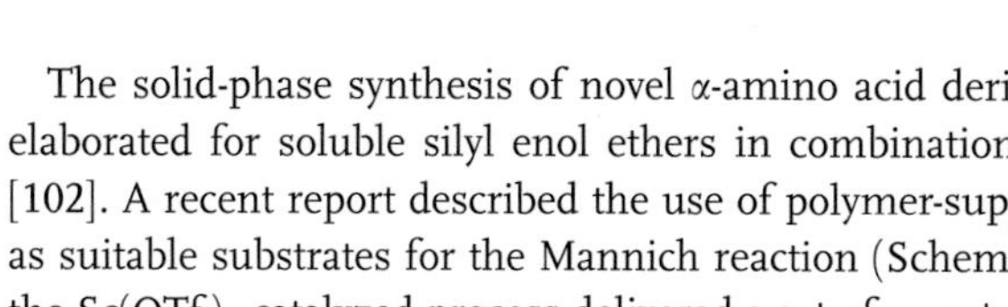

Scheme 18.33. Mannich reactions on solid support to yield β-amino alcohols.

The solid-phase synthesis of novel α-amino acid derivatives (**119**) has also been elaborated for soluble silyl enol ethers in combination with immobilized imines [102]. A recent report described the use of polymer-supported imino acetates (**118**) as suitable substrates for the Mannich reaction (Scheme 18.34) [103]. Accordingly, the Sc(OTf)$_3$-catalyzed process delivered a set of unnatural aspartic acid derivatives (**119**) which were cleaved from the resin by transesterification using NaOMe. Under similar conditions polymer-supported acylhydrazones reacted with ketene silyl acetals and the corresponding adducts underwent cyclative cleavage upon treatment with NaOMe to yield substituted pyrazolones [104].

Scheme 18.34. Solid-phase synthesis of aspartic acid derivatives.

18.8.2
Solid-phase Synthesis of β-Lactams via Enolate Additions to Imines

The synthesis of β-lactams is conveniently achieved by [2 + 2] cycloaddition of ketenes with imines [105], or, alternatively, by addition of enolates to imines with subsequent ring closure [106]. The latter strategy has been adapted to solid-phase synthesis by Enders and coworkers, who utilized the T1-triazene linker system [107]. The resin was loaded with alanine methylester to give **120** (Scheme 18.35) [108]. The precursor **120** was then transferred into its dianion using an excess of LiHMDS and subsequently treated with imines derived from aniline and aromatic aldehydes. Traceless cleavage of the resulting β-lactams was accomplished in a two-step process including formation of the diazonium salt followed by decomposition of the diazonium intermediate to yield the target molecules **121**. With R = aryl or heteroaryl, a range of products (eight examples) was obtained in good yields (53–71%) with high purities (88–98%) and excellent diastereoselectivity in favor of the trans-configurated product (90 to ≥96% de).

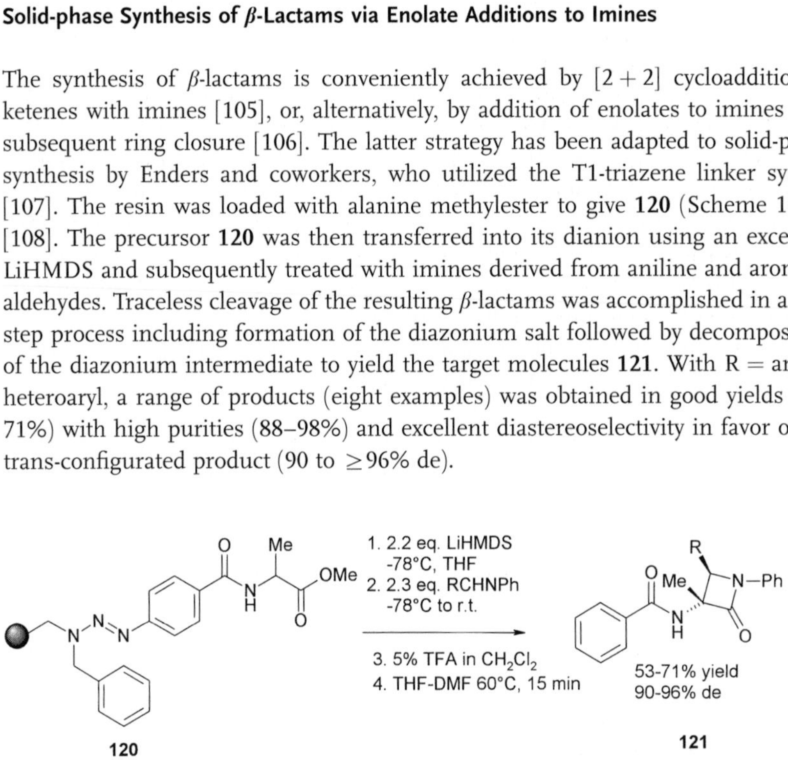

Scheme 18.35. Solid-phase synthesis of β-lactams employing a traceless linker strategy.

A related process was reported involving *in situ*-generated Ti enolates and imines immobilized on soluble PEG resins [109].

18.9
Nitro-aldol Reactions

The nitro-aldol reaction (Henry reaction) represents a useful tool for the introduction of a nitrogen functionality onto an organic scaffold. The base-promoted addition of C–H acidic nitroalkanes to aldehydes has been realized in solution phase by utilizing polymer-supported reagents. For instance, immobilized 1,5,7-triaza-bizyclo[4.4.0]dec-1-ene (TBD) has been reported to catalyze the aforementioned process [110]. The use of polymeric TBD dramatically simplified work-up, and product isolation was reduced to a simple filtration process. However, only a limited number of examples were reported that lacked data on experimental details and product purity.

Other polymeric bases which catalyze the Henry reaction are polymeric ylides [65], ion-exchange resins such as Amberlyst A21 [111], or basic La^{3+} polymers [112]. Again, the scope of these methods is rather limited to a small set of examples, and an assessment regarding the suitability of these reagents for library synthesis is therefore exceedingly difficult.

Polymer-supported reagents were also successfully used to promote the 1,4-addition of nitroalkanes to Michael acceptors. Accordingly, KF on basic alumina was found to catalyze the addition of C–H acidic nitro-compounds to a range of α,β-unsaturated ketones and esters (Scheme 18.36) [113, 114]. The reactions were driven to completion by using a large excess of the nitroalkane. Finally, the adducts **124** were isolated after filtration and evaporation of the excess nitroalkane in good yields (39–100% for 14 examples). The best results were obtained if the Michael acceptors were unsubstituted in the β-position.

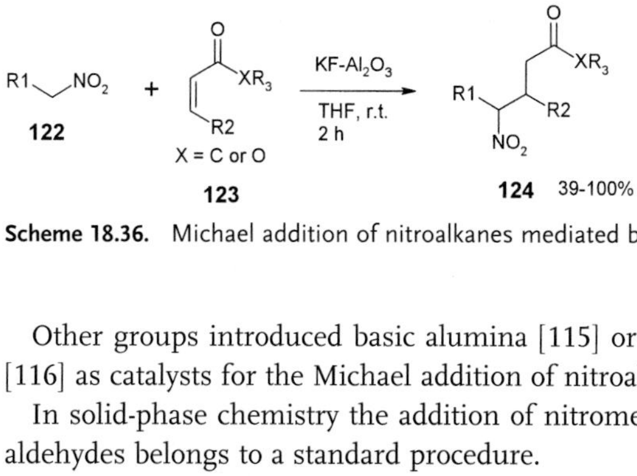

Scheme 18.36. Michael addition of nitroalkanes mediated by KF on alumina.

Other groups introduced basic alumina [115] or Amberlyst ion-exchange resins [116] as catalysts for the Michael addition of nitroalkanes.

In solid-phase chemistry the addition of nitromethane to immobilized aromatic aldehydes belongs to a standard procedure.

A typical example of the Henry reaction on solid support is illustrated in Scheme 18.37 [117]. The isoxazole aldehyde **125** loaded onto the polystyrene resin via a Cl-trityl linker was treated with a large excess of nitromethane and triethyl-amine in a THF/EtOH solvent mixture. The addition proceeded at rt within 3 h

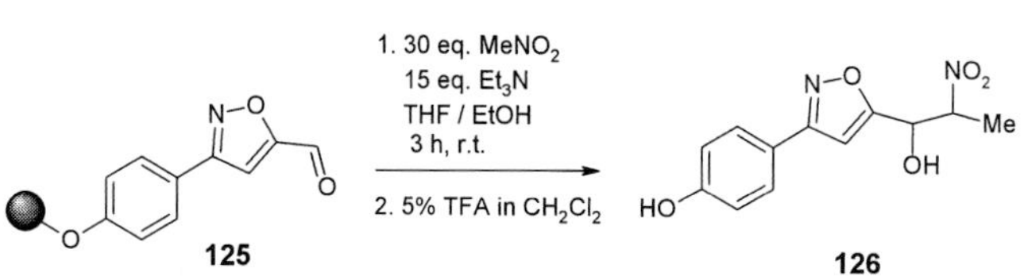

Scheme 18.37. Henry reaction on solid support.

and the resulting Henry adduct was cleaved from the resin to produce phenol **126** with a 96% yield and with 95% purity. Practically identical conditions were reported for other polymeric substrates [118].

18.10
The Baylis–Hillman Reaction

The Baylis–Hillman reaction comprises the addition of aldehydes to activated alkenes mediated by tertiary amines such as 1,4-diazabicyclo[2.2.2]octane (DABCO) or 3-quinuclidinol (3-QDL). The resulting allylic alcohols display a dense array of functional groups and thereby serve as versatile templates for further modifications.

Several protocols for the Baylis–Hillman reaction on solid phase have been described. In a typical example reported by Jung and coworkers, acryloin-loaded Cl-trityl resin was treated with a large excess of *p*-trifluoromethylbenzaldehyde and 10 equiv. DABCO in a 1:1 mixture of DMSO/CHCl₃ (Scheme 18.38) [119].

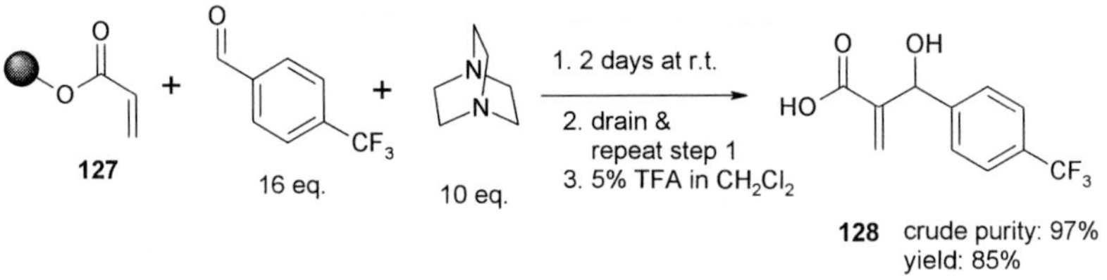

Scheme 18.38. Baylis–Hillman reaction on solid support.

The mixture was maintained at rt for 2 days and the procedure was repeated to achieve complete conversion. After acidolysis with TFA, the Baylis–Hillman product **128** was obtained with a crude purity of 97% and with an 85% yield after preparative HPLC. Similar protocols have been disclosed by other groups. Künzer and coworkers used dimethylformamide (DMF) as solvent and 3-QDL as base to promote the Baylis–Hillman addition [120], while Kulkarni and Ganesan reported that a 3:1 DMF/MeCN solvent mixture in combination with DABCO and catalytic amounts of La(OTf)₃ were beneficial to the process [121]. The major limitations for the solid-phase versions of the Baylis–Hillman transformation arise from incomplete reactions and a narrow range of aldehydes that are able to react smoothly under the aforementioned conditions. Large excesses of the reagents or prolonged reaction times with repeated incubations did not dramatically improve the situation.

Preferably, aromatic aldehydes with electron-withdrawing substituents work well in these reactions while less reactive aromatic or aliphatic aldehydes gave sluggish reactions. A slightly different protocol for the polymer-supported Baylis–Hillman reaction was described by Reiser and coworkers [122]. Herein, acrolein immobilized on a soluble PEG resin via an ester linkage was reacted with a range of aldehydes using ethanol as solvent and 3-QDL as base. The reaction with a diverse set of aldehydes (six examples), including aliphatic aldehydes, could be driven to completion since the soluble polymer guaranteed solution-like kinetics. However, it was also reported that the ester linkage to the PEG polymer was partly cleaved under the reaction conditions, leading to a lower yield of the desired product.

In conclusion, the Baylis–Hillman reaction on solid phase is an established process; however, owing to the inherent limitations, a careful selection of the polymeric linker as well as the aldehyde set used in this reaction should be addressed prior to multicomponent library synthesis.

Finally, an interesting three-component version of the Baylis–Hillman reaction was reported on solid support. Thus, resin **127** was treated with aldehydes and sulfonamides in the presence of DABCO in dioxane at 70 °C. This cocktail gave rise to the formation of functionalized sulfonamides **129** which represent useful intermediates for further manipulations (Scheme 18.39) [123]. The versatility of this chemistry has been shown for aromatic and heteroaromatic aldehydes in combination with arylsulfonamides. The best results were obtained if electron-deficient aldehydes were employed in this process.

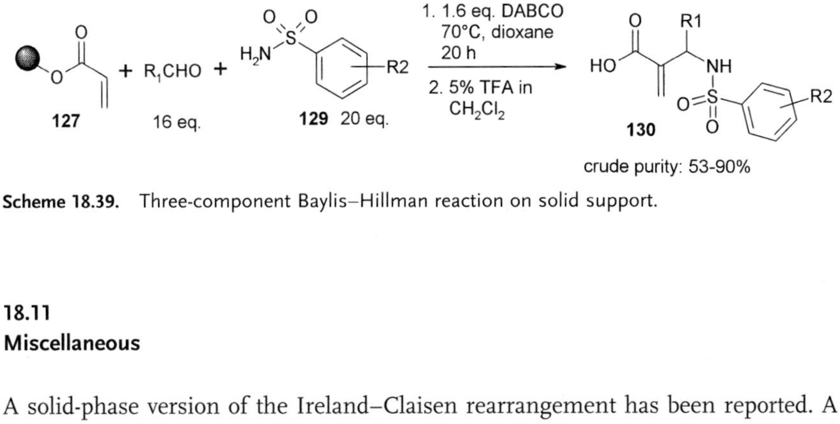

Scheme 18.39. Three-component Baylis–Hillman reaction on solid support.

18.11
Miscellaneous

A solid-phase version of the Ireland–Claisen rearrangement has been reported. A key feature of this concept was the formation of silyl ketene acetals on solid support employing polymeric silyl triflate **132** (Scheme 18.40) [124]. The resulting enolate intermediate (**133**) cleanly underwent an Ireland–Claisen rearrangement

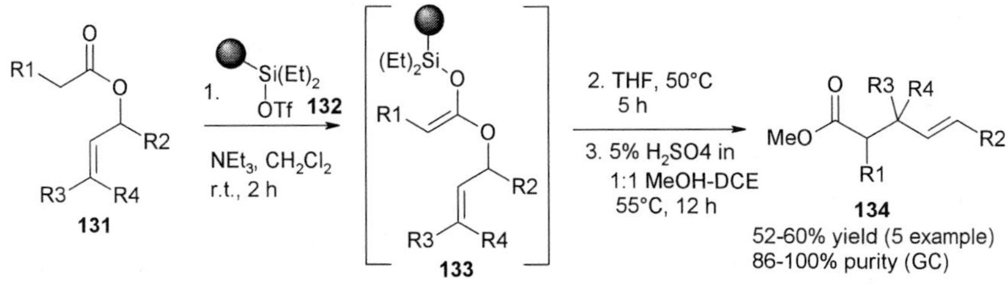

Scheme 18.40. Ireland–Claisen rearrangement on solid support.

in THF at 50 °C to yield the corresponding silyl esters which were cleaved from the resin by transesterification. The reaction worked well for a range of allylic esters, however, esters derived from acetic acid failed to give the desired rearrangement product (**134**). This result is presumably due – to a large extent – to C-silylation in the enolization–immobilization event.

An elegant approach to substituted phenols via a cyclative cleavage strategy was reported by Katritzky and coworkers (Scheme 18.41) [125]. Merrifield resin loaded with 3-hydroxypyridine was alkylated with bromoacetone to give pyridinium salt (**135**). The latter was reacted with α,β-unsaturated ketones under basic conditions to yield Robinson anellation product **136**, which aromatized with concomitant cleavage from the resin. A range of chalcones (12 examples) were successfully utilized in this reaction and the phenols **137** were isolated after aqueous work-up and column chromatography in good yields with high purities.

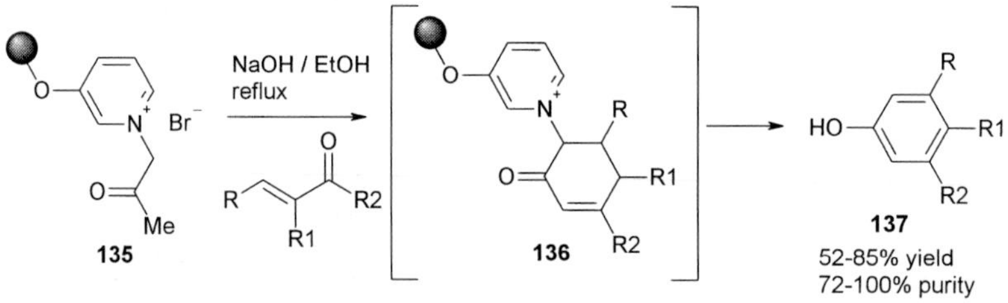

Scheme 18.41. Solid-phase synthesis of substituted phenols.

Another example demonstrates that functionalized organometallics can conveniently be handled under anhydrous conditions in solid-phase organic synthesis (Scheme 18.42) [126]. The amino acid derivative **138** loaded on Wang resin was treated with an excess of LDA under a nitrogen atmosphere to generate the corresponding lithium enolate. Subsequent transmetalation with ZnBr$_2$ resulted in the formation of the Zn enolate, which underwent intramolecular carbozincation to yield organozinc intermediate **139**. The latter was quenched with reactive electrophiles (I$_2$ or H$^+$) and cleavage from the resin led to the formation of proline derivatives (**140**). As concluded from liquid chromatography/mass spectrometry (LC/MS), the desired products have been formed virtually quantitatively (no yields were reported) as single diastereomers.

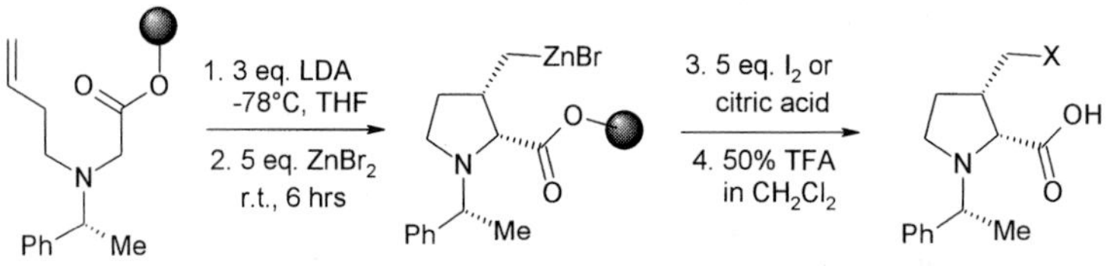

Scheme 18.42. Solid-phase synthesis of substituted proline derivatives.

This carbozincation protocol (depicted in Scheme 18.42) offers an innovative approach to substituted proline derivatives; however, the scope of this process is limited with respect to diversification. Thus, the extension of this methodology to a broader range of substrates as well as the adaptation of the reaction protocol to automation clearly remains a challenge.

References

1 a) For solid phase, see: R. E. SAMMELSON, M. J. KURTH, *Chem. Rev.* **2001**, *101*, 137–202; b) for polymer-supported reagents, see: S. V. LEY, I. R. BAXENDALE, R. N. BREAM, P. S. JACKSON, A. G. LEACH, D. A. LONGBOTTOM, M. NESI, J. S. SCOTT, R. I. STORER, S. TAYLOR, *J. Chem. Soc., Perkin Trnas. 1*, **2000**, 3815–4195.

2 T. RUHLAND, H. KÜNZER, *Tetrahedron Lett.* **1996**, *37*, 2757–2760.

3 M. J. KURTH, L. A. AHLBERG RANDALL, C. CHEN, C. MELANDER, R. B. MILLER, *J. Org. Chem.* **1994**, *59*, 5862.

4 a) K. C. NICOLAOU, D. VOURLOUMIS, T. LI, J. PASTOR, N. WINSSINGER, Y. HE, S. NINKOVIC, F. SARABIA, H. VALLBERG, F. ROSCHANGAR, N. P. KING, M. R. V. FINLAY, P. GIANNAKAKOU, P. VERDIER-PINARD, E. HAMEL, *Angew. Chem.* **1997**, *109*, 2181–2187; b) K. C. NICOLAOU, N. WINSSINGER, J. PASTOR, S. NINKOVIC, F. SARABIA, Y. HE, D. VOURLOUMIS, Z. YANG, T. LI, P. GIANNAKAKOU, E. HAMEL, *Nature*, **1997**, *387*, 268–272.

5 a) S. P. HOLLINSHEAD, *Tetrahedron Lett.* **1996**, *37*, 9157–9160; b) S. P. HOLLINSHEAD, WO 98/08813; c) for additional examples, see: P. GROSCHE, A. HÖLTZEL, T. B. WALK, A. W. TRAUTWEIN, G. JUNG, *Synthesis* **1999**, *11*, 1961–1970.

6 C. CHIU, T. TANG, J. W. ELLINGBOE, *J. Comb. Chem.* **1999**, *1*, 73–77.

7 A. L. MARZINZIK, WO 98/155532.

8 A. L. MARZINZIK, E. R. FELDER, *J. Org. Chem.* **1998**, *63*, 723–727.

9 C. GENNARI, S. CECCARELLI, U. PIARULLI, K. ABOUTAYAB, M. DONGHI, I. PATERSON, *Tetrahedron* **1998**, *54*, 14999–15016.

10 I. PATERSON, M. DONGHI, K.

GERLACH, *Angew. Chem. Int. Ed.* **2000**, *112*, 3453–3457.

11 M. REGGELIN, V. BRENIG, *Tetrahedron Lett.* **1996**, *37*, 6851–6852.

12 M. REGGELIN, V. BRENIG, R. WELCKER, *Tetrahedron Lett.* **1998**, *39*, 4801–4804.

13 M. REGGELIN, V. BRENIG, C. ZUR, *Org. Lett.* **2000**, *2*, 531–533.

14 a) A. V. PURANDARE, S. NATARAJAN, *Tetrahedron Lett.* **1997**, *38*, 8777–8780; b) C. W. PHOON, C. ABELL, *Tetrahedon Lett.* **1998**, *39*, 2655–2658.

15 H. DANDA, M. M. HANSEN, C. H. HEATCHCOCK, *J. Org. Chem.* **1990**, *55*, 173–181.

16 T. MUKAIYAMA, H. IWAKIRI, *Chem. Lett.* **1985**, 1363–1366.

17 a) S. KOBAYASHI, R. AKIYAMA, T. FURUTA, M. MORIWAKI, *Molecules Online* **1998**, *2*, 35–39; b) L. YU, D. CHEN, J. LI, P. G. WANG, *J. Org. Chem.* **1997**, *62*, 3575–3581.

18 M. KAWAI, M. ONAKA, Y. IZUMI, *Bull. Chem. Soc. Jpn.* **1988**, *61*, 1237–1245.

19 T. P. LOH, X.-R. LI, *Tetrahedron* **1999**, *55*, 10789–10802.

20 B. C. RANU, R. CHAKRABORTY, *Tetrahedron* **1993**, *49*, 5333–5338.

21 B. C. RANU, M. SAHA, S. BHAR, *Synthetic Commun.* **1997**, *27*, 3065–3077.

22 a) S. KOBAYASHI, S. NAGAYAMA, *J. Am. Chem. Soc.* **1998**, *120*, 2985–2986; b) S. KOBAYASHI, S. NAGAYAMA, *J. Am. Chem. Soc.* **1998**, *120*, 4554; c) S. NAGAYAMA S. KOBAYASHI, *Angew. Chem. Int. Ed.* **2000**, *39*, 567–569.

23 F. HAUNERT, M. H. BOLLI, B. HINZEN, S. V. LEY, *J. Chem. Soc., Perkin Trans. 1* **1998**, 2235–2237.

24 S. KIYOOKA, Y. KIDO, Y. KANEKO, *Tetrahedron Lett.* **1994**, *35*, 5243–5246.

25 A. MANDOLI, D. PINI, S. ORLANDI, F. MAZZINI, P. SALVADORI, *Tetrahedron Asymmetry* **1998**, *9*, 1479–1482.

26 A. FUJII, M. SODEOKA, *Tetrahedron Lett.* **1999**, *40*, 8011–8014.

27 S. KOBAYASHI, I. HACHIYA, M. YASUDA, *Tetrahedron Lett.* **1996**, *37*, 5569–5572.

28 S. KOBAYASHI, T. WAKABAYASHI, M. YASUDA, *J. Org. Chem.* **1998**, *63*, 4868–4869.

29 A. GRAVEN, M. GRØTLI, M. MELDAL, *J. Chem. Soc., Perkin Trans. 1*, **2000**, 955–962.

30 E. D. BERGMANN, R. CORETT, *J. Org. Chem.* **1958**, *23*, 1507–1510.

31 I. RODRIGUEZ, S. IBORRA, A. CORMA, F. REY, J. JORDÁ, *Chem. Commun.* **1999**, 593–594.

32 a) P. HODGE, E. KHOSHDEL, J. WATERHOUSE, *J. Chem. Soc., Perkin Trans I*, **1983**, 2205–2209; b) M. INAGAKI, J. HIRATAKE, Y. YAMAMOTO, J. ODA, *Bull. Chem. Soc. Jpn.*, **1987**, *60*, 4121–4126; c) R. ALVAREZ, M. A. HOURDIN, C. CAVÉ, J. D'ANGELO, P. CHAMINADE, *Tetrahedron Lett.* **1999**, *40*, 7091–7094.

33 S. V. LEY, A. MASSI, *J. Comb. Chem.* **2000**, *2*, 104–107.

34 a) S. V. LEY, D. M. MYNETT, W.-J. KOOT, *Synlett*, **1995**, 1017–1020; b) H.-J. GUTKE, D. SPITZNER, *Tetrahedron*, **1999**, *55*, 3931–3936.

35 E. DOMINGUEZ, M. J. O'DONNELL, W. L. SCOTT, *Tetrahedron Lett.* **1998**, *39*, 2167–2170.

36 C. CHIU, Z. TANG, J. W. ELLINGBOE, *J. Comb. Chem.* **1999**, *1*, 73–77.

37 H. MOLINARI, F. MONTANARI, S. QUICI, P. TUNDO, *J. Am. Chem. Soc.* **1979**, *101*, 3920–3927.

38 G. L. KAD, V. SINGH, A. KHURANA, S. CHAUDHARY, J. SINGH, *Synth. Commun.* **1999**, *29*, 3439–3442.

39 a) B. J. BACKES, J. A. ELLMAN, *J. Am. Chem. Soc.* **1994**, *116*, 11171–11172; b) J. A. ELLMAN, *Acc. Chem. Res.* **1996**, *29*, 132–143.

40 J. A. PORCO JR, T. GEEGAN, W. DEVONPORT, O. W. GOODING, K. HEISLER, J. W. LABADIE, B. NEWCOMB, C. NGUYEN, P. VAN EIKEREN, J. WONG, P. WRIGHT, *Mol. Diversity* **1996**, *2*, 197–206.

41 G. W. KENNER, J. R. MCDERMOTT, R. C. SHEPPARD, *J. Chem. Soc., Chem. Commun.* **1971**, 636–637.

42 Z. ZHU, B. MCKITTRICK, *Tetrahedron Lett.* **1998**, *39*, 7479.

43 S. HANESSIAN, F. XIE, *Tetrahedron Lett.* **1998**, *39*, 737–740.

44 a) G. GELBARD, *Synthesis* **1977**, 113–116; b) K. SHIMO, S. WAKAMATSU, *J. Org. Chem.* **1963**, *28*, 504–506.

45 J. M. MILLER, K.-H. SO, *J. Chem. Soc., Chem. Comm.* **1978**, 466–467.

46 J. YAMAWAKI, T. ANDO, *Chem Lett.* **1979**, 755–758.

47 L. T. TIETZE, A. STEINMETZ, F. BALKENHOHL, *Bioorg. Med. Chem. Lett.* **1997**, *7*, 1303–1306.

48 A. L. MARZINIK, E. R. FELDER, *Tetrahedron Lett.* **1996**, *37*, 1003–1006.

49 A. L. MARZINIK, E. R. FELDER, *Molecules*, **1997**, *2*, 17–30.

50 H. STEPHENSEN, F. ZARAGOZA, *Tetrahedron Lett.* **1999**, *40*, 5799–5802.

51 a) O. W. GOODING, S. BAUDART, T. L. DEEGAN, K. HEISLER, J. W. LABADIE, W. S. NEWCOMB, J. A. PORCO JR, P. VAN EIKEREN, *J. Comb. Chem.* **1999**, *1*, 113–122; b) T. REDEMANN, H. BANDEL, G. JUNG, *Mol. Diversity* **1998**, *4*, 191–197.

52 E. T. M. WOLTERS, G. I. TESSER, R. J. F. NIVARD, *J. Org. Chem.* **1974**, *39*, 3388–3392.

53 J. M. J. FRÉCHET, M. D. DE SMET, M. J. FARRALL, *J. Org. Chem.* **1979**, *44*, 1774–1778.

54 L. F. TIETZE, T. HIPPE, A. STEINMETZ, *Chem. Commun.* **1998**, 793–794.

55 L. F. TIETZE, A. STEINMETZ, *Synlett* **1996**, 667–668.

56 P. M. WORSTER, C. R. MCARTHUR, C. C. LEZNOFF, *Angew. Chem.* **1979**, *91*, 255.

57 C. R. MCARTHUR, P. M. WORSTER, J.-L. JIANG, C. C. LEZNOFF, *Can. J. Chem.* **1982**, *60*, 1837–1841.

58 H.-S. MOON, N. E. SCHORE, M. J. KURTH, *J. Org. Chem.* **1992**, *57*, 6088–6089.

59 H.-S. MOON, N. E. SCHORE, M. J. KURTH, *Tetrahedron Lett.* **1994**, *35*, 8915–8918.

60 A. R. COLWELL, L. R. DUCKWALL, R. BROOKS, S. P. MCMANUS, *J. Org. Chem.* **1981**, *46*, 3097–3102.

61 S. M. ALLIN, S. SHUTTELWORTH, *Tetrahedron Lett.* **1996**, *37*, 8023–8026.

62 K. BURGESS, D. LIM, *Chem. Commun.* **1997**, 785–786.

63 S. HANESSIAN, J. MA, W. WANG, *Tetrahedron Lett.* **1999**, *40*, 4631–4634.

64 S. CHEN, K. D. JANDA, *Tetrahedron Lett.* **1998**, *39*, 3943–3946.

65 F. PALACIOS, D. APARICIO, J. M. DE LOS SANTOS, A. BACEIREDO, G. BERTRAND, *Tetrahedron* **2000**, *56*, 663–669.

66 R. CHINCHILLA, P. MAZÓN, C. NÁJERA, *Tetrahedron Asymmetry*, **2000**, *11*, 3277–3281.

67 M. J. O'DONNELL, C. W. LUGAR, R. S. POTTORF, C. ZHOU, *Tetrahedron Lett.* **1997**, *38*, 7163–7166.

68 M. J. O'DONNELL, C. ZHOU, W. L. SCOTT, *J. Am. Chem. Soc.* **1996**, *118*, 6070–6071.

69 B. SAUVAGNAT, F. LAMATY, R. LAZARO, J. MARTINEZ, *Tetrahedron Lett.* **1998**, *39*, 821–824.

70 W. L. SCOTT, C. ZHOU, Z. FANG, M. O'DONNELL, *Tetrahedron Lett.* **1997**, *38*, 3695–3698.

71 M. J. O'DONNELL, F. DELGADO, R. S. POTTORF, *Tetrahedron* **1999**, *55*, 6347–6362.

72 a) B. J. COHEN, M. A. KRAUS, A. PATCHORNIK *J. Am. Chem. Soc.* **1977**, *99*, 4165–4167; b) B. J. COHEN, M. A. KRAUS, A. PATCHORNIK, *J. Am. Chem. Soc.* **1981**, *103*, 7620–7629.

73 K. H. BLEICHER, J. R. WAREING, *Tetrahedron Lett.* **1998**, *39*, 4587–4590.

74 K. C. NICOLAOU, G.-Q. CAO, J. A. PFEFFERKORN, *Angew. Chem. Int. Ed.* **2000**, *39*, 739–743.

75 F. BALKENHOHL, C. VON DEM BUSSCHE-HÜNNEFELD, A. LANSKY, C. ZECHEL, *Angew. Chem.* **1996**, *108*, 2436–2488.

76 D. R. CODY, S. H. DEWITT, J. C. HODGES, B. D. ROTH, M. C. SCHROEDER, C. J. STANKOVIC, W. H. MOOS, M. R. PAVIA, F. S. KIELY, WO 94/08711.

77 B. A. KULKARNI, A. GANESAN, *Angew. Chem.* **1997**, *109*, 2565–2566.

78 For the synthesis of 4-hydroxy-quinolin-2-(1*H*)-ones, see: B. A. KULKARNI, A. GANESAN, *Chem. Commun.* **1998**, 785–786.

79 a) J. I. CROWLEY, H. RAPOPORT, *J. Am. Chem. Soc.* **1970**, *92*, 6363–6365; b) J. I. CROWLEY, H. RAPOPORT, *J. Org. Chem.* **1980**, *45*, 3215–3227.

80 J. MATTHEWS, R. A. RIVERO, *J. Org. Chem.* **1998**, *63*, 4808–4810.

81 T. T. ROMOFF, L. MA, Y. WANG, *Synlett* **1998**, 1341–1342.

82 L. WEBER, P. IAIZA, G. BIRINGER, P. BARBIER, *Synlett* **1998**, 1156–1158.

83 B. A. KULKARNI, A. GANESAN, *Tetrahedron Lett.* **1998**, *39*, 4369–4372.

84 E. ANGELETTI, C. CANEPA, G. MARTINETTI, P. VENTURELLO, *J. Chem. Soc., Perkin Trans. I* **1989**, 105–107.

85 R. W. HEIN, M. J. ASTLE, J. R. SHELTON, *J. Org. Chem.* **1961**, *26*, 4874–4878.

86 I. RODRIGUEZ, S. IBORRA, A. CORMA, F. REY, J. JORDÁ, *Chem. Commun.* **1999**, 593–594.

87 a) J. A. CABELLO, J. M. CAMPELO, A. GARCIA, D. LUNA, J. M. MARINAS, *J. Org. Chem.* **1984**, *49*, 5195; b) S. CHALAIS, P. LASZLO, A. MATHY, *Tetrahedron Lett.* **1982**, *23*, 4927.

88 a) D. VILLEMIN, *Chem. Commun.* **1983**, 1092–1093; b) J. YAMAWAKI, T. KAWATE, T. ANDO, T. HANAFUSA, *Bull. Chem. Soc. Jpn.* **1983**, *56*, 1885–1886.

89 G. A. TAYLOR, *J. Chem. Soc., Perkin Trans. I* **1981**, 3132.

90 S. TADESSE, A. BHANDARI, M. A. GALLOP, *J. Comb. Chem.* **1999**, *1*, 184–187.

91 A. BHANDARI, B. LI, M. A. GALLOP, *Synthesis* **1999**, 1951–1960.

92 M. F. GORDEEV, D. V. PATEL, J. WU, E. M. GORDON, *Tetrahedron Lett.* **1996**, *37*, 4643–4646.

93 a) B. T. WATSON, G. E. CHRISTIANSEN, *Tetrahedron Lett.* **1998**, *39*, 6087–6090; b) A. SVENSSON, K.-E. BERGQUIST, T. FEX, J. KIHLBERG, *Tetrahedron Lett.* **1998**, *39*, 7193–9196; c) Y. XIA, Z.-Y. YANG, A. BROSSI, K.-H. LEE, *Org. Lett.* **1999**, *1*, 2113–2115.

94 B. C. HAMPER, D. M. SNYDERMAN, T. J. OWEN, A. M. SCATES, D. C. OWSLEY, A. S. KESSELRING, R. C. CHOTT, *J. Comb. Chem.* **1999**, *1*, 140–150.

95 a) L. F. TIETZE, A. STEINMETZ, *Angew.*

Chem., Int. Ed. **1996**, *35*, 651–652; b) L. F. Tietze, T. Hippe, A. Steinmetz, *Synlett* **1996**, 1043–1044; c) B. C. Hamper, K. Z. Gan, T. J. Owen, *Tetrahedron Lett.* **1999**, *40*, 4973–4976.

96 a) S. Kobayashi, S. Nagayama, *J. Org. Chem.* **1996**, *61*, 2256; b) S. Kobayashi, S. Nagayama, *J. Am. Chem. Soc.* **1996**, *118*, 8977; c) S. Kobayashi, S. Nagayama, *Synlett* **1997**, 653–654; d) L. Yu, D. Chen, J. Li, P. G. Wang, *J. Org. Chem.* **1997**, *62*, 3575–3581.

97 N. Tanaka, Y. Masaki, *Synlett* **2000**, 406–408.

98 E. N. Prabhakaran, J. Iqbal, *J. Org. Chem.* **1999**, *64*, 3339–3341.

99 B. Bhatia, M. M. Reddy, J. Iqbal, *J. Chem. Soc., Chem. Commun.* **1994**, 713–714.

100 a) S. Kobayashi, I. Hachiya, S. Suzuki, M. Moriwaki, *Tetrahedron Lett.* **1996**, *37*, 2809–2812; b) S. Kobayashi, M. Moriwaki, *Tetrahedron Lett.* **1997**, *38*, 4251–4254.

101 S. Kobayashi, M. Moriwaki, R. Akiyama, S. Suzuki, I. Hachiya, *Tetrahedron Lett.* **1996**, *37*, 7783–7786.

102 S. Kobayashi, Y. Aoki, *Tetrahedron Lett.* **1998**, *39*, 7345–7348.

103 S. Kobayashi, R. Akiyama, H. Kitagawa, *J. Comb. Chem.* **2000**, *2*, 438–440.

104 a) S. Kobayashi, T. Furuta, K. Sugita, O. Okitsu, H. Oyamada, *Tetrahedron Lett.* **1999**, *40*, 1341–1344; b) S. Kobayashi, H. Oyamada, WO 99/46238.

105 G. I. Georg, V. T. Ravikumar in: The Organic Chemistry of β-Lactams, VCH, New York 1993, pp. 295–368.

106 D. J. Hart, D. C. Ha, *Chem. Rev.* **1989**, *89*, 1447–1465.

107 S. Bräse, D. Enders, J. Köbberling, F. Avemaria, *Angew. Chem.* **1998**, *110*, 3614–3616; *Angew. Chem., Int. Ed.* **1998**, *37*, 3413–3415.

108 S. Schunk, D. Enders, *Org. Lett.* **2000**, *2*, 907–910.

109 a) V. Molteni, R. Annunziata, M. Cinquini, F. Cozzi, M. Benaglia, *Tetrahedron Lett.* **1998**, *39*, 1257; b) R. Annunziata, M. Benaglia, M. Cinquini, F. Cozzi, *Chem. Eur. J.* **2000**, *6*, 133–138.

110 D. Simoni, R. Rondanin, M. Morini, R. Baruchello, F. P. Invidiata, *Tetrahedron Lett.* **2000**, *41*, 1607–1610.

111 G. Rosini, R. Ballini, M. Petrini, *Synthesis* **1986**, 46–48.

112 T. Saiki, Y. Aoyama, *Chem. Lett.* **1999**, 197–798.

113 D. E. Bergbreiter, J. J. Lalonde, *J. Org. Chem.* **1987**, *52*, 1601–1603.

114 J. H. Clark, G. Cork, *J. Chem. Soc., Perkin Trans. I* **1983**, 2253–2258.

115 B. C. Ranu, S. Bhar, *Tetrahedron* **1992**, *48*, 1327–1332.

116 a) R. Ballini, M. Petrini, G. Rosini, *Synthesis* **1987**, 711–713; b) R. Ballini, P. Marzialia, A. Mozzicafreddo, *J. Org. Chem.* **1996**, *61*, 3209–3211.

117 S. Batra, S. K. Rastogi, B. Kundu, A. Patra, A. P. Bhaduri, *Tetrahedron Lett.* **2000**, *41*, 5971.

118 a) X. Beebe, N. E. Schore, M. J. Kurth, *J. Am. Chem. Soc.* **1992**, *114*, 10061–10062; b) X. Beebe, N. E. Schore, M. J. Kurth, *J. Org. Chem.* **1995**, *60*, 4196–4203; c) X. Beebe, C. L. Chiappari, M. M. Olmstead, M. J. Kurth, *J. Org. Chem.* **1995**, *60*, 4204–4212; d) J. Rademann, M. Meldal, K. Bock, *Chem. Eur. J.* **1999**, *5*, 1218–1225.

119 a) H. Richter, G. Jung, *Mol. Diversity* **1998**, *3*, 191–194; b) H. Richter, T. Walk, A. Höltzel, G. Jung, *J. Org. Chem.* **1999**, *64*, 1362–1365.

120 O. Prien, K. Rölfing, M. Thiel, H. Künzer, *Synlett* **1997**, 325–326.

121 B. A. Kulkarni, A. Ganesan, *J. Comb. Chem.* **1999**, *1*, 373–378.

122 R. Räcker, K. Döring, O. Reiser, *J. Org. Chem.* **2000**, *65*, 6932–6939.

123 H. Richter, G. Jung, *Tetrahedron Lett.* **1998**, *39*, 2729–2730.

124 Y. Hu, J. A. Porco Jr, *Tetrahedron Lett.* **1999**, *40*, 3289–3292.

125 A. R. Katritzky, S. A. Belyakov, Y. Fang, J. S. Kiely, *Tetrahedron Lett.* **1998**, *39*, 8051–8054.

126 P. Karoyan, A. Triolo, R. Nannicini, D. Giannotti, M. Altamura, G. Chassaing, E. Perrotta, *Tetrahedron Lett.* **1999**, *40*, 71–74.

19

Solid-phase Palladium Catalysis for High-throughput Organic Synthesis

Yasuhiro Uozumi and Tamio Hayashi

19.1
Introduction

Palladium-mediated organic transformations have emerged as a powerful tool in the domain of synthetic organic chemistry [1]. Recently, high-throughput organic synthesis by solid-phase chemistry has been gaining in popularity owing to the ease of purification of the products [2]. Palladium-catalyzed reactions have also found widespread utility in the preparation of small molecule libraries, especially for medicinal screening purposes. This chapter surveys solid-phase synthetic reactions [3] by palladium catalysis.

19.2
Carbon–Carbon and Carbon–Nitrogen Bond-forming Reactions of Aryl and Alkenyl Halides

19.2.1
Cross-coupling Reactions

Palladium complexes catalyze the reaction of organometallic reagents (R-m) with aryl or alkenyl halides and related compounds (R′–X) to give cross-coupling products (R–R′) (Scheme 19.1). It is generally accepted that the catalytic cycle of the reaction proceeds via oxidative addition of R′–X to palladium(0) and subsequent ligand exchange of the resulting LnPd(R′)X with R-m to give, as the key intermediate, the unsymmetrical diorganometal LnPd(II)(R)R′. From this intermediate, the product R–R′ is released by reductive elimination to leave a LnPd(0) species that undergoes the next catalytic cycle. Transfer of an alkyl group from R-m to LnPd(R′)X, the so-called transmetalation step, takes place with organometallic compounds of Mg, Zn, B, Al, Sn, Si, Zr, etc.

Among the many organometallic reagents used for cross-coupling, organoboron reagents and organotin reagents have been extensively examined for solid-phase applications. Palladium-catalyzed coupling of aryl halides with a terminal alkyne in the presence of a copper salt has also been studied on solid phase.

$$R\text{-}m + R'\text{-}X \xrightarrow{\text{[M]}} R\text{-}R' + m\text{-}X$$

M = Ni, Pd
m = Mg, Zn, Al, Si, Zr, Sn, B, etc.
R' = aryl, alkenyl
X = Cl, Br, I, OSO_2CF_3, $OPO(OR)_2$, etc.

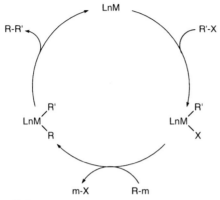

Scheme 19.1. Reaction pathway of cross-coupling reactions.

19.2.1.1 Reactions of Aryl and Alkenyl Halides with Organoboron Reagents

The palladium-catalyzed cross-coupling of aryl halides with organoboron reagents (the so-called Suzuki–Miyaura coupling) has become an indispensable carbon–carbon bond-forming reaction in fine organic synthesis [4]. One of the earliest examples of its solid-phase application was using the aryl bromide **1** supported on polystyrene resin with a sulfonamide linker (Scheme 19.2) [5]. The Suzuki–Miyaura coupling of the bromide **1** with aryl boronic acids or alkyl 9-BBN in aqueous tetrahydrofuran (THF) took place in the presence of $Pd(Ph_3P)_4$ with Na_2CO_3 as the base at 65 °C. Nucleophilic cleavage of the resulting resin-bound products **2** and **3** gave the corresponding coupling products in excellent yields.

Couplings of various halogenobenzoates (**4–9**), the *m*-iodobenzyl ester **10**, and the *p*-, *m*-, *o*-bromobenzaldehyde acetals **12–14** bound to the polystyrene (PS) matrix resin are summarized in Tables 19.1, 19.2, and 19.3, respectively. Reactions of the resin-supported *p*-iodobenzoate **4** with arylboronic acids were catalyzed by $Pd_2(dba)_3$ (dba = dibenzylideneacetone) at room temperature in aqueous dimethylformamide (DMF) to give good to excellent yields of the corresponding biaryl carboxylic acids after hydrolysis (Table 19.1, entries 1–3). The iodobenzoate **4** also reacted with pinacol alkenylborates and tributylborane in the presence of $Pd(Ph_3P)_4$ (Table 19.1, entries 4 and 5) [6]. Coupling of the *p*-bromobenzoate **5** with various electron-rich as well as electron-deficient arylboronic acids was examined with $Pd(Ph_3P)_4$ in aqueous dimethoxyethane (DME) (Table 19.1, entries 6–10) [7]. Base-mediated methanolysis of the resulting resin gave methyl 4-(4'-

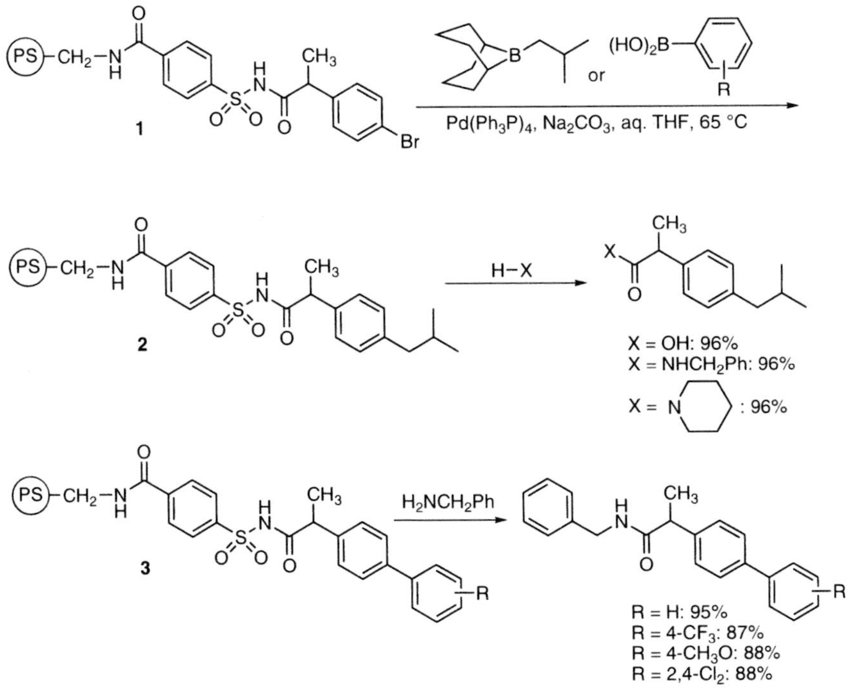

Scheme 19.2. Solid-phase Suzuki–Miyaura coupling: preparation of substituted arylacetic acid derivatives.

substituted aryl)benzoates in excellent yields. The bromobenzoates **7–9** having ortho substituents underwent the Suzuki–Miyaura coupling under the same reaction conditions (Table 19.1, entries 12–14).

The iodobenzene **10** connected to the PS resin by benzyl benzoate linkage reacted with a wide variety of arylboronic acids, including heteroaromatic reagents in aqueous dioxane at 100 °C (Table 19.2) [8]. Palladium acetate catalyzed the coupling to give the *m*-arylbenzyl alcohols (**11**) after methanolysis in satisfactory yields. Palladium-catalyzed cross-coupling of the cyclic acetals of *o*-, *m*-, and *p*-bromobenzaldehydes bound to the PS resin (**12–14**) was also examined with various arylboron reagents in aqueous DME in the presence of Na₂CO₃ to produce various biaryl carboxaldehydes after acid hydrolysis of the acetal linker (Table 19.3) [9]. It is worth noting that the biaryl coupling products having substituents at their 2,6- as well as 2,2′-positions were readily obtained despite their steric hindrance (Table 19.3, entries 3 and 12). Thus, the coupling of mesitylboronic acid with the *p*-bromobenzene **12** and *o*-carbamoylphenylboronic acid with the *o*-bromobenzene **14** gave 4-mesitylbenzaldehyde and 2-(2-carbamoylphenyl)benzaldehyde, respectively.

Tab. 19.1. Coupling of resin-bound aryl iodide with boron reagents.

Entry	Aryl halide	Boron reagent	Product	Yield (%)
1	**4**	(HO)$_2$B—		72
2		(HO)$_2$B—		91
3		(HO)$_2$B—		73
4		B—OSiMe$_3$		68
5		(n-Bu)$_3$B		56
6	**5**	R = H		95
7		R = Me		95
8		R = OMe		>95
9		R = F		91
10		R = NO$_2$		95

Tab. 19.1. *(continued)*

Entry	Aryl halide	Boron reagent	Product	Yield (%)
11	**6**	$(HO)_2B$—⬡—OMe		90
12	**7**			>95
13	**8**			>95
14	**9**			>95

Entries 1–5, ref. 6; entries 6–14, ref. 7. Conditions: $Pd_2(dba)_3$
(5–10 mol%), K_2CO_3 (2 equiv.), DMF, rt, then CF_3COOH/CH_2Cl_2
(for entries 1–3); $Pd(Ph_3P)_4$ (5–10 mol%), K_2CO_3 (2 equiv.), DMF,
rt, then CF_3COOH/CH_2Cl_2 (for entries 4 and 5); $Pd(Ph_3P)_4$
(5 mol%), 2 M Na_2CO_3, DME, reflux, then NaOMe, MeOH (for
entries 6–14).

Tab. 19.2. Solid-phase Suzuki–Miyaura coupling with base labile linker.

+ Ar–B(OH)$_2$

Entry	Ar–B(OH)$_2$	Conditions	Yield (%)
1	(HO)$_2$B—⟨phenyl⟩	A	95
2	(HO)$_2$B—⟨phenyl⟩—Me	A	95
3	(HO)$_2$B—⟨o-Me-phenyl⟩	A	91
4		B	96
5	(HO)$_2$B—⟨phenyl⟩—OMe	A	95
6	(HO)$_2$B—⟨o-MeO-phenyl⟩	A	89
7	(HO)$_2$B—⟨phenyl⟩—CF$_3$	A	90
8	(HO)$_2$B—⟨phenyl⟩—CH$_2$OH	A	70
9	(HO)$_2$B—⟨phenyl⟩—CHO	A	56
10	(HO)$_2$B—⟨phenyl⟩—COOH	B	No reaction

Tab. 19.2. *(continued)*

Entry	Ar–B(OH)$_2$	Conditions	Yield (%)
11	(HO)$_2$B— [phenyl with NH$_2$]	A	51
12	(HO)$_2$B— [phenyl with CF$_3$]	A	89
13	(HO)$_2$B— [phenyl with CF$_3$, Cl, Cl]	A	69
14		B	80
15	(HO)$_2$B— [phenyl with BocNH]	A	89
16		B	91
17	(HO)$_2$B— [thiophene S]	A	66
18		B	86
19	(HO)$_2$B— [thiophene S]	A	79
20	(HO)$_2$B— [naphthalene]	A	76

Ref. 8: conditions: A, Ar–B(OH)$_2$ (4 equiv.), K$_2$CO$_3$ (9 equiv.),
Pd(OAc)$_2$ (10 mol%), dioxane/H$_2$O = 6/1, 100 °C, 24 h; B,
Ar–B(OH)$_2$ (8 equiv.), K$_2$CO$_3$ (18 equiv.), Pd(OAc)$_2$ (20 mol%),
dioxane/H$_2$O = 6/1, 100 °C, 24 h.

Tab. 19.3. Coupling of resin-bound aryl iodide with boron reagents.

Entry	Aryl halide	Boron reagent	Product	Yield (%)
1	**12**	(HO)₂B—		>95
2		(HO)₂B— (Me)		60
3		(HO)₂B— (Me, Me, Me)		45
4		(HO)₂B— (OMe)		>95
5		Et₂NC(O)O (HO)₂B—		>95
6		(HO)₂B— S		65
7		(HO)₂B— S		>95

Tab. 19.3. *(continued)*

Entry	Aryl halide	Boron reagent	Product	Yield (%)
8	13 Br	(HO)$_2$B—		>95
9		(HO)$_2$B— OMe		>95
10		(HO)$_2$B— S		85
11	14 Br	(HO)$_2$B—		>95
12		Et$_2$NC(O)O (HO)$_2$B—		>95
13		(HO)$_2$B— NO$_2$		87
14		(HO)$_2$B— S		>95
15		BocHN		90

Ref. 9: conditions: boron reagent (3 equiv.), Pd(Ph$_3$P)$_4$ (5 mol%),
2 M Na$_2$CO$_3$ (8 equiv.), DME, reflux, 24 h; then dioxane, 3 M HCl,
80 °C, 24 h.

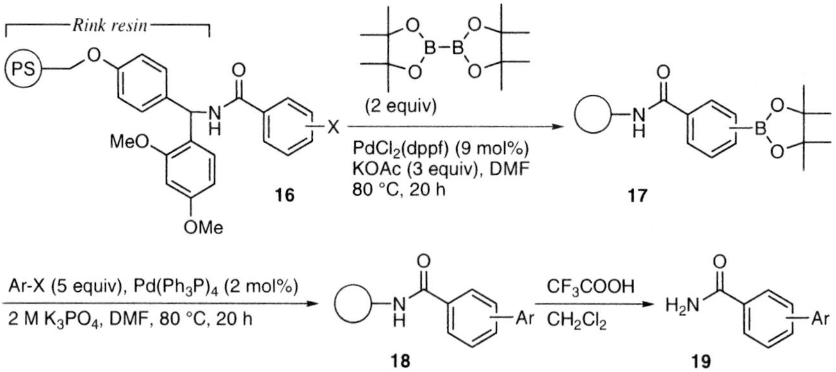

Scheme 19.3. Sequential transformation of polymer-supported haloamides into boronates and biaryls by use of Miyaura's arylboronate formation and Suzuki–Miyaura coupling.

The Suzuki–Miyaura coupling using resin-supported boron reagents has also been investigated (Scheme 19.3 and Table 19.4) [10]. The resin-supported aryl-boron reagents (**17**) were prepared by palladium-catalyzed introduction of pinacol borate on solid support. Thus, *p*-, *m*-, and *o*-iodobenzene Rink amide resins (**16**) were treated with 2 equiv. of bis-(pinacolato)diboron in DMF in the presence of $PdCl_2(dppf)$ to give **17** in high yields. The coupling reactions of the resulting borates with an excess of aryl halides (Ar–X in Table 19.4) were catalyzed by $Pd(Ph_3P)_4$ in DMF upon heating to give excellent yields of the corresponding biaryl carboxamides **18**, which were readily converted to **19** (Table 19.4).

One advantage of solid-phase coupling was demonstrated in the preparation of tetrasubstituted alkenes (Scheme 19.4) [11]. The *Z*-bisborate **20** prepared by palladium-catalyzed diboration was treated with an alkyl halide R′–X in the presence of catalytic $PdCl_2(Ph_3P)_2$ and KOH as base in aqueous DMF to give a mixture of the mono- and dialkylated olefins (**21** and **22**). The reaction mixture was then taken onto the coupling with a Rink *p*-iodobenzamide without further addition of the palladium catalyst. The monoalkylated intermediate bearing the unreacted borate group underwent solid-phase coupling to form **23**, leaving the dialkylated alkene **22** in the solution phase. After removal of the dialkylated alkene **22** by filtration, the resulting resin-bound tetrasubstituted alkene **23** was detached from the resin to give the *p*-alkenylbenzamide **24** in excellent purity.

Solid-phase cross-coupling of the alkyl 9-BBN was used to introduce the silyl traceless linker on solid support (Scheme 19.5) [12]. Palladium-catalyzed solid-phase coupling of the alkyl 9-BBN **26** having an anisyl-(dimethyl)silyl group gave the PS-alkyl-(anisyl)dimethylsilane **27**, which was readily converted into the chlorosilane **28** by treatment with HCl. The alkylchlorosilane resin reacted with the lithiated pyridine **29** to give **30**, the intermediate for the pyridine-fused benzazepines **31** being released in traceless fashion on subsequent treatment with a fluoride reagent.

Tab. 19.4. Transformation of haloamides into polymer-supported boronates and biaryls.

Starting haloamide	Boronate	Yield (%)	Ar–X	Product	Yield (%)
		92–95			95
					80
					82
					72
					90
	94				92
		71 (80 °C, 40 h)			48

Ref. 10: yields are based on incorporation of halobenzoyl group on the resin.

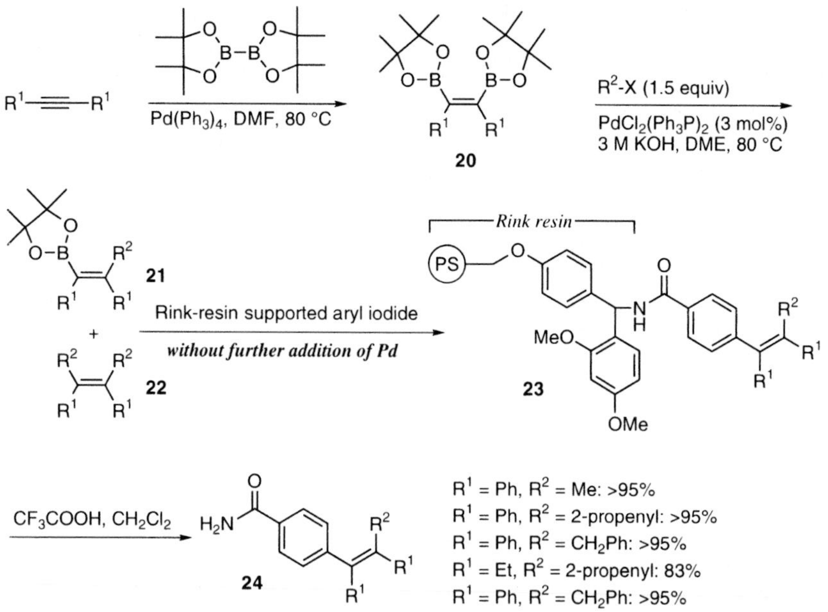

Scheme 19.4. Synthesis of tetrasubstituted ethylenes on solid support via resin capture.

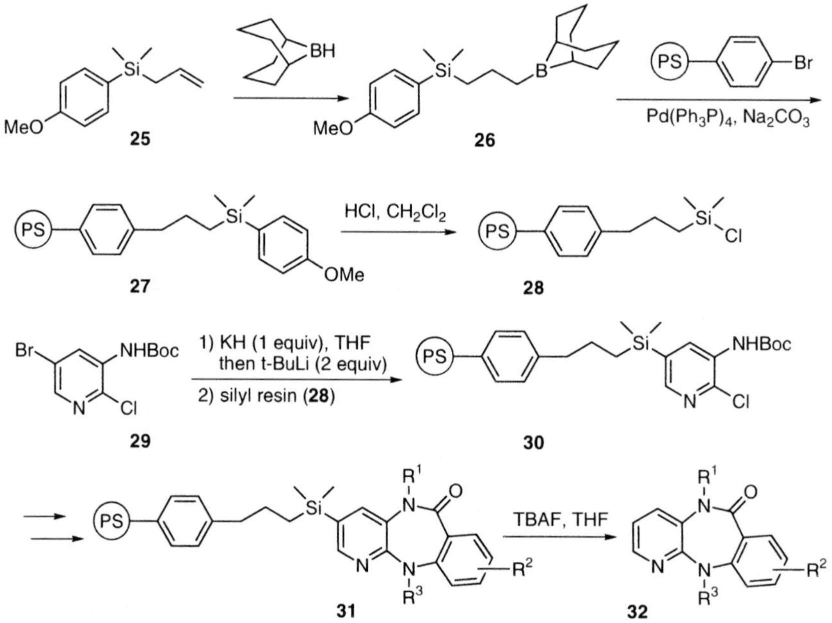

R^1 = Ph, R^2 = Me: >95%
R^1 = Ph, R^2 = 2-propenyl: >95%
R^1 = Ph, R^2 = CH$_2$Ph: >95%
R^1 = Et, R^2 = 2-propenyl: 83%
R^1 = Ph, R^2 = CH$_2$Ph: >95%

Scheme 19.5. Introduction of a traceless silicon linker onto polymer support using Suzuki–Miyaura coupling.

The solid-phase synthesis of the prostaglandin E (PGE) and prostaglandin F (PGF) families was carried out via palladium-catalyzed coupling of alkyl 9-BBN as the key step (Scheme 19.6) [13]. The solid-supported bromocyclopentene **34** and the (bromopropenyl)cyclopentene **38** were substituted with alkyl groups by coupling with butyl or pentyl 9-BBN to give the key intermediates of PGE and PGF derivatives (for example **36**), respectively.

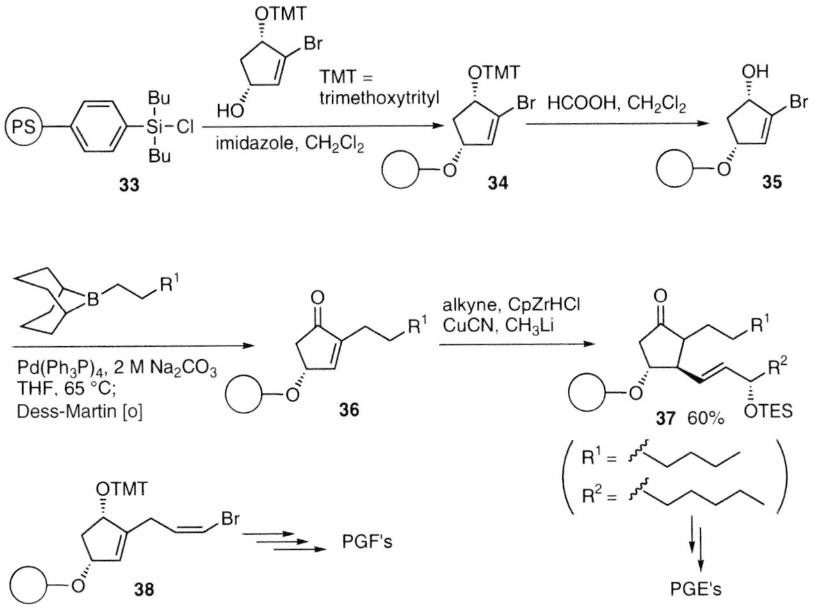

Scheme 19.6. Solid-phase synthesis of PGE and -F series.

Another interesting application of the Suzuki–Miyaura coupling is the solid-supported pinacolate boron linker (Scheme 19.7) [14]. The boron pinacolate prepared from the arylboronic acid **39** and the pinacol carboxylic acid **40** was connected to polystyrene-poly(ethylene glycol) resin (TentaGel) to give **41**. After condensation with **42**, the resulting aryl iodide–aryl borate **43** underwent intramolecular cross-coupling to release the biaryl macrocyclic peptide **44**.

19.2.1.2 Reactions of Aryl and Alkenyl Halides with Organotin Reagents

In addition to the Suzuki–Miyaura coupling, another widely investigated coupling reaction is palladium-catalyzed cross-coupling of aryl and alkenyl halides with organotin reagents (the so-called Stille coupling) [15]. Scope and limitation studies on the coupling of halogenoaryl carboxylic amides supported on Rink, Wang, and PS resin are summarized in Table 19.5 [16, 17]. The Stille coupling of *o*-, *m*-, and *p*-halogenobenzamides with alkenyl and aryl-(tributyl)stannanes (including heteroaromatics) was carried out in aprotic solvents using palladium complexes of triarylphosphine or arsenic ligands to give good to excellent yields of the corresponding cross-coupling products (Table 19.5).

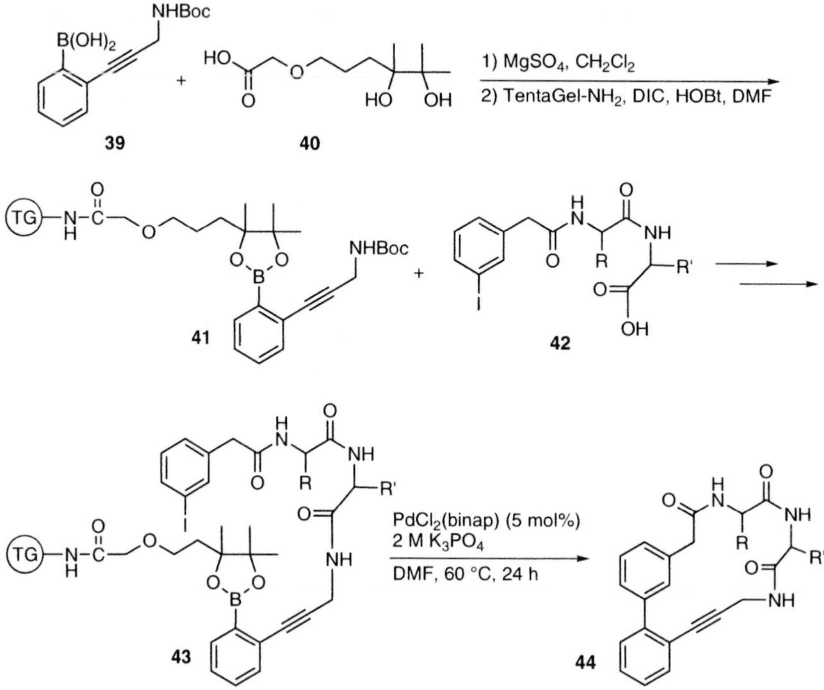

Scheme 19.7. Solid-phase synthesis of macrocyclic β-turn mimics with boronate linker.

The resin-supported arylstannane **45** was also examined for Stille coupling with bromobenzenes (Table 19.6) [18]. Thus, the reaction of the PS resin-supported *p*-tributylstannylbenzoate **45** with 3 equiv. of aryl bromide took place in 1-methyl-2-pyrrolidinone (NMP) at 90 °C to give the coupling products on solid support. A palladacycle catalyst **46** was used for the coupling in the presence of LiCl to promote the reaction more efficiently. Methanolysis of the resulting resin gave the corresponding methyl *p*-arylbenzoates in 80–95% yields.

The solid-phase synthesis of the library of the highly functionalized fused ring system shown in Scheme 19.8 was achieved via the Stille coupling of an alkenyl bromide with a series of alkenylstannanes, resulting in conjugate dienes and a subsequent Diels–Alder reaction (Scheme 19.8) [19].

19.2.1.3 Reactions of Aryl Halides with Terminal Alkynes

The palladium-catalyzed coupling of an aryl halide with a terminal alkyne promoted by CuI and base, the so-called Sonogashira reaction [20], is recognized as the most powerful method for the preparation of aryl acetylene derivatives. The solid-phase Sonogashira reaction was examined for the coupling of the resin-supported *m*-iodobenzylbenzoate **54** and (*m*-iodobenzyloxy)acetamide **55** with various terminal alkynes, as shown in Table 19.7 [21]. The coupling was carried out with an excess of alkynes using a PdCl$_2$(Ph$_3$P)$_2$/CuI/Et$_3$N system to give the corre-

Tab. 19.5. Coupling of polymer-supported aryl iodides with vinyl/aryl stannanes.

$$Ar–X + (alkyl)_3SnR \ (R = vinyl \ or \ aryl) \longrightarrow Ar–R \ (R = vinyl \ or \ aryl)$$

Entry	Aryl iodide	Stannane	Product[a]	Yield (%)[b]
1		$(Bu)_3Sn$	Ar^1	89
2		$(Bu)_3Sn$	Ar^1	91
3	= Rink resin	$(Bu)_3Sn$	Ar^1	85
4		$(Bu)_3Sn$—Ph	Ar^1—Ph	89
5		$(Me)_3Sn$	Ar^1	90
6		$(Bu)_3Sn$	Ar^2	92
7	= Wang resin	$(Bu)_3Sn$	Ar^2	88
8	X = Br	$(Bu)_3Sn$	Ar^3	95
9	X = I	$(Bu)_3Sn$	Ar^3	88
10	X = I	$(Bu)_3Sn$	Ar^3	71
11	X = Br	$(Bu)_3Sn$	Ar^4	94
12	X = I	$(Bu)_3Sn$	Ar^4	91
13	X = I	$(Bu)_3Sn$	Ar^4	87
14	X = I	$(Bu)_3Sn$	Ar^4	86
15	X = I	$(Bu)_3Sn$	Ar^4	89
16	X = Br	$(Bu)_3Sn$	Ar^5	93
17	X = Br	$(Bu)_3Sn$	Ar^5	88

Tab. 19.5. (continued)

Entry	Aryl iodide	Stannane	Product[a]	Yield (%)[b]
18		X = Br (Bu)$_3$Sn—	Ar^5—	84
19		X = Br (Bu)$_3$Sn—	Ar^5—	91
20		X = I (Bu)$_3$Sn—	Ar^5—	95
21		X = I (Bu)$_3$Sn—	Ar^5—	95
22		X = I (Bu)$_3$Sn—	Ar^5—	80
23		X = I (Bu)$_3$Sn—	Ar^5—	93
24	(PS)—CH$_2$O—C(O)—Br (pyridyl)	(Bu)$_3$Sn—	Ar^6—	96
25		(Bu)$_3$Sn—	Ar^6—	48
26	(PS)—CH$_2$O—C(O)—(furyl)—Br	(Bu)$_3$Sn—	Ar^7—	75
27		(Bu)$_3$Sn—	Ar^7—	68

Conditions: entries 1–7 (ref. 16): Pd$_2$(dba)$_3$ (5 mol%), Ph$_3$As (20
mol%), NMP, 45 °C; entries 8, 9, 11–15, 17–24, and 26 (ref. 17):
stannane (3 equiv.), Pd(Ph$_3$P)$_4$ (5 mol%), DMF, 60 °C, 24 h;
entries 10, 16, 25, and 27: Pd$_2$(dba)$_3$ (5 mol% Pd), (2-furyl)$_3$P,
DMF, 60 °C, 48 h.

[a] Entries 1–5: product cleaved from the resin with 5% TFA/DCM
(entries 1–5), 90% TFA (entries 6–7), or aq. LiOH THF-MeOH-
H$_2$O (entries 8–27). Ar^1 = 4-H$_2$NCOC$_6$H$_4$, Ar^2 = 4-
HOCOCH(CH$_3$)NH-COC$_6$H$_4$, Ar^3 = 2-HOCOC$_6$H$_4$, Ar^4 = 3-
HOCOC$_6$H$_4$, Ar^5 = 4-HOCOC$_6$H$_4$, Ar^6 = (3-hydroxycarbonyl)-5-
pyridyl, Ar^7 = (2-hydroxycarbonyl)-5-furyl.

[b] Entries 1–24: isolated yield; entries 25–27: determined by HPLC
and ^{1}H NMR.

Tab. 19.6. Coupling of polymer-supported aryl stannane with aryl halides.

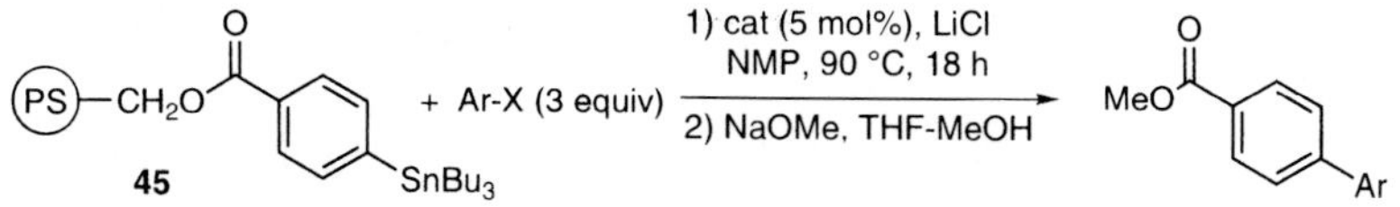

Entry	Aryl halide	Yield (%)
1	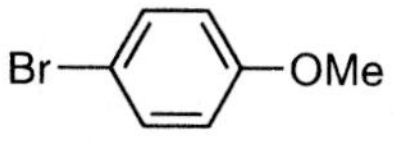 Br—⟨ ⟩—OMe	>95
2	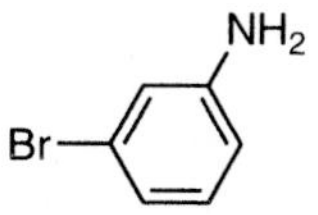 Br—⟨ ⟩ NH$_2$	>95
3	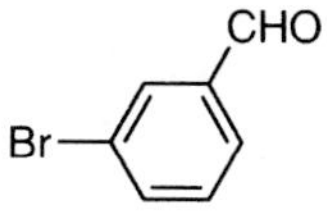 Br—⟨ ⟩ CHO	>95
4	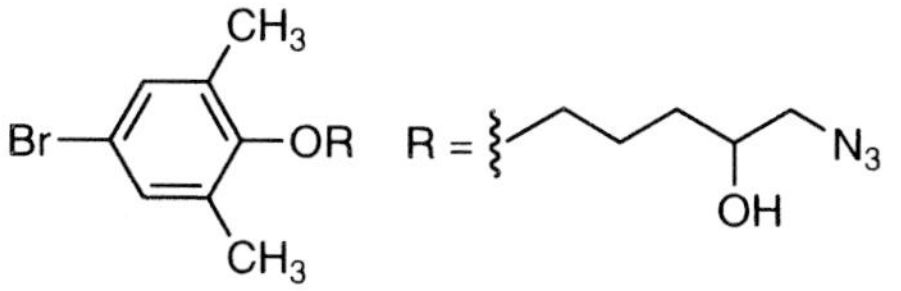 Br—⟨ ⟩—OR (CH$_3$, CH$_3$) R = azide chain with OH	>95
5	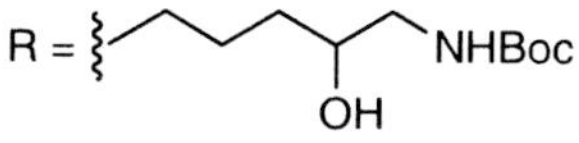 R = NHBoc chain with OH	80
6	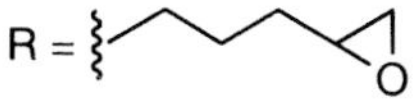 R = epoxide chain	>95

Ref. 18

cat = **46**

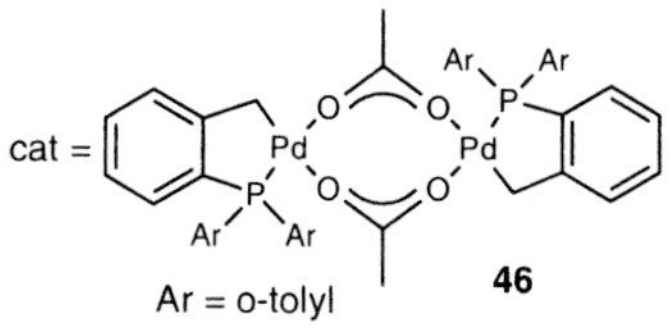

Ar = o-tolyl

sponding ethynylbenzenes in nearly quantitative yields. The ethynylbenzyl alcohols **56** or the (ethynylbenzyloxy)acetamide **57** were released from the resulting resin by methanolysis of the benzoate linker or acid hydrolysis of the Rink acetamide linker, respectively.

The Sonogashira reaction has often been used as a key step for the solid-phase preparation of indole derivatives. Thus, for example, the solid-phase coupling of *o*-iodoacetanilide **58** with phenylacetylene using PdCl$_2$(Ph$_3$P)$_2$, CuI, and base, followed by alkaline hydrolysis of the ester linkage, gave the indole 5-carboxylic acid

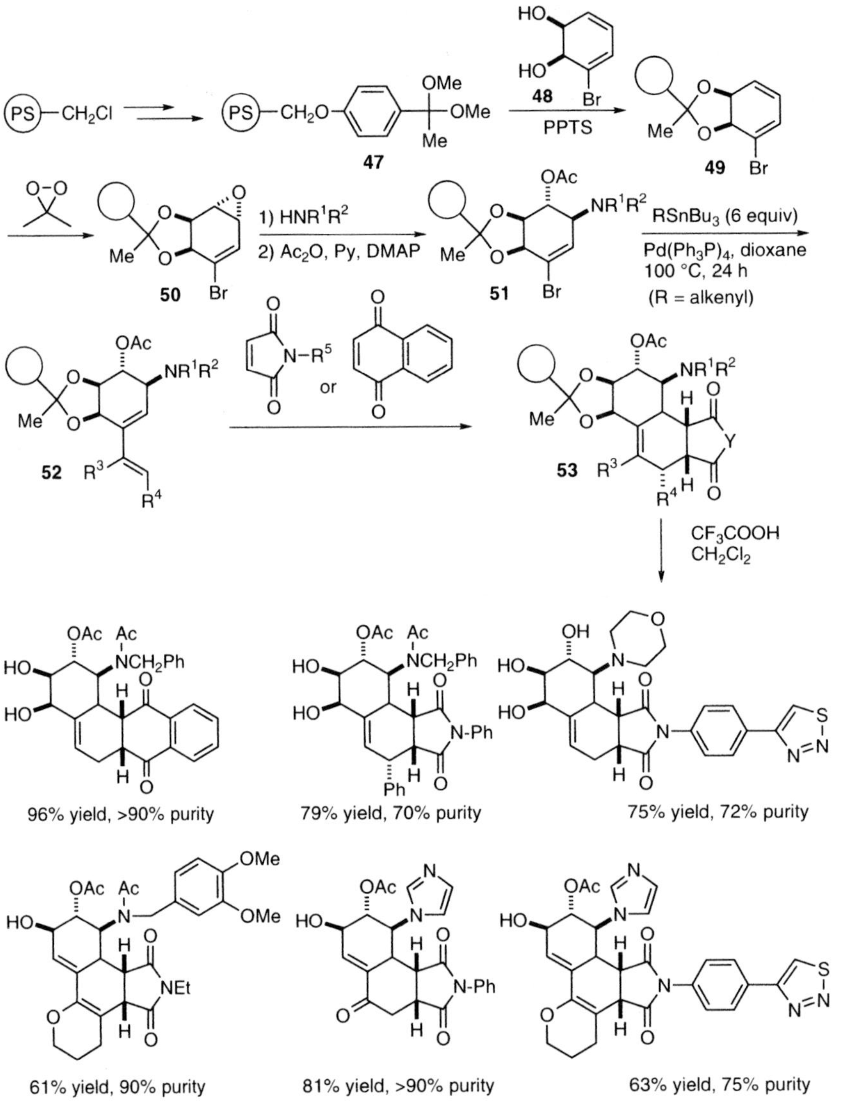

Scheme 19.8. Solid-phase synthesis of highly functionalized fused ring systems.

59 in 72% yield in one pot, thereby showing that the Sonogashira coupling and subsequent intramolecular cyclization to form the indole ring proceeds on solid phase (Scheme 19.9) [22].

The Rink resin-supported *o*-iodo(*N*-methanesulfonyl)anilide **60** gave the 2-substituted indoles **61** which underwent, after deprotection of the *N*-mesyl group, a solid-phase Mannich reaction to afford the (2-alkyl-3-aminomethyl)indole 5-carboxamides **64** (Scheme 19.10) [23].

Tab. 19.7. Palladium-catalyzed coupling of solid-supported aryl iodides with terminal acetylenes.

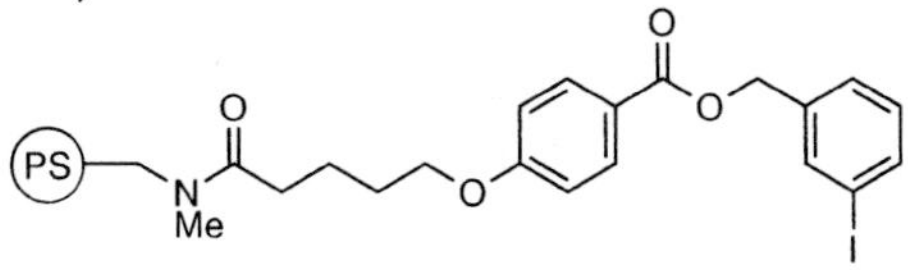

54

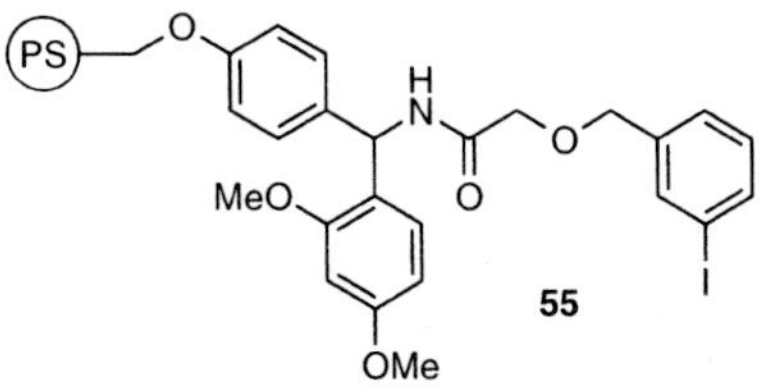

55

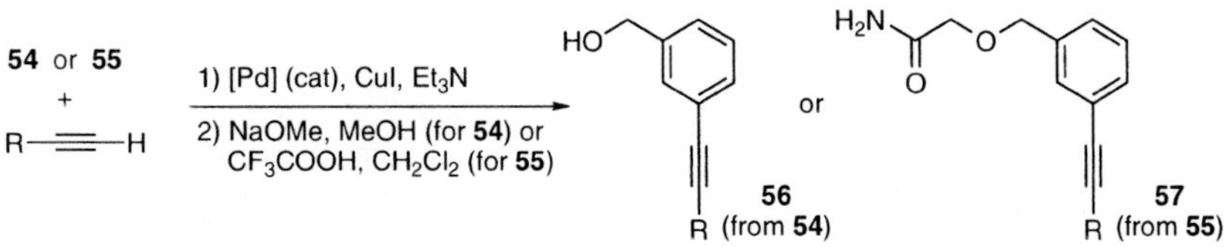

54 or 55
+
R——H

1) [Pd] (cat), CuI, Et₃N

2) NaOMe, MeOH (for **54**) or
CF₃COOH, CH₂Cl₂ (for **55**)

56
R (from **54**)

or

57
R (from **55**)

Entry	Aryl iodide	Acetylene	Conditions	Yield (%)
1	54	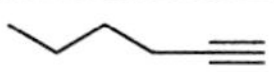	A	95
2	54	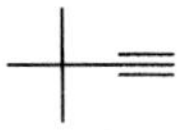	A	95
3	54	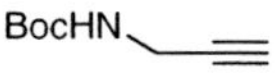 BocHN	A	96
4	54	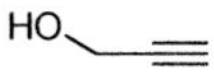 HO	A	93
5	54	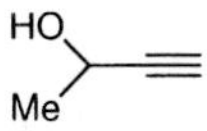 HO / Me	A	95
6	54	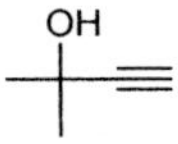 OH	A	95
7	54	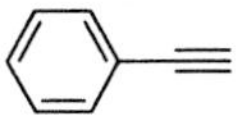	A	95

Tab. 19.7. *(continued)*

Entry	*Aryl iodide*	*Acetylene*	*Conditions*	*Yield (%)*
8	**55**	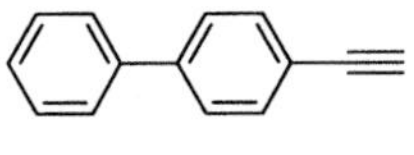	B	79
9	**55**	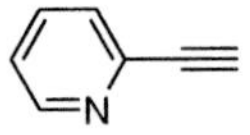	C	95
10	**55**	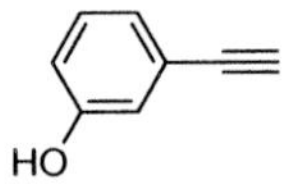	C	83
11	**55**	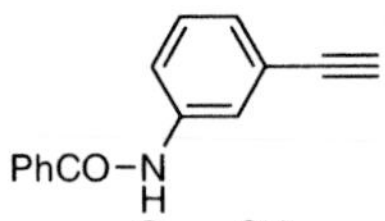	C	95
12	**55**	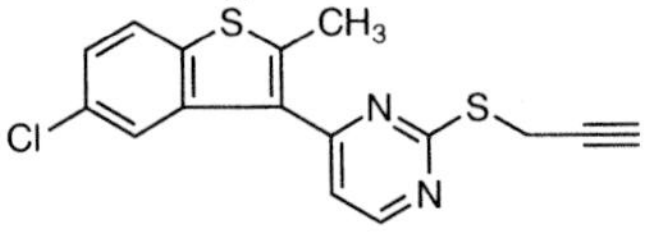	C	91

Ref. 21 conditions: A, acetylene (4.0 equiv.), $PdCl_2(Ph_3P)_2$ (10 mol%), CuI (20 mol%), Et_3N/dioxane = 1/2, rt, 24 h; B, acetylene (8.0 equiv.), $PdCl_2(Ph_3P)_2$ (20 mol%), CuI (40 mol%), Et_3N/dioxane = 1/2, rt, 24 h; C, acetylene (4.0 equiv.), $PdCl_2(Ph_3P)_2$ (10 mol%), CuI (20 mol%), Et_3N/dioxane = 1/2, 50 °C, 24 h.

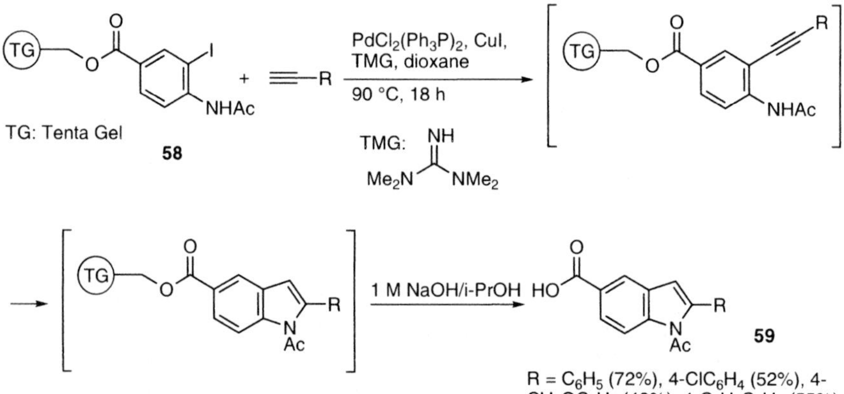

R = C_6H_5 (72%), 4-ClC_6H_4 (52%), 4-$CH_3OC_6H_4$ (48%), 4-$C_3H_7C_6H_4$ (55%), $(CH_2)_3OH$ (95%), $CH_2SC_6H_5$ (81%)

Scheme 19.9. Solid-phase synthesis of indoles using Pd-catalyzed acetylene coupling.

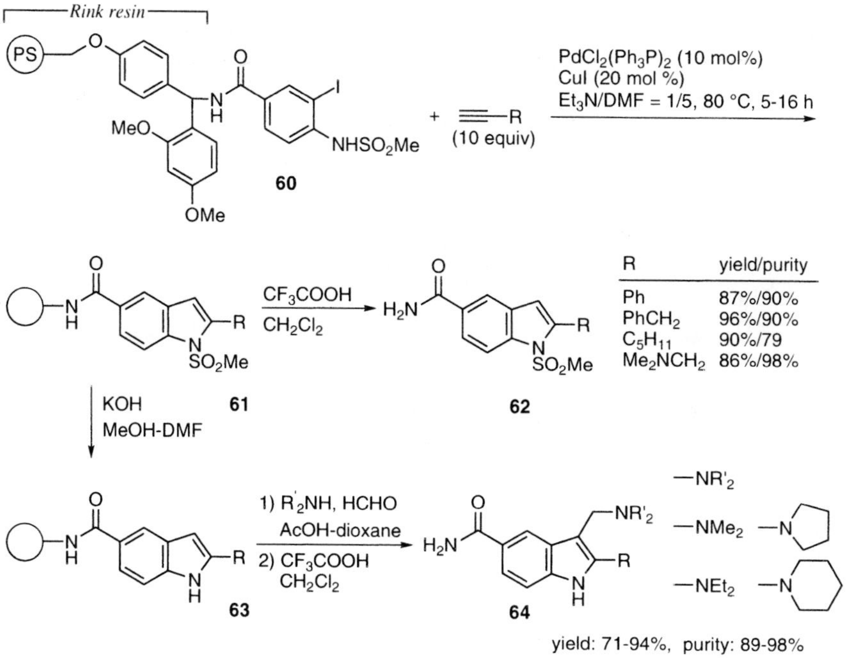

R	yield/purity
Ph	87%/90%
PhCH$_2$	96%/90%
C$_5$H$_{11}$	90%/79
Me$_2$NCH$_2$	86%/98%

yield: 71-94%, purity: 89-98%

Scheme 19.10. Solid-phase Mannich substitution of indoles.

Similarly, the benzofurans **66** as well as the (2,3,6-substituted)indoles **70** were also synthesized on solid phase (Schemes 19.11 and 19.12) [24, 25].

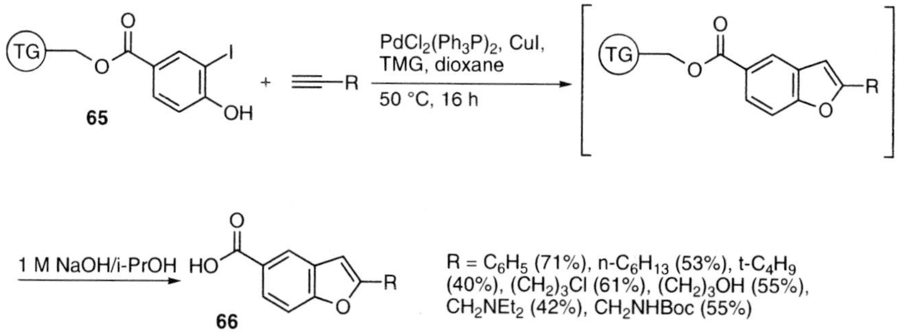

$R = C_6H_5$ (71%), n-C$_6$H$_{13}$ (53%), t-C$_4$H$_9$ (40%), (CH$_2$)$_3$Cl (61%), (CH$_2$)$_3$OH (55%), CH$_2$NEt$_2$ (42%), CH$_2$NHBoc (55%)

Scheme 19.11. Solid-phase synthesis of benzofurans using Pd-catalyzed acetylene coupling.

An unsubstituted terminal acetylene group was introduced on a resin-supported aromatic ring by Sonogashira coupling of (trimethylsilyl)acetylene followed by carbon–silicon bond cleavage with tetrabutylammonium fluoride (Scheme 19.13) [26]. The resulting arylacetylene **73** was subjected to the copper-mediated Mannich reaction on solid support with various aldehydes and piperazines to afford a library of arylpropynylamines **74** in excellent chemical yields and purities.

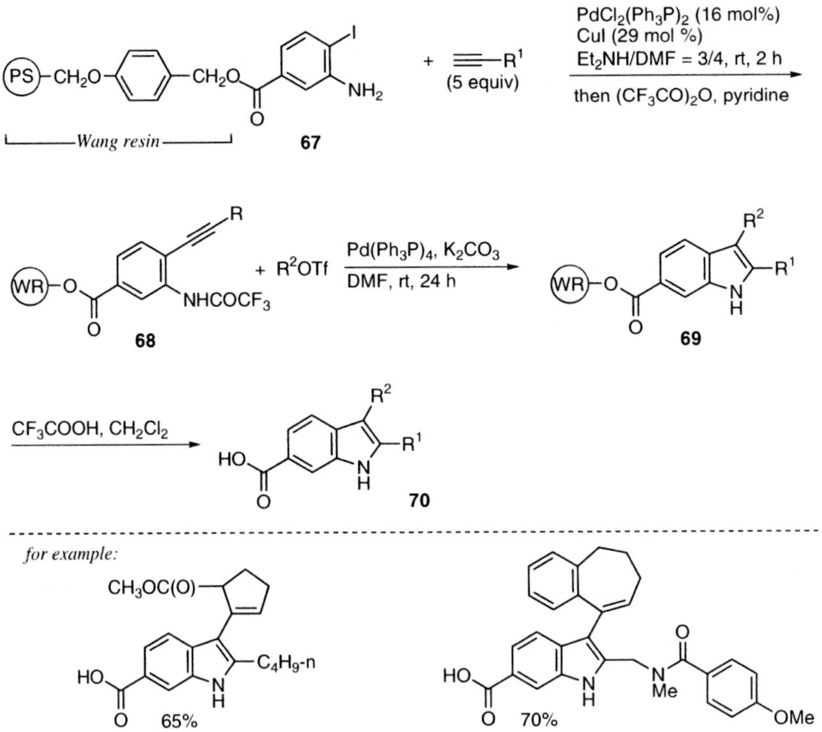

Scheme 19.12. Solid-phase synthesis of trisubstituted indoles.

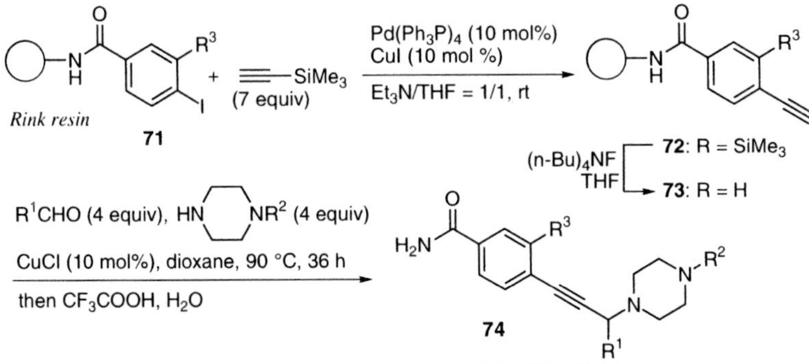

R^1	R^2	R^3	yield (%)	purity (%)
4-CH$_3$C$_6$H$_4$	(E)-PhCH=CHCH$_2$	H	71	95
3-FC$_6$H$_4$	2,3-(CH$_3$)$_2$C$_6$H$_3$	H	85	>95
3-PhOC$_6$H$_4$	2-FC$_6$H$_4$	H	75	>95
1-naphthyl	cyclohexyl	H	89	94
4-CH$_3$C$_6$H$_4$	PhCH$_2$	H	90	>95
4-PhC$_6$H$_4$	2-FC$_6$H$_4$	H	88	>95
2-CH$_3$C$_6$H$_4$	2-ClC$_6$H$_4$	H	88	89
2-naphthyl	PhCH$_2$	CH$_3$	73	82
Ph	Ph	CH$_3$	70	85

Scheme 19.13. Combination of Sonogashira and Mannich reactions.

A solid-supported terminal alkyne was also examined for the Sonogashira coupling (Scheme 19.14) [27]. Thus, propiolic acid was connected to PS oxyamino resin prepared from the phthalimide **75** via the Gabriel synthesis to give the solid-supported propiolic amide **76**. The palladium-catalyzed coupling of the supported propiolic amide **76** with the halogenonucleoside **77** proceeded under the standard Sonogashira conditions to give the nucleoside hydroxamic acid **79**.

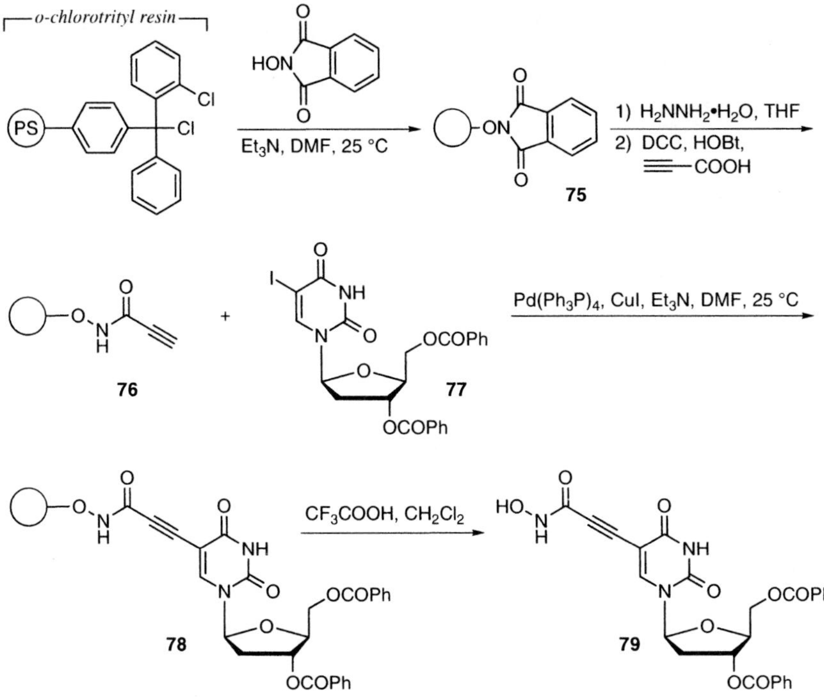

Scheme 19.14. Solid-phase synthesis of nucleoside hydroxamic acids.

19.2.1.4 Solid-phase Palladium-catalyzed Cross-coupling Using Aryl and Benzylzinc Reagents

It has been well established that cross-coupling of aryl halides (pseudo-halides) with aryl and benzylzinc reagents are also catalyzed by palladium–phosphine complexes, as are those with organoboron and organostannane reagents. Solid-phase cross-coupling of arylzinc reagents with supported aryl halides was catalyzed by a palladium complex of bis-(diphenylphosphino)ferrocene (dppf) in THF to give biaryls in good to excellent yields. Representative results are summarized in Table 19.8 [28]. Reactions of supported *p*-bromobenzoate with 2 equiv. of *o*-, *m*-, and *p*-substituted arylzinc reagents bearing an electron-withdrawing or electron-donating group took place at room temperature to give the corresponding biaryls in good chemical yields (Table 19.8, entries 1–7). *Meta*- and *o*-bromobenzoate also underwent coupling on resin support under the same reaction conditions to give the

Tab. 19.8. Coupling of polymer-supported aryl bromides with arylzinc reagents.

Entry	Aryl halide	Aryl zinc	Product	Yield (%)
1		BrZn—⟨⟩—Y Y = H		76
2		Y = OMe		82
3		Y = F		84
4		Y = CN		61
5		BrZn Y=OMe		80
6		Y = F		84
7				75
8				85
9				75
10				71
11				55

Ref. 28   dppf =

biaryl 2-carboxylic ester and the 3-carboxylic ester, respectively (Table 19.8, entries 9 and 10). Heteroaromatic biaryl derivatives were obtained from solid-supported bromopyridine or thiophenylzinc (Table 19.8, entries 8 and 11).

It is noteworthy that zinc-mediated cross-coupling was used for the preparation of diarylmethanes (Scheme 19.15) [29]. Thus, the coupling of the Rink amide of the o-, m-, and p-iodobenzoyl substrates with 10 equiv. of 4-cyanobenzylzinc and 2,6-dichlorobenzylzinc reagents was catalyzed by a catalyst generated from Pd$_2$(dba)$_3$ and tri(2-furyl)phosphine to give the corresponding benzylphenylcarboxamides **80** and **81** in high yields. The synthetic sequences shown in Scheme 19.16, where the zinc-mediated aryl–aryl and aryl–benzyl couplings were performed on resin support, gave Ar–Ar–CH$_2$–Ar products in high purity.

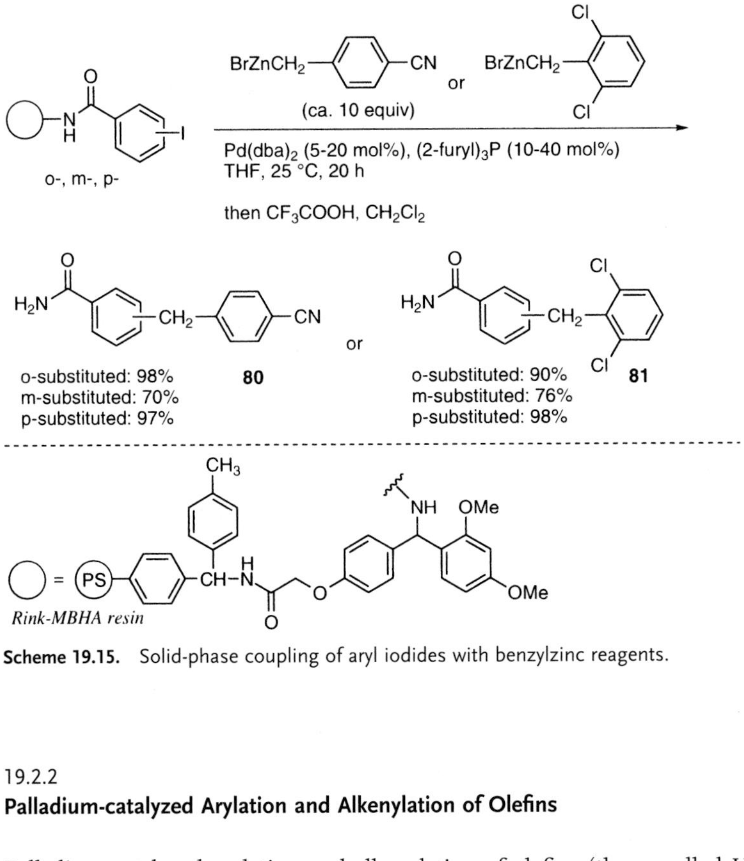

Scheme 19.15. Solid-phase coupling of aryl iodides with benzylzinc reagents.

19.2.2
Palladium-catalyzed Arylation and Alkenylation of Olefins

Palladium-catalyzed arylation and alkenylation of olefins (the so-called Heck reaction) [30] are versatile means for making a carbon–carbon bond. Aryl and alkenyl halides (pseudo-halides) (R–X) are employed for the Heck reaction as alkylation agents for the olefinic substrates.

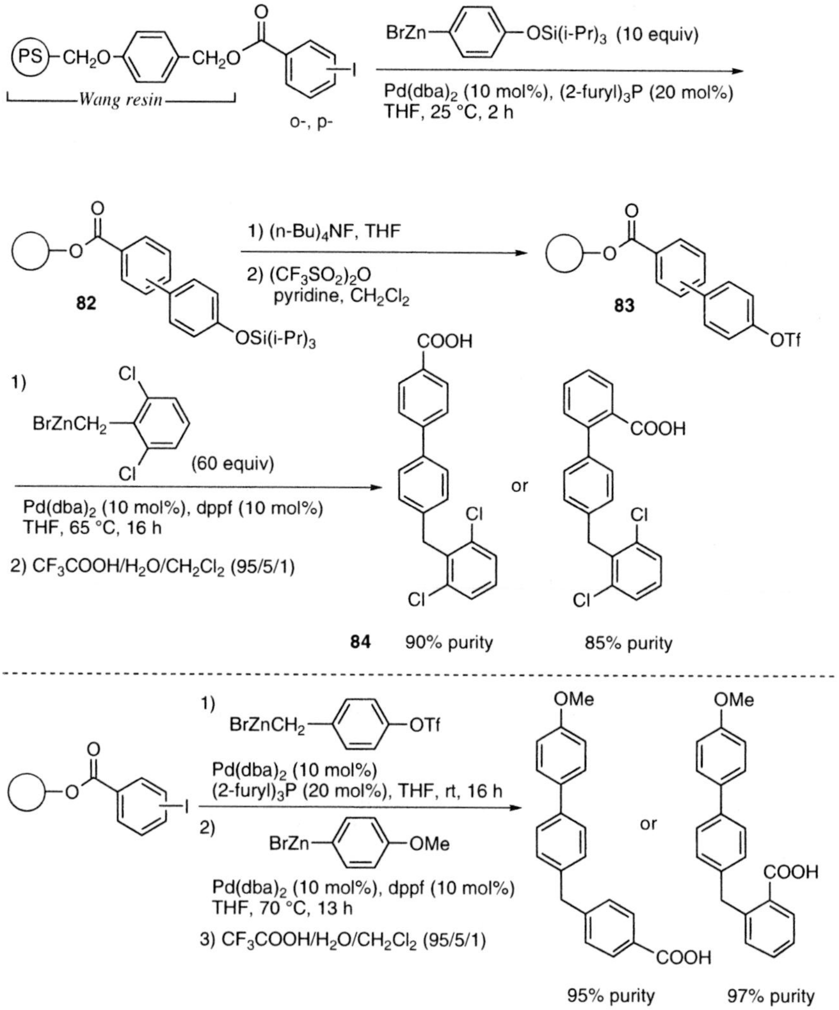

Scheme 19.16. Multiple coupling process of aryl- and benzylzinc reagents with supported aryl halides.

The Heck reaction is generally thought to proceed via the reaction pathway shown in Scheme 19.17. Oxidative addition of R–X to palladium(0) gives LnPd(R)X. Coordination of an olefin substrate to LnPd(R)X gives Pd(II) (η^2-olefin)R. Insertion of the η^2-olefin ligand into the Pd–R bond gives a σ-alkylpalladium intermediate which subsequently undergoes β-hydride elimination to give the Heck product and LnPd(H)X.

The intermolecular Heck reaction of resin-supported aryl iodides with olefin substrates was examined on various polymer supports. Representative results are

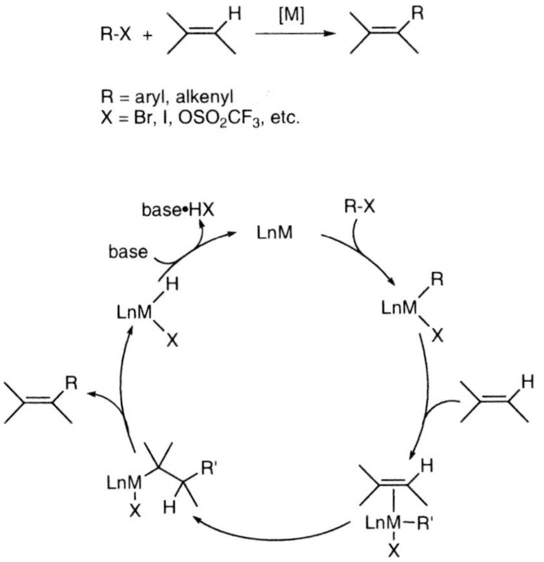

Scheme 19.17. Pathway of Heck reaction.

summarized in Table 19.9, where conjugated olefins were used to exhibit high re-activity [31, 32]. The Heck reaction of the Wang resin-supported 4-iodobenzoate with ethyl acrylate was catalyzed by Pd(OAc)$_2$ in DMF to give the ethyl cinnamate product in 91% yield (Table 19.9, entry 1). The unsymmetrical stilbene was ob-tained from methoxystyrene in 90% yield (Table 19.9, entry 2). Alkenylation of io-dobenzoates supported on PS–polyethylene glycol (PEG) resin and peptide amide linker (PAL) resin took place at 37 °C in the presence of Pd(OAc)$_2$-PPh$_3$ in aque-ous DMF (Table 19.9, entries 4–7).

The Heck reaction of a supported styrene with solution-phase aryl iodides was also examined under essentially the same conditions (Table 19.10). Thus, the Wang resin-supported styrene carboxylate reacted with iodobenzene, bromonaphthalene, bromothiophene, and bromopyridine in DMF on heating to give the corresponding β-arylstyrene-4-carboxylic acids upon release from the resin support. A palladium complex generated *in situ* by mixing Pd$_2$(dba)$_3$ and tri-(2-furyl)phosphine (Table 19.9, conditions B) exhibited high catalytic activity for the reaction with aryl bro-mides (Table 19.9, entries 2, 4, and 5).

Aryl halides having allylamine or acrylamide substituents at their ortho posi-tions have been well documented to undergo an intramolecular Heck reaction to form indole ring systems. The intramolecular indole ring construction has been applied to the solid-phase synthesis of libraries of indole derivatives. Thus, the 2-bromoaniline **85**, which was connected at its 5-position to PS–PEG resin by a Wang-type linker, was converted to the *N*-acyl-*N*-allylanilines **86** using the standard fluorenylmethoxycarbonyl (Fmoc) method. The supported substrates **86** were sub-

Tab. 19.9. Solid-phase Heck reaction of supported aryl halides.

Entry	Ar–X	Alkene (alkyne)	Product	Yield (%)
1		COOEt		91
2		OMe		90
3		COOEt		>95
4		CONH₂		>95
5		CN		54
6		CHO		73

Entries 1–3, ref. 31; entries 4–7, ref. 32. Conditions: entries 1 and 3, Pd(OAc)₂, Et₃N, Bu₄NCl, DMF, 80–90 °C, 16 h; entry 2, Pd₂(dba)₃, (o-tol)₃P, DMF, 100 °C, 20 h; entries 4–7, Pd(OAc)₂, PPh₃, Et₃N, Bu₄NCl, DMF/H₂O (9:1), 37 °C, 4 h.

jected to the palladium-catalyzed Heck reaction. The resulting resin was treated with CF$_3$COOH to give *N*-acyl-3-alkyl-6-hydroxyindoles (**87**) in good to excellent yields and purities (Scheme 19.18) [33]. An acrylamide group showed good reactivity for the intramolecular Heck reaction (Scheme 19.19) [34]. Thus, cyclization of Rink resin-supported (2-iodo)acrylanilides (**90**) was catalyzed by Pd(OAc)$_2$-PPh$_3$

Tab. 19.10. Solid-phase Heck reaction of supported styrene with aryl halides.

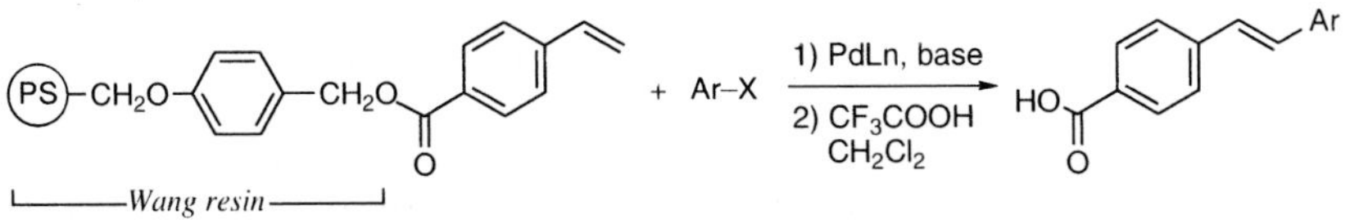

Entry	Ar–X	Conditions	Product	Yield (%)
1	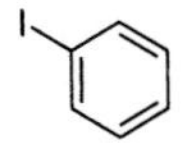 (iodobenzene)	A	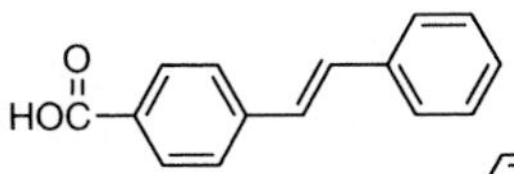 (product)	81
2	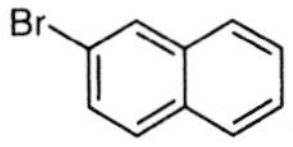 (bromonaphthalene)	A	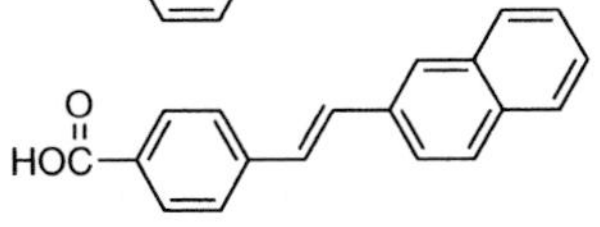 (product)	NR
3		B		64
4	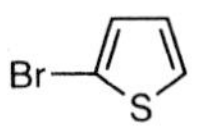 (bromothiophene)	B	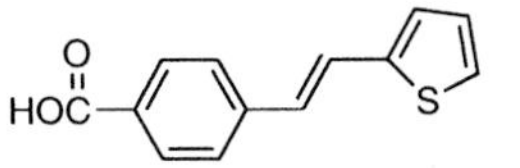 (product)	76
5	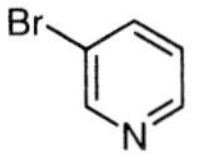 (bromopyridine)	A	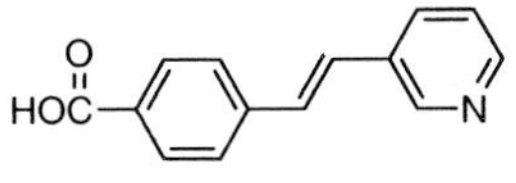 (product)	NR
6		B		87

Ref. 31: condition A, Pd(OAc)$_2$, Et$_3$N, Bu$_4$NCl, DMF, 80–90 °C, 16 h; B, Pd$_2$(dba)$_3$, (*o*-tol)$_3$P, DMF, 100 °C, 20 h.

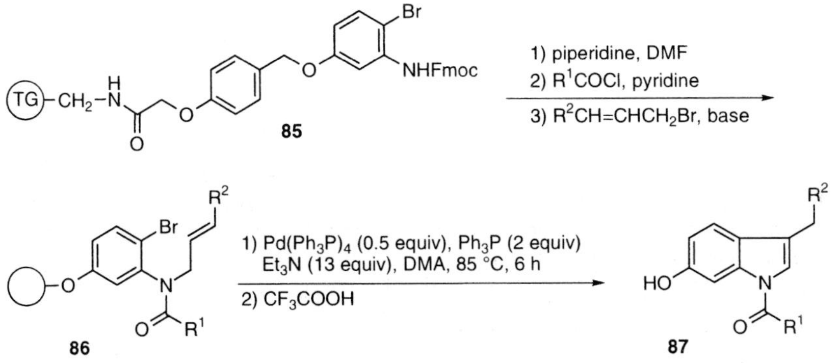

R^1	R^2	yield (%)	purity (%)
Et	H	90	88
Et	Ph	94	93
i-Pr	Me	92	91
Ph	H	70	48
Ph	Ph	65	55

TG = TentaGel
Fmoc = 9-fluorenylmethoxycarbonyl

Scheme 19.18. Solid-phase synthesis of indoles.

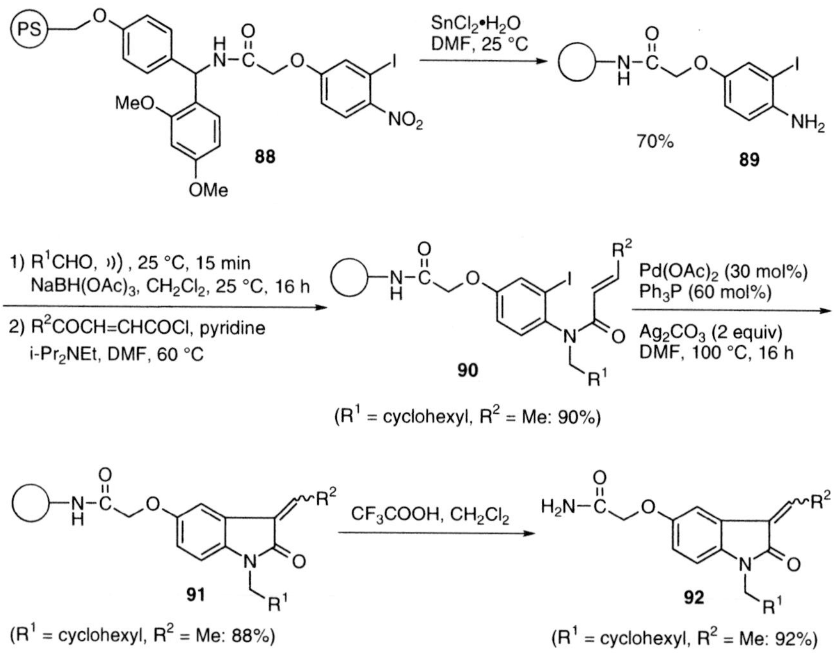

Scheme 19.19. Solid-phase synthesis of oxindoles.

entry	R^1	R^2	yield (%)	purity (%)	E/Z
1	H	H	65	10	—
2	H	Me	91	65	3/1
3	H	Ph	92	76	5.5/1
4	cyclo-Hex	Me	92	70	2.7/1
5	cyclo-Hex	Ph	90	71	5.8/1

in DMF to give the oxindoles **91**. N-Allyl-2-iodoanilines (**94**) which were supported on Rink amide resin by their N-alkyl side-chains also underwent Heck cyclization in the presence of Pd-PPh$_3$ species in aqueous DMF at 80 °C to afford 3-(resin-connected alkyl)indoles (**95**) (Scheme 19.20) [35]. Cleavage of the resin support under acidic conditions gave high yields of the indoles **96**, which bear 3-(amino-carbonyl)methyl and N-benzyl groups. Preparation of an oligopeptide library at the 5-position of the indole skeleton was achieved by use of resin-supported 5-carboxy indole **95** as a scaffold (Scheme 19.20, bottom).

A macrocyclic peptide was synthesized on solid support via Heck cyclization (Scheme 19.21) [36]. An oligopeptide chain (**98**) bearing aryl iodide and acrylamide groups was prepared by the standard Fmoc method on PS resin. Heck reaction of the oligopeptide **98** took place at 37 °C in aqueous DMF in the presence of the Pd(OAc)$_2$-PPh$_3$ catalyst to give, after deprotection and release from resin support,

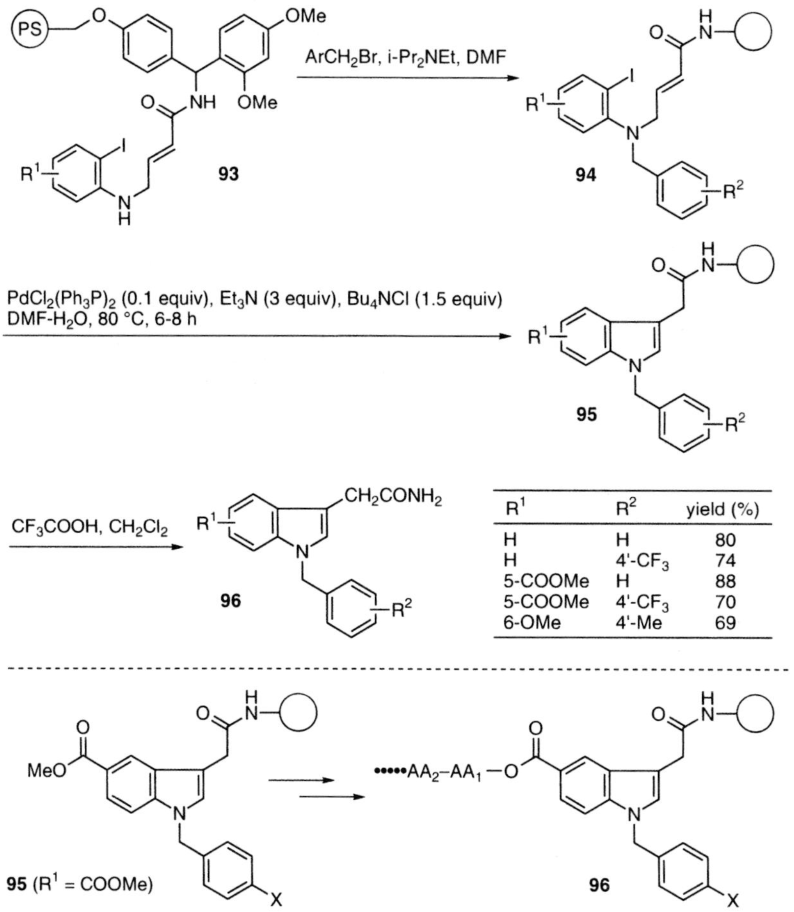

R^1	R^2	yield (%)
H	H	80
H	4'-CF_3	74
5-COOMe	H	88
5-COOMe	4'-CF_3	70
6-OMe	4'-Me	69

Scheme 19.20. Solid-phase synthesis of indoles (2).

the macrocyclic peptide **100** in 30% overall yield based on the loading of the starting resin. A benzazepine skeleton was constructed on solid support by intramolecular Heck reaction (Scheme 19.22) [37]. The Wang resin-supported (*N*-butenyl)-2-iodobenzamides **101** (R = CH_3 or CH_2Ph) underwent Heck cyclization with $Pd(OAc)_2$-PPh_3 to give **102**. Acidic cleavage of the Wang ester, followed by treatment with diazomethane, gave the benzazepines **103** in high yields.

19.2.3
Amination of Aryl Halides

Palladium-catalyzed amination of aromatic halides [38] has become a powerful tool in solid-phase organic synthesis. Thus, various Rink resin-supported aryl bromides were coupled with aniline derivatives to give *N*-aryl anilines in quantitative yields, as seen in Table 19.11, entries 1, 2, 4, and 5 [39]. The aromatic ami-

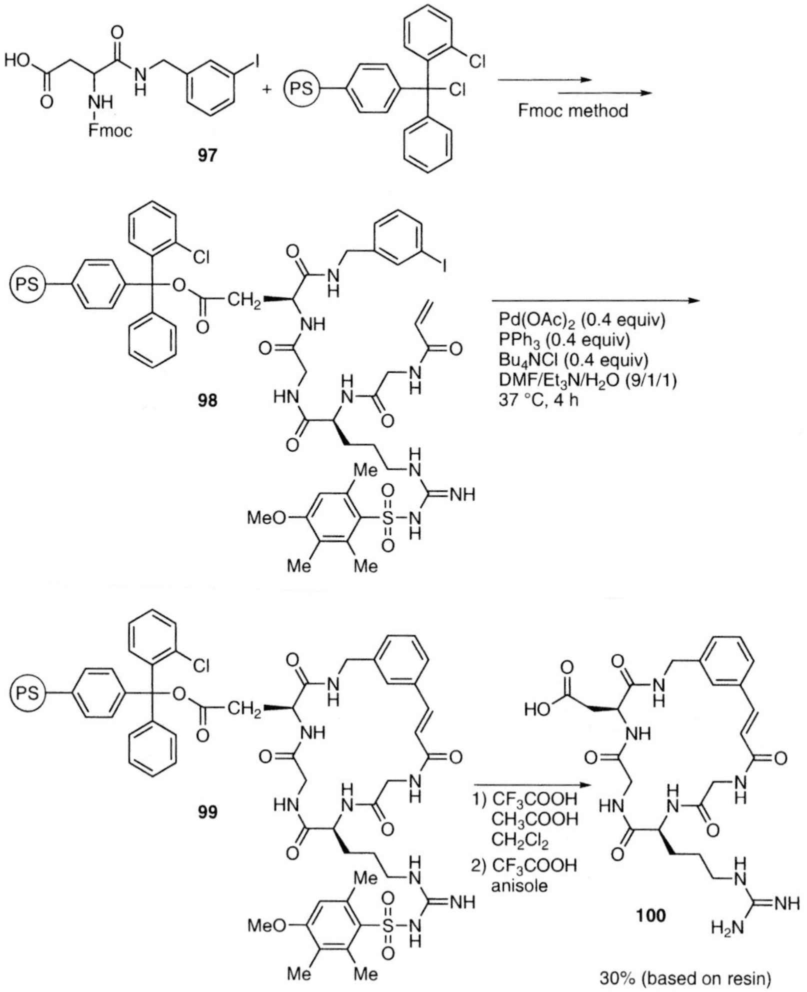

Scheme 19.21. Palladium-catalyzed formation of macrocyclic peptide on solid support.

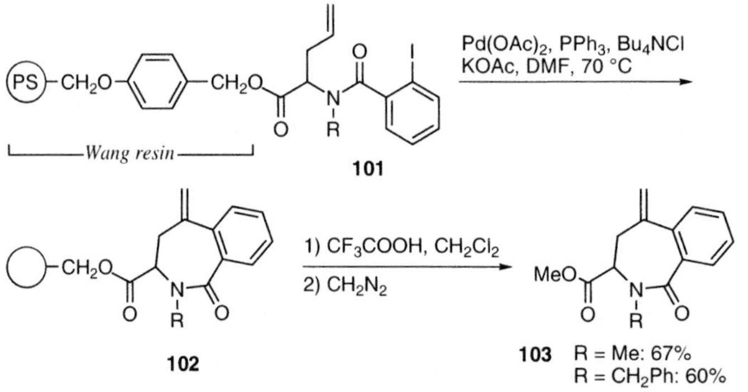

Scheme 19.22. Solid-phase synthesis of benzazepines via intermolecular Heck reaction.

Tab. 19.11. Palladium-catalyzed amination of resin-bound aryl bromide.

Entry	Aryl halide	Amine	Condition	Product	Yield (%)
	PS Rink resin				
1			A		100% conversion
2			A		100% conversion
3			A	no reaction	
4			A		100% conversion
5			A		100% conversion
	TG RAM resin				
6		HN⟩	B		81

Tab. 19.11. *(continued)*

Entry	Aryl halide	Amine	Condition	Product	Yield (%)
7			B		49
8			C		93
9			C		99
10			C		89

Entries 1–5, ref. 39; entries 6–10, ref. 40. Conditions: A, amine
(3 equiv.), Pd$_2$(dba)$_3$ (5 mol%), (o-tol)$_3$P, NaOBu-t (10–20 equiv.),
toluene, 100 °C; B, amine (10 equiv.), Pd$_2$(dba)$_3$ (20 mol%),
(o-tol)$_3$P (80 mol%), NaOt-Bu (10–20 equiv.), toluene, 100 °C; C,
amine (10 equiv.), Pd$_2$(dba)$_3$ (20 mol%), BINAP (80 mol% P),
NaOBu-t (10–20 equiv.), toluene, 100 °C

nations were carried out with the Pd$_2$(dba)$_3$/tri-(o-tolyl)phosphine catalyst system
and NaOt-Bu in toluene at 100 °C to give complete conversion of the substrates.
Resin-bound o-bromides showed little activity, presumably owing to their steric
hindrance (Table 19.11, entry 3). PS–PEG Rink amide (TG RAM) resin-bound p-
bromobenzamide was also examined for coupling with piperidine and pyrrolidine
to give the N-arylpiperidine and N-arylpyrrolidine in 81% and 49% yields, respectively, under essentially the same conditions (Table 19.11, entries 6 and 7) [40]. It
has been documented that primary and secondary aliphatic amines result in significant reduction of the bromide using (o-tol)$_3$P and that the improved conditions
with 2,2'-bis(diphenylphosphino)-1,1'-binaphthyl (BINAP) (Table 19.11, conditions
C) decrease this side reaction. The yield of N-arylpyrrolidine (49%) increased to
93% with BINAP (Table 19.11, compare entry 7 with 8). The use of BINAP as a
ligand also allowed for the successful coupling of primary amines. Thus, benzylamine reacted with the TG RAM-supported p- and m-bromobenzamide with
the Pd/BINAP catalyst to give 99% and 89% yields, respectively, of the p- and m-
(benzylamino)benzamide (Table 19.11, entries 9 and 10).

19.2.4
Miscellaneous Reactions [41]

19.2.4.1 **Heteroannulation**
It has been reported that annulation of a 2-iodoaniline with an internal alkyne takes place in the presence of a palladium catalyst to give a 2,3-disubstituted indole in one step (Larock annulation) [42]. The annulation of 4-carboxamide-2-iodoanilines (**104**) supported on Rink resin with an excess amount of disubstituted alkynes was catalyzed by Pd(OAc)$_2$-PPh$_3$ to give the indoles **105** (Table 19.12, route A) [43]. Cleavage of the resin moiety from **105** by trifluoroacetic acid gave the 2,3,5-trisubstituted indoles **108** in excellent yields (Table 19.12, entries 1–5). The 2-iodoaniline **106** bound to the resin support at its N1 position by the THP linker reacted with alkynes under palladium-catalyzed conditions to give N-resin-bound indole **107** (Table 19.12, route B). Acidic cleavage of the N–THP linkage gave high

Tab. 19.12. Solid-phase Larock heteroannulation.

Entry	Route	Condition	X	R	R1	R2	Yield (%)
1	A	i	CONH$_2$	H	Pr	Pr	91
2	A	i	CONH$_2$	H	Me	t-Bu	87
3	A	i	CONH$_2$	H	Me	Ph	86
4	A	ii	CONH$_2$	COCH$_3$	Pr	Pr	95
5	A	i	CONH$_2$	COCH(CH$_3$)$_2$	Me	t-Bu	75
6	B	iii	H	H	Ph	SiMe$_3$	73
7	B	iii	H	H	Me	t-Bu	55

Entries 1–5, ref. 41; entries 6–7, ref. 42. Condition i, alkyne
(10–15 equiv.), Pd(OAc)$_2$ (10 mol%), Ph$_3$P (20 mol%), LiCl
(1 equiv.), K$_2$CO$_3$ (5 equiv.), DMF, 80 °C; ii, alkyne (10–15 equiv.),
Pd(OAc)$_2$ (10 mol%), Ph$_3$P (20 mol%), Bu$_4$NCl (1 equiv.), KOAc
(5 equiv.), DMF, 80 °C; iii, alkyne (excess), PdCl$_2$(Ph$_3$P)$_2$
(20 mol%), tetramethylguanidine (10 equiv.), DMF, 110 °C.

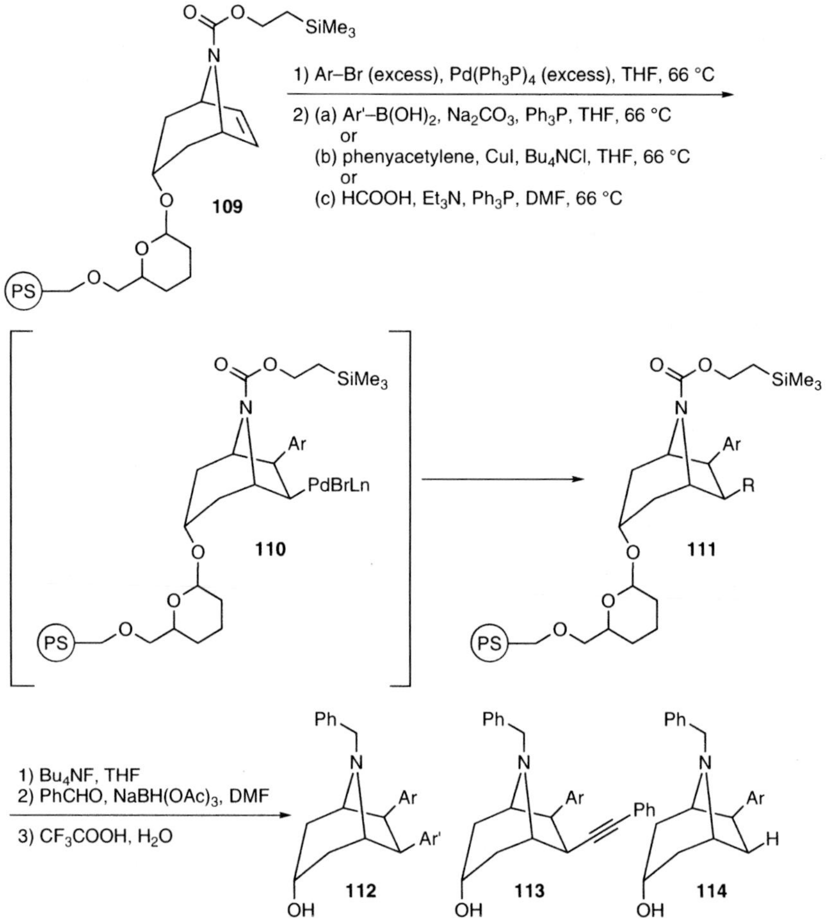

Scheme 19.23. Solid-phase synthesis of tropane derivatives via palladium-mediated three-component coupling.

yields of the 2,3-disubstituted indole **108** (X = H) (Table 19.12, entries 6 and 7) [44].

19.2.4.2 Insertion Cross-coupling Sequence (Dialkylation of Tropene)

Three-component couplings of the resin-supported tropene **109**, an aryl bromide, and an arylboronic acid or phenylacetylene were promoted by palladium(0) to give vicinal disubstituted tropanes (**111**) (Scheme 19.23) [45]. Thus, the reaction of **109** with an aryl bromide took place in THF with palladium(0) to give the σ-alkylpalladium intermediate **110**, which coupled with an arylboronic acid or phenylacetylene successively to give **111** (R = Ar or CCPh). Tropane **112** or **113** was obtained from **111** through deprotection, reductive N-alkylation, and acidic cleavage of resin. The monosubstituted tropane **114** was obtained similarly by reductive cleavage of the σ-alkylpalladium bond of the intermediate **110** with formic acid.

19.2.4.3 Coupling Reactions on Various Solid Supports

The Heck reaction, the Suzuki–Miyaura, Sonogashira, and Stille couplings with aryl iodide were examined on various resin supports (Table 19.13) [46]. Thus, aryl

Tab. 19.13. Palladium-catalyzed various coupling with a traceless linker.

Resin	Substrate	Coupling conditions	Cleavage conditions	Product	Yield (%)
TentaGel		Substrate (6 equiv.) Pd(OAc)₂ (20 mol%)	A		83
Polystyrene		NaOAc (3 equiv.) Bu₄NBr (1 equiv.)	B		85
ArgoPore		DMA, 100 °C, 24 h	B		96
TentaGel	OMe	Substrate (10 equiv.) Pd(Ph₃P)₄ (2 mol%)	A		74
TentaGel		K₃PO₄ (2 equiv.) aq. DMA, 80 °C, 24 h	B		86
TentaGel	B(OH)₂		C		93
Polystyrene			C		67
ArgoPore			A		60
TentaGel		Substrate (6 equiv.) PdCl₂(Ph₃P)₂ (10 mol%)	C		50
Polystyrene		CuI (20 mol%)	C		92
ArgoPore		Dioxane, Et₃N, rt, 24 h	C		86
TentaGel		Substrate (5 equiv.) Pd₂(dba)₃ (10 mol%)	C		90
Polystyrene	SnBu₃	Ph₃As (40 mol%)	C		79
ArgoPore		Dioxane, 60 °C, 24 h	C		80

Ref. 46: conditions A, Cu(OAc)₂ (0.5 equiv.), MeOH, pyridine (10 equiv.), rt, 2 h; B, Cu(OAc)₂ (0.5 equiv.), *n*-propylamine, rt, 2 h; C, NBS (2 equiv.), CH₂Cl₂, rt, 45 min.

iodide was connected to PS–PEG resin (TentaGel), standard PS resin, and macroporous PS resin (ArgoPore) with a hydrazine linker. The supported aryl iodides (**115**) were subjected to the coupling reactions under the conditions listed in Table 19.13 to give **116**. The resulting resins (**116**) were subsequently subjected to linker cleavage conditions A, B, or C (see Table 19.13) to give the substituted aromatics in a traceless fashion.

19.3
Solid-phase Reactions by Way of π-Allylpalladium Intermediates

Substitution reactions of allylic substrates with nucleophiles have been shown to be catalyzed by certain palladium complexes. The catalytic cycle of the reactions involves π-allylpalladium as the key intermediate (Scheme 19.24). Oxidative addition of the allylic substrate to a palladium(0) species forms a π-allylpalladium(II) complex, which undergoes attack of a nucleophile on the π-allyl moiety to give an allylic substitution product.

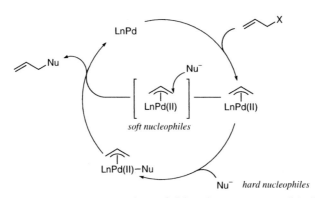

Scheme 19.24.   Reaction pathway of allylic substitution via a π-allylpalladium intermediate.

19.3.1
Cleavage of Allyl Ester Linkers

A carboxylic acid moiety connected to a polymer resin by an allyl ester linker was released under palladium-catalyzed allylic substitution conditions. Thus, an allyl ester group of the PS resin-supported tripeptide **117** was cleaved reductively by tin hydride in the presence of a palladium-PPh$_3$ catalyst to release the peptide in high yield. Carbon–oxygen bonds of supported allyl esters **118–120** were also readily cleaved by morpholine by way of π-allylpalladium intermediates (Scheme 19.25) [47].

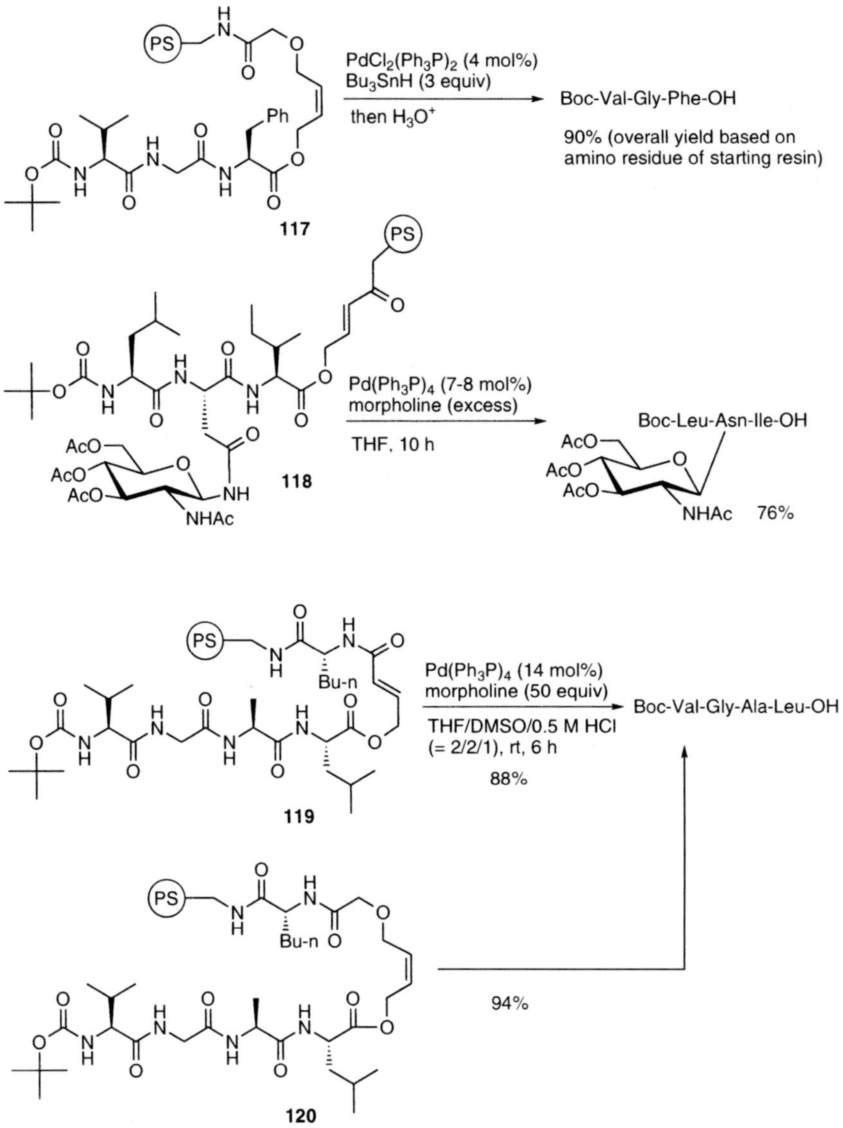

Scheme 19.25. Palladium-catalyzed cleavage of allylic anchoring groups.

An allyl ester of a resin-bound carboxylic acid was activated with palladium(0) to form the Pd(η^3-allyl)(OC(O)-resin) species **124** which readily undergoes attack by a nucleophile to provide functionalization and release of the allyl moiety **123** in one step (Table 19.14) [48]. Thus, the resin carboxylate ester **122** bearing a conjugated diene moiety prepared by solid-phase ruthenium-mediated metathesis reacted with an active methylene compound (Table 19.14, entries 1–4) or morpholine (Table

Tab. 19.14. Preparation of conjugated dienes via Ru-catalyzed cross-metathesis and Pd-catalyzed allylic substitution on solid support.

Entry	NuH or NuNa	Product	Yield (%)
1			54
2			66
3			59
4			67
5			86

Ref. 48.

19.14, entry 5) in the presence of a palladium-phosphine catalyst to give the diene **123** in high yield.

Treatment of compound **125** bearing an amino group at the homoallylic position with palladium-dppe catalyst gave the exo-methylenepyrrole **126** via formation of a π-allylpalladium intermediate and subsequent intramolecular nucleophilic attack of an amino group (Scheme 19.26) [49].

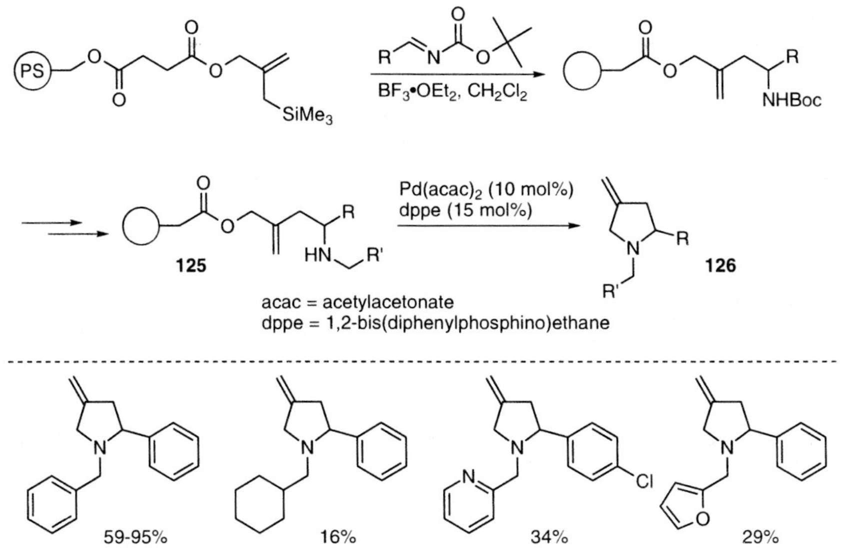
PS
SiMe3
BF3•OEt2, CH2Cl2
NHBoc
Pd(acac)2 (10 mol%)
dppe (15 mol%)
125
126
acac = acetylacetonate
dppe = 1,2-bis(diphenylphosphino)ethane
59-95%
16%
34%
29%

Scheme 19.26. Solid-phase synthesis of pyrrolidines via palladium-catalyzed cyclization cleavage.

19.3.2
N-Allylation via π-Allylpalladium Intermediates

The reaction of an allyl ester with a nitrogen nucleophile bound to the PS–PEG resin gave the *N*-allylation product (Scheme 19.27) [50]. Thus, the reaction of 2-methoxycarbonylmethyl-2-propen-1-ol (**128**) with the TentaGel-bound benzylamine **127** in the presence of $Pd(PPh_3)_4$ gave the *N*-allylation product **129**. After esterification of the allylic alcohol of **129**, the resulting allyl acetate **130** was subjected to palladium-catalyzed allylic substitution, again with various nitrogen nucleophiles. A resin-supported π-allylpalladium intermediate generated *in situ* underwent nucleophilic attack by primary, secondary, tertiary, and cyclic amines to give the corresponding allylic amines (**131**) on solid support. The *N*-(2-aminomethyl-2-propenyl)-*N*-benzylglycine derivatives **132** were released from the resin **131** by alkaline hydrolysis in moderate to high yields.

19.3.3
Insertion–π-Allylic Substitution System

Solid-phase synthesis of the (2-alkenyl)indoline derivatives **134** has been achieved in one pot by the reaction of the immobilized aryl halides **133** and conjugated dienes which proceeded through a palladium-catalyzed insertion–π-allylic substitution sequence (Scheme 19.28) [51]. Thus, the Rink resin-supported aryl iodide **133** was added to palladium(0) oxidatively to form the arylpalladium intermediate **135**. The arylpalladium intermediate **135** reacted with the diene to give the π-allylpalla-

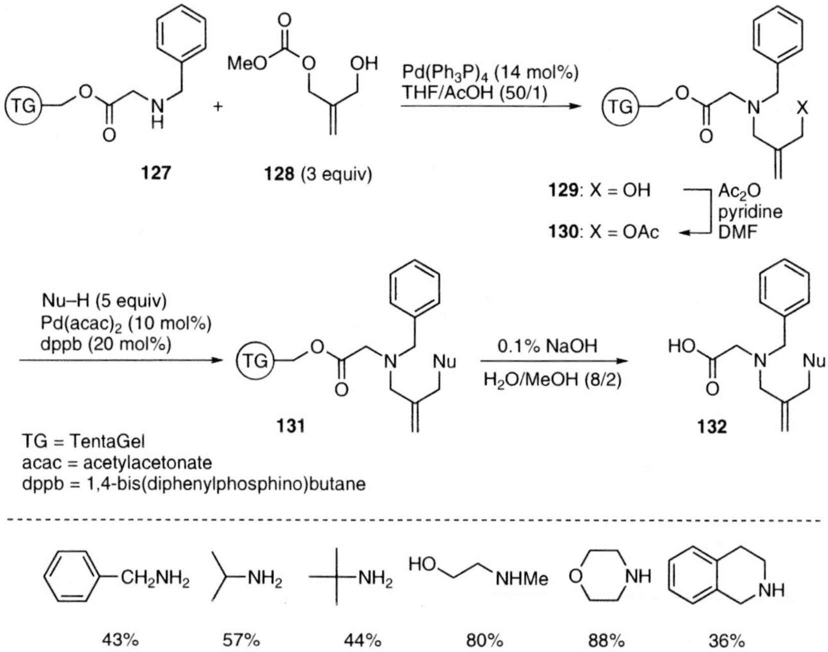

Scheme 19.27. Solid-phase preparation of *N*-benzylglycine derivatives.

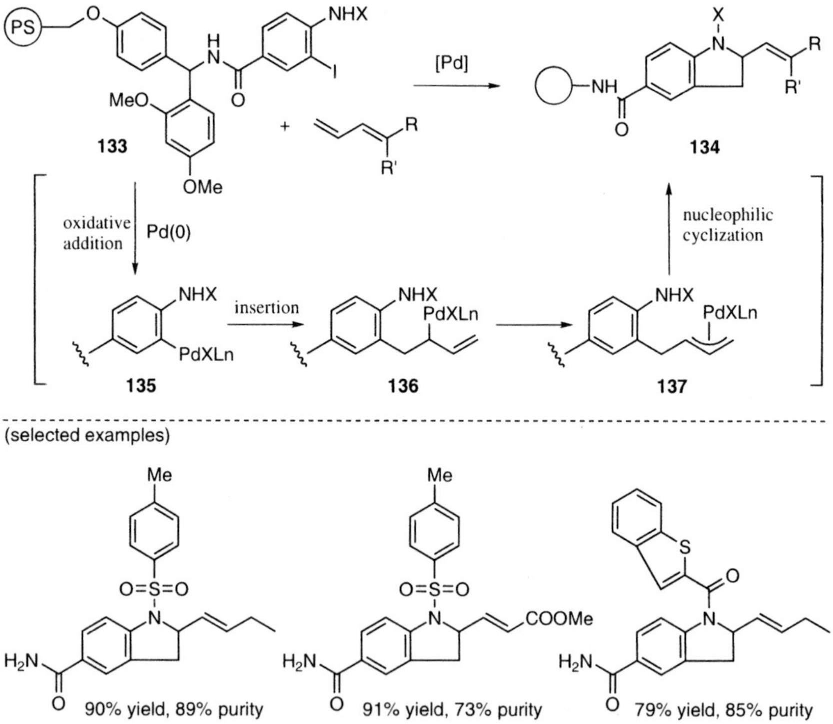

TG = TentaGel
acac = acetylacetonate
dppb = 1,4-bis(diphenylphosphino)butane

127   128 (3 equiv)

129: X = OH
130: X = OAc

Nu–H (5 equiv)
Pd(acac)₂ (10 mol%)
dppb (20 mol%)

131

132

43%   57%   44%   80%   88%   36%

Scheme 19.28. Synthesis of indoles via palladium-catalyzed annulation.

133

134

oxidative
addition | Pd(0)

insertion

135   136   137

(selected examples)

90% yield, 89% purity   91% yield, 73% purity   79% yield, 85% purity

nucleophilic
cyclization

dium **137** via the alkylpalladium **136**. The π-allylpalladium should readily undergo intramolecular nucleophilic attack of nitrogen atom at the ortho position to form the 2-(alkenyl)indoline **134**.

Three-component coupling of an aryl halide, 1,5-hexadiene, and the Rink-supported piperidine **138** was catalyzed by palladium to give the *N*-(6-aryl-2-hexenyl)-piperidine **141** via the insertion–π-allylic substitution pathway (Scheme 19.29) [52]. The alkylpalladium intermediate **140** generated in solution phase underwent a β-elimination–insertion process which was terminated by the formation of thermodynamically stable π-allylpalladium **143**. The resulting π-allylpalladium complex **143** reacted with piperidine on the resin supports to give the *N*-alkylated piperidines **139** in high yield.

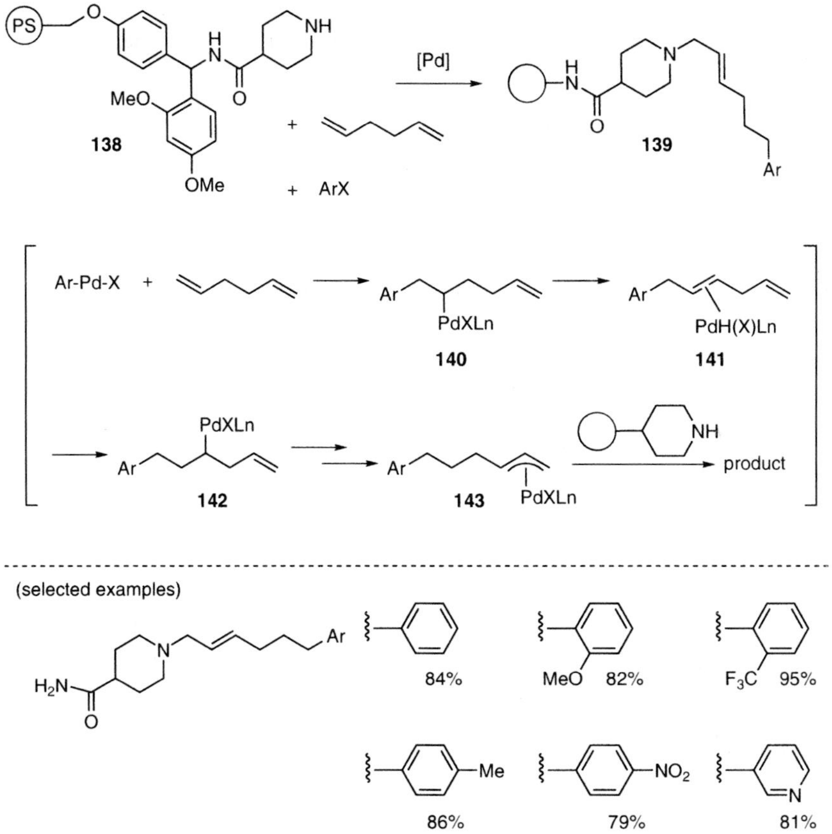

Scheme 19.29. Palladium-catalyzed three-component coupling.

19.4
Palladium Catalysis with Solid-supported Complexes

Homogeneous transition metal catalysts are widely used for a variety of organic transformations. High-throughput synthesis by solution-phase catalysis has also been recognized as a useful methodology with the advent of efficient methods for compound purification. One approach employs supported catalysts that can be readily removed by filtration. Several reviews have covered the synthetic use of solid-supported reagents, including transition metal complexes [53]. A number of support-bound palladium complexes, in particular palladium–phosphine complexes, have been designed and prepared to combine the advantages of both homogeneous and heterogeneous catalysts in one system [54]. This class of resin-bound palladium catalysts would solve the basic problems of homogeneous catalysts, namely the separation and recycling of the catalysts. These palladium complex catalysts are also advantageous in that contamination of the ligand residue in the products is avoided.

19.4.1
Preparation of Solid-supported Palladium Complexes and Their Use in Palladium Catalysis

Standard procedures for the preparation of polymer-supported catalysts usually entail surface modification of commercially available polymer resins, e.g. polystyrene–divinylbenzene (PS–DVB) or chloromethylated PS–DVB resin. Thus, the reaction of chloromethylated polystyrene with an excess of lithium diphenylphosphide gave the (diphenylphosphino)methylated polystyrene **145** in quantitative yield (Scheme 19.30). The palladium(0) complex **146** was obtained by the treatment of **145** with Pd(PPh$_3$)$_4$. The reaction of **145** with PdCl$_2$ (or PdCl$_2$(cod)) gave the resin-bound palladium(II) complex **147** which was readily converted to **146** by reduction with hydrazine in the presence of PPh$_3$. The physical properties of the resin matrix and the loading value of the phosphine residue are dependent on the crosslinking value (DVB, %) and the yield of the chloromethylation step, respec-

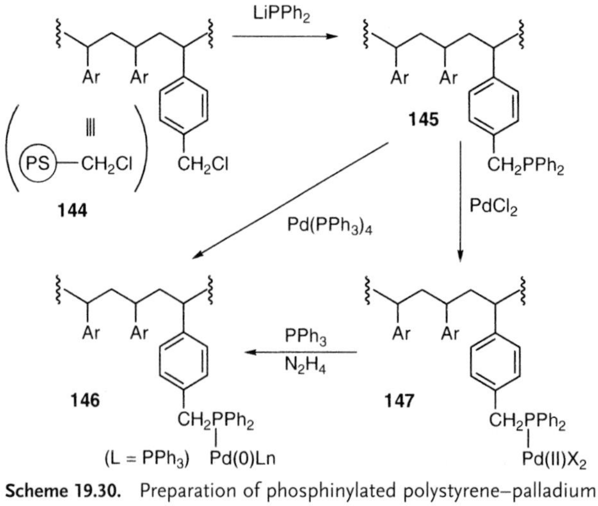

Scheme 19.30. Preparation of phosphinylated polystyrene–palladium complexes.

tively. The resin-bound palladium–phosphine complex **146** catalyzed nucleophilic allylic substitution via π-allylpalladium intermediates [55], telomerization of dienes [56], the Heck reaction [57], the Suzuki–Miyaura coupling [58], etc.

The bisphosphines **148** and **150** bearing alkyl substituents on their phosphorus atoms were supported on PS resin by the nucleophilic substitution of the chloromethyl groups on the resin to give **149** and **151**, respectively (Scheme 19.31) [59]. A palladium complex of **149** showed moderate catalytic activity to promote the Heck reaction of iodobenzene with methyl acrylate.

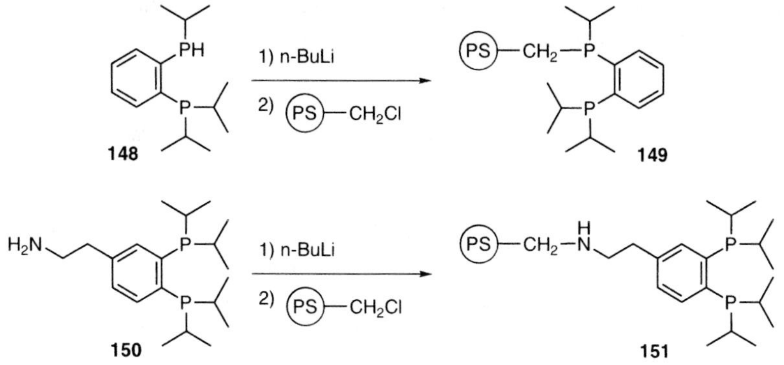

Scheme 19.31. Various ligands bound to polystyrene support.

The biarylphosphines **152** also reacted with the chloromethylated PS resin under basic conditions to give the PS-supported biarylphosphines **153** (Scheme 19.32) [60]. The resin-bound biaryl-(dialkyl)phosphines **153** were the ligands designed for use in the palladium-catalyzed amination and Suzuki–Miyaura coupling of aryl halides, especially those of aryl chlorides, whereas the use of electron-rich phosphine ligands allowed for an increase in the scope of the aryl halide substrate [61].

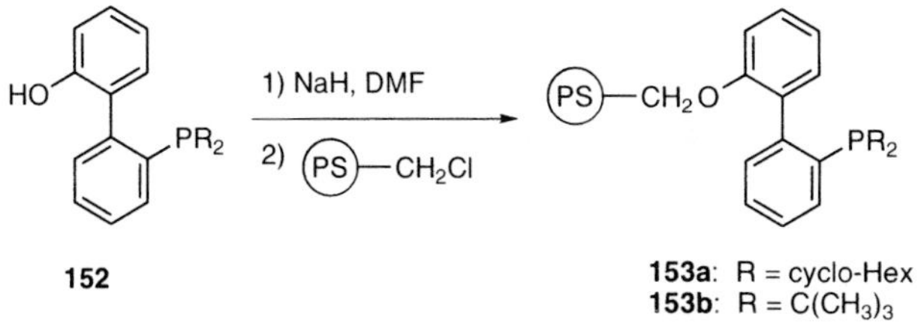

152

153a: R = cyclo-Hex
153b: R = C(CH$_3$)$_3$

Scheme 19.32. Various ligands bound to polystyrene support (2).

The polymer-supported carbene complexes of palladium **155** were prepared by the nucleophilic substitution of the bromomethylated Wang resin with **154** under basic reaction conditions (Scheme 19.33) [62]. The catalytic activity of **155** for the Heck reaction of aryl bromides with acrylates or styrene was found to exhibit high turnover numbers (TON) up to 5000. The supported carbene complexes **155** were air-stable and recyclable catalysts.

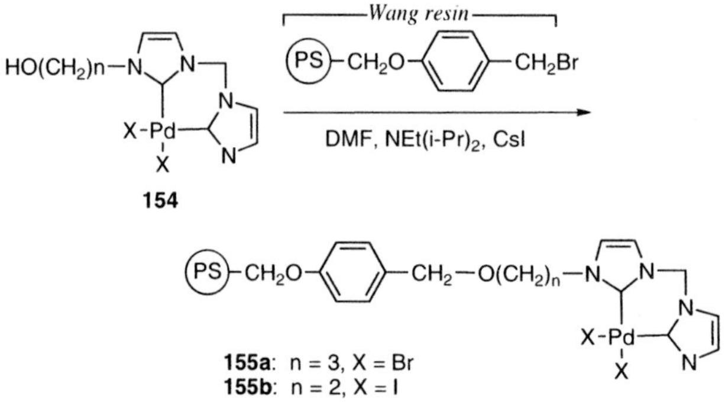

Scheme 19.33. Various ligands bound to polystyrene support (3).

Polymerization of ligand monomers is a useful tool for preparing polymer-supported ligands. The crosslinked polystyrene-bound ferrocenyl bisphosphine ligand **157** was prepared by the copolymerization of styrene, divinylbenzene, and 1,1′-bis-(diphenylphosphino)-2-vinylferrocene (**156**) (Scheme 19.34) [63]. The loading density of the catalyst on the support was readily controlled by the ratio of the monomers used.

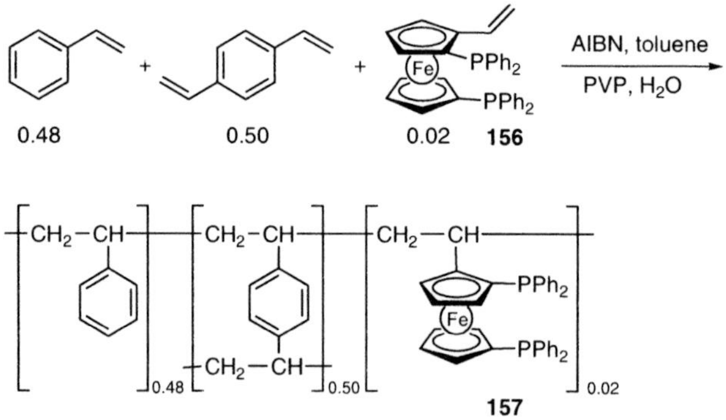

Scheme 19.34. Preparation of polymer-bound ferrocenylphosphine.

Carbonylative intramolecular Stille coupling to form macrocyclic molecules was investigated with a palladium complex of the polymer-bound ferrocenyl phosphine **157** (Scheme 19.35). One of the major problems encountered in the intramolecular macrocyclization is the formation of linear oligomers via an intermolecular pathway. "Site isolation" of the catalytic sites on a polymer backbone has been achieved with relatively low loading density of the catalyst to suppress the intermolecular reactions. Thus, ester **158** bearing an alkenylstannane and an alkenyl triflate gave high yields of the corresponding keto lactone **159** with the Pd(0)/**157** complex and LiCl under carbon monoxide, whereas only moderate yields of the macrocycles were obtained under homogeneous conditions using Pd(PPh$_3$)$_4$ or PdCl$_2$(dppf).

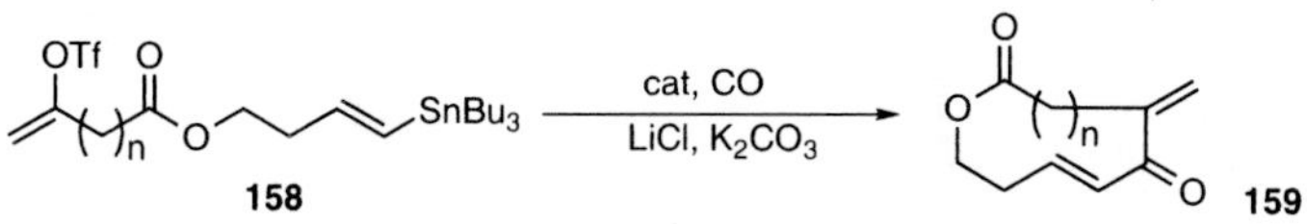

cat	n = 6	n = 7	n = 8
Pd(Ph$_3$P)$_4$	—	59	60
PdCl$_2$(dppf)	55	—	—
Pd(dba)$_2$/**157**	78	76	70

Scheme 19.35. Palladium-catalyzed macrocyclization.

The polypyrrole-bound mono- and bisphosphines **162** and **163** were prepared as their *P*-borane complexes from the corresponding monomers **160** and **161** via FeCl$_3$-induced or electrochemical polymerization conditions (Scheme 19.36) [64]. These phosphine–borane complexes reacted with palladium(II) without prede-complexation to give the polypyrrole-bound palladium(0)–phosphine complexes, where the borane on the phosphorus atom served as a reducing agent of palladium(II). The resulting immobilized polypyrrole palladium(0)–phosphine complexes catalyzed the Heck reaction and the π-allylic substitution of allyl acetates.

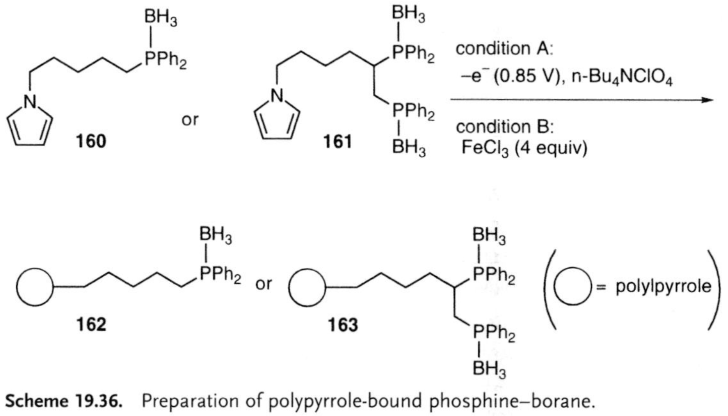

Scheme 19.36. Preparation of polypyrrole-bound phosphine–borane.

Ring-opening methathesis polymerization of the norbornene monomer **164** having a 2-endo-*N,N*-di-(2-pyridyl)carbamide group was carried out via a "living polymerization" using the Schrock catalyst (Scheme 19.37) [65]. The resulting living polymer chains were crosslinked using 1,4,4a,5,8,8a-hexahydro-1,4,5,8-exo-endo-dimethanonaphthalene (**165**) to give the bispyridyl ligand **166**. Its palladium complex **167** generated by treatment of **166** with H_2PdCl_4 catalyzed the Heck reaction of aryl bromides and even aryl chlorides. Thus, the reaction of chlorobenzene with styrene in the presence of 0.003 mol% of the palladium species **167** and tetrabutylammonium bromide in dimethylacetamide at 140 °C gave an 89% isolated yield of *trans*-stilbene where the TON observed reached 23,600.

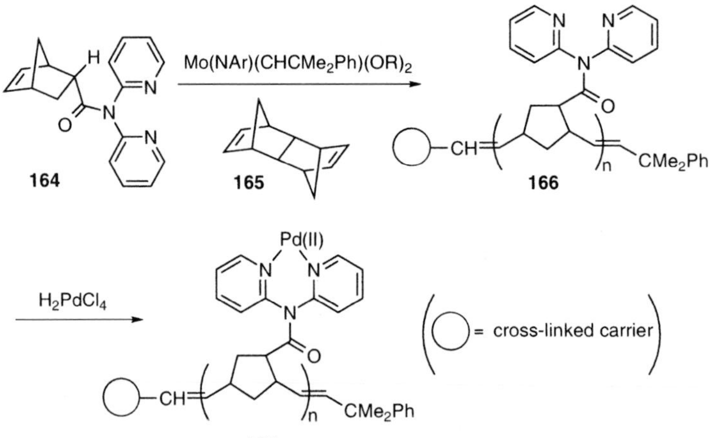

Scheme 19.37. Ring-opening metathesis polymerization (ROMP) of monomer ligands.

Polyaminoamide (PAMAM) dendrimers of generation 0–4 on silica [66] and carbosilane dendrimers [67] were used as solid support for immobilization of the palladium catalysts. Thus, for example, (diphenylphosphino)methyl groups were introduced on the terminal nitrogen of PAMAM chains by treatment of **168** with paraformaldehyde and diphenylphosphine (Scheme 19.38). Treatment of the resulting dendrimer bearing diphenylphosphino groups with $PdMe_2$(tmeda) gave the chelate complex **169**, which showed good catalytic activity in the Heck reaction.

The triarylphosphine moiety was incorporated into the PS–PEG resin by a solid-phase amide-forming reaction (Scheme 19.39) [68]. Thus, a mixture of the PEG–PS amino resin, 2 equiv. of 4-(diphenylphosphino)benzoic acid, 1-(3-dimethylaminopropyl)-3-ethylcarbodiimide hydrochloride (EDCI), and 1-hydroxybenzotriazole hydrate (HOBt) in DMF was agitated to give the PS–PEG resin-supported phosphine **171**. The complete consumption of the amino residue of the PEG chain was conveniently monitored by the Kaiser test. Formation of the palladium–phosphine complex **172** on the resin was performed by mixing $[PdCl(\pi\text{-}C_3H_5)]_2$ and **171** in an appropriate organic solvent at ambient temperature for 10 min. The PS–PEG resin-supported complex **172** exhibited high catalytic activity in water due to its amphiphilic property. Allylic substitution [68], Heck reaction [69], carbonylation [70], and Suzuki–Miyaura [70] coupling took place in a single aqueous medium at room temperature by use of **172**.

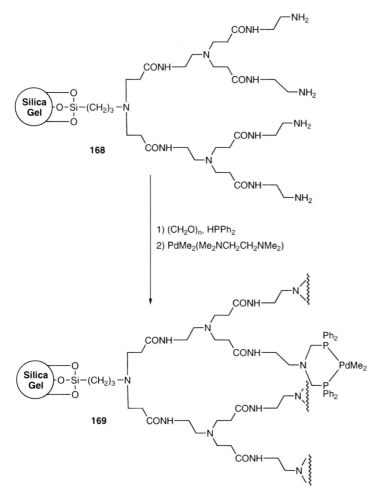

Scheme 19.38. Dendrimer-bound palladium–phosphine complex.

Amphiphilic polymer-supported phosphine ligands were also prepared on poly(N-isopropyl)acrylamide (PNIPAM) resin (Scheme 19.40) [71]. The palladium complex of the PNIPAM–phosphine, formed from reaction of **174** or **175** with Pd(dba)$_2$, showed high catalytic activity both in organic solvents and in water to promote π-allylic substitution of allyl carbonates and the Sonogashira reaction of aryl iodides.

19.4.2
Solid-supported Chiral Palladium Catalysts

Asymmetric reactions catalyzed by transition metal complexes containing optically active ligands have attracted great interest because of their synthetic utility. A vast

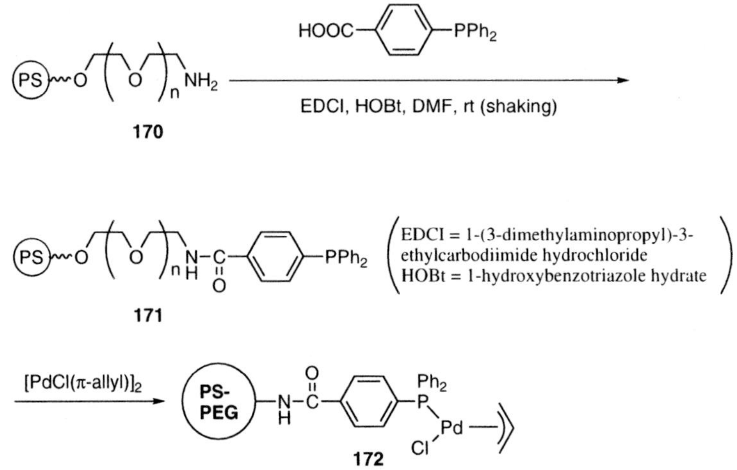

Scheme 19.39. PS–PEG resin-supported amphiphilic palladium–phosphine complexes.

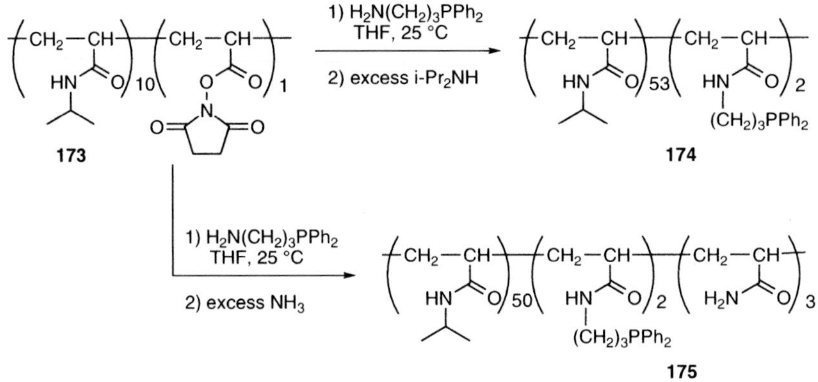

Scheme 19.40. PNIPAM-supported amphiphilic phosphine ligands.

amount of research has been reported to date on asymmetric reactions using homogeneous catalyst systems in which activity and stereoselectivity can be tuned by varying the ligand structure. Recently, immobilization of the enantioselective catalysts has been recognized as one of the most promising strategies for achieving highly stereoselective catalysis under heterogeneous conditions [72]. Several examples of chiral ligands supported on polymer resin, which have found utility in asymmetric palladium catalysis, are shown in Scheme 19.41. Palladium complexes of the resin-supported 2-diphenylphosphino-2′-substituted-1,1′-binaphthyl (MOP) ligand **176** [73] and pyridinoxazoline **179** [74] catalyzed allylic substitution with good to high stereoselectivity. The PS-supported BINAP **177** [75] was applied to a palladium(II)-catalyzed aldol reaction of a silyl enolate [76]. A novel chiral ligand, (3R,9aS)-(2-aryl-3-(2-diphenylphosphino)-phenyl)-tetrahydro-1H-imidazo[1,5-a]indole-1-one was designed, prepared, and immobilized on an amphiphilic polystyrene-poly(ethylene glycol) graft copolymer (PS–PEG) resin (**178**) [77]. A palladium complex of the PS–PEG resin-supported ligand **178** catalyzed the allylic substitution of both cyclic and acyclic allylic esters in water with high enantioselectivity (up to 98% ee). The PS–PEG-supported Pd complex was readily recovered by simple filtration and reused without loss of catalytic activity or enantioselectivity.

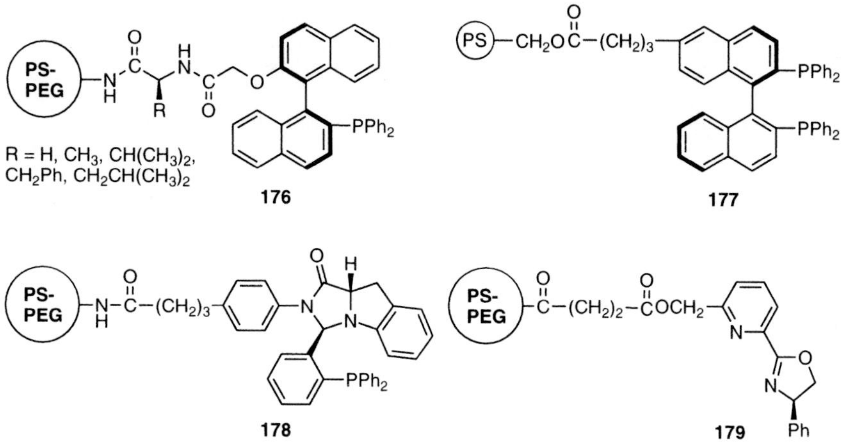

Scheme 19.41. Resin-supported chiral ligands.

References

1 For reviews, see: a) J. Tsuji, *Palladium Reagents and Catalysts.* John Wiley and Sons, Chichester 1995; b) R. F. Heck, *Palladium Reagents in Organic Synthesis.* Academic Press, New York 1985.

2 For recent reviews, see: a) B. A. Bunin, *Combinatorial Index.* Academic Press, London 1998; b) D. Obrecht, J. M. Villalgordo, *Solid-Supported Combinatorial and Parallel Synthesis of Small-Molecular-Weight Compound Libraries,* Tetrahedron Organic Chemistry Series Vol. 17. Elsevier, Oxford 1998.

3 For a recent review, see: F. Z. Dörwald, *Organic Synthesis on Solid Phase,* Wiley-VCH, Weinheim 2000.

4 a) A. Suzuki, *Pure Appl. Chem.* **1985**, *57*, 1749; b) N. Miyaura, A. Suzuki, *Chem. Rev.* **1995**, *95*, 2457.

5 B. J. Backes, J. A. Ellman, *J. Am. Chem. Soc.* **1994**, *116*, 11171–11172.

6 J. W. Guiles, S. G. Johnson, W. V. Murray, *J. Org. Chem.* **1996**, *61*, 5169–5171.

7 R. Frenette, R. W. Friesen, *Tetrahedron Lett.* **1994**, *35*, 9177–9180.

8 S. Wendeborn, S. Berteina, W. K.-D. Brill, A. D. Mesmaeker, *Synlett* **1998**, 671–675.

9 S. Chamoin, S. Houldsworth, C. G. Kruse, W. I. Bakker, V. Snieckus, *Tetrahedron Lett.* **1998**, *39*, 4179–4182.

10 S. R. Piettre, S. Baltzer, *Tetrahedron Lett.* **1997**, *38*, 1197–1200.

11 a) S. D. Brown, R. W. Armstrong, *J. Am. Chem. Soc.* **1996**, *118*, 6331–6332; b) S. D. Brown, R. W. Armstrong, *J. Org. Chem.* **1997**, *62*, 7076–7077.

12 a) F. X. Woolard, J. Paetsch, J. A. Ellman, *J. Org. Chem.* **1997**, *62*, 6102–

6103; b) C. VANIER, A. WAGNER, C. MIOSKOWSKI, *Tetrahedron Lett.* **1999**, *40*, 4335–4338.

13 L. A. THOMPSON, F. L. MOORE, Y.-C. MOON, J. A. ELLMAN, *J. Org. Chem.* **1998**, *63*, 2066–2067.

14 W. LI, K. BURGESS, *Tetrahedron Lett.* **1999**, *40*, 6527–6530.

15 J. K. STILLE, *Angew. Chem. Int. Ed. Engl.* **1986**, *25*, 508–524.

16 M. S. DESHPANDE, *Tetrahedron Lett.* **1994**, *35*, 5613–5614.

17 S. CHAMOIN, S. HOUDSWORTH, V. SNIECKUS, *Tetrahedron Lett.* **1998**, *39*, 4175–418.

18 M. S. BRODY, M. G. FINN, *Tetrahedron Lett.* **1999**, *40*, 415–418.

19 S. WENDEBORN, A. D. MESMAEKER, *Synlett* **1998**, 865–867.

20 K. SONOGASHIRA, Y. TOHDA, *Tetrahedron Lett.* **1975**, 4467–4470.

21 S. BERTEINA, S. WENDEBORN, W. K.-D. BRILL, A. D. MESMAEKER, *Synlett* **1998**, 676–678.

22 M. C. FAGNOLA, I. CANDIANI, G. VISENTIN, W. CABRI, F. ZARINI, N. MONGELLI, A. BEDESCHI, *Tetrahedron Lett.* **1997**, *38*, 2307–2310.

23 H.-C. ZHANG, K. K. BRUMFIELD, L. JAROSKOVA, B. E. MARYANOFF, *Tetrahedron Lett.* **1998**, *39*, 4449–4452.

24 D. FANCELLI, M. C. FAGNOLA, D. SEVERINO, A. BEDESCHI, *Tetrahedron Lett.* **1997**, *38*, 2311–2314.

25 M. D. COLLINI, J. W. ELLINGBOE, *Tetrahedron Lett.* **1997**, *38*, 7963–7966.

26 A. B. DYATKIN, R. A. RIVERO, *Tetrahedron Lett.* **1998**, *39*, 3647–3650.

27 S. I. KHAN, M. W. GRINSTAFF, *Tetrahedron Lett.* **1998**, *39*, 8031–8034.

28 S. MARQUAIS, M. ARLT, *Tetrahedron Lett.* **1996**, *37*, 5491–5494.

29 M. ROTTLÄNDER, P. KNOCHEL, *Synlett* **1997**, 1084–1086.

30 a) R. F. HECK, *Org. React.* **1982**, *27*, 345; b) R. F. HECK in: Comprehensive Organic Synthesis, Vol. 4. TROST, B. M., FLEMING, I. (eds), Pergamon Press, New York 1991, 833.

31 K.-L. YU, M. S. DESHPANDE, D. M. VYAS, *Tetrahedron Lett.* **1994**, *35*, 8919–8922.

32 M. HIROSHIGE, J. R. HAUSKE, P. ZHOU, *Tetrahedron Lett.* **1995**, *36*, 4567–4570.

33 W. YUN, R. MOHAN, *Tetrahedron Lett.* **1996**, *37*, 7189–7192.

34 V. ARUMUGAM, A. ROUTLEDGE, C. ABELL, S. BALASUBRAMANIAN, *Tetrahedron Lett.* **1997**, *38*, 6473–6476.

35 a) H.-C. ZHANG, B. E. MARYANOFF, *J. Org. Chem.* **1997**, *62*, 1804–1809.

36 K. AKAJI, Y. KISO, *Tetrahedron Lett.* **1997**, *38*, 5185–5188.

37 G. L. BOLTON, J. C. HODGES, *J. Comb. Chem.* **1999**, *1*, 130–133.

38 a) S. L. BUCHWALD, A. S. GURAM, R. A. RENNELS, *Angew. Chem. Int. Ed. Engl.* **1995**, *34*, 1348–1350; b) J. F. HARTWIG, J. LOUIE, *Tetrahedron Lett.* **1995**, *36*, 3609.

39 Y. D. WARD, V. FARINA, *Tetrahedron Lett.* **1996**, *37*, 6993–6996.

40 C. A. WILLOUGHBY, K. T. CHAPMAN, *Tetrahedron Lett.* **1996**, *37*, 7181–7184.

41 For examples of miscellaneous synthetic applications of palladium-catalyzed coupling reactions, see: a) P. A. TEMPEST, R. W. ARMSTRONG, *J. Am. Chem. Soc.*, **1997**, *119*, 7607–7608 (Suzuki–Miyaura coupling, Stille coupling); b) M. A. LAGO, T. T. NGUYEN, P. BHATNAGAR, *Tetrahedron Lett.* **1998**, *39*, 3885–3888 (Suzuki–Miyaura coupling); c) Y. HAN, A. GIROUX, C. LÉPINE, F. LALIBERTÉ, Z. HUANG, H. PERRIER, C. I. BAYLY, R. N. YOUNG, *Tetrahedron*, **1999**, *55*, 11669–11685 (Suzuki–Miyaura coupling); d) M. LARHED, G. LINDEBERG, A. HALLBERG, *Tetrahedron Lett.* **1996**, *37*, 8219–8222 (Stille coupling); e) F. W. FORMAN, I. SUCHOLEIKI, *J. Org. Chem* **1995**, *60*, 523–528 (Stille coupling); f) P. R. L. MALENFANT, J. M. J. FRÉCHET, *Chem. Commun.* **1998**, 2657–2568 (Stille coupling); g) J. C. NELSON, J. K. YOUNG, J. S. MOORE, *J. Org. Chem* **1996**, *61*, 8160–8168 (Sonogashira reaction); h) S. I. KHAN, M. W. GRINSTAFF, *J. Am. Chem. Soc.* **1999**, *121*, 4704–4705 (Sonogashira reaction); i) S. HUANG, J. M. TOUR, *J. Am. Chem. Soc.* **1999**, *121*, 4908–4909

(Sonogashira reaction); j) F. Homsi, K. Nozaki, T. Hiyama, *Tetrahedron Lett.* **2000**, *41*, 5869–5872 (Si-mediated coupling); k) Y. Han, A. Roy, A. Giroux, *Tetrahedron Lett.* **2000**, *41*, 5447–5451 (Zn-mediated coupling); l) H. A. Dondas, R. Grigg, W. S. MacLachlan, D. T. MacPherson, J. Markandu, V. Sridharan, S. Suganthan, *Tetrahedron Lett.* **2000**, *41*, 967–970 (Heck reaction); m) S. Ma, D. Duan, Z. Shi, *Org. Lett.* **2000**, *2*, 1419–1422 (Heck-type reaction).

42 R. C. Larock, E. K. Yum, *J. Am. Chem. Soc.* **1991**, *113*, 6689–6690.

43 H.-C. Zhang, K. K. Brumfield, B. E. Maryanoff, *Tetrahedron Lett.* **1997**, *38*, 2439–2442.

44 A. L. Smith, G. I. Stevenson, C. J. Swain, J. L. Castro, *Tetrahedron Lett.* **1998**, *39*, 8317–8320.

45 J. S. Koh, J. A. Ellman, *J. Org. Chem.* **1996**, *61*, 4494–4495.

46 F. Stieber, U. Grether, H. Waldmann, *Angew. Chem. Int. Ed.* **1999**, *38*, 1073–1077.

47 a) F. Guibé, O. Dangles, G. Balavoine, *Tetrahedron Lett.* **1989**, *30*, 2641–2644; b) H. Kunz, B. Dombo, *Angew. Chem. Int. Ed. Engl.* **1988**, *27*, 711–713; c) P. L.-Williams, G. Jou, F. Albericio, E. Giralt, *Tetrahedron Lett.* **1991**, *32*, 4207–4210; d) O. Seitz, H. Kunz, *J. Org. Chem.* **1997**, *62*, 813–826.

48 S. C. Schürer, S. Blechert, *Synlett.* **1997**, 166–168.

49 R. C. D. Brown, M. Fisher, *Chem. Commun.* **1999**, 1547–1548.

50 Z. Flegelova, M. Patek, *J. Org. Chem.* **1996**, *61*, 6735–6738.

51 Y. Wang, T.-N. Huang, *Tetrahedron Lett.* **1998**, *39*, 9605–9608.

52 Y. Wang, T.-N. Huang, *Tetrahedron Lett.* **1999**, *40*, 5837–5840.

53 For reviews, see: a) D. C. Bailey, S. H. Langer, *Chem. Rev.* **1981**, *81*, 109–148; b) S. J. Shuttleworth, S. M. Allin, P. K. Sharma, *Synthesis* **1997**, 1217–1239; c) K. Burgess, A. M. Porte, *Advance Catalytic Processes* **1997**, *2*, 69–82; d) S. J. Shuttleworth, S. M. Allin, R. D. Wilson,

D. Nasturica, *Synthesis* **2000**, 1035–1074.

54 Development of nano-particles of palladium crystallites encapsulated inside polymer resin have become a major challenge; their synthetic use is, however, still limited to hydrogenation of unsaturated compounds. For examples, see: a) D. E. Bergbreiter, B. Chen, T. J. Lynch, *J. Org. Chem.* **1983**, *48*, 4179–4186; b) M. Zecca, R. Fisera, G. Palma, S. Lora, M. Hronec, M. Kralik, *Chem. Eur. J.* **2000**, *6*, 1980–1986.

55 B. M. Trost, E. Keinan, *J. Am. Chem. Soc.* **1978**, *100*, 7779–7781.

56 K. Kaneda, H. Kurosaki, M. Terasawa, T. Imanaka, S. Teranishi, *J. Org. Chem.* **1981**, *46*, 2356–2362.

57 C.-M. Andersson, K. Karabelas, A. Hallberg, *J. Org. Chem.* **1985**, *50*, 3891–3895.

58 a) S.-B. Jang, *Tetrahedron Lett.* **1997**, *38*, 1793–1796; b) I. Fenger, C. L. Drian, *Tetrahedron Lett.* **1998**, *39*, 4287–4290; c) K. Inada, N. Miyaura, *Tetrahedron* **2000**, *56*, 8661–8664.

59 a) M. A. Fox, D. A. Chandler, P.-W. Wang, *Macromolecules* **1991**, *24*, 4626–4636; b) P.-W. Wang, M. A. Fox, *J. Org. Chem.* **1994**, *59*, 5358–5364.

60 C. A. Parrish, S. L. Buchwald, *J. Org. Chem.* **2001**, *66*, 3820–3827.

61 A. F. Littke, C. Dai, G. C. Fu, *J. Am. Chem. Soc.* **2000**, *122*, 4020.

62 J. Schwarz, V. P. W. Böhm, M. G. Gardiner, M. Grosche, W. A. Herrmann, W. Hieringer, G. Raudaschl-Sieber, *Chem. Eur. J.* **2000**, *6*, 1773–1780.

63 J. K. Stille, H. Su, D. H. Hill, P. Schneider, M. Tanaka, D. L. Morrison, L. S. Hegedus, *Organometallics* **1991**, *10*, 1993–2000.

64 N. Riegel, C. Darcel, O. Stéphan, S. Jugé, *J. Organomet. Chem.* **1998**, *567*, 219–233.

65 M. R. Buchmeiser, K. Wurst, *J. Am. Chem. Soc.* **1999**, *121*, 11101–11107.

66 a) H. Alper, P. Arya, S. C. Bourque, G. R. Jefferson, L. E. Manzer, *Can. J. Chem.* **2000**, *78*, 920–924; b) similar dendrimer-bound phosphine ligands

anchored on PS resin have been prepared for rhodium-catalyzed reactions, see: P. Arya, G. Panda, N. V. Rao, H. Alper, S. C. Bourque, L. E. Manzer, *J. Am. Chem. Soc.* **2001**, *123*, 2889–2890.

67 E. B. Eggeling, N. J. Hovestad, J. T. B. H. Jastrzebski, D. Vog, G. V. Koten, *J. Org. Chem.* **2000**, *65*, 8857–8865.

68 a) Y. Uozumi, H. Danjo, T. Hayashi, *Tetrahedron Lett.* **1997**, *38*, 3557; b) H. Danjo, D. Tanaka, T. Hayashi, Y. Uozumi, *Tetrahedron* **1999**, *55*, 14341.

69 Y. Uozumi, T. Watanabe, *J. Org. Chem.* **1999**, *64*, 6921.

70 Y. Uozumi, H. Danjo, T. Hayashi, *J. Org. Chem.* **1999**, *64*, 3384.

71 D. E. Bergbreiter, Y.-S. Liu, *Tetrahedron Lett.* **1997**, *38*, 7843.

72 For a review, see: D. E. De Vos, I. F. J. Vankelecom, P. A. Jacobs (eds) *Chiral Catalyst Immobilization and Recycling.* Wiley-VCH, Weinheim 2000.

73 Y. Uozumi, H. Danjo, T. Hayashi, *Tetrahedron Lett.* **1998**, *39*, 8303.

74 K. Hallman, E. Macedo, K. Nordström, C. Moberg, *Tetrahedron Asymmetry* **1999**, *10*, 4037–4046.

75 a) D. J. Bayston, J. L. Fraser, M. R. Ashton, A. D. Baxter, M. E. Polywka, E. Moses, *J. Org. Chem.* **1998**, *63*, 3137–3140; b) D. J. Bayston, J. L. Fraser, M. R. Ashton, A. D. Baxter, M. E. Polywka, E. Moses, *Speciality Chemicals* **1998**, *18*, 224–226.

76 A. Fujii, M. Sodeoka, *Tetrahedron Lett.* **1999**, *40*, 8011–8014.

77 Y. Uozumi, K. Shibatomi, *J. Am. Chem. Soc.* **2001**, *123*, 2919–2920.

20

Olefin Metathesis and Related Processes for CC Multiple Bond Formation

Florencio Zaragoza

20.1
Introduction

Olefin metathesis refers to a reaction in which two alkenes exchange their alkylidene fragments (Scheme 20.1) [1–5]. This reaction has been applied to the preparation of compound libraries in solution and has also been used for solid-phase synthesis, thus enabling its application to automated parallel synthesis. Examples of metathesis on solid phase include the chemical transformation of resin-bound intermediates as well as the cleavage of final products from the support.

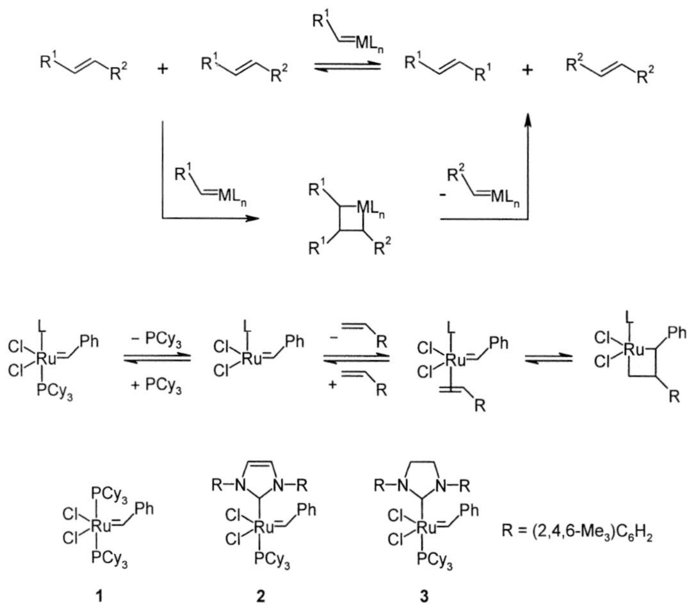

Scheme 20.1. The mechanism of olefin metathesis, dissociative mechanism for ruthenacyclobutane formation [27], and some useful ruthenium-based catalysts.

Olefin metathesis can be catalyzed both by heterogeneous catalysts (mainly supported transition metal oxides) and by soluble transition metal complexes (mainly tungsten, molybdenum, or ruthenium carbene complexes); the process has been known and exploited industrially since the early 1960s [6, 7]. However, it was not until soluble and highly effective catalysts were discovered by Schrock, Grubbs, and others that this reaction could be performed under sufficiently mild conditions to enable its systematic application to organic synthesis. Ruthenium carbene complex **1** [8], and the newer, more effective complexes **2** [9–12] and **3** [13–18] (Scheme 20.1) are particularly well suited for organic synthesis because of their high stability and chemoselectivity. These and related complexes tolerate a broad selection of functional groups and mediate olefin metathesis even in the presence of air and protic solvents, including water [19–21]. New catalysts with improved properties are continuously being developed [22–26].

The mechanism by which carbene complexes catalyze olefin metathesis is shown in Scheme 20.1. A reversible $2 + 2$ cycloaddition of the carbene complex to the alkene yields a metallacyclobutane, which upon cycloreversion can yield either the products of olefin metathesis or the starting materials. Because all of these transformations are reversible, equilibrium toward the desired products must be shifted either by continuous removal of one of the products (e.g. ethylene) or by using substrates for which olefin metathesis cannot be reversed (e.g. strained cycloalkenes).

In the case of ruthenium carbene complexes such as **1**, one of the phosphine ligands dissociates from the complex during catalysis [27]. It is still being debated whether this dissociation takes place before or after coordination to the alkene. In the case of complex **3**, it has been shown that olefin metathesis is initiated by a dissociative mechanism, as shown in Scheme 20.1 [27].

Although the currently available catalysts are exceedingly useful for many applications, a series of drawbacks still limits their scope, and there is ample room for improvement. Hopefully, some of these weaknesses will be overcome as our understanding of the precise mechanism of carbene complex-mediated olefin metathesis deepens and new types of catalysts emerge. One of the problems of current metathesis catalysts is that they generally yield internal alkenes as mixtures of *E*- and *Z*-isomers. Moreover, because the required carbene complexes can also react with soft nucleophiles other than alkenes, care must be taken to remove all traces of amines, pyridines, imidazoles, or other potential ligands from the solvent and the reactants in order to avoid deactivation of the catalyst. For the same reason, olefin metathesis proceeds best in solvents of low nucleophilicity [$CH_2Cl_2 >$ toluene $>$ tetrahydrofuran (THF)]. Electron-rich alkenes (enamines, enol ethers) do not usually undergo metathesis because donor-substituted carbene complexes are formed as intermediates, which are no longer electrophilic enough to act as metathesis catalysts. With some of the most recently developed catalysts (such as **3**, Scheme 20.1), however, even enol ethers may undergo metathesis [17]. The reaction rate of olefin metathesis sharply decreases in the series terminal alkene $>$ internal, disubstituted alkene $>$ trisubstituted alkene [2, 14], and 1,2-disubstituted *cis*-alkenes are usually more reactive than the corresponding *trans*-alkenes.

Most of the currently used catalysts (e.g. **1**, **2**, and **3**) are ruthenium *benzylidene* complexes. One of the reasons for choosing the benzylidene ligand is that styrene does not generally undergo cross-metathesis efficiently when using ruthenium carbene complexes as catalysts [28]. For this reason, only small amounts of product resulting from a cross-metathesis with styrene will result, even if large amounts of catalyst are used (see, for example, Scheme 20.20).

Olefin metathesis reactions have been grouped into different categories (Scheme 20.2). These include self-metathesis (reaction of one alkene with itself), cross-metathesis (reaction of two different alkenes with each other), ring-opening metathesis polymerization (ROMP, polymerization of a strained, cyclic alkene), and ring-closing metathesis (RCM, cyclization of a diene). All these types of olefin metathesis can be conducted on insoluble supports, and can thus be adapted to automated, parallel solid-phase synthesis. For combinatorial chemistry in solution, mainly cross-metathesis and ROMP have been used.

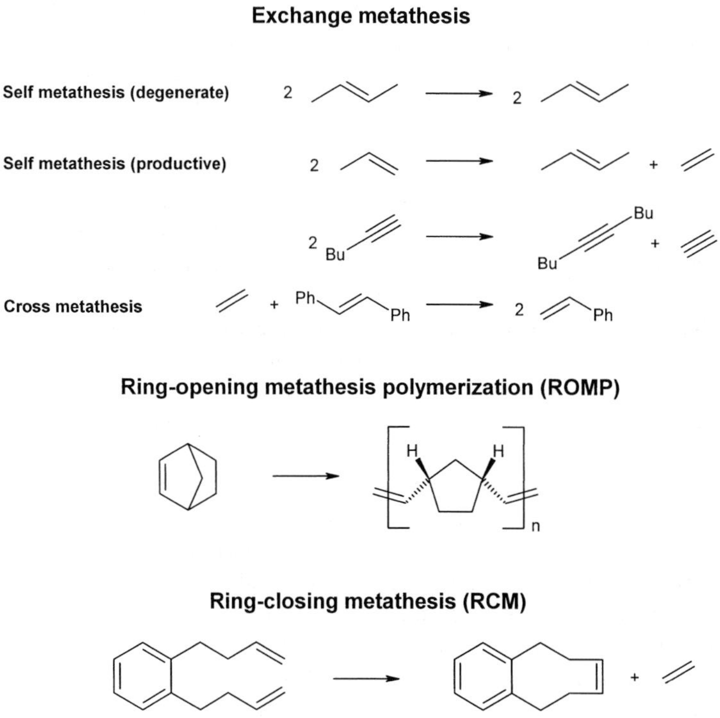

Exchange metathesis

Scheme 20.2. Categories of olefin metathesis reactions [2].

Alkynes are also suitable substrates for catalytic olefin metathesis (Scheme 20.3). Treatment of a mixture of an alkene and an alkyne with a metathesis catalyst can lead to the clean formation of dienes [29, 30]. Because alkynes usually react faster than alkenes with carbene complexes, the formation of dienes from alkenes and alkynes is assumed to proceed via the initial addition of the catalyst to the alkyne, followed by cycloreversion and reaction of the resulting vinylcarbene complex with the alkene (Scheme 20.3). Treatment of alkynes with carbyne complexes can bring about alkyne metathesis [1, 31], which presumably also proceeds via reversible 2 + 2 cycloadditions (Scheme 20.3). Some of these intriguing transformations have also been performed on insoluble supports, and might be suitable for parallel synthesis.

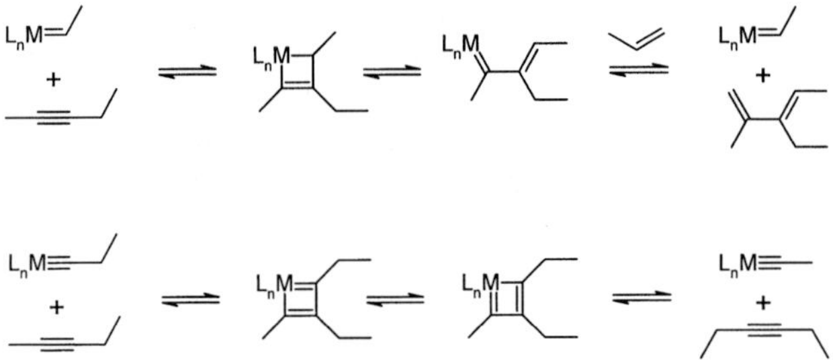

Scheme 20.3. Metathesis of alkynes.

20.2
Olefin Metathesis in Solution

For the preparation of compound libraries in solution using olefin metathesis, mainly two strategies have been employed. These are the cross-metathesis of mixtures of terminal alkenes, to yield mixtures of internal, disubstituted alkenes, and the oligomerization by ring-opening metathesis of strained, cyclic alkenes (Scheme 20.4). Ring-closing metathesis in solution has been used mainly for the preparation of small arrays of compounds or of single compounds.

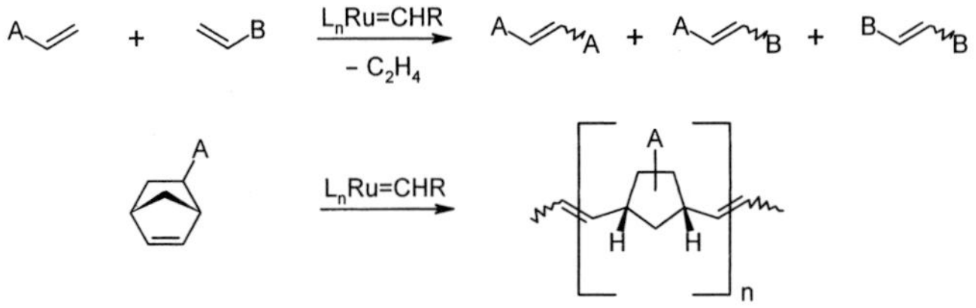

Scheme 20.4. Strategies for the preparation of compound mixtures using olefin metathesis.

20.2.1
Scope and Limitations of Olefin Metathesis in Solution

Cross-metathesis of two different terminal alkenes in solution only rarely gives high yields of one product (for recent advances in selective cross-metathesis, see

[14, 28, 32–34]). Usually, mixtures of the products of cross-metathesis and of self-metathesis are obtained, each of them as mixtures of *E*- and *Z*-isomers. Unfortunately, some alkenes show a high tendency to undergo self-metathesis (to form symmetric "dimers"; see, for example, [35, 36]), whereas other alkenes (acrylonitrile, styrenes) undergo self-metathesis only slowly or not at all. For this reason, cross-metathesis of mixtures of different olefins will not always yield the statistically expected amounts of internal olefins. This feature can cause problems during the deconvolution of such compound libraries because potent ligands formed only in low quantities are usually difficult to identify by deconvolution.

A further problem of cross-metathesis in solution is that a purification step will usually be required to remove the catalyst. With the recent development of immobilized catalysts (e.g. **4–6**) [37–41], however, this problem has been reduced. Unfortunately, all of the immobilized ruthenium carbene complexes described so far (Scheme 20.5) lose activity rather quickly; this fact might be due to the inherent instability of these complexes and to the fact that during catalysis detachment of the metal from the support can readily occur.

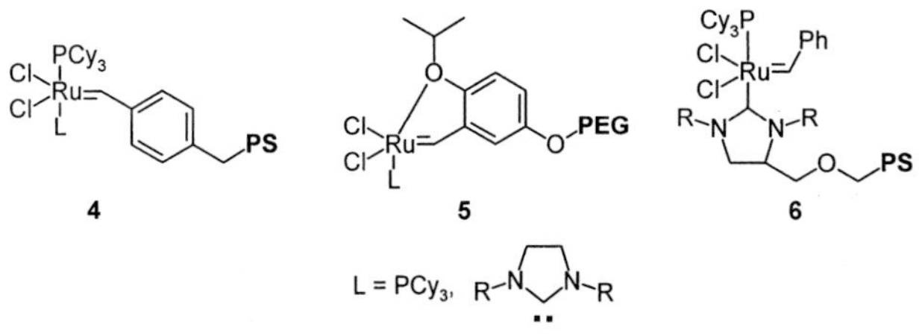

Scheme 20.5. Support-bound ruthenium carbene complexes, useful as insoluble metathesis catalysts [24, 38–41]. PS, crosslinked polystyrene; PEG, poly(ethylene glycol).

All known metathesis catalysts, being essentially electrophilic reagents, react with nucleophiles such as amines, nitrogen-containing heterocycles, and thiols. Accordingly, alkenes containing these functional groups (which are often important for the interaction of small molecules with proteins) cannot be used as building blocks for library preparation, unless these functional groups are effectively masked.

20.2.2
Examples of Library Preparation by Cross-metathesis in Solution

One of the first examples of the preparation of compound libraries by cross-metathesis was reported by Boger and coworkers [42–44], who dimerized mixtures of alkenoyl iminodiacetamides by cross-metathesis in solution (Scheme 20.6). The aim of this work was to identify new agonists or antagonists for biochemical signal transduction processes which involve the dimerization or oligomerization of pro-

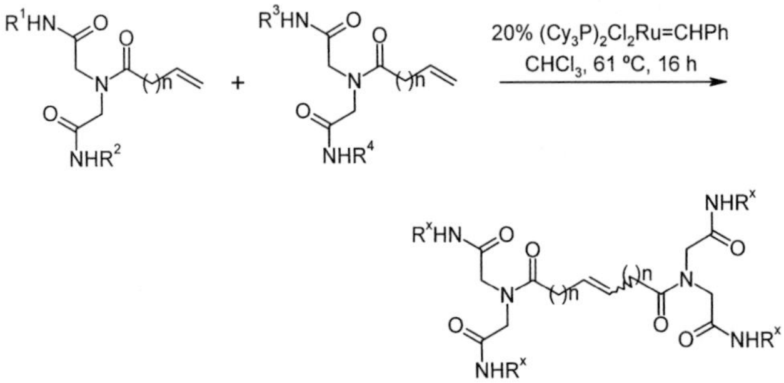

Scheme 20.6. Preparation of libraries of iminodiacetamides by cross-metathesis.

teins [e.g. tyrosin/serine/threonine kinase receptors, cytokine receptors, tumor necrosis factor (TNF) receptors].

During optimization of the chemistry they found that 3-butenamides ($n = 1$; Scheme 20.6) did not undergo metathesis at all, and 4-pentenamides ($n = 2$) only reacted sluggishly under the conditions of cross-metathesis. Longer ω-alkenoyl amides, however, cleanly yielded the expected internal alkenes [44]. The libraries were usually purified by column chromatography.

Similarly, Benner and coworkers [45] prepared mixtures of internal alkenes by cross-metathesis of mixtures of terminal olefins. The resulting libraries of alkenes were oxidized to the corresponding diols or epoxides. The mixtures of diols were the starting monomers for "receptor-assisted combinatorial synthesis", in which these diols were to be dimerized reversibly to borate esters in the presence of a receptor. Under conditions of dynamic equilibrium, enhanced concentrations of those borate esters with highest affinity to the receptor are to be expected [45].

The authors observed during the optimization of the metathesis reaction that certain alkenes (Scheme 20.7) failed to undergo cross-metathesis and others only reacted sluggishly, depending on the functional groups present in these alkenes. In particular, nitrogen-containing alkenes did not undergo metathesis – this might be due to complexation with the catalyst.

A further example of target-accelerated combinatorial synthesis has been reported by Nicolaou et al. [46]. With the aim of finding new vancomycin dimers with improved antibiotic activity, various alkenylated derivatives of vancomycin were subjected to conditions of olefin metathesis in the presence of derivatives of L-Lys-D-Ala-D-Ala, the peptide to which vancomycin strongly binds and thereby inhibits the cell wall growth of bacteria (Scheme 20.8). Cross-metathesis was performed in aqueous solution at 23 °C in the presence of a phase-transfer catalyst ($C_{12}H_{25}NMe_3Br$) and with $(Cy_3P)_2Cl_2Ru=CHPh$ (0.2 equiv.) (**1**) as metathesis catalyst. In this instance, it was observed that addition of the target peptide in fact led to increased concentrations of those dimers which were also the more potent antibiotics.

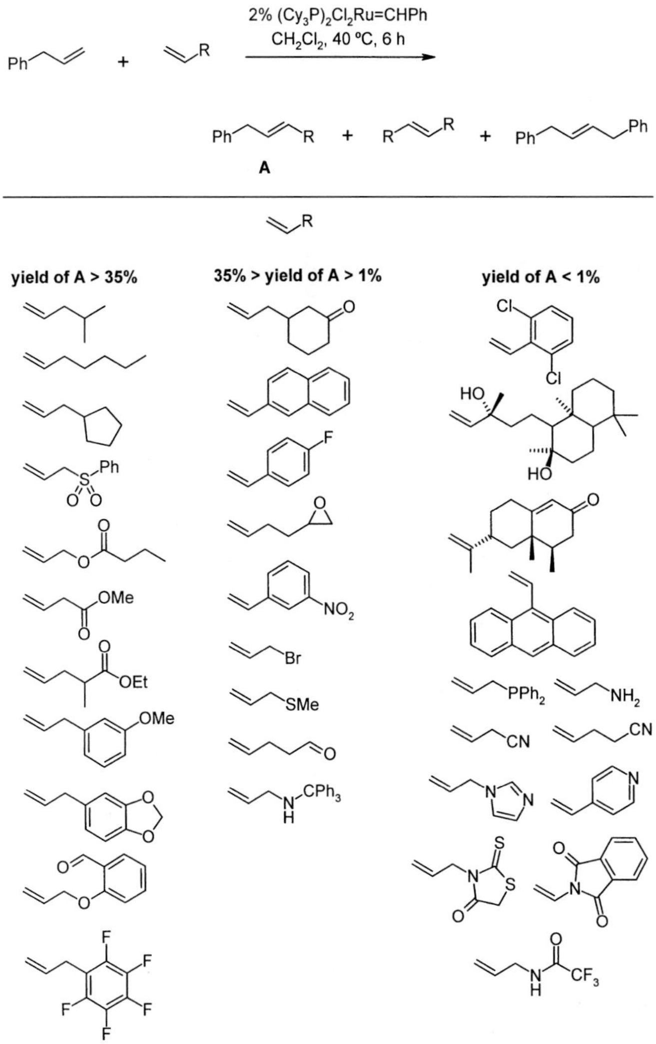

Scheme 20.7. Suitability of alkenes for cross-metathesis [45].

Brändli and Ward [47] prepared mixtures of internal, disubstituted alkenes by equilibration of internal olefins (oleic acid derivatives). Their synthesis was performed either in dichloromethane or without any solvent, and proceeded satisfactorily with as little as 0.1% of $(Cy_3P)_2Cl_2Ru=CHPh$ **(1)** if no solvent was used. With gas chromatography–mass spectrometry (GC/MS) and ^{13}C-NMR (nuclear magnetic resonance) spectroscopy the authors were able to identify all ten expected products (each as E/Z mixture) of the equilibration of two different, unsymmetrical alkenes (Scheme 20.9).

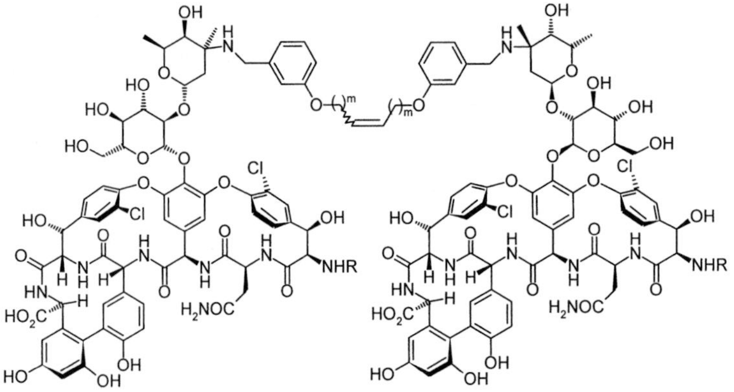

Scheme 20.8. Vancomycin dimers prepared by cross-metathesis (m = 1, 2, 3, 7; R = H, β-Ala, L-Asn, D-LeuNMe, γ-Abu, L-Arg, L-Phe [46]).

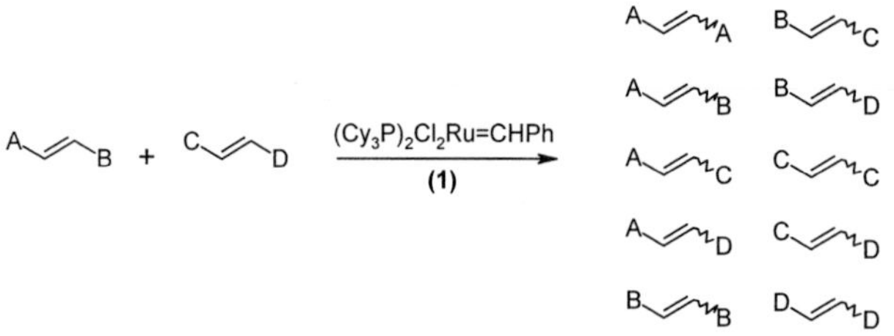

Scheme 20.9. Equilibration of internal alkenes by cross-metathesis [47].

20.2.3

Examples of Library Preparation by Ring-closing Metathesis in Solution

Ring-closing metathesis is increasingly being used for the preparation of conformationally constrained analogs of peptides. Most of the examples reported, however, only describe the synthesis of single compounds or of small arrays of compounds. These syntheses are only rarely based on easily available dienes, and are therefore not always suitable for the preparation of large compound libraries. Moreover, unlike cross-metathesis, ring-closing metathesis is an intramolecular reaction which does not increase the number of products or their diversity. Hence, ring-closing metathesis only allows the conversion of one library into another, without changing the total number of products within this library.

Cyclic peptides are an important tool for the identification of turns within a peptide which are critical for its biological activity. In analogy to the "Ala-scan", in which all the amino acids of a peptide are sequentially replaced by alanine to identify those amino acids which are crucial for biological activity, a "loop scan" (Scheme 20.10 [48]) may be used to locate possible turns within a peptide and to identify conformationally constrained analogs (**7–10**) of the original peptide (**11**).

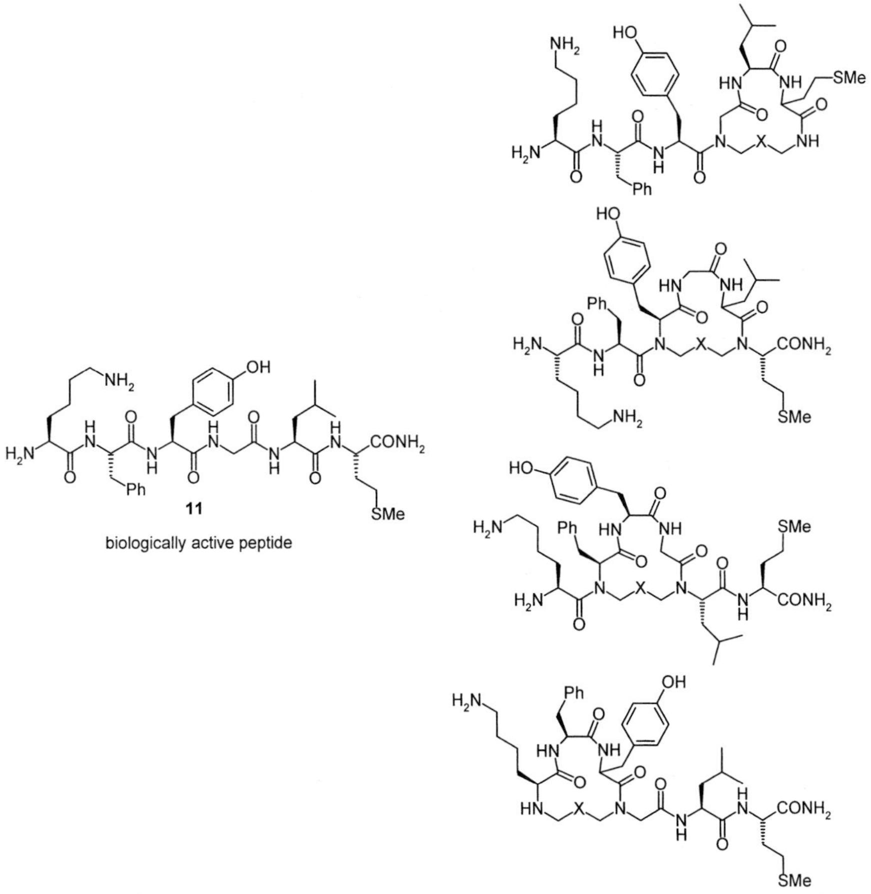

Scheme 20.10. Illustrative example of "loop scan". Four cyclic analogs (**7–10**) of the original peptide are prepared and their biological activity is compared with the activity of the original peptide (**11**). X = variable spacer.

Some new strategies for the preparation of cyclic peptides by ring-closing metathesis are presented below to illustrate the scope of these cyclizations.

Liskamp and coworkers have investigated the cyclization of N-alkenylated peptides (**12**) by ring-closing metathesis (Scheme 20.11) [48–50]. The peptides were prepared by standard solid-phase synthesis, and the N-alkenylation was effected during the assembly of the peptide by N-sulfonylation with 2-nitrobenzenesulfonyl chloride, followed by N-alkenylation under Mitsunobu conditions and sulfonamide cleavage by treatment with mercaptoethanol/1,8-diazabicyclo[5.4.0]undec-7-ene (DBU). Ring-closing metathesis could be performed either in solution or on solid phase, but in solution higher yields were usually obtained [50]. Cyclization experiments showed that the length of the N-alkenyl group was crucial for ring closure. *N*-Allyl peptides (**12**) could only be cyclized to yield eight-member rings (**13**). Larger ring sizes required the use of *N*-homoallyl or *N*-(4-penten-1-yl) peptides. The cyclization of tripeptides (to form a 15-member ring, e.g. (**14**)) was particularly difficult, and only proceeded satisfactorily with *N*-(4-penten-1-yl) substitution (Scheme 20.11).

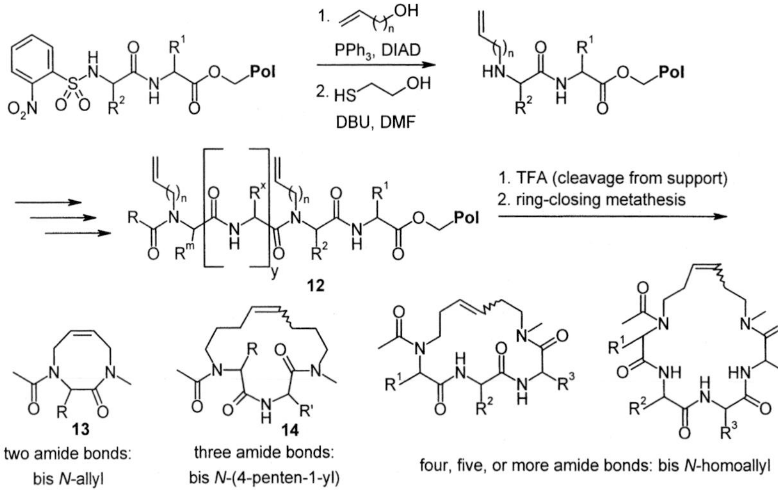

Scheme 20.11. Preparation of cyclic peptides by ring-closing metathesis in solution, and lengths of the *N*-alkenyl substituent required for ring formation [49].

Other recent examples of the preparation of cyclic peptide analogs by ring-closing metathesis in solution include the cyclic sulfamides **15** [51], cyclic sulfonamides **16** [37], and siloxanes **17** [52] (Scheme 20.12). The last were synthetic intermediates for the preparation of diols such as **18**, which were used as building blocks for the solid-phase synthesis of peptide analogs [52]. Further examples of

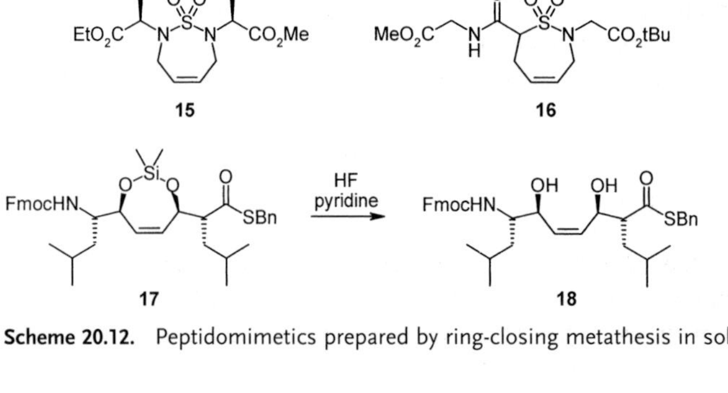

Scheme 20.12. Peptidomimetics prepared by ring-closing metathesis in solution.

solution-phase synthesis of peptide mimetics by ring-closing metathesis have been reported [2, 53–56].

20.2.4
Examples of Library Preparation by Ring-opening Metathesis Polymerization in Solution

Functionalized oligomers and polymers are of interest for a variety of applications. These include their use in chromatography as the stationary phase [57] for the separation of metals [58] or soluble receptors, and as carriers for the controlled release of drugs [59]. Oligomers functionalized with biologically relevant molecules such as amino acids or carbohydrates can also be used to mimic various biopolymers (proteins, DNA) or the surface of a cell. Such biopolymer mimetics are useful tools for studying the interaction of cell surfaces with biopolymers.

Ring-opening metathesis polymerization (ROMP), in which a strained, cyclic alkene is polymerized with the aid of a metathesis catalyst, offers several features which make this reaction particularly attractive for the preparation of functionalized oligomers [58, 60–62]. ROMP can be conducted as a living polymerization because the rate of initiation can be faster than the rate of propagation. This feature enables the preparation of oligomers with well-defined length and narrow molecular weight distribution. Because the oligomers persist as active carbene complexes even when one monomer has been consumed, ROMP also enables the preparation of block copolymers, in which various different monomers are polymerized sequentially.

Kiessling and coworkers have used ROMP for the preparation of carbohydrate-functionalized oligomers, which were used as ligands for various carbohydrate-binding proteins (concanavalin A [63], P-selectin [64], L-selectin [65]). Initially, ROMP was performed with norbornenes that were already covalently linked to a carbohydrate. However, better results were later, obtained by preparing activated oligomers by ROMP, which were then derivatized with the carbohydrate (Scheme 20.13).

Maynard et al. [67] used ROMP of *exo*-5-norbornene-2-carboxylic acid derivatives for the preparation of oligomers displaying the peptide sequences Gly–Arg–Gly–Asp and Ser–Arg–Asn, which play an important role in the binding of extracellular matrix proteins to cell-surface integrins. Both homopolymers and copolymers were prepared and characterized (Scheme 20.14). Polymers substituted with these peptides are being considered for use in the treatment of cancer [67].

20.3
Olefin Metathesis on Solid Phase

In solid-phase synthesis, the metathesis of alkenes has been used both for the chemical transformation of support-bound intermediates as well as for the cleavage of products from the support. Although these techniques have not yet been

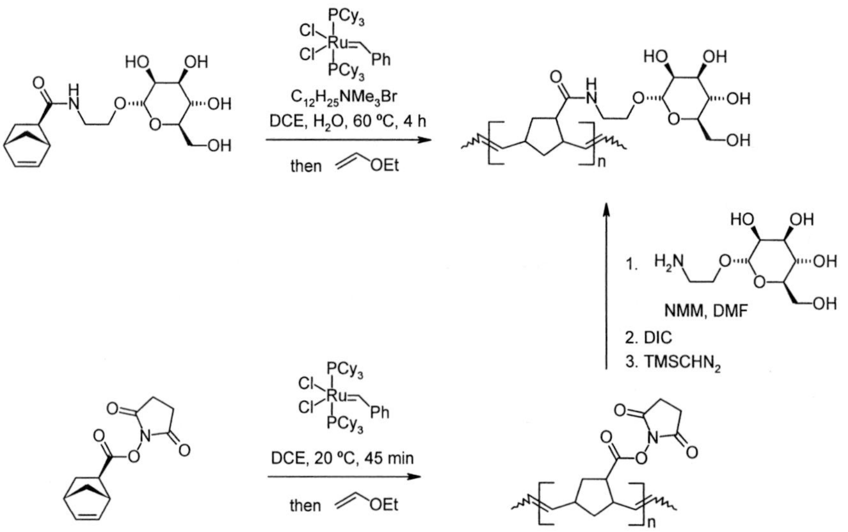

Scheme 20.13. Strategies for the preparation of carbohydrate-functionalized oligomers by ROMP [65, 66]. DCE, 1,2-dichloroethane.

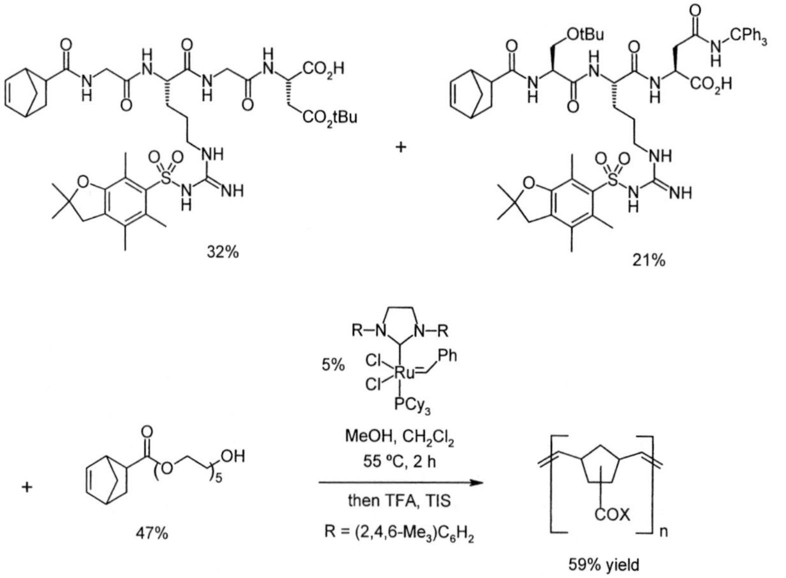

Scheme 20.14. Preparation of peptide-functionalized oligomers by ROMP [67].

extensively used for the preparation of large libraries by parallel synthesis, solid-phase chemistry is generally well suited for this purpose, and some of the reactions described below can probably be used for the preparation of compound libraries.

20.3.1
Cleavage from the Support by Olefin Metathesis

20.3.1.1 Scope and Limitations

With the discovery of highly efficient and robust soluble catalysts which mediate olefin metathesis under mild reaction conditions even in the presence of water and air, the use of alkenes as linkers for solid-phase synthesis became a realistic option. The use of alkenes as linkers is an attractive alternative to other types of linkers because alkenes are inert toward a broad range of reaction conditions, and because they provide for a reliable fixation of intermediates to the support.

Various strategies for the cleavage of compounds from insoluble supports by olefin metathesis have been described (Scheme 20.15). Support-bound dienes can yield either terminal alkenes or cycloalkenes, depending on how the diene is bound to the resin. Terminal alkenes can also be prepared by cross-metathesis of resin-bound internal alkenes with ethylene [68].

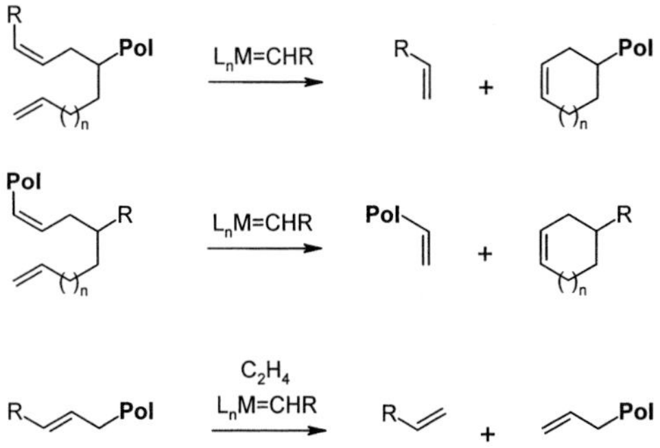

Scheme 20.15. Strategies for the cleavage of alkenes from insoluble supports by olefin metathesis.

Occasionally, carbene complex-mediated cleavage reactions give only low yields. When an additional olefin was added to the reaction mixture, however, better yields could be obtained [69]. This effect was attributed to the irreversible fixation of the carbene complex to the support when little or no amounts of terminal alkenes were present in the reaction mixture (Scheme 20.16).

Later studies [71] suggest that the irreversible fixation of the catalyst to the support is not necessarily detrimental to the yield of the cleavage reaction if spacers of sufficient flexibility are used. Thus, diene **19** (Scheme 20.17) could not be cleaved from the support, and even in the presence of 1-octene only traces of the desired product were obtained. The more flexible diene **20**, on the other hand, underwent smoothly RCM smoothly in the absence of any additional alkene, to give the expected cyclic product (**21**) in high yield. The fact that the support was colored after cleavage and catalyzed olefin metathesis suggests that carbene complexes were indeed covalently bound to the support. The flexibility of the spacer enables the metal fragment to migrate from one attachment point to the next, so that catalytic

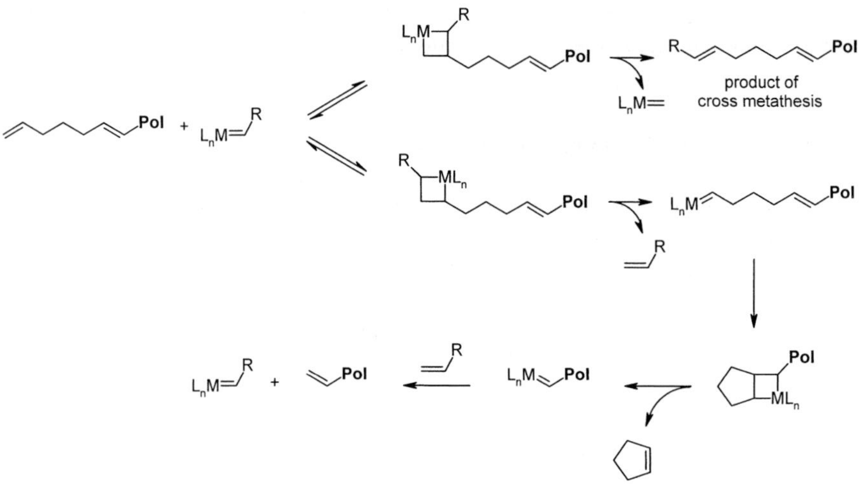

Scheme 20.16. Mechanism of the cleavage of dienes from supports by RCM [70].

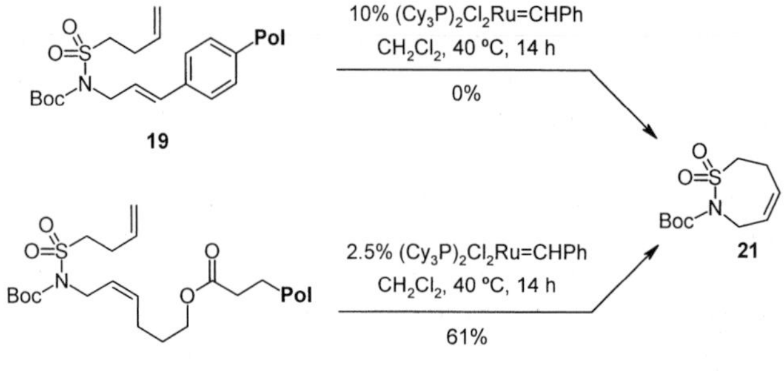

Scheme 20.17. Dependence of cleavage yields on the flexibility of spacers and on the type of alkene used as linker [71].

amounts of the carbene complex are sufficient to achieve complete metathesis of all attachment sites. Another reason for the resistance of **19** toward carbene complex-mediated cleavage may be the fact that **19** is a styrene derivative. Styrenes usually react more slowly with ruthenium carbene complexes than unconjugated, internal *cis*-alkenes.

One problem which is inherent to olefin metathesis-induced cleavage is the elution of catalyst-derived byproducts together with the final product. The currently known metathesis catalysts (mainly ruthenium carbene complexes) decompose slowly during the metathesis reaction to yield various ruthenium complexes, which do not remain attached to the support. These impurities have to be removed by chromatographic purification of the products. However, large libraries of compounds for direct biological screening cannot always be purified, and cleavage by olefin metathesis will only be of limited use for the preparation of such libraries unless more stable metathesis catalysts or selective scavengers for metal-containing byproducts become available.

20.3.1.2 Examples of Cleavage from the Support by Olefin Metathesis

Knerr and Schmidt [72, 73] have used a metathesis-based cleavage strategy for the solid-phase synthesis of oligosaccharides (Scheme 20.18). Cleavage by treatment with Grubbs' catalyst yielded O-allyl glycosides (22), which represent versatile, protected intermediates for further synthetic manipulations [73].

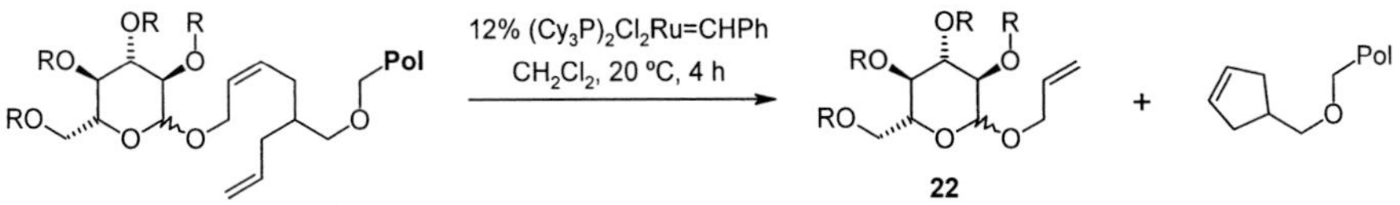

Scheme 20.18. Synthesis of allyl glycosides by RCM-mediated cleavage from a polymeric support [72, 73].

A similar strategy has been described by Peters and Blechert [74], in which RCM of a support-bound diene was used to release styrenes from a polystyrene-based, insoluble support. Linkers of this type can also be cleaved by cross-metathesis with ethylene [68].

Several groups have investigated the preparation of cyclic compounds by RCM with simultaneous cleavage from the support [71, 75, 76]. One recent example, reported by Piscopio et al. [77], is shown in Scheme 20.19. The substrate (23) for olefin metathesis was prepared in one step by an Ugi reaction. The product (24), a Freidinger lactam, was designed to mimic β-turns, which play a pivotal role in the

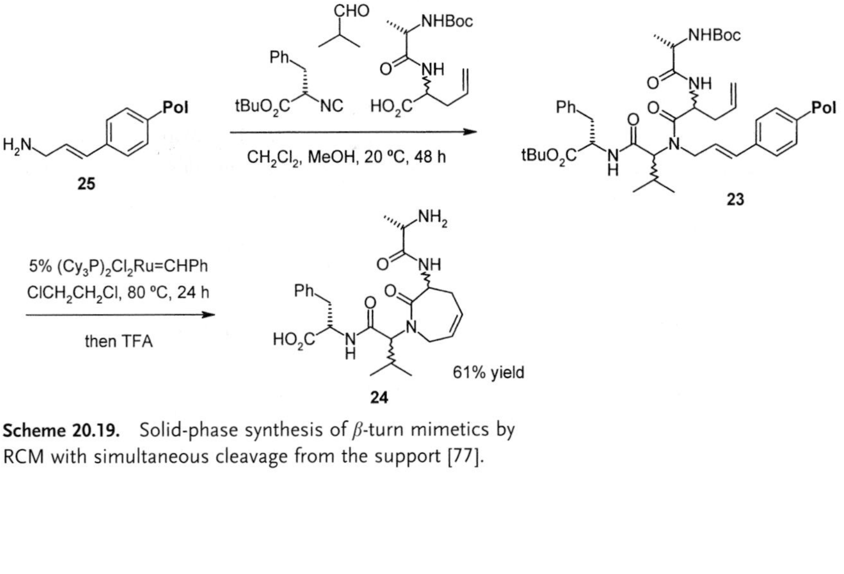

Scheme 20.19. Solid-phase synthesis of β-turn mimetics by RCM with simultaneous cleavage from the support [77].

interaction of proteins. Because styrene derivative **25** was chosen as linker, cleavage required prolonged heating for a long time. An unconjugated *cis*-alkene would probably allow milder cleavage conditions (cf. Schemes 20.17 and 20.18).

Nicolaou et al. [78, 79] have used ring-closing metathesis with simultaneous cleavage from the support in an elegant solid-phase synthesis of epothilone analogs (Scheme 20.20). Epothilones are a group of natural products which promote the polymerization of α- and β-tubulin subunits, and which show higher cytotoxicity than taxol [80]. These interesting biological properties have prompted several research groups to develop syntheses for these compounds and analogs thereof [80, 81].

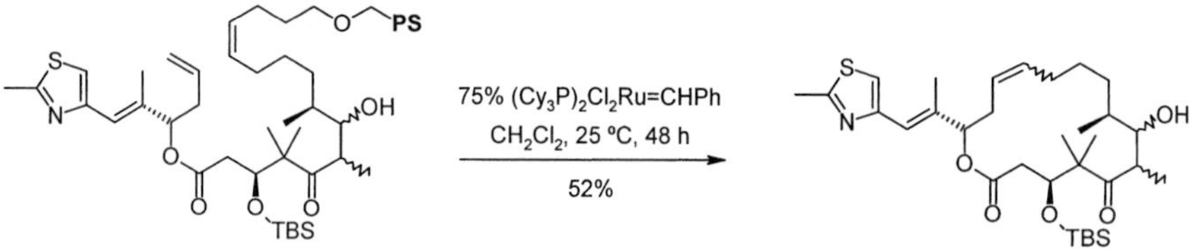

Scheme 20.20. Solid-phase synthesis of epothilone analogs [78, 79].

In Nicolaou's solid-phase synthesis of epothilone analogs, a Merrifield resin (PS–CH$_2$Cl) with low loading (0.3 mmol g^{-1}) was used. After release from the support the products were purified by preparative thin layer chromatography. More than 100 epothilone analogs have been prepared using this methodology, and their biological evaluation gave detailed insight into the structure–activity relationship of this family of natural products [79].

20.3.2
Ring-closing Metathesis on Solid Phase

20.3.2.1 **Scope and Limitations**
Ring-closing metathesis (RCM), being a reversible process, is best suited to the preparation of unstrained cyclic compounds. In most of the reported examples of RCM on solid phase [2], five- or six-member rings were generated. Other ring sizes are also accessible, but careful optimization of the reaction conditions are often necessary. Macrocyclizations, for instance, usually require the use of supports with low loading to avoid self-metathesis (Scheme 20.21).

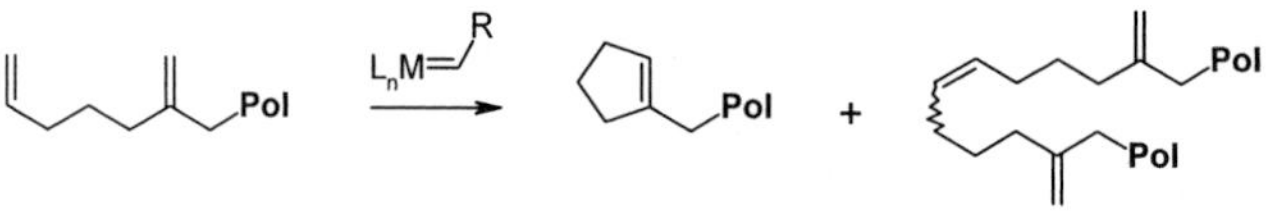

Scheme 20.21. Ring-closing metathesis on solid phase, and self-metathesis as a potential side reaction.

20.3.2.2 Examples of Ring-closing Metathesis on Solid Phase

Several examples of the solid-phase synthesis of nitrogen-containing heterocycles have been reported (Scheme 20.22) (for an example performed on soluble poly-(ethylene glycol), see [82]). Heating and substantial amounts of ruthenium carbene complex are usually required to attain complete conversion of the starting diene (e.g. 25). Eneyne 28 is an interesting example of an intramolecular ene–yne coupling, which gives ready access to highly substituted dienes (29), which in turn are suitable starting materials for Diels–Alder reactions [83].

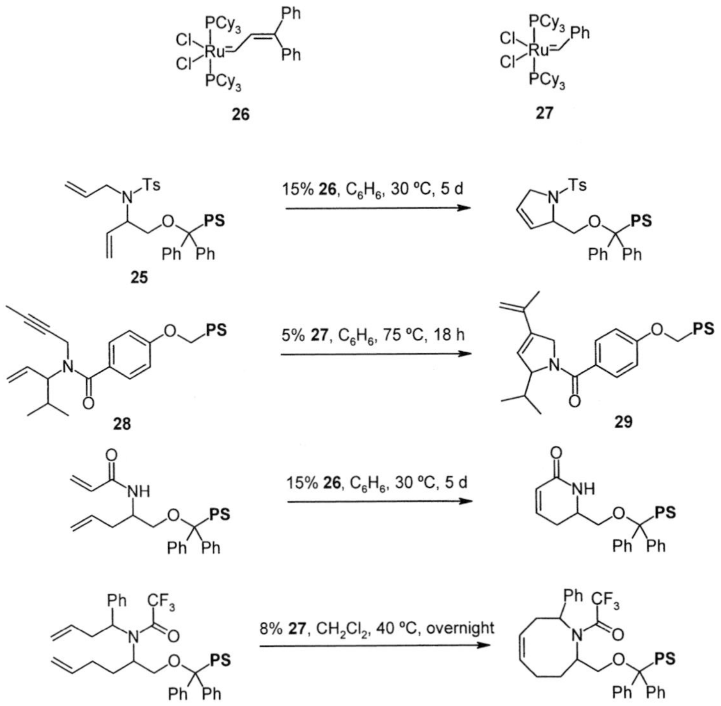

Scheme 20.22. Examples of the preparation of nitrogen-containing heterocycles by RCM on solid phase [83–85]. PS, crosslinked polystyrene.

Various groups have investigated the preparation of cyclic peptides by RCM on solid phase. In Section 20.2.3, the work of Liskamp and coworkers concerning the cyclization of peptide-derived dienes was presented. These cyclizations generally give higher yields in solution than on solid phase [48–50]. Another example on crosslinked polystyrene is shown in Scheme 20.23.

With the aim of finding efficient routes to structurally complex, polycyclic compounds, Lee et al. [86] recently developed the synthesis shown in Scheme 20.24. With an Ugi reaction followed by an intramolecular Diels–Alder reaction and an

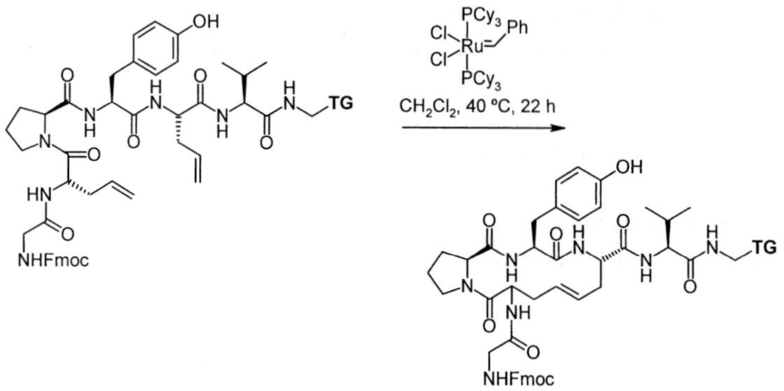

Scheme 20.23. Preparation of cyclic peptides by RCM on solid phase [55]. TG, Tentagel.

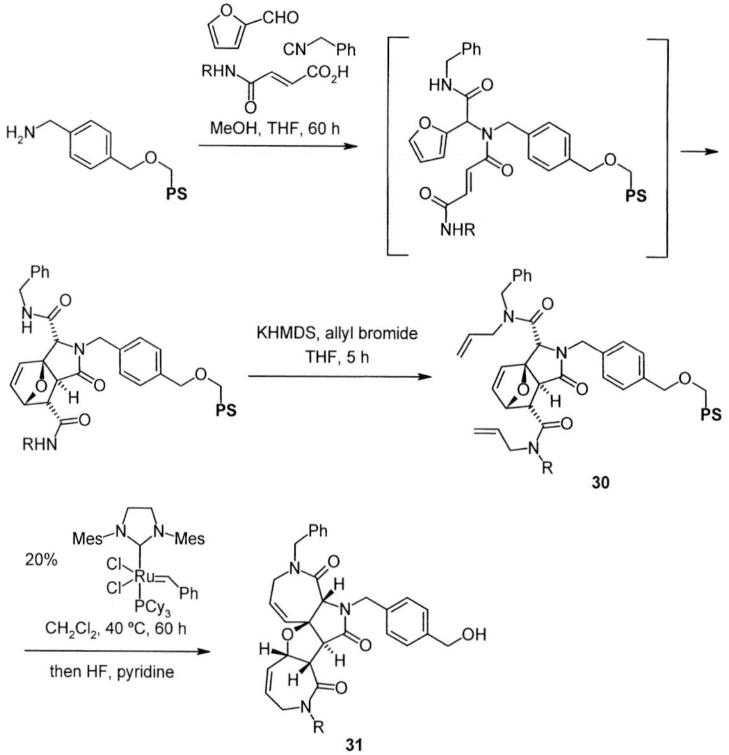

Scheme 20.24. Synthesis of polycyclic structures using a ring-opening ring-closing metathesis cascade [86].

allylation, strained triene (**30**) was generated, which upon treatment with a metathesis catalysts underwent a ring-opening/ring-closing cascade to yield highly substituted, tetracyclic compounds (**31**). A valuable feature of this synthesis is the ready availability of some of the four building blocks. One drawback of this reaction sequence is the low selectivity of the allylation reaction, which necessitates the protection of all nucleophilic functional groups.

20.3.3
Cross- and Self-metathesis on Solid Phase

20.3.3.1 Scope and Limitations
Cross-metathesis should in principle enable the efficient preparation of unsymmetrical, acyclic alkenes on solid phase. Unfortunately, this reaction does not always proceed as expected, mainly because only few types of terminal alkenes smoothly undergo cross-metathesis (see Scheme 20.7). Alkenes bearing "interesting" functional groups (hydrogen bond donors and acceptors) sometimes react only sluggishly or not at all, leading either to complete consumption of the catalyst (formation of unreactive carbene complexes) and/or to formation of large amounts of the products of self-metathesis (e.g. **33**) (Scheme 20.25).

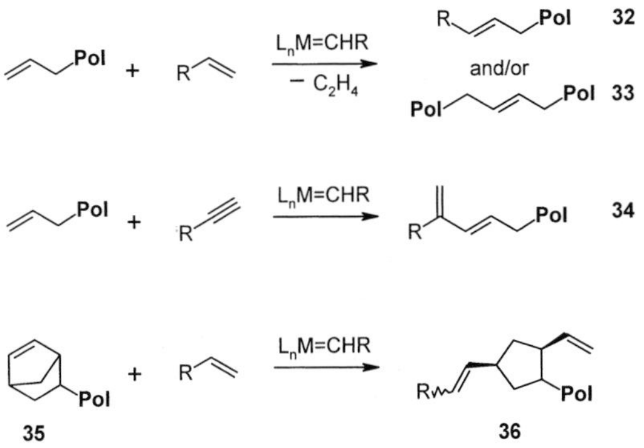

Scheme 20.25. Strategies for performing cross-metathesis on solid phase.

An interesting variant of cross-metathesis is the so-called ring-opening cross-metathesis. Strained, cyclic alkenes (e.g. norbornene (**35**), cyclobutene) react rapidly and irreversibly with metathesis catalysts to yield a new carbene complex, which can react with a second alkene to yield the product of cross-metathesis (**36**). This reaction has also been performed successfully on solid phase (Scheme 20.25).

20.3.3.2 Examples of Cross- and Self-metathesis on Solid Phase
Some illustrative preparations of internal alkenes by cross-metathesis on solid phase are shown in Scheme 20.26. Allylsilanes (**37**) appear to be well suited for this reaction [33, 87], and substituted allylsilanes (**38**), which are valuable synthetic intermediates, can be easily prepared by cross-metathesis (Scheme 20.26). Nicolaou et al. used cross-metathesis of support-bound, alkenyl-substituted ketophospho-

nates (**39**) with ω-alkenols (**40**) in a solid-phase synthesis of muscone analogs [88] (Scheme 20.26). Unsaturated α-amino acid derivatives (**42**) have also been prepared on solid phase by cross-metathesis [35] (Scheme 20.26). The low loading of the starting resin (0.07 mmol g^{-1}) was required to suppress self-metathesis.

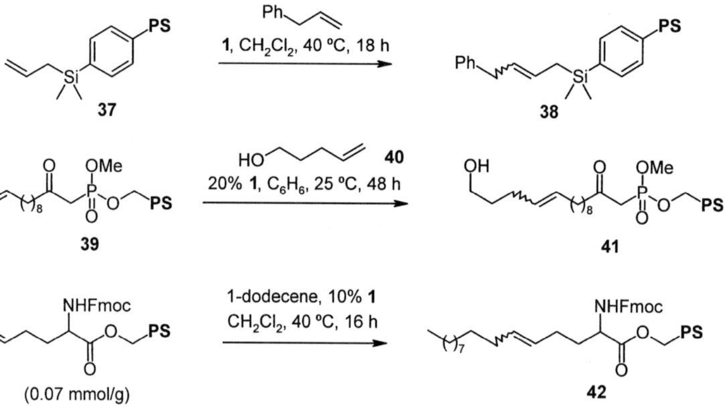

Scheme 20.26. Cross-metathesis on solid phase [35, 87, 89]. PS, crosslinked polystyrene; **1**, (Cy$_3$P)$_2$Cl$_2$Ru=CHPh.

The combination of ring-opening with cross-metathesis ("ring-opening cross-metathesis") has been realized on insoluble supports by Cuny and coworkers [90–92]. Support-bound norbornene derivative **43** was treated with styrenes **44** (which do not undergo self-metathesis efficiently) and a ruthenium carbene complex, to yield regioisomeric mixtures of highly substituted cyclopentenes **45** (Scheme 20.27). Substituents on styrene which were tolerated included *tert*-butyl, alkoxy, acyloxy, and hydroxy.

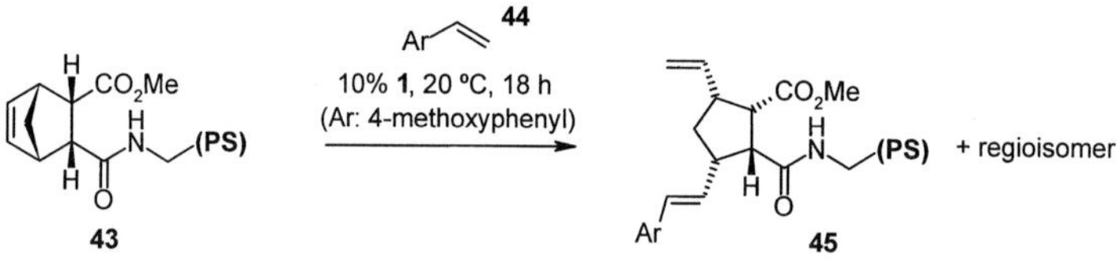

Scheme 20.27. Ring-opening cross metathesis on solid phase [92]. (PS), crosslinked polystyrene with spacer; **1**, (Cy$_3$P)$_2$Cl$_2$Ru=CHPh.

The carbene complex-mediated coupling of alkynes with alkenes to yield 1,3-dienes [13, 16, 93] is one of the most surprising metathesis reactions. Despite a number of potential side reactions (polymerization of the alkyne, self-metathesis of the alkene), high yields of dienes can be obtained. This reaction can also be conducted on insoluble supports, with either the alkene or the alkyne attached to the resin (Scheme 20.28). The resulting dienes can be further transformed by 2 + 4 cycloaddition with suitable dienophiles to yield substituted cyclohexenes [94].

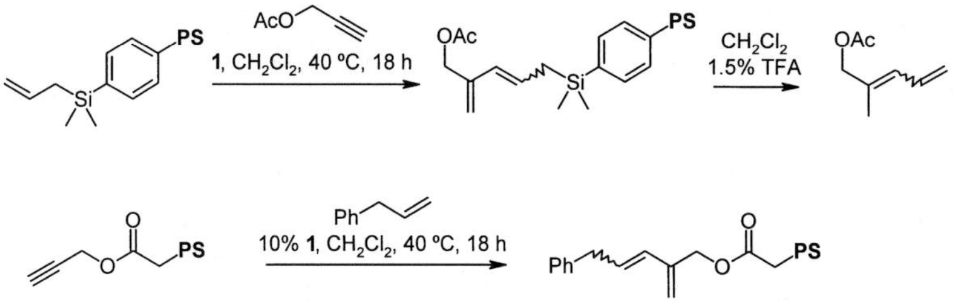

Scheme 20.28. Examples of alkene/alkyne cross-metathesis on solid phase [29, 95]. PS, crosslinked polystyrene; **1**, $(Cy_3P)_2Cl_2Ru=CHPh$.

Self-metathesis of support-bound N-alkenoylated peptides was used by Conde-Friboes et al. [96] for the preparation of symmetric peptidomimetics (Scheme 20.29). Peptides were prepared on crosslinked polystyrene by standard fluorenyl-methoxycarbonyl (Fmoc) chemistry and then acylated with an ω-alkenoic acid. In accordance with similar results of Boger and Chai [44] (Section 20.2.2), neither 3-butenamides nor 4-pentenamides underwent self-metathesis efficiently. With longer alkenoic acids, however, the peptide dimers **46** were formed in high yield and purity as mixtures of E/Z isomers.

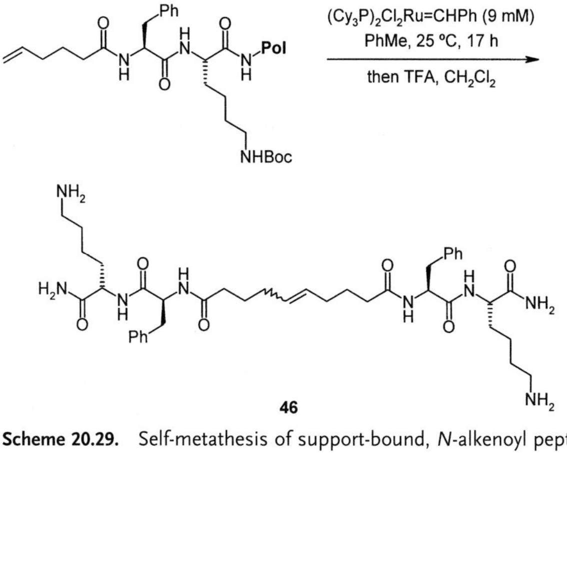

46

Scheme 20.29. Self-metathesis of support-bound, N-alkenoyl peptides [96].

20.4
Conclusion

With the development of highly efficient and selective catalysts for olefin metathesis in recent years, this reaction has become a valuable tool for organic chemists. Cross-metathesis and ROMP in solution can now be performed with functionalized alkenes, and can offer interesting new possibilities for the preparation of compound libraries. In particular, the selective cross-metathesis of different alkenes, for which the underlying principles are now slowly emerging, has huge potential and could become a process with an impact similar to those of the Wittig or the Diels–Alder reactions if its scope and limitations are clearly understood. Solid-phase synthesis has also greatly benefited from these new catalysts, and new cleavage strategies and other methodologies based on carbene complex-mediated olefin metathesis on solid phase have been developed successfully.

The properties of the currently available catalysts are, however, far from ideal. Because of their limited stability, large amounts of catalyst are often required to drive reactions to completion. This feature can lead to significant amounts of metal-derived impurities in the crude products. Moreover, most carbene complexes suitable for olefin metathesis are also highly sensitive toward amines, azoles, and other nucleophiles, which severely limits the choice of functional groups tolerated in the starting materials. This facet is particularly problematic for the preparation of libraries of biologically active compounds because nucleophilic functional groups are often of crucial importance for biological activity. Future research should aim to overcome these limitations of current catalysts by further enhancing their selectivity and stability.

References

1 A. Fürstner, *Angew. Chem. Int. Ed.* **2000**, *39*, 3012–3043.

2 F. Zaragoza, *Metal Carbenes in Organic Synthesis.* Wiley-VCH, Weinheim 1999.

3 R. H. Grubbs, S. Chang, *Tetrahedron* **1998**, *54*, 4413–4450.

4 K. J. Ivin, J. C. Mol, *Olefin Metathesis and Metathesis Polymerization.* Academic Press, London 1997.

5 M. Schuster, S. Blechert, *Angew. Chem. Int. Ed.* **1997**, *36*, 2037–2056.

6 H. S. Eleuterio, *J. Mol. Catal.* **1991**, *65*, 55–61.

7 R. Streck, *J. Mol. Catal.* **1992**, *76*, 359–372.

8 P. Schwab, R. H. Grubbs, J. W. Ziller, *J. Am. Chem. Soc.* **1996**, *118*, 100–110.

9 M. Scholl, T. M. Trnka, J. P. Morgan, R. H. Grubbs, *Tetrahedron Lett.* **1999**, *40*, 2247–2250.

10 J. Huang, E. D. Stevens, S. P. Nolan, J. L. Petersen, *J. Am. Chem. Soc.* **1999**, *121*, 2674–2678.

11 A. Fürstner, O. R. Thiel, L. Ackermann, H. J. Schanz, S. P. Nolan, *J. Org. Chem.* **2000**, *65*, 2204–2207.

12 A. Briot, M. Bujard, V. Gouverneur, S. P. Nolan, C. Mioskowski, *Org. Lett.* **2000**, *2*, 1517–1519.

13 R. Stragies, U. Voigtmann, S. Blechert, *Tetrahedron Lett.* **2000**, *41*, 5465–5468.

14 A. K. Chatterjee, R. H. Grubbs, *Org. Lett.* **1999**, *1*, 1751–1753.

15 M. Scholl, S. Ding, C. W. Lee, R. H. Grubbs, *Org. Lett.* **1999**, *1*, 953–956.

16 J. A. Smulik, S. T. Diver, *Tetrahedron Lett.* **2001**, *42*, 171–174.

17 J. D. Rainier, J. M. Cox, S. P. Allwein, *Tetrahedron Lett.* **2001**, *42*, 179–181.

18 C. W. Bielawski, R. H. Grubbs, *Angew. Chem. Int. Ed.* **2000**, *39*, 2903–2906.

19 W. Apichatachutapan, L. J. Mathias, *J. Appl. Pol. Sci.* **1998**, *67*, 183–190.

20 D. M. Lynn, B. Mohr, R. H. Grubbs, *J. Am. Chem. Soc.* **1998**, *120*, 1627–1628.

21 D. M. Lynn, S. Kanaoka, R. H. Grubbs, *J. Am. Chem. Soc.* **1996**, *118*, 784–790.

22 J. Louie, R. H. Grubbs, *Angew. Chem. Int. Ed.* **2001**, *40*, 247–249.

23 D. M. Lynn, B. Mohr, R. H. Grubbs, L. M. Henling, M. W. Day, *J. Am. Chem. Soc.* **2000**, *122*, 6601–6609.

24 S. B. Garber, J. S. Kingsbury, B. L. Gray, A. H. Hoveyda, *J. Am. Chem. Soc.* **2000**, *122*, 8168–8179.

25 M. S. Sanford, L. M. Henling, M. W. Day, R. H. Grubbs, *Angew. Chem. Int. Ed.* **2000**, *39*, 3451–3453.

26 H. Katayama, H. Urushima, T. Nishioka, C. Wada, M. Nagao, F. Ozawa, *Angew. Chem. Int. Ed.* **2000**, *39*, 4513–4515.

27 M. S. Sanford, M. Ulman, R. H. Grubbs, *J. Am. Chem. Soc.* **2001**, *123*, 749–750.

28 H. E. Blackwell, D. J. O'Leary, A. K. Chatterjee, R. A. Washenfelder, D. A. Bussmann, R. H. Grubbs, *J. Am. Chem. Soc.* **2000**, *122*, 58–71.

29 M. Schuster, S. Blechert, *Tetrahedron Lett.* **1998**, *39*, 2295–2298.

30 R. Stragies, M. Schuster, S. Blechert, *Angew. Chem. Int. Ed.* **1997**, *36*, 2518–2520.

31 A. Fürstner, G. Seidel, *Angew. Chem. Int. Ed.* **1998**, *37*, 1734–1736.

32 W. E. Crowe, Z. J. Zhang, *J. Am. Chem. Soc.* **1993**, *115*, 10998–10999.

33 W. E. Crowe, D. R. Goldberg, Z. J. Zhang, *Tetrahedron Lett.* **1996**, *37*, 2117–2120.

34 W. E. Crowe, D. R. Goldberg, *J. Am. Chem. Soc.* **1995**, *117*, 5162–5163.

35 S. C. G. Biagini, S. E. Gibson, S. P. Keen, *J. Chem. Soc. Perkin Trans. 1* **1998**, 2485–2499.

36 S. E. Gibson, V. C. Gibson, S. P. Keen, *Chem. Commun.* **1997**, 1107–1108.

37 D. D. Long, A. P. Termin, *Tetrahedron Lett.* **2000**, *41*, 6743–6747.

38 S. C. Schürer, S. Gessler, N. Buschmann, S. Blechert, *Angew. Chem. Int. Ed.* **2000**, *39*, 3898–3900.

39 Q. Yao, *Angew. Chem. Int. Ed.* **2000**, *39*, 3896–3898.

40 L. Jafarpour, S. P. Nolan, *Org. Lett.* **2000**, *2*, 4075–4078.

41 M. Ahmed, T. Arnauld, A. G. M. Barrett, D. C. Braddock, P. A. Procopiou, *Synlett* **2000**, 1007–1009.

42 D. L. Boger, W. Y. Chai, R. S. Ozer, C. M. Andersson, *Bioorg. Med. Chem. Lett.* **1997**, *7*, 463–468.

43 D. L. Boger, W. Chai, Q. Jin, *J. Am. Chem. Soc.* **1998**, *120*, 7220–7225.

44 D. L. Boger, W. Y. Chai, *Tetrahedron* **1998**, *54*, 3955–3970.

45 T. Giger, M. Wigger, S. Audétat, S. A. Benner, *Synlett* **1998**, 688–691.

46 K. C. Nicolaou, R. Hughes, S. Y. Cho, N. Winssinger, C. Smethurst, H. Labischinski, R. Endermann, *Angew. Chem. Int. Ed.* **2000**, *39*, 3823–3828.

47 C. Brändli, T. R. Ward, *Helv. Chim. Acta* **1998**, *81*, 1616–1621.

48 J. F. Reichwein, B. Wels, J. A. W. Kruijtzer, C. Versluis, R. M. J. Liskamp, *Angew. Chem. Int. Ed.* **1999**, *38*, 3684–3687.

49 J. F. Reichwein, C. Versluis, R. M. J. Liskamp, *J. Org. Chem.* **2000**, *65*, 6187–6195.

50 J. F. Reichwein, R. M. J. Liskamp, *Eur. J. Org. Chem.* **2000**, 2335–2344.

51 J. M. Dougherty, D. A. Probst, R. E. Robinson, J. D. Moore, T. A. Klein, K. A. Snelgrove, P. R. Hanson, *Tetrahedron* **2000**, *56*, 9781–9790.

52 T. M. Gierasch, M. Chytil, M. T. Didiuk, J. Y. Park, J. J. Urban, S. P. Nolan, G. L. Verdine, *Org. Lett.* **2000**, *2*, 3999–4002.

53 Y. Gao, P. Lane-Bell, J. C. Vederas, *J. Org. Chem.* **1998**, *63*, 2133–2143.

54 R. M. Williams, J. Liu, *J. Org. Chem.* **1998**, *63*, 2130–2132.

55 S. J. Miller, H. E. Blackwell, R. H. Grubbs, *J. Am. Chem. Soc.* **1996**, *118*, 9606–9614.

56 A. S. Ripka, R. S. Bohacek, D. H. Rich, *Bioorg. Med. Chem. Lett.* **1998**, *8*, 357–360.

57 M. R. Buchmeiser, N. Atzl, G. K. Bonn, *J. Am. Chem. Soc.* **1997**, *119*, 9166–9174.

58 F. Sinner, M. R. Buchmeiser, R. Tessadri, M. Mupa, K. Wurst, G. K. Bonn, *J. Am. Chem. Soc.* **1998**, *120*, 2790–2797.

59 A. Godwin, M. Hartenstein, A. H. E. Müller, S. Brocchini, *Angew. Chem. Int. Ed.* **2001**, *40*, 594–597.

60 S. C. G. Biagini, V. C. Gibson, M. R. Giles, E. L. Marshall, M. North, *Chem. Commun.* **1997**, 1097–1098.

61 V. C. Gibson, E. L. Marshall, M. North, D. A. Robson, P. J. Williams, *Chem. Commun.* **1997**, 1095–1096.

62 C. Fraser, R. H. Grubbs, *Macromolecules* **1995**, *28*, 7248–7255.

63 M. Kanai, K. H. Mortell, L. L. Kiessling, *J. Am. Chem. Soc.* **1997**, *119*, 9931–9932.

64 D. D. Manning, L. E. Strong, X. Hu, P. J. Beck, L. L. Kiessling, *Tetrahedron* **1997**, *53*, 11937–11952.

65 E. J. Gordon, J. E. Gestwicki, L. E. Strong, L. L. Kiessling, *Chem. Biol.* **2000**, *7*, 9–16.

66 L. E. Strong, L. L. Kiessling, *J. Am. Chem. Soc.* **1999**, *121*, 6193–6196.

67 H. D. Maynard, S. Y. Okada, R. H. Grubbs, *Macromolecules* **2000**, *33*, 6239–6248.

68 R. B. Andrade, O. J. Plante, L. G. Melean, P. H. Seeberger, *Org. Lett.* **1999**, *1*, 1811–1814.

69 J. J. N. Veerman, J. H. van Maarseveen, G. M. Visser, C. G. Kruse, H. E. Schoemaker, H. Hiemstra, F. P. J. T. Rutjes, *Eur. J. Org. Chem.* **1998**, 2583–2589.

70 F. Zaragoza, *Organic Synthesis on Solid Phase*; Wiley-VCH, Weinheim 2000.

71 R. C. D. Brown, J. L. Castro, J.-D.

Moriggi, *Tetrahedron Lett.* **2000**, *41*, 3681–3685.

72 L. Knerr, R. R. Schmidt, *Eur. J. Org. Chem.* **2000**, 2803–2808.

73 L. Knerr, R. R. Schmidt, *Synlett* **1999**, 1802–1804.

74 J. U. Peters, S. Blechert, *Synlett* **1997**, 348–350.

75 J. Pernerstorfer, M. Schuster, S. Blechert, *Chem. Commun.* **1997**, 1949–1950.

76 J. H. van Maarseveen, J. A. J. den Hartog, V. Engelen, E. Finner, G. Visser, C. G. Kruse, *Tetrahedron Lett.* **1996**, *37*, 8249–8252.

77 A. D. Piscopio, J. F. Miller, K. Koch, *Tetrahedron* **1999**, *55*, 8189–8198.

78 K. C. Nicolaou, N. Winssinger, J. Pastor, S. Ninkovic, F. Sarabia, Y. He, D. Vourloumis, Z. Yang, T. Li, P. Giannakakou, E. Hamel, *Nature* **1997**, *387*, 268–272.

79 K. C. Nicolaou, D. Vourloumis, T. Li, J. Pastor, N. Winssinger, Y. He, S. Ninkovic, F. Sarabia, H. Vallberg, F. Roschangar, N. P. King, M. R. V. Finlay, P. Giannakakou, P. Verdier-Pinard, E. Hamel, *Angew. Chem. Int. Ed.* **1997**, *36*, 2097–2103.

80 K. C. Nicolaou, N. P. King, M. R. V. Finlay, Y. He, F. Roschangar, D. Vourloumis, H. Vallberg, F. Sarabia, S. Ninkovic, D. Hepworth, *Bioorg. Med. Chem.* **1999**, *7*, 665–697.

81 J. D. Winkler, J. M. Holland, J. Kasparec, P. H. Axelsen, *Tetrahedron* **1999**, *55*, 8199–8214.

82 S. Varray, C. Gauzy, F. Lamaty, R. Lazaro, J. Martinez, *J. Org. Chem.* **2000**, *65*, 6787–6790.

83 D. A. Heerding, D. T. Takata, C. Kwon, W. F. Huffman, J. Samanen, *Tetrahedron Lett.* **1998**, *39*, 6815–6818.

84 M. Schuster, J. Pernerstorfer, S. Blechert, *Angew. Chem. Int. Ed.* **1996**, *35*, 1979–1980.

85 J. Pernerstorfer, M. Schuster, S. Blechert, *Synthesis* **1999**, 138–144.

86 D. Lee, J. K. Sello, S. L. Schreiber, *Org. Lett.* **2000**, *2*, 709–712.

87 M. Schuster, N. Lucas, S. Blechert, *Chem. Commun.* **1997**, 823–824.

88 K. C. Nicolaou, J. Pastor, N. Winssinger, F. Murphy, *J. Am. Chem. Soc.* **1998**, *120*, 5132–5133.

89 J. Barluenga, F. Aznar, S. Barluenga, A. Martín, S. García Granda, E. Martín, *Synlett* **1998**, 473–474.

90 G. D. Cuny, J. R. Cao, J. R. Hauske, *Tetrahedron Lett.* **1997**, *38*, 5237–5240.

91 J. Cao, G. D. Cuny, J. R. Hauske, *Mol. Diversity* **1998**, *3*, 173–179.

92 G. D. Cuny, J. Cao, A. Sidhu, J. R. Hauske, *Tetrahedron* **1999**, *55*, 8169–8178.

93 A. Kinoshita, N. Sakakibara, M. Mori, *J. Am. Chem. Soc.* **1997**, *119*, 12388–12389.

94 S. C. Schürer, S. Blechert, *Synlett* **1999**, 1879–1882.

95 S. C. Schürer, S. Blechert, S. *Synlett* **1998**, 166–168.

96 K. Conde-Frieboes, S. Andersen, J. Breinholt, *Tetrahedron Lett.* **2000**, *41*, 9153–9156.